Masterclass in Neuroendocrinology

Volume 17

Masterclass in Neuroendocrinology is published in collaboration with the International Neuroendocrine Federation (INF) https://www.inf-neuroendocrinology.org/. This series aims to illustrate and promote the use of the latest and most impactful technologies in basic and clinical research in the exciting field of neuroendocrinology. Neuroendocrinology is defined as the study of the control of endocrine function by the brain and the actions of hormones on the brain.

The series is made up of volumes that highlight different topics in neuroendocrinology that are of current interest and encompass the study of both normal and abnormal function, and the developmental origins of disease. They include the study of neural networks in the brain that regulate and form neuroendocrine systems, and also include the study of behaviours and mental states that are influenced or regulated by hormones. They also necessarily include the study of peripheral physiological systems that are regulated by neuroendocrine mechanisms. Some recent examples include: Developmental Neuroendocrinology, Neuroendocrine Clocks and Calendars, and Neuroanatomy of Neuroendocrine Systems. Each book

- is edited by leading experts in the field
- is written by a team of internationally respected researchers
- is focused on an area of neuroendocrinology research
- demonstrates the present state of knowledge and understanding in the field of neuroendocrinology and highlights the translation of such knowledge to improve human health and well-being
- includes assessments of different experimental approaches both
- *in vivo* and
- *in vitro*, and of how the resulting data are interpreted.

The Series is aimed at a wide audience, as neuroendocrinology integrates neuroscience and endocrinology. It is intended for established scientists, clinicians and early-career researchers.

Founding Series Co-Editors: William E. Armstrong and John A. Russell

The series is indexed in Scopus.

Valery Grinevich • Rui F. Oliveira
Editors

Evolutionary and Comparative Neuroendocrinology

Springer

Editors
Valery Grinevich
Department of Neuropeptide Research in Psychiatry, Central Institute of Mental Health, German Center for Mental Health (DZPG), Medical Faculty Mannheim
University of Heidelberg
Mannheim, Germany

Rui F. Oliveira
ISPA - University Institute for Psychological Social and Life Sciences
Lisbon, Portugal

GIMM - Gulbenkian Institute for Molecular Medicine
Oeiras, Portugal

ISSN 2662-2068 ISSN 2662-2076 (electronic)
Masterclass in Neuroendocrinology
ISBN 978-3-031-80208-9 ISBN 978-3-031-80209-6 (eBook)
https://doi.org/10.1007/978-3-031-80209-6

Cover illustration: Illustration by deblik, based on an image from © Nablys/stock.adobe.com

This Springer imprint is published by the registered company Springer Nature Switzerland AG
The registered company address is: Gewerbestrasse 11, 6330 Cham, Switzerland

Series Preface

This Series began as a joint venture between the International Neuroendocrine Federation and Wiley-Blackwell, and now is continuing with Springer-Nature as publisher for the Federation. The broad aim of the Series is to provide established researchers, trainees, and students with authoritative, up-to-date accounts of the present state of knowledge, and prospects for the future, across a range of topics in the flourishing field of neuroendocrinology. The Series is aimed at a wide audience as neuroendocrinology integrates the disciplines of neuroscience and endocrinology. We define neuroendocrinology as the study of how endocrine function is controlled by the brain and the actions of hormones on the brain. It encompasses the study of normal and abnormal function, and the developmental origins of disease. It includes investigation of the neural networks in the brain that regulate and form neuroendocrine systems and also includes the study of behaviors and mental states that are influenced or regulated by hormones. In addition, neuroendocrinology encompasses the understanding and study of peripheral physiological systems that are regulated by neuroendocrine mechanisms. While neuroendocrinology embraces many issues of concern to human health and well-being, research in reductionist animal models is usually required to fully understand these issues.

Contemporary research in neuroendocrinology involves a wide range of techniques and technologies, from the subcellular and systems level to the whole-organism level. A particular aim of the Series is to provide expert advice and discussion about experimental or technical protocols in neuroendocrinology research and to further advance the field by giving information and advice about novel techniques, technologies, and interdisciplinary approaches.

To achieve our aims, each book focuses on a particular theme in neuroendocrinology. For each book we recruit editors, who are leaders in their field, to engage an international team of experts to contribute chapters in their individual areas of expertise. The mission of each contributor is to provide an update of current knowledge and recent discoveries and to discuss new approaches, "gold-standard" protocols, translational possibilities, and future prospects. Authors are asked to write for a broad audience, to use references selectively, and to consider the use of video clips and explanatory text boxes; each chapter is peer-reviewed and has a glossary.

The Masterclass Series is open-ended; books in the Series published to date are:

Neurophysiology of Neuroendocrine Neurons (2014, ed. WE Armstrong & JG Tasker)
Neuroendocrinology of Stress (2015, ed. JA Russell & MJ Shipston)
Molecular Neuroendocrinology: From Genome to Physiology (2016, ed. D Murphy & H Gainer)
Computational Neuroendocrinology (2016, ed. DJ MacGregor & G Leng)
Neuroendocrinology of Appetite (2016, ed. SL Dickson & JG Mercer)
The GnRH Neuron and its Control (2018, ed. AE Herbison & TM Plant)
Model Animals in Neuroendocrinology (2019, ed. M Ludwig & G Levkowitz).

The first books of the series published by Springer Nature are:

- *Neurosecretion: Secretory Mechanisms* (2020, ed. JR Lemos & G Dayanithi)
- *Developmental Neuroendocrinology* (2020, ed. S Wray & S Blackshaw)
- *Neuroendocrine Clocks and Calendars* (2020, ed. FJP Ebling & HD Piggins)
- *Glial-Neuronal Signaling in Neuroendocrine Systems* (2021, ed. JG Tasker, JS Bains, & JA Chowen)
- *Neuroanatomy of Neuroendocrine Systems* (2022, ed. V Grinevich & A Dobolyi)
- *Neuroendocrine-Immune System Interactions* (2023, ed. JP Konsman & TM Reyes)
- *Cardiovascular Neuroendocrinology* (2023, ed. T Cunningham & G Yosten)
- *Neuroendocrine Regulation of mammalian Pregnancy and Lactation (2024, ed. PJ Brunton & DR Grattan)*
- *Neuroendocrinology of Behavior and Emotions (2024, ed. HK Caldwell & HE Albers)*

Feedback and suggestions are welcome.
International Neuroendocrine Federation—https://www.inf-neuroendocrinology.org/

Edinburgh, UK
Dunedin, New Zealand

Mike Ludwig
Rebecca Campbell

Preface

The evolution of multicellular animals created the demand for integrated physiological responses across distant populations or tissues, which has been implemented by long distance cell–cell communication through: (1) chemical signals (hormones) released systemically into the extracellular space, where they diffuse over large distances (endocrine system) or (2) coupled electro-chemical signaling of specialized cells (neurons) that have the capacity of generating electrical signals that travel along cell body projections (axons and dendrites) that establish direct intercellular contacts (synapses) with other neurons, where chemical signals (neurotransmitters) are used for short range communication, hence forming a network of coupled cells (nervous system). These two intercellular communication systems offer complementary ways of coordinating multicellular organisms since endocrine communication involves a slower propagation of the signal that is broadcast across the whole organism affecting all cells that express receptors for a given hormone when it is released, whereas neuronal communication conveys faster signaling but only across synaptically connected cells. Interestingly, some neurons, known as neurosecretory or neuroendocrine cells, also produce hormones and have a dual capacity of cell–cell signaling, and together with their target tissues form the neuroendocrine system. In evolutionary terms endocrine communication predates neuronal communication and is present in all animals without nervous systems (e.g., sponges and placozoa), and neurosecretion is observed in all animals that possess a nervous system (e.g., cnidarians). In bilaterians neurosecretion tends to be organized in axes that include central populations of neurosecretory cells in the brain (e.g., hypothalamus in vertebrates) that control the activity of endocrine glands spatially close to the brain (e.g., anterior pituitary in vertebrates and corpora cardiaca in insects), which in turn regulate peripheral glands or organs. The ubiquity of neuroendocrine systems in metazoans results from the need to integrate the endocrine and the nervous system in the regulation of homeostatic mechanisms and behavior.

This volume presents the current state of knowledge in comparative and evolutionary neuroendocrinology that aims help identify general principles and variations in the evolution of neuroendocrine signaling. For this purpose, the book is organized in three parts.

Part I addresses general principles of neuroendocrine signaling and its evolution. Chapter 1 by Krabichler and Grinevich sets the stage for the whole book by laying down the evolution of neuropeptide signaling from autocrine communication in unicellular organisms to paracrine, endocrine, and neuroendocrine cell–cell communication in metazoans. Chapter 2 by Elphick focuses on the evolution of neuropeptides of the hypothalamic-pituitary axis and their ancestral role in invertebrates. Chapter 3 examines volume transmission in the brain. Rodriguez Cortes, Chachlaki, and Prevot discuss the role volume transmission as a complementary mode of cell–cell communication to wired synaptic transmission in the nervous system. Finally, Chap. 4 by Santiago-Andres, Fiordelisio, del-Rio, and Cubillos describes the appearance of the pituitary gland and the formation of the hypothalamus-pituitary system, which regulates a wide range of physiological processes to maintain homeostasis, as a key event in vertebrate evolution.

Part II expands the theme of the evolution of the hypothalamic-pituitary system by addressing the multiple hypothalamic-pituitary-endocrine gland axes in vertebrates. Chapter 5 by Herbison traces the evolution of GnRH neurons from early chordates and examines their organization and function among vertebrates. The next two chapters are focused on the hypothalamic-pituitary-thyroid system. Chapter 6 by Singru, Basu, and Singh examines the evolution of thyrotropin-releasing hormone (TRH) neurons across invertebrates and vertebrates and the role they play in metabolism regulation, whereas Chap. 7 by Lazcano, Joseph-Bravo, and Orozco provides a comparative analysis of the hypothalamic-pituitary-thyroid axis across vertebrates. Chapter 8 moves on to the analysis of hypothalamic-pituitary growth regulation. Vélez and Unniappan provide a comparative review of growth-hormone-releasing hormone (GHRH) regulatory role of growth and other functions in vertebrates. Chapter 9 by Leal, Muriach, and Cerdá-Reverter examines the evolution of the melanocortin system and discusses alternative hypotheses that have been proposed to explain the diversity of the various melanocortin partners. Chapter 10 discusses the evolution of the corticotropin-releasing hormone (CRH). Deussing shows how its signaling system has been evolutionarily co-opted for the regulation of multiple physiological processes and how it became restricted to the control of the hypothalamic-pituitary-adrenal (HPA) axis in mammals. Chapter 11 by Zhang, Hernandez, and Eiden ends Part II with a discussion on the co-evolution of PACAP and VIP and their receptors.

Part III theme is the evolution of neurohypophyseal peptides in vertebrates and their physiological and behavioral effects. Chapter 12 by Leng examines the evolution of magnocellular neurons as multifunctional neurosecretory cells whose peptides can act both as endocrine signals in peripheral tissues and as central neuromodulators acting in a paracrine fashion on other neurons. Chapter 13 addresses the evolution of AVP-like peptides and their roles in osmoregulation. Lin, Adem, Murphy, and Greenwood discuss the structural and functional adaptations of the hypothalamus-kidney axis and the mechanisms that enable water-stressed mammals to cope with osmotic challenge. The three final chapters of this section are focused on the evolution of oxytocin and its role in the regulation of social behavior across vertebrates.

Chapter 14 by Sartorius, Theofanopoulou, and Quintana examines the evolutionary history of the oxytocin receptors and provides the tools for phylogenetic analysis. Their results suggest that oxytocin receptor orthologs are present since the origin of vertebrates and have been under positive selection in the marsupial lineage. Chapter 15 by Kareklas, Sorigue, and Oliveira describes the evolutionary history of nonapeptide ligands and their receptors in fish and discusses their role in the broad variation in social behavior observed in teleosts. Chapter 16 by Saverio and Canoine examines the role of oxytocin-like peptides in bird behavior and describes novel methods used to quantify these peptides.

After a golden period of comparative and evolutionary studies in neuroendocrinology, recent decades have observed a decrease in this trend with most studies focused on mammals, mainly rodents, and with a strong emphasis toward questions with potential for biomedical translational research. This collection from experts in the field provides a comprehensive view of the field that unveils the added value of comparative research that we hope will help to stimulate others to develop new ideas and prompt comparative and evolutionary research in neuroendocrinology.

Mannheim, Baden-Württemberg, Germany — Valery Grinevich
Lisbon, Portugal — Rui F. Oliveira

Contents

About the Editors

Valery Grinevich is known for his studies on neuropeptide signaling in the brain. He obtained his MD degree in 1992 at the Kursk State Medical University (USSR/Russia) with specialization in neurology and psychiatry. His PhD thesis was defended in 1996 on a subject of evolutionary and comparative neuroanatomy of oxytocin and vasopressin systems in vertebrates at the Russian Academy of Sciences in Saint-Petersburg (supervisor: Prof. Andrey L. Polenov). Thereafter, as a winner of Russian President Mr. Boris Yeltsin's award, he did his first postdoc with Profs. Fernand Labrie and George Pelletier in Quebec on hypothalamic mechanisms of the stress response. Next, he joined the team of Greti Aguilera at NIH to work on neuro-immuno-endocrine interactions. In 2001, as an awardee of the Alexandre von Humboldt fellowship, he joined Prof. Gustav Jirikowski's group in Jena for two years, focusing on morphological alterations of human hypothalamic neurons afflicted by various pathologies. In 2003 he was appointed as a Full University Professor at the Moscow State Medical University, but resigned this offer, starting his next postdoc training at the Max Plank Institute for Medical Research Heidelberg with the father of modern neuroscience and neuroendocrinology, Prof. Peter H. Seeburg. In 2008, he established his own group in Seeburg's Department, achieving cell-type specific tagging of OT neurons by viral vectors and deciphered a novel mechanism of neuropeptide signaling in the brain via axonal release. In 2012, as a winner of the Chica and Heinz Schaller Research Foundation Award, he became the Head of an independent Group on Neuropeptides allocated at the German Cancer Research Center (DKFZ) and the University of Heidelberg. From May 1st 2019 he has been promoted for a Full University Professor (W3) position at the University of Heidelberg and a Head of Department of Neuropeptide Research in Psychiatry at the Central Institute of Mental Health. From March 2020, he is Affiliate Professor at the Center for Neuroinflammation at the Neuroscience Institute, Georgia State University, Atlanta, USA. In 2022, Valery Grinevich was awarded by the Synergy ERC and the German-Israel Excellence Research Program (DIP) grants, both of which he is coordinating.

Rui F. Oliveira is Full Professor and Chair of the Neuroscience and Behavior at ISPA—University Institute for Psychological Social and Life sciences, and Head of the Laboratory for Integrative Behavioral Biology at the Gulbenkian Institute of Science, in Lisbon, Portugal. He obtained his PhD in Biology in 1996 with a study on the neuroendocrinology of social dominance in cichlid fish, and after a postdoc in the laboratory of Prof. Peter McGregor at the University of Nottingham to study animal communication networks, he became a faculty member at ISPA, where he served as Dean from 2010 to 2020. He is also an active science communicator, having organized a science exhibition about the brain for a general audience ("Brain: wider than the sky," 2019, Calouste Gulbenkian Museum, Lisbon), which attracted over 85,000 visitors, and recorded a Gulbenkian Masterclass video series on the brain. He has also been an active member of the research community, having served in different bodies of several scientific societies (Portuguese Ethological Society, Association for the Study of Animal Behavior, Society for Behavioral Neuroendocrinology), including as President of the Society for Social Neuroscience. His main research interests are focused on the evolutionary neurobiology of social behavior, and his research agenda is centered on an integrative approach to the study of social behavior, which combines the study of proximate causes (gene modules, hormones, neural circuits, cognitive processes) and ultimate effects (evolutionary consequences). In particular, he aims to understand how brain and behavior can be shaped by the social environment and how the cognitive, neural, and genetic mechanisms underlying plasticity in the expression of social behavior have evolved. During his career, he made significant contributions to different areas of behavioral neuroendocrinology, working mainly with teleost fish as study models. First, he focused on the social modulation of androgens and its adaptive function. He then investigated the neuroendocrine mechanisms involved in the evolution of reproductive plasticity in fish, in particular the occurrence of alternative reproductive tactics within males. He then became interested in the vasotocin/oxytocin nonapeptide family as a neuroendocrine toolkit for the evolution of social behavior, and he has been investigating the role of oxytocin on the development and evolution of social behavior in zebrafish and cichlid fishes, and how this system may have been co-opted to regulate also interspecific interactions, using the cleaner wrasse marine mutualism as a model system.

Part I

General Principles of Neuropeptide Signaling in Invertebrates and Vertebrates

Evolution of Neuropeptide Signaling: From a Single Cell to Mammals

1

Quirin Krabichler and Valery Grinevich

Nothing in biology makes sense except in the light of evolution

— Theodosius Dobzhansky (1900–1975), geneticist and evolutionist

Abstract

Neuropeptides are small protein-like molecules produced and released by various cells in the brain, including neurosecretory and other neuronal cell types. They function both as hormones, acting on distant organs, and as neuromodulators or neurotransmitters, affecting cells expressing the respective G protein-coupled receptors within the brain. Since the early days of neuroendocrinology, oxytocin (OT) and vasopressin (AVP) have been the most extensively studied neuropeptides, due to both historical reasons (their early identification, isolation, and chemical synthesis) and their crucial roles in physiology and survival of animals including humans. However, they are just two of the 200–300 different neuropeptides found in the brain of every mammalian species. To better understand any neuropeptide system, it is of great importance to regard its evolutionary history. Even though it seems that the comparative neuroscience approach, which was once so fruitful, has been in decline over the past decades, we argue that it is more important than ever to reconsider it. And indeed, much has recently been learned about how neuropeptide systems evolved. In this chapter, we intend to first put the history of neuropeptide research into an evolutionary perspective. We will argue that it was a

Q. Krabichler (✉) · V. Grinevich (✉)
Department of Neuropeptide Research in Psychiatry, Central Institute of Mental Health, German Center for Mental Health (DZPG), Medical Faculty Mannheim, University of Heidelberg, Mannheim, Germany
e-mail: quirin.krabichler@zi-mannheim.de; valery.grinevich@zi-mannheim.de

V. Grinevich, R. F. Oliveira (eds.), *Evolutionary and Comparative Neuroendocrinology*, Masterclass in Neuroendocrinology 17,
https://doi.org/10.1007/978-3-031-80209-6_1

comparative evolutionary approach which facilitated major breakthroughs in the research of the best studied neuropeptides OT and AVP, and will provide a short summary of the history of OT and AVP research with some focus on the broad spectrum of model organisms that were used in this decade-long endeavor. Afterwards, we will provide an extensive review on what is known about the evolution of neuropeptidergic signaling in general. Finally, we will turn back to AVP and OT and review what is known about the phylogenetic origin and evolution of these important neuropeptides, which form highly complex interactive networks in the central nervous system that govern homeostasis as well as many behaviors essential for evolutionarily driven propagation of species.

Keywords

Evolution · Invertebrates · Vertebrates · Mammals · Neuropeptides · Oxytocin · Vasopressin · Receptors

Abbreviations

AAV	Adeno-associated virus
AN	Accessory magnocellular nuclei
AVP	Arginine vasopressin
AVT	Arginine vasotocin
GABA	γ-Aminobutyric acid
GFP	Green fluorescent protein
HPA	Hypothalamic–pituitary axis
OT	Oxytocin
OXTR	Oxytocin receptor
PON	Preoptic nucleus
PVN	Paraventricular nucleus
SNARE	SNAp (Soluble *N*-ethylmaleimide-Sensitive Factor Attachment proteins) REceptors
SON	Supraoptic nucleus
TAVT	Tunicate vasotocin-like peptide
V1aR	Vasopressin receptor 1a
V1bR	Vasopressin receptor 1b
V2R	Vasopressin receptor 2

1.1 Introduction

1.1.1 Neuropeptides in a Nutshell, and a Reappraisal of Their Discovery

Neuropeptides are small protein-like molecules, which are produced by and released from a variety of cells in the brain, including neurosecretory cells and many other neuronal cell types (Hökfelt et al. 2000; Burbach 2011). They function as hormones by acting on distant organs, and at the same time as neuromodulators or neurotransmitters by acting on cells expressing the respective G protein-coupled receptors within the brain (Burbach 2011). Their wide variety of functions in the brain and body make them indispensable for the proper physiological functioning and thus survival of animals (Leng and Russell 2019; Grinevich and Ludwig 2021). It is currently known that there are many hundreds of different neuropeptides—for example, every mammalian species has approximately 200–300 of different neuropeptides in the brain (Kim et al. 2011; Wang et al. 2015; Russo 2017). However, two of them, arginine vasopressin (AVP) and oxytocin (OT), have always stood out—albeit especially due to historical reasons, because they were the first to be identified, isolated, and chemically synthetized. The history of their discovery is very interesting in the scope of this chapter, as it is replete with comparative neuroscience hinting at the evolutionary conservation of neuropeptides. It is always worth to look back to reconsider how the past giants of neuroscience managed to make their discoveries. Therefore, before delving into our advanced knowledge on neuropeptide evolution in general, as well as AVP/OT evolution in particular, we shall first dare a brief reappraisal of the early history of neuropeptide research with a special focus on the comparative angle.

1.1.2 The History of Neuropeptide Research: A Comparative Endeavor

The end of the nineteenth and beginning of the twentieth century saw a true scientific revolution in the field of physiology, which had only relatively recently been unofficially founded thanks to Claude Bernard (1813–1878), one of the "fathers of physiology." He had developed the theory that the internal vital conditions, the *milieu de l'intérieur* of the body, were more crucial for survival than the external physical conditions, and—importantly—that the body somehow acted to *maintain* the optimal interior milieu (Holmes 1986). This notion is of course known today as *homeostasis*, but back then it was a novel idea which laid the foundation for the subsequent scientific revolution in "regulatory physiology."

Seeking to answer the question of how distant parts of the body were able to communicate with one another to "regulate" the interior milieu, physiologists focused their attention on the pituitary gland: In 1895, Oliver and Schäfer reported that hypophyseal extracts from sheep, when intravenously injected into dogs, increase blood pressure—the *pressor* effect (Oliver and Schäfer 1895). The same was true when using bovine pituitary extract and injecting it into cats, mice, or rats (Schäfer and Vincent 1899), or even into frogs (Herring 1904). Soon, bovine posterior pituitary extracts were also found to generate a strong uterine-contracting—*oxytocic*—effect in cats (Dale 1906), a lactation-inducing effect on mammary glands of goats (Ott and Scott 1910) and cats (Schäfer and Mackenzie 1911), and an inhibitory—*antidiuretic*—effect on urine excretion in the kidneys of humans (Von den Velden 1913).

However, the nature of the chemical compounds underlying these effects were still elusive, and it would take 15 more years to purify and clearly distinguish at

Box 1.1: First Historical Remark

It is now well-established that AVP is always antidiuretic. A second earlier name of this peptide was "antidiuretic hormone (ADH)" due to its capacity to decrease urine excretion. However, scientists at the beginning of the twentieth century frequently reported *diuretic* actions of partially purified "pressor" fractions of posterior pituitary extracts injected intravenously into experimental animals (Kamm et al. 1928). When the separation of AVP from OT fractions was improved, it started to become clear that (1) the diuresis was not due to AVP, and (2) the water excretion was accompanied by a relatively even stronger NaCl excretion, which produced an overall increase in blood osmolarity (Kuschinsky and Bundschuh 1939). This so-called **natriuresis**—strong sodium chloride excretion in the kidney—was ultimately proven to be **mediated by OT** action in the kidney (Verbalis et al. 1991; see also Chap. 13 in this book).

least two separate compounds with either pressor and diuretic-antidiuretic activities or oxytocic activities, and for the first time assign them their names under which they are known today: *vasopressin* and *oxytocin* (Kamm et al. 1928). After a further two decades of laborious research by many chemists (see, e.g., du Vigneaud 1935), Vincent du Vigneaud and his team succeeded in chemically identifying and synthesizing the two nonapeptides (i.e., consisting of nine amino acids) OT and AVP as originally isolated from bovine pituitaries (du Vigneaud et al. 1954a, b), for which he received the Nobel Prize in Chemistry in 1955 (du Vigneaud 1956). In parallel, Roger Acher and his team in Paris identified the structure of bovine AVP (Acher and Chauvet 1953, 1954), which had an identical structure.

Importantly, these discoveries were made in a variety of mammalian species, and often by cross-reacting pituitary extract from one species in another. Early on, this indicated the presence of evolutionarily conserved mechanisms in the physiology of the pituitary hormones. Another important question in addition to the structure and identity of the "pituitary hormones" AVP and OT was the site of their biosynthesis. In the late nineteenth and early twentieth century that was generally assumed to be the pituitary itself. This paradigm was finally shifted by pivotal research done in fish; In 1928, Ernst Scharrer postulated based on his histological material that teleost fish possessed "specific vacuolized nerve cells of the hypothalamic preoptic nuclei", which possessed "endocrine activity related to pituitary functions" (Scharrer 1928). This led to the birth of the concept of "neurosecretion" (Scharrer and Scharrer 1937). Also the important question of how the active substances reached the pituitary was first resolved in teleost fish, where a direct axonal projection from magnocellular neurons in the hypothalamic supraoptic nucleus (SON) to the posterior pituitary (neurohypophysis) was shown, which contained a granular substance representing AVP and OT (Palay 1945). This was later confirmed to also be the case in mammals (Bargmann 1949) and other vertebrates such as snakes (Scharrer 1951), in which, in addition to the SON, a separate paraventricular nucleus (PVN) was also found to project to the posterior pituitary.

Until the 1970s, only the *hormonal* functions of pituitary extracts on, e.g., the vascular system or distant organs such as the kidney were appreciated. The idea that hormones might act *on the brain itself* was mostly unheard of. This changed when David de Wied, who first coined the term "*neuropeptide*" in 1971 (cited in: Klavdieva 1995), demonstrated behavioral effects of AVP via central nervous system action in rats (de Wied 1969, 1971; Lande et al. 1971; de Wied 1973; reviewed in Klavdieva 1996). Soon afterwards, neuropeptide-transporting axonal projections from magnocellular PVN and SON neurons throughout the brain were shown (Buijs 1978; Buijs et al. 1978, 1980; de Vries and Buijs 1983), which possessed true synapses in their target regions (Buijs and Swaab 1979). This indicated that behavioral phenotypes observed after AVP and OT infusions into different brain areas (see the early review by De Wied 1980) were resembling natural functions via central projections (Buijs et al. 1980; De Wied 1980; de Vries and Buijs 1983). These discoveries were followed by immunohistochemical studies in a variety of species across the vertebrate phylum. Various different homologous forms of the mammalian OT and AVP were discovered (in particular isotocin/vasotocin in fish, and mesotocin/vasotocin in amphibians, reptiles, and birds), and the distribution of these magnocellular neuroendocrine cells in hypothalamus, plus sometimes also their pituitary-targeting and extrahypothalamic axonal projections were studied (for details, see final part of this chapter).

Box 1.2: Second Historical Remark

For the major part of the twentieth century, only the hormonal functions of OT and AVP were recognized. Only after David de Wied published a set of experimental studies, in which he clearly demonstrated behavioral effects of AVP action directly in the brain, the dual nature of neuropeptides as both hormones and neurotransmitters became widely accepted. However, this development could have been vastly accelerated if scientists had been aware of the solitary voice of Austrian-US-American physiologist Wilhelm Raab (1895–1970); in a series of elegant experiments in dogs (Raab 1926a, b, 1928, 1930), Raab demonstrated already in the 1920s that intracranial injections of very small doses of pituitary extract into the hypothalamic third ventricle had a profound effect on decreasing blood fat levels. Importantly, Raab demonstrated by means of various control experiments that the observed effects were not mediated by hormonal action through the blood, but had to occur via central nervous action. For example, subcutaneous injections of the extract did not produce the effect. Furthermore, the blood-fat-decreasing effect upon intracranial infusion of the pituitary extract into the third ventricle was abolished by inactivating various intracerebral pathways, for example by destruction or pharmacological silencing of the ventral hypothalamus, by transection of the cervical spinal cord, or by sectioning of the sympathetic nervous pathways innervating the guts. Thus, Wilhelm Raab was probably the first scientist in history to recognize the neurotransmitter function of pituitary hormones. Unfortunately, however, even though Wilhelm Raab was a distinguished scientist and medical doctor (Surawicz 1997), this particular aspect of his work was never recognized, and thus its implications were overlooked.

1.1.3 The Genetic Revolution in Neuroscience and Neuroendocrinology

The "genetic revolution" of the late 1980s and early 1990s led to a quantum leap in our understanding of the neurobiology of neuropeptide systems (Murphy et al. 2012). With respect to OT and AVP, this was spear-headed by the successful genetic cloning of the AVP receptors Avpr1a (=V1aR), Avpr2 (=V2R), and Avpr1b (=V1bR) as well as the single OT receptor OXTR (Birnbaumer et al. 1992; Kimura et al. 1992; Lolait et al. 1992; Morel et al. 1992; de Keyzer et al. 1994; Sugimoto et al. 1994; Thibonnier et al. 1994; Rozen et al. 1995). Knowing the genetic code of the receptors, it became possible to study the distribution of their expression in the brain by means of the mRNA in situ hybridization technique in different species (Young 1999; Freeman and Young 2016; Powell et al. 2023). Interestingly, whereas the receptors are phylogenetically conserved (Hoyle 1999), their expression patterns in the brain are not; even closely related species, e.g. from the same genus, can have vast differences in the expression levels of AVP and OT receptors in particular brain areas, which strongly correlate with different behavioral lifestyles (Insel and Shapiro 1992; Young 1999; Beery et al. 2008; Campbell et al. 2009; Albers 2015; Jurek and Neumann 2018; Wigger et al. 2004).

Identification of the genes involved enabled the development of new tools including viral vectors for gene transfer to study neuropeptide systems at the genetic level, such as the artificial overexpression of neuropeptide genes in specific brain regions (Murphy et al. 2012). For example, in an elegant study the group of Larry J. Young were probing whether artificially changing the expression levels of Avp1a receptors in a particular brain region in one species to resemble the expression levels in another species with different behavior, could be sufficient to induce the adoption of a similar behavior as the latter's (Lim et al. 2004). To tackle this question, they used two vole species as models: The prairie vole (*Microtus ochrogaster*) is very social and preferentially lives monogamously with only one sexual partner, whereas its closely related cousin, the meadow vole (*M. pennsylvanicus*), is asocial and sexually promiscuous as was discovered by Sue Carter and Lowell L. Getz after his 25 years of field work (Carter and Getz 1993). Importantly, the meadow vole has considerably lower V1aR expression levels in ventral pallidum of the forebrain compared to the prairie vole. When Larry Young and his team used viral vectors to artificially overexpress V1aR in the meadow vole ventral pallidum, they found increased partner preference behavior in thusly treated animals, resembling the monogamous social behavior of prairie voles. We can conclude from these and many analogous findings (Johnson and Young 2015) that the distribution of receptor expression in neuropeptide systems is highly plastic, possibly providing a neurobiological means for fast evolutionary adaptation of behavior and physiology.

The growing possibilities to generate transgenic animals led to major breakthroughs early on. For example, the high degree of phylogenetic conservation of essential genes and gene regulatory elements of neuropeptide systems became evident in an impressive experiment: Scientists cloned a 46 kb segment of the teleost fish *Fugu rubripes* genome, which covered the isotocin and vasotocin gene locus (the fish equivalents of OT and AVP, see also further below) and introduced it into the rat genome (Venkatesh et al. 1997). They found that in these transgenic rats, the fish isotocin was specifically expressed in OT neurons. Furthermore, isotocin expression level was upregulated in response to the same physiological changes that also increased OT expression. While the initial study did not find expression of the fish AVP equivalent vasotocin in their transgenic rats, a subsequent study showed that fish vasotocin was also expressed under control of AVP promoter and gene regulatory elements in transgenic mice when additional downstream sequences to the vasotocin gene were included (Gilligan et al. 2003). Taken together, these studies demonstrated that despite the long evolutionary divergence between Fugu and rat lineages of over 400 million years, the regulatory mechanisms of the isotocin and OT genes as well of the vasotocin and AVP genes have been preserved (Murphy et al. 1998).

The advent of adeno-associated virus (AAV) vectors started another small genetic revolution on its own; in the 2010s, it became possible to use AAVs with specific promoters to exclusively target and manipulate specific neuropeptide-expressing neurons. This facilitated the study of their fine-detailed neuroanatomy as well as the behavioral functions of particular circuits by means of activity manipulations such as optogenetics. One major breakthrough in this respect was achieved in 2012 with the generation of a novel AAV vector containing an evolutionary-conserved fragment of the mouse OT promoter (Knobloch et al. 2012). Many different genetic tools can be inserted as transgenes into this vector (e.g., GFP or optogenetic channelrhodopsins), and after injection into the brain will only be expressed in OT neurons with active OT promoter, but not in other cell types such as AVP neurons. Since then, a plethora of publications from research groups around the world has been using these AAVs to study OT circuits in the brain of many different species including rodents and primates (Grinevich et al. 2016; Grinevich and Neumann 2020).

Quite recently, the emergence of single-cell transcriptomics has further revolutionized neuroendocrinology: For example, the use of this technique has revealed that the seemingly homogenous OT/AVP neurons are in fact composed of a highly diverse combination of cell types (Romanov et al. 2017; Romanov and Harkany 2022; Berkhout et al. 2024). These findings might challenge our current understanding of the functional organization of the OT/AVP systems. However, the genetic and developmental control as well as the functions of these different sub-cell-types is not understood, and knowledge is limited to rodent species.

We want to reiterate that many of the early great discoveries, which advanced our understanding of neuropeptide signaling, were made in a variety of different species. As mentioned above, findings of major importance were made in frogs, fish, reptiles, birds, and various mammalian species. In the recent two decades however, an ever-growing focus has been on rodent models, as the early genome sequencing of the mouse led to the generation of advanced genetic tools specific for experiments in rodents. We would like to argue that in order to facilitate future scientific innovations and further expand our in-depth understanding of OT/AVP as well as other neuropeptide systems, refocusing our attention on broader, comparative approaches in line with evolutionary neurobiology will be of paramount significance (Odekunle and Elphick 2020). As famously said by Theodosius Dobzhansky (https://www.britannica.com/biography/Theodosius-Dobzhansky): "Nothing in biology makes sense except in the light of evolution" (Dobzhansky 1964, 1973). Much is already known about the evolution of neuropeptides, but a lot is yet to be understood.

The following major part of this chapter will be dedicated to what is currently known about the question how neuropeptides, these astounding cellular messenger systems, have evolved. We will first provide an overview on the evolutionary origin of peptidergic signaling in general. Subsequently, we will explore what is known about the evolutionary trajectory of OT and AVP as major and highly interesting neuropeptide representatives.

1.2 The Phylogenetic Origins of Neuropeptide Signaling in Metazoa and Beyond

Already very early in the history of neuropeptide research it became evident that neuropeptides in different species must have evolved from common ancestors and maintained a high degree of phylogenetic conservation. For example, the conserved physiological effects of *mammalian* pituitary extract on the organs of other, non-mammalian vertebrates, for example frogs (Herring 1904) and birds (Paton and Watson 1912), were experimentally demonstrated in the early twentieth century. This implied that (at least among vertebrates) the then known pituitary hormones had very similarly structured homologs in even distantly related species. Another example of early evidence includes the difficulties to isolate AVP and OT as purified compounds from pituitary extracts implying that they must be chemically and structurally very similar and must have evolved from a single common ancestor neuropeptide.

Today, thanks to the great advances in molecular genetics, including its "children" evolutionary genomics, evolutionary transcriptomics, and evolutionary proteomics, our understanding of neuropeptide evolution is clearer than ever before—though many mysteries still remain. A few of the most basic questions that come to mind are: Why was there any need for neuropeptides instead of just relying on "normal" neurotransmitters such as glutamate, GABA, and acetylcholine? When and how in the phylogeny of animals did neuropeptides first emerge?

As already pointed out, neuropeptides are chemical compounds consisting of short sequences of amino acids, which are released by cells and then sensed by specific cell receptors. Arguably, specificity is the main defining characteristic of neuropeptides, which may have allowed them to be positively selected for in evolution. By means of changes to the peptide sequence or chemistry, going hand in hand with appropriate changes in the receptor sequences, highly specific signaling mechanisms could evolve. These considerations immediately suggest that the emergence of multicellularity must be the key, which necessitated distinct means of cell–cell communication. Indeed, with the evolution of multicellular animals (Metazoa) about 833–650 million years ago (dos Reis et al. 2015), it became a necessity for cells within individual animals to "communicate" with each other over various distances, especially if these cells had specialized functions which needed to be coordinated. Therefore, early representatives of Metazoa, which includes the extant clades Porifera (sponges), Ctenophora (comb jellies), and Parahoxozoa (Placozoa, Cnidaria, and Bilateria; Ryan et al. 2010; Simion et al. 2017; Whelan et al. 2017), must have been among the first animals to evolve mechanisms for paracrine cell–cell communication (see Fig. 1.1b).

When we look at the extant root of Metazoa, the sponges Porifera are probably the most basal representatives alive. Phylogenomically however, the identity of either Porifera or Ctenophora as sister taxa to all other extant animals is still not

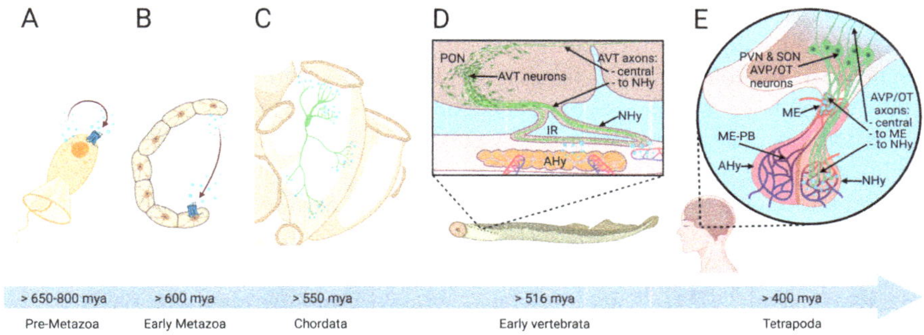

Fig. 1.1 Major evolutionary cornerstones in the phylogeny of neuropeptide signaling. (**a**) In unicellular pre-Metazoa such as choanoflagellates, autocrine peptidergic signaling had evolved between 650 and 800 million years ago (mya). (**b**) In early multicellular metazoan animals, paracrine signaling between individual cells evolved. (**c**) Tunicates, early Chordata, have since approximately 550 mya possessed neuropeptides with some similarity to vertebrate ones, such as a tunicate vasotocin-like (TAVT) peptide. It is produced by specialized neurons in the cerebral ganglion, which project nerves to innervate different body parts, especially the oral and atrial siphons, which have important functions for feeding, excretion, and reproduction. (**d**) Early vertebrates, represented by the Cyclostomata, evolved approx. 516 mya, and were the first to possess "proper" vasotocin. It is produced by specialized neuroendocrine cells in the preoptic nucleus of the hypothalamus, which mostly terminate in the sac-like neurohypophysis (NHy). There, vasotocin and other neuropeptides are released into the surrounding tissue and diffuse to the separate adenohypophysis (AHy) and to a plexus of blood capillaries near the caudal wall of the NHy. (**e**) Tetrapoda evolved approx. 400 mya and developed a highly specialized hypothalamic–pituitary axis (HPA). Their hypothalamus has a median eminence (ME), which is directly connected to the AHy via a portal blood system (ME-PB) that transports neuropeptides and hormones from the hypothalamus to the AHy. In addition, their NHy is highly vascularized with blood capillaries, on which peptidergic axons directly terminate. They possess both AVP and OT. Note that the special HPA system of teleost fish, which instead of a ME-PB possess direct hypothalamic innervation of the AHy, is not touched on here (for more information on this, see, e.g., the excellent review by Trudeau and Somoza 2020). Abbreviations: AHy, adenohypophysis; AVP, arginine vasopressin; AVT, arginine vasotocin; IR, infundibular recess; ME, median eminence; ME-PB, median eminence portal blood system; NHy, neurohypophysis; OT, oxytocin; PON, preoptic nucleus. Phylogenetic divergence times were extracted from the following papers Trudeau and Somoza (2020); Ros-Rocher et al. (2021); Nanglu et al. (2023a, b). Created in https://BioRender.com, publication license JA27TGN3TJ

fully resolved, a discussion known as the "sponge-sister/ctenophore-sister" debate (Steenwyk et al. 2023). Some evidence has been favoring the sponge-sister hypothesis (Simion et al. 2017), though more and more evidence has accumulated in favor of the ctenophore-sister hypothesis (Ryan et al. 2013; Moroz et al. 2014; Schultz et al. 2023). From a neurological point of view, sponges as the outgroup makes more sense. They possess neither neurons nor synapses nor indeed muscles (Steenwyk et al. 2023). Furthermore, they do not possess any homologues to bilaterian neuropeptides (Yañez-Guerra et al. 2022; Najle et al. 2023). However, recent phylogenomic analyses suggest the potential presence of precursor "proneuropeptides" (Musser et al. 2021; Hayakawa et al. 2022; Yañez-Guerra et al. 2022).

In contrast to Porifera, the comb jellies (Ctenophora) evolved a unique nervous system: Many of their neurons form a *syncytial* nerve net devoid of any synapses, though some kinds of chemical synapses appear to exist between the fused nerve net cells and outlying effector cells (Burkhardt et al. 2023). Ctenophora lack many "canonical" neurotransmitters (such as GABA, acetylcholine, dopamine, noradrenaline, adrenaline, octopamine, histamine, or glycine), and their main neurotransmitter is L-Glutamate even for neuromuscular transmission (Moroz et al. 2014). In addition, they appear to lack homologous forms of any of the many dozens of neuropeptides that are ubiquitous among Bilateria, even among distant Bilaterian phyla such as Arthropoda and Chordata (Moroz et al. 2014). In contrast, Ctenophora do possess a whole range of unique neuropeptides not found in any other taxon (Sachkova et al. 2021). Thus, they have evolved a unique combination of neurotransmission mechanisms including their very own neuropeptides, which apparently play an important role even in a syncytial-nerve-net-based nervous system.

Placozoa, the probable sister clade to Cnidaria and Bilateria (though see Laumer et al. 2018), are millimeter-small multicellular animals which have neither muscles nor neurons, but they nevertheless coordinate their own behavior via paracrine signaling by a variety of small neuropeptides (Senatore et al. 2017; Varoqueaux et al. 2018). Single-cell transcriptomics has revealed a high variety of peptidergic cell types in Placozoa (Sebé-Pedrós et al. 2018; Najle et al. 2023). Indeed, approximately two thirds of all their cell types express one or several peptides (Najle et al. 2023). In addition, their peptidergic cells co-express a high number of G protein-coupled receptors in unique combinations, and many of them are autapomorphic to Placozoa with no orthologs in other animal phyla (Najle et al. 2023).

Finally, the phylogenetically more advanced Cnidaria and even the basal Bilateria possess for the first time "true" neurons, as well as clearly identifiable orthologs to the neurotransmitter and neuropeptide systems of advanced Bilateria such as vertebrates (Steger et al. 2022).

Results from the investigation of the non-Cnidarian/non-Bilaterian Metazoa representatives Porifera, Ctenophora, and Placozoa teach us that the evolutionary origins of neuropeptides clearly *predate* the evolution of the nervous system as we know it from extant Bilateria animals (including most animals alive today) (Burkhardt and Jékely 2021). Thus, peptidergic cell–cell communication must have first evolved at least in the last common ancestor of Metazoa, though it must have been a solely paracrine mechanism (Jékely 2021). Accumulating evidence indicates that peptidergic signaling may go back even further, to Premetazoan single-cell life forms. Much of the cellular machinery for neurosecretion in general, especially the molecular elements of functional small synaptic vesicles such as SNAREs (e.g., synaptobrevin) and vesicular neurotransmitter transporters (e.g., the vesicular glutamate and GABA transporters VGLUT and VGAT, and the vesicular acetyl choline transporter VACHT), go back far beyond Metazoa and even their closest extant unicellular cousins Choanoflagellates, to unicellular Holozoa (a clade including many unicellular and all multicellular Eukaryotes except Fungi) or even to Pre-Holozoan origins (Göhde et al. 2021). Similarly, the evolution of neuropeptide signaling is likely to have preceded the evolution of Metazoa. For example, all extant Metazoa (Porifera, Ctenophora, Placozoa, and Cnidaria+Bilateria) share conserved enzymatic machinery for the processing of the precursor peptides forms of neuropeptides (Yañez-Guerra et al. 2022). Thus, even very early single-celled pre-animalian eukaryotes may have first evolved neuropeptides as an autocrine mechanism, potentially as a means for signal amplification or sustained activation (Jékely 2021) (see Fig. 1.1a).

In summary, recent research into the earliest phylogenetic origins of neuropeptide systems of animals has shown that they first evolved in the simplest and single-celled forms of animal or even pre-animalian life, where they probably served various functions of autocrine and later (with the onset of multicellularity) of paracrine cell–cell communication. The early appearance in the history of life shows that neuropeptide systems have always been essential, if not defining, machineries of animal cells. Their nature as peptides, which can be directly produced via transcription and translation from DNA, probably facilitated their (and their receptors') quick adaptive radiation toward numerous different manifestations with specialized physiological functions.

The neuropeptides OT and AVP are of particular interest not only because of their highly important functions in the body and nervous system of many animals, because of which they have become two of the most-studied neuropeptides, but also regarding their evolution. They are chemically and structurally highly similar and differ by only two amino acids, because of which they are often referred to as "sister peptides" (Legros 2001; Stoop et al. 2015; Althammer et al. 2021), but they are expressed in separate only slightly overlapping hypothalamic neuronal populations (Soumier et al. 2022) and have different, in fact often opposite behavioral functions (De Wied 1980; Legros 2001; Neumann and Landgraf 2012; Stoop 2012; Stoop et al. 2015; Neugebauer et al. 2020; Oliveira et al. 2021). These are conveyed by specialized receptors differentially expressed throughout the body and brain (Ostrowski et al. 1994; Ostrowski 1998; Vaccari et al. 1998; Young 1999; Freeman et al. 2014a; Freeman and Young 2016; Rokicki et al. 2022; Powell et al. 2023).

For the next part of this chapter, we will focus on the evolution of the neuropeptides OT and AVP, considering their phylogenetic origins, their evolution across the animal kingdom, the evolution of their neuroanatomy, and evolutionary aspects of their connectivity and behavioral functions.

1.3 Evolution of the Peptides Oxytocin and Vasopressin

1.3.1 The Genetics of OT and AVP

The genes of AVP and OT (and all their related forms) do not only consist of the code for these short amino acid sequences. Rather, they each consist of three protein-coding exons with introns in between. After initial transcription into RNA, the intron parts are spliced out and the resulting mRNA is translated into a protein. AVP and OT have similar exons: The first exon contains a signal peptide domain and the sequence for the neuropeptide itself, the second and third exons together constitute the neurophysin domains (NP-I in case of OT, NP-II in case of AVP) (Ivell and Richter 1984; de Bree and Burbach 1998). Translation produces one coherent, large precursor protein (prepropeptide). Its signal peptide serves for its transfer into the Golgi apparatus, where it is packaged into neurosecretory vesicles for transport along axons or dendrites. Within the vesicles, it is enzymatically cleaved into the neuropeptide plus its separate neurophysin and both are subjected to further post-translational processing by specific enzymes (Altstein and Gainer 1988). The neurophysins appear to act as carrier proteins, though their additional physiological functions remain unclear until today. In the following part, we are focusing on the amino acid structures of the neuropeptides themselves, but not the neurophysins. Genomic analyses of the whole gene loci including neurophysins and introns yield more detailed information on the evolution of these neuropeptide systems (Elphick and Rowe 2009; Wallis 2012; Mirabeau and Joly 2013) (see Chap. 2 in this book for more details).

1.3.2 Pre-bilaterian Origins of OT/AVP-Like Neuropeptides and Their Radiation in Invertebrates

OT/AVP-like neuropeptide signaling emerged over 600 million years ago (Donaldson and Young 2008; Grinevich et al. 2016) and became a ubiquitous and vital machinery of almost all animal taxa which have survived to this day. In Fig. 1.2, we visualize a simplified phylogenetic tree of animals while considering the presence (or absence) and structural differences of OT/AVP-like peptides in extant taxa. When looking at the amino acid sequence of OT/AVP-variants across the animal kingdom, it is immediately striking that some positions are relatively variable, whereas others are strongly conserved (Gruber 2014). Taking into account the average "nonapeptide" (nine amino acids) layout, the landmark of OT/AVP-like neuropeptides are the cysteine amino acids at #1 and #6, which form a disulfide bond and thus create a ring structure essential for the biological activity (du Vigneaud 1956). Positions #2 and #3 are generally conserved to be hydrophobic or aromatic amino acids such as isoleucine (I), phenylalanine (F), and tyrosine (Y), leucine (L) or valine (V). Positions #4 and #5 are generally conserved to be hydrophilic amino acids such as glutamine (Q), asparagine (N), or arginine (R). Position #7 is mostly proline (P) and #9 is almost always glycine (G). Nevertheless, subtle changes in the

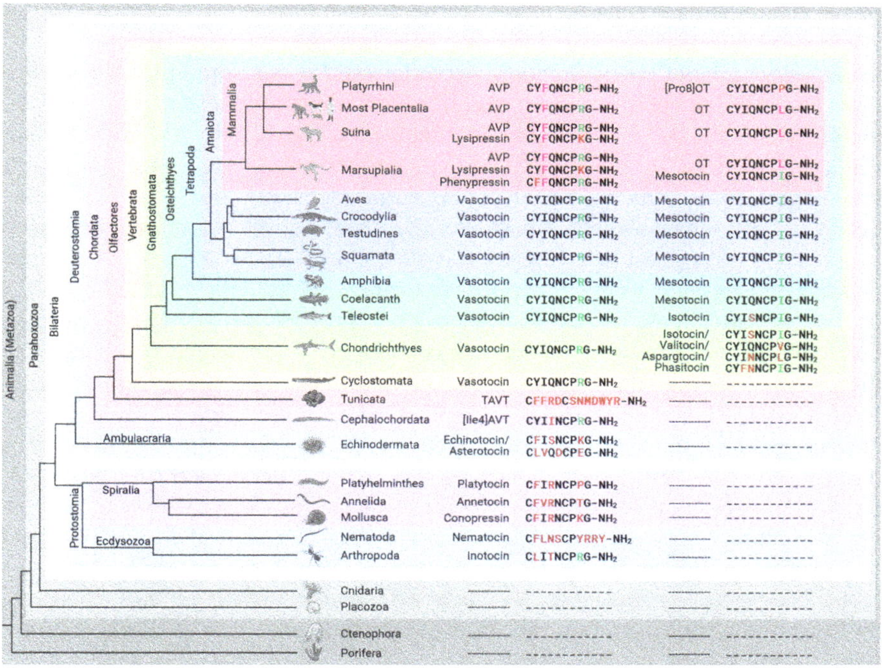

Fig. 1.2 Phylogenetic tree and AVP- and OT-like neuropeptides of animals. Amino acid sequences of representative AVP-like (middle) and OT-like (right) peptides are shown as single letter codes for each extant taxon, with their amidation group $-NH_2$ at the end. Abbreviations: AVP, arginine vasopressin; OT, Oxytocin; TAVT, Tunicate vasotocin-like peptide; [Ile4]AVT, [Ile4]Vasotocin. The Phylogenetic tree was drawn based on Simion et al. (2017); Delsuc et al. (2018); Schultz et al. (2023); Yu et al. (2024) (tree branch lengths are arbitrary). The peptide sequences were extracted from both review articles (Hamann et al. 1992; de Bree and Burbach 1998; Hoyle 1998; Gimpl and Fahrenholz 2001; Takuwa-Kuroda et al. 2003; Hyodo et al. 2004; Donaldson and Young 2008; Gruber 2014; Liutkeviciute et al. 2016; Banerjee et al. 2017; Odekunle et al. 2019; Odekunle and Elphick 2020; Kawada et al. 2021) as well as from primary literature: Platyrrhini (Lee et al. 2011; Ren et al. 2014); Suina (Ferguson 1969); Marsupialia (Chauvet et al. 1980, 1984, 1987; Rouillé et al. 1988; Bathgate et al. 1990; Parry et al. 2000); Coelacanth (Gwee et al. 2008); Chondrichthyes (Acher et al. 1972, 1999; Chauvet et al. 1994; Hyodo et al. 2004; Gwee et al. 2008, 2009); Cyclostomata (Lane et al. 1988; Gwee et al. 2009; Sower 2018); Tunicata (Kawada et al. 2008; Ukena et al. 2008); Cephalochordata (Gwee et al. 2009); Echinodermata (Elphick and Rowe 2009; Semmens et al. 2016; Odekunle et al. 2019); Annelida, Mollusca (Oumi et al. 1996; Hoyle 1998; Gruber 2014; Kobayashi et al. 2022); Platyhelminthes (Kobayashi et al. 2022); Nematoda, Arthropoda (Gruber and Muttenthaler 2012; Gruber 2014; Koto et al. 2019; Fetter-Pruneda et al. 2021; Kobayashi et al. 2022). Created in https://BioRender.com, publication license SR270Q20AH

neuropeptide amino acid sequences can have significant effects on receptor binding and potency, and thus probably reciprocally affected receptor evolution.

The original form of the AVP/OT sister neuropeptides was a "proto-AVP", whereas the OT-like form evolved later in evolution (see below). The last common ancestor of Bilateria must already have possessed such a "proto-AVP" including its associated neurophysin, as variants of it are found throughout extant representatives of the two bilaterian clades *Protostomia* and *Deuterostomia* (Mirabeau and Joly 2013; Odekunle and Elphick 2020) (see Fig. 1.2). Also, a recently defined sister taxon to Proto- and Deuterostomia within Bilateria, Xenacoelomorpha, already features an AVP-like neuropeptide in addition to many other conserved peptide systems (Thiel et al. 2018). Non-bilaterian extant animal phyla however, which include Cnidaria, Placozoa, Porifera, and Ctenophora, do not possess any clear AVP or OT homologue. From which original pre-Bilaterian substrate the "proto-AVP" first evolved is still not resolved and may never be known, as we are missing the necessary links to reconstruct its evolution. It is clear, however, that a variety of different neuropeptide systems were already present in ancestral Bilateria and have remained fixed in extant taxa (Mirabeau and Joly 2013; Thiel et al. 2018). A whole range of vertebrate-homologous neuroendocrine systems are for example found in basal Protostomia, such as the nematode *Caenorhabditis elegans* (Istiban et al. 2024). Interestingly, whole-genome sequencing in a range of phylogenetically basal species has revealed some indications of how AVP/OT may have first evolved: Extant basal Proto- and Deuterostomia such as sea urchins (Echinodermata) have been found to possess families of *non*-AVP-like peptides associated with neurophysin-like domains, which had previously been thought to be exclusive to AVP/OT-like neuropeptides (Elphick and Rowe 2009; Elphick 2010). Thus, various neurophysin-associated peptide systems must have evolved in pre-Bilaterian animals from some proto-neurophysin-associated one. One of those became the "AVP-like" system which is still present in all Bilateria.

In extant Protostomia, which consist of Spiralia and Ecdysozoa, a clear AVP-like neuropeptide is ubiquitous, even though no OT-like neuropeptide is yet present (see Fig. 1.2). In Bilaterians of the Ecdysozoa (Arthropoda, Nematoda, and Onychophora), two AVP-like forms are widely found. Some arthropods, especially insects, possess inotocin (Liutkeviciute et al. 2016). The insect inotocin was first discovered in the locust (Proux et al. 1987). Later, it was also detected in a beetle and in a parasitic wasp (Stafflinger et al. 2008), and recently in various species of ants (Gruber and Muttenthaler 2012; Koto et al. 2019; Fetter-Pruneda et al. 2021). Interestingly, the functions of inotocin have been experimentally elucidated in several ant species: It has clear "antidiuretic"-like action for water balancing via synthesis of anti-desiccation cuticula (Koto et al. 2019), while also partaking in modulation of social foraging behaviors (Fetter-Pruneda et al. 2021). Thus, the single neuropeptide inotocin appears to simultaneously mediate AVP- as well as OT-like functions. Curiously, some eusocial hymenopteran species, such as the honey bee, appear to be able to live entirely without inotocin, as their genome does not contain genes encoding this neuropeptide (Gruber and Muttenthaler 2012). The protostomian cousins of arthropods, nematodes (roundworms) possess a slightly different AVP-like peptide, nematocin (Beets et al. 2012; Garrison et al. 2012; Beets et al. 2023). It is involved in reproductive behaviors (Garrison et al. 2012) as well as salt-tasting behavior (Beets et al. 2012), once again combining the two functional domains of OT and AVP in vertebrates.

In Protostomia of the Spiralia (Platyhelminthes, Annelida, and Mollusca), various slightly differing AVP-like peptides exist (see Fig. 1.2). In flat worms (Platyhelminthes) a gene for an AVP-like peptide was recently identified in a marine planarian and named Platytocin (Kobayashi et al. 2022), after other attempts in non-marine flat worms had been unsuccessful (Lockard et al. 2017). In their cousin group Annelida, which is considered to be evolutionarily more advanced, the OT/AVP-like peptide Annetocin has long been known to exist (Hoyle 1998, 1999). Annetocin signaling in Annelida (similar to OT in vertebrates) has a role in the neural control of reproductive behaviors, especially in egg-laying, which was demonstrated in earthworms as well as in leeches (Oumi et al. 1996). In flat worms, Platytocin seems to have similar functions, as it also induces egg-laying (Kobayashi et al. 2022). Interestingly however, it has also "anti-diuretic hormone-like" functions in these marine planarians, which live in environments of changing salinity. When treated with platytocin, planarians survived significantly longer in hyperosmic salt water, and endogenous platytocin expression increased with medium osmolality (Kobayashi et al. 2022). Thus, once again the single peptide nematocin can have both homeostatic and reproductive functions, combining OT and AVP roles. In Mollusca, conopressin is present as an AVP-like peptide in gastropod snails (Cruz et al. 1987; Martinez-Padrón et al. 1992). Early on, it was shown in a hermaphrodite snail to be involved in both male and female reproductive behaviors including ejaculation and egg production/laying (van Kesteren et al. 1995a, b). Curiously, in cephalopods such as the octopus, a double-peptide system of cephalotocin and octopressin exists (Reich 1992; Takuwa-Kuroda et al. 2003). Even more interestingly, octopressin appears to convey reproductive functions (Takuwa-Kuroda et al. 2003), whilst cephalotocin appears to be involved in osmoregulation (Sakamoto et al. 2015). Thus, cephalotocin/octopressin may represent a fascinating case of evolutionary convergence, as they may have independently evolved "division of labor" between reproductive and homeostatic functions, similar as the vertebrate pair OT/AVP.

Deuterostomia consist of Ambulacraria, including Echinodermata, and Chordata, including Cephalochordata, Tunicata, and Vertebrata (reviewed in: Donaldson and Young 2008; Gruber 2014; Odekunle and Elphick 2020). The invertebrate Deuterostomia are interesting targets for studying the evolution of neuropeptide systems, as they bridge the gap between vertebrates and well-studied Protostomia (such as *C. elegans* and some insects). Echinodermata such as sea urchins and starfish have been shown to possess several dozens of neuropeptide precursors including the AVP homologues echinotocin and asterotocin (Elphick and Rowe 2009; Rowe and Elphick 2012; Semmens et al. 2016; Odekunle et al. 2019) (see Fig. 1.2). In starfish, asterotocin as well as its receptor is expressed in a variety of different organs and body parts, and asterotocin axons are often in close apposition to receptor expressing cells. For example, the receptor is expressed in the mucosa of the stomach, which is innervated by asterotocin processes. In vivo experiments with asterotocin injection into the body have suggested a critical involvement in feeding behaviors (Odekunle et al. 2019) (see Fig. 1.3).

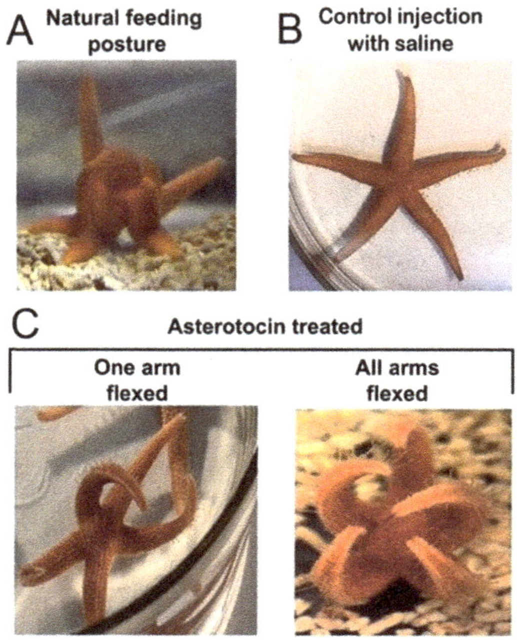

Fig. 1.3 Asterotocin, an AVP/OT homologue in starfish (Echinodermata), induced a feeding-like body posture when injected into a starfish. (**a**) Natural feeding posture of starfish with flexed arms. (**b**) Control injections with vehicle did not change arm posture. (**c**) Asterotocin injections produced flexed arm postures resembling the natural feeding posture. Adapted from Odekunle et al. (2019), originally published under CC-BY 4.0

1.3.3 AVP-Like Neuropeptides in Basal Chordates and Primitive Vertebrates

The "non-vertebrate" Chordata is represented by only two extant clades, namely Tunicata (also known as Urochordata) and Cephalochordata (also known as lancelets or amphioxi). Vertebrates and Tunicates are monophyletic and together form the taxon Olfactores, whereas Lancelets are their sister group (Delsuc et al. 2018). The paraphyletic Cephalochordata and Tunicata both do not yet possess a brain resembling the vertebrate one; accordingly, they do not possess any distinguishable anterior pituitary, posterior pituitary, or hypothalamus structures. Interestingly, however, both of them already possess a variety of vertebrate-like "anterior pituitary" hormones (such as a growth-hormone-like protein and kisspeptin) and "hypophysiotropic hypothalamic" hormones (such as gonadotropin-releasing hormone (GnRH)-like peptides) (reviewed in: Trudeau and Somoza 2020).

Regarding AVP/OT-like neuropeptides in cephalochordates, the genome sequence of lancelets suggests the presence of a vasotocin-like precursor ([Ile4] arginine vasotocin), whose derived neuropeptide structure has been predicted (see Fig. 1.2), and also the presence of two receptors for this neuropeptide (Gwee et al. 2009; Elphick 2010; Mirabeau and Joly 2013), However, experimental data on pharmacological properties and physiological or behavioral functions are still missing (Odekunle and Elphick 2020).

Similar to cephalochordates, Tunicata do not possess a vertebrate-like brain, but instead a cerebral ganglion and a neural gland. In these, "tunicate arginine vasotocin-like" neuropeptides (TAVT) have been identified and characterized in several species (Kawada et al. 2008; Ukena et al. 2008) (see Figs. 1.1c and 1.2). Their gene loci have a similar general composition as in vertebrates and encode a precursor peptide, which is reminiscent of other OT/AVP family precursors and consists of three segments: a signal peptide, a vasotocin-like sequence flanked by a C-terminal amidation signal (for posttranslational enzymatic processing to enhance the peptide's stability and/or receptor-binding potency), and a neurophysin domain. However, in contrast to most OT/AVP-like peptides, the TAVT consist of 14 amino acids instead of 9 (Ukena et al. 2008). Recently, transgenic tunicates (the Ascidian *Ciona intestinalis*) were generated, which either express the GFP variant Venus driven by the TAVT promoter (see Fig. 1.4), or have a knock-out of TAVT (Kawada et al. 2021). Interestingly, the TAVT-knock-out *Ciona* showed a reduction in body length compared to the wildtype, suggesting a role for the tunicate TAVT in growth. Moreover, all TAVT neurons were found to reside in the cerebral ganglion of the adult tunicate, from where several nerves emerged. These were shown to innervate different body parts, especially the oral and atrial siphons, which have important functions for feeding, excretion, and reproduction. Along these lines, another recent study showed that TAVT is involved in ovulation and oocyte maturation in tunicates (Matsubara et al. 2019).

The phylogenetically most primitive extant vertebrates, Cyclostomata ("round mouths"; also called Agnatha, for "jawless"), which include hagfish (Myxini) and lampreys (Petromyzontiformes), already possess a brain with a similar *Bauplan* as the tetrapod brain. In contrast to basal chordates Tunicata and Cephalochordata (see above), they already possess a hypothalamus and are the first group to have evolved a true pituitary (Trudeau and Somoza 2020). Their pituitary, however, is still primitive compared with "higher" vertebrates; it consists of a sac-like neurohypophysis bulging out from the hypothalamus and containing a ventricular "infundibular recess," and separate from it a homogenous (no distinguishable *pars distalis* and *pars intermedia*) glandular adenohypophysis embedded in connective tissue (Tsuneki and Gorbman 1975; Nozaki and Gorbman 1983). Even though there is neither a direct innervation of the adenohypophysis (Goossens et al. 1977) as in teleost fish, nor a median eminence and associated portal blood system as in tetrapods to convey hypothalamo-pituitary communication (Tsuneki and Gorbman 1975; Sower 2018), Cyclostomata do possess a functional hypothalamic–pituitary axis (HPA) (see reviews by Sower 2018; Trudeau and Somoza 2020; and also Chap. 4 in this book). However, the neuropeptides released into the neurohypophysis sac reach the adenohypophysis by means of passive diffusion. In the same way, they reach blood capillaries which form a vascular plexus close to the posterior wall of the neurohypophysis sac (Nozaki 2013) (see Fig. 1.1d).

Cyclostomata do not possess any OT-like peptide (Goossens et al. 1977), but they have a form of vasotocin (Lane et al. 1988) (see Fig. 1.2), which, like AVP in advanced vertebrates, is produced by magnocellular neurosecretory cells of the hypothalamus. Most of vasotocin cells reside in a homologue of the SON, the

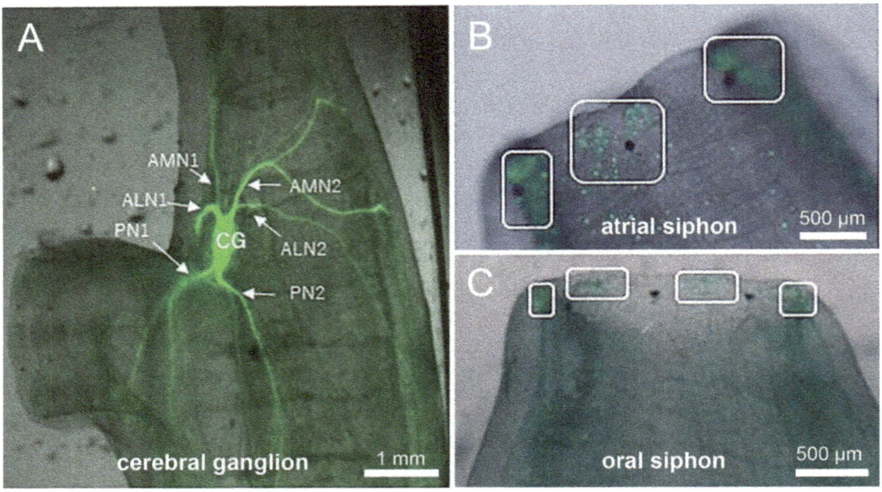

Fig. 1.4 Transgenic tunicate (*Ciona intestinalis*) with GFP expression driven by the endogenous TAVT promoter. (**a**) All green-fluorescent TAVT neurons were concentrated in the cerebral ganglion (CG), from where six fluorescent nerves emerge (AMN, anterior medial nerves; ALN, anterior lateral nerves; PN, posterior nerves), which innervate the body. (**b**, **c**) TAVT axons terminate in various organs of the tunicate, such as the oral and the atrial siphon. Modified from Kawada et al. (2021); CC BY license

preoptic nucleus (PON; see Figs. 1.1d and 1.5a) (Goossens et al. 1977; Lamanna et al. 2023), but a number of them appear to also be found in more caudal parts of the hypothalamus (Nozaki and Gorbman 1983). In the PON, the lamprey possesses numerous vasotocin neurons lining the third ventricle and often squeezing in between ependymal cells, similar to other fish, amphibian, and reptiles (see Fig. 1.5). In addition to their connection to the third ventricle, these cells also project axons to the homologue of the posterior pituitary (the neurohypophysis; see Fig. 1.1d) (Hoheisel et al. 1978; Nozaki and Gorbman 1983). In contrast to advanced vertebrates, only relatively few extrahypothalamic vasotocin fibers were found in the brain of hagfish (Nozaki and Gorbman 1983).

Importantly, in contrast to other vertebrates, Cyclostomata do not yet possess any OT-like peptide (Goossens et al. 1977; Nozaki and Gorbman 1983; Lane et al. 1988; Lamanna et al. 2023) as they miss the entire OT gene locus. Indeed, the genome of lampreys and other cyclostomata, such as hagfish, is in many ways quite different from their sister group of jawed vertebrates, Gnathostomata. For example, the sea lamprey (*Petromyzon marinus*) possesses 99 chromosome pairs in its germline (Smith et al. 2018). However, the absence of the OT gene locus is not due to some secondary loss as it is just as absent in the ancestors of cyclostomata (Smith et al. 2018; Lamanna et al. 2023).

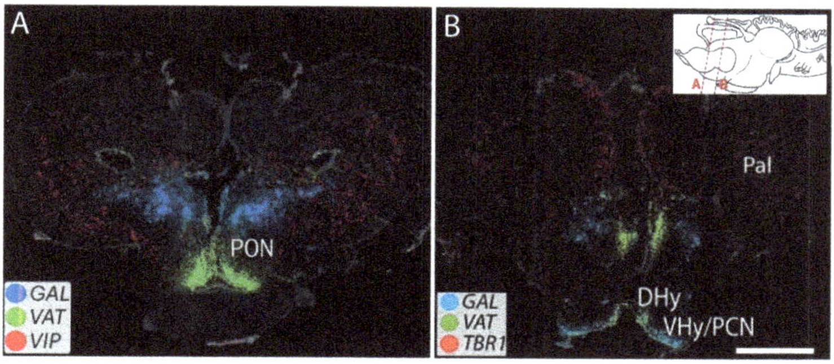

Fig. 1.5 Vasotocin gene expression (green) in the adult lamprey brain, shown by single-molecule RNA fluorescence in situ hybridization (smRNA-FISH) in two coronal sections (levels indicated in inset). (**a**) Dense vasotocin (here abbreviated as VAT)-expressing cells are found in the ventral hypothalamus just dorsal to the optic chiasm. (**b**) Further vasotocin cells are found in more caudal parts. Apart from vasotocin, three other markers are also shown (VIP, GAL, and TBR1). Abbreviations: DHy, dorsal hypothalamus; GAL, galanin; Pal, pallium; PCN, postoptic commissure nucleus; PON, preoptic nucleus; TBR1, T-box brain protein 1; VAT, vasotocin; VHy, ventral hypothalamus; VIP, vasoactive intestinal peptide. Scale bar, 500 µm. Modified from Lamanna et al. (2023), originally published under CC-BY 4.0

1.3.4 The Emergence of OT in Vertebrates as a Result of Whole-Genome Duplications

When and how in evolution did OT-like peptides emerge? As mentioned, jawless vertebrates (Cyclostomata) do not yet possess any such form. It is now well-established that a OT-like peptide first evolved in the last common ancestor of all jawed vertebrates or Gnathostomata, which include all cartilaginous and bony fish, the latter of which not only contain teleost fish, but also all tetrapods including mammals. Most likely, the dual AVP and OT-like peptides evolved in a common ancestor of jawed vertebrates through tandem duplication of the ancestral vasotocin gene (Gwee et al. 2009; Theofanopoulou et al. 2021; Sartorius et al. 2024). Furthermore, at least two whole-genome duplication events occurred during the evolution of early vertebrates (Kuraku et al. 2023), one before the Cyclostomata/Gnathostomata split and another one in the early Gnathostomata line (Yu et al. 2024). Many original vertebrate innovations (e.g., the basic brain layout of hindbrain, midbrain, and forebrain) originated in the ancestor of extant vertebrates, as they are already present in cyclostomes such as the Lamprey (Lamanna et al. 2023). As Cyclostomata do not yet have dual nonapeptides, the duplication of vasotocin into vasotocin and oxytocin-like peptides must have happened during a whole-genome duplication event in the ancestral line of Gnathostomata. Subsequently, the double versions genetically diverged and got fixed as independent neuropeptides systems, which evolved to serve different functions.

Chondrichthyes (e.g., sharks and dogfish) are the most basal extant vertebrates to possess the "double-nonapeptides" corresponding to OT and AVP, and both nonapeptide archetypes are ubiquitous in extant Chondrichthyes (reviewed by

Kuraku et al. 2023). However, their OT variants experienced frequent amino acid changes during their phylogenetic evolution, leading to a high structural variability in extant taxa (Acher et al. 1972; Chauvet et al. 1994; Acher et al. 1999; Hyodo et al. 2004). Importantly, the chondrichthyan vasotocin retained its classical functions in, e.g., osmoregulation, whereas their OT-like peptides do not react to osmotic challenges with water of different salinities, indicating other roles (Hyodo et al. 2004).

By the time bony fish (Osteichthyes) emerged, the structure of the OT-like isotocin had been fixed and remained relatively stable throughout the phylogenetic evolution of tetrapods, with only one amino acid change between the isotocin of ray-finned fish (Actinopterygii) and mesotocin of tetrapods (though of course more in the neurophysin genes). When looking at the brain of teleost fish such as the trout, vasotocin and isotocin neurons are found in the PON of the hypothalamus, whose several subdivisions have been suggested to be homologous to both PVN and SON of vertebrates (Herget et al. 2014). Their neuroanatomical layout is thus already quite similar as in tetrapods, in that their vasotocin and isotocin neurons project to the pituitary as well as to central regions such as the telencephalon and thalamus, as was shown in carp (Saito et al. 2004) (see Fig. 1.6) and zebra fish (Akinrinade et al. 2023). An important difference to tetrapods consists in the lack in teleost fish of a median eminence and associated portal blood system, which in tetrapods conveys endocrine communication between hypothalamus and adenohypophysis (=anterior pituitary) (see review by Trudeau and Somoza 2020) (see also Fig. 1.1e). Interestingly, vasotocin in teleost fish appears to have an important role in female reproductive biology, as its knock-out causes reduced oocyte numbers and changes in oocyte maturation, female courtship behavior, and overall reproductive success in zebra fish (Ramachandran et al. 2023).

Fig. 1.6 Intermingled isotocin (red) and vasotocin (green) neurons in the PON of a bony fish, the rainbow trout (sagittal section). A single isotocin neuron was intracellularly filled (black) and its arborizing axonal projections found to innervate pituitary (Pit), but also extrahypothalamic ventral thalamic (Thv) and forebrain regions (Tel). Adapted from Saito et al. (2004), with permission

1.3.5 The Evolution of the OT/AVP Systems in Tetrapods, from Anamniotes to Early Amniotes

In anamniote tetrapods, represented by amphibians, vasotocin and mesotocin magnocellular neurons are still found in a single hypothalamic structure, the PON, similar to fish (see Fig. 1.7). They are tightly connected to the lumen of the third ventricle and also project to both pituitary and extrahypothalamic areas including spinal cord (Conway and Gainer 1987; Jokura and Urano 1987). Furthermore, amphibians are the most basal animals to have the true "*tetrapod Bauplan*" of the HPA that includes a median eminence and associated portal blood system (see Fig. 1.1e).

When comparing the basal tetrapods amphibians with basal amniotes, e.g. Squamata, an interesting feature strikes the eye: Squamata, such as geckos (Stoll and Voorn 1985; Thepen et al. 1987) and snakes (Silveira et al. 2002), possess for the first time two major segregated magnocellular nuclei containing the OT/AVP-like neurons, the PVN and the SON (Knobloch and Grinevich 2014). In addition, in

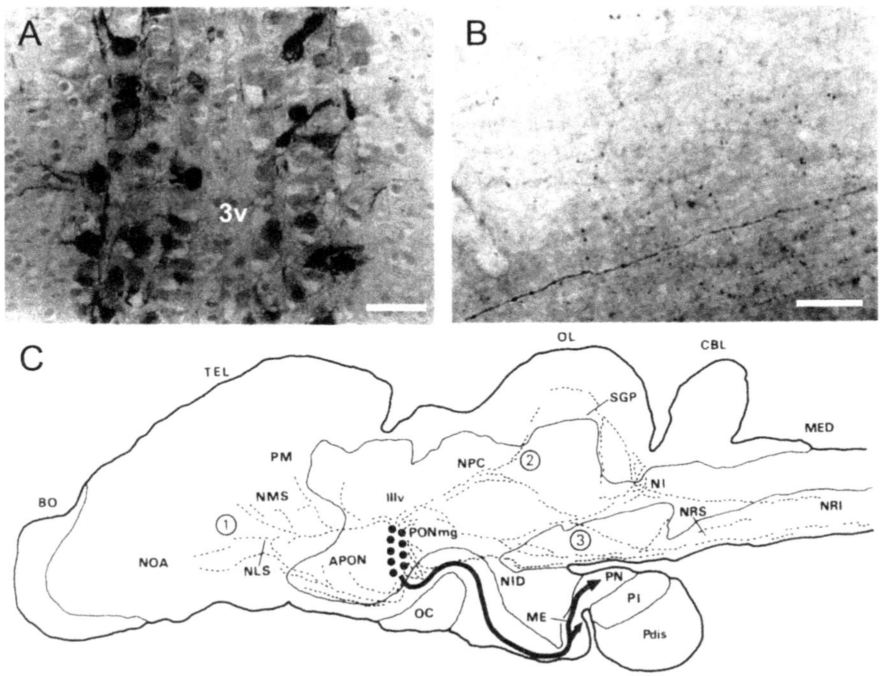

Fig. 1.7 Vasotocin system in an amphibian, the toad *Bufo japonicus* (filled circles). (**a**) Vasotocin-immunoreactive magnocellular neurons in the PON in close proximity to the third ventricle (3v). (**b**) Vasotocin axon in the spinal cord, indicating the brainwide extrahypothalamic outputs. (**c**) Schematic diagram, illustrating the distribution of magnocellular vasotocin-immunoreactive perikarya (dots) and their local and distant projections (broken lines), vasotocin fibers are assembled into three projection groups: (1) projection to the telencephalon; (2) to the thalamus and the tectum; and (3) to the brain stem. Adapted from Jokura and Urano (1987)

most amniotes, a smaller number of magnocellular OT/AVP-like peptide neurons also resides in so-called accessory nuclei (AN) outside the main nuclei of PVN/SON (Rhodes et al. 1981; Grinevich and Polenov 1994). It is not clear why amniotes evolved two main hypothalamic centers of magnocellular AVP/OT neurons, PVN and SON, plus several smaller AN. This layout appears quite different from the single isotocin-containing PON in cyclostomata, fish, amphibia, and fish. The polycentric OT system may constitute an epiphenomenon as developmental gene expression studies suggest that the more dorsal magnocellular portion of the fish PON is homologous to the amniote PVN, while its more posterior and ventral parvocellular portion could correspond to the amniote SON (Herget et al. 2014). This suggests that the ancestral PON of pre-amniotes may have compartmentalized into two main structures, which develop from one embryonic anlage due to cell migration. Indeed, developmental studies suggest that in the mouse the cells of PVN/SON form together in the supraoptic/paraventricular (spv) area at around embryonic day E.12 and subsequently migrate to their final destinations (Wang and Lufkin 2000; Karim and Sloper 1980; Acampora et al. 1999). In this respect, the AN may represent magnocellular OT neurons which got "stuck" along their migration paths during development. Interestingly, in marsupial mammals, such as Echidna and Platypus, OT neurons form rather dense aggregations between the SON and PVN, which contain similar numbers of neurons as the main nuclei (Ashwell et al. 2006). This may reflect a momentary picture of migrating OT neurons during ontogenesis having been preserved in adulthood. Alternatively, it could also reflect adult neurogenesis of OT neurons, a possibility which cannot be excluded even in advanced mammalian species as pigs during lactation (Rankin et al. 2003), or in case of the sister peptide AVP, in rats (Zhang et al. 2024).

Regarding early amniota, such as the gecko, it is furthermore interesting to note that mesotocin staining and extrahypothalamic projections are still quite weak compared with vasotocin staining and projections (Thepen et al. 1987) (see Fig. 1.8). This may suggest that when their ancestors diverged from the lines leading to "advanced" amniotes, mesotocin had not yet evolved its major behavioral functions and vasotocin itself was still mediating most behavioral modulations.

1.3.6 The Progressive Behavioral "Empowerment" of the Oxytocin-Like Nonapeptides in Advanced Amniotes

In more advanced amniotes, such as birds, the OT-homologue mesotocin is strongly expressed in the brain and forms a network of numerous extrahypothalamic axonal projections throughout the brain. In chicken (Robinzon et al. 1988; Kamkrathok et al. 2017) and more recently in song birds (Haakenson et al. 2022), the presence of elaborate mesotocinergic extrahypothalamic axonal projections has been reported, as well as the presence of mesotocin neurons in several extrahypothalamic regions. In the zebra finch and the canary, numerous axonal mesotocin terminals were found in many parts of the brain, including periaqueductal gray, ventral

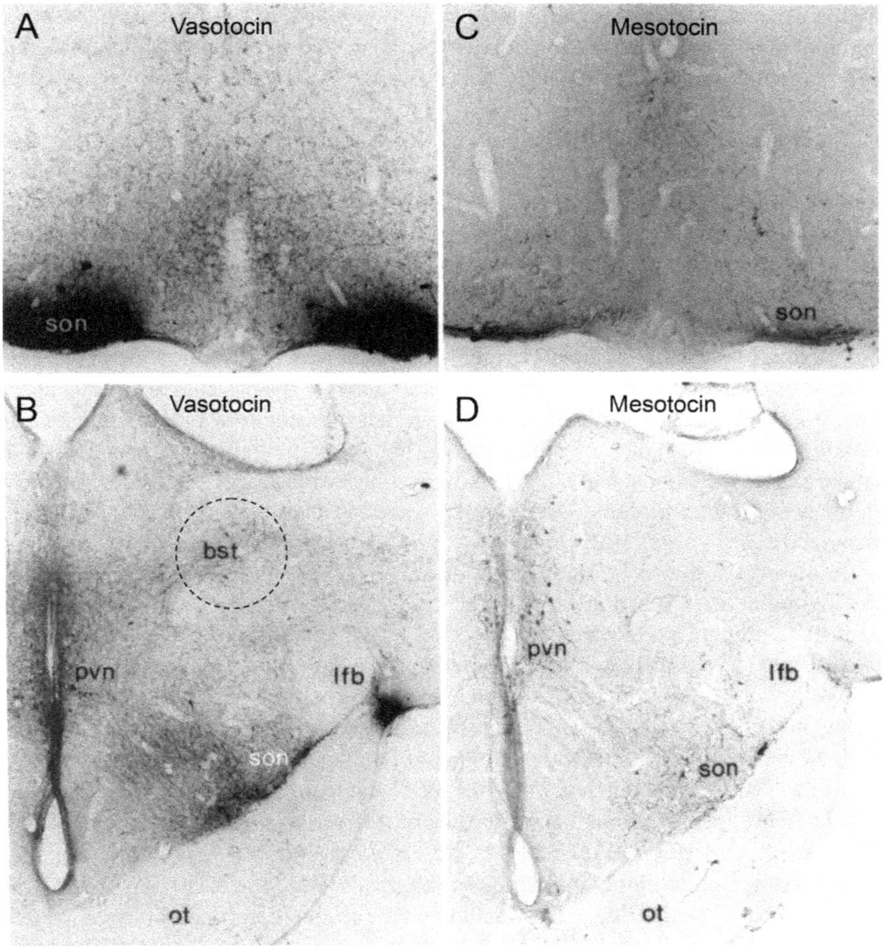

Fig. 1.8 Vasotocin and mesotocin in the hypothalamus of a basal amniote, the gecko. (**a**, **b**) The majority of vasotocin immunoreactive neurons are located in the SON (**a**) and PVN (**b**), while sparse neurons are detected outside the main nuclei with preferential location in the region of the bed nuclei of stria terminalis (bst, encircled); (**b**). Furthermore, dense and strongly labeled vasotocinergic projecting axons can be seen (**a**, **b**). (**c**, **d**) Mesotocin staining revealed comparatively few magnocellular neurons in the SON (**c**) and PVN (**d**). Also, labeled mesotocinergic projecting axons were scarcer (**d**). Abbreviations not previously defined: lfb, lateral forebrain bundle; ot, optic tract. No scalebars were available. Adapted from Thepen et al. (1987)

tegmental area, hypothalamus, lateral septum, and the avian cortical song control nucleus HVC ("higher vocal center"). In addition, mesotocin-expressing neurons were found throughout the mesopallium, an important cortical area involved in song control as well as in multisensory processing (Haakenson et al. 2022). Curiously, numerous mesotocin neurons were even found in the cerebellum. This indicates quite a distinct organization of the mesotocin system in birds compared with reptiles and mammals, characterized by a very broad distribution of mesotocin peptidergic cells in many parts of the brain as well as wide-spread distribution of their axons through the brain (Fig. 1.9).

Such anatomical peculiarities in the organization of the avian mesotocin system may reflect multifaceted roles of this neuropeptide in reproduction and other behaviors of birds. Indeed, a number of pharmacological studies with intraventricular, intranasal, or intracerebral applications of mesotocin or its antagonist have shown that it facilitates distinct aspects of bird social behaviors (Goodson et al. 2009, 2012; Duque et al. 2018, 2020). For example, intracerebral infusion of mesotocin in gregarious zebra finches significantly increased the time the birds spent together with large groups and familiar social partners (Goodson et al. 2009). Interestingly, however, in contrast to OT in mammals (cf. paragraphs on the prairie vole in this chapter), it has not yet been conclusively shown that mesotocin in birds is involved in initiation or maintenance of monogamous behavior (though some anecdotal evidence exists, see Goodson et al. 2012), even though monogamy is a frequent feature in many species of birds but very rarely evolved in mammals (Lukas and Clutton-Brock 2013).

In mammals, the AVP and OT peptides evolved very early in their phylogenetic tree by amino acid substitutions from the pre-mammalian mesotocin and vasotocin forms (see Fig. 1.2). However, the ancestral mesotocin form of OT has been maintained in extant basal mammals, the marsupials (Bathgate et al. 1990; Parry et al. 2000). In some cases, mutations of the OT gene occurred in advanced mammals, producing a novel form of OT; several species of new world monkeys possess an OT with a one amino acid substitution compared to other primates and rodents, [P8]OT (Lee et al. 2011). The same holds true for AVP, which mutated several times in certain side branches of the mammalian evolutionary tree, for example, once again, in new world monkeys (Ren et al. 2014). Nevertheless, the OT and AVP peptides have remained quite highly conserved in mammals, which in fact holds true for all exons of the AVP and OT gene loci, including the carrier proteins neurophysin I and II.

At the anatomical level, the general layout of hypothalamic OT/AVP nuclei as well as their efferent extrahypothalamic projections are evolutionarily conserved in all classes of vertebrates (Knobloch, Grinevich, 2014). In mammals, similar extrahypothalamic projections from OT and AVP neurons were reported in various rodents (Ross et al. 2009; Knobloch et al. 2012; Li et al. 2024), carnivores (Herbeck et al. 2018), and primates including humans (Rogers et al. 2018; Lefevre et al. 2024). Thus, axonal neuropeptide release seems to be highly conserved and the most efficient pathway to reach neurons expressing receptors of respective neuropeptides and in turn to modulate brain-region specific behaviors (Knobloch et al. 2012; Eliava et al. 2016; Hung et al. 2017; Hasan et al. 2019; Ino et al. 2022; Iwasaki et al. 2023; Liu et al. 2023).

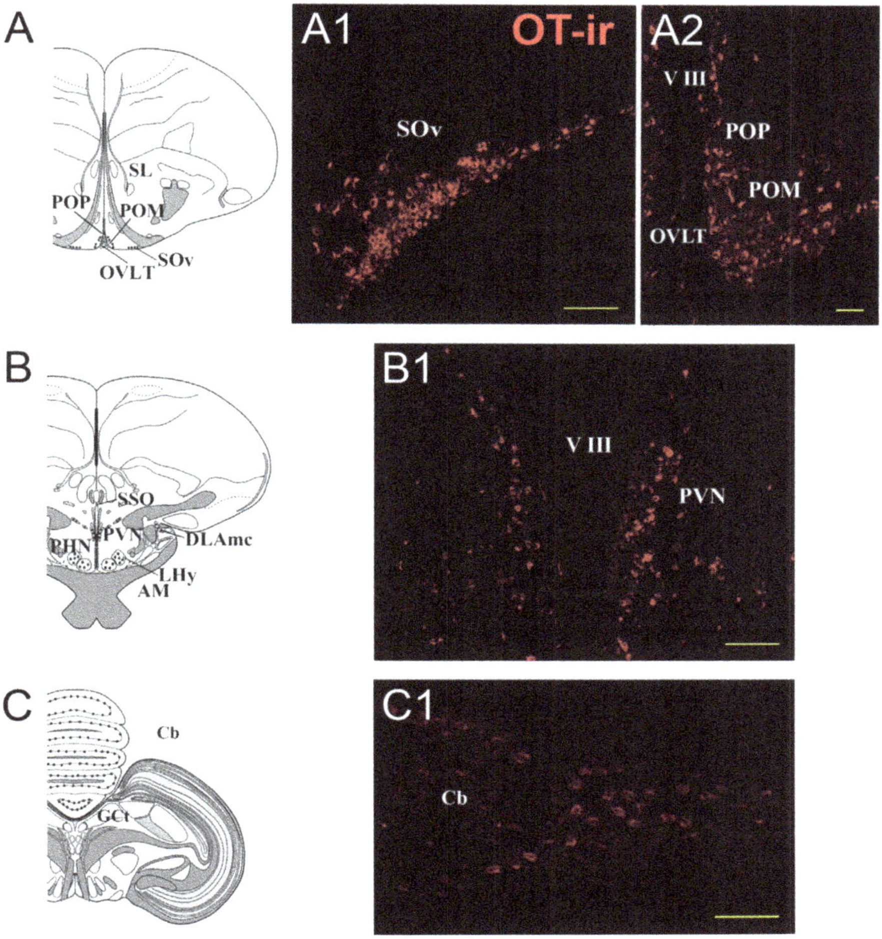

Fig. 1.9 The mesotocin system in the brain of the Thai chicken (*Gallus domesticus*). (**a**–**c**) Schematic diagrams show the distribution of mesotocin-immunoreactive cell bodies throughout the chicken brain. (**a1, b1, c1**) Examples of stained mesotocin neurons distributed in the preoptic and supraoptic area (**a1**), the hypothalamus (**b, b1**), and cerebellum (**c1**). Abbreviations not previously defined: AM, nucleus anterior medialis hypothalami; Cb, cerebellum; DLAmc, nucleus dorsolateralis anterior thalami, pars lateralis; GCt, substantia grisea centralis; LHy, regio lateralis hypothalami (lateral hypothalamic area); OVLT, organum vasculosum laminae terminalis; PHN, nucleus periventricularis hypothalami; POM, nucleus preopticus medialis; POP, nucleus preopticus periventricularis; PVN, nucleus paraventricularis magnocellularis; PVO, organum paraventriculare; ROT, nucleus rotundus; SCNm, nucleus suprachiasmaticus, pars medialis; SL, nucleus septalis lateralis; SOv, nucleus supraopticus, pars ventralis; SSO, organum subseptale (subseptal organ). Scalebars: **a1, b1, c1**, 100 μm; **a2**, 200 μm. Adapted from Kamkrathok et al. (2017), with permission

1.3.7 OT and AVP Receptors in Mammals

It remains enigmatic how the mammalian OT/AVP neurons and extrahypothalamic projections can, on the one hand, be so highly conserved, and, on the other hand, modulate such an enormous variety of different behaviors. One answer to this question may lie in the receptors. In contrast to the rather similar distribution of OT and AVP axonal projections across mammals (see above), OT and AVP receptors can have very distinct expression patterns in different species. For example, in highly "olfactory" mammals, such as rodents, OTR expression predominates in olfactory-related areas, while in "more visual" mammals, such as marmosets, OTR is more densely expressed in visual brain centers than in olfactory areas (Schorscher-Petcu et al. 2009; Freeman and Young 2016). For more details, see (Grinevich et al. 2016) and Fig. 1.10.

OTR expression pattern may differ even in closely related species exhibiting very similar social strategies. For example, dense OTR binding in the nucleus accumbens (the key brain structure for processing rewarding and reinforcing stimuli) has been reported in common marmosets (Schorscher-Petcu et al. 2009), but not in titi monkeys (Freeman and Young 2016), although both species are highly social and monogamous. Such cases are telling of the complex laws governing the evolution of OTR receptors. In contrast to common marmosets, titi monkeys for example have a high expression of AVPR1a in the nucleus accumbens and inexplicably dense OTR expression in the hippocampal dentate gyrus (Freeman et al. 2014b; Freeman and Young 2016). It is still unknown how these behaviorally similar species evolved such different receptor distributions for these sister peptides and what the functional implications are at the neural network level. Thus, it would be important to study whether blocking OTRs or AVPR1a in the nucleus accumbens prevents pair-bond formation, as observed in prairie voles (Young et al. 2001), which would help to clarify whether species-dependent peptidergic mechanisms of monogamy convergently evolved in various mammalian branches, such as voles, beavers, wolves, and marmosets (Carter and Perkeybile 2018).

Similar findings regarding high variability of neuropeptide receptors in closely related species exist with respect to the AVP receptors (Hammock and Levitt 2013). For example, monogamous prairie voles display numerous V1aR receptors in the ventral pallidum, unlike their non-monogamous relatives, the meadow voles (Young 1999; Lim et al. 2004; Lim and Young 2006). Introducing V1aR into the ventral pallidum of meadow voles enables male voles to form partner preferences (Lim et al. 2004). This suggests that both species-specific genetic differences and early life experiences can influence OT/AVP receptor distribution in the brain, thereby affecting how these receptors modulate behaviors in adulthood (Hammock 2015).

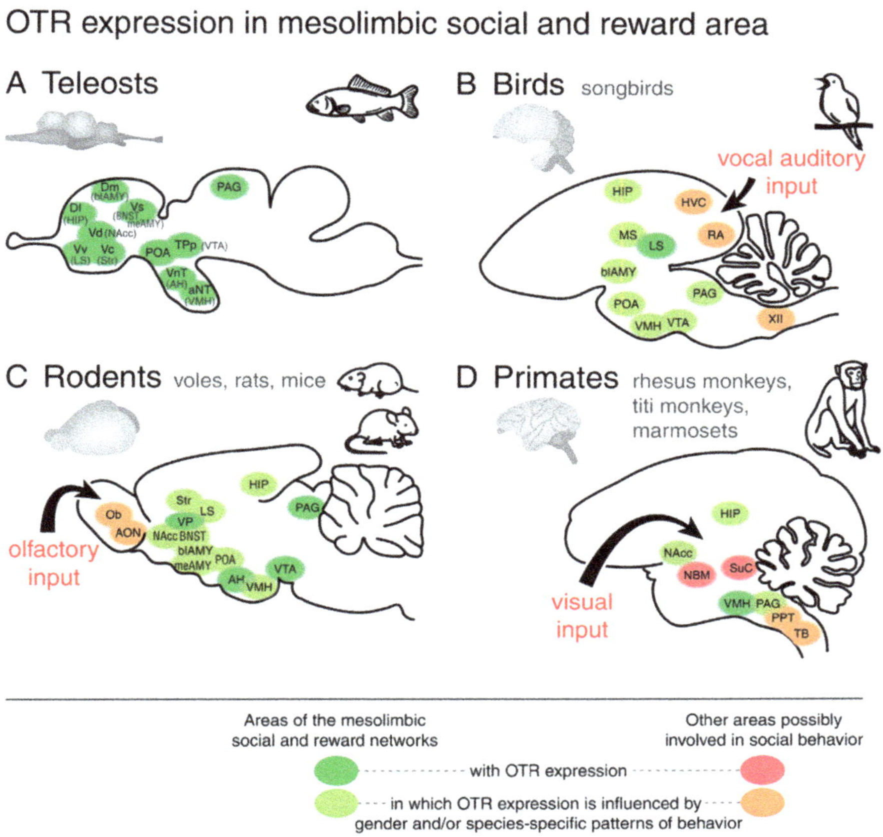

Fig. 1.10 Expression of OTRs within the mesolimbic and reward brain networks of four classes of vertebrates: teleosts, songbirds, rodents, and primates. Areas of the mesolimbic social decision-making network include nucleus accumbens (NAcc), ventral pallidum (VP), lateral septum (LS), preoptic area (POA), ventromedial nucleus of hypothalamus (VMH), basolateral amygdala (blAMY), medial amygdala (meAMY), bed nucleus of the stria terminalis (BNST), anterior hypothalamus (AH), hippocampus (HIP), ventral tegmental area (VTA), periaqueductal gray (PAG), and striatum (Str). Teleost areas of the network (**a**) include anterior tuberal nucleus (aNT), lateral pallial (dorsal) part of the telencephalon (Dl), medial pallial (dorsal) part of the telencephalon (Dm), periventricular nucleus of the posterior tuberculum (TPp), central part of subpallial (ventral) telencephalon (Vc), dorsal nucleus of subpallial (ventral) telencephalon (Vd), ventral tuberal nucleus (VnT), supracommissural nucleus of subpallial (ventral) telencephalon (Vs), and ventral part of subpallial (ventral) telencephalon (Vv). Other areas involved in social behavior depicted here include: medial septum (MS), high vocal center (HVC), robust nucleus of the arcopallium (RA), and motor nucleus of the XII cranial nerve (XII) in panel (**b**); olfactory bulb (Ob) and anterior olfactory nucleus (AON) in panel (**c**); nucleus basalis of Meynert (NBM), superior colliculus (SuC), pedunculopontine tegmental nucleus (PPT), and trapezoid body (TB) in panel (**d**). Reproduced from Grinevich et al. (2016), with permission

1.4 Conclusions

In our chapter, we have crystalized key literature on the evolution of peptidergic signaling from the genetic and anatomical to the behavioral level. We have primarily focused on the best studied neuropeptide systems of OT and AVP, which can be considered model systems for studying the evolution of the functional organization of neuropeptide systems, analogous to the other 200 neuropeptides in the brain. We have also illustrated progressive evolution of general peptidergic mechanisms from unicellular pre-Metazoan species, via primitive metazoan pre-Bilaterians, to highly specialized peptidergic systems integrated into the CNS of vertebrates (Fig. 1.1). The latter allowed peptides to elaborate their functions far beyond neuroendocrine effects on reproduction and metabolism, namely, to coordinate complex forms of adaptive behaviors in vertebrates, especially in mammals. The deeper our understanding, the closer we may get to possibilities for targeted circuit-specific therapies based on neuropeptides in different representatives of the phylogenetic tree. New techniques for manipulation and monitoring activity of peptidergic circuits are rapidly developing for laboratory rodents (Wu et al. 2022; Ino et al. 2022; Althammer 2023; Qian et al. 2023; Iwasaki et al. 2023), and many of these might be adapted for non-canonical species of reptiles, birds, and non-rodent mammals. Nature has granted us a glimpse into the vast richness of neuropeptide signaling, which has been produced during the course of hundreds of millions of years of evolution. It may be time to return to the cultivation of the comparative approach of evolutionary neuroscience to learn more comprehensively about mechanisms of neuropeptidergic control of both classic neuroendocrine functions and a wide range of complex behaviors, ranging from pain and fear to empathy and social cooperation.

Acknowledgments The authors would like to dedicate this chapter to the memory of Larry J. Young (1967–2024), who among other scientific achievements, greatly contributed to evolutionary and comparative neuroendocrinology, especially to peptidergic mechanisms underlying pair-bonding and monogamy. This work was supported by the Synergy European Research Council (ERC) grant "OxytocINspace" 101071777 and German-Israeli Cooperation Program (DIP) GR3619-1 to VG.

Key Literature

Fetter-Pruneda et al. (2021): This paper reported that an OT/AVP-related neuropeptide modulates social foraging behavior in ants, challenging the dogma that these socially organized insects lack the expression of OT/AVP homologues.

du Vigneaud (1956): This paper was based on a speech by Vincent du Vigneaud, the discoverer of the chemical structure and synthesis of OT and AVP, upon receiving the Nobel Prize for Chemistry in 1955.

Sartorius et al. (2024): Based on complex analyses of OT pathway genes, the authors reported that the majority are "modern," from around the emergence of vertebrates (540–530 million years ago), while a minority are "ancient," dating back to the evolution of cellular organisms (3500–1100 mya).

Theofanopoulou et al. (2021): The authors performed extensive genomic analyses on 35 vertebrate species and confirmed the origin of OT and AVP/vasotocin as well as their receptors from gene or whole-genome duplication events. Based on their data, they proposed a universal nomenclature for OT and vasotocin ligands and receptor families.

Yañez-Guerra et al. (2022): The authors identified two neuropeptide precursors with broad evolutionary conservation, being present in bilaterians, cnidarians, ctenophores, sponges, and unicellular (non-metazoan) choanoflagellates. Thus, secreted neuropeptide homologs that derive from longer precursors have pre-metazoan origins and evolved even before the appearance of neurons.

Knobloch et al. (2012): In this paper, the authors achieved cell type-specific targeting of OT neurons in rats using an AAV equipped with evolutionarily conserved 2.6 kb sequence of the mouse OT gene promoter.

Göhde et al. (2021): This paper found that most of the core neurosecretory vesicle proteome was already present in unicellular organisms, underlining the ancestral molecular machinery of neurosecretory vesicles, synapses, and neurons.

Lamanna et al. (2023): This paper established a spatially resolved cell type atlas of the entire brain of the sea lamprey (a basal jawless vertebrate) based on extensive single-cell RNA-seq and in situ sequencing data. This revealed shared neural features between the brains of jawless and jawed vertebrates, as well as many which only later evolved in jawed vertebrates.

References

Acampora D, Postiglione MP, Avantaggiato V et al (1999) Progressive impairment of developing neuroendocrine cell lineages in the hypothalamus of mice lacking the Orthopedia gene. Genes Dev 13:2787–2800

Acher R, Chauvet J (1953) La structure de la vasopressine de boeuf. Biochim Biophys Acta 12:487–488. https://doi.org/10.1016/0006-3002(53)90173-5

Acher R, Chauvet J (1954) La structure de la vasopressine de boeuf. Biochim Biophys Acta 14:421–429. https://doi.org/10.1016/0006-3002(54)90202-4

Acher R, Chauvet J, Chauvet M-T (1972) Phylogeny of the neurohypophysial hormones: two new active peptides isolated from a cartilaginous fish, *Squalus acanthias*. Eur J Biochem 29:12–19. https://doi.org/10.1111/j.1432-1033.1972.tb01951.x

Acher R, Chauvet J, Chauvet M-T, Rouille Y (1999) Unique evolution of neurohypophysial hormones in cartilaginous fishes: Possible implications for urea-based osmoregulation. J Exp Zool 284:475–484. https://doi.org/10.1002/(SICI)1097-010X(19991001)284:5<475::AID-JEZ2>3.0.CO;2-9

Akinrinade I, Kareklas K, Teles MC et al (2023) Evolutionarily conserved role of oxytocin in social fear contagion in zebrafish. Science 379:1232–1237. https://doi.org/10.1126/science.abq5158

Albers HE (2015) Species, sex and individual differences in the vasotocin/vasopressin system: Relationship to neurochemical signaling in the social behaviour neural network. Front Neuroendocrinol 36:49–71. https://doi.org/10.1016/j.yfrne.2014.07.001

Althammer F (2023) Heralding a new era of oxytocinergic research: New tools, new problems? J Neuroendocrinol 35:e13333. https://doi.org/10.1111/jne.13333

Althammer F, Eliava M, Grinevich V (2021) Chapter 3—Central and peripheral release of oxytocin: Relevance of neuroendocrine and neurotransmitter actions for physiology and behaviour. In: Swaab DF, Kreier F, Lucassen PJ et al (eds) Handbook of clinical neurology. Elsevier, pp 25–44

Altstein M, Gainer H (1988) Differential biosynthesis and posttranslational processing of vasopressin and oxytocin in rat brain during embryonic and postnatal development. J Neurosci 8:3967–3977. https://doi.org/10.1523/JNEUROSCI.08-11-03967.1988

Ashwell KWS, Lajevardi S, Cheng G, Paxinos G (2006) The hypothalamic supraoptic and paraventricular nuclei of the echidna and platypus. Brain Behav Evol 68:197–217. https://doi.org/10.1159/000094358

Banerjee P, Joy KP, Chaube R (2017) Structural and functional diversity of nonapeptide hormones from an evolutionary perspective: A review. Gen Comp Endocrinol 241:4–23. https://doi.org/10.1016/j.ygcen.2016.04.025

Bargmann W (1949) Über die neurosekretorische Verknüpfung von Hypothalamus und Hypophyse. Klin Wochenschr 27:617–622. https://doi.org/10.1007/BF01484668

Bathgate RAD, Sernia C, Gemmell RT (1990) Mesotocin in the brain and plasma of an Australian marsupial, the brushtail possum (*Trichosurus vulpecula*). Neuropeptides 16:121–127. https://doi.org/10.1016/0143-4179(90)90123-G

Beery AK, Lacey EA, Francis DD (2008) Oxytocin and vasopressin receptor distributions in a solitary and a social species of tuco-tuco (Ctenomys haigi and Ctenomys sociabilis). J Comp Neurol 507:1847–1859. https://doi.org/10.1002/cne.21638

Beets I, Janssen T, Meelkop E et al (2012) Vasopressin/oxytocin-related signaling regulates gustatory associative learning in C. elegans. Science 338:543–545. https://doi.org/10.1126/science.1226860

Beets I, Zels S, Vandewyer E et al (2023) System-wide mapping of peptide-GPCR interactions in C. elegans. Cell Rep 42. https://doi.org/10.1016/j.celrep.2023.113058

Berkhout JB, Poormoghadam D, Yi C et al (2024) An integrated single-cell RNA-seq atlas of the mouse hypothalamic paraventricular nucleus links transcriptomic and functional types. J Neuroendocrinol 36:e13367. https://doi.org/10.1111/jne.13367

Birnbaumer M, Seibold A, Gilbert S et al (1992) Molecular cloning of the receptor for human antidiuretic hormone. Nature 357:333–335. https://doi.org/10.1038/357333a0

Buijs RM (1978) Intra- and extrahypothalamic vasopressin and oxytocin pathways in the rat: Pathways to the limbic system, medulla oblongata and spinal cord. Cell Tissue Res 192:423–435. https://doi.org/10.1007/BF00212323

Buijs RM, Swaab DF (1979) Immuno-electron microscopical demonstration of vasopressin and oxytocin synapses in the limbic system of the rat. Cell Tissue Res 204:355–365. https://doi.org/10.1007/BF00233648

Buijs RM, Swaab DF, Dogterom J, van Leeuwen FW (1978) Intra- and extrahypothalamic vasopressin and oxytocin pathways in the rat. Cell Tissue Res 186:423–433. https://doi.org/10.1007/BF00224932

Buijs RM, Velis DN, Swaab DF (1980) Ontogeny of vasopressin and oxytocin in the fetal rat: Early vasopressinergic innervation of the fetal brain. Peptides 1:315–324. https://doi.org/10.1016/0196-9781(80)90009-1

Burbach JPH (2011) What are neuropeptides? In: Merighi A (ed) Neuropeptides: methods and protocols. Humana, Totowa, NJ, pp 1–36

Burkhardt P, Jékely G (2021) Evolution of synapses and neurotransmitter systems: The divide-and-conquer model for early neural cell-type evolution. Curr Opin Neurobiol 71:127–138. https://doi.org/10.1016/j.conb.2021.11.002

Burkhardt P, Colgren J, Medhus A et al (2023) Syncytial nerve net in a ctenophore adds insights on the evolution of nervous systems. Science 380:293–297. https://doi.org/10.1126/science.ade5645

Campbell P, Ophir AG, Phelps SM (2009) Central vasopressin and oxytocin receptor distributions in two species of singing mice. J Comp Neurol 516:321–333. https://doi.org/10.1002/cne.22116

Carter CS, Getz LL (1993) Monogamy and the prairie vole. Sci Am 268:100–106. https://doi.org/10.1038/scientificamerican0693-100

Carter CS, Perkeybile AM (2018) The monogamy paradox: what do love and sex have to do with it? Front Ecol Evol 6. https://doi.org/10.3389/fevo.2018.00202

Chauvet MT, Hurpet D, Chauvet J, Acher R (1980) Phenypressin (Phe2-Arg8-vasopressin), a new neurohypophysial peptide found in marsupials. Nature 287:640–642. https://doi.org/10.1038/287640a0

Chauvet J, Hurpet D, Michel G et al (1984) Two multigene families for marsupial neurohypophysial hormones? Identification of oxytocin, mesotocin, lysipressin and arginine vasopressin in the North American opossum (Didelphis virginiana). Biochem Biophys Res Commun 123:306–311. https://doi.org/10.1016/0006-291X(84)90413-3

Chauvet J, Rouille Y, Chauvet MT, Acher R (1987) Evolution of marsupials traced by their neurohypophyseal hormones: Microidentification of mesotocin and arginine vasopressin in two Australian families, Dasyuridae and Phascolarctidae. Gen Comp Endocrinol 67:399–408. https://doi.org/10.1016/0016-6480(87)90195-X

Chauvet J, Rouille Y, Chauveau C et al (1994) Special evolution of neurohypophysial hormones in cartilaginous fishes: asvatocin and phasvatocin, two oxytocin-like peptides isolated from the spotted dogfish (Scyliorhinus caniculus). Proc Natl Acad Sci USA 91:11266–11270. https://doi.org/10.1073/pnas.91.23.11266

Conway KM, Gainer H (1987) Immunocytochemical studies of vasotocin, mesotocin, and neurophysins in the Xenopus hypothalamo-neurohypophysial system. J Comp Neurol 264:494–508. https://doi.org/10.1002/cne.902640405

Cruz LJ, de Santos V, Zafaralla GC et al (1987) Invertebrate vasopressin/oxytocin homologs. Characterization of peptides from Conus geographus and Conus straitus venoms. J Biol Chem 262:15821–15824. https://doi.org/10.1016/S0021-9258(18)47661-2

Dale HH (1906) On some physiological actions of ergot. J Physiol 34:163–206

de Bree FM, Burbach JPH (1998) Structure–function relationships of the vasopressin prohormone domains. Cell Mol Neurobiol 18:173–191. https://doi.org/10.1023/A:1022564803093

de Keyzer Y, Auzan C, Lenne F et al (1994) Cloning and characterization of the human V3 pituitary vasopressin receptor. FEBS Lett 356:215–220. https://doi.org/10.1016/0014-5793(94)01268-7

Delsuc F, Philippe H, Tsagkogeorga G et al (2018) A phylogenomic framework and timescale for comparative studies of tunicates. BMC Biol 16:39. https://doi.org/10.1186/s12915-018-0499-2

de Vries GJ, Buijs RM (1983) The origin of the vasopressinergic and oxytocinergic innervation of the rat brain with special reference to the lateral septum. Brain Res 273:307–317. https://doi.org/10.1016/0006-8993(83)90855-7

de Wied D (1969) Effects of peptide hormones on behaviour. In: Ganong WF, Martini L (eds) Frontiers in neuroendocrinology. Oxford University Press, New York, pp 97–140

de Wied D (1971) Long term effect of vasopressin on the maintenance of a conditioned avoidance response in rats. Nature 232:58–60. https://doi.org/10.1038/232058a0

de Wied D (1973) The role of the posterior pituitary and its peptides on the maintenance of conditioned avoidance behaviour. In: Lissák K (ed) Hormones and brain function. Plenum, New York, pp 391–397

De Wied D (1980) Behavioural actions of neurohypophysial peptides. Proc R Soc Lond B Biol Sci 210:183–194. https://doi.org/10.1098/rspb.1980.0127

Dobzhansky T (1964) Biology, molecular and organismic. Am Zool 4:443–452. https://doi.org/10.1093/icb/4.4.443

Dobzhansky T (1973) Nothing in biology makes sense except in the light of evolution. Am Biol Teach 75:87–91. https://doi.org/10.2307/4444260

Donaldson ZR, Young LJ (2008) Oxytocin, vasopressin, and the neurogenetics of sociality. Science 322:900–904. https://doi.org/10.1126/science.1158668

dos Reis M, Thawornwattana Y, Angelis K et al (2015) Uncertainty in the timing of origin of animals and the limits of precision in molecular timescales. Curr Biol 25:2939–2950. https://doi.org/10.1016/j.cub.2015.09.066

Duque JF, Leichner W, Ahmann H, Stevens JR (2018) Mesotocin influences pinyon jay prosociality. Biol Lett 14:20180105. https://doi.org/10.1098/rsbl.2018.0105

Duque JF, Rasmussen T, Rodriguez A, Stevens JR (2020) The role of mesotocin on social bonding in pinyon jays. Ethology 126:165–175. https://doi.org/10.1111/eth.12990

du Vigneaud V (1935) The chemistry of the hormones from a structural standpoint. Sci Mon 40:138–145

du Vigneaud V (1956) Trail of sulfur research: from insulin to oxytocin. Science 123:967–974. https://doi.org/10.1126/science.123.3205.967

du Vigneaud V, Gish DT, Katsoyannis PG (1954a) A synthetic preparation possessing biological properties associated with arginine-vasopressin. J Am Chem Soc 76:4751–4752. https://doi.org/10.1021/ja01647a089

du Vigneaud V, Ressler C, Swan JM et al (1954b) The synthesis of oxytocin. J Am Chem Soc 76:3115–3121. https://doi.org/10.1021/ja01641a004

Eliava M, Melchior M, Knobloch-Bollmann HS et al (2016) A new population of parvocellular oxytocin neurons controlling magnocellular neuron activity and inflammatory pain processing. Neuron 89:1291–1304. https://doi.org/10.1016/j.neuron.2016.01.041

Elphick MR (2010) NG peptides: A novel family of neurophysin-associated neuropeptides. Gene 458:20–26. https://doi.org/10.1016/j.gene.2010.03.004

Elphick MR, Rowe ML (2009) NGFFFamide and echinotocin: structurally unrelated myoactive neuropeptides derived from neurophysin-containing precursors in sea urchins. J Exp Biol 212:1067–1077. https://doi.org/10.1242/jeb.027599

Ferguson DR (1969) The genetic distribution of vasopressins in the peccary (*Tayassu angulatus*) and warthog (*Phacochoerus aethiopicus*). Gen Comp Endocrinol 12:609–613. https://doi.org/10.1016/0016-6480(69)90181-6

Fetter-Pruneda I, Hart T, Ulrich Y et al (2021) An oxytocin/vasopressin-related neuropeptide modulates social foraging behaviour in the clonal raider ant. PLoS Biol 19:e3001305. https://doi.org/10.1371/journal.pbio.3001305

Freeman SM, Young LJ (2016) Comparative perspectives on oxytocin and vasopressin receptor research in rodents and primates: translational implications. J Neuroendocrinol 28. https://doi.org/10.1111/jne.12382

Freeman SM, Inoue K, Smith AL et al (2014a) The neuroanatomical distribution of oxytocin receptor binding and mRNA in the male rhesus macaque (Macaca mulatta). Psychoneuroendocrinology 45:128–141. https://doi.org/10.1016/j.psyneuen.2014.03.023

Freeman SM, Walum H, Inoue K et al (2014b) Neuroanatomical distribution of oxytocin and vasopressin 1a receptors in the socially monogamous coppery titi monkey (*Callicebus cupreus*). Neuroscience 273:12–23. https://doi.org/10.1016/j.neuroscience.2014.04.055

Garrison JL, Macosko EZ, Bernstein S et al (2012) Oxytocin/vasopressin-related peptides have an ancient role in reproductive behaviour. Science 338:540–543. https://doi.org/10.1126/science.1226201

Gilligan P, Brenner S, Venkatesh B (2003) Neurone-specific expression and regulation of the pufferfish isotocin and vasotocin genes in transgenic mice. J Neuroendocrinol 15:1027–1036. https://doi.org/10.1046/j.1365-2826.2003.01090.x

Gimpl G, Fahrenholz F (2001) The oxytocin receptor system: structure, function, and regulation. Physiol Rev 81:629–683. https://doi.org/10.1152/physrev.2001.81.2.629

Göhde R, Naumann B, Laundon D et al (2021) Choanoflagellates and the ancestry of neurosecretory vesicles. Philos Trans R Soc B Biol Sci 376:20190759. https://doi.org/10.1098/rstb.2019.0759

Goodson JL, Schrock SE, Klatt JD et al (2009) Mesotocin and nonapeptide receptors promote estrildid flocking behaviour. Science 325:862–866. https://doi.org/10.1126/science.1174929

Goodson JL, Kelly AM, Kingsbury MA (2012) Evolving nonapeptide mechanisms of gregariousness and social diversity in birds. Horm Behav 61:239–250. https://doi.org/10.1016/j.yhbeh.2012.01.005
Goossens N, Dierickx K, Vandesande F (1977) Immunocytochemical demonstration of the hypothalamo-hypophysial vasotocinergic system of Lampetra fluviatilis. Cell Tissue Res 177:317–323. https://doi.org/10.1007/BF00220307
Grinevich V, Ludwig M (2021) The multiple faces of the oxytocin and vasopressin systems in the brain. J Neuroendocrinol 33:e13004. https://doi.org/10.1111/jne.13004
Grinevich V, Neumann ID (2020) Brain oxytocin: how puzzle stones from animal studies translate into psychiatry. Mol Psychiatry 1–15. https://doi.org/10.1038/s41380-020-0802-9
Grinevich VV, Polenov AL (1994) The evolution of the nonapeptidergic neurosecretory formations of the hypothalamus in vertebrate animals. J Evol Biochem Physiol Zh Evol Biokhim Fiziol 30:270–292
Grinevich V, Knobloch-Bollmann HS, Eliava M et al (2016) Assembling the puzzle: pathways of oxytocin signaling in the Brain. Biol Psychiatry 79:155–164. https://doi.org/10.1016/j.biopsych.2015.04.013
Gruber CW (2014) Physiology of invertebrate oxytocin and vasopressin neuropeptides. Exp Physiol 99:55–61. https://doi.org/10.1113/expphysiol.2013.072561
Gruber CW, Muttenthaler M (2012) Discovery of defense- and neuropeptides in social ants by genome-mining. PLoS One 7:e32559. https://doi.org/10.1371/journal.pone.0032559
Gwee P-C, Amemiya CT, Brenner S, Venkatesh B (2008) Sequence and organization of coelacanth neurohypophysial hormone genes: Evolutionary history of the vertebrate neurohypophysial hormone gene locus. BMC Evol Biol 8:93. https://doi.org/10.1186/1471-2148-8-93
Gwee P-C, Tay B-H, Brenner S, Venkatesh B (2009) Characterization of the neurohypophysial hormone gene loci in elephant shark and the Japanese lamprey: origin of the vertebrate neurohypophysial hormone genes. BMC Evol Biol 9:47. https://doi.org/10.1186/1471-2148-9-47
Haakenson CM, Balthazart J, Madison FN, Ball GF (2022) The neural distribution of the avian homologue of oxytocin, mesotocin, in two songbird species, the zebra finch and the canary: A potential role in song perception and production. J Comp Neurol 530:2402–2414. https://doi.org/10.1002/cne.25338
Hamann D, Hunt N, Ivell R (1992) The chicken vasotocin gene. J Neuroendocrinol 4:505–513. https://doi.org/10.1111/j.1365-2826.1992.tb00198.x
Hammock EAD (2015) Developmental perspectives on oxytocin and vasopressin. Neuropsychopharmacology 40:24–42. https://doi.org/10.1038/npp.2014.120
Hammock E, Levitt P (2013) Oxytocin receptor ligand binding in embryonic tissue and postnatal brain development of the C57BL/6J mouse. Front Behav Neurosci 7. https://doi.org/10.3389/fnbeh.2013.00195
Hasan MT, Althammer F, Silva da Gouveia M et al (2019) A fear memory engram and its plasticity in the hypothalamic oxytocin system. Neuron 103:133–146.e8. https://doi.org/10.1016/j.neuron.2019.04.029
Hayakawa E, Guzman C, Horiguchi O et al (2022) Mass spectrometry of short peptides reveals common features of metazoan peptidergic neurons. Nat Ecol Evol 6:1438–1448. https://doi.org/10.1038/s41559-022-01835-7
Herbeck YE, Gulevich RG, Eliava M et al (2018) Domestication: neuroendocrine mechanisms of Canidae-human bonds. In: Ludwig M, Levkowitz G (eds) Model animals in neuroendocrinology. Wiley, pp 313–334
Herget U, Wolf A, Wullimann MF, Ryu S (2014) Molecular neuroanatomy and chemoarchitecture of the neurosecretory preoptic-hypothalamic area in zebrafish larvae. J Comp Neurol 522:1542–1564. https://doi.org/10.1002/cne.23480
Herring PT (1904) The action of pituitary extracts on the heart and circulation of the frog. J Physiol 31:429–437
Hoheisel G, Rühle H-J, Sterba G (1978) The reticular formation of lampreys (Petromyzonidae) — A target area for exohypothalamic vasotocinergic fibres. Cell Tissue Res 189:331–345. https://doi.org/10.1007/BF00209281

Hökfelt T, Broberger C, Xu Z-QD et al (2000) Neuropeptides — an overview. Neuropharmacology 39:1337–1356. https://doi.org/10.1016/S0028-3908(00)00010-1
Holmes FL (1986) Claude Bernard, The "Milieu Intérieur", and regulatory physiology. Hist Philos Life Sci 8:3–25
Hoyle CHV (1998) Neuropeptide families: evolutionary perspectives. Regul Pept 73:1–33. https://doi.org/10.1016/S0167-0115(97)01073-2
Hoyle CHV (1999) Neuropeptide families and their receptors: evolutionary perspectives. Brain Res 848:1–25. https://doi.org/10.1016/S0006-8993(99)01975-7
Hung LW, Neuner S, Polepalli JS et al (2017) Gating of social reward by oxytocin in the ventral tegmental area. Science 357:1406–1411. https://doi.org/10.1126/science.aan4994
Hyodo S, Tsukada T, Takei Y (2004) Neurohypophysial hormones of dogfish, *Triakis scyllium*: structures and salinity-dependent secretion. Gen Comp Endocrinol 138:97–104. https://doi.org/10.1016/j.ygcen.2004.05.009
Ino D, Tanaka Y, Hibino H, Nishiyama M (2022) A fluorescent sensor for real-time measurement of extracellular oxytocin dynamics in the brain. Nat Methods 19:1286–1294. https://doi.org/10.1038/s41592-022-01597-x
Insel TR, Shapiro LE (1992) Oxytocin receptor distribution reflects social organization in monogamous and polygamous voles. Proc Natl Acad Sci USA 89:5981–5985. https://doi.org/10.1073/pnas.89.13.5981
Istiban MN, De Fruyt N, Kenis S, Beets I (2024) Evolutionary conserved peptide and glycoprotein hormone-like neuroendocrine systems in *C. elegans*. Mol Cell Endocrinol 584:112162. https://doi.org/10.1016/j.mce.2024.112162
Ivell R, Richter D (1984) Structure and comparison of the oxytocin and vasopressin genes from rat. Proc Natl Acad Sci 81:2006–2010. https://doi.org/10.1073/pnas.81.7.2006
Iwasaki M, Lefevre A, Althammer F et al (2023) An analgesic pathway from parvocellular oxytocin neurons to the periaqueductal gray in rats. Nat Commun 14:1066. https://doi.org/10.1038/s41467-023-36641-7
Jékely G (2021) The chemical brain hypothesis for the origin of nervous systems. Philos Trans R Soc B Biol Sci 376:20190761. https://doi.org/10.1098/rstb.2019.0761
Johnson ZV, Young LJ (2015) Neurobiological mechanisms of social attachment and pair bonding. Curr Opin Behav Sci 3:38–44. https://doi.org/10.1016/j.cobeha.2015.01.009
Jokura Y, Urano A (1987) Extrahypothalamic Projection of Immunoreactive Vasotocin Fibers in the Brain of the Toad, Bufo japonicus. Zool Sci 4:675–681
Jurek B, Neumann ID (2018) The oxytocin receptor: from intracellular signaling to behaviour. Physiol Rev 98:1805–1908. https://doi.org/10.1152/physrev.00031.2017
Kamkrathok B, Porter TE, El Halawani ME, Chaiseha Y (2017) Distribution of mesotocin-immunoreactive neurons in the brain of the male native Thai chicken. Acta Histochem 119:804–811. https://doi.org/10.1016/j.acthis.2017.10.004
Kamm O, Aldrich TB, Grote IW et al (1928) The active principles of the posterior lobe of the pituitary gland. I. The demonstration of the presence of two active principles. II. The separation of the two principles and their concentration in the form of potent solid preparations. J Am Chem Soc 50:573–601. https://doi.org/10.1021/ja01389a050
Karim MA, Sloper JC (1980) Histogenesis of the supraoptic and paraventricular neurosecretory cells of the mouse hypothalamus. J Anat 130:341–347
Kawada T, Sekiguchi T, Itoh Y et al (2008) Characterization of a novel vasopressin/oxytocin superfamily peptide and its receptor from an ascidian, *Ciona intestinalis*. Peptides 29:1672–1678. https://doi.org/10.1016/j.peptides.2008.05.030
Kawada T, Shiraishi A, Matsubara S et al (2021) Vasopressin promoter transgenic and vasopressin gene-edited ascidian, Ciona intestinalis type A (Ciona robusta): innervation, gene expression profiles, and phenotypes. Front Endocrinol 12. https://doi.org/10.3389/fendo.2021.668564
Kim Y, Bark S, Hook V, Bandeira N (2011) NeuroPedia: neuropeptide database and spectral library. Bioinformatics 27:2772–2773. https://doi.org/10.1093/bioinformatics/btr445
Kimura T, Tanizawa O, Mori K et al (1992) Structure and expression of a human oxytocin receptor. Nature 356:526–529. https://doi.org/10.1038/356526a0

Klavdieva MM (1995) The history of neuropeptides I. Front Neuroendocrinol 16:293–321. https://doi.org/10.1006/frne.1995.1011
Klavdieva MM (1996) The history of neuropeptides III. Front Neuroendocrinol 17:155–179. https://doi.org/10.1006/frne.1996.0004
Knobloch HS, Grinevich V (2014) Evolution of oxytocin pathways in the brain of vertebrates. Front Behav Neurosci 8. https://doi.org/10.3389/fnbeh.2014.00031
Knobloch HS, Charlet A, Hoffmann LC et al (2012) Evoked axonal oxytocin release in the central amygdala attenuates fear response. Neuron 73:553–566. https://doi.org/10.1016/j.neuron.2011.11.030
Kobayashi A, Hamada M, Yoshida M et al (2022) Vasopressin-oxytocin-type signaling is ancient and has a conserved water homeostasis role in euryhaline marine planarians. Sci Adv 8:eabk0331. https://doi.org/10.1126/sciadv.abk0331
Koto A, Motoyama N, Tahara H et al (2019) Oxytocin/vasopressin-like peptide inotocin regulates cuticular hydrocarbon synthesis and water balancing in ants. Proc Natl Acad Sci USA 116:5597–5606. https://doi.org/10.1073/pnas.1817788116
Kuraku S, Kaiya H, Tanaka T, Hyodo S (2023) Evolution of vertebrate hormones and their receptors: insights from non-osteichthyan genomes. Annu Rev Anim Biosci 11:163–182. https://doi.org/10.1146/annurev-animal-050922-071351
Kuschinsky G, Bundschuh HE (1939) Über die Diuretische Wirkung von Hinter-Lappen-Präparaten und Ihre Beziehung zur Antidiuretischen Wirkung des Vasopressins. Klin Wochenschr 18:207–208. https://doi.org/10.1007/BF01777622
Lamanna F, Hervas-Sotomayor F, Oel AP et al (2023) A lamprey neural cell type atlas illuminates the origins of the vertebrate brain. Nat Ecol Evol 7:1714–1728. https://doi.org/10.1038/s41559-023-02170-1
Lande S, Witter A, de Wied D (1971) Pituitary peptides: an octapeptide that stimulates conditioned avoidance acquisition in hypophysectomized rats. J Biol Chem 246:2058–2062. https://doi.org/10.1016/S0021-9258(19)77188-9
Lane TF, Sower SA, Kawauchi H (1988) Arginine vasotocin from the pituitary gland of the lamprey (*Petromyzon marinus*): Isolation and amino acid sequence. Gen Comp Endocrinol 70:152–157. https://doi.org/10.1016/0016-6480(88)90104-9
Laumer CE, Gruber-Vodicka H, Hadfield MG et al (2018) Support for a clade of Placozoa and Cnidaria in genes with minimal compositional bias. eLife 7:e36278. https://doi.org/10.7554/eLife.36278
Lee AG, Cool DR, Grunwald WC et al (2011) A novel form of oxytocin in New World monkeys. Biol Lett 7:584–587. https://doi.org/10.1098/rsbl.2011.0107
Lefevre A, Meza J, Miller CT (2024) Long-range projections of oxytocin neurons in the marmoset brain. J Neuroendocrinol. https://doi.org/10.1111/jne.13397
Legros J-J (2001) Inhibitory effect of oxytocin on corticotrope function in humans: are vasopressin and oxytocin ying–yang neurohormones? Psychoneuroendocrinology 26:649–655. https://doi.org/10.1016/S0306-4530(01)00018-X
Leng G, Russell JA (2019) The osmoresponsiveness of oxytocin and vasopressin neurones: Mechanisms, allostasis and evolution. J Neuroendocrinol 31:e12662. https://doi.org/10.1111/jne.12662
Li H, Jiang T, An S et al (2024) Single-neuron projectomes of mouse paraventricular hypothalamic nucleus oxytocin neurons reveal mutually exclusive projection patterns. Neuron. https://doi.org/10.1016/j.neuron.2023.12.022
Lim MM, Young LJ (2006) Neuropeptidergic regulation of affiliative behaviour and social bonding in animals. Horm Behav 50:506–517. https://doi.org/10.1016/j.yhbeh.2006.06.028
Lim MM, Wang Z, Olazábal DE et al (2004) Enhanced partner preference in a promiscuous species by manipulating the expression of a single gene. Nature 429:754–757. https://doi.org/10.1038/nature02539
Liu Y, Li A, Bair-Marshall C et al (2023) Oxytocin promotes prefrontal population activity via the PVN-PFC pathway to regulate pain. Neuron 111:1795–1811.e7. https://doi.org/10.1016/j.neuron.2023.03.014

Liutkeviciute Z, Koehbach J, Eder T et al (2016) Global map of oxytocin/vasopressin-like neuropeptide signalling in insects. Sci Rep 6:39177. https://doi.org/10.1038/srep39177

Lockard MA, Ebert MS, Bargmann CI (2017) Oxytocin mediated behaviour in invertebrates: An evolutionary perspective. Dev Neurobiol 77:128–142. https://doi.org/10.1002/dneu.22466

Lolait SJ, O'Carroll A-M, McBride OW et al (1992) Cloning and characterization of a vasopressin V2 receptor and possible link to nephrogenic diabetes insipidus. Nature 357:336–339. https://doi.org/10.1038/357336a0

Lukas D, Clutton-Brock TH (2013) The evolution of social monogamy in mammals. Science 341:526–530. https://doi.org/10.1126/science.1238677

Martinez-Padrón M, González GC, Lederis K, Lukowiak K (1992) Localization of vasopressin-like immunoreactivity in the CNS of *Aplysia californica*. J Neurocytol 21:672–678. https://doi.org/10.1007/BF01191728

Matsubara S, Shiraishi A, Osugi T et al (2019) The regulation of oocyte maturation and ovulation in the closest sister group of vertebrates. eLife 8:e49062. https://doi.org/10.7554/eLife.49062

Mirabeau O, Joly J-S (2013) Molecular evolution of peptidergic signaling systems in bilaterians. Proc Natl Acad Sci USA 110:E2028–E2037. https://doi.org/10.1073/pnas.1219956110

Morel A, O'Carroll A-M, Brownstein MJ, Lolaft SJ (1992) Molecular cloning and expression of a rat Via arginine vasopressin receptor. Nature 356:523–526. https://doi.org/10.1038/356523a0

Moroz LL, Kocot KM, Citarella MR et al (2014) The ctenophore genome and the evolutionary origins of neural systems. Nature 510:109–114. https://doi.org/10.1038/nature13400

Murphy D, Si-Hoe S-L, Brenner S, Venkatesh B (1998) Something fishy in the rat brain: molecular genetics of the hypothalamo-neurohypophysial system. BioEssays 20:741–749. https://doi.org/10.1002/(SICI)1521-1878(199809)20:9<741::AID-BIES7>3.0.CO;2-J

Murphy D, Konopacka A, Hindmarch C et al (2012) The hypothalamic-neurohypophyseal system: from genome to physiology. J Neuroendocrinol 24:539–553. https://doi.org/10.1111/j.1365-2826.2011.02241.x

Musser JM, Schippers KJ, Nickel M et al (2021) Profiling cellular diversity in sponges informs animal cell type and nervous system evolution. Science 374:717–723. https://doi.org/10.1126/science.abj2949

Najle SR, Grau-Bové X, Elek A et al (2023) Stepwise emergence of the neuronal gene expression program in early animal evolution. Cell 186:4676–4693.e29. https://doi.org/10.1016/j.cell.2023.08.027

Nanglu K, Cole SR, Wright DF, Souto C (2023a) Worms and gills, plates and spines: the evolutionary origins and incredible disparity of deuterostomes revealed by fossils, genes, and development. Biol Rev 98:316–351. https://doi.org/10.1111/brv.12908

Nanglu K, Lerosey-Aubril R, Weaver JC, Ortega-Hernández J (2023b) A mid-Cambrian tunicate and the deep origin of the ascidiacean body plan. Nat Commun 14:3832. https://doi.org/10.1038/s41467-023-39012-4

Neugebauer V, Mazzitelli M, Cragg B et al (2020) Amygdala, neuropeptides, and chronic pain-related affective behaviours. Neuropharmacology 170:108052. https://doi.org/10.1016/j.neuropharm.2020.108052

Neumann ID, Landgraf R (2012) Balance of brain oxytocin and vasopressin: implications for anxiety, depression, and social behaviours. Trends Neurosci 35:649–659. https://doi.org/10.1016/j.tins.2012.08.004

Nozaki M (2013) Hypothalamic-pituitary-gonadal endocrine system in the hagfish. Front Endocrinol 4. https://doi.org/10.3389/fendo.2013.00200

Nozaki M, Gorbman A (1983) Immunocytochemical localization of somatostatin and vasotocin in the brain of the Pacific hagfish, Eptatretus stouti. Cell Tissue Res 229:541–550. https://doi.org/10.1007/BF00207697

Odekunle EA, Elphick MR (2020) Comparative and evolutionary physiology of vasopressin/oxytocin-type neuropeptide signaling in invertebrates. Front Endocrinol 11. https://doi.org/10.3389/fendo.2020.00225

Odekunle EA, Semmens DC, Martynyuk N et al (2019) Ancient role of vasopressin/oxytocin-type neuropeptides as regulators of feeding revealed in an echinoderm. BMC Biol 17:60. https://doi.org/10.1186/s12915-019-0680-2

Oliveira VEDM, Lukas M, Wolf HN et al (2021) Oxytocin and vasopressin within the ventral and dorsal lateral septum modulate aggression in female rats. Nat Commun 12:2900. https://doi.org/10.1038/s41467-021-23064-5

Oliver G, Schäfer EA (1895) On the physiological action of extracts of pituitary body and certain other glandular organs. J Physiol 18:277–279. https://doi.org/10.1113/jphysiol.1895.sp000565

Ostrowski NL (1998) Oxytocin receptor mRNA expression in rat brain: implications for behavioural integration and reproductive success. Psychoneuroendocrinology 23:989–1004. https://doi.org/10.1016/S0306-4530(98)00070-5

Ostrowski NL, Lolait SJ, Young WS 3rd (1994) Cellular localization of vasopressin V1a receptor messenger ribonucleic acid in adult male rat brain, pineal, and brain vasculature. Endocrinology 135:1511–1528. https://doi.org/10.1210/en.135.4.1511

Ott I, Scott JC (1910) The action of infundibulin upon the mammary secretion. Proc Soc Exp Biol Med 8:48–49. https://doi.org/10.3181/00379727-8-27

Oumi T, Ukena K, Matsushima O et al (1996) Annetocin, an annelid oxytocin-related peptide, induces egg-laying behaviour in the earthworm, Eisenia foetida. J Exp Zool 276:151–156. https://doi.org/10.1002/(SICI)1097-010X(19961001)276:2<151::AID-JEZ8>3.0.CO;2-N

Palay SL (1945) Neurosecretion. VII. The preoptico-hypophysial pathway in fishes. J Comp Neurol 82:129–143. https://doi.org/10.1002/cne.900820202

Parry LJ, Bathgate RAD, Ivell R (2000) Mammalian mesotocin: cDNA sequence and expression of an oxytocin-like gene in a macropodid marsupial, the tammar wallaby. Gen Comp Endocrinol 118:187–199. https://doi.org/10.1006/gcen.2000.7464

Paton DN, Watson A (1912) The actions of pituitrin, adrenalin and barium on the circulation of the bird. J Physiol 44:413–424

Powell JM, Inoue K, Wallace KJ et al (2023) Distribution of vasopressin 1a and oxytocin receptor protein and mRNA in the basal forebrain and midbrain of the spiny mouse (Acomys cahirinus). Brain Struct Funct 228:413–431. https://doi.org/10.1007/s00429-022-02581-z

Proux JP, Miller CA, Li JP et al (1987) Identification of an arginine vasopressin-like diuretic hormone from Locusta migratoria. Biochem Biophys Res Commun 149:180–186. https://doi.org/10.1016/0006-291X(87)91621-4

Qian T, Wang H, Wang P et al (2023) A genetically encoded sensor measures temporal oxytocin release from different neuronal compartments. Nat Biotechnol 1–14. https://doi.org/10.1038/s41587-022-01561-2

Raab W (1926a) Das hormonal-nervöse Regulationssystem des Fettstoffwechsels. Z Für Gesamte Exp Med 49:179–269. https://doi.org/10.1007/BF02622485

Raab W (1926b) Wärmeregulation und Fettstoffwechsel. Z Für Gesamte Exp Med 53:317–331. https://doi.org/10.1007/BF02625159

Raab W (1928) Pituitrin-Fettstoffwechselwirkung und vegetatives Nervensystem. Z Für Gesamte Exp Med 62:366–372. https://doi.org/10.1007/BF02623943

Raab W (1930) The action of pituitrin, pitressin and pitocin on the blood phosphatides. Endocrinology 14:150–156. https://doi.org/10.1210/endo-14-3-150

Ramachandran D, Sharma K, Saxena V et al (2023) Knock-out of vasotocin reduces reproductive success in female zebrafish, Danio rerio. Front Endocrinol 14. https://doi.org/10.3389/fendo.2023.1151299

Rankin SL, Partlow GD, McCurdy RD et al (2003) Postnatal neurogenesis in the vasopressin and oxytocin-containing nucleus of the pig hypothalamus. Brain Res 971:189–196. https://doi.org/10.1016/S0006-8993(03)02350-3

Reich G (1992) A new peptide of the oxytocin/vasopressin family isolated from nerves of the cephalopod *Octopus vulgaris*. Neurosci Lett 134:191–194. https://doi.org/10.1016/0304-3940(92)90514-8

Ren D, Chin KR, French JA (2014) Molecular variation in AVP and AVPR1a in New World monkeys (primates, platyrrhini): evolution and implications for social monogamy. PLoS One 9:e111638. https://doi.org/10.1371/journal.pone.0111638

Rhodes CH, Morriell JI, Pfaff DW (1981) Immunohistochemical analysis of magnocellular elements in rat hypothalamus: distribution and numbers of cells containing neurophysin, oxytocin, and vasopressin. J Comp Neurol 198:45–64. https://doi.org/10.1002/cne.901980106

Robinzon B, Koike TI, Neldon HL, Kinzler SL (1988) Distribution of immunoreactive mesotocin and vasotocin in the brain and pituitary of chickens. Peptides 9:829–833. https://doi.org/10.1016/0196-9781(88)90129-5

Rogers CN, Ross AP, Sahu SP et al (2018) Oxytocin- and arginine vasopressin-containing fibers in the cortex of humans, chimpanzees, and rhesus macaques. Am J Primatol 80:e22875. https://doi.org/10.1002/ajp.22875

Rokicki J, Kaufmann T, de Lange A-MG et al (2022) Oxytocin receptor expression patterns in the human brain across development. Neuropsychopharmacology 47:1550–1560. https://doi.org/10.1038/s41386-022-01305-5

Romanov RA, Harkany T (2022) Neuronal heterogeneity in the paraventricular nucleus of the hypothalamus as revealed by single-cell RNA-seq. Curr Opin Endocr Metab Res 25:100366. https://doi.org/10.1016/j.coemr.2022.100366

Romanov RA, Zeisel A, Bakker J et al (2017) Molecular interrogation of hypothalamic organization reveals distinct dopamine neuronal subtypes. Nat Neurosci 20:176–188. https://doi.org/10.1038/nn.4462

Ros-Rocher N, Pérez-Posada A, Leger MM, Ruiz-Trillo I (2021) The origin of animals: an ancestral reconstruction of the unicellular-to-multicellular transition. Open Biol 11:200359. https://doi.org/10.1098/rsob.200359

Ross HE, Cole CD, Smith Y et al (2009) Characterization of the oxytocin system regulating affiliative behaviour in female prairie voles. Neuroscience 162:892–903. https://doi.org/10.1016/j.neuroscience.2009.05.055

Rouillé Y, Chauvet M-T, Chauvet J, Acher R (1988) Dual duplication of neurohypophysial hormones in an Australian marsupial: Mesotocin, oxytocin, lysine vasopressin and arginine vasopressin in a single gland of the Northern bandicoot (Isoodon macrourus). Biochem Biophys Res Commun 154:346–350. https://doi.org/10.1016/0006-291X(88)90691-2

Rowe ML, Elphick MR (2012) The neuropeptide transcriptome of a model echinoderm, the sea urchin *Strongylocentrotus purpuratus*. Gen Comp Endocrinol 179:331–344. https://doi.org/10.1016/j.ygcen.2012.09.009

Rozen F, Russo C, Banville D, Zingg HH (1995) Structure, characterization, and expression of the rat oxytocin receptor gene. Proc Natl Acad Sci USA 92:200–204. https://doi.org/10.1073/pnas.92.1.200

Russo AF (2017) Overview of neuropeptides: awakening the senses? Headache J Head Face Pain 57:37–46. https://doi.org/10.1111/head.13084

Ryan JF, Pang K, Mullikin JC et al (2010) The homeodomain complement of the ctenophore Mnemiopsis leidyi suggests that Ctenophora and Porifera diverged prior to the ParaHoxozoa. EvoDevo 1:9. https://doi.org/10.1186/2041-9139-1-9

Ryan JF, Pang K, Schnitzler CE et al (2013) The genome of the ctenophore Mnemiopsis leidyi and its implications for cell type evolution. Science 342:1242592. https://doi.org/10.1126/science.1242592

Sachkova MY, Nordmann E-L, Soto-Àngel JJ et al (2021) Neuropeptide repertoire and 3D anatomy of the ctenophore nervous system. Curr Biol 31:5274–5285.e6. https://doi.org/10.1016/j.cub.2021.09.005

Saito D, Komatsuda M, Urano A (2004) Functional organization of preoptic vasotocin and isotocin neurons in the brain of rainbow trout: central and neurohypophysial projections of single neurons. Neuroscience 124:973–984. https://doi.org/10.1016/j.neuroscience.2003.12.038

Sakamoto T, Ogawa S, Nishiyama Y et al (2015) Osmotic/ionic status of body fluids in the euryhaline cephalopod suggest possible parallel evolution of osmoregulation. Sci Rep 5:14469. https://doi.org/10.1038/srep14469

Sartorius AM, Rokicki J, Birkeland S et al (2024) An evolutionary timeline of the oxytocin signaling pathway. Commun Biol 7:1–13. https://doi.org/10.1038/s42003-024-06094-9
Schäfer EA, Mackenzie K (1911) The action of animal extracts on milk secretion. Proc R Soc Lond Ser B Contain Pap Biol Character 84:16–22. https://doi.org/10.1098/rspb.1911.0042
Schäfer EA, Vincent S (1899) The physiological effects of extracts of the pituitary body. J Physiol 25:87–97
Scharrer E (1928) Die Lichtempfindlichkeit Blinder Elritzen. (Untersuchungen über das Zwischenhirn der Fische I.). Z Für Vgl Physiol 7:1–38. https://doi.org/10.1007/BF00341151
Scharrer E (1951) Neurosecretion. X. A relationship between the paraphysis and the paraventricular nucleus in the garter snake (Thamnophis sp.). Biol Bull 101:106–113. https://doi.org/10.2307/1538506
Scharrer E, Scharrer B (1937) Über Drüsen-Nervenzellen und Neurosekretorische Organe bei Wirbellosen und Wirbeltieren. Biol Rev 12:185–216. https://doi.org/10.1111/j.1469-185X.1937.tb01229.x
Schorscher-Petcu A, Dupré A, Tribollet E (2009) Distribution of vasopressin and oxytocin binding sites in the brain and upper spinal cord of the common marmoset. Neurosci Lett 461:217–222. https://doi.org/10.1016/j.neulet.2009.06.016
Schultz DT, Haddock SHD, Bredeson JV et al (2023) Ancient gene linkages support ctenophores as sister to other animals. Nature 618:110–117. https://doi.org/10.1038/s41586-023-05936-6
Sebé-Pedrós A, Chomsky E, Pang K et al (2018) Early metazoan cell type diversity and the evolution of multicellular gene regulation. Nat Ecol Evol 2:1176–1188. https://doi.org/10.1038/s41559-018-0575-6
Semmens DC, Mirabeau O, Moghul I et al (2016) Transcriptomic identification of starfish neuropeptide precursors yields new insights into neuropeptide evolution. Open Biol 6:150224. https://doi.org/10.1098/rsob.150224
Senatore A, Reese TS, Smith CL (2017) Neuropeptidergic integration of behaviour in Trichoplax adhaerens, an animal without synapses. J Exp Biol 220:3381–3390. https://doi.org/10.1242/jeb.162396
Silveira PF, Breno MC, Martín del Río MP, Mancera JM (2002) The distribution of vasotocin and mesotocin immunoreactivity in the brain of the snake, *Bothrops jararaca*. J Chem Neuroanat 24:15–26. https://doi.org/10.1016/S0891-0618(02)00016-9
Simion P, Philippe H, Baurain D et al (2017) A large and consistent phylogenomic dataset supports sponges as the sister group to all other animals. Curr Biol 27:958–967. https://doi.org/10.1016/j.cub.2017.02.031
Smith JJ, Timoshevskaya N, Ye C et al (2018) The sea lamprey germline genome provides insights into programmed genome rearrangement and vertebrate evolution. Nat Genet 50:270–277. https://doi.org/10.1038/s41588-017-0036-1
Soumier A, Habart M, Lio G et al (2022) Differential fate between oxytocin and vasopressin cells in the developing mouse brain. iScience 25:103655. https://doi.org/10.1016/j.isci.2021.103655
Sower SA (2018) Landmark discoveries in elucidating the origins of the hypothalamic-pituitary system from the perspective of a basal vertebrate, sea lamprey. Gen Comp Endocrinol 264:3–15. https://doi.org/10.1016/j.ygcen.2017.10.016
Stafflinger E, Hansen KK, Hauser F et al (2008) Cloning and identification of an oxytocin/vasopressin-like receptor and its ligand from insects. Proc Natl Acad Sci USA 105:3262–3267. https://doi.org/10.1073/pnas.0710897105
Steenwyk JL, Li Y, Zhou X et al (2023) Incongruence in the phylogenomics era. Nat Rev Genet 24:834–850. https://doi.org/10.1038/s41576-023-00620-x
Steger J, Cole AG, Denner A et al (2022) Single-cell transcriptomics identifies conserved regulators of neuroglandular lineages. Cell Rep 40. https://doi.org/10.1016/j.celrep.2022.111370
Stoll CJ, Voorn P (1985) The distribution of hypothalamic and extrahypothalamic vasotocinergic cells and fibers in the brain of a lizard, Gekko gecko: Presence of a sex difference. J Comp Neurol 239:193–204. https://doi.org/10.1002/cne.902390206
Stoop R (2012) Neuromodulation by oxytocin and vasopressin. Neuron 76:142–159. https://doi.org/10.1016/j.neuron.2012.09.025

Stoop R, Hegoburu C, van den Burg E (2015) New opportunities in vasopressin and oxytocin research: a perspective from the amygdala. Annu Rev Neurosci 38:369–388. https://doi.org/10.1146/annurev-neuro-071714-033904

Sugimoto T, Saito M, Mochizuki S et al (1994) Molecular cloning and functional expression of a cDNA encoding the human V1b vasopressin receptor. J Biol Chem 269:27088–27092. https://doi.org/10.1016/S0021-9258(18)47129-3

Surawicz B (1997) Wilhelm Raab. Clin Cardiol 20:310–311. https://doi.org/10.1002/clc.4960200325

Takuwa-Kuroda K, Iwakoshi-Ukena E, Kanda A, Minakata H (2003) Octopus, which owns the most advanced brain in invertebrates, has two members of vasopressin/oxytocin superfamily as in vertebrates. Regul Pept 115:139–149. https://doi.org/10.1016/S0167-0115(03)00151-4

Theofanopoulou C, Gedman G, Cahill JA et al (2021) Universal nomenclature for oxytocin–vasotocin ligand and receptor families. Nature 592:747–755. https://doi.org/10.1038/s41586-020-03040-7

Thepen T, Voorn P, Stoll CJ et al (1987) Mesotocin and vasotocin in the brain of the lizard Gekko gecko. An immunocytochemical study. Cell Tissue Res 250:649–656. https://doi.org/10.1007/BF00218959

Thibonnier M, Auzan C, Madhun Z et al (1994) Molecular cloning, sequencing, and functional expression of a cDNA encoding the human V1a vasopressin receptor. J Biol Chem 269:3304–3310. https://doi.org/10.1016/S0021-9258(17)41863-1

Thiel D, Franz-Wachtel M, Aguilera F, Hejnol A (2018) Xenacoelomorph neuropeptidomes reveal a major expansion of neuropeptide systems during early bilaterian evolution. Mol Biol Evol 35:2528–2543. https://doi.org/10.1093/molbev/msy160

Trudeau VL, Somoza GM (2020) Multimodal hypothalamo-hypophysial communication in the vertebrates. Gen Comp Endocrinol 293:113475. https://doi.org/10.1016/j.ygcen.2020.113475

Tsuneki K, Gorbman A (1975) Ultrastructure of the anterior neurohypophysis and the pars distalis of the lamprey, *Lampetra tridentata*. Gen Comp Endocrinol 25:487–508. https://doi.org/10.1016/0016-6480(75)90160-4

Ukena K, Iwakoshi-Ukena E, Hikosaka A (2008) Unique form and osmoregulatory function of a neurohypophysial hormone in a urochordate. Endocrinology 149:5254–5261. https://doi.org/10.1210/en.2008-0607

Vaccari C, Lolait SJ, Ostrowski NL (1998) Comparative distribution of vasopressin V1b and oxytocin receptor messenger ribonucleic acids in brain. Endocrinology 139:5015–5033. https://doi.org/10.1210/endo.139.12.6382

van Kesteren RE, Smit AB, de Lange RP et al (1995a) Structural and functional evolution of the vasopressin/oxytocin superfamily: vasopressin-related conopressin is the only member present in Lymnaea, and is involved in the control of sexual behaviour. J Neurosci 15:5989–5998. https://doi.org/10.1523/JNEUROSCI.15-09-05989.1995

van Kesteren RE, Tensen CP, Smit AB et al (1995b) A novel G protein-coupled receptor mediating both vasopressin- and oxytocin-like functions of Lys-conopressin in Lymnaea stagnalis. Neuron 15:897–908. https://doi.org/10.1016/0896-6273(95)90180-9

Varoqueaux F, Williams EA, Grandemange S et al (2018) High cell diversity and complex peptidergic signaling underlie placozoan behaviour. Curr Biol 28:3495–3501.e2. https://doi.org/10.1016/j.cub.2018.08.067

Venkatesh B, Si-Hoe SL, Murphy D, Brenner S (1997) Transgenic rats reveal functional conservation of regulatory controls between the Fugu isotocin and rat oxytocin genes. Proc Natl Acad Sci USA 94:12462–12466. https://doi.org/10.1073/pnas.94.23.12462

Verbalis JG, Mangione MP, Stricker EM (1991) Oxytocin produces natriuresis in rats at physiological plasma concentrations. Endocrinology 128:1317–1322. https://doi.org/10.1210/endo-128-3-1317

Von den Velden R (1913) Die Nierenwirkung von Hypophysenextrakten beim Menschen. Berl Klin Wochenschr 45:2083–2087

Wallis M (2012) Molecular evolution of the neurohypophysial hormone precursors in mammals: Comparative genomics reveals novel mammalian oxytocin and vasopressin analogues. Gen Comp Endocrinol 179:313–318. https://doi.org/10.1016/j.ygcen.2012.07.030

Wang W, Lufkin T (2000) The murine Otp homeobox gene plays an essential role in the specification of neuronal cell lineages in the developing hypothalamus. Dev Biol 227:432–449. https://doi.org/10.1006/dbio.2000.9902

Wang Y, Wang M, Yin S et al (2015) NeuroPep: a comprehensive resource of neuropeptides. Database 2015:bav038. https://doi.org/10.1093/database/bav038

Whelan NV, Kocot KM, Moroz TP et al (2017) Ctenophore relationships and their placement as the sister group to all other animals. Nat Ecol Evol 1:1737–1746. https://doi.org/10.1038/s41559-017-0331-3

Wigger A, Sánchez MM, Mathys KC et al (2004) Alterations in central neuropeptide expression, release, and receptor binding in rats bred for high anxiety: critical role of vasopressin. Neuropsychopharmacology 29:1–14. https://doi.org/10.1038/sj.npp.1300290

Wu Z, Lin D, Li Y (2022) Pushing the frontiers: tools for monitoring neurotransmitters and neuromodulators. Nat Rev Neurosci 23:257–274. https://doi.org/10.1038/s41583-022-00577-6

Yañez-Guerra LA, Thiel D, Jékely G (2022) Premetazoan origin of neuropeptide signaling. Mol Biol Evol 39:msac051. https://doi.org/10.1093/molbev/msac051

Young LJ (1999) Oxytocin and vasopressin receptors and species-typical social behaviours. Horm Behav 36:212–221. https://doi.org/10.1006/hbeh.1999.1548

Young LJ, Lim MM, Gingrich B, Insel TR (2001) Cellular mechanisms of social attachment. Horm Behav 40:133–138. https://doi.org/10.1006/hbeh.2001.1691

Yu D, Ren Y, Uesaka M et al (2024) Hagfish genome elucidates vertebrate whole-genome duplication events and their evolutionary consequences. Nat Ecol Evol 8:519–535. https://doi.org/10.1038/s41559-023-02299-z

Zhang L, Zetter M, Hernández VS, Hernández-Pérez OR, Jáuregui-Huerta F, Krabichler Q, Grinevich V (2024) Morphological signatures of neurogenesis and neuronal migration in hypothalamic vasopressinergic magnocellular nuclei of the adult rat. Int. J. Mol. Sci. 25(13), Article 13. https://doi.org/10.3390/ijms25136988

Evolution of Hypothalamic–Pituitary Neuropeptides and Their Receptors: Insights from Invertebrates

2

Maurice R. Elphick

Abstract

Identification of the hypothalamic–pituitary neuropeptides vasopressin (VP), oxytocin (OT), gonadotropin-releasing hormone (GnRH), thyrotropin-releasing hormone (TRH), corticotropin-releasing hormone (CRH), somatostatin, and growth hormone-releasing hormone (GHRH) in mammals provided the molecular foundations of neuroendocrinology. Furthermore, discovery of G-protein coupled receptors that mediate effects of these neuropeptides facilitated investigation of their phylogenetic distribution and evolution. Orthologs of VP/OT-type, GnRH-type, TRH-type, CRH-type, and somatostatin-type receptors have been identified in protostome and/or deuterostome invertebrates. Identification and characterization of ligands for these receptors have provided interesting insights into the evolution of the structures and functions of hypothalamic–pituitary neuropeptides. Roles of VP/OT-type neuropeptides as ancient regulators of reproductive processes, osmoregulation, and feeding have been revealed in a variety of invertebrates. Novel TRH-type amidated neuropeptides that have a core ELF/EFL motif have been discovered in protostomes, with functional studies on the nematode *Caenorhabditis elegans* revealing a role in regulation of growth. Corazonin-type and allatostatin-C-type neuropeptides have been identified as paralogs of GnRH-type and somatostatin-type neuropeptides, respectively, that have been variously retained in protostome and deuterostome invertebrates but lost in vertebrates. Furthermore, functional studies have revealed that allatostatin-C-type and somatostatin-type neuropeptides have opposing myoexcitatory and myoinhibitory roles, respectively, in an echinoderm—the starfish *Asterias*

M. R. Elphick (✉)
Centre for Evolutionary & Functional Genomics, School of Biological & Behavioural Sciences, Queen Mary University of London, London, UK
e-mail: m.r.elphick@qmul.ac.uk

V. Grinevich, R. F. Oliveira (eds.), *Evolutionary and Comparative Neuroendocrinology*, Masterclass in Neuroendocrinology 17,
https://doi.org/10.1007/978-3-031-80209-6_2

rubens. Contrasting with the other hypothalamic–pituitary neuropeptides that have an urbilaterian origin, GHRH-type neuropeptides and their cognate receptors originated more recently in the vertebrate lineage. Looking ahead, facilitated by genome sequencing, there are exciting opportunities to investigate the evolution and comparative physiology of hypothalamic–pituitary neuropeptides in an ever-growing diversity of invertebrate taxa.

Keywords

Evolution · Neuropeptide · Hypothalamic · Invertebrate · Phylogeny · Vasopressin · Oxytocin · Gonadotropin-releasing hormone · Thyrotropin-releasing hormone · Corticotropin-releasing hormone · Somatostatin · Growth hormone-releasing hormone (GHRH)

Abbreviations

ACP	Adipokinetic hormone/corazonin-related peptide
AKH	Adipokinetic hormone
ArSS1	*Asterias rubens* somatostatin-like peptide 1
ArSS2	*Asterias rubens* somatostatin-like peptide 2
ASTC	Allatostatin-C
CRH	Corticotropin-releasing hormone
CRZ	Corazonin
DH44	Diuretic hormone 44
ELH	Egg-laying hormone
GHRH	Growth hormone-releasing hormone
GnRH	Gonadotropin-releasing hormone
GPCR	G-protein coupled receptor
OT	Oxytocin
TRH	Thyrotropin-releasing hormone
UNC	Urocortin
VP	Vasopressin

2.1 Introduction

The first neuropeptides to be purified and sequenced were vasopressin and oxytocin, structurally related hormones that are synthesized by hypothalamic neurons and released into the circulatory system from the posterior lobe of the pituitary gland in mammals. The importance of the molecular identification of vasopressin and oxytocin was recognized in the award of the 1955 Nobel Prize for Chemistry to Vincent du Vigneaud (Du Vigneaud 1964). In the same year, Geoffrey Harris' monograph *Neural Control of the Pituitary Gland* presented the growing evidence that secretion of hormones from the anterior (endocrine) lobe of the pituitary gland in mammals is regulated by neurohormones released by hypothalamic neurons (Harris 1955), providing the foundations for a new branch of physiology—neuroendocrinology. Molecular identification of these hypothalamic releasing factors revealed that they are peptides, including gonadotropin-releasing hormone (GnRH), thyrotropin-releasing hormone (TRH), corticotropin-releasing hormone (CRH), somatostatin, and growth hormone-releasing hormone (GHRH). The importance of the discovery of these neurohormones was recognized in the award of the 1977 Nobel Prize for Physiology or Medicine to Roger Guillemin and Andrew Schally (Guillemin 1992; Schally 1992). Then, facilitated by advances in molecular biology in the 1980s and 1990s, the G-protein coupled receptors (GPCRs) that mediate the effects of the hypothalamic–pituitary neuropeptides were discovered (Civelli et al. 2013). Furthermore, investigation of hypothalamic–pituitary neuropeptides in non-mammalian vertebrates has provided insights into their evolution and comparative physiology (Norris and Carr 2013).

The aim of this chapter is to review and discuss important insights into the evolution and comparative physiology of hypothalamic–pituitary neuropeptides that have been provided by research on invertebrates, which do not have a hypothalamic–pituitary system but do have, with a few exceptions, a nervous system. Recognizing that students of mammalian/vertebrate neuroendocrinology may not be familiar with the variety and relationships of invertebrate taxa, I will first provide a brief introduction to animal phylogeny (Holland 2011). The animal kingdom comprises over thirty phyla and the majority of these are bilaterians—animals that exhibit bilateral symmetry during at least one stage of their life cycle and that have a common "urbilaterian" ancestor. There are also a few non-bilaterian phyla, including Cnidaria (e.g., sea anemones), Placozoa, Porifera (sponges), and Ctenophora (comb jellies), but these will not be discussed here because the hypothalamic–pituitary neuropeptides discussed in this chapter are only present in bilaterians. The Bilateria comprise two clades: the deuterostomes and the protostomes. The deuterostomes include three phyla, the Chordata (including vertebrates, urochordates, and cephalochordates), the Hemichordata (e.g., acorn worms), and the Echinodermata (e.g., starfish), although there is also evidence for a fourth phylum—the Xenacoelomorpha. The protostomes comprise at least twenty-two phyla, which sub-divide into two clades—the Lophotrochozoa and the Ecdysozoa. However, for the purposes of this chapter, I will mainly focus on four protostome phyla—(1) the Arthropoda (e.g., the insect *Drosophila melanogaster*) and the Nematoda (e.g., *Caenorhabditis elegans*), both of which are ecdysozoans and (2) the Annelida (e.g., the polychaete *Platynereis dumerilii*) and the Mollusca (e.g., the sea hare *Aplysia californica*). With this animal phylogeny as a framework, as shown in Fig. 2.1, I will now go on to discuss each of the hypothalamic–pituitary neuropeptides from an invertebrate perspective. I will start with vasopressin and oxytocin, hypothalamic neuropeptides released from the posterior pituitary gland in mammals, and then I will proceed onto the hypothalamic releasing factors that were discovered on account of their effects as regulators of hormone release from the anterior pituitary gland: GnRH, TRH, CRH, somatostatin, and GHRH.

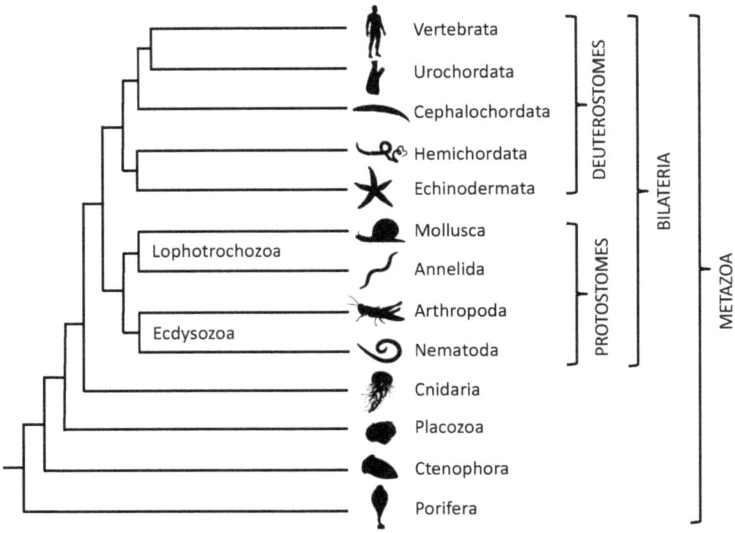

Fig. 2.1 Phylogenetic tree showing relationships of selected animal (metazoan) phyla, which include both bilaterian and non-bilaterian phyla. The Bilateria comprise two clades: (1) the deuterostomes, which include the phyla Chordata (sub-phyla Vertebrata, Urochordata, and Cephalochordata), Hemichordata, and Echinodermata and (2) the protostomes, which include over twenty phyla that sub-divide into two clades—the Lophotrochozoa (e.g., Mollusca and Annelida) and the Ecdysozoa (e.g., Arthropoda and Nematoda). The non-bilaterians include phyla that have nervous systems (Ctenophora and Cnidaria) and phyla that lack nervous systems (Porifera and Placozoa). The branch lengths in the tree are arbitrary. This diagram is adapted from a figure published previously (Elphick et al. 2018)

2.2 Vasopressin/Oxytocin-Type Neuropeptide Signaling

Research on vasopressin and oxytocin was initiated by the discovery that extracts of the pituitary gland contain substances that have a variety of physiological effects in mammals, including (1) an anti-diuretic effect (reduction in urine production), (2) an increase in blood pressure, (3) stimulation of lactation, and (4) stimulation of uterine contraction. Subsequently, it was discovered that the anti-diuretic and vasopressive effects were attributable to one substance (vasopressin) and the effects on lactation and the uterus were attributable to another substance (oxytocin). Furthermore, purification and sequencing of vasopressin and oxytocin revealed that they are structurally related neuropeptides, with vasopressin having the structure CYFQNCPRG-NH_2 (with a disulphide bridge between the cysteines) and oxytocin having the structure CYIQNCPLG-NH_2 (with a disulphide bridge between the cysteines). The GPCRs that mediate the effects of vasopressin and oxytocin have been identified in mammals, with three receptor types (V1a, V1b, V2) mediating effects of vasopressin and a single oxytocin receptor mediating effects of oxytocin. Furthermore, investigation of the evolution of vasopressin/oxytocin (VP/OT)-type neuropeptide signaling in vertebrates indicates that these two neuropeptide types evolved by duplication of a gene encoding a common ancestral peptide that would have been present in the common ancestor of jawed vertebrates and agnathans (e.g., lampreys) (Odekunle and Elphick 2020; Bichet et al. 2019; Elphick et al. 2018).

Insights from invertebrates into the evolution and comparative physiology of VP/OT-type neuropeptides were reviewed recently (Odekunle and Elphick 2020). Readers are referred to this article for a comprehensive overview of the primary literature up until 2020, but here I will present a short overview and an update on some more recent studies. Evidence that VP/OT-type neuropeptides occur in invertebrates was first obtained using antibodies to these neuropeptides, which revealed the presence of immunoreactive substances in the nervous systems of invertebrates (Rémy et al. 1979). However, definitive evidence that the evolutionary origin of VP/OT-type peptides can be traced back to the common ancestor of the Bilateria was first obtained with the sequencing of VP/OT-type peptides in an insect, the *Locusta migratoria*, and in molluscs, the cone snails *Conus geographus* and *Conus striatus* (Schooley et al. 1987; Cruz et al. 1987). Furthermore, sequencing of cDNAs/genes encoding the precursors of these peptides revealed the presence of a neurophysin domain that is a characteristic feature of VP/OT-type precursors in vertebrates (Van Kesteren et al. 1992a, b). The GPCRs that mediate the effects of these VP/OT-type neuropeptides were also identified in insects and molluscs (Van Kesteren et al. 1995; Stafflinger et al. 2008). Furthermore, VP/OT-type neuropeptides have also been identified in a variety of other invertebrates, including the urochordate *Styela clava*, the cephalochordate *Branchiostoma floridae*, the hemichordate *Saccoglossus kowalevskii*, the echinoderm *Strongylocentrotus purpuratus*, the annelid *Eisenia fetida*, and the nematode *C. elegans* (Ukena et al. 2008; Elphick 2010; Elphick and Rowe 2009; Oumi et al. 1994; Beets et al. 2012; Garrison et al. 2012). Investigation of the physiological roles of VP/OT-type neuropeptides in invertebrates has revealed both evolutionary conservation and diversification of their functions. For example, relevant to the effects of vasopressin on diuresis in vertebrates, the VP/OT-type neuropeptide inotocin stimulates diuresis in the beetle *Tribolium castaneum* (phylum Arthropoda) (Aikins et al. 2008). In accordance with roles of oxytocin in regulation of reproductive physiology and behavior in mammals, VP/OT-type neuropeptides have been found to have reproduction-associated roles in invertebrates. For example, in the earthworm *Eisenia fetida* (phylum Annelida), the VP/OT-type neuropeptide annetocin induces egg-laying related behaviors (Oumi et al. 1996), whilst in the nematode *C. elegans* VP/OT-type neuropeptide signaling is required for normal mating behavior in male animals (Garrison et al. 2012).

Broader perspectives on VP/OT-type neuropeptide function have been obtained as research has been extended to wider range of phyla. Echinoderms are unique amongst bilaterians in acquiring a radially symmetrical body plan as adult animals and therefore they provide an unusual anatomical context for exploration of neuropeptide function. Interestingly, a VP/OT-type neuropeptide known as asterotocin (CLVQDCPEG-NH_2, with a disulphide bridge between the cysteines) causes relaxation of neuromuscular preparations in the starfish *Asterias rubens* (phylum Echinodermata) (Odekunle et al. 2019). This contrasts with the myoexcitatory effects of VP/OT-type neuropeptides in vertebrates and several invertebrates, including other echinoderms; for example, the VP/OT-type neuropeptide echinotocin causes contraction of tube foot and esophagus preparations in the sea urchin *Echinus esculentus* (Elphick and Rowe 2009). Furthermore, asterotocin has a striking behavioral effect on starfish, triggering "fictive feeding" where eversion of the stomach out of the mouth (Fig. 2.2) and induction of a change in body posture resemble their extra-oral feeding behavior on prey (e.g., mussels) (Odekunle et al. 2019). These findings indicate that, in addition to ancient roles in regulation of reproduction and diuresis, VP/OT-type neuropeptides are evolutionarily ancient regulators of feeding behavior and, accordingly, inhibitory effects of oxytocin-type neuropeptides on feeding behavior in vertebrates have been reported (Olszewski et al. 2022). More recently, the effects of a VP/OT-type neuropeptide (holotocin) have been investigated in another echinoderm species—the sea cucumber *Apostichopus japonicus*. This revealed that holotocin causes contraction of the longitudinal body wall muscles in vitro, consistent with the myoexcitatory effect of echinotocin in sea urchins but contrasting with the myoinhibitory effect of asterotocin in starfish. Furthermore, consistent with the in vitro effect of holotocin, injection of this peptide in vivo causes changes in body shape in sea cucumbers (Cong et al. 2023). Thus, in both starfish and sea cucumbers, VP/OT-type neuropeptides have striking effects on body posture/shape.

Another phylum in which VP/OT-type neuropeptide signaling has been investigated for the first time recently is the Platyhelminthes, which are lophotrochozoan protostomes. Analysis of genome sequence data from parasitic species and the freshwater planarian *Schmidtea mediterranea* did not reveal the presence of genes encoding VP/OT-type neuropeptides. However, a gene encoding a VP/OT-type neuropeptide known as platytocin (CFIRNCPPG-NH_2, with a disulphide bridge between the cysteines) was identified in *Macrostomum lignano*, a free-living marine flatworm that lives in the intertidal zone. Furthermore, investigation of the effects of platytocin on marine flatworm species indicates physiological roles in regulation of water homeostasis and reproduction, providing further evidence that these are evolutionarily ancient and conserved roles of VP/OT-type neuropeptides in the Bilateria (Kobayashi et al., 2022).

Lastly, extending earlier studies that revealed the diuretic action of the VP/OT-type neuropeptide inotocin in an insect, the beetle *T. castaneum* (Aikins et al. 2008), more recent studies on other beetle species have revealed a potential roles for inotocin as a regulator of reproductive and parental behavior (Nagy et al. 2021; Potticary et al. 2023). Furthermore, experimental studies on several ant species have revealed evidence of roles of inotocin in regulation of social foraging behavior and metabolic processes associated with locomotion, appetite, and water retention (Fetter-Pruneda et al. 2021; Liutkeviciute et al. 2018; Koto et al. 2019).

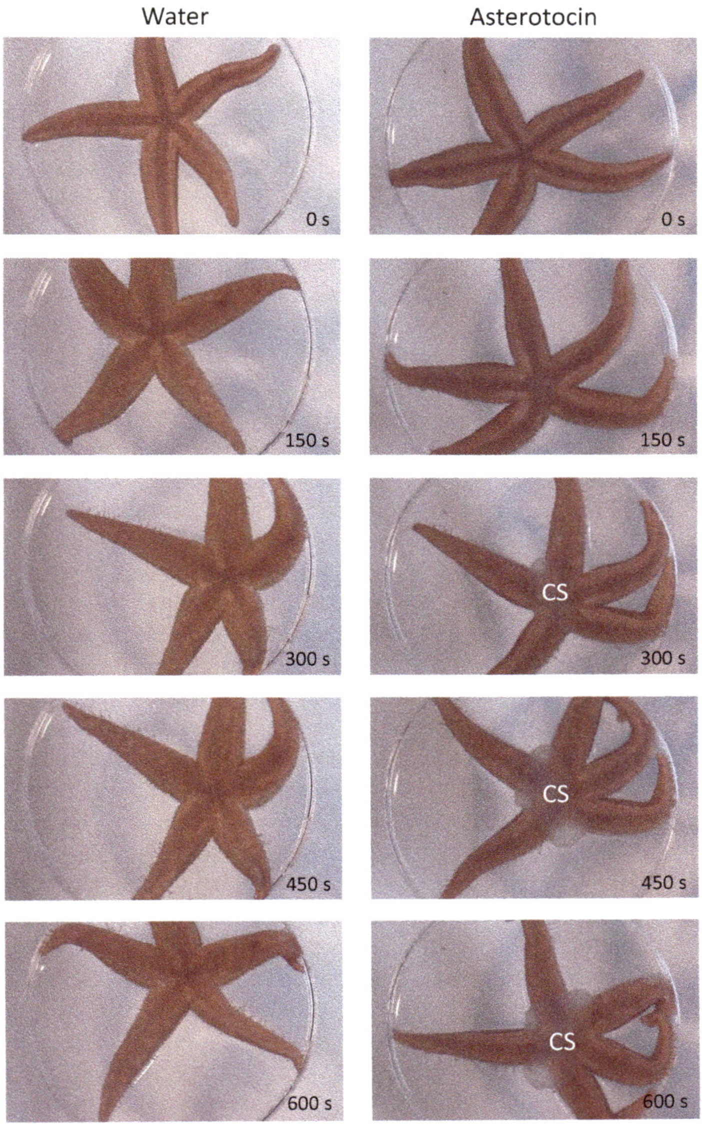

Fig. 2.2 The vasopressin/oxytocin-type neuropeptide asterotocin triggers eversion of the cardiac stomach in the starfish *Asterias rubens*. Starfish were injected with 10 μl of water (control; left panel) or 10 μl of 1 mM asterotocin (right panel). In starfish injected with asterotocin, eversion of the cardiac stomach (CS) occurred during a period of 10 min (600 s) after injection. In starfish injected with water, cardiac stomach eversion is not observed. Asterotocin also causes changes in the posture of starfish (not shown) that resemble the "humped" posture that starfish adopt when feeding on their prey. Thus, asterotocin induces "fictive feeding" in starfish, indicating that it acts as a neural mediator of the extra-oral feeding behavior of starfish, where the cardiac stomach is everted out of the mouth and over the soft, digestible tissue of prey such as mussels. The photographs in this figure are taken from video recordings of experiments reported previously (Odekunle et al. 2019)

2.3 GnRH-Type Neuropeptide Signaling

Purification and sequencing of mammalian GnRH revealed that it is a decapeptide with the structure pQHWSYGLRPG-NH_2 (where pQ = pyroglutamate). Subsequently, a second GnRH-type peptide (pQHWSHGWYPG-NH_2) was identified in humans, and GPCRs that mediate effects of GnRH-type peptides in mammals were identified (Cheng and Leung 2005). Furthermore, investigation of the occurrence and actions of GnRH-type signaling in non-mammalian vertebrates has revealed evolutionarily conserved roles in regulation of reproductive maturation and function, as well as other physiological processes (Ogawa et al. 2022). The first definitive evidence for the existence of GnRH-type neuropeptides in invertebrates was the identification of the peptides pQHWSDYFKPG-NH_2 and pQHWSLCHAPG-NH_2 in the sea-squirt *Chelyosoma productum* (phylum Urochordata) (Powell et al. 1996). More recently, a neuropeptide (YSYSYGFAP-NH_2) that acts as a ligand for two GnRH-type receptors in the cephalochordate *Branchiostoma floridae* has been identified (Yañez Guerra and Zandawala 2023). Thus, the discovery of GnRH-type signaling systems in invertebrate chordates has provided valuable insights into the evolution of GnRH-type signaling in the phylum Chordata. However, it was the identification of the ligands for GnRH-type receptors in protostomes, in particular arthropods and nematodes, that really transformed our understanding of the evolution of GnRH-type signaling. It was discovered that adipokinetic hormone (AKH; pQLNFTPNWGT-NH_2), a neuropeptide first discovered in insects in 1976, is the ligand for a GnRH-type receptor in *Drosophila* (Staubli et al. 2002). Likewise, an AKH-like neuropeptide in the nematode *C. elegans* was identified as the ligand for a GnRH-type receptor in this species (Lindemans et al. 2009). In retrospect, there are structural similarities between AKH and GnRH, with both peptides being decapeptides and both having the same N-terminal (pyroglutamate) and C-terminal (amide) post-translational modifications. However, the overall sequence identity between AKH and GnRH was too low to provide a basis for a hypothesis that they might be evolutionarily related, and it was only through the discovery that their receptors are closely related that it became evident that GnRH-type peptides in vertebrates and AKH-type peptides in arthropods are orthologs. Thus, two separate fields of neuropeptide research were united in a trans-phyletic embrace (Zandawala et al. 2018).

The discovery of the hitherto unknown relationship between GnRH and AKH was complicated, however, by the existence of two other AKH-like neuropeptides in insects and other arthropods: corazonin (CRZ) and AKH/corazonin-related peptide (ACP). Identification of the receptors for CRZ and ACP revealed that their receptors are related to vertebrate GnRH-type receptors and insect AKH-type receptors. However, the phylogenetic distribution and evolutionary origin of the AKH, ACP, and CRZ was unclear. It was first proposed that AKH, ACP, and CRZ are GnRH-related neuropeptides that uniquely evolved in the arthropod or protostome lineages (Hauser and Grimmelikhuijzen 2014). However, phylogenetic analysis revealed that CRZ-type receptors are also present in deuterostomes (e.g., hemichordates, cephalochordates), but not vertebrates (Roch et al. 2011). This indicated that CRZ-type signaling originated in the common ancestor of Bilateria but was lost in the lineage leading to the vertebrates (Mirabeau and Joly 2013). However, definitive proof of this hypothesis required the identification of neuropeptide ligands for CRZ-type receptors in invertebrate deuterostomes. Importantly, this was accomplished with the discovery of neuropeptides that act as ligands for a GnRH-type receptor and a CRZ-type receptor in an echinoderm, the starfish *A. rubens*. Thus, ArGnRH (pQI-HYKNPGWGPG-NH_2) acts as the ligand for the GnRH-type receptor ArGnRHR and ArCRZ (HNTFTMGGQNRWKAG-NH_2) acts as a ligand for the CRZ-type receptor ArCRZR (Tian et al. 2016). Furthermore, it was proposed that the short peptide FTYTHTW-NH_2 is a ligand for CRZ-type receptors in the cephalochordate *B. floridae* (Zandawala et al. 2018; Tian et al. 2016) and, importantly, this hypothesis has recently been confirmed experimentally (Yañez Guerra and Zandawala 2023). These findings have enabled reconstruction of the evolution of GnRH-type and CRZ-type neuropeptide signaling and a basis for revision of the nomenclature for the variety of GnRH/AKH/ACP/CRZ-related neuropeptides that have been identified in invertebrates. Thus, it was concluded that GnRH-type and CRZ-type neuropeptides are paralogs that evolved by gene duplication in the common ancestor of the Bilateria. GnRH-type neuropeptides and their receptors appear to have been retained in most, and possibly all, bilaterian phyla. However, a second gene duplication in the arthropod lineage gave rise to two paralogous GnRH-related neuropeptides, AKH and ACP, and their cognate receptors. In contrast to the conservation of GnRH-type signaling in bilaterians, there has been loss of CRZ-type signaling in vertebrates, urochordates, and nematodes (*C. elegans*) (Fig. 2.3) (Zandawala et al. 2018).

Turning now to the physiological roles of GnRH-related neuropeptides in invertebrates, evidence of evolutionarily ancient roles in regulation of reproductive processes has been obtained, consistent with the prototypical action of GnRH in vertebrates. For example, GnRH-type neuropeptides induce gamete release in the urochordate *Ciona intestinalis* (Terakado 2001). However, with the discovery that AKH is a GnRH-type neuropeptide in insects, it was clear that the roles of GnRH-type neuropeptides in invertebrates extend beyond reproductive physiology. Thus, AKH and/or ACP as co-orthologs of GnRH in insects are involved in regulation of a variety of physiological processes; for example, both AKH and ACP cause elevation of hemolymph carbohydrate and lipid levels and reduce food intake in the cricket *Gryllus bimaculatus* (Zhou et al. 2018). In the nematode *C. elegans*, silencing of genes encoding a GnRH/AKH-type neuropeptide or its receptor was found to cause a delay in egg-laying in the nematode *C. elegans*, indicative of a reproduction-associated role physiologically (Lindemans et al. 2009). Furthermore, more recent studies have revealed that there is an expanded family of genes encoding GnRH/AKH-related peptides and their receptors in *C. elegans* and other nematodes, with experimental investigations indicating that a subset of these signaling pathways are involved in regulation of sleep and wakefulness in *C. elegans* (Van der Auwera et al. 2020; Beets et al. 2023).

Turning to lophotrochozoans, recently the receptor for a GnRH/AKH-type neuropeptide (pQIHFSPDWGT-NH_2) in the mollusc *A. californica* has been

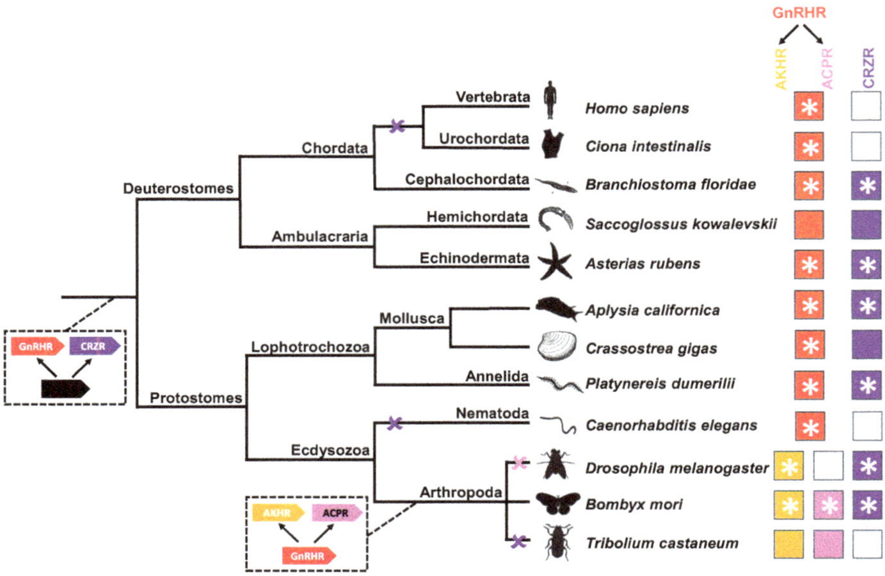

Fig. 2.3 Phylogenetic distribution and evolution of paralogous GnRH-type and corazonin-type neuropeptide receptors in the Bilateria. Analysis of the phylogenetic distribution of GnRH-type receptors (red) and corazonin-type receptors (purple) and their neuropeptide ligands indicates these are paralogous signaling systems that originated following duplication of genes encoding an ancestral receptor in an urbilaterian common ancestor of deuterostomes and protostomes. However, the corazonin-type signaling system has been lost in vertebrates, urochordates, nematodes, and some arthropods. Duplication of a GnRH-type receptor gene in a common ancestor of the arthropods has given rise to the paralogous adipokinetic hormone-type (AKH; yellow) and AKH/corazonin-related peptide-type (ACP; pink) receptors. On the right side of the phylogenetic tree, the empty squares (white) indicate loss of a signaling system and colored squares indicate presence of a signaling system in selected taxa, based on analysis of transcriptome/genome sequence data. Inclusion of an asterisk in a colored square indicates that the neuropeptide ligand for the receptor has been identified and tested experimentally. Loss of a signaling system is represented by X in branches of the phylogenetic tree. This figure has been adapted (with updates) from a diagram that was published previously (Tian et al. 2016)

characterized experimentally (Tsai et al. 2024), extending an earlier study that reported the effects of this peptide, which include inhibition of feeding, an increase in defecation, a reduction in body mass, and a reduction in gonadal mass and oocyte diameter (Johnson et al. 2014). GnRH/AKH-type neuropeptide signaling systems have also been identified in the annelid *Platynereis dumerilii* and analysis of the expression of a gene encoding the peptide pQLTQSLGWGSAGSS-NH_2 revealed that it is up-regulated in sexually-mature animals in comparison with immature animals, suggesting a reproduction-associated role (Andreatta et al. 2020). However, further studies are needed to gain more detailed insights into the physiological roles of GnRH/AKH-type neuropeptides in annelids.

Turning to invertebrate deuterostomes, in addition to aforementioned studies on urochordates (Terakado 2001), the expression and action of the GnRH-type neuropeptide ArGnRH have been investigated in the echinoderm *A. rubens*. This revealed a widespread pattern of expression in the central nervous system, digestive system, and body wall-associated appendages. Accordingly, investigation of the in vitro pharmacological effects of ArGnRH revealed that it is myotropic and causes contraction of three different neuromuscular organs: (1) the cardiac stomach, which is everted out of the mouth when starfish feed, (2) tube feet, body wall appendages that enable locomotion in starfish, and (3) the apical muscle, a strand of muscle that runs along the aboral inner surface of the arms in starfish (Tian et al. 2017). More recently, expression of ArGnRH in the reproductive system of *A. rubens* was investigated using immunohistochemistry with specific antibodies to this neuropeptide, but it was not detected. Accordingly, when tested on in vitro preparations of gonads from male and female starfish, no effects of ArGnRH were observed (Piñon Gonzalez et al. 2024).

In conclusion, whilst evidence of reproductive roles of GnRH-type neuropeptides in invertebrates has been reported, their functions clearly extend beyond reproductive biology as regulators of diverse physiological processes. Likewise, investigation of the physiological roles of the paralogous CRZ-type neuropeptide signaling system, which has been lost in vertebrates, has revealed diverse roles in both reproductive and non-reproductive processes (Tian et al. 2017; Tsai et al. 2010; Tayler et al. 2012; Kubrak et al. 2016).

2.4 TRH-Type Neuropeptide Signaling

The molecular identification of TRH in mammals revealed this to be one of the smallest known neuropeptides, comprising just three residues, pGlu-His-Pro-NH_2, with both N-terminal pyroglutamylation and C-terminal amidation as post-translational modifications. The effects of TRH as a regulator of pituitary hormone release and as neurotransmitter/neuromodulator in other regions of the brain are mediated by two receptors—TRH-R1 and TRH-R2 (Trubacova et al. 2022). The structure of TRH is conserved throughout the vertebrates and it has an evolutionarily conserved role as a regulator of the release of thyrotropin and/or other pituitary hormones (Galas et al. 2009).

During the four decades following the discovery of TRH in 1969, TRH-like peptides were not found in invertebrates and therefore it was thought that TRH may be unique to vertebrates. However, facilitated by application of genome sequencing methods to a growing variety of invertebrate taxa, in 2012, the discovery of a precursor of TRH-like peptides in the sea urchin *Strongylocentrotus purpuratus* was reported. This protein comprises peptides with amino acid sequences QYPGG, QWPGG, QFPAG, and QFPGG, which were predicted to be post-translationally modified to form the mature tetrapeptides pQYPG-NH_2, pQWPG-NH_2, pQFPA-NH_2, and pQFPG-NH_2 (Rowe and Elphick 2012). Subsequently, related precursor proteins have been identified in other echinoderms (Zandawala et al. 2017), and GPCRs that are orthologs of vertebrate TRH receptors have also been identified in echinoderms (Mirabeau and Joly 2013). However, experimental proof that the

TRH-like tetrapeptides (e.g., pQFPG-NH_2 in *S. purpuratus*) act as ligands for these receptors has yet to be reported. There have, however, been reports on the expression of the TRH-type precursor in *S. purpuratus*, revealing that it is expressed by just two cells in the pluteus larvae in this species (Wood et al. 2018).

Importantly, discovery of the TRH-type precursor in *S. purpuratus* also facilitated identification of the peptide FSEFLG-NH_2 as a putative TRH-type neuropeptide in the annelid *Platynereis dumerilii*. Furthermore, experimental testing of synthetic FSEFLG-NH_2 revealed that it acts as a ligand for a *Platynereis* GPCR that is an ortholog of vertebrate TRH-type receptors. Thus, the first experimental evidence for the existence of a TRH-type signaling system in an invertebrate was obtained (Bauknecht and Jekely 2015). Discovery of a TRH-type signaling system in *Platynereis* subsequently facilitated discovery of TRH-type signaling in the nematode *C. elegans*, including a protein that is the precursor of two pentapeptides (GRELF-NH_2 and ANELF-NH_2), both of which act as ligands for a GPCR that is an ortholog of vertebrate TRH receptors. Furthermore, investigation of the physiological roles of this TRH-type signaling system revealed that it promotes growth in *C. elegans*. In particular, expression of the TRH-type precursor in pharyngeal motor neurons is required for animals to attain a normal body size, and knockdown of the TRH-type receptor in pharyngeal muscle cells causes a reduction in growth. The discovery of this growth-promoting function of TRH-type signaling in *C. elegans* is of interest from a comparative perspective with respect to the roles of TRH in mammals in promoting growth by stimulating release of pituitary thyrotropin (Van Sinay et al. 2017). Thus, it appears that TRH may have an evolutionarily ancient and conserved role as a regulator of growth in the Bilateria. It will be of interest, therefore, to investigate the physiological roles of the TRH-type signaling system in other invertebrates. From this perspective, an obvious experimental system would be the insect *Drosophila*; however, analysis of the phylogenetic distribution of the TRH-type signaling system has revealed that it has been lost in *Drosophila*. Nevertheless, TRH-type signaling has not been lost in all arthropods and precursors of peptides that have a C-terminal EFL-NH_2 motif (EFLamides), in common with the *C. elegans* TRH-type peptides, have been identified in several arthropods, including some insect species (Veenstra and Simo 2020; Kotwica-Rolinska et al. 2020). Accordingly, experimental studies on the locust *Locusta migratoria* have demonstrated that two EFLamides act as ligands for a TRH-type receptor in this species. Interestingly, anatomical analysis of the EFLamide precursor in *L. migratoria* has revealed that it is expressed by a single pair of bilaterally symmetrical neurons in the protocerebrum of the brain. However, nothing is known about the physiological roles of EFLamides in the locust, although, informed by the pattern of neuronal EFLamide expression, it has been postulated that ELFamide-expressing neurons may modulate input onto the locust navigation system (Veenstra and Simo 2020). Null mutants for the ELFamide precursor have been generated in another insect species, the linden bug *Pyrrhocoris apterus*, but no effects on development, reproduction, lifespan, diapause induction, or circadian rhythmicity were observed (Kotwica-Rolinska et al. 2020). Thus, further studies are needed to gain insights into the physiological roles of TRH-type signaling in insects and other arthropods.

Discovery of the TRH-type signaling system in annelids, nematodes, and arthropods has also enabled recognition that a neuropeptide known as fulicin (FNEFV-NH_2), which was first identified in the mollusc *Achatina fulica* in 1991 (Ohta et al. 1991), is a homolog of EFLamides (Van Sinay et al. 2017). Fulicin has a variety of modulatory effects on in vitro preparations from *A. fulica*, including the penis retractor muscle, buccal muscle, ventricular muscle of the heart, and central neurons. Furthermore, the fulicin precursor gives rise to one copy of fulicin and nine other C-terminally amidated neuropeptides comprising four to six residues and expression of this precursor is detected in the parietal ganglia of the central nervous system and in the atrium of the heart (Yasuda-Kamatani et al. 1995; Satake et al. 1999). Hitherto, interest in fulicin may have been limited to those with a specific focus on molluscan neurobiology. The discovery that fulicin is a homolog of TRH may attract broader interest and further investigation of the physiological roles of TRH-type signaling molluscs.

2.5 CRH-Type Neuropeptide Signaling

The structure of CRH was first determined in 1981 as a C-terminally amidated 41-residue peptide. This revealed that CRH belongs to a family of related peptides that include sauvagine, a vasoactive peptide isolated from the skin of the frog *Phyllomedusa sauvagii*, and urotensin-I, a vasoactive peptide isolated from the urophysis of teleost fish. Furthermore, CRH-like peptides known as urocortin 1–3 (UCN1–3) have subsequently been identified in mammals and two GPCRs that are activated by CRH and/or UCN1–3 have been characterized (Hauger et al. 2003).

Precursors of CRH-type peptides have also been identified in deuterostome invertebrates, including the urochordate *C. intestinalis*, the cephalochordate *B. floridae*, and the echinoderms *A. rubens* and *Ophionotus victoriae* (Lovejoy and Barsyte-Lovejoy 2010; Mirabeau and Joly 2013; Semmens et al. 2016; Zandawala et al. 2017). However, little is known about the physiological roles of CRH-type neuropeptides in deuterostome invertebrates. Nevertheless, a detailed analysis of the expression pattern of the CRH-type neuropeptide ArCRH in *A. rubens* has been reported, providing a neuroanatomical basis for further investigation of the functions of this peptide in starfish (Cai et al. 2023).

A key insight into the evolutionary origin of CRH-type neuropeptide signaling was the discovery of a CRH-like neuropeptide that acts as an egg-laying hormone (ELH) in gastropod molluscs (Chiu et al. 1979; Mirabeau and Joly 2013). Subsequently, a diuretic hormone (DH44) in insects which shares sequence similarity with vertebrate CRH-type peptides was identified (Kataoka et al. 1989). Accordingly, the receptors that mediate the effects of DH44 in insects are closely related to vertebrate CRH-type receptors (Hector et al. 2009). Furthermore, a variety of other functions of DH44 in insects have been revealed, including roles in regulation of nutrient sensing, feeding, gut activity, immunity, wakefulness, and circadian activity (Lee et al. 2023). Thus, CRH-type signaling systems have been identified in both ecdysozoan and lophotrochozoan protostomes and, accordingly, the evolutionary origin of CRH-type neuropeptide signaling can be traced back to a common ancestor of the Bilateria (Mirabeau and Joly 2013; Cardoso et al. 2014). However, aside from ELH in gastropod molluscs and DH44 in insects, little is known about the physiological roles of CRH-type neuropeptides in other protostomes. Thus, there remains much to be learnt about the evolution and comparative physiology of CRH-type neuropeptide signaling in invertebrates.

2.6 Somatostatin-Type Neuropeptide Signaling

The discovery of somatostatin was facilitated by the discovery that hypothalamic extracts contain substances that can either stimulate (GHRH; see below) or inhibit (somatostatin) the release of growth hormone from the anterior lobe of the pituitary gland in mammals. Purification and sequencing of somatostatin revealed that it is a tetradecapeptide with a disulphide bridge between cysteine residues at positions 3 and 14. A paralog of somatostatin named cortistatin was discovered subsequently, enabled by cDNA sequencing of transcripts expressed in cortical regions of the mammalian brain. Furthermore, five GPCRs that mediate the effects of somatostatin and cortistatin have been discovered in mammals (Gunther et al. 2018).

Important advances in our knowledge and understanding of the evolutionary history of somatostatin-type signaling have been made recently with the discovery and functional characterization of somatostatin-related neuropeptides in a deuterostome invertebrate, the starfish *Asterias rubens* (phylum Echinodermata) (Zhang et al. 2020, 2022b), as discussed in more detail below. However, as a prelude, it is first necessary to discuss the discovery of allatostatin-C, a closely related neuropeptide that has been characterized in protostome invertebrates, and in particular in insects. Allatostatins are neurohormones that inhibit synthesis of juvenile hormone by the corpora allata, which are insect endocrine glands analogous to the pituitary gland in vertebrates. Three neuropeptide types have been found to act as allatostatins in insects—allatostatin-A, allatostatin-B, and allatostatin-C. However, it is only allatostatin-C that exhibits structural similarity with somatostatin, including the presence of a disulphide bridge between two cysteine residues. Furthermore, discovery of the GPCR that mediates the effects of allatostatin-C provided definitive evidence that allatostatin-C-type signaling is closely related to somatostatin-type signaling (Verlinden et al. 2015). Thus, phylogenetic analysis of GPCRs has revealed that allatostatin-C type receptors in protostomes are positioned in the same clade as vertebrate somatostatin-type receptors (Mirabeau and Joly 2013). Informed by this finding, it has been proposed that the allatostatin-C-type signaling system in protostomes is orthologous to the somatostatin-C-type signaling system in vertebrates. Furthermore, embracement of this concept has led to publication of papers that refer to allatostatin-C-type signaling as "the insect somatostatin pathway" (Zhang et al. 2022a). However, research on a deuterostome invertebrate, the starfish *A. rubens*, indicates that the concept of an "insect somatostatin pathway" may be flawed, as discussed below.

Analysis of neural transcriptome sequence data from *A. rubens* revealed the presence of two transcripts encoding precursor proteins that contain a somatostatin/allatostatin-C-like neuropeptide located in the C-terminal region of the proteins. Firstly, a precursor containing a neuropeptide with the amino acid sequence KCIGRFQPFSMPC, which was initially named ArSS but then subsequently renamed ArSS1, following discovery of the precursor of a second somatostatin/allatostatin-C-like neuropeptide. Secondly, a precursor containing a neuropeptide with the amino acid sequence RAKNARCMADFWKGRGLVDC, which was named ArSS2. Analysis of the sequence of ArSS1 revealed that it contains the core sequence Phe-Gln-Pro, which is similar to the Phe-Asn-Pro motif found in allatostatin-C-type peptides in the insect *D. melanogaster* and the nematode *C. elegans* (Fig. 2.4a). Furthermore, cluster-based analysis of the sequences of the precursors of allatostatin-C/somatostatin-type neuropeptides in bilaterians revealed a relationship between the ArSS1 precursor and precursors of SS1-type peptides in other echinoderms and protostome allatostatin-C-type precursors. Conversely, analysis of the sequence of ArSS2 revealed that it contains the core sequence Phe-Trp-Lys, which is identical to the core sequence of somatostatin and cortistatin in mammals (Fig. 2.4a). Accordingly, cluster-based analysis of the sequences of the precursors of allatostatin-C/somatostatin-type neuropeptides in bilaterians revealed a relationship between the ArSS2 precursor and precursors of SS2-type peptides in other echinoderms and vertebrate somatostatin/cortistatin-type precursors. Based on these findings, it was concluded that SS1-type neuropeptides in starfish and other echinoderms are orthologs of protostome allatostatin-C-type neuropeptides and SS2-type neuropeptides in starfish and other echinoderms are orthologs of vertebrate somatostatin/cortistatin-type neuropeptides (Zhang et al. 2020, 2022b).

The discovery that both allatostatin-C-type and somatostatin-type neuropeptides occur in echinoderms has changed our understanding of the evolutionary relationship of vertebrate somatostatin-type and protostome allatostatin-C-type neuropeptides. Thus, it indicates that somatostatin-type neuropeptides and allatostatin-C-type neuropeptides are paralogs, not orthologs as proposed previously. Furthermore, it is proposed that the gene duplication event that gave rise to paralogous somatostatin-type neuropeptides and allatostatin-C-type neuropeptides occurred in a common ancestor of the Bilateria. Then differential loss/retention of genes encoding these two neuropeptide types occurred in different animal lineages. Thus, there was loss of allatostatin-C-type neuropeptides and retention of somatostatin-type neuropeptides in the vertebrate/chordate lineage. Conversely, there was loss of somatostatin-type neuropeptides and retention of allatostatin-C-type neuropeptides in the protostome lineage (Fig. 2.4b). Furthermore, the retention of both allatostatin-C-type and somatostatin-type neuropeptides may be unique to echinoderms (Zhang et al. 2020, 2022b). This then raises the question as to why both neuropeptide types have been retained in echinoderms but differentially lost in vertebrates and protostomes. Addressing this issue has been facilitated by functional characterization of ArSS1 and ArSS2 in *A. rubens*. Analysis of the transcriptome/genome of *A. rubens* revealed the presence of three genes encoding GPCRs (ArSSR1–3) that are closely related to vertebrate somatostatin-type receptors. Furthermore, consistent with its

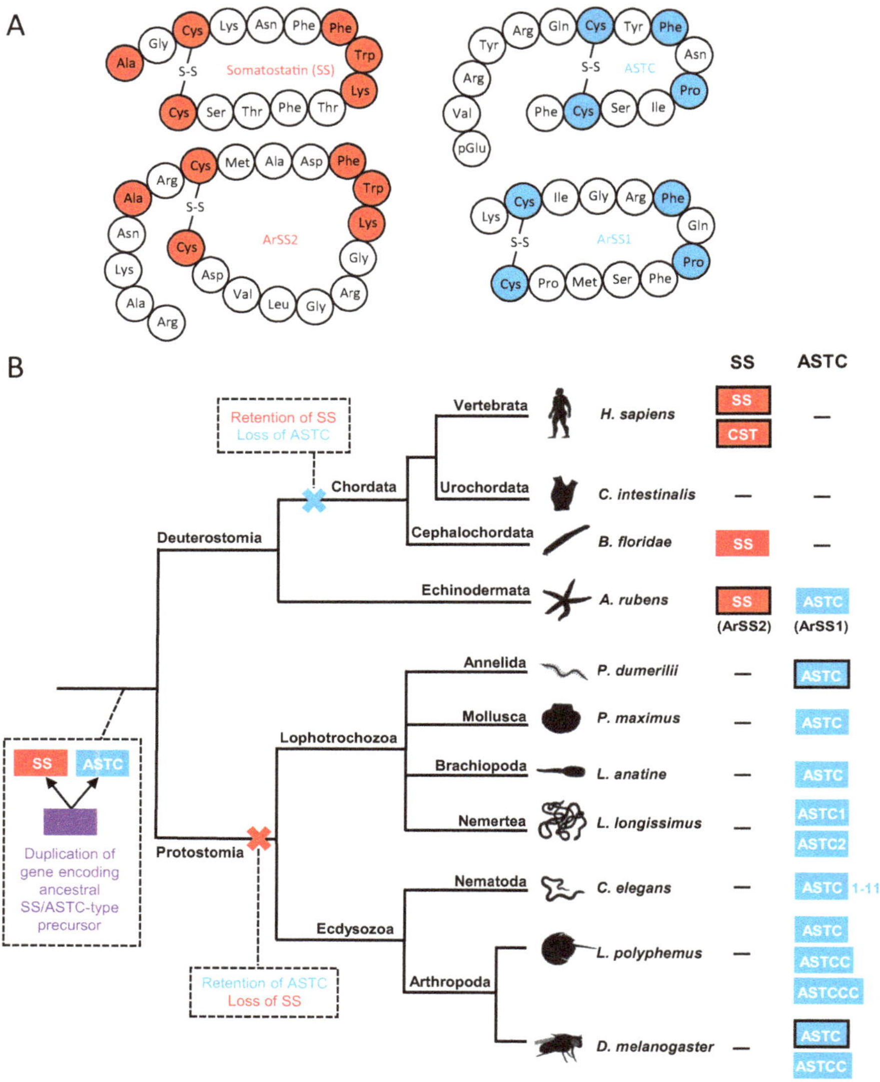

Fig. 2.4 Somatostatin-type and allatostatin-C-type neuropeptides are paralogous and originated following duplication of a gene encoding an ancestral neuropeptide in Urbilateria—the common ancestor of deuterostomes and protostomes. (**a**) Comparison of the structures of human somatostatin (SS), *Drosophila* allatostatin-C (ASTC), and the SS/ASTC-type neuropeptides ArSS1 and ArSS2 in the starfish *Asterias rubens*. ArSS2 shares sequence similarity with SS (as shown in red), whereas ArSS1 shares sequence similarity with ASTC (as shown in blue). (**b**) Phylogenetic diagram showing the occurrence of SS/ASTC-type signaling systems in selected bilaterian taxa. Following a gene duplication event that gave rise to SS-type (red) and ASTC-type (blue) neuropeptides in Urbilateria, there has been differential retention of these neuropeptides in different taxa. In chordates SS-type neuropeptides have been retained (SS and cortistatin (CST)), but ASTC-type neuropeptides have been lost (blue X). Conversely, in protostomes ASTC-type neuropeptides have

structural similarity with somatostatin/cortistatin-type neuropeptides, heterologous expression of ArSSR1–3 in mammalian cells revealed that ArSS2 acts as a ligand for all three receptors. In contrast, ArSS1 does not act as a ligand for ArSSR1–3 and ArSS1 receptor(s) remain to be identified (Zhang et al. 2020). It is noteworthy, however, that receptors closely related to protostome allatostatin-C-type receptors have not been found in echinoderms and therefore it is hypothesized that SS1-type neuropeptides in echinoderms exert their physiological effects via as yet undiscovered receptor mechanisms (Zhang et al. 2022b). Although the receptors that mediate effects of ArSS1 in starfish have yet to be discovered, the physiological roles of ArSS1 have been investigated. Informed by analysis of the expression of ArSS1 in *A. rubens*, in vitro pharmacological experiments revealed that ArSS1 is myoexcitatory, causing contraction of two neuromuscular organs—the cardiac stomach and locomotory appendages known as tube feet. Interestingly, ArSS2 has opposing effects, causing relaxation of in vitro preparations of the cardiac stomach and tube feet. Furthermore, it is noteworthy that the myoinhibitory effect of ArSS2 in starfish is consistent with the inhibitory actions of somatostatin as a regulator of physiological processes in mammals, including inhibition of growth hormone release. Thus, it is postulated that the ancestral role of an allatostatin-C/somatostatin-type neuropeptide in a common ancestor of the Bilateria was to act as an inhibitory regulator of physiological processes. Then following gene duplication to give rise to the paralogous allatostatin-C-type and somatostatin-type neuropeptides, this inhibitory regulatory role was: (1) retained by somatostatin-type neuropeptides in vertebrates/chordates, but with loss of allatostatin-C-type neuropeptide due to functional redundancy and (2) retained by allatostatin-C-type neuropeptides in protostomes, but with loss of somatostatin-type neuropeptides due to functional redundancy. Retention of both somatostatin-type and allatostatin-C-type neuropeptides in echinoderms may therefore be explained by acquisition of excitatory regulatory roles for allatostatin-C-type neuropeptides in this phylum, possibly in association with the evolution of novel receptors (Zhang et al. 2022b). Further insights into this issue will be obtained if the receptor for ArSS1 in *A. rubens* and/or for SS1-type neuropeptides in other echinoderms is discovered. In addition, it will also be of interest to investigate if SS1-type and SS2-type neuropeptides have excitatory and inhibitory effects, respectively, in other echinoderms.

A key conclusion from the molecular characterization of allatostatin-C-type and somatostatin-type neuropeptides in starfish was the absence/loss of these signaling systems in vertebrates/chordates and protostomes, respectively. However, there have recently been reports of identification of somatostatin-type neuropeptides in protostomes and it is important, therefore, to examine these findings. A long-term

Fig. 2.4 (continued) been retained, but SS-type neuropeptides have been lost (red X). However, both SS-type and ASTC-type neuropeptides have been uniquely retained in the phylum Echinodermata, where they have opposing myoinhibitory (ArSS2) and myoexcitatory (ArSS1) effects in the starfish *A. rubens*. Species in which the receptor for the SS/ASTC-type neuropeptide has been identified and tested experimentally are highlighted on the right side of the diagram by inclusion of black edging around the red or blue boxes. Part **b** of this figure has been adapted from a diagram published previously (Zhang et al. 2022b)

programme of research targeted toward identification of molecular components of venoms in predatory marine cone snails belonging to the genus *Conus* has identified a variety of bioactive peptides (Olivera 2002), including the VP/OT-type peptides known as conopressins (see Sect. 2.2 above). Thus, neuropeptides that normally regulate internal physiological processes appear to have been repurposed for their external effects on prey animals. Recently, novel components of *Conus* venoms have been identified which comprise peptides that share some structural similarity with somatostatins and which hence have been named "consomatins" (Ramiro et al. 2022). Furthermore, homologs of consomatins have been identified in non-venomous molluscs and annelids (Koch et al. 2022). However, there are two features of this family of protostome peptides which suggest that they are not in fact orthologs of the somatostatin-type neuropeptides that have been characterized in vertebrates and echinoderms. Firstly, analysis of the sequences of consomatins and related peptides in other protostomes reveals that there are only four residues between two conserved cysteine residues that form a disulphide bridge. This contrasts with vertebrate somatostatin-type peptides that have ten residues between two conserved cysteine residues that form a disulphide bridge and echinoderm somatostatin-type peptides (SS2-type) that have twelve residues between two conserved cysteine residues that form a disulphide bridge (Fig. 2.4). However, consomatins also share sequence similarity with urotensin-II-type peptides, a family of peptides that are closely related to somatostatins. Furthermore, and importantly, urotensin-II-type peptides are similar to consomatins in also having four residues between two cysteine residues that form a disulphide bridge (Koch et al. 2022). This suggests that consomatins could be homologs of urotensin-II-type peptides, rather than somatostatins. Further evidence in support of this hypothesis is the identification of a receptor for consomatin-related peptides in the annelid *Capitella teleta*. Phylogenetic analysis revealed that this receptor is not positioned in a clade that comprises vertebrate somatostatin receptors and the *A. rubens* SS2 receptors ArSSR1–3. Instead, the *Capitella* receptor is positioned in a clade which, together with vertebrate urotensin-II-type receptors, is positioned outside a clade that comprises somatostatin/allatostatin-C-type receptors (Koch et al. 2022). Collectively, these findings suggest that consomatins and related peptides in other molluscs and annelids are in fact members of the urotensin-II family, and not members of the somatostatin family. This conclusion is consistent with the proposition that there has been loss of somatostatin-type peptides in the protostome lineage (Zhang et al. 2022b).

Lastly, an interesting finding emerging from analysis of neuropeptide signaling in nematodes has been the discovery of expanded families of genes encoding allatostatin-C-type neuropeptide precursors and allatostatin-C/somatostatin-type receptors. Thus, in *C. elegans* eleven genes encoding allatostatin-C-type precursors have been identified and twelve genes encoding allatostatin-C/somatostatin-type neuropeptide receptors have been identified, whilst ligand–receptor pairings have been determined experimentally for two of these receptors (Beets et al. 2023; Zhang et al. 2020, 2022b). However, the functional significance of the expansion of allatostatin-C-type signaling in *C. elegans* and other nematodes remains to be determined. Opportunities to address this issue will be facilitated by recent mapping of the neuropeptidergic connectome in *C. elegans* and development of experimental methods for functional visualization of neuropeptide signaling in this species (Ripoll-Sanchez et al. 2023; Randi et al. 2023).

2.7 GHRH-Type Neuropeptide Signaling

GHRH (also known as somatoliberin or somatocrinin) was the last of the hypothalamic releasing factors to be discovered. It was isolated from a human pancreatic tumor that caused acromegaly and was revealed to be a forty-four-residue peptide, which shares sequence similarity with the glucagon-type family of peptide hormones that include glucagon and secretin. Accordingly, the receptor that mediates the physiological effects of GHRH belongs to the secretin-type family of GPCRs, which have a large N-terminal extracellular region that is involved in ligand binding (Mayo et al. 2003).

Analysis of the phylogenetic distribution of the glucagon-type family of GPCRs, including the GHRH receptor, has revealed that it appears to only occur in chordates. This contrasts with other secretin family receptors such as CRH-type receptors (see above), calcitonin-type receptors, and pigment dispersing factor-type receptors that are found in deuterostome and protostome invertebrates (Mirabeau and Joly 2013). Thus, it appears that GHRH, other glucagon-type peptides and their cognate receptors may have evolved uniquely in the vertebrate/chordate lineage.

2.8 Perspectives

Reviewing the phylogenetic distribution and evolution of the hypothalamic–pituitary neuropeptides highlights how inappropriate the nomenclature of these peptides is in the context of invertebrates that do not even have a hypothalamic–pituitary system. However, even in the context of mammalian physiology, it is clear that the names vasopressin, oxytocin, GnRH, TRH, CRH, somatostatin, and GHRH do not appropriately reflect the diversity of roles of these neuropeptides. Their names were, of course, assigned to them based on a single biological role, either as regulators of blood pressure or parturition (vasopressin and oxytocin, respectively) or as regulators of pituitary hormone release (GnRH, TRH, CRH, somatostatin, and GHRH). If these neuropeptides were first discovered now in the post-genomic era, they would likely be given much more generic names. So, should their names be changed? In my opinion, no. There is such a long and extensive history of publications on these neuropeptides that to change their names now would cause confusion and might also diminish recognition of many decades of research on their properties and functions. For the same reason, in my opinion, the names given to the hypothalamic–pituitary neuropeptides in mammals should generally be used for their invertebrate orthologs, with the primary motivation for doing so being to help ensure that research on neuropeptides in invertebrates gains appropriate attention and recognition. An exception to this recommendation though would be where an alternative name has already been assigned to a neuropeptide prior to the discovery of its relationship with a mammalian neuropeptide (e.g., DH44 and ELH, as homologs of CRH). Regarding VP/OT-type neuropeptides in invertebrates, a tradition for a nomenclature that combines the relationship with VP or OT and the source taxon has emerged (e.g., annetocin, inotocin, nematocin, and echinotocin). Some may regard this as rather whimsical, but naming of genes/proteins is often somewhat arbitrary or even intentionally comical; for example the *Drosophila* learning mutants that were given names of vegetables (Waddell and Quinn 2001).

The remarkable advances in our knowledge of the evolution of hypothalamic–pituitary neuropeptides in the twenty-first century have largely been enabled by genome sequencing. This began at the turn of the century with the sequencing of the genomes of the nematode *C. elegans* and the insect *D. melanogaster* (The *C. elegans* Sequencing Consortium 1998; Adams et al. 2000). However, as genome sequences have been determined for an ever-growing diversity of taxa, this has reaffirmed what zoologists always knew—there is no such thing as a "model invertebrate." And this perspective applies as much to hypothalamic–pituitary neuropeptides as any other aspect of biology. Thus, for example, the loss of genes encoding VP/OT-type and CRH-type neuropeptides in *D. melanogaster* and *C. elegans*, respectively, has necessitated investigation of other species to obtain comparative/evolutionary perspectives on these signaling molecules. Furthermore, the opportunities for advances in comparative neuroendocrinology emerge at an ever-increasing pace through international projects targeted toward sequencing of the genomes of all species on earth (https://www.earthbiogenome.org/). Looking ahead, the possibilities for obtaining new insights into the physiological roles of hypothalamic–pituitary neuropeptides in invertebrates are seemingly endless. But what a long way we have come so far, as we approach the one hundredth anniversary of the 1928 meeting of American Chemical Society where Oliver Kamm first reported the isolation of the pituitary hormones vasopressin and oxytocin.

Key Literature

Cruz et al. (1987): One of two papers that were the first to report the molecular identification of vasopressin/oxytocin-type peptides in invertebrates.

Mirabeau and Joly (2013): First comprehensive genomic analysis of the occurrence of neuropeptide signaling systems in bilaterian animals, which revealed the phylogenetic distribution of hypothalamic–pituitary neuropeptides and/or their receptors.

Schooley et al. (1987): One of two papers that were the first to report the molecular identification of vasopressin/oxytocin-type peptides in invertebrates.

Staubli et al. (2002): First paper to report the functional identification of a receptor for adipokinetic hormone (AKH) in an insect, which provided the basis for discovery of the evolutionary relationship (orthology) between AKH and GnRH.

Tian et al. (2016): First paper to report experimental identification of a neuropeptide ligand for a corazonin-type receptor in a deuterostome, revealing the urbilaterian evolutionary origin of corazonin-type neuropeptides as paralogs of GnRH-type neuropeptides.

Van Sinay et al. (2017): First paper to report the identification and functional characterization of a TRH-type signaling system in an invertebrate, the nematode *C. elegans*, revealing evidence of an evolutionarily ancient and conserved role for TRH-type neuropeptides as regulators of growth.

Zhang et al. (2022b): First paper to report the evolutionary origin of somatostatin-type neuropeptides as paralogs of allatostatin-C-type neuropeptides.

References

Adams MD, Celniker SE, Holt RA, Evans CA, Gocayne JD, Amanatides PG, Scherer SE, Li PW, Hoskins RA, Galle RF, George RA, Lewis SE, Richards S, Ashburner M, Henderson SN, Sutton GG, Wortman JR, Yandell MD, Zhang Q, Chen LX, Brandon RC, Rogers YH, Blazej RG, Champe M, Pfeiffer BD, Wan KH, Doyle C, Baxter EG, Helt G, Nelson CR, Gabor GL, Abril JF, Agbayani A, An HJ, Andrews-Pfannkoch C, Baldwin D, Ballew RM, Basu A, Baxendale J, Bayraktaroglu L, Beasley EM, Beeson KY, Benos PV, Berman BP, Bhandari D, Bolshakov S, Borkova D, Botchan MR, Bouck J, Brokstein P, Brottier P, Burtis KC, Busam DA, Butler H, Cadieu E, Center A, Chandra I, Cherry JM, Cawley S, Dahlke C, Davenport LB, Davies P, de Pablos B, Delcher A, Deng Z, Mays AD, Dew I, Dietz SM, Dodson K, Doup LE, Downes M, Dugan-Rocha S, Dunkov BC, Dunn P, Durbin KJ, Evangelista CC, Ferraz C, Ferriera S, Fleischmann W, Fosler C, Gabrielian AE, Garg NS, Gelbart WM, Glasser K, Glodek A, Gong F, Gorrell JH, Gu Z, Guan P, Harris M, Harris NL, Harvey D, Heiman TJ, Hernandez JR, Houck J, Hostin D, Houston KA, Howland TJ, Wei MH, Ibegwam C et al (2000) The genome sequence of *Drosophila melanogaster*. Science 287:2185–2195

Aikins MJ, Schooley DA, Begum K, Detheux M, Beeman RW, Park Y (2008) Vasopressin-like peptide and its receptor function in an indirect diuretic signaling pathway in the red flour beetle. Insect Biochem Mol Biol 38:740–748

Andreatta G, Broyart C, Borghgraef C, Vadiwala K, Kozin V, Polo A, Bileck A, Beets I, Schoofs L, Gerner C, Raible F (2020) Corazonin signaling integrates energy homeostasis and lunar phase to regulate aspects of growth and sexual maturation in *Platynereis*. Proc Natl Acad Sci U S A 117:1097–1106

Bauknecht P, Jékely G (2015) Large-scale combinatorial deorphanization of *Platynereis* neuropeptide GPCRs. Cell Rep 12:684–693

Beets I, Janssen T, Meelkop E, Temmerman L, Suetens N, Rademakers S, Jansen G, Schoofs L (2012) Vasopressin/oxytocin-related signaling regulates gustatory associative learning in *C. elegans*. Science 338:543–545

Beets I, Zels S, Vandewyer E, Demeulemeester J, Caers J, Baytemur E, Courtney A, Golinelli L, Hasakiogullari I, Schafer WR, Vertes PE, Mirabeau O, Schoofs L (2023) System-wide mapping of peptide-GPCR interactions in *C. elegans*. Cell Rep 42:113058

Bichet D, Bouvier M, Chini B, Gimpl G, Guillon G, Kimura T, Knepper M, Lolait S, Manning M, Mouillac B, O'Carroll A-M, Serradeil-Le Gal C, Soloff M, Verbalis JG, Wheatley M, Zingg HH (2019) Vasopressin and oxytocin receptors (version 2019.4) in the IUPHAR/BPS Guide to Pharmacology Database. IUPHAR/BPS Guide to Pharmacology CITE, 2019(4). https://doi.org/10.2218/gtopdb/F66/2019.4

Cai W, Egertová M, Zampronio CG, Jones AM, Elphick MR (2023) Molecular identification and cellular localization of a corticotropin-releasing hormone-type neuropeptide in an echinoderm. Neuroendocrinology 113:231–250

Cardoso JC, Felix RC, Bergqvist CA, Larhammar D (2014) New insights into the evolution of vertebrate CRH (corticotropin-releasing hormone) and invertebrate DH44 (diuretic hormone 44) receptors in metazoans. Gen Comp Endocrinol 209:162–170

Cheng CK, Leung PC (2005) Molecular biology of gonadotropin-releasing hormone (GnRH)-I, GnRH-II, and their receptors in humans. Endocr Rev 26:283–306

Chiu AY, Hunkapiller MW, Heller E, Stuart DK, Hood LE, Strumwasser F (1979) Purification and primary structure of the neuropeptide egg-laying hormone of *Aplysia californica*. Proc Natl Acad Sci U S A 76:6656–6660

Civelli O, Reinscheid RK, Zhang Y, Wang Z, Fredriksson R, Schioth HB (2013) G protein-coupled receptor deorphanizations. Annu Rev Pharmacol Toxicol 53:127–146

Cong X, Liu H, Zheng Y, Chen M (2023) A putative role of vasopressin/oxytocin-type neuropeptide in osmoregulation and feeding inhibition of *Apostichopus japonicus*. Int J Mol Sci 24

Cruz LJ, de Santos V, Zafaralla GC, Ramilo CA, Zeikus R, Gray WR, Olivera BM (1987) Invertebrate vasopressin/oxytocin homologs. Characterization of peptides from *Conus geographus* and *Conus straitus* venoms. J Biol Chem 262:15821–15824

Du Vigneaud V (ed) (1964) A trail of sulfa research: from insulin to oxytocin—Nobel Lecture, December 12, 1955. Nobel Lectures, Chemistry 1942-1962. Elsevier, Amsterdam

Elphick MR (2010) NG peptides: A novel family of neurophysin-associated neuropeptides. Gene 458:20–26

Elphick MR, Rowe ML (2009) NGFFFamide and echinotocin: structurally unrelated myoactive neuropeptides derived from neurophysin-containing precursors in sea urchins. J Exp Biol 212:1067–1077

Elphick MR, Mirabeau O, Larhammar D (2018) Evolution of neuropeptide signalling systems. J Exp Biol 221:jeb151092

Fetter-Pruneda I, Hart T, Ulrich Y, Gal A, Oxley PR, Olivos-Cisneros L, Ebert MS, Kazmi MA, Garrison JL, Bargmann CI, Kronauer DJC (2021) An oxytocin/vasopressin-related neuropeptide modulates social foraging behavior in the clonal raider ant. PLoS Biol 19:e3001305

Galas L, Raoult E, Tonon MC, Okada R, Jenks BG, Castano JP, Kikuyama S, Malagon M, Roubos EW, Vaudry H (2009) TRH acts as a multifunctional hypophysiotropic factor in vertebrates. Gen Comp Endocrinol 164:40–50

Garrison JL, Macosko EZ, Bernstein S, Pokala N, Albrecht DR, Bargmann CI (2012) Oxytocin/vasopressin-related peptides have an ancient role in reproductive behavior. Science 338:540–543

Guillemin R (1992) Peptides in the brain. The new endocrinology of the neuron—Nobel Lecture, December 8, 1977. In: Lindsten J (ed) Nobel lectures, physiology or medicine 1971-1980. World Scientific, Singapore

Gunther T, Tulipano G, Dournaud P, Bousquet C, Csaba Z, Kreienkamp HJ, Lupp A, Korbonits M, Castano JP, Wester HJ, Culler M, Melmed S, Schulz S (2018) International Union of Basic and Clinical Pharmacology. CV. Somatostatin receptors: structure, function, ligands, and new nomenclature. Pharmacol Rev 70:763–835

Harris GW (1955) Neural control of the pituitary gland. Edward Arnold, London, UK

Hauger RL, Grigoriadis DE, Dallman MF, Plotsky PM, Vale WW, Dautzenberg FM (2003) International Union of Pharmacology. XXXVI. Current status of the nomenclature for receptors for corticotropin-releasing factor and their ligands. Pharmacol Rev 55:21–26

Hauser F, Grimmelikhuijzen CJ (2014) Evolution of the AKH/corazonin/ACP/GnRH receptor superfamily and their ligands in the Protostomia. Gen Comp Endocrinol 209:35–49

Hector CE, Bretz CA, Zhao Y, Johnson EC (2009) Functional differences between two CRF-related diuretic hormone receptors in *Drosophila*. J Exp Biol 212:3142–3147

Holland PWH (2011) The animal kingdom: a very short introduction. Oxford University Press, Oxford

Johnson JI, Kavanaugh SI, Nguyen C, Tsai PS (2014) Localization and functional characterization of a novel adipokinetic hormone in the mollusk, *Aplysia californica*. PLoS One 9:e106014

Kataoka H, Troetschler RG, Li JP, Kramer SJ, Carney RL, Schooley DA (1989) Isolation and identification of a diuretic hormone from the tobacco hornworm, *Manduca sexta*. Proc Natl Acad Sci U S A 86:2976–2980

Koch TL, Ramiro IBL, Florez Salcedo P, Engholm E, Jensen KJ, Chase K, Olivera BM, Bjorn-Yoshimoto WE, Safavi-Hemami H (2022) Reconstructing the origins of the somatostatin and allatostatin-C signaling systems using the accelerated evolution of biodiverse cone snail toxins. Mol Biol Evol 39:msac075

Koto A, Motoyama N, Tahara H, McGregor S, Moriyama M, Okabe T, Miura M, Keller L (2019) Oxytocin/vasopressin-like peptide inotocin regulates cuticular hydrocarbon synthesis and water balancing in ants. Proc Natl Acad Sci U S A 116:5597–5606

Kotwica-Rolinska J, Kristofova L, Chvalova D, Pauchova L, Provaznik J, Hejnikova M, Sehadova H, Lichy M, Vaneckova H, Dolezel D (2020) Functional analysis and localisation of a thyrotropin-releasing hormone-type neuropeptide (EFLa) in hemipteran insects. Insect Biochem Mol Biol 122:103376

Kubrak OI, Lushchak OV, Zandawala M, Nässel DR (2016) Systemic corazonin signalling modulates stress responses and metabolism in *Drosophila*. Open Biol 6:160152

Lee G, Jang H, Oh Y (2023) The role of diuretic hormones (DHs) and their receptors in *Drosophila*. BMB Rep 56:209–215

Lindemans M, Liu F, Janssen T, Husson SJ, Mertens I, Gade G, Schoofs L (2009) Adipokinetic hormone signaling through the gonadotropin-releasing hormone receptor modulates egg-laying in *Caenorhabditis elegans*. Proc Natl Acad Sci U S A 106:1642–1647

Liutkeviciute Z, Gil-Mansilla E, Eder T, Casillas-Perez B, Di Giglio MG, Muratspahic E, Grebien F, Rattei T, Muttenthaler M, Cremer S, Gruber CW (2018) Oxytocin-like signaling in ants influences metabolic gene expression and locomotor activity. FASEB J 32:fj201800443

Lovejoy DA, Barsyte-Lovejoy D (2010) Characterization of a corticotropin-releasing factor (CRF)/diuretic hormone-like peptide from tunicates: insight into the origins of the vertebrate CRF family. Gen Comp Endocrinol 165:330–336

Mayo KE, Miller LJ, Bataille D, Dalle S, Goke B, Thorens B, Drucker DJ (2003) International Union of Pharmacology. XXXV. The glucagon receptor family. Pharmacol Rev 55:167–194

Mirabeau O, Joly JS (2013) Molecular evolution of peptidergic signaling systems in bilaterians. Proc Natl Acad Sci U S A 110:E2028–E2037

Nagy NA, Nemeth Z, Juhasz E, Poliska S, Racz R, Kiss J, Kosztolanyi A, Barta Z (2021) Inotocin, a potential modulator of reproductive behaviours in a biparental beetle, *Lethrus apterus*. J Insect Physiol 132:104253

Norris DO, Carr JA (2013) Vertebrate endocrinology. Elsevier, Amsterdam

Odekunle EA, Elphick MR (2020) Comparative and evolutionary physiology of vasopressin/ oxytocin-type neuropeptide signaling in invertebrates. Front Endocrinol (Lausanne) 11:225

Odekunle EA, Semmens DC, Martynyuk N, Tinoco AB, Garewal AK, Patel RR, Blowes LM, Zandawala M, Delroisse J, Slade SE, Scrivens JH, Egertová M, Elphick MR (2019) Ancient role of vasopressin/oxytocin-type neuropeptides as regulators of feeding revealed in an echinoderm. BMC Biol 17:60

Ogawa S, Yamamoto N, Hagio H, Oka Y, Parhar IS (2022) Multiple gonadotropin-releasing hormone systems in non-mammalian vertebrates: Ontogeny, anatomy, and physiology. J Neuroendocrinol 34:e13068

Ohta N, Kubota I, Takao T, Shimonishi Y, Yasuda-Kamatani Y, Minakata H, Nomoto K, Muneoka Y, Kobayashi M (1991) Fulicin, a novel neuropeptide containing a D-amino acid residue isolated from the ganglia of *Achatina fulica*. Biochem Biophys Res Commun 178:486–493

Olivera BM (2002) Conus venom peptides: reflections from the biology of clades and species. Annu Rev Ecol Syst 33:25–47

Olszewski PK, Noble EE, Paiva L, Ueta Y, Blevins JE (2022) Oxytocin as a potential pharmacological tool to combat obesity. J Neuroendocrinol 34:e13106

Oumi T, Ukena K, Matsushima O, Ikeda T, Fujita T, Minakata H, Nomoto K (1994) Annetocin: An oxytocin-related peptide isolated from the earthworm, *Eisenia foetida*. Biochem Biophys Res Commun 198:393–399

Oumi T, Ukena K, Matsushima O, Ikeda T, Fujita T, Minakata H, Nomoto K (1996) Annetocin, an annelid oxytocin-related peptide, induces egg-laying behavior in the earthworm, *Eisenia foetida*. J Exp Zool 276:151–156

Piñon Gonzalez VM, Feng Y, Egertová M, Elphick MR (2024) Neuropeptide expression and action in the reproductive system of the starfish *Asterias rubens*. J Comp Neurol 532(1):e25585
Potticary AL, Cunningham CB, McKinney EC, Moore PJ, Belay AT, Moore AJ (2023) Insect homolog of oxytocin/vasopressin associated with parenting of males but not females in a subsocial beetle. Evolution 77:2029–2038
Powell JF, Reska-Skinner SM, Prakash MO, Fischer WH, Park M, Rivier JE, Craig AG, Mackie GO, Sherwood NM (1996) Two new forms of gonadotropin-releasing hormone in a protochordate and the evolutionary implications. Proc Natl Acad Sci U S A 93:10461–10464
Ramiro IBL, Bjorn-Yoshimoto WE, Imperial JS, Gajewiak J, Salcedo PF, Watkins M, Taylor D, Resager W, Ueberheide B, Brauner-Osborne H, Whitby FG, Hill CP, Martin LF, Patwardhan A, Concepcion GP, Olivera BM, Safavi-Hemami H (2022) Somatostatin venom analogs evolved by fish-hunting cone snails: From prey capture behavior to identifying drug leads. Sci Adv 8:eabk1410
Randi F, Sharma AK, Dvali S, Leifer AM (2023) Neural signal propagation atlas of *Caenorhabditis elegans*. Nature 623:406–414
Rémy C, Girardie J, Dubois MP (1979) Vertebrate neuropeptide-like substances in the suboesophageal ganglion of two insects: *Locusta migratoria* R. and F. (Orthoptera) and *Bombyx mori* L. (Lepidoptera) immunocytological investigation. Gen Comp Endocrinol 37:93–100
Ripoll-Sanchez L, Watteyne J, Sun H, Fernandez R, Taylor SR, Weinreb A, Bentley BL, Hammarlund M, Miller DM 3rd, Hobert O, Beets I, Vertes PE, Schafer WR (2023) The neuropeptidergic connectome of *C. elegans*. Neuron 111:3570–3589 e5
Roch GJ, Busby ER, Sherwood NM (2011) Evolution of GnRH: diving deeper. Gen Comp Endocrinol 171:1–16
Rowe ML, Elphick MR (2012) The neuropeptide transcriptome of a model echinoderm, the sea urchin *Strongylocentrotus purpuratus*. Gen Comp Endocrinol 179:331–344
Satake H, Yasuda-Kamatani Y, Takuwa K, Nomoto K, Minakata H, Nagahama T, Nakabayashi K, Matsushima O (1999) Characterization of a cDNA encoding a precursor polypeptide of a D-amino acid-containing peptide, achatin-I and localized expression of the achatin-I and fulicin genes. Eur J Biochem 261:130–136
Schally AV (1992) Aspects of hypothalamic regulation of the pituitary gland with major emphasis on its implications for the control of reproductive processes—Nobel Lecture, December 8, 1977. In: Lindsten J (ed) Nobel Lectures, physiology or medicine 1971-1980. World Scientific, Singapore
Schooley DA, Miller CA, Proux JP (1987) Isolation of two arginine vasopressin-like factors from ganglia of *Locusta migratoria*. Arch Insect Biochem Physiol 5:157–166
Semmens DC, Mirabeau O, Moghul I, Pancholi MR, Wurm Y, Elphick MR (2016) Transcriptomic identification of starfish neuropeptide precursors yields new insights into neuropeptide evolution. Open Biol 6:150224
Stafflinger E, Hansen KK, Hauser F, Schneider M, Cazzamali G, Williamson M, Grimmelikhuijzen CJ (2008) Cloning and identification of an oxytocin/vasopressin-like receptor and its ligand from insects. Proc Natl Acad Sci USA 105:3262–3267
Staubli F, Jorgensen TJ, Cazzamali G, Williamson M, Lenz C, Sondergaard L, Roepstorff P, Grimmelikhuijzen CJ (2002) Molecular identification of the insect adipokinetic hormone receptors. Proc Natl Acad Sci U S A 99:3446–3451
Tayler TD, Pacheco DA, Hergarden AC, Murthy M, Anderson DJ (2012) A neuropeptide circuit that coordinates sperm transfer and copulation duration in *Drosophila*. Proc Natl Acad Sci U S A 109:20697–20702
Terakado K (2001) Induction of gamete release by gonadotropin-releasing hormone in a protochordate, *Ciona intestinalis*. Gen Comp Endocrinol 124:277–284
The *C. elegans* Sequencing Consortium (1998) Genome sequence of the nematode *C. elegans*: a platform for investigating biology. Science 282:2012–2018
Tian S, Zandawala M, Beets I, Baytemur E, Slade SE, Scrivens JH, Elphick MR (2016) Urbilaterian origin of paralogous GnRH and corazonin neuropeptide signalling pathways. Sci Rep 6:28788

Tian S, Egertová M, Elphick MR (2017) Functional characterization of paralogous gonadotropin-releasing hormone-type and corazonin-type neuropeptides in an echinoderm. Front Endocrinol (Lausanne) 8:259

Trubacova R, Drastichova Z, Novotny J (2022) Biochemical and physiological insights into TRH receptor-mediated signaling. Front Cell Dev Biol 10:981452

Tsai PS, Sun B, Rochester JR, Wayne NL (2010) Gonadotropin-releasing hormone-like molecule is not an acute reproductive activator in the gastropod, *Aplysia californica*. Gen Comp Endocrinol 166:280–288

Tsai PS, Sanders KN, Dai X, Plachetzki D (2024) Authentication of a lophotrochozoan adipokinetic hormone receptor in a Gastropod, *Aplysia californica*. Gen Comp Endocrinol 345:114393

Ukena K, Iwakoshi-Ukena E, Hikosaka A (2008) Unique form and osmoregulatory function of a neurohypophysial hormone in a urochordate. Endocrinology 149:5254–5261

Van der Auwera P, Frooninckx L, Buscemi K, Vance RT, Watteyne J, Mirabeau O, Temmerman L, De Haes W, Fancsalszky L, Gottschalk A, Raizen DM, Nelson MD, Schoofs L, Beets I (2020) RPamide neuropeptides NLP-22 and NLP-2 act through GnRH-like receptors to promote sleep and wakefulness in *C. elegans*. Sci Rep 10:9929

Van Kesteren RE, Smit AB, De With ND, Van Minnen J, Dirks RW, Van der Schors RC, Joosse J (1992a) A vasopressin-related peptide in the mollusc *Lymnaea stagnalis*: peptide structure, prohormone organization, evolutionary and functional aspects of Lymnaea conopressin. In: Joosse J, Buijs RM, Tilders FJH (eds) Progress in Brain Research. Elsevier

Van Kesteren RE, Smit AB, Dirks RW, De With ND, Geraerts WPM, Joosse J (1992b) Evolution of the vasopressin/oxytocin superfamily: characterization of a cDNA encoding a vasopressin-related precursor, preproconopressin, from the mollusc *Lymnaea stagnalis*. Proc Natl Acad Sci USA 89:4593–4597

Van Kesteren RE, Tensen CP, Smit AB, Van Minnen J, Van Soest PF, Kits KS, Meyerhof W, Richter D, Van Heerikhuizen H, Vreugdenhil E, Geraerts WPM (1995) A novel G protein-coupled receptor mediating both vasopressin- and oxytocin-like functions of Lys-conopressin in *Lymnaea stagnalis*. Neuron 15:897–908

Van Sinay E, Mirabeau O, Depuydt G, Van Hiel MB, Peymen K, Watteyne J, Zels S, Schoofs L, Beets I (2017) Evolutionarily conserved TRH neuropeptide pathway regulates growth in *Caenorhabditis elegans*. Proc Natl Acad Sci U S A 114:E4065–E4074

Veenstra JA, Simo L (2020) The TRH-ortholog EFLamide in the migratory locust. Insect Biochem Mol Biol 116:103281

Verlinden H, Gijbels M, Lismont E, Lenaerts C, Vanden Broeck J, Marchal E (2015) The pleiotropic allatoregulatory neuropeptides and their receptors: A mini-review. J Insect Physiol 80:2–14

Waddell S, Quinn WG (2001) Flies, genes, and learning. Annu Rev Neurosci 24:1283–1309

Wood NJ, Mattiello T, Rowe ML, Ward L, Perillo M, Arnone MI, Elphick MR, Oliveri P (2018) Neuropeptidergic systems in pluteus larvae of the sea urchin *Strongylocentrotus purpuratus*: neurochemical complexity in a “simple” nervous system. Front Endocrinol (Lausanne) 9:628

Yañez Guerra LA, Zandawala M (2023) Discovery of paralogous GnRH and corazonin signaling systems in an invertebrate chordate. Genome Biol Evol 15:evad108

Yasuda-Kamatani Y, Nakamura M, Minakata H, Nomoto K, Sakiyama F (1995) A novel cDNA sequence encoding the precursor of the D-amino acid-containing neuropeptide fulicin and multiple alpha-amidated neuropeptides from *Achatina fulica*. J Neurochem 64:2248–2255

Zandawala M, Moghul I, Yañez Guerra LA, Delroisse J, Abylkassimova N, Hugall AF, O'Hara TD, Elphick MR (2017) Discovery of novel representatives of bilaterian neuropeptide families and reconstruction of neuropeptide precursor evolution in ophiuroid echinoderms. Open Biol 7:170129

Zandawala M, Tian S, Elphick MR (2018) The evolution and nomenclature of GnRH-type and corazonin-type neuropeptide signaling systems. Gen Comp Endocrinol 264:64–77

Zhang Y, Yañez Guerra LA, Egertová M, Zampronio CG, Jones AM, Elphick MR (2020) Molecular and functional characterization of somatostatin-type signalling in a deuterostome invertebrate. Open Biol 10:200172

Zhang C, Kim AJ, Rivera-Perez C, Noriega FG, Kim YJ (2022a) The insect somatostatin pathway gates vitellogenesis progression during reproductive maturation and the post-mating response. Nat Commun 13:969

Zhang Y, Yañez-Guerra LA, Tinoco AB, Escudero Castelán N, Egertová M, Elphick MR (2022b) Somatostatin-type and allatostatin-C-type neuropeptides are paralogous and have opposing myoregulatory roles in an echinoderm. Proc Natl Acad Sci U S A 119:e2113589119

Zhou YJ, Fukumura K, Nagata S (2018) Effects of adipokinetic hormone and its related peptide on maintaining hemolymph carbohydrate and lipid levels in the two-spotted cricket, *Gryllus bimaculatus*. Biosci Biotechnol Biochem 82:274–284

Volume Transmission of Neurotransmitters and Neuropeptides in the Brain

3

Betty Rodriguez Cortes, Konstantina Chachlaki, and Vincent Prevot

Abstract

The brain performs a multitude of astonishing functions, ranging from fundamental processes such as regulating eating and sleeping to more complex phenomena, such as self-consciousness, believed to have evolved particularly in primates. In vertebrates, these functions are partly modulated by well-established wired transmission, known as synaptic transmission. However, this mode of operation has proven insufficient to explain more complex emergent processes. The cerebrospinal fluid (CSF) and interstitial fluid bathe the entire brain, providing access to various cell types not only in the brain but also in the spinal cord. Exploring volume transmission may offer insights into aspects of brain function that cannot be elucidated solely through wired transmission studies. In this chapter, we will delve into various instances of volume transmission in vertebrates, exploring its potential therapeutic applications in humans. Additionally, we will explore the evolutionary implications of volume transmission, providing insights into its development and significance across phyla.

Keywords

Cerebrospinal fluid · Ependymal cells · Tanycytes · Neuropeptides

B. R. Cortes · K. Chachlaki · V. Prevot (✉)
Univ. Lille, CHU Lille, Inserm, Laboratory of Development and Plasticity of the Neuroendocrine Brain, Lille Neuroscience & Cognition Research Center, UMR_S1172, Lille, France
e-mail: vincent.prevot@inserm.fr

V. Grinevich, R. F. Oliveira (eds.), *Evolutionary and Comparative Neuroendocrinology*, Masterclass in Neuroendocrinology 17,
https://doi.org/10.1007/978-3-031-80209-6_3

Abbreviations

3V	Third ventricle of the brain
AMPA	α-amino-3-hydroxy-5-methyl-4-isoxazolepropionic acid
AP	Area postrema
ARC/ME	Anatomical complex of arcuate nucleus/median eminence
ARF6	ADP-ribosylation factor 6
AVP	Arginine vasopressin, a small peptide expressed in the brain
BBB	Blood-brain barrier
CA	Cerebral aqueduct
cGMP	Cyclic guanosine monophosphate
CO	Carbone monoxide
CO2	Carbon dioxide
CSF	Cerebrospinal fluid
Ctβ	Cholera toxin β subunit
CVO	Circumventricular organs
DMH	Dorsomedial hypothalamus
ECS	Extracellular space
eNOS or NOS3	Endothelial nitric oxide synthase or nitric oxide synthase 3
Fe2+	Element iron in its +2-oxidation state
GABA	Gamma-aminobutyric acid
GnRH	Gonadotropin-releasing hormone
Hb	Hemoglobin
i.c.v.	Intracerebroventricular
iNOS or NOS2	Inducible nitric oxide synthase or nitric oxide synthase 2
ISF	Interstitial space fluid
IT	Intrathecal
kDa	Kilodalton, unit of molecular weight
LV	Lateral ventricle
LVT	Local volume transmission
MCH	Melanin-concentrating hormone
MCH1R	Melanin-concentrating hormone receptor 1
MCHR	Melanin-concentrating hormone receptor
MCHR2	Melanin-concentrating hormone receptor 2
MePO	Medial preoptic area
MVb	Multi vesicular bodies
nm	Nanometer, unit of length
nM	Nanomolar, unit of concentration
NMDAR	*N*-methyl-D-aspartate receptor
nNOS or NOS1	Neuronal nitric oxide synthase or nitric oxide synthase 1
NO	Nitric oxide
OT	Oxytocin
OVLT	Organum vasculosum of the lamina terminalis

PDZ	Type of domain of 80–110 amino acids
pM	Picomolar, a unit of concentration
PSD95	Postsynaptic density protein 95
PVN	Paraventricular nucleus of the hypothalamus
ROR2	Receptor tyrosine kinase-like orphan receptor-2
SCO	Sub-commissural organ
SFO	Subfornical organ
VMH	Ventromedial hypothalamus
VT	Volume transmission
Wnt5a	Wingless-type MMTV integration site family member 5A
WT	Wired transmission

3.1 Introduction

Pioneering experiments by Legendre and Pieron in 1913 demonstrated that injecting cerebrospinal fluid (CSF) from a sleep-deprived dog into the cisterna magna of a normal dog induced sleep in the recipient for 2–6 h (Pieron and Legendre 1910, 1913). In contrast, recipient dogs injected with CSF from well-rested dogs remained awake (Pappenheimer et al. 1967).

In 1973, CSF from 24-h fasted or fed/satiated sheep (donors) was collected from the lateral ventricles and injected into the lateral or third ventricle (LV or 3V, respectively) of either 24-h fasted or fed sheep (recipients). Recipient-fed sheep ate more following injections of CSF from fasted donors compared to when they were injected with CSF from equally "satiated" donors (Martin et al. 1973).

The concept of volume transmission (VT) of brain-derived molecules was explicitly proposed in 1986 by Agnati et al. (1986). Unlike synaptic transmission, which is discrete and dependent on neuronal connections, VT operates more diffusely and can affect non-neuronal cells since it does not rely on synapses.

In the 1980s, techniques such as receptor autoradiography and immunohistochemistry allowed the mapping of neurotransmitter locations. It was observed that the density of positive nerve terminals for a particular neurotransmitter and the density of its receptors were not always compatible. The presence of neurotransmitters outside of synapses under various experimental conditions suggested that spillover from the synaptic cleft into the surrounding extracellular space (ECS) was a source of nonsynaptic communication, known as local volume transmission (Rice and Cragg 2008).

We will review both classical and recent examples of volume transmission, highlighting its potential for widespread brain communication. Unlike synaptic transmission, which is relatively fast, volume transmission operates more slowly and can involve various cell types, including neurons, tanycytes, ependymal glial cells in the walls of the third ventricle, astrocytes, oligodendrocytes, and other non-neuronal cells, regardless of anatomical connectivity.

3.2 Transmission via the CSF

The physiological relevance of neuronal communication through CSF transmission has not yet been systemically investigated; for example, in the CSF of rats one can find neurotransmitters such as dopamine (ng/mL) 14.3 ± 7.0, noradrenaline (ng/mL) 31.6 ± 7.9, adrenaline (ng/mL) 0.89 ± 1.77, serotonin (ng/mL) 14.0 ± 20.3, melatonin (ng/mL) 0.411 ± 0.202, phosphoethanolamine (ng/mL) 1431 ± 298, GABA (ng/mL) 33.4 ± 54.5, and histamine (ng/mL) 3.43 ± 5.78 (Tsutsumi et al. 2020). What are the actual target(s) and functionality(ies) of all of these neurotransmitters?

Studies mapping the movement of a 76 kDa protein (iduronate-2-sulfatase, I2S) into the CSF show that after intrathecal administration (IT) in non-human primates, even large proteins can diffuse and reach the brain in significant amounts (Papisov et al. 2013; Yang et al. 2013). Also, both intracerebroventricular (icv) and lumbar intrathecal administration of recombinant I2S in dogs and non-human primates resulted in the wide distribution of this protein both in neurons and oligodendrocytes (Calias et al. 2012; Papisov et al. 2013). The fact that I2S entered the CSF and from the CSF into different brain cells illustrates the potential of proteins in the CSF to be internalized by various cell types. However, it is important to note that this does not constitute an example of VT (volume transmission). In VT, one must consider the receptor affinities, the abundance of the receptors at sites bordering the ventricles, and whether the concentration of the particular neurotransmitter is sufficient to initiate cellular signaling in neurons or other types of cells.

It is thought that volume transmission involving neuropeptide release into the CSF is especially favored as a hypothalamic output, given the anatomical arrangement of the entire tuberal region of the hypothalamus, where both neuronal dendrites and axons extend toward the ventricular surface (Alpár et al. 2019; Chen et al. 2017; Leak and Moore 2012; Taub et al. 2021). Nerve endings can be found very close to the ventricular surface (Leak and Moore 2012; Noble et al. 2018), and even more compelling, neuronal activity correlates with surges in ventricular neuropeptide levels (Caraty and Skinner 2008; Noble et al. 2018).

To determine which brain regions communicate with the CSF, Leak and Moore (2012) describe the distribution of retrograde neuronal labeling in the rat brain following ventricular injection of Cholera toxin, β subunit (CTβ). Within 2–10 days, researchers reported extensive labeling of neuronal perikarya in specific nuclear groups in the telencephalon, thalamus, hypothalamus, and brainstem, many at a considerable distance from the ventricles. This finding reinforces the idea that ventricular CSF can serve as a channel for volume transmission (Leak and Moore 2012). The study identifies potential sources of ventricular innervation, including proximal (periventricular regions) and distal hypothalamic areas (lateral, posterior, paraventricular, and ventromedial hypothalamus), as well as extrahypothalamic regions (cholinergic territories and dorsal raphe).

3.3 Ependymal Cells and Tanycytes: Key Actors in Volume Transmission

Based on the evidence presented in the previous section, one might initially conceive that axons are situated on the outer surface of the cerebral ventricles. However, this notion seems improbable. Instead, axons are typically found within the subependymal space, positioned opposite the basal surface of ciliated ependymal cells (Alpár et al. 2019; Franke et al. 1978), although exceptions exist (Taub et al. 2021). Local movement of CSF depends on ependymal cell ciliary beating in ventricular regions. However, adding more layers of complexity, the ciliary beating and CSF flow vary inside the third ventricle and appear to be regulated by the circadian clock (Faubel et al. 2016). Additionally, one study identified three distinct types of ependymal cells in the third ventricle; these cells were classified based on their cilia beating frequency as type I (>60 Hz), type II (30–60 Hz), or type III (<30 Hz) (Liu et al. 2014).

Ciliary beating and CSF flow vary inside the third ventricle according to the ATP/ADP ratio, as a proxy of the energy expenditure, as well as with the Ca^{2+} concentration. A decrease in the glucose concentration (from 25 to 0.1 mM) induces a reversible increase in ciliary beating frequency in vitro (Meste et al. 2015), tentatively influencing volume transmission.

Recent single-cell RNA sequencing assays (Campbell et al. 2017; Chen et al. 2017; Romanov et al. 2017) have unveiled a fascinating aspect of ependymal cells. These cells not only express a variety of receptors, including NMDA, AMPA, GABAA, muscarinic acetylcholine and monoamine receptors, but also have the capacity to produce various neuromodulators such as lipids, peptides, amines, histamine, nucleotides, and different glycoproteins.

This discovery suggests the possibility of synaptic-like innervation of spatially segregated ependymal cells (Alpár et al. 2018), which could directly modulate or initiate signaling cascades within the ventricular system. These processes may regulate ciliary beating, influencing the pace and direction of CSF flow. Moreover, they may initiate and/or amplify long-range ventricular volume transmission by triggering communication processes among ependymal cells, neurons, astrocytes, or other cell types responsive to the modulators released by these cells (Alpár et al. 2019).

Supporting this notion, experiments conducted in mice have demonstrated that the release of melanin-concentrating hormone (MCH) by neurons in the lateral hypothalamic area can regulate ciliary beating in the ventral region of the third ventricle. However, this regulation is not observed in the dorsal ventricular area or at the entrance to the cerebral aqueduct (CA). This regulation is achieved through the activation of the MCHR1 receptor expressed in ependymal cells (Conductier et al. 2013a, b) (Fig. 3.1).

Besides multiciliated ependymocytes, tanycytes are positioned strategically to influence CSF composition, making them significant contributors to volume transmission. These specialized ependymoglial cells, equipped with one or two non-beating cilia, line the CSF within circumventricular organs (CVOs) such as the organum vasculosum of the lamina terminalis (OVLT), the subfornical organ (SFO),

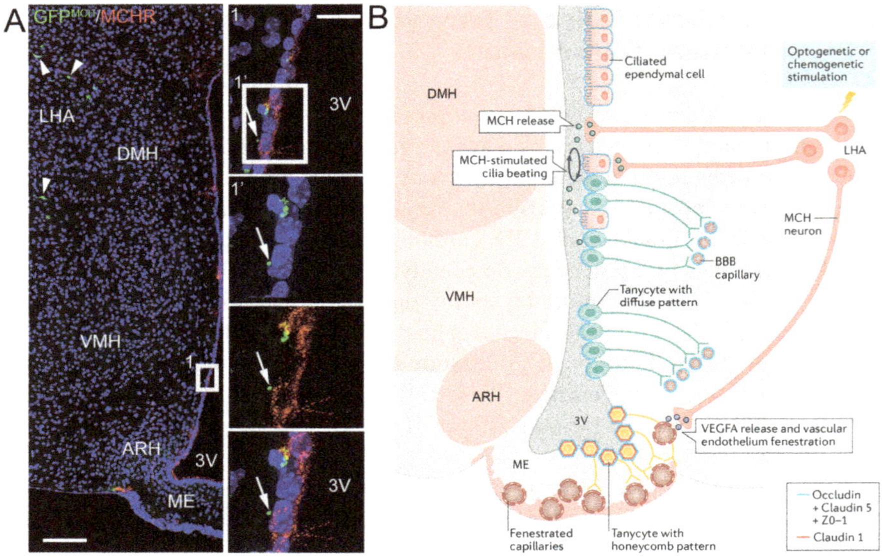

Fig. 3.1 MCH neurons form synaptoid contacts with MCHR1-expressing ciliated ependymal cells lining the wall of the third ventricule in the tuberal region of the hypothalamus to control cilia beating and CSF flow. (**a**) Representative frontal section from showing cell bodies of MCH neurons (green, arrowheads) in the lateral hypothalamic area (LHA) and synaptoid-like MCH buttons (arrows) morphologically associated with ependymal cells of the wall of the third ventricle (3V). 1 and 1′ insets show higher magnification of the GFP and MCHR1 immunoreactivities along the wall of the 3V in the transition between the arcuate nucleus of the hypothalamus (ARH) and the ventromedial nucleus of the hypothalamus (VMH). DMH, dorsomedial nucleus of the hypothalamus; ME, median eminence. (**b**) Schematics showing that MCH neurons in the LHA send processes toward the wall of the 3V, which is lined by ciliated ependymocytes, the blood-brain barrier (BBB), and fenestrated capillaries (which are located in the tuberal region of the hypothalamus). Chemogenetic or optogenetic activation of MCH neurons releases MCH into the cerebrospinal fluid (CSF). MCH regulates CSF flow by promoting the beating of cilia in ependymocytes that express MCH receptor 1 (MCHR1). This increased CSF flow accelerates volume transmission of CSF-borne metabolic signals. In parallel, MCH neurons induce fenestration (thereby increasing the permeability of the BBB) of blood vessels arising from the median eminence (ME) that run close to the ventricular wall, where they interact with tanycyte processes, via the release of vascular endothelial growth factor A (VEGFA). The color key indicates the distribution of tight junction proteins in the cell types forming the BBB in this hypothalamic region, namely tanycytes and vascular endothelial cells in brain capillaries. Adapted with permission from Al-Massadi et al. (2021)

the area postrema (AP), and the anatomical complex of arcuate nucleus of the hypothalamus (ARH)/median eminence (ME). In the tuberal region of the hypothalamus, tanycytes border the dorsomedial and ventromedial nuclei (DMH and VMH, respectively).

Tanycytes have drawn early attention due to their unique location and extensive interactions within the brain (Rodríguez et al. 2005; Prevot et al. 2018). Their soma directly contacts the CSF, while their processes extend into the brain parenchyma, engaging with neurons, astrocytes, and blood vessels across various nuclei in the mediobasal hypothalamus. They notably interface with permeable fenestrated capillaries in the OVLT, SFO, ME, and AP. These observations have led to hypotheses suggesting tanycytes as communication bridges between the bloodstream, CSF, and neural circuits in the hypothalamus.

In the tuberal hypothalamic region, tanycytes facilitate communication between the CSF and the perivascular space of fenestrated portal capillaries. They also connect the CSF with blood vessels protected by the blood-brain barrier (BBB), astrocytes and neurons in key hypothalamic circuits (Prevot et al. 2021). Early research indicates a polarized expression of proteins involved in endocytosis and transcytosis in tanycytes, suggesting bidirectional movement of molecules between the CSF, hypothalamic parenchyma, and portal circulation (Rodríguez et al. 2005; Prevot et al. 2018).

Further examination through transmission electron microscopy reveals intricate cellular structures within tanycytes, including coated pits, vesicles, vacuoles, and multivesicular bodies, indicating their involvement in the transport of bioactive molecules between CSF, portal blood, and parenchyma (Peruzzo et al. 2004). Tanycytes likely play a crucial role in facilitating the entry of circulating metabolic signaling molecules into the CSF from the periphery, enabling far-reaching communication within the brain via volume transmission (Prevot et al. 2021).

Supporting evidence indicates that tanycytes express numerous neurotransmitter and metabolic signal receptors or transporters, which facilitate activity-dependent release and bidirectional communication between the CSF and pituitary portal blood circulation (Prevot et al. 2018). Furthermore, their expression may be modulated by circadian rhythms (Rodríguez-Cortés et al. 2022). While tanycytes have been recognized as molecular shuttles for various hormones, including ghrelin, leptin, GLP1 analogs, liver FGF21 and glucose (Nampoothiri et al. 2022), their counterparts lining the lateral and dorsal walls of the third ventricle have received less attention even though they appear to perceive and translate CSF-borne information for neurons (Lhomme et al. 2021; Barahona et al. 2024; Bolborea et al. 2020).

A compelling study by Nualart et al. (2023) suggests that tanycytic-like ependymal cells located in the subcommissural organ (SCO), a circumventricular organ situated at the caudal extent of the third ventricle, may influence the ciliary beating of adjacent ependymal cells. This modulation occurs through the release of secretory granules containing spondin, which interacts with ependymal cilia, increasing their stiffness. Such SCO-spondin release was seen to occur in hyperglycemic conditions (Nualart et al. 2023).

3.4 Local Volume Transmission

3.4.1 Structural Changes of the Interstitial Space Fluid

Local volume transmission occurs as a result of the movement of neuroactive substances through the interstitial space fluid (ISF). Current models of neuronal plasticity greatly stress the importance of transient electrical and biochemical events associated with the excitation process. However, there is compelling evidence that activation is accompanied by other important physical phenomena. Microstructural changes in tissues have been observed in response to the changing physiological context, initially in electron microscopic studies (Hatton 1997; Theodosis et al. 2008) and subsequently through in vitro diffusion measurements (Piet et al. 2004). For example, within the magnocellular neurosecretory system, astrocytes can retract their processes from around synapses to influence transmitter spillover and excitability of magnocellular neurons during dehydration, parturition, and lactation (Oliet et al. 2001; Panatier et al. 2006; see for review Clasadonte and Prevot 2018). Similarly, it has been established that fluctuating levels of gonadal steroids during the reproductive cycle have the capacity to alter the tanycytic sheath, thus regulating "neuroendocrine synapse" formation by hypothalamic neurons that release gonadotropin-releasing hormone (GnRH) both in rodents (King and Letourneau 1994; Prevot et al. 1999) and birds (Yamamura et al. 2004), and to promote structural plasticity in neuronal circuits controlling sexual behavior in mice (Inoue et al. 2019). Interestingly, with the development of advanced magnetic resonance imaging (MRI) techniques such as diffusion and spectroscopy, tissue structure can now be probed and imaged on a microscopic scale in vivo, including in humans. One approach consists in assessing the water diffusion coefficient (Fig. 3.2). This technique provides information about the cellular structure of tissue by detecting microscopic obstacles to diffusion of water molecules, such as membranes and macromolecules. Indeed, the extracellular space is not empty but contains different proteins of different sizes, electrostatic charges, etc.; the movement, direction, and speed of the molecules through the ECF will be influenced by all these factors (Syková and Nicholson 2008). A longitudinal study of healthy young female volunteers under oral contraception has shown that diffusion of water molecules in the hypothalamus changes during the artificial menstrual cycle and diffusivity is maximal at the pill-free period (Baroncini et al. 2010). This result strongly suggests the existence of sex steroid hormone-driven changes in local volume transmission in the hypothalamus of young adult women, a phenomenon that may be key to the control of reproductive competence in humans.

3.4.2 Nitric Oxide: The Prototypic Volume Neurotransmitter

Nitric oxide (NO) serves as the quintessential neurotransmitter for volume transmission. NO is a biological intracellular messenger endogenously produced by three NO synthase (NOS) isoforms: neuronal (nNOS or NOS1), endothelial (eNOS or NOS3), and inducible (iNOS or NOS2) NOS, all of which are responsible for the enzymatic formation of NO through a reaction that converts L-arginine and oxygen into citrulline and NO. The production of brain-derived NO is primarily triggered by the activation of the NMDA subtype of glutamate receptor. This occurs through their interaction via PDZ domains and the formation of a ternary complex involving the scaffolding protein PSD95 (for review, see Chachlaki and Prevot 2020). The

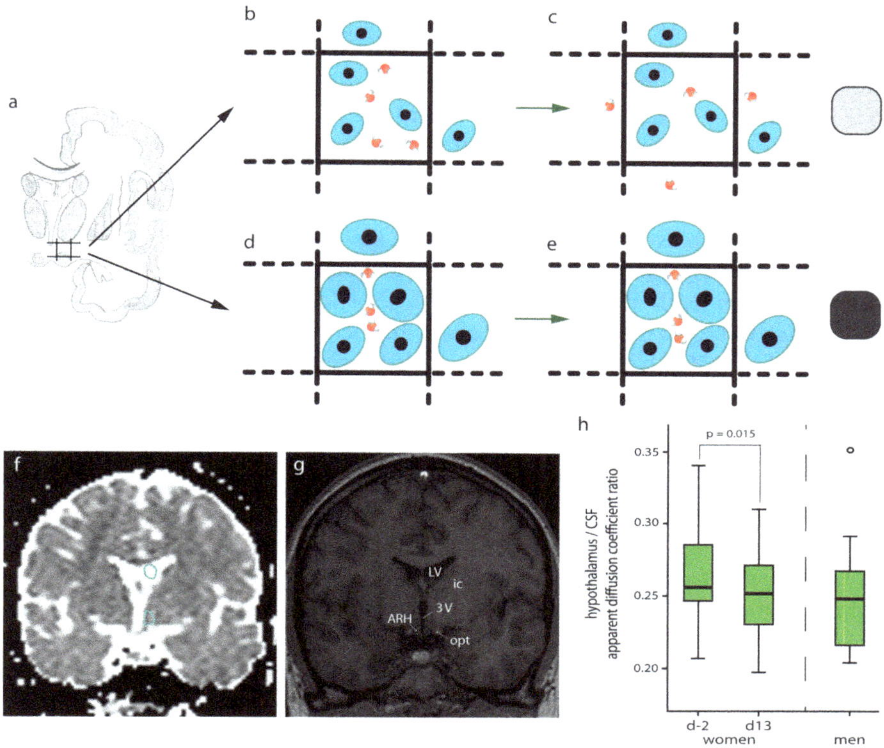

Fig. 3.2 Measurement of Brownian motion of water molecules using diffusion MRI. (**a**) Anatomical specification of the voxel (sampling volume) chosen in this example. In a given tissue, when the space between cells is relatively large (**b**), water molecules are free to move from one voxel to another (**c**). Upon application of the second gradient the spin of hydrogen nuclei of displaced water molecules are not in phase. This results in a high apparent diffusion coefficient (ADC) that is coded as hypersignal displayed in clear gray. Conversely, when the space between the cells is relatively small (**d**), water molecules are densely packed allowing the spins of hydrogen nuclei to be in phase. This results in a low ADC that is coded as hyposignal displayed in dark gray (**e**). (**f**) ADC mapping of the brain in coronal view with highlighted regions of interest (CSF of the lateral ventricles and mediobasal hypothalamus) where ADC is measured. (**g**) T1-weighted MRI in the same plane identifying anatomical landmarks (LV, lateral ventricle; ic, internal capsule; 3V, third ventricle; ARH, arcuate nucleus of the hypothalamus; opt, optic tract). (**h**) Box plot representation of the ratio of hypothalamic ADC/cerebrospinal fluid (CSF) ADC values measured in women (n = 10) at two stages of their artificial menstrual cycle (d-2: 2 days before use of oral contraception in the pill-free interval; d13, 13 days after initiating oral contraception) and in men (n = 10). Adapted from Baroncini et al. (2010) with permission

influx of Ca^{+2} through the activated NMDA receptors is largely responsible for activating nNOS and subsequently producing NO (Garthwaite et al. 1988).

NO is a molecule that is unique due to its physicochemical properties, including its lack of chemical specialization. The three properties that contribute most to its chemistry include its small size, uncharged nature, and the presence of an unpaired electron. Given its relatively small size and uncharged state, NO has a high diffusion constant, allowing it to rapidly diffuse and travel long distances in a short period. However, due to the presence of its third characteristic, i.e., the unpaired electron, NO is searching to acquire chemical stability by binding to either other unpaired electrons (other radical species), or transition metals, confining thus its diffusion field in proximity to the production source. For example, in the blood, NO cannot exist in meaningful amounts since its radical nature favors covalent binding to transition metals such as ferrous (Fe^{2+}), and thus the majority of NO is rapidly scavenged by different ferroheme proteins, such as hemoglobin (Hb). However, in the tissue, this high reactivity of NO with Fe^{2+} is a "biological advantage," allowing for the binding of NO to the ferrous iron of another heme-containing enzyme, the guanylate cyclase (GC). Unlike Hb, GC has an extremely low affinity to oxygen, even in the presence of enormous amounts, enabling GC to sense minimal NO, leading to several hundred-fold activation of its guanosine 3′,5′-cyclic monophosphate (cGMP) synthesis activity, rendering GC the only known and highly selective NO receptor, mediating most of NO's physiological effects (Chachlaki and Prevot 2020)

Exploring the mechanism of NO synthesis within the postsynaptic density, it becomes clear that NO primarily acts as a synapse-specific signal. Based on the assumption that each NMDAR forms a complex with one nNOS enzyme, a postsynaptic structure can generate approximately 40 molecules of NO per second, resulting in a concentration of about 60 pM in the postsynaptic source. This concentration is expected to decrease dramatically within a distance of 400 nm, corresponding to the synapse's outer boundaries. According to the Garthwaite mathematical model (Garthwaite 2016, 2019) at the level of the single synapse, NO receptor activation would result in about 10 nM cGMP, a concentration too low to induce any physiological effects. Hence, NO is most probably operating in an activity-dependent manner, becoming effective only with repeated NO pulses. In fact, repeated NO pulses of approximately 10–20 Hz, which is a stimulation frequency used to optogenetically promote kisspeptin-induced GnRH neuronal activity and/or secretion, can provide the summated cGMP formation capable of transmitting physiological effects (e.g., ten pulses of NO release per second could result in 100 nM cGMP). Importantly, a large number of NO molecules in a tissue volume can give rise to non-negligible NO concentrations outside the structure, allowing NO to act away from its point of synthesis and engage target cells regardless of their anatomical connectivity. This local volume transmission can occur in regions where a great number of NOS-expressing cells synchronize their activity (Fig. 3.3).

One of the most well-described examples of such a phenomenon occurs in the organum vasculosum of the lamina terminalis (OVLT) and the medial preoptic area (MePO) of the hypothalamus, where nNOS-expressing neurons are densely packed, with NO sources being less than 50 μm away from each other (Bellefontaine et al. 2014) (Fig. 3.3). Peripheral hormones such as estrogen and leptin or even neuropeptides such as kisspeptin can "synchronize" the activation status of the nNOS enzyme in the area, enabling sufficient build-up of intercellular NO concentration, allowing

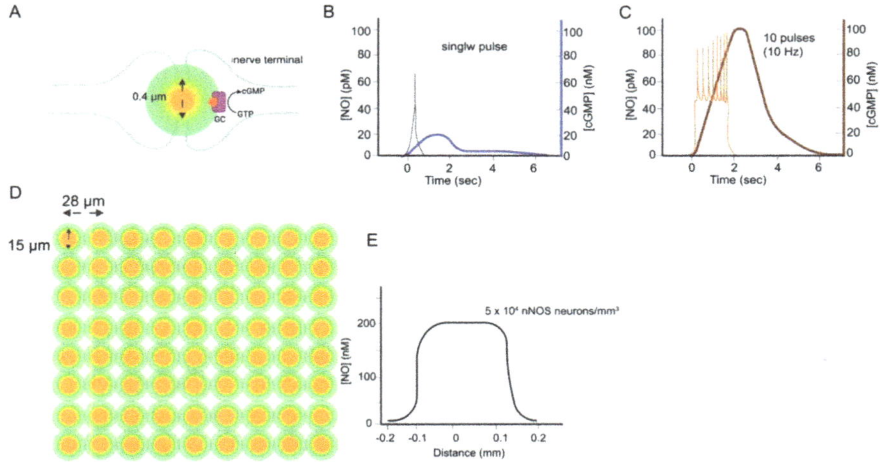

Fig. 3.3 Nitric oxide volume transmission. (**a**) Within an individual synapse NO acts as a retrograde messenger operating in an activity-dependent manner. NO production will result in the activation of guanylate cyclase in a 1:1 ratio, leading to the conversion of GTP to cGMP. (**b**) With a rate of synthesis of 40 molecules per second, the NO levels within the source (orange circle) will arrive at about 60 pM, dropping about ten times in the outer boundaries of the synapse. This single NO pulse will result in the production of approximately 10 nM of cGMP, a concentration that is too low to have a physiological effect. (**c**) However, rapid, multiple pulses of NO can result in the accumulation of cGMP in a concentration range able to initiate downstream physiological actions. For example, ten repeated NO pulses of ~10–20 Hz are capable of eliciting the production of about 100 nM cGMP. (**d**) Supramaximal cGMP responses can also be elicited when multiple NO sources are simultaneously releasing NO, which is then able to act via volume transmission. This could be the case in the area of the organum vasculosum of the lamina terminalis (OV), where the packing density of nNOS neurons (diameter of about 15 μm) comes to about 5×10^4 nNOS neurons/mm^3. According to the proposed mathematical model, in such an area, nNOS cells will have a distance of about 28 μm between them, hypothesizing that they are similarly packed (orange spheres in a $9 \times 9 \times 9$ conformation). Hence, in the presence of activation signals such as leptin, estrogen, and kisspeptin activation of the nNOS enzyme will lead to NO being emitted from the neuronal soma (spheres) synchronously. (**e**) Based on the activation and deactivation kinetics of the involved enzymes in the hypothalamic region, this synchronous activation of such a densely packed population could lead to about 200 nM of NO being released. Adapted from Bellefontaine et al. (2014) and Chachlaki et al. (2017)

NO to affect neighboring GnRH cells via volume transmission (Bellefontaine et al. 2014; Chachlaki et al. 2022; Delli et al. 2023, 2024). The physiological effects of NO, such as the ones on GnRH neurons, will depend not only on the dynamics of NO production, but also on its degradation rate. The high rates of NO inactivation, described in the rat cerebellum to result in a NO half-life of about 5 ms (Hall and Garthwaite 2006), are further conferring to the generation of transient NO pulses, and as such a dynamic alteration of GnRH neuronal activity (or other potential targets).

3.4.3 Gliotransmission

In the brain, including the hypothalamus, gliotransmission, that is the active information transfer from glia to neurons, is now recognized to have wide implications on brain function. This signaling involves calcium-dependent vesicular transmitter release, including ATP, D-serine, taurine, glutamate, and GABA, that results in a variety of regulatory effects at neuronal synapses (Parpura and Zorec 2010), including in the hypothalamus (Panatier 2009; Clasadonte and Prevot 2018). Gliotransmission also involves molecules not released by vesicles such as prostaglandins, which have potent postsynaptic effects on neuronal firing (see for review Clasadonte and Prevot 2018).

3.4.4 Chemotrophic Factors

Chemotropic molecules play a crucial role in establishing gradients within tissues, originating from their source, which are essential for proper embryonic development. Interestingly, these same molecules not only guide cell migration and circuit formation during embryogenesis but also regulate postnatal development and adult physiology. For example, semaphorin 3A, which is crucial for the migration of GnRH neurons from the nose to the brain during embryogenesis (Hanchate et al. 2012; Vanacker et al. 2020), is released by endothelial cells in the median eminence (ME), where GnRH neurons project. This release contributes to the cyclic plasticity of the Sema3A receptor Neuropilin-1-expressing GnRH axons in the adult hypothalamus, thereby influencing reproductive physiology (Giacobini et al. 2014). During early postnatal development, GnRH neurons provide another example of local volume transmission. These neurons utilize the same molecule family as glial cells for gliotransmission, specifically releasing prostaglandin species such as PGD2. These prostaglandins attract DP1-expressing progenitor cells to the vicinity of GnRH neurons. Once anchored to the cell body of GnRH neurons, these progenitor cells undergo differentiation into astrocytes. This process is crucial for enabling the maturation of synaptic contacts reaching these neurons, thereby facilitating their wiring for reproduction (Pellegrino et al. 2021).

3.5 Neuropeptide-Mediated Volume Transmission

3.5.1 Suprachiasmatic-Nucleus-Derived Peptides

In the 1980s, elegant experiments by René Drucker-Colín et al. demonstrated that transplants of the biological clock, the suprachiasmatic nucleus (SCN), could restore circadian activity rhythms in rats whose own SCN had been ablated (Drucker-Colín et al. 1984). Importantly, the graft was deposited into the third ventricle (3V). Although the exact nature of the coupling signal from the grafted SCN to the host brain remains unknown, it is presumed that functional recovery required the re-establishment of appropriate synaptic connections. Similar results were later obtained in hamsters by Rae Silver (Silver et al. 1996).

To further investigate this question, Rae Silver isolated SCN tissue within a semipermeable polymeric capsule before transplantation in hamsters. This approach prevented neural outgrowth but allowed the diffusion of humoral signals into the brain ventricles' CSF. She and her collaborators demonstrated that the transplanted SCN could sustain activity rhythms using an unknown diffusible signal, as measured by locomotor activity (Drucker-Colín et al. 1984; Silver et al. 1996). These effects likely involve more than one neurotransmitter from the SCN, with volume transmission playing a role in restoring diurnal locomotor activity post-transplant. One possible neurotransmitter is vasopressin (AVP), which exhibits a consistent and robust rhythm in the CSF (Kalsbeek et al. 2010; Schwartz and Reppert 1985). In mice, AVP release in the CSF could occur via vasopressinergic fibers extending from the SCN into contact with the CSF (Taub et al. 2021).

Besides AVP, SCN neurons are known to release other signaling molecules such as transforming growth factor alpha (Kramer et al. 2001), prokineticin 2 (Cheng et al. 2002; Prosser et al. 2007), cardiotropin-like cytokine (Kraves and Weitz 2006), and gastrin-releasing peptide (Maywood et al. 2011). It is possible that SCN neurons' CSF-contacting protrusions also sense CSF-borne signals, as the CSF composition changes across the circadian cycle (Fame et al. 2023).

A 2021 study intriguingly discovered a portal system connecting the capillary beds of the SCN with the organum vasculosum laminae terminalis (OVLT), a circumventricular organ (CVO) that relays peripheral blood-borne information from fenestrated capillaries to distinct neuronal circuits. These circuits control various functions under circadian regulation, such as anticipatory thirst (Gizowski et al. 2016), ovulation (Everett et al. 1949), osmoregulation (Gizowski et al. 2018; Trudel and Bourque 2010), and fever and sickness behavior (Guerrero-Vargas et al. 2014; Soto-Tinoco et al. 2016, 2020). Yao et al. (2021) speculate that this pathway could be a source of circadian control over OVLT functions via volume transmission.

Additionally, SCN AVP neurons target the ME/ARH hypothalamic region, where they communicate with tanycytes to allow circulating glucose to periodically access glucose-sensing neurons in the ARH and thus regulate circadian changes in glycemia (Rodríguez-Cortés et al. 2022). While the authors suggest this phenomenon occurs via SCN neuronal fibers releasing AVP, the possibility of volume transmission cannot be excluded, as injecting AVP receptor blockers into the ventricle blunts SCN-mediated control of tanycytic function (Rodríguez-Cortés et al. 2022).

3.5.2 Oxytocin

Oxytocin, a nine amino acid CNS neuropeptide, was discovered by Sir Henry Dale in 1906 when he found that extracts from the human posterior pituitary gland contracted the uterus of a pregnant cat (reviewed in Viero et al. 2010). Oxytocin (OT) has been observed to signal via local volume transmission: Long OT axons may activate several parts of different circuits, even in more than one brain area, by releasing OT from *en passant* boutons that diffuse in the interstitial space fluid (ISF) modulating neurons expressing the oxytocin receptors (OTR) in distant brain regions (Grinevich and Ludwig 2021). Hereby, OT neurons could potentially control neurons with different functions simultaneously (Chini et al. 2017; Ludwig and Leng 2006). The amount of OT released by hypothalamic neurons by via local volume transmission is sufficient to produce OT effects at distant locations in the brain activating micro-ensembles of OT-sensitive neurons and modulating region-specific behaviors in the organism (Chini et al. 2017). However, the spillover of OT competes with both specific and unspecific peptidases in the extracellular space that break down OT after release (Ebstein et al. 2012).

OT has also been detected in the CSF of many species, and their levels vary considerably both diurnally and between species (Amico et al. 1983, 1990; Devarajan and Rusak 2004; Grinevich and Ludwig 2021). Early experiments showed that oxytocin levels in the CSF were sufficiently high to induce contraction of the rabbit uterus (Friedman and Friedman 1932). However, the physiological function of CSF-borne OT remains yet to be uncovered. Oxytocin levels in the CSF are regulated independently from the blood levels in humans. Indeed, the levels of this neurotransmitter fluctuate with the time of the day in the CSF but not in the blood (Amico et al. 1983). Conversely, the rise in OT levels in the blood are not mirrored by increased levels of the OT CSF as seen in lactating Rhesus monkeys during suckling (Amico et al. 1990) or in humans (Takeda et al. 1985). These data appear to indicate that the central levels of oxytocin in CSF and in the plasma are controlled independently.

Intriguingly, clinical trials are underway to assess the effects of administering oxytocin through the nose (intranasally) to infants with Prader-Willi syndrome, aimed at enhancing their neurodevelopment (Tauber and Diene 2021). These trials registered under the NCT03114371 reference on Clinicaltrials.gov draw upon promising results from preclinical studies conducted in mice (Meziane et al. 2015). Additionally, research conducted in macaques revealed interesting insights into the bioavailability of OT in different body compartments upon its intranasal administration (Monte et al. 2014). It was found that this method led to changes in oxytocin levels in the CSF but not in the bloodstream. This differs from another administration method using a nebulizer, which increased oxytocin levels in both the CSF and blood compartments. These findings suggest potential differences in how oxytocin is distributed and metabolized based on the method of administration. Further exploration of these differences in OT volume transmission could shed light on optimizing oxytocin therapy for various conditions.

3.5.3 GnRH

GnRH is the master decapeptide of the hypothalamic–pituitary–gonadal (HPG) axis regulating reproduction in vertebrates (Fig. 3.4). GnRH stimulates the synthesis and release of pituitary gonadotropins, follicle-stimulating hormone, and luteinizing hormone to control gametogenesis and sex steroid production; this hormone was originally considered and named for its important reproductive function (Amoss et al. 1971), however recent evidence suggests that GnRH could also be involved in cognition, metabolism, odor recognition and behavior (reviewed in Prevot et al. 2023).

Pioneering work in sheep and primates demonstrated that at the median eminence of the hypothalamus, each pulse of GnRH release into the pituitary portal circulation corresponds to an episode of increase in GnRH concentrations in the CSF, suggesting that CSF-borne GnRH could act remotely on GnRH receptor (GnRHR)-expressing sites by volume transmission (Fig. 3.4) (Skinner et al. 1997). Intriguingly, in addition to CSF-borne GnRH, GnRH-immunoreactive fibers are also found in the extra hypothalamic areas not directly involved in reproduction and expressing the GnRHR (Fig. 3.5) (Manfredi-Lozano et al. 2022). Even if the

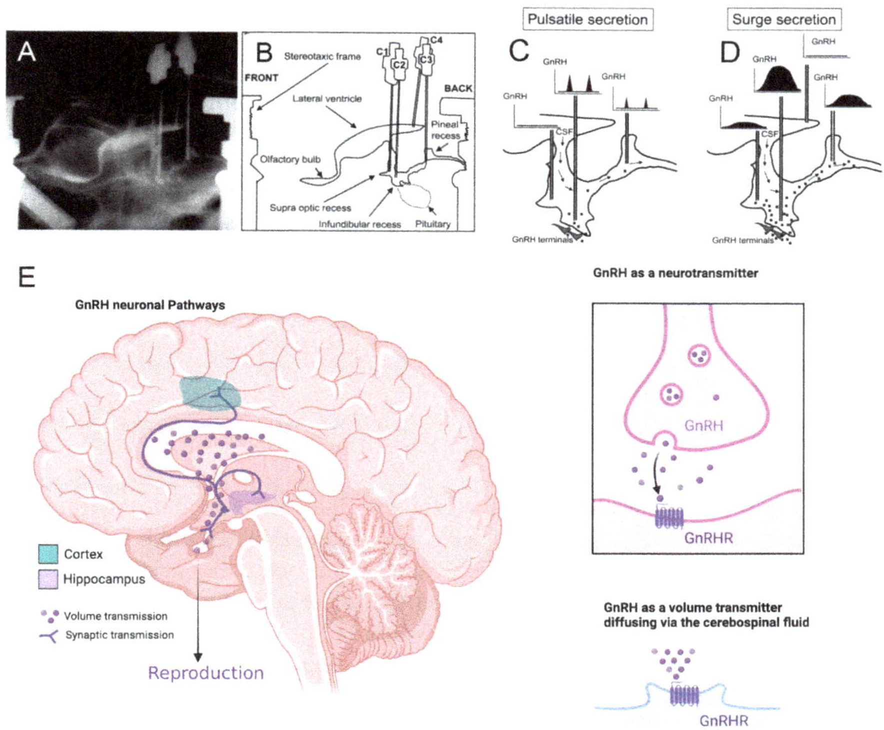

Fig. 3.4 GnRH volume transmission. (**a–d**) Microdialysis experiments conducted in sheep demonstrated that each episode of GnRH release in the pituitary portal blood system was accompanied by an equivalent episode of CSF-borne GnRH in the third ventricle, just above the median eminence (**a–d**). Intriguingly, the further the microdialysis probes were implanted from the median eminence, the more the amplitude of the GnRH pulse or surge was attenuated (**c**, **d**). Overall, this suggests that CSF-borne GnRH could act remotely on GnRHR-expressing sites by volume transmission. Adapted from Caraty and Skinner (2008). (**e**) In addition to volume transmission, neuroendocrine GnRH neurons controlling reproduction also send neuronal processes to neocortical brain regions where GnRHR-expressing neurons are located and could therefore deliver GnRH in a synaptic manner to them. Adapted from Prevot et al. (2023)

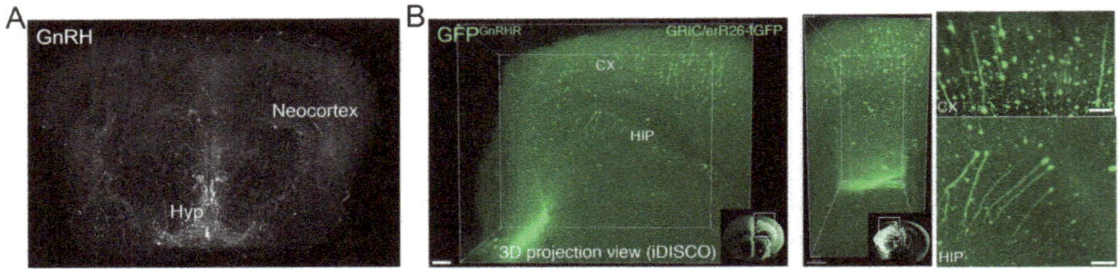

Fig. 3.5 GnRH neurons send projections to GnRH-receptor-expressing neocortical brain areas. (**a**) GnRH immunoreactivity in a transparentized adult mouse brain. Hyp, hypothalamus. (**b**) GFP expression in neurons of the cortex (Cx) and the hippocampus (HIP) in a transparentized adult mouse brain from *Gnrhr*::*Cre*;*Tau-Gfp*$^{\text{loxP-STOP-loxP}}$ mice (GRIC/erR26-tGFP)

respective contribution of synaptic vs. volumic GnRH transmission in modulating cognitive processes remains to be clarified in normal physiology, peripheral delivery of pulsatile GnRH, which very likely accesses its target brain areas via volume transmission, is able to improve cognitive performances in both mouse and humans with pathological rhythms of GnRH secretion (Fig. 3.6) (Manfredi-Lozano et al. 2022). In contrast, continuous infusion of GnRH, which desensitizes GnRH receptors and suppresses the GnRH axis, results in marked cognitive impairments in wildtype mice (Manfredi-Lozano et al. 2022). This latter condition resembles the secondary effects observed in patients with prostate cancer and endometriosis who undergo prolonged treatment with GnRH agonists. These patients often experience cognitive decline and an elevated risk of developing Alzheimer's disease (see for review Prevot et al. 2023).

3.5.4 MCH

Fish and amphibians can change their skin color acutely in response to environmental changes and stress. This rapid change occurs within minutes and is orchestrated by the migration of melanosome cells. When melanosomes are dispersed, the skin appears dark; when they concentrate in the perinuclear area, the skin becomes light (Kawauchi et al. 1983; Pissios et al. 2006; Westerfield et al. 1980).

First identified in the early 1980s from chum salmon pituitaries, the melanin-concentrating hormone (MCH) owes its name to its ability to concentrate melanin (Conductier et al. 2013b; Rance and Baker 1979). In the brains of mammals, however, MCH is a neuropeptide produced primarily by neurons in the lateral hypothalamic area (LHA) and zona incerta (ZI). It increases appetite and food intake, even in already obese animals (Gomori et al. 2003). As is common with neuropeptide systems, the functions of MCH are much more diverse than just regulating food intake. It also plays roles in sleep (Izawa et al. 2019), energy balance (Shimada et al. 1998), glucose homeostasis (Al-Massadi et al. 2019), thermoregulation (Ahnaou et al. 2011), learning and memory (Sita et al. 2016), reward, and anxiety (Borowsky et al. 2002).

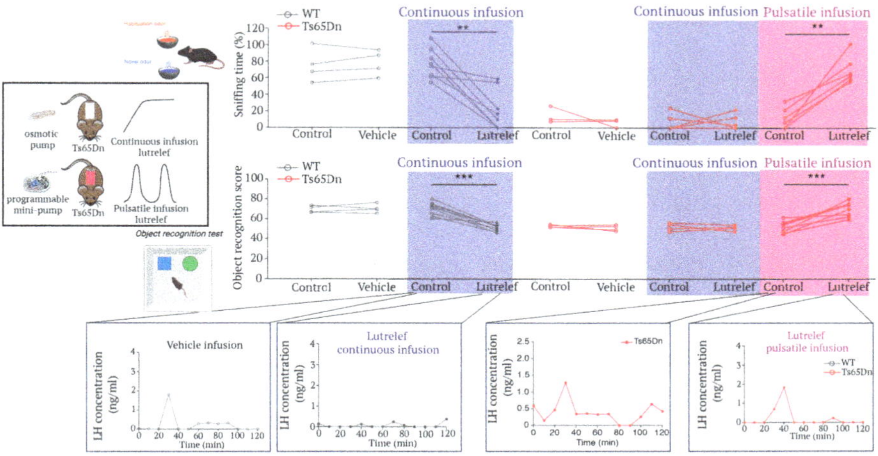

Fig. 3.6 Pulsatile GnRH treatment rescues olfaction and cognition in a mouse model of Down syndrome, whereas continuous GnRH administration impairs these non-reproductive functions in wildtype mice. Wildtype mice or a mouse model of Down syndrome (Ts65Dn mice) were implanted either with a minipump delivering continuous GnRH levels or with a programmable minipump delivering a pattern of GnRH release to mimic in Ts65Dn mice the one seen in wildtype animals. While 15-day continuous GnRH infusion impaired the ability of wildtype mice to recognize new odors and new objects, 15-day pulsatile GnRH therapy rescued olfaction and cognition in Ts65Dn mice. From Manfredi-Lozano et al. (2022) with permission

Studies involving the injection of MCH and MCH receptor (MCHR) antagonists or agonists intracerebroventricularly (i.c.v.) suggest that volume transmission plays a role in signaling. Several studies (Al-Massadi et al. 2019; Gomori et al. 2003; Imbernon et al. 2013) have shown that the release of MCH into the CSF can reach targets that are not synaptically connected (Al-Massadi et al. 2021).

Recently, in vivo optogenetic approaches using a dual viral vector method have revealed that selective activation of putative CSF-projecting MCH neurons increases food intake. Conversely, food intake was reduced following the immunosequestration of MCH present in the CSF (Noble et al. 2018). Notably, MCH is present in the CSF, and confocal analysis has confirmed the presence of synaptophysin immunoreactive (ir) elements co-localized with MCH-ir axon terminals in close proximity to ciliated ependymal cells lining the cerebral aqueduct and the lateral, third, and fourth ventricles (Noble et al. 2018), as initially reported by Nahon and collaborators (Fig. 3.1) (Conductier et al. 2013a; b). Importantly, the ependymal cells that make up the ventricular wall express the MCH receptor MCH1R (Fig. 3.1).

Consistently, the selective activation of MCH neurons increases MCH release into the CSF (Noble et al. 2018), which can modulate the movement of the CSF by regulating ciliary beating in ependymocytes (Conductier et al. 2013a; b). Additionally, data suggest that the release of MCH into the CSF is under circadian control, with a potential role in promoting the initial feeding bout during the nocturnal feeding cycle (Bluet-Pajot et al. 1995; Noble et al. 2018).

3.6 Evolution of Volume Transmission

As with any biological adaptation, neural systems and neurotransmitters are products of non-teleological evolution, meaning they lack a specific goal (Moroz et al. 2021). The resulting adaptations can be beneficial or disadvantageous depending on the context. Additionally, there are exaptations—changes that neither increase nor decrease an organism's fitness. Tracing the evolution of these traits and their impact on the global fitness of an organism can sometimes be challenging.

Interestingly, 85% of biological evolution on Earth (over 3 billion years) occurred without neurons as we know them. Therefore, the origin(s) of neurons is a rare event (Moroz et al. 2021). "Natura non facit saltus" (Nature does not make leaps), and many examples of volume transmission (VT) can be traced back to lower organisms. The basic machinery for the regulated vesicular release of peptides and monoamines is present in cells ranging from protozoa to endocrine, neuroendocrine, and neuronal cells in all multicellular organisms (Agnati et al. 1995).

As Schmitt (1984) pointed out, neurons evolved from cells with axon-like elongations that transported and secreted molecules into the interstitial fluid near target cells. Accordingly, many examples of nonsynaptic or para-synaptic release sites exist in invertebrates, suggesting some type of VT exists in these organisms (Agnati et al. 1995). Nonsynaptic release sites are common across the animal kingdom. One interesting example is found in the neuroendocrine bag cells in the abdominal ganglia of the mollusc Aplysia. Morphological analysis shows that the axons of these bag cells only reach connective tissue surrounding the ganglia (Haskins et al. 1981). Upon activation, peptides are released from these axons, diffuse into the ganglion, and regulate the activity of ganglionic neurons (Haskins et al. 1981). Further studies by Arshavsky et al. (1988) demonstrated that even after synaptic contacts were ablated, locomotor command neurons in isolated pedal ganglia could still regulate other locomotor neurons, indicating that command neurons operate via VT to organize networks for locomotion (Agnati et al. 1995).

Placozoa provides a close example of an existing organism that uses VT and integrates behaviors without canonical synapses (Moroz et al. 2021). These simple animals exhibit complex behaviors (Armon et al. 2018; Fortunato and Aktipis 2019; Heyland et al. 2014; Romanova et al. 2021; Senatore et al. 2017; Smith et al. 2015, 2019; Ueda et al. 1999; Varoqueaux et al. 2018) with only six recognized cell types (Smith et al. 2014) organized in three layers. They have no tissues or organs but display complex and rapid behaviors, including social interactions and ultrafast contractions (Heyland et al. 2014; Romanova et al. 2020; Senatore et al. 2017; Smith et al. 2015), despite the absence of synapses and gap junctions (Schierwater and Eitel 2015; Smith et al. 2014). The genome of Trichoplax adhaerens and its kin lacks traditional gap junction proteins (Eitel et al. 2018; Srivastava et al. 2008). Most cells in Placozoa are secretory, and their extensive system of extracellular spaces, surrounded by secretory vesicles, allows the release of complex molecular packages (Bobrie et al. 2011; Hoshino et al. 2013; Morton et al. 2018).

Both low molecular weight and peptide transmitters are present in Placozoa, targeting multiple receptors and effectors, thereby modulating and coordinating behaviors without synapses (Arendt 2021). Secretory and receptive components can function at large distances using VT (Borroto-Escuela et al. 2015; Syková 2004). Synapses and additional synaptic cleft components evolved later, after early neurons (Moroz and Kohn 2016).

Placozoans are a powerful model for studying alternative integrative systems, where chemical gradients and oscillations form a dynamic internal microenvironment with multiple chemical pacemakers and integrators of novel functions and behaviors. Another example is found in Cnidaria, where nitric oxide synthase (NOS) is widely expressed in non-neuronal cells across species. In Hydra, non-neuronal NO signaling controls regeneration, feeding behavior, and chemosensation (Colasanti et al. 1997, 2009; Cristino et al. 2008). Additionally, in Aplysia, the egg-laying hormone (ELH) acts as a neurotransmitter that diffuses long distances to its targets via VT (Mayeri et al. 1985).

Nonsynaptic communication allows groups of neurons to communicate without direct contact. Studies in Aplysia (*L. stagnalis*), rats, and cockroaches (*P. americana*) suggest that nonsynaptic release sites are present in these organisms' central nervous systems (Golding and Bayraktaroglu 1984). In *L. stagnalis*, about 15 different types of peptides have been identified in central ganglia (Bonilla 1977; Schot et al. 1981). Nonsynaptic release sites contain various secretory vesicles, even if some small electron-lucent vesicles are resorption structures formed by membrane recapture after exocytosis (Buma and Roubos 1983, 1986; Golding and Bayraktaroglu 1984).

Many theories explain nervous system evolution, with VT-mediated chemical signals linking signal-secreting sender cells to receptor-expressing receiver cells. Both chemical and synaptic transmission can wire complex neuronal networks with specific connections. The idea that VT evolved before synaptic transmission has been proposed by Grundfest, Horridge, and others (Mackie 1990; Moroz 2009).

3.7 Conclusion and Perspective

The pharmacological relevance of volume transmission in the brain remains largely underexplored, representing a promising frontier in neuropsychiatric research. The CSF, abundant in various neurotransmitters including dopamine, noradrenaline, serotonin, and GABA, among others, awaits systematic investigation of its physiological role (Tsutsumi et al. 2020). Additionally, CO and CO2, akin to NO, may serve as a significant diffusible signal in volume transmission, potentially modulating protein configuration via pH alteration in the extracellular matrix (Orengo et al. 1999).

Advancing our understanding of volume transmission holds significant therapeutic promise and implications for drug design. Notably, many neuropsychiatric medications, including antidepressants and anxiolytics, are orally administered and subsequently diffuse into the brain through volume transmission pathways. This underscores the medical importance of comprehensively investigating volume transmission mechanisms, as drug efficacy and distribution may be impacted by this mode of transmission.

Tanycytes, with their ability to facilitate the shuttling of peptide-derived drugs from blood to CSF, exemplify the potential clinical significance of volume transmission research (Duquenne et al. 2024). For instance, studies have demonstrated the crucial role of tanycytic transport in mediating the effects of liraglutide, an anti-diabetic drug and GLP1R agonist, in the hypothalamus (Imbernon et al. 2022). This suggests that CSF-borne peripheral hormones can influence the activity of hypothalamic and extrahypothalamic neural cells expressing their receptors via volume transmission, enabling long-range communication between the brain and the rest of the body (Prevot et al. 2018).

Finally, further investigation is warranted to elucidate the mechanisms underlying local volume transmission in the interstitial fluid within the brain, particularly regarding signal inactivation, clearance, and cellular uptake. These endeavors promise to uncover novel therapeutic targets and enhance our understanding of various neurological disorders.

Acknowledgements This work was supported by the European Research Council (ERC) Synergy Grant-2019-WATCH-810331, ERC-2023-PoC UPGRADE (No 101123221), the H2020 miniNO (No 847941), and a fellowship to BRC by the Foundation pour la Recherche Medical SPF202209015962.

Bibliography

Agnati LF, Fuxe K, Zoli M, Ozini I, Toffano G, Ferraguti F (1986) A correlation analysis of the regional distribution of central enkephalin and β-endorphin immunoreactive terminals and of opiate receptors in adult and old male rats. Evidence for the existence of two main types of communication in the central nervous system: The volume transmission and the wiring transmission. Acta Physiol Scand 128(2):201–207. https://doi.org/10.1111/J.1748-1716.1986.TB07967.X

Agnati LF, Zoli M, Strömberg I, Fuxe K (1995) Intercellular communication in the brain: Wiring versus volume transmission. Neuroscience 69(3):711–726. https://doi.org/10.1016/0306-4522(95)00308-6

Ahnaou A, Dautzenberg FM, Huysmans H, Steckler T, Drinkenburg WHIM (2011) Contribution of melanin-concentrating hormone (MCH1) receptor to thermoregulation and sleep stabilization: Evidence from MCH1 (−/−) mice. Behav Brain Res 218(1):42–50. https://doi.org/10.1016/j.bbr.2010.11.019

Amoss M, Burgus R, Blackwell R, Vale W, Fellows R, Guillemin R. Purification, amino acid composition and N-terminus of the hypothalamic luteinizing hormone releasing factor (LRF) of ovine origin. Biochem Biophys Res Commun. 1971;44(1):205–210. https://doi.org/10.1016/s0006-291x(71)80179-1

Al-Massadi O, Quiñones M, Clasadonte J, Hernandez-Bautista R, Romero-Picó A, Folgueira C, Morgan DA, Kalló I, Heras V, Senra A, Funderburk SC, Krashes MJ, Souto Y, Fidalgo M, Luquet S, Chee MJ, Imbernon M, Beiroa D, García-Caballero L et al (2019) MCH regulates SIRT1/FoxO1 and reduces POMC neuronal activity to induce hyperphagia, adiposity, and glucose intolerance. Diabetes 68(12):2210–2222. https://doi.org/10.2337/db19-0029

Al-Massadi O, Dieguez C, Schneeberger M, López M, Schwaninger M, Prevot V, Nogueiras R (2021) Multifaceted actions of melanin-concentrating hormone on mammalian energy homeostasis. Nat Rev Endocrinol 17(12):745–755. https://doi.org/10.1038/s41574-021-00559-1

Alpár A, Zahola P, Hanics J, Hevesi Z, Korchynska S, Benevento M, Pifl C, Zachar G, Perugini J, Severi I, Leitgeb P, Bakker J, Miklosi AG, Tretiakov E, Keimpema E, Arque G, Tasan

RO, Sperk G, Malenczyk K et al (2018) Hypothalamic CNTF volume transmission shapes cortical noradrenergic excitability upon acute stress. EMBO J 37(21):e100087. https://doi.org/10.15252/embj.2018100087

Alpár A, Benevento M, Romanov RA, Hökfelt T, Harkany T (2019) Hypothalamic cell diversity: Non-neuronal codes for long-distance volume transmission by neuropeptides. Curr Opin Neurobiol 56:16–23. https://doi.org/10.1016/j.conb.2018.10.012

Amico JA, Tenicela R, Johnston J, Robinson AG (1983) A time-dependent peak of oxytocin exists in cerebrospinal fluid but not in plasma of humans. J Clin Endocrinol Metab 57(5):947–951. https://doi.org/10.1210/JCEM-57-5-947

Amico JA, Challinor SM, Cameron JL (1990) Pattern of oxytocin concentrations in the plasma and cerebrospinal fluid of lactating rhesus monkeys (Macaca mulatta): Evidence for functionally independent oxytocinergic pathways in primates. J Clin Endocrinol Metab 71(6):1531–1535. https://doi.org/10.1210/JCEM-71-6-1531

Arendt D (2021) Elementary nervous systems. Philos Trans R Soc Lond Ser B Biol Sci 376(1821):20200347. https://doi.org/10.1098/rstb.2020.0347

Armon S, Bull MS, Aranda-Diaz A, Prakash M (2018) Ultrafast epithelial contractions provide insights into contraction speed limits and tissue integrity. Proc Natl Acad Sci USA 115(44):E10333–E10341. https://doi.org/10.1073/pnas.1802934115

Arshavsky YI, Deliagina TG, Gelfand IM, Orlovsky GN, Panchin YV, Pavlova GA, Popova LB (1988) Non-synaptic interaction between neurons in molluscs. Comp Biochem Physiol Part C Comp Pharmacol 91(1):199–203. https://doi.org/10.1016/0742-8413(88)90187-9

Balland E, Dam J, Langlet F, Caron E, Steculorum S, Messina A, Rasika S, Falluel-Morel A, Anouar Y, Dehouck B, Trinquet E, Jockers R, Bouret SG, Prévot V (2014) Hypothalamic tanycytes are an ERK-gated conduit for leptin into the brain. Cell Metab 19(2):293–301. https://doi.org/10.1016/J.CMET.2013.12.015

Barahona MJ, Ferrada L, Vera M, Nualart F (2024) Tanycytes release glucose using the glucose-6-phosphatase system during hypoglycemia to control hypothalamic energy balance. Mol Metab 101940. https://doi.org/10.1016/j.molmet.2024.101940

Baroncini M, Jissendi P, Catteau-Jonard S, Dewailly D, Pruvo JP, Francke JP, Prevot V (2010) Sex steroid hormones-related structural plasticity in the human hypothalamus. NeuroImage 50:428–433

Bellefontaine N, Chachlaki K, Parkash J, Vanacker C, Colledge W, d'Anglemont de Tassigny X, Garthwaite J, Bouret SG, Prevot V (2014) Leptin-dependent neuronal NO signaling in the preoptic hypothalamus facilitates reproduction. J Clin Invest 124(6):2550–2559. https://doi.org/10.1172/JCI65928

Bluet-Pajot MT, Presse F, Voko Z, Hoeger C, Mounier F, Epelbaum J, Nahon JL (1995) Neuropeptide-E-I antagonizes the action of melanin-concentrating hormone on stress-induced release of adrenocorticotropin in the rat. J Neuroendocrinol 7(4):297–303. https://doi.org/10.1111/j.1365-2826.1995.tb00761.x

Bobrie A, Colombo M, Raposo G, Théry C (2011) Exosome secretion: molecular mechanisms and roles in immune responses. Traffic 12(12):1659–1668. https://doi.org/10.1111/j.1600-0854.2011.01225.x

Bolborea M, Pollatzek E, Benford H, Sotelo-Hitschfeld T, Dale N (2020) Hypothalamic tanycytes generate acute hyperphagia through activation of the arcuate neuronal network. Proc Natl Acad Sci USA 117(25):14473–14481. https://doi.org/10.1073/PNAS.1919887117/-/DCSUPPLEMENTAL

Bonilla E (1977) Staining of transverse tubular system of skeletal muscle by tannic acid-glutaraldehyde fixation. J Ultrastruct Res 2:162–165. https://doi.org/10.1016/s0022-5320(77)90028-4

Borowsky B, Durkin MM, Ogozalek K, Marzabadi MR, DeLeon J, Heurich R, Lichtblau H, Shaposhnik Z, Daniewska I, Blackburn TP, Branchek TA, Gerald C, Vaysse PJ, Forray C (2002) Antidepressant, anxiolytic and anorectic effects of a melanin-concentrating hormone-1 receptor antagonist. Nat Med 8(8):825–830. https://doi.org/10.1038/nm741

Borroto-Escuela DO, Agnati LF, Bechter K, Jansson A, Tarakanov AO, Fuxe K (2015) The role of transmitter diffusion and flow versus extracellular vesicles in volume transmission in the brain neural-glial networks. Philos Trans R Soc Lond Ser B Biol Sci 370(1672):20140183. https://doi.org/10.1098/rstb.2014.0183

Buma P, Roubos EW (1983) Calcium dynamics, exocytosis, and membrane turnover in the ovulation hormone-releasing caudo-dorsal cells of Lymnaea stagnalis. Cell Tissue Res 233(1):143–159. https://doi.org/10.1007/BF00222239

Buma P, Roubos EW (1986) Ultrastructural demonstration of nonsynaptic release sites in the central nervous system of the snail Lymnaea stagnalis, the insect Periplaneta americana, and the rat. Neuroscience 17(3):867–879. https://doi.org/10.1016/0306-4522(86)90051-5

Calias P, Papisov M, Pan J, Savioli N, Belov V, Huang Y, Lotterhand J, Alessandrini M, Liu N, Fischman AJ, Powell JL, Heartlein MW (2012) CNS penetration of intrathecal-lumbar idursulfase in the monkey, dog and mouse: implications for neurological outcomes of lysosomal storage disorder. PLoS One 7(1):e30341. https://doi.org/10.1371/JOURNAL.PONE.0030341

Campbell JN, Macosko EZ, Fenselau H, Pers TH, Lyubetskaya A, Tenen D, Goldman M, Verstegen AMJ, Resch JM, McCarroll SA, Rosen ED, Lowell BB, Tsai LT (2017) A molecular census of arcuate hypothalamus and median eminence cell types. Nat Neurosci 20(3):484–496. https://doi.org/10.1038/nn.4495

Caprile T, Hein S, Rodríguez, S., Montecinos, H., & Rodríguez, E. (2003) Reissner fiber binds and transports away monoamines present in the cerebrospinal fluid. Mol Brain Res 110(2):177–192. https://doi.org/10.1016/S0169-328X(02)00565-X

Caraty A, Skinner DC (2008) Gonadotropin-releasing hormone in third ventricular cerebrospinal fluid: Endogenous distribution and exogenous uptake. Endocrinology 149(10):5227–5234. https://doi.org/10.1210/EN.2007-1636

Casoni F, Malone SA, Belle M, Luzzati F, Collier F, Allet C, Hrabovszky E, Rasika S, Prevot V, Chédotal A, Giacobini P (2016) Development of the neurons controlling fertility in humans: New insights from 3D imaging and transparent fetal brains. Development (Cambridge, England) 143(21):3969–3981. https://doi.org/10.1242/dev.139444

Chachlaki K, Garthwaite J, Prevot V (2017) The gentle art of saying NO: how nitric oxide gets things done in the hypothalamus. Nat Rev Endocrinol 13(9):521–535. https://doi.org/10.1038/nrendo.2017.69

Chachlaki K, Prevot V (2020) Nitric oxide signalling in the brain and its control of bodily functions. Br J Pharmacol 177(24):5437–5458. https://doi.org/10.1111/BPH.14800

Chachlaki K, Messina A, Delli V, Leysen V, Maurnyi C, Huber C, Ternier G, Skrapits K, Papadakis G, Shruti S, Kapanidou M, Cheng X, Acierno J, Rademaker J, Rasika S, Quinton R, Niedziela M, L'Allemand D, Pignatelli D et al (2022) NOS1 mutations cause hypogonadotropic hypogonadism with sensory and cognitive deficits that can be reversed in infantile mice. Sci Transl Med 14(665):eabh2369. https://doi.org/10.1126/scitranslmed.abh2369

Chen C-C, Fernald RD (2008) GnRH and GnRH receptors: Distribution, function and evolution. J Fish Biol 73(5):1099–1120. https://doi.org/10.1111/j.1095-8649.2008.01936.x

Chen R, Wu X, Jiang L, Zhang Y (2017) Single-cell RNA-Seq reveals hypothalamic cell diversity. Cell Rep 18(13):3227–3241. https://doi.org/10.1016/j.celrep.2017.03.004

Cheng MY, Bullock CM, Li C, Lee AG, Bermak JC, Belluzzi J, Weaver DR, Leslie FM, Zhou Q-Y (2002) Prokineticin 2 transmits the behavioural circadian rhythm of the suprachiasmatic nucleus. Nature 417(6887):405–410. https://doi.org/10.1038/417405a

Chini B, Verhage M, Grinevich V (2017) The action radius of oxytocin release in the mammalian CNS: from single vesicles to behavior. Trends Pharmacol Sci 38(11):982–991. https://doi.org/10.1016/J.TIPS.2017.08.005

Clasadonte J, Prevot V (2018) The special relationship: glia-neuron interactions in the neuroendocrine hypothalamus. Nat Rev Endocrinol 14:25–44

Colasanti M, Venturini G, Merante A, Musci G, Lauro GM (1997) Nitric oxide involvement in Hydra vulgaris very primitive olfactory-like system. J Neurosci Off J Soc Neurosci 17(1):493–499. https://doi.org/10.1523/JNEUROSCI.17-01-00493.1997

Colasanti M, Mazzone V, Mancinelli L, Leone S, Venturini G (2009) Involvement of nitric oxide in the head regeneration of Hydra vulgaris. Nitric Oxide Biol Chem 21(3–4):164–170. https://doi.org/10.1016/j.niox.2009.07.003

Collden G, Balland E, Parkash J, Caron E, Langlet F, Prevot V, Bouret SG (2015) Neonatal overnutrition causes early alterations in the central response to peripheral ghrelin. Mol Metab 4(1):15–24. https://doi.org/10.1016/J.MOLMET.2014.10.003

Conductier G, Brau F, Viola A, Langlet F, Ramkumar N, Dehouck B, Lemaire T, Chapot R, Lucas L, Rovère C, Maitre P, Hosseiny S, Petit-Paitel A, Adamantidis A, Lakaye B, Risold PY, Prévot V, Meste O, Nahon JL, Guyon A (2013a) Melanin-concentrating hormone regulates beat frequency of ependymal cilia and ventricular volume. Nat Neurosci 16(7):845–847. https://doi.org/10.1038/NN.3401

Conductier G, Martin A, Risold P-Y, Jégo S, Lavoie R, Lafont C, Mollard P, Adamantidis A, Nahon J-L (2013b) Control of ventricular ciliary beating by the melanin concentrating hormone-expressing neurons of the lateral hypothalamus: a functional imaging survey. Front Endocrinol 4. https://doi.org/10.3389/fendo.2013.00182

Cristino L, Guglielmotti V, Cotugno A, Musio C, Santillo S (2008) Nitric oxide signaling pathways at neural level in invertebrates: Functional implications in cnidarians. Brain Res 1225:17–25. https://doi.org/10.1016/j.brainres.2008.04.056

Decoster L, Trova S, Zucca S, Gouveia A, Bulk J, Ternier G, Lhomme T, Legrand A, Boehm U, Hrabovszky E, Rácz G, Luzzati F, Peretto P, Steculorum S, Schriever S, Bernecker M, Pfluger P, Bovetti S, Rasika S et al (2023) An olfactory bulb GnRH neuronal population translates social relevant odors into reproductive behavior in male mice. https://doi.org/10.21203/rs.3.rs-3115610/v1

Delli V, Dehame J, Franssen D, Rasika S, Parent A-S, Prevot V, Chachlaki K (2023) Male minipuberty involves the gonad-independent activation of preoptic nNOS neurons. Free Radic Biol Med 194:199–208. https://doi.org/10.1016/j.freeradbiomed.2022.11.040

Delli V, Seux C-A, Dehame J, Nair S, Lhomme T, Chachlaki K (2024) The preoptic Kisspeptin/nNOS/GnRH (KiNG) neuronal network regulates rhythmic LH release through a dual activation-inhibition mechanism. bioRxiv. https://doi.org/10.1101/2024.01.15.575688

Devarajan K, Rusak B (2004) Oxytocin levels in the plasma and cerebrospinal fluid of male rats: Effects of circadian phase, light and stress. Neurosci Lett 367(2):144–147. https://doi.org/10.1016/J.NEULET.2004.05.112

Drucker-Colín R, Aguilar-Roblero R, García-Hernández F, Fernández-Cancino F, Rattoni FB (1984) Fetal suprachiasmatic nucleus transplants: Diurnal rhythm recovery of lesioned rats. Brain Res 311(2):353–357. https://doi.org/10.1016/0006-8993(84)90099-4

Duquenne M, Folgueira C, Bourouh C, Millet M, Silva A, Clasadonte J, Imbernon M, Fernandois D, Martinez-Corral I, Kusumakshi S, Caron E, Rasika S, Deliglia E, Jouy N, Oishi A, Mazzone M, Trinquet E, Tavernier J, Kim YB et al (2021) Leptin brain entry via a tanycytic LepR–EGFR shuttle controls lipid metabolism and pancreas function. Nat Metab 3(8):1071–1090. https://doi.org/10.1038/s42255-021-00432-5

Duquenne M, Deligia E, Folgueira C, Bourouh C, Caron E, Pfrieger F, Schwaninger M, Nogueiras R, Annicotte JS, Imbernon M, Prevot V (2024). Tanycytic transcytosis inhibition disrupts energy balance, glucose homeostasis and cognitive function in male mice. Mol Metab 87: 101996. https://doi.org/10.1016/j.molmet.2024.101996

Ebstein RP, Knafo A, Mankuta D, Chew SH, Lai PS (2012) The contributions of oxytocin and vasopressin pathway genes to human behavior. Horm Behav 61(3):359–379. https://doi.org/10.1016/J.YHBEH.2011.12.014

Eitel M, Francis WR, Varoqueaux F, Daraspe J, Osigus H-J, Krebs S, Vargas S, Blum H, Williams GA, Schierwater B, Wörheide G (2018) Comparative genomics and the nature of placozoan species. PLoS Biol 16(7):e2005359. https://doi.org/10.1371/journal.pbio.2005359

Everett JW, Sawyer CH, Markee JE (1949) A neurogenic timing factor in control of the ovulatory discharge of luteinizing hormone in the cyclic rat. Endocrinology 44(3):234–250. https://doi.org/10.1210/endo-44-3-234

Fame RM, Kalugin PN, Petrova B, Xu H, Soden PA, Shipley FB, Dani N, Grant B, Pragana A, Head JP, Gupta S, Shannon ML, Chifamba FF, Hawks-Mayer H, Vernon A, Gao F, Zhang Y, Holtzman MJ, Heiman M et al (2023) Defining diurnal fluctuations in mouse choroid plexus and CSF at high molecular, spatial, and temporal resolution. Nat Commun 14(1):1–20. https://doi.org/10.1038/s41467-023-39326-3

Faubel R, Westendorf C, Bodenschatz E, Eichele G (2016) Cilia-based flow network in the brain ventricles. Science 353(6295):176–178. https://doi.org/10.1126/science.aae0450

Fortunato A, Aktipis A (2019) Social feeding behavior of Trichoplax adhaerens. Front Ecol Evol 7:19. https://doi.org/10.3389/fevo.2019.00019

Franke WW, Schmid E, Osborn M, Weber K (1978) Different intermediate-sized filaments distinguished by immunofluorescence microscopy. Proc Natl Acad Sci USA 75(10):5034–5038. https://doi.org/10.1073/pnas.75.10.5034

Friedman GS, Friedman MH (1932) An examination of cerebrospinal fluid for oxytocic activity as tested by the rabbit uterine fistula preparation. Am J Physiol 103(1):244–254. https://doi.org/10.1152/AJPLEGACY.1932.103.1.244

Garthwaite J (2016) From synaptically localized to volume transmission by nitric oxide. J Physiol 594(1):9–18. https://doi.org/10.1113/JP270297

Garthwaite J (2019) NO as a multimodal transmitter in the brain: Discovery and current status. Br J Pharmacol 176(2):197–211. https://doi.org/10.1111/BPH.14532

Garthwaite J, Charles SL, Chess-Williams R (1988) Endothelium-derived relaxing factor release on activation of NMDA receptors suggests role as intercellular messenger in the brain. Nature 336(6197):385–388. https://doi.org/10.1038/336385a0

Giacobini P, Parkash J, Campagne C, Messina A, Casoni F, Vanacker C, Langlet F, Hobo B, Cagnoni G, Gallet S, Hanchate NK, Mazur D, Tanaguchi M, Mazzone M, Verhaagen J, Ciofi P, Bouret SG, Tamagnone L, Prevot V (2014) Brain Endothelial Cells Control Fertility through Ovarian-Steroid-Dependent Release of Semaphorin 3A. PLoS Biol 12(3):e1001808. https://doi.org/10.1371/journal.pbio.1001808

Gizowski C, Zaelzer C, Bourque CW (2016) Clock-driven vasopressin neurotransmission mediates anticipatory thirst prior to sleep. Nature 537(7622):685–688. https://doi.org/10.1038/nature19756

Gizowski C, Zaelzer C, Bourque CW (2018) Activation of organum vasculosum neurons and water intake in mice by vasopressin neurons in the suprachiasmatic nucleus. J Neuroendocrinol. https://doi.org/10.1111/jne.12577

Gobron S, Creveaux I, Meiniel R, Didier R, Herbet A, Bamdad M, El Bitar F, Dastugue B, Meiniel A (2000) Subcommissural organ/Reissner's fiber complex: Characterization of SCO-spondin, a glycoprotein with potent activity on neurite outgrowth. Glia 32(2):177–191. https://doi.org/10.1002/1098-1136(200011)32:2<177::AID-GLIA70>3.0.CO;2-V

Golding DW, Bayraktaroglu E (1984) Exocytosis of secretory granules—A probable mechanism for the release of neuromodulators in invertebrate neuropiles. Experientia 40(11):1277–1280. https://doi.org/10.1007/BF01946673

Gomori A, Ishihara A, Ito M, Mashiko S, Matsushita H, Yumoto M, Ito M, Tanaka T, Tokita S, Moriya M, Iwaasa H, Kanatani A (2003) Chronic intracerebroventricular infusion of MCH causes obesity in mice. Am J Physiol Endocrinol Metab 284(3):E583–E588. https://doi.org/10.1152/ajpendo.00350.2002

Gonzalez-Santander R (2008) Electron-microscopic study of the secretion of the ependymal cells in the domestic cat (ependymin-β cells). Acta Anat 103(3):266–277. https://doi.org/10.1159/000145024

Grinevich V, Ludwig M (2021) The multiple faces of the oxytocin and vasopressin systems in the brain. J Neuroendocrinol 33(11):e13004. https://doi.org/10.1111/JNE.13004

Guerrero-Vargas NN, Salgado-Delgado R, Basualdo MDC, García J, Guzmán-Ruiz M, Carrero JC, Escobar C, Buijs RM (2014) Reciprocal interaction between the suprachiasmatic nucleus and the immune system tunes down the inflammatory response to lipopolysaccharide. J Neuroimmunol 273(1–2):22–30. https://doi.org/10.1016/j.jneuroim.2014.05.012

Hall CN, Garthwaite J (2006) Inactivation of nitric oxide by rat cerebellar slices. J Physiol 577(Pt 2):549–567. https://doi.org/10.1113/jphysiol.2006.118380

Hanchate NK, Giacobini P, Lhuillier P, Parkash J, Espy C, Fouveaut C, Leroy C, Baron S, Campagne C, Vanacker C, Collier F, Cruaud C, Meyer V, Garcia-Pinero A, Dewailly D, Cortet-

Rudelli C, Gersak K, Metz C, Chabrier G, Pugeat M, Young J, Hardelin JP, Prevot V, Dode C (2012) SEMA3A, a Gene Involved in Axonal Pathfinding, Is Mutated in Patients with Kallmann Syndrome. PLoS Genet 8(8):e1002896. https://doi.org/10.1371/journal.pgen.1002896

Haskins JT, Price CH, Blankenship JE (1981) A light and electron microscopic investigation of the neurosecretory bag cells of Aplysia. J Neurocytol 10(5):729–747. https://doi.org/10.1007/BF01262650

Hatton GI (1997) Function-related plasticity in hypothalamus. Annu Rev Neurosci 20:375–397

Heyland A, Croll R, Goodall S, Kranyak J, Wyeth R (2014) Trichoplax adhaerens, an enigmatic basal metazoan with potential. Methods Mol Biol (Clifton, N.J.) 1128:45–61. https://doi.org/10.1007/978-1-62703-974-1_4

Hoshino D, Kirkbride KC, Costello K, Clark ES, Sinha S, Grega-Larson N, Tyska MJ, Weaver AM (2013) Exosome secretion is enhanced by invadopodia and drives invasive behavior. Cell Rep 5(5):1159–1168. https://doi.org/10.1016/j.celrep.2013.10.050

Imbernon M, Beiroa D, Vázquez MJ, Morgan DA, Veyrat-Durebex C, Porteiro B, Díaz-Arteaga A, Senra A, Busquets S, Velásquez DA, Al-Massadi O, Varela L, Gándara M, López-Soriano F, Gallego R, Seoane LM, Argiles JM, López M, Davis RJ et al (2013) Central melanin-concentrating hormone influences liver and adipose metabolism via specific hypothalamic nuclei and efferent autonomic/JNK1 pathways. Gastroenterology 144(3):636–649.e6. https://doi.org/10.1053/j.gastro.2012.10.051

Imbernon M, Saponaro C, Helms HCC, Duquenne M, Fernandois D, Deligia E, Denis RGP, Chao DHM, Rasika S, Staels B, Pattou F, Pfrieger FW, Brodin B, Luquet S, Bonner C, Prevot V (2022) Tanycytes control hypothalamic liraglutide uptake and its anti-obesity actions. Cell Metab 34(7):1054–1063.e7. https://doi.org/10.1016/j.cmet.2022.06.002

Inoue S, Yang R, Tantry A, Davis CH, Yang T, Knoedler JR, Wei Y, Adams EL, Thombare S, Golf SR, Neve RL, Tessier-Lavigne M, Ding JB, Shah NM (2019) Periodic remodeling in a neural circuit governs timing of female sexual behavior. Cell 179:1393–1408 e1316

Izawa S, Chowdhury S, Miyazaki T, Mukai Y, Ono D, Inoue R, Ohmura Y, Mizoguchi H, Kimura K, Yoshioka M, Terao A, Kilduff TS, Yamanaka A (2019) REM sleep-active MCH neurons are involved in forgetting hippocampus-dependent memories. Science 365(6459):1308–1313. https://doi.org/10.1126/science.aax9238

Kalsbeek A, Fliers E, Hofman MA, Swaab DF, Buijs RM (2010) Vasopressin and the output of the hypothalamic biological clock. J Neuroendocrinol 22(5):362–372. https://doi.org/10.1111/J.1365-2826.2010.01956.X

Kawauchi H, Kawazoe I, Tsubokawa M, Kishida M, Baker BI (1983) Characterization of melanin-concentrating hormone in chum salmon pituitaries. Nature 305(5932):321–323. https://doi.org/10.1038/305321a0

King JC, Letourneau RJ (1994) Luteinizing hormone-releasing hormone terminals in the median eminence of rats undergo dramatic changes after gonadectomy, as revealed by electron microscopic image analysis. Endocrinology 134:1340–1351

Kramer A, Yang F-C, Snodgrass P, Li X, Scammell TE, Davis FC, Weitz CJ (2001) Regulation of daily locomotor activity and sleep by hypothalamic EGF receptor signaling. Science 294(5551):2511–2515. https://doi.org/10.1126/science.1067716

Kraves S, Weitz CJ (2006) A role for cardiotrophin-like cytokine in the circadian control of mammalian locomotor activity. Nat Neurosci 9(2):212–219. https://doi.org/10.1038/nn1633

Langlet F, Levin BE, Luquet S, Mazzone M, Messina A, Dunn-Meynell AA, Balland E, Lacombe A, Mazur D, Carmeliet P, Bouret SG, Prevot V, Dehouck B (2013) Tanycytic VEGF-A boosts blood-hypothalamus barrier plasticity and access of metabolic signals to the arcuate nucleus in response to fasting. Cell Metab 17(4):607–617. https://doi.org/10.1016/j.cmet.2013.03.004

Leak RK, Moore RY (2012) Innervation of ventricular and periventricular brain compartments. Brain Res 1463:51–62. https://doi.org/10.1016/j.brainres.2012.04.055

Lhomme T, Clasadonte J, Imbernon M, Fernandois D, Sauve F, Caron E, Lima NDS, Heras V, Martinez-Corral I, Mueller-Fielitz H, Rasika S, Schwaninger M, Nogueiras R, Prevot V (2021) Tanycytic networks mediate energy balance by feeding lactate to glucose-insensitive POMC neurons. J Clin Invest 131(18). https://doi.org/10.1172/JCI140521

Liu T, Jin X, Prasad RM, Sari Y, Nauli SM (2014) Three types of ependymal cells with intracellular calcium oscillation are characterized by distinct cilia beating properties. J Neurosci Res 92(9):1199–1204. https://doi.org/10.1002/jnr.23405

Ludwig M, Leng G (2006) Dendritic peptide release and peptide-dependent behaviours. Nat Rev Neurosci 7(2):126–136. https://doi.org/10.1038/nrn1845

Mackie GO (1990) The elementary nervous system revisited. Am Zool 30(4):907–920. https://doi.org/10.1093/icb/30.4.907

Manfredi-Lozano M, Leysen V, Adamo M, Paiva I, Rovera R, Pignat J-M, Timzoura FE, Candlish M, Eddarkaoui S, Malone SA, Silva MSB, Trova S, Imbernon M, Decoster L, Cotellessa L, Tena-Sempere M, Claret M, Paoloni-Giacobino A, Plassard D et al (2022) GnRH replacement rescues cognition in Down syndrome. Science 377(6610):eabq4515. https://doi.org/10.1126/science.abq4515

Martin KAC, Spühler IA (2013) The fine structure of the dopaminergic innervation of area 10 of macaque prefrontal cortex. Eur J Neurosci 37(7):1061–1071. https://doi.org/10.1111/EJN.12124

Martin FH, Seoane JR, Baile CA (1973) Feeding in satiated sheep elicited by intraventricular injections of CSF from fasted sheep. Life Sci 13(2):177–184. https://doi.org/10.1016/0024-3205(73)90193-8

Mayeri E, Rothman BS, Brownell PH, Branton WD, Padgett L (1985) Nonsynaptic characteristics of neurotransmission mediated by egg-laying hormone in the abdominal ganglion of Aplysia. J Neurosci 5(8):2060–2077. https://doi.org/10.1523/JNEUROSCI.05-08-02060.1985

Maywood ES, Chesham JE, O'Brien JA, Hastings MH (2011) A diversity of paracrine signals sustains molecular circadian cycling in suprachiasmatic nucleus circuits. Proc Natl Acad Sci USA 108(34):14306–14311. https://doi.org/10.1073/pnas.1101767108

McCaslin AFH, Chen BR, Radosevich AJ, Cauli B, Hillman EMC (2011) In vivo 3D morphology of astrocyte-vasculature interactions in the somatosensory cortex: Implications for neurovascular coupling. J Cereb Blood Flow Metab 31(3):795–806. https://doi.org/10.1038/JCBFM.2010.204

Meiniel O, Meiniel A (2007) The complex multidomain organization of SCO-spondin protein is highly conserved in mammals. Brain Res Rev 53(2):321–327. https://doi.org/10.1016/j.brainresrev.2006.09.007

Meste O, Brau F, Guyon A (2015) Robust estimation of the motile cilia beating frequency. Med Biol Eng Comput 53(10):1025–1035. https://doi.org/10.1007/s11517-015-1345-0

Meziane H, Schaller F, Bauer S, Villard C, Matarazzo V, Riet F, Guillon G, Lafitte D, Desarmenien MG, Tauber M, Muscatelli F (2015) An early postnatal oxytocin treatment prevents social and learning deficits in adult mice deficient for Magel2, a gene involved in Prader-Willi syndrome and autism. Biol Psychiatry 78:85–94

Mirzadeh Z, Kusne Y, Duran-Moreno M, Cabrales E, Gil-Perotin S, Ortiz C, Chen B, Garcia-Verdugo JM, Sanai N, Alvarez-Buylla A (2017) Bi- and uniciliated ependymal cells define continuous floor-plate-derived tanycytic territories. Nat Commun 8(1):1–12. https://doi.org/10.1038/ncomms13759

Monte OD, Noble PL, Turchi J, Cummins A, Averbeck BB (2014) CSF and blood oxytocin concentration changes following intranasal delivery in macaque. PLoS One 9(8):103677. https://doi.org/10.1371/journal.pone.0103677

Moroz LL (2009) On the independent origins of complex brains and neurons. Brain Behav Evol 74(3):177–190. https://doi.org/10.1159/000258665

Moroz LL, Kohn AB (2016) Independent origins of neurons and synapses: Insights from ctenophores. Philos Trans R Soc Lond Ser B Biol Sci 371(1685):20150041. https://doi.org/10.1098/rstb.2015.0041

Moroz LL, Romanova DY, Kohn AB (2021) Neural versus alternative integrative systems: Molecular insights into origins of neurotransmitters. Philos Trans R Soc B Biol Sci 376(1821):20190762. https://doi.org/10.1098/rstb.2019.0762

Morton DJ, Kuiper EG, Jones SK, Leung SW, Corbett AH, Fasken MB (2018) The RNA exosome and RNA exosome-linked disease. RNA (New York, N.Y.) 24(2):127–142. https://doi.org/10.1261/rna.064626.117

Mullier A, Bouret SG, Prevot V, Dehouck B (2010) Differential distribution of tight junction proteins suggests a role for tanycytes in blood-hypothalamus barrier regulation in the adult mouse brain. J Comp Neurol 518(7):943–962. https://doi.org/10.1002/CNE.22273

Muñoz RI, Kähne T, Herrera H, Rodríguez S, Guerra MM, Vío K, Hennig R, Rapp E, Rodríguez E (2019) The subcommissural organ and the Reissner fiber: Old friends revisited. Cell Tissue Res 375(2):507–529. https://doi.org/10.1007/s00441-018-2917-8

Myung J, Wu D, Simonneaux V, Lane TJ (2018) Strong circadian rhythms in the choroid plexus: implications for sleep-independent brain metabolite clearance. J Exp Neurosci 12. https://doi.org/10.1177/1179069518783762/FORMAT/EPUB

Nampoothiri S, Nogueiras R, Schwaninger M, Prevot V (2022) Glial cells as integrators of peripheral and central signals in the regulation of energy homeostasis. Nat Metab 4:813–825

Noble EE, Hahn JD, Konanur VR, Hsu TM, Page SJ, Cortella AM, Liu CM, Song MY, Suarez AN, Szujewski CC, Rider D, Clarke JE, Darvas M, Appleyard SM, Kanoski SE (2018) Control of feeding behavior by cerebral ventricular volume transmission of melanin-concentrating hormone. Cell Metab 28(1):55–68.e7. https://doi.org/10.1016/j.cmet.2018.05.001

Nualart F, Hein S (2001) Biosynthesis and molecular biology of the secretory proteins of the subcommissural organ. Microsc Res Tech 52(5):468–483. https://doi.org/10.1002/1097-0029(20010301)52:5<468::AID-JEMT1033>3.0.CO;2-U

Nualart F, Cifuentes M, Ramírez E, Martínez F, Barahona MJ, Ferrada L, Saldivia N, Bongarzone ER, Thorens B, Salazar K (2023) Hyperglycemia increases SCO-spondin and Wnt5a secretion into the cerebrospinal fluid to regulate ependymal cell beating and glucose sensing. PLoS Biol 21(9):e3002308. https://doi.org/10.1371/journal.pbio.3002308

Oliet SH, Piet R, Poulain DA (2001) Control of glutamate clearance and synaptic efficacy by glial coverage of neurons. Science 292:923–926

Orengo CA, Todd AE, Thornton JM (1999) From protein structure to function. Curr Opin Struct Biol 9(3):374–382. https://doi.org/10.1016/S0959-440X(99)80051-7

Panatier A (2009) Glial cells: indispensable partners of hypothalamic magnocellular neurones. J Neuroendocrinol 21:665–672

Panatier A, Theodosis DT, Mothet JP, Touquet B, Pollegioni L, Poulain DA, Oliet SH (2006) Glia-derived D-serine controls NMDA receptor activity and synaptic memory. Cell 125:775–784

Papisov MI, Belov VV, Gannon KS (2013) Physiology of the intrathecal bolus: The leptomeningeal route for macromolecule and particle delivery to CNS. Mol Pharm 10(5):1522–1532. https://doi.org/10.1021/MP300474M

Pappenheimer JR, Miller TB, Goodrich CA (1967) Sleep-promoting effects of cerebrospinal fluid from sleep-deprived goats. Proc Natl Acad Sci USA 58(2):513–517. https://doi.org/10.1073/PNAS.58.2.513

Parpura V, Zorec R (2010) Gliotransmission: Exocytotic release from astrocytes. Brain Res Rev 63(1-2):83–92. https://doi.org/10.1016/j.brainresrev.2009.11.008

Pasquettaz R, Kolotuev I, Rohrbach A, Gouelle C, Pellerin L, Langlet F (2021) Peculiar protrusions along tanycyte processes face diverse neural and nonneural cell types in the hypothalamic parenchyma. J Comp Neurol 529(3):553–575. https://doi.org/10.1002/CNE.24965

Pellegrino G, Martin M, Allet C, Lhomme T, Geller S, Franssen D, Mansuy V, Manfredi-Lozano M, Coutteau-Robles A, Delli V, Rasika S, Mazur D, Loyens A, Tena-Sempere M, Siepmann J, Pralong FP, Ciofi P, Corfas G, Parent AS, Ojeda SR, Sharif A, Prevot V (2021) GnRH neurons recruit astrocytes in infancy to facilitate network integration and sexual maturation. Nat Neurosci 24(12):1660–1672. https://doi.org/10.1038/s41593-021-00960-z

Peruzzo B, Pastor FE, Blázquez JL, Amat P, Rodríguez EM (2004) Polarized endocytosis and transcytosis in the hypothalamic tanycytes of the rat. Cell Tissue Res 317(2):147–164. https://doi.org/10.1007/S00441-004-0899-1/FIGURES/7

Pieron H, Legendre R (1910) Le problème des facteurs du sommeil. Résultats d'injections vasculaires et intracérébrales de liquides insomniques 68:1077–1079

Pieron H, Legendre R (1913) Recherches sur le besoin de sommeil consecutif a une vielle prolongee 14:235–262
Piet R, Vargova L, Sykova E, Poulain DA, Oliet SH (2004) Physiological contribution of the astrocytic environment of neurons to intersynaptic crosstalk. Proc Natl Acad Sci USA 101:2151–2155
Pissios P, Bradley RL, Maratos-Flier E (2006) Expanding the scales: the multiple roles of MCH in regulating energy balance and other biological functions. Endocr Rev 27(6):606–620. https://doi.org/10.1210/er.2006-0021
Prevot V, Croix D, Bouret S, Dutoit S, Tramu G, Stefano GB, Beauvillain JC (1999) Definitive evidence for the existence of morphological plasticity in the external zone of the median eminence during the rat estrous cycle: implication of neuro-glio-endothelial interactions in gonadotropin-releasing hormone release. Neuroscience 94:809–819
Prevot V, Dehouck B, Sharif A, Ciofi P, Giacobini P, Clasadonte J (2018) The versatile tanycyte: a hypothalamic integrator of reproduction and energy metabolism. Endocr Rev 39(3):333–368. https://doi.org/10.1210/er.2017-00235
Prevot V, Nogueiras R, Schwaninger M (2021) Tanycytes in the infundibular nucleus and median eminence and their role in the blood-brain barrier. Handb Clin Neurol 180:253–273
Prevot V, Tena-Sempere M, Pitteloud N (2023) New horizons: gonadotropin-releasing hormone and cognition. J Clin Endocrinol Metab 108:2747–2758
Prosser HM, Bradley A, Chesham JE, Ebling FJP, Hastings MH, Maywood ES (2007) Prokineticin receptor 2 (Prokr2) is essential for the regulation of circadian behavior by the suprachiasmatic nuclei. Proc Natl Acad Sci USA 104(2):648–653. https://doi.org/10.1073/pnas.0606884104
Rance TA, Baker BI (1979) Possible hypothalamic origin of the teleost melanin-concentrating hormone [proceedings]. J Endocrinol 81(2):166P
Rice ME, Cragg SJ (2008) Dopamine spillover after quantal release: Rethinking dopamine transmission in the nigrostriatal pathway. Brain Res Rev 58(2):303–313. https://doi.org/10.1016/J.BRAINRESREV.2008.02.004
Rodríguez EM, Blázquez JL, Pastor FE, Peláez B, Peña P, Peruzzo B, Amat P (2005) Hypothalamic tanycytes: a key component of brain–endocrine interaction. Int Rev Cytol 247:89–164. https://doi.org/10.1016/S0074-7696(05)47003-5
Rodríguez-Cortés B, Hurtado-Alvarado G, Martínez-Gómez R, León-Mercado LA, Prager-Khoutorsky M, Buijs RM (2022) Suprachiasmatic nucleus-mediated glucose entry into the arcuate nucleus determines the daily rhythm in blood glycemia. Curr Biol 32(4):796–805.e4. https://doi.org/10.1016/j.cub.2021.12.039
Romanov RA, Zeisel A, Bakker J, Girach F, Hellysaz A, Tomer R, Alpár A, Mulder J, Clotman F, Keimpema E, Hsueh B, Crow AK, Martens H, Schwindling C, Calvigioni D, Bains JS, Máté Z, Szabó G, Yanagawa Y et al (2017) Molecular interrogation of hypothalamic organization reveals distinct dopamine neuronal subtypes. Nat Neurosci 20(2):176–188. https://doi.org/10.1038/nn.4462
Romanova DY, Heyland A, Sohn D, Kohn AB, Fasshauer D, Varoqueaux F, Moroz LL (2020) Glycine as a signaling molecule and chemoattractant in Trichoplax (Placozoa): Insights into the early evolution of neurotransmitters. Neuroreport 31(6):490–497. https://doi.org/10.1097/WNR.0000000000001436
Romanova DY, Varoqueaux F, Daraspe J, Nikitin MA, Eitel M, Fasshauer D, Moroz LL (2021) Hidden cell diversity in Placozoa: Ultrastructural insights from Hoilungia hongkongensis. Cell Tissue Res 385(3):623–637. https://doi.org/10.1007/s00441-021-03459-y
Rose CD, Pompili D, Henke K, Van Gennip JLM, Meyer-Miner A, Rana R, Gobron S, Harris MP, Nitz M, Ciruna B (2020) SCO-spondin defects and neuroinflammation are conserved mechanisms driving spinal deformity across genetic models of idiopathic scoliosis. Curr Biol 30(12):2363–2373.e6. https://doi.org/10.1016/j.cub.2020.04.020
Schang A-L, Ngô-Muller V, Bleux C, Granger A, Chenut M-C, Loudes C, Magre S, Counis R, Cohen-Tannoudji J, Laverrière J-N (2011) GnRH receptor gene expression in the developing rat hippocampus: transcriptional regulation and potential roles in neuronal plasticity. Endocrinology 152(2):568–580. https://doi.org/10.1210/en.2010-0840

Schierwater B, Eitel M (2015) Placozoa. In: Wanninger A (ed) Evolutionary developmental biology of invertebrates 1: introduction, non-bilateria, Acoelomorpha, Xenoturbellida, Chaetognatha. Springer, Vienna, pp 107–114. https://doi.org/10.1007/978-3-7091-1862-7_5

Schmitt FO (1984) Molecular regulators of brain function: A new view. Neuroscience 13(4):991–1001. https://doi.org/10.1016/0306-4522(84)90283-5

Schot LP, Boer HH, Swaab DF, Van Noorden S (1981) Immunocytochemical demonstration of peptidergic neurons in the central nervous system of the pond snail Lymnaea stagnalis with antisera raised to biologically active peptides of vertebrates. Cell Tissue Res 216(2):273–291. https://doi.org/10.1007/BF00233620

Schwartz WJ, Reppert SM (1985) Neural regulation of the circadian vasopressin rhythm in cerebrospinal fluid: A pre-eminent role for the suprachiasmatic nuclei. J Neurosci 5(10):2771–2778. https://doi.org/10.1523/JNEUROSCI.05-10-02771.1985

Senatore A, Reese TS, Smith CL (2017) Neuropeptidergic integration of behavior in Trichoplax adhaerens, an animal without synapses. J Exp Biol 220(Pt 18):3381–3390. https://doi.org/10.1242/jeb.162396

Shimada M, Tritos NA, Lowell BB, Flier JS, Maratos-Flier E (1998) Mice lacking melanin-concentrating hormone are hypophagic and lean. Nature 396(6712):670–674. https://doi.org/10.1038/25341

Silver R, LeSauter J, Tresco PA, Lehman MN (1996) A diffusible coupling signal from the transplanted suprachiasmatic nucleus controlling circadian locomotor rhythms. Nature 382(6594):810–813. https://doi.org/10.1038/382810a0

Sita LV, Diniz GB, Canteras NS, Xavier GF, Bittencourt JC (2016) Effect of intrahippocampal administration of anti-melanin-concentrating hormone on spatial food-seeking behavior in rats. Peptides 76:130–138. https://doi.org/10.1016/j.peptides.2015.12.007

Skinner DC, Caraty A, Malpaux B, Evans NP (1997) Simultaneous measurement of gonadotropin-releasing hormone in the third ventricular cerebrospinal fluid and hypophyseal portal blood of the ewe. Endocrinology 138(11):4699–4704. https://doi.org/10.1210/endo.138.11.5494

Skrapits K, Sárvári M, Farkas I, Göcz B, Takács S, Rumpler É, Váczi V, Vastagh C, Rácz G, Matolcsy A, Solymosi N, Póliska S, Tóth B, Erdélyi F, Szabó G, Culler MD, Allet C, Cotellessa L, Prévot V et al (2021) The cryptic gonadotropin-releasing hormone neuronal system of human basal ganglia. eLife 10:e67714. https://doi.org/10.7554/eLife.67714

Smith CL, Varoqueaux F, Kittelmann M, Azzam RN, Cooper B, Winters CA, Eitel M, Fasshauer D, Reese TS (2014) Novel cell types, neurosecretory cells, and body plan of the early-diverging metazoan Trichoplax adhaerens. Curr Biol 24(14):1565–1572. https://doi.org/10.1016/j.cub.2014.05.046

Smith CL, Pivovarova N, Reese TS (2015) Coordinated feeding behavior in Trichoplax, an animal without synapses. PLoS One 10(9):e0136098. https://doi.org/10.1371/journal.pone.0136098

Smith CL, Reese TS, Govezensky T, Barrio RA (2019) Coherent directed movement toward food modeled in Trichoplax, a ciliated animal lacking a nervous system. Proc Natl Acad Sci USA 116(18):8901–8908. https://doi.org/10.1073/pnas.1815655116

Soto-Tinoco E, Guerrero-Vargas NN, Buijs RM (2016) Interaction between the hypothalamus and the immune system. Exp Physiol 101(12):1463–1471. https://doi.org/10.1113/EP085560

Soto-Tinoco E, Santacruz E, Basualdo-Sigales MDC, Guerrero-Vargas NN, Buijs RM (2020) Time-of-day-dependent gating of the liver-spinal axis initiates an anti-inflammatory reflex in the rat. eNeuro 7(6). https://doi.org/10.1523/ENEURO.0463-20.2020

Srivastava M, Begovic E, Chapman J, Putnam NH, Hellsten U, Kawashima T, Kuo A, Mitros T, Salamov A, Carpenter ML, Signorovitch AY, Moreno MA, Kamm K, Grimwood J, Schmutz J, Shapiro H, Grigoriev IV, Buss LW, Schierwater B et al (2008) The Trichoplax genome and the nature of placozoans. Nature 454(7207):955–960. https://doi.org/10.1038/nature07191

Syková E (2004) Extrasynaptic volume transmission and diffusion parameters of the extracellular space. Neuroscience 129(4):861–876. https://doi.org/10.1016/j.neuroscience.2004.06.077

Syková E, Nicholson C (2008) Diffusion in brain extracellular space. Physiol Rev 88(4):1277–1340. https://doi.org/10.1152/PHYSREV.00027.2007/ASSET/IMAGES/LARGE/Z9J0040824840013.JPEG

Taber KH, Hurley RA (2014) Volume transmission in the brain: Beyond the synapse. J Neuropsychiatry Clin Neurosci 26(1):1. https://doi.org/10.1176/APPI.NEUROPSYCH.13110351/ASSET/IMAGES/LARGE/IVF3B.JPEG

Takeda S, Kuwabara Y, Mizuno M (1985) Effects of pregnancy and labor on oxytocin levels in human plasma and cerebrospinal fluid. Endocrinol Jpn 32(6):875–880

Taub A, Carbajal Y, Rimu K, Holt R, Yao Y, Hernandez AL, Lesauter J, Silver R (2021) Arginine vasopressin-containing neurons of the suprachiasmatic nucleus project to CSF. eNeuro 8(2). https://doi.org/10.1523/ENEURO.0363-20.2021

Tauber M, Diene G (2021) Prader-Willi syndrome: Hormone therapies. Handb Clin Neurol 181:351–367

Theodosis DT, Poulain DA, Oliet SH (2008) Activity-dependent structural and functional plasticity of astrocyte-neuron interactions. Physiol Rev 88:983–1008

Trudel E, Bourque CW (2010) Central clock excites vasopressin neurons by waking osmosensory afferents during late sleep. Nat Neurosci 13(4):467–474. https://doi.org/10.1038/nn.2503

Tsutsumi H, Yonemitsu K, Sasao A, Ohtsu Y, Furukawa S, Nishitani Y (2020) Cerebrospinal fluid neurotransmitter levels and central nervous system depression in a rat drug overdose model. Toxicol Mech Methods 30(2):139–145. https://doi.org/10.1080/15376516.2019.1672122

Ueda T, Koya S, Maruyama YK (1999) Dynamic patterns in the locomotion and feeding behaviors by the placozoan *Trichoplax adhaerence*. Biosystems 54(1):65–70. https://doi.org/10.1016/S0303-2647(99)00066-0

Vanacker C, Trova S, Shruti S, Casoni F, Messina A, Croizier S, Malone S, Ternier G, Hanchate NK, Rasika S, Bouret SG, Ciofi P, Giacobini P, Prevot V (2020) Neuropilin-1 expression in GnRH neurons regulates prepubertal weight gain and sexual attraction. EMBO J 39(19):e104633. https://doi.org/10.15252/embj.2020104633

Varoqueaux F, Fasshauer D (2017) Getting nervous: an evolutionary overhaul for communication. Annu Rev Genet 51(1):455–476. https://doi.org/10.1146/annurev-genet-120116-024648

Varoqueaux F, Williams EA, Grandemange S, Truscello L, Kamm K, Schierwater B, Jékely G, Fasshauer D (2018) High cell diversity and complex peptidergic signaling underlie placozoan behavior. Curr Biol 28(21):3495–3501.e2. https://doi.org/10.1016/j.cub.2018.08.067

Veening JG, de Jong T, Barendregt HP (2010) Oxytocin-messages via the cerebrospinal fluid: Behavioral effects; a review. Physiol Behav 101(2):193–210. https://doi.org/10.1016/J.PHYSBEH.2010.05.004

Viero C, Shibuya I, Kitamura N, Verkhratsky A, Fujihara H, Katoh A, Ueta Y, Zingg HH, Chvatal A, Sykova E, Dayanithi G (2010) REVIEW: oxytocin: crossing the bridge between basic science and pharmacotherapy. CNS Neurosci Ther 16(5):e138–e156. https://doi.org/10.1111/j.1755-5949.2010.00185.x

Vizi ES, Fekete A, Karoly R, Mike A (2010) Non-synaptic receptors and transporters involved in brain functions and targets of drug treatment. Br J Pharmacol 160(4):785–809. https://doi.org/10.1111/J.1476-5381.2009.00624.X

Westerfield DB, Pang PKT, Burns JM (1980) Some characteristics of melanophore-concentrating hormone (MCH) from teleost pituitary glands. Gen Comp Endocrinol 42(4):494–499. https://doi.org/10.1016/0016-6480(80)90215-4

Yamamura T, Hirunagi K, Ebihara S, Yoshimura T (2004) Seasonal morphological changes in the neuro-glial interaction between gonadotropin-releasing hormone nerve terminals and glial endfeet in Japanese quail. Endocrinology 145:4264–4267

Yang L, Kress BT, Weber HJ, Thiyagarajan M, Wang B, Deane R, Benveniste H, Iliff JJ, Nedergaard M (2013) Evaluating glymphatic pathway function utilizing clinically relevant intrathecal infusion of CSF tracer. J Transl Med 11(1). https://doi.org/10.1186/1479-5876-11-107

Yao Y, Taub AB, LeSauter J, Silver R (2021) Identification of the suprachiasmatic nucleus venous portal system in the mammalian brain. Nat Commun 12(1):5643. https://doi.org/10.1038/s41467-021-25793-z

Yoon NA, Diano S (2021) Hypothalamic glucose-sensing mechanisms. Diabetologia 64(5):985–993. https://doi.org/10.1007/s00125-021-05395-6

The Evolution of the Pituitary Gland

4

Yorgui Santiago-Andres, Tatiana Fiordelisio, Diana del-Rio, and Juan F. Cubillos

Abstract

From an evolutionary perspective, the neuroendocrine system is a robust, complex, and sophisticated communication mechanism by which organisms coordinate and integrate signals to regulate a wide range of physiological processes, maintain homeostasis, and enable an adaptive response. One of the key events of the evolution of vertebrates was the appearance of the pituitary gland and the formation of the hypothalamus–pituitary system. This chapter discusses the molecular evolution of the structure and function of hormones and receptors, the complexity of the vasculature, as well as changes in the structural characteristics, diversity of composition, and zonal distribution of endocrine cells in the gland. The diversification of hypothalamic–pituitary communication mechanisms in vertebrates is reviewed, ranging from passive diffusion to neuroglandular and neurovascular pathways, such as the hypophyseal portal system. This chapter explores the increasing complexity in the integration of the regulation of the endocrine axes in vertebrates and their degree of independence in operation, with emphasis on the physiological implications and specific adaptations of the dynamics of the axes regulating growth, metabolism, and reproduction. Finally,

Y. Santiago-Andres · J. F. Cubillos
Laboratorio de Neuroendocrinología, Facultad de Ciencias, Universidad Nacional Autónoma de México, Coyoacán, CDMX 04510, Mexico
e-mail: sayorgui@ciencias.unam.mx; jcubillos@ciencias.unam.mx

T. Fiordelisio (✉) · D. del-Rio
Laboratorio de Neuroendocrinología, Facultad de Ciencias, Universidad Nacional Autónoma de México, Coyoacán, CDMX 04510, Mexico

Laboratorio Nacional de Soluciones Biomiméticas para Diagnóstico y Terapia, Facultad de Ciencias, Universidad Nacional Autónoma de México, Coyoacán, CDMX 04510, Mexico
e-mail: tfiorde@ciencias.unam.mx; delrio@ciencias.unam.mx

V. Grinevich, R. F. Oliveira (eds.), *Evolutionary and Comparative Neuroendocrinology*, Masterclass in Neuroendocrinology 17,
https://doi.org/10.1007/978-3-031-80209-6_4

pulsatile hormone secretion is addressed as an adaptive feature of high-level signaling that involves a system of internal glandular regulation. This results from the interplay between intercellular communication elements, highly plastic cellular networks, and the vascular system.

Keywords

Neuroendocrine evolution · Adenohypophysis · Hypothalamus–pituitary communication · Vertebrate pituitary · Hormonal pulse · Neuroglandular · Neurovascular · Endocrine evolution

List of Abbreviations

ACP	Corazonin-related peptide
ACTH	Adrenocorticotropic hormone
AKH	Adipokinetic hormone
AVP	Arginine-vasopressin
AVT	Arginine-vasotocin
CRH	Corticotropin-releasing hormone
CRZ	Corazonin
DA	Dopamine
DIT	Diiodotyrosine
Dyn	Dynorphin
ELISA	Enzyme-linked immunosorbent assay
ER	Estrogen receptor
FSH	Follicle-stimulating hormone
GH	Growth hormone
GHRH	Growth hormone-releasing hormone
GnRH	Gonadotropin-releasing hormone
HP system	Hypothalamus–pituitary system
HPG axis	Hypothalamic–pituitary–gonadal axis
HPT axis	Hypothalamic–pituitary–thyroid axis
IGF-1	Insulin-like growth factor 1
Kiss	Kisspeptin
lGpH	Lamprey glycoprotein hormone
LH	Luteinizing hormone
MIT	Monoiodotyrosine
MSH	Melanocyte-stimulating hormone
MT	Mesotocin
NKB	Neurokinin B
OT	Oxytocin
PACAP	Pituitary adenylate cyclase-activating peptide
PRL	Prolactin
PVN	Paraventricular nucleus

SON	Supraoptic nucleus
T_3	Triiodothyronine
T_4	Thyroxine
TPO	Thyroperoxidase
TRH	Thyrotropin-releasing hormone
TSH	Thyroid-stimulating hormone
TSN	Thyrostimulin

4.1 Introduction

> Constancy is constructed. To understand it, then, we must look at its processes, not only at its constituents, and be ready to meet them on as many levels as their complexity justifies.
>
> Susan Oyama, 2000

Cellular communication is the building block of all animal life. It consists of the intercellular transfer of information—that is, a set of organized data—emitted in the form of signals which, when received and integrated, evoke a biological response. The evolution of multicellularity and the compartmentalization of functions require the coordinated action of different systems that, despite relatively large distances between transmitting and receiving cells, allow the transmission, collection, and integration of information, and the most appropriate response to adapt to variation (internal or external) and to ensure the internal environment is kept stable and constant; in other words, to maintain homeostasis. The adaptive response of organisms has developed two systems for internal communication and interaction with the outside world: the nervous system and the endocrine system, which work together in the neuroendocrine system.

From an evolutionary point of view, chordates have structures that perform both endocrine and sensory functions, such as Hatschek's pit and the neural gland, which may have given rise to the neuroendocrine system (Heyland et al. 2012; Fontaine et al. 2022). In vertebrates, the most important neuroendocrine system is the hypothalamus–pituitary (HP) system, which regulates multiple physiological functions (growth, metabolism, reproduction, osmoregulation, stress response, among others) by means of chemical messengers or bioregulators (hormones and a wide variety of biologically active compounds) transported through the blood, forming axes of communication between the hypothalamus, the pituitary gland, and several target organs. The hypothalamus was long thought to be the leading component of the HP system, relegating the pituitary to a subordinate role in the regulation of vital functions and homeostasis, but it is now clear that its contribution to the HP system is much more complex, far from a simple hierarchical relationship, with the vascular system as a co-protagonist (Le Tissier et al. 2017).

In protochordates, which do not possess endocrine organs as such, primitive neurons release hormone molecules that reach their effector sites via the extracellular medium. Furthermore, in these organisms, no part of the genome appears to encode the ligands and receptors required for signaling as seen in the HP system, although general mechanisms linking the neuronal and endocrine systems can be observed. As we will explore in this chapter, the HP system is an important adaptation and a crucial element in vertebrate evolution, which has led to divergences in hypothalamic and pituitary structures, mechanisms of communication, hormones and receptors, structures and functions of enormous physiological relevance for the organisms and their ecological niches.

4.2 The Origin of the Pituitary Gland

The pituitary gland is an organ found in all extant vertebrates, made up of three main hormone-producing structures: the *pars nervosa*, the *pars intermedia*, and the *pars distalis*. The *pars nervosa* is a tissue composed mainly of projections of hypothalamic neurons that produce oxytocin-like (OT-like) and arginine-vasopressin-like (AVP-like) hormones; these peptides are stored and released into capillary beds directly connected to the venous drainage of the bloodstream. The *pars intermedia* consists of a single cell type, the melanotropes, which secrete melanocyte-stimulating hormones (α-, β-, and γ-MSH). The *pars distalis* contains a variety of cell types, whose diversity and distribution throughout the gland has changed in the course of evolution (Fig. 4.1).

The endocrine cells of the *pars distalis* synthesize and secrete hormones into a complex of capillary beds: gonadotropes secrete luteinizing hormone (LH) and follicle-stimulating hormone (FSH), somatotropes produce growth hormone (GH), lactotropes synthesize prolactin (PRL), thyrotropes release thyroid-stimulating hormone (TSH), and corticotropes produce adrenocorticotropic hormone (ACTH). Two other endocrine cell types have been identified in the *pars distalis* of fish: somatolactotropes and proto-glycotropes. Teleosts (infraclass Teleostei), which comprise about 95% of all extant species of fishes, possess a sixth hormone-producing cell type called somatolactotropes that secrete the hormone somatolactin, which is distinct from both GH and PRL (Kaneko 1996); jawless fish of class Hyperoartia, a basal vertebrate lineage that includes lampreys, on the other hand, lacks thyrotropes and gonadotropes, and instead possesses proto-glycotropes that synthesize two glycoprotein hormones termed lamprey glycoprotein hormone (lGpH) and thyrostimulin (TSN), which are homologous to LH/FSH and TSH in jawed vertebrates, respectively (Marquis et al. 2017).

Although the three main units of the pituitary gland and their cell types are mostly conserved throughout living vertebrates, the precise origin of this organ is still subject to research. In evolutionary biology, the origin of a structure is traced by phylogenetic reconstructions of all species in which the structure in question can be found, in order to determine the molecular, cellular, and functional changes it has undergone as a sequence over time. This method also allows us to hypothesize about the morphology and ecology of the common ancestor of the species being studied, and whether the trait itself evolved from a pre-existing element or as an evolutionary novelty. For example, living birds have feathers that play a role in flight, sexual behavior, and thermoregulation. However, fossil data from related species of

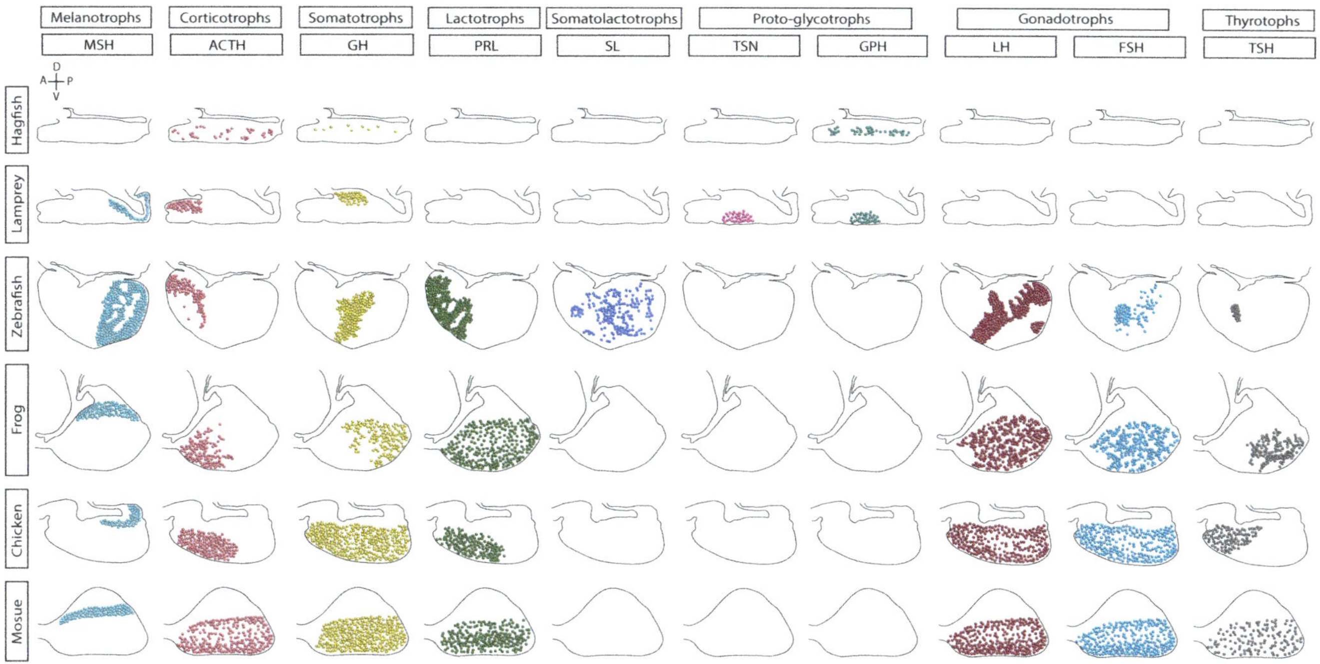

Fig. 4.1 Diversity and distribution of secretory cells in the *pars distalis* of the pituitary gland. Sagittal map of pituitary glands representative of the main groups of vertebrates. In jawless fish (lamprey and hagfish) and teleosts (zebrafish) there is a zonal distribution of endocrine cells. In amphibians (frog) and reptiles (chicken), the distribution of some cell types is restricted to the anterior or the posterior portion of the gland, but are widely found across the entire *pars distalis* in other groups. No zonal distribution of cells is observed in mammals (mouse) except for MSH-producing cells. MSH, melanocyte-stimulating hormones; ACTH, adrenocorticotropic hormone; GH, growth hormone; PRL, prolactin; SL, somatolactin; TSN, thyrostimulin; GpH, glycoprotein hormone; LH, luteinizing hormone; FSH, follicle-stimulating hormone; TSH, thyroid-stimulating hormone; A, anterior or rostral; P, posterior or caudal; D, dorsal; V, ventral. Modified from Santiago-Andres et al. (2020) with permission

feathered dinosaurs suggest that the function of feathers in maintaining body temperature preceded that of flight. We also know from extant reptiles that the development of feathers arose from the same ontogenetic process that gives rise to scales in crocodiles, modified at the gene network level, resulting in feathers rather than scales. We apply the same type of reasoning to the origin and evolution of the pituitary gland, based on evidence about morphological and functional differences in the gland itself and the conformation of neuroendocrine systems across taxa.

The pituitary gland is present in all groups of vertebrates (subphylum Vertebrata) that are divided into jawless fish (all extant members in superclass Cyclostomi), whose only living members are hagfish and lampreys, and jawed vertebrates (infraphylum Gnathostomata), further classified into cartilaginous fish (class Chondrichthyes; sharks, rays, and chimeras), bony fish (superclass Osteichthyes; ray-finned fish and lobe-finned fish), and tetrapods (superclass Tetrapoda), represented by amphibians, reptiles (including birds), and mammals. All vertebrates belong to phylum Chordata, which also includes tunicates (subphylum Tunicata) and lancelets (subphylum Cephalochordata). As the two living lineages closest to vertebrates, the study of tunicates and lancelets has provided some insights into the possible origins of the vertebrate pituitary gland. However, this evidence must be interpreted with caution, as similarity in function does not necessarily indicate homology: convergent solutions to similar environmental pressures in unrelated species are a common occurrence in evolution, as seen in the independent origins of wings in bats, birds, and insects (Box 4.1).

Box 4.1: Searching for Common Origin of the Pituitary Gland: Steps in Homology Assessment

In its most general definition, homology refers to traits shared by two or more species due to common ancestry. In this sense, the search for homologies across the living world makes sense in the light of evolutionary theory, and since shared traits of different species may accumulate variation as a consequence of their independent evolutionary history and ecology after divergence, the identification of homologous features is not always a straightforward process. According to Wagner (1989), the identification of homologies requires the assessment of some basic biological properties:

1. Conservation: Homologous features are sufficiently independent of the rest of the body that can be anatomically delimited. Homologous traits are like building blocks that may present different states of character due to different ways of living between species (and changes in function and morphology), but conserve certain elements that the variation moves around an identifiable plan. For example, the three main structures of the vertebrate pituitary gland are conserved and are easily identifiable, despite species variation in patterns of reproduction, growth, or the environment in

(continued)

which they develop. The conservation property also correlates with the similar position of the trait compared to other building blocks of the organisms (the pituitary is always close to, and below, the hypothalamus) and the components of a trait have a conserved correspondence of relative position (the *pars intermedia* of the pituitary always resides between the *pars nervosa* and the *pars distalis*).

2. Individuality: During embryonic ontogeny and trait formation, features that are homologous possess an important degree of independence from other structures under development. The quasi-autonomy of development is reflected at the different levels of organization such as genetic network dynamics, epigenetics, and cell–cell interactions. One prediction of this property of homologous traits is that the perturbation of their development (through genetic changes or environmental modifications) will not significantly affect the normal pathway of other traits. Of course, there is always some degree of interaction between adjacent developmental units but "the lack of genetic and developmental individuality may be the main reason for dubious homologies" (Wagner 1989). Classic experiments on the pituitary have demonstrated that removal or transplantation of the gland to another part of the body has no effect on the function of other organs.
3. Uniqueness: Since homologous features are well defined and separate from other structures and possess a certain level of complexity and integration, it is also possible to identify that homologies are not repetitive structures in the organisms being compared. This uniqueness is also a direct consequence of the fact that the common ancestor of the species acquired the trait only once in evolution, which was then passed down to its descendants. This is the case for almost every trait in biology, not just the pituitary gland, and "hence, it appears that the probability of evolving a new character is sufficiently low that most characters originate only once" (Wagner 1989). However, we need to be cautious since convergence of characters is also a common event in evolution.
4. Ontogenetic networks: The growing body of information from the comparative study of the development and evolution of species (also called Evo-Devo) has shown that homologous structures have a close relationship in the formation of their components throughout development. This interaction creates a feedback in which the development of one element becomes the "bed" for a second step in the development of the whole structure. For example, the expression of the gene Pit-1, a transcription factor, is a definitive gene for pituitary identity and its expression regulates the fate of cells differentiating in the gland. The existence of ontogenetic networks underlying traits also gives rise to relationships of dependence between different subunits during development. For instance, in mammals, the successful proliferation and migration of gonadotropes depends heavily on the establishment of a pre-existing network of corticotropes.

(continued)

Together, these properties provide a robust hypothesis in the research of homologous traits between species that must be considered when studying and evaluating origins of the vertebrate pituitary gland. Finally, we would like to emphasize the lack of correspondence when analyzing homology at different levels of organization (genetic, ontogenetic, and anatomical), since evolution is also a contingent process that generates variation with different patterns at different levels (Tessmar-Raible et al. 2007). Advances in genetic techniques and the analysis of gene expression have pointed out another level of homology termed *deep homology*, in which non-homologous structures with similar functions (such as the different types of "eyes" in animals) share common regulatory genes or genetic networks that directs their formation during development (for example, Pax6 expression is involved in eye development in both fruit flies and mice). The study of deep homology in the origin of the pituitary gland is still a black box.

Tunicates, also called sea squirts or ascidians, are barrel-shaped animals that feed by filtering plankton and algae using a system of oral and atrial siphons for water intake and expulsion, respectively. During the larval stage, tunicates possess a notochord (a defining trait of all chordates) which protects the nerve cords and acts as a site for muscle attachment. These animals also have a neural gland which is located parallel to the cerebral ganglion (the ascidian brain). Several functions have been proposed for this neural gland, including blood volume regulation, involvement in digestion, lymph production and excretion, and endocrine communication (Deyts et al. 2006). The neural gland has been proposed as a homologous structure to the pituitary gland because of its proximity to the cerebral ganglion, which has been shown to have endocrine activity and because it derives from the neuroectoderm (Sekiguchi et al. 2007). The possible expression of pituitary hormones (Fig. 4.1) in the tunicate neural gland has been explored using genome sequencing, although no genes for pituitary hormones are present in the ascidian genome (Campbell et al. 2004; Kano 2010).

The search for a regulatory gene network for endocrine cell specification and hormone expression in the tunicate genome showed that not all genes relevant to pituitary function are present in tunicates, while others are expressed at different times in the life history, suggesting that no true homology exists between the pituitary and any of the structures found in tunicates, but that several elements of the vertebrate neuroendocrine system were already present in the ancestor of all chordates (Kano 2010). In addition, using different techniques of microscopy, gene expression, and phylogenetic analysis, Deyts et al. (2006) challenged the idea that the neural gland is homologous to the pituitary and proposed that it functions as an osmoregulatory system homologous to the periventricular organs in vertebrates. These results contradict previous data from immunolabeling of pituitary hormones with antibodies that may cross-react with other molecules in the tunicate cells. More recently, the ciliated funnel has also been proposed as a pituitary homolog because of the expression of Pitx, a gene involved in organogenesis and cell differentiation of this gland, but more research is needed to fully establish the role of Pitx in the development of the ciliated funnel.

Lancelets, also known as amphioxi, are benthic, fish-shaped, filter-feeding animals. Like tunicates, lancelets possess a notochord below the neural cord, as well as a ciliated groove on the dorsal surface of the buccal cavity termed Hatschek's pit, which aids feeding and has been proposed as a pituitary homolog on anatomical, functional, and genetic grounds. In juvenile and adult organisms, Hatschek's pit is located below the neural cord and is connected to the cerebral vesicle (which has been proposed to correspond to the vertebrate diencephalon) by an infundibulum-like extension, although in larval stages these structures are separated from each other and have no direct link between them, either by innervation or through a shared circulatory system as seen in the vertebrate pituitary gland's link to the hypothalamus (Gorbman et al. 1999; Kano 2010; Zhang and Ji 2022). It has been reported that this structure contains granulated secretory cells, suggesting possible endocrine activity. The administration of ovine PRL and LH to lancelets has also been shown to increase sex steroid production, although attempts to identify pituitary hormones in amphioxi have been unsuccessful (Nozaki and Gorbman 1992; Uchida et al. 2010). However, recent analysis of immunoreactivity to several antibodies in lancelets has revealed the expression of genes with known endocrine function in vertebrates, such as Pit-1, substance P, kisspeptin (Kiss), LH, GH-like protein, met-enkephalin, the aromatase enzyme that converts androgens into estrogens, gpa2 and gpb5 (two glycoproteins that are thought to be related to the alpha- and beta-subunits of vertebrate FSH, LH, and TSH) (Zhang and Ji 2022). Thirdly, the analysis of the lancelet genome has revealed that the genes for all other pituitary hormones appear to be absent but genes for gonadotropin-releasing hormone-like (GnRH-like) peptides, GnRH receptors, Kiss receptor, TSN, steroidogenic enzymes, and estrogen receptors are present, although their pattern of expression in time and space needs further investigation (Holland et al. 2008; Kano 2010; Trudeau and Somoza 2020; Zhang and Ji 2022).

Despite wide differences in life histories among vertebrates, the pituitary gland is a well-defined organ throughout all of them, suggesting that variation in the gland is limited as a consequence of its fundamental role in organismal fitness. Nevertheless, although tunicates and lancelets are the closest living relatives of vertebrates, they do not possess a true pituitary gland. The protochordate neural complex (neural gland and ciliated funnel) and Hatschek's pit have been proposed as homologous structures to the pituitary gland, although discrepancies exist at the molecular, functional, and developmental levels of these structures that invite caution about interpreting homology, raising the possibility that the pituitary gland is an evolutionary novelty that was not present before the origin of vertebrates, even if several of the components were present in the last common ancestor of vertebrates and protochordates (Campbell et al. 2004; Gerhart and Kirschner 2007).

4.3 Evolution of the Neurohypophysis

The posterior pituitary, or neurohypophysis, is an anatomical extension of the hypothalamus that consists mainly of the axonal terminals of magnocellular neurons located in the supraoptic (SON) and paraventricular (PVN) nuclei of the hypothalamus. In all but the most basal vertebrates, the neurons of these nuclei secrete homologs of the hormones OT and AVP into the venous sinusoids of the posterior pituitary region, where they enter the systemic circulation and are able to act upon target tissues in a classical endocrine manner. This structure also contains pituicytes, which are specialized glial cells that assist in the storage and release of hormones. The magnocellular neuroendocrine system is involved in a broad range of biological functions, including reproduction, social behavior, osmoregulation, and thermoregulation.

The anatomical connection between the hypothalamus and the posterior pituitary was first described by Santiago Ramón y Cajal in 1894. Years later, Ernst Scharrer observed colloid-like inclusions in the magnocellular neurons of the SON of the hypothalamus of the Eurasian minnow, a teleost. He correctly interpreted this as evidence of neurosecretion, supported by the extensive vascularization of the area and its connection to the pituitary gland. The idea of neurosecretion, a novel concept at the time, was controversial for several decades, and it was accepted in large part due to the extensive work of Ernst and Berta Scharrer. The Scharrers described neurosecretory organs in all extant classes of vertebrates, as well as in annelids, mollusks, and arthropods (Scharrer and Scharrer 1937).

The biological functions of the posterior pituitary hormones began to be discovered at the end of the nineteenth century, when George Oliver and Edward Albert Schäfer were the first to describe the vasopressor effect of pituitary extracts (Oliver and Schäfer 1895). Shortly afterwards, William Henry Howell demonstrated that the effects of the extracts were solely attributable to the posterior pituitary portion. Prior to his experiments, it was assumed that the posterior pituitary was a rudimentary organ with no physiological function (Howell 1898). In the following years, studies of the properties of these posterior pituitary extracts would provide further clues about the biological functions of the neurohypophysis: Henry Hallet Dale (1906) described the contraction-inducing properties on the uterus of monkeys, cats, and rabbits; Isaac Ott and John C. Scott (1910) reported the induction of milk secretion in nursing goats, which became known as the milk-ejection effect; later, in 1913, F. Farini and R. von den Velden independently discovered anti-diuretic effects of this extract (Dale 1906; Ott and Scott 1910; Lévêque 1974).

While the research of these and other scientists helped to establish the broad effects of posterior pituitary hormones, it was the Nobel Prize-winning work of Vincent du Vigneaud, who isolated and synthesized OT and AVP for the first time, that allowed for the distinction of two separate hormones in posterior pituitary extracts and, consequently, allowed the separate exploration of their chemical composition and biological function (Du Vigneaud 1956).

The hormonal components of the magnocellular neurosecretory system have been described in a wide range of animals belonging to all extant classes of

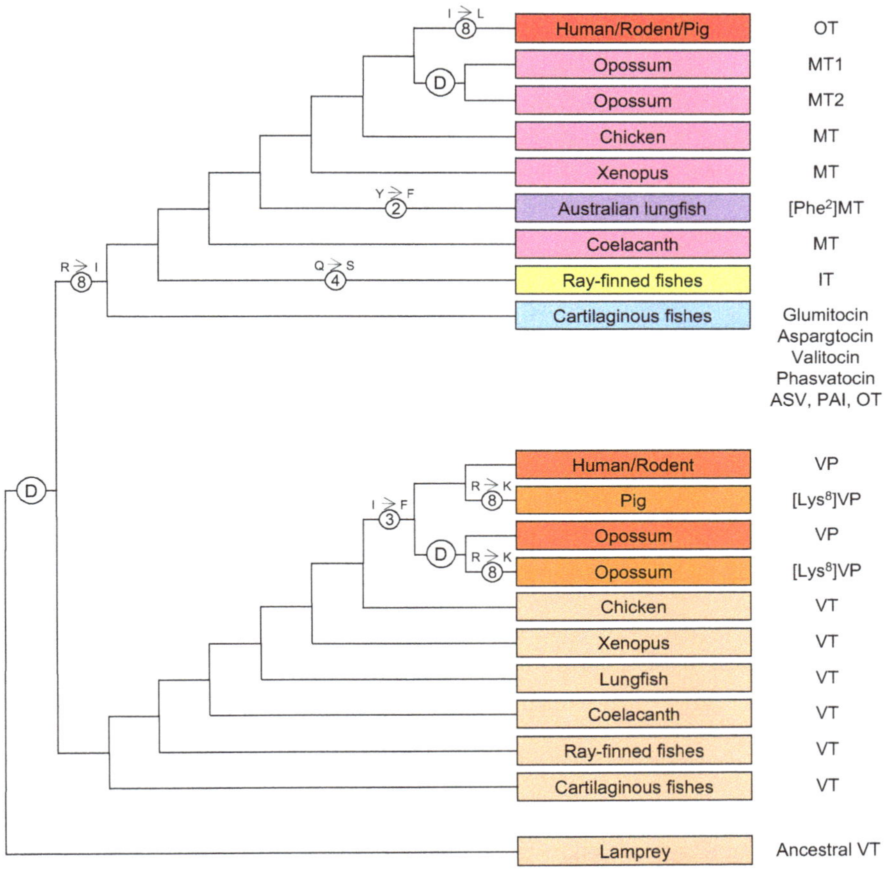

Fig. 4.2 The molecular evolution of the neurohypophysis nonapeptides, vasopressin and oxytocin family. Gene duplication events in the phylogeny are represented with the letter "D" and numbers within the circles refer to the position of the amino acid residue that has been substituted. OT, oxytocin; MT, mesotocin; IT, isotocin; VP, vasopressin; [Lys^{8}]VP, lysipressin; VT, vasotocin; ASV, asvatocin; PAI, phasitocin. Reprinted from Gwee et al. (2008), under the terms of the Creative Commons Attribution License (https://creativecommons.org/licenses/by/2.0)

vertebrates. All jawed vertebrates possess two types of neurohypophysial hormones: one belonging to the arginine-vasotocin (AVT)/AVP family and one belonging to the OT-like family, forming an AVP/OT superfamily of hormones. All of the molecules in this superfamily are nonapeptides that differ from each other by one to three amino acid residues. Evidence from extant jawless fish (hagfish and lampreys) suggests that both lines of vertebrate neurohypophysial hormones originated from an ancestral duplication event of the provasotocin gene, as hagfish and lampreys possess a single AVP-like posterior pituitary hormone (Fig. 4.2) (Acher 1996).

All non-mammalian vertebrates express a conserved AVT as their neurohypophysial hormone of the AVT/AVP family (Fig. 4.2), with the vast majority of mammals producing AVP instead. The amino acid sequence of AVP differs from that of AVT by a single residue. There is some variation in this hormone family within mammals, with swine (family Suidae), hippopotamuses, and some marsupials in the families Macropodidae, Didelphidae, and Peramelidae expressing lysipressin instead of AVP, and some macropods expressing phenypressin. The similar pharmacological properties of these mammalian AVP variants have led to the conclusion that they are products of neutral evolution (Acher 1996).

Peptides of the AVT/AVP family play important roles in the maintenance of osmolarity and fluid balance. It is notable that all non-mammalian species of vertebrates studied to date present AVT (or, in the case of cyclostomes, the ancestral form of AVT), while mammals exhibit a higher degree of variation in the sequence of their AVP-like peptides. The conservation of AVT in the vertebrate lineage suggests it has been subjected to intense evolutionary pressure, likely in relation to the maintenance of osmotic homeostasis in the aquatic habit that is ancestral to the vertebrate clade.

OT is present in all placental mammals, as well as in some marsupials and in one species of chimera—the ratfish (Fig. 4.2). The presence of OT in ratfish is considered to be an instance of convergent evolution, rather than an ancestral trait of the jawed vertebrate clade. Meanwhile, most marsupials, all non-mammalian vertebrates, and lobe-finned fish possess mesotocin (MT) instead, which differs from OT by one amino acid residue. All bony fishes have isotocin, the amino acid sequence of which is shared with that of MT except for one substitution. The cartilaginous fishes show a remarkable variability in the OT-like hormone family compared to other vertebrates: besides OT, glumitocin, aspargtocin, valitocin, phasvatocin, asvatocin, and phasitocin have all been described in these fishes (Caldwell and Young 2006).

Contrary to what happens in the AVT/AVP family of hormones, OT-like peptides show much higher variation, especially in cartilaginous fishes, which reflects a relaxed evolutionary pressure. It has been proposed that the diversity in chondrichthyan OT-like peptides is due to the high importance of urea and methylamines in maintaining osmotic balance in these animals, which lessens their reliance on ionic osmoregulation and relieves the pressure on the neurohormones that govern it (Wright and Wood 2015). While OT-like peptides have roles in osmoregulation and sexual behavior across vertebrates, OT plays a central role in placental mammals, where it is required for processes such as lactation and uterine contractions. The higher importance of OT for the mammalian life history helps explain the relative invariability of OT-like peptides in this class (Acher 1996).

Finally, homologs of vertebrate neurohypophysial hormones, associated with neurosecretory organs, have been described in several invertebrate phyla, namely annelids (annetocin in earthworms), mollusks (cephalotocin and octopressin in octopuses and conopressin in cones), nematodes (nematocin), and arthropods (inotocin in several groups of insects and the grasshopper diuretic hormone), with a further OT-like tetradecapeptide described in the tunicate *Styela plicata*. In all these cases, only one member of the OT/AVP superfamily is present in each taxon (Scharrer and Scharrer 1937; Caldwell and Young 2006; Donaldson and Young

2008; Ukena et al. 2008). It has been proposed that the vertebrate magnocellular neuroendocrine system may share a common origin with the insect *pars intercerebralis* as well as neurosecretory brain centers in annelids, dating back to the ancestor of bilaterians and with vasotocinergic neurons as the ancestral cell type (Tessmar-Raible et al. 2007).

4.4 Evolution of the Adenohypophysis Structure

The anterior pituitary gland, or adenohypophysis, includes the *pars intermedia* and *pars distalis* with the secretory cell types represented in Fig. 4.1. As in the case of the neurohypophysis, these hormone-producing cells are regulated by the activity of neurons from different nuclei of the hypothalamus, which release hormones to the pituitary gland, forming a neuroendocrine complex known as the hypothalamus–pituitary system (Fig. 4.3). In the following sections, we describe the evolution of the communication in the HP system across vertebrates and compare the dynamics of the neuroendocrine axes that regulate the three main components of organismal fitness: growth, metabolism, and reproduction.

The evolution of the adenohypophysis is the diversification of communication between the hypothalamus and the anterior pituitary gland. In contrast to the unchanging arrangement of the vertebrate neurohypophysis described above, where projections from hypothalamic neurons terminate directly in the neurohypophysis, with local control over hormone storage and release, the communication between the adenohypophysis and the hypothalamus has changed drastically in vertebrate evolution.

At the level of the pituitary gland, one of the main evolutionary trends is the diversification of cell types. In hagfish, for example, three cell types have been identified which produce three different hormones: corticotropes, somatotropes, and proto-glycotropes, the latter with the potential to regulate both reproduction and metabolism through a single hormone. In lampreys, two more hormones—MSH, produced by melanotropes, and TSN, produced by proto-glycotropes—are present (Nozaki 2013). In other fish species, as well as in all other vertebrates, the six cell types described in mammals have been identified and the hormones LH, FSH, and TSH appear to have replaced the function of GpH and TSN, although at the molecular level, their corresponding genes are highly similar and are the result of several gene duplication events followed by gene neofunctionalization (Uchida et al. 2010). Furthermore, by analyzing the distribution of pituitary secretory cell types across vertebrates, we find that all hormone-producing cell types in fishes are compartmentalized in specific regions of the gland, with almost no contact between cell clusters. Such compartmentalization is conserved for some cell types in lungfish and amphibians, while others are found distributed throughout the adenohypophysis. Meanwhile, the distribution of cell types in the anterior pituitary gland of reptiles and mammals is highly heterogeneous, but presents some density tendencies in the ventro-dorsal distribution of cells. The only exception to this trend are melanotropes, which conserve a limited distribution in the *pars intermedia* of all vertebrates.

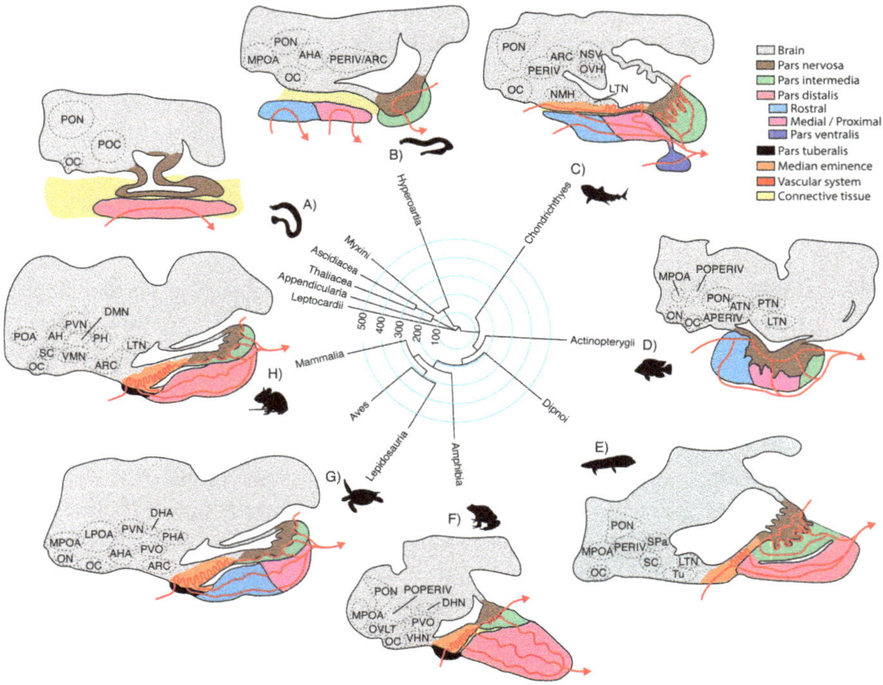

Fig. 4.3 Anatomical organization of the pituitary gland in vertebrates. The phylogenetic tree shows the evolutionary relationships and divergence times in millions of years between the major groups of vertebrates, extracted from the TimeTree database (Kumar et al. 2022). A sagittal schematic of the anatomical boundaries of the pituitary in representative living species is shown in different colors. The vascular system is depicted in red, showing a simplification of the hypothalamus–pituitary communication and the formation of the hypophyseal portal system in tetrapods. (**a**) Hagfish (Myxini); (**b**) lamprey (Hyperoartia); (**c**) shark (Chondrichthyes); (**d**) tilapia (Actinopterygii); (**e**) lungfish (Dipnoi); (**f**) bullfrog (Amphibia); (**g**) turtle (Reptilia), and (**h**) mouse (Mammalia). The phylum Chordata includes other lineages that are closely related to vertebrates (see main text): Leptocardii (lancelets), Appendicularia (larvaceans), Thaliacea (salps, doliolids, and pyrosomes), and Ascidiacea (tunicates). The regionalization of the hypothalamus in most of the species is according to Norris and Carr (2020): APERIV, anterior periventricular nucleus; AHA, anterior hypothalamic area; AHN, anterior hypothalamic nucleus; ARC, arcuate nucleus; ATN, anterior tuberal nucleus; DHA, dorsal hypothalamic area; DHN, dorsal hypothalamic nucleus; DMN, dorsomedial nucleus; LTN, nucleus lateralis tuberis; LPOA, lateral preoptic area; MHN, medial hypothalamic nucleus; MPOA, medial preoptic area; NSV, nucleus sacci vasculosi; ON, optic nerve; OC, optic chiasm; OVH, organum vasculosum hypothalami; OVLT, organum vasculosum lateralis terminae; PERIV, periventricular nucleus; PHA, posterior hypothalamic area; PHN, posterior hypothalamic nucleus; POC, postoptic commissure; PON, preoptic nucleus; POPERIV, periventricular preoptic nucleus; PTN, posterior tuberal nucleus; PVN, paraventricular nucleus; PVO, paraventricular organ; SC, suprachiasmatic nucleus; SON, supraoptic nucleus; SPa, subparaventricular hypothalamic area; Tu, tuberal hypothalamic region; VHN, ventral hypothalamic nucleus; VMN, ventromedial nucleus. Modified from Santiago-Andres et al. (2020) with permission

One of the main aspects of the hypothalamus–pituitary communication that has been subject to changes in vertebrate evolution is the site at which the axonal terminals of neurons release and deliver neurohormones to the pituitary secretory cells. In cyclostomes, the *pars distalis* and the hypothalamus are separated by connective tissue, and neurons terminate in the *pars nervosa*, where neurosecretion happens (Fig. 4.3). Because of this, it has been suggested that the secreted hypothalamic factors of cyclostomes reach their target cell by means of passive diffusion. The same hypothalamus–pituitary interface has been described in chondrichthyans, sturgeons, and gars. Other vertebrates use two ways of non-passive neurohormone delivery to the pituitary gland: a neuroglandular and a neurovascular mode of communication. In neuroglandular communication, the axonal terminals of hypothalamic neurons are present within the pituitary and make direct contact with their target cells in a "synaptic-like" configuration. This is the main mode of communication of the HP system in teleost fishes (Fig. 4.4). In the neurovascular mode of communication, there is no direct contact between the hypothalamus and the anterior pituitary, but axons from hypothalamic neurons terminate in a novel structure called the median eminence (Fig. 4.3), where the bloodstream forms a complex arrangement of capillaries known as the hypophyseal portal system (see Box 4.2 for a detailed explanation). Once released into the capillaries, hypothalamic hormones travel to their targets via the hypophyseal portal system. This type of interaction governs the communication of the mammalian HP system.

Limited anatomical observations showing the hypothalamic regulation of the pituitary gland in most teleost and mammal species to be exclusively neuroglandular or neurovascular, respectively, led to early suggestions of a trend in the evolution of these communication systems. This was refuted, however, by evidence showing the coexistence of both systems to varying degrees in other species of vertebrates (Trudeau and Somoza 2020). For example, the chondrichthyan lineage, which originated before teleosts in vertebrate evolution (see the phylogeny in Fig. 4.3), presents a complex median eminence with two distinct neurohemal regions that preferentially irrigate the rostral or posterior portion of the pituitary gland, limiting the delivery of some hypothalamic hormones to specific cell populations. However, the intermediate lobe of chondrichthyans is not completely separate from the neurohypophysis and has been termed the neurointermediate lobe, where direct regulation of melanotropes by the brain has been reported. Furthermore, members of class Actinopterygii, a large group of fishes including teleosts, show great diversity in the neuroglandular communication. A median eminence with nerve fibers in close association to the vascular system, as well as axons permeating the pituitary and making direct contact with the endocrine cells, has been reported in non-teleostean actinopterygians. More recently, it has been reported that, even in teleosts, not all nerve fibers reach pituitary cells directly, but some hypothalamic axons terminate at the basal membrane separating the neurohypophysis and the adenohypophysis, where they release their contents, which then diffuse into the nearby capillaries of the adenohypophysis (Golan et al. 2015). Finally, tetrapod and lungfish species exhibit a well-developed median eminence, with no apparent direct innervation from the hypothalamus to the anterior pituitary gland, including the intermediate lobe, which

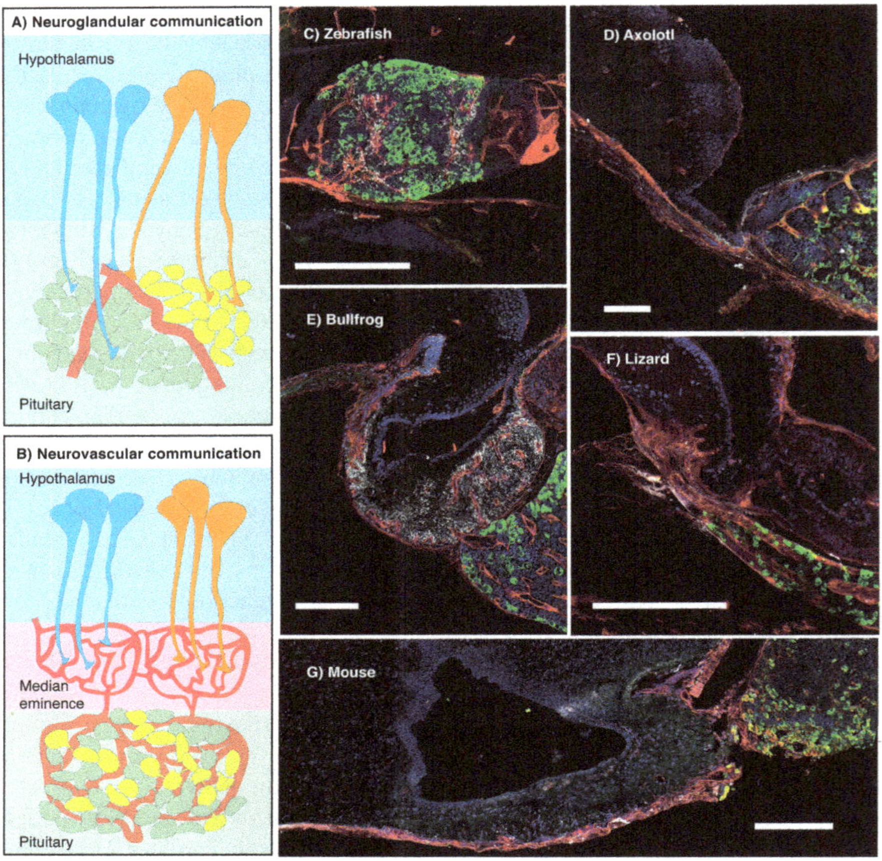

Fig. 4.4 Types of communication in the hypothalamus–pituitary system. (**a**, **b**) Schematic representation of neuroglandular and neurovascular communication. (**c**–**g**) Confocal images of immunohistochemistry preparations showing gonadotropes, GnRH neurons, and the vasculature in green, white, and red color, respectively. The anterior region of the pituitary gland is at the left in each picture, with the dorsal region at the top. (**c**) The neuroglandular communication is shown in immunolabeled samples of zebrafish pituitary gland (teleost, *Danio rerio*), the sagittal section shows the β subunit of LH-containing gonadotropes and the terminal nerves of GnRH neurons unfolding throughout the pituitary with some capillaries crossing the pituitary but not the hypothalamus. In contrast, neuroglandular communication is the most common pattern in tetrapods, shown here in representative species of the major groups within the superclass (**d**, axolotl, *Ambystoma mexicanum*; **e**, bullfrog, *Lithobates catesbeianus*; **f**, lizard, *Sceloporus aeneus*; and **g**, mouse, *Mus musculus*). In this case, no clear neuron invasion of the pituitary by nerve fibers is observed, with the nerves ending at a complex bed of capillaries in the median eminence. However, GnRH neuronal terminals maintain spatial proximity with gonadotropes in the bullfrog and gonadotropes of the *pars tuberalis* maintain a neuroglandular-like communication in the mouse. In addition, gonadotrope distribution in zebrafish is restricted to the rostral *pars distalis* but it occupies the entire *pars distalis* in the other species with continuous cell–cell contact across the gland. Scale bars represent 200 μm. DAPI, a specific nucleus marker, is shown in blue as a counterstain. Modified from Santiago-Andres et al. (2020) with permission

is clearly separated from the neurohypophysis in these groups. Hypothalamic neurons are, however, found in close anatomical proximity to pituitary cells of the *pars tuberalis* and rostral adenohypophysis, as seen in Fig. 4.4. This raises the question as to whether mammalian neurohormones could be transmitted to endocrine cells of these regions by passive diffusion or in a neuroglandular manner, as happens in other vertebrates.

The emergence of the median eminence and the hypophyseal portal system is strongly associated with the distribution of the secretory cell types in the anterior pituitary gland, as well as with the predominance of the neuroglandular regulation of its activity. Where pituitary cells are restricted to specific areas of the gland, we find a direct innervation of the pituitary endocrine cells with no median eminence. On the contrary, a scattered distribution of pituitary cell types is accompanied by a highly developed median eminence without neuroglandular regulation. An exception to this are the cells of the *pars tuberalis* of tetrapods, a structure that envelops the hypophyseal portal system of the infundibulum and is populated by thyrotropes and gonadotropes. These endocrine cells influence secretion from the *pars distalis* in relation to seasonal and photoperiod changes (Wittkowski et al. 1999; Fekete 2022).

Box 4.2: Hypophyseal Portal System

The portal circulation is a particular type of circulatory system in which blood is collected in a plexus, a network of capillary blood vessels, before passing through one or more long vessels onto another capillary plexus and then rejoining the systemic circulation. The fundamental structural elements that make up a portal system are: a feeding artery, primary capillary plexus, portal vessels, secondary capillary plexus, and draining vein. In addition to particular portal systems, such as the placental portal system in placental mammals, there are three known portal systems in vertebrates: the hepatic, renal, and hypophyseal portal systems. The hypophyseal portal system, also known as the pituitary or hypothalamic–hypophyseal portal system, enables the efficient and direct delivery of concentrated neurochemical signals, such as hypothalamic releasing and inhibitory hormones, from the median eminence and pituitary stalk to the endocrine cell types of the anterior pituitary.

This system originates from a pair of branches of the internal carotid artery—the superior and inferior hypophyseal arteries—which form two primary capillary plexuses connected by portal vessels to the same secondary capillary plexus in the anterior pituitary. The superior hypophyseal artery divides to form a primary capillary plexus that penetrates and surrounds the internal and external median eminence zones, and then converges into long portal vessels that carry blood down the pituitary stalk to the anterior pituitary capillary bed. On the other hand, the inferior hypophyseal artery forms another primary plexus located far below the pituitary stalk, in the posterior pituitary gland, where other neurons secrete their factors directly and are

(continued)

carried by short portal vessels to the secondary plexus. The secondary capillary bed thus coalesces into the lateral and inferior hypophyseal veins, through which blood containing pituitary hormones drains into the cavernous and posterior intravenous sinus, thus accessing the systemic circulation.

One of the fundamental elements that make the hypophyseal portal system functional are fenestrated capillaries, which facilitate diffusion into and out of the portal circulation, and the presence of pericytes as the perivascular cells of the extensive fenestrated capillary network. The pituitary vasculature is highly dynamic and it is thought that the endothelium-pericyte-endocrine cellular unit may act in a coordinated manner to generate hormonal pulses.

4.5 Evolution of the Adenohypophysis Function

Each pituitary hormone-producing cell type is primarily regulated by one hypothalamic factor. As an intermediate regulator, the pituitary gland induces the activity of target tissues or peripheral endocrine glands; the peripheral targets then generate a feedback loop that regulates, usually in a negative fashion, the future release of hormones at the hypothalamic and pituitary levels, creating a dynamic balance of hormone concentrations in the body. In mammals, the regulation of physiological processes in different organs by the HP unit depends on specific ligands that form axes of communication: the growth hormone-releasing hormone (GHRH)-GH-liver, GnRH-LH/FSH-gonads, thyrotropin-releasing hormone (TRH)-TSH-thyroid, dopamine (DA)-PRL-mammary glands, and corticotropin-releasing hormone (CRH)-ACTH-adrenal axes. Moreover, in mammals, each axis operates quasi-independently of the others, giving the HP system a high level of efficiency and responsiveness to physiological perturbations without compromising the overall function of the organism. This well-delimited function of the HP system is not necessarily observed in all vertebrate species. In many organisms, the functions of the individual axes can be intertwined; in other cases, a single factor can induce the activity of more than one axis, or axes can be absent. The most extreme example of this variation in function is the PRL axis, which has the capacity to regulate metabolism, growth, development, reproduction, behavior (from migration to parental care), lactation, immunity, and osmoregulation. As detailed below, three of the most important elements determining organismal fitness are growth, metabolism, and reproduction, all of which are under neuroendocrine control and show significant adaptations in vertebrate evolution.

4.5.1 The Hypothalamus–Pituitary System in Growth

In mammals, GH secretion by somatotropes is triggered by the pulsatile release of GHRH from neurons of the hypothalamic arcuate nucleus. In addition to stimulating GH secretion, GHRH is also important for GH synthesis and somatotrope proliferation through the activity of Pit-1. About half of the anterior pituitary mass is composed of somatotropes. It is also known that, at the hypothalamic level, somatostatin is the main inhibitory factor affecting GH synthesis and secretion. However, the complexity of the somatostatin-GHRH relationship in somatotrope activity goes further due to a dual effect of somatostatin, where on the one hand it inhibits GH secretion, while on the other it potentiates the next episode of GHRH action. Meanwhile, the growth-promoting effect of GH in the liver and other target tissues is mediated by the induction of gene expression including hepatic insulin-like growth factor 1 (IGF-1), which in turn stimulates bone growth and induces cell growth, differentiation, and metabolism.

The GH effect on its target organ is sex-specific and results in sexual dimorphism, a major feature of the ecological and behavioral differences between males and females in many vertebrate species. Studies in murine models have shown that somatotrope differentiation is not GHRH-dependent in the early stages of development, but the consequences of GHRH deficiency are observed later in postnatal life. In addition, the differential patterns of gene expression, hormone secretion, sexual maturation, and sexual dimorphism traits may be the result of previous induction or imprinting of the GH axis during development. This is the case for effects of ontogenetic gonadal steroid exposure on the hypothalamic and pituitary phenotype in rats (Chowen et al. 2004): under normal conditions, females secrete GH in small and continuous pulses that maintain a stable concentration of GH in the circulating blood over time (30–60 ng/mL). In contrast, males secrete GH in bigger pulses (200 ng/mL) every 3.5 h with intervals of relatively null blood GH concentrations (Waxman and O'Connor 2006). These patterns of GH secretion have been experimentally manipulated in young and adult male mice using androgens and estrogens, and the results show that there is a switch in gene expression and function at each level of the axis that produces a female-like phenotype, revealing exquisite regulation across the GH axis and linkage between growth and reproduction.

Furthermore, changes in hepatic or hypothalamic function have an impact on pituitary GH secretion and the organism. For example, an excess of GHRH will result in an increase in somatotropic secretion as well as cell number (hyperplasia) and size (hypertrophy), in some cases causing a condition known as acromegaly in humans. Conversely, GHRH deficiency results in a decrease in the number of somatotropes (hypoplasia), loss of GH secretion, delayed puberty, and dwarfism. However, these mechanisms of GH regulation in mammals demand further investigation in other vertebrates to elucidate some of the variation of life histories that may have the same regulatory mechanisms as mammals. Internal biological constraints affect growth in many vertebrates. For example, in fishes, unlike in mammals, the growth process continues throughout adult life. In other species, sexual maturation is only possible after an optimal body size has been reached; in vertebrates with female mate choice, males compete with other males only after achieving a specific umbral of body size or secondary sexual features. Other factors affecting the rate of growth in species are external, as many species are dependent on temperature, photoperiod, and resource availability, establishing a trade-off between growth and reproduction. Both the pulsatile secretion of GH and its action on the hepatic tissue have been observed in other vertebrates after being initially described in mammals, and IGF-1 inhibits the activity of somatotropes and GHRH neurons in fish, amphibians, and birds. In a wide variety of teleosts, a circadian variation of GH has been characterized, with hormonal peaks during the day that increase in amplitude during the period of darkness.

As pointed out previously, GH, PRL, and somatolactin have very similar protein structures, with the potential to activate the same receptors and activity of other axes in some vertebrates. Analysis of vertebrate phylogeny and molecular variations of these three pituitary hormones has suggested that GH was the first hormone to appear in vertebrate evolution, as GH is present in hagfish and lampreys, while PRL and somatolactin are absent. Then, after genomic duplication, PRL appears in the evolution of the pituitary gland, as evidenced by its shared presence among jawed vertebrates and the structural and functional similarity between GH and PRL. The hypothesis that GH appeared before PRL and somatolactin in evolutionary history is in line with the so-called one-to-four rule of vertebrate gene evolution, where a first round of genome duplication occurred in the common ancestor of all deuterostomes, with a second duplication event in the ancestor of all vertebrates. A third genome duplication event affected exclusively teleost fishes, which may have given rise to the somatolactin hormone in this group of vertebrates. The discovery of a GH-like hormone in lancelets with osmoregulatory and growth function, together with the osmoregulatory role of GH reported in teleosts, supports the notion that the original function of PRL was in osmoregulation, and that GH played more distinct regulatory roles earlier in evolution (Lee et al. 2007; Li et al. 2017).

The emergence of PRL in vertebrate evolution is associated with a specialization of the GH axis, with many of the ancestral roles of GH in metabolic regulation being filled by PRL in tetrapods and, most markedly, in mammals. Consistent with this, PRL and thyroid hormones stimulate and accelerate growth in non-mammalian vertebrates, but thyroid hormones also have an inhibitory effect in amphibians and the ammocete (the larval stage of lampreys), interrupting metamorphosis. In vivo and in vitro experiments using the European eel have demonstrated that GH may stimulate the production of thyroid hormones and that these hormones are able to inhibit the activity of somatotropes. However, there is evidence that this feedback between the axes is not a regular pattern among vertebrates, as it exhibits species-related variation in teleosts and mammals. Finally, an increase in somatotrope activity has been observed to correlate with sexual maturity and gonadal function in other vertebrates in a sex-specific manner, as is the case in mammals.

The presence of GH early in vertebrate evolution (Fig. 4.1) indicates the importance of growth regulation in the organism. The conserved function of GH was tested experimentally in the 1960s using radioimmunoassays and other chemical assays to compare the capacity of mammalian antibodies to cross-react with purified GH from fish, amphibians, reptiles, birds, and different mammals, and by testing GH action with direct injections into several vertebrate species. More recently, studies of the molecular identity of GH have confirmed the very limited variation that this hormone has undergone during evolution due to its specific and conserved function in promoting growth.

The correlation between food intake control and GH has been observed in teleosts, lizards, turtles, chickens, sheep, dogs, and humans: GH increases appetite, and food deprivation increases GH concentrations. An example of the ecological relevance of this association is the cost–benefit relation of many animal behaviors (Jönsson and Björnsson 2002; Weidner et al. 2020): foraging activity in the rainbow trout is a plastic strategy that depends on risk of predation; in the absence of predators, trout do not exhibit anti-predator behaviors such as hiding and can feed freely. However, in the presence of predators, this fish can opt to hide rather than feed. Nevertheless, with an experimentally-increased GH concentration, foraging behavior increases even in the presence of predators.

Although early experiments doubted the effect of GHRH on GH expression and secretion in non-mammalian vertebrates, it was soon discovered that GHRH, the GHRH receptor, and the GH receptor are all expressed in fishes including lampreys, and that the GHRH stimulatory effect on somatotropes is a general mechanism in vertebrates (Lee et al. 2007). Other hypothalamic regulators of GH secretion have been identified in vertebrates, although their importance has varied throughout evolution. Pituitary adenylate cyclase-activating peptide (PACAP), like GHRH, is able to induce GH release in several species of fishes, amphibians, and birds, but its role in mammals is controversial because some unvalidated reports suggest a stimulatory activity. TRH is also capable of stimulating secretion by somatotropes in vertebrates, where it shows considerable variation: a strong stimulatory effect of both TRH and GnRH has been observed in some teleost species (tilapia, goldfish, and the common carp), with no apparent effect in others (catfish, eel, and turbot). The GHRH-like effect of TRH on somatotropes has also been reported in amphibians, reptiles, birds, and mammals. The stimulatory effects of CRH and neuropeptide Y have been demonstrated in teleost species, but their effects in other vertebrates remain to be investigated.

In contrast to the conserved function of GHRH in vertebrates, somatostatin appears to have a more flexible and context-dependent effect on GH secretion (Gahete et al. 2009). For instance, exogenous administration of somatostatin inhibits GH secretion in turbot, rainbow trout, tilapia, goldfish, and salmon. However, somatostatin has no direct effect on GH secretion in amphibians (grass frog, bullfrog, and clawed toad) and reptiles (terrapin turtles), although it has an apparent inhibitory effect on TRH release and, indirectly, on GH secretion through TRH action. Studies in birds and mammals have demonstrated an inhibitory effect of somatostatin on GH secretion.

Serotonin is yet another factor with opposing effects in the species studied. In goldfish, serotonin inhibits somatotrope activity, whereas it lacks a clear effect on GH secretion in chicken and it has a stimulatory role in mammals. GH cells in goldfish are directly innervated by norepinephrine-releasing adrenergic neurons, which act directly and negatively on GH secretion, but norepinephrine does not exhibit such a strong effect in mammals and birds. Finally, ghrelin has been shown to stimulate the release of GH, and IGF-1 has an inhibitory role in all species studied so far, ranging from teleosts to mammals.

In summary, there is an apparent pattern in the reduction of factors regulating GH activity and growth along vertebrate evolution. In fish, there are more factors involved in somatrope activity, including hypothalamic factors besides GHRH, and feedback loops with target tissues, and environmental factors. In amphibians and reptiles, the main hypothalamic regulators of GH secretion are somatostatin, GHRH, and PACAP, but in mammals PACAP has no apparent role, restricting the activity of the GH axis to the somatostatin-GHRH-IGF-1 interplay. This pattern reflects the trend for increased internalization of biological processes in vertebrate evolution, where said processes tend to become more determined by internal mechanisms and less so by direct environmental factors. In this case, an integrated HP system receives signals from the internal and external environments and coordinates multiple regulatory axes that function quasi-independently from each other, reducing the number of factors directly affecting their activity in the process (Gahete et al. 2009).

4.5.2 The Hypothalamus–Pituitary System in Metabolism

The hypothalamic–pituitary–thyroid (HPT) axis is an evolutionarily conserved neuroendocrine system present in all vertebrates, where it is most often composed of TRH-releasing parvocellular neurons in the paraventricular nucleus of the hypothalamus, TSH-secreting thyrotropes in the anterior pituitary, and the thyroid gland. The HPT axis is involved in multiple biological processes across the taxa, including regulation of metabolism, thermogenesis, development, growth, reproduction, and molting.

The thyroid gland of all vertebrates is composed of functional units called thyroid follicles, which can concentrate iodine from the plasma and synthesize thyroid hormones. Thyroid follicles comprise a spherically arranged layer of follicular cells surrounding an acellular, protein-rich fluid compartment termed the thyroid colloid. The follicular cells are able to import iodide (I^-) from the bloodstream and export it to the colloid-filled lumen of the follicle, where it becomes oxidized by the enzyme thyroperoxidase (TPO). The follicular cells also synthesize a tyrosine-rich glycoprotein named thyroglobulin, which they secrete to the colloid. An enzymatic reaction catalyzed by TPO iodinates the tyrosyl residues in the thyroglobulin protein, forming two thyroid hormone precursors: monoiodotyrosine (MIT) and diiodotyrosine (DIT). Iodinated forms of thyroglobulin can accumulate in the follicular lumen, serving as a reservoir of iodine for future use. Neighboring MIT and DIT molecules attached to thyroglobulin can then be coupled together by TPO-linkage of one MIT and one DIT gives rise to triiodothyronine (T_3), while two linked DITs form thyroxine (T_4). Stimulation with TSH triggers endocytosis of the colloid by the follicular cells, with the resulting vesicles fusing with lysosomes inside the cell. Proteolytic cleavage separates T_3 and T_4 from the thyroglobulin molecule, after which they are released to the bloodstream as active thyroid hormones. Across all living vertebrates, T_3 appears to be more biologically active than T_4, and can also be generated through deiodination of T_4, both in the thyroid and in peripheral tissues (Rousset et al. 2015).

The thyroid follicles can be located in one of two general arrangements in vertebrates: scattered follicles in the pharyngeal region, with occasional presence of extrapharyngeal thyroid follicles, and organized follicles forming one or two discrete glands encased in connective tissue, associated with rich vascularization by blood and lymphatic vessels and innervation. A diffuse arrangement is considered to be the ancestral organization of the thyroid and is present in cyclostomes as well as in most teleosts. Thyroid follicles arranged in lobes are present in chondrichthyans, sarcopterygians—including tetrapods—and some teleostean groups (e.g., tunas, anglerfish, and pufferfishes) (Chanet and Meunier 2014).

The function of the HPT axis in mammals is tightly interlinked with other neuroendocrine axes, with permissive or cooperative roles in growth, sexual maturation, and gonadal development and function, among others, and the axis is a key controller of metabolism and thermogenesis. Thyroid hormones are also necessary for hair replacement and molting in mammals and stimulate cell proliferation and maturation in most tissues. Insufficient thyroid hormone production during mammalian development has severe consequences for the structure and function of the central nervous system (Pascual and Aranda 2013).

In birds, the other extant clade of homeothermic animals besides mammals, the HPT axis is also necessary for thermogenesis. It has been suggested that this function of the axis first evolved in the ancestors of amniotes, from which both birds and mammals are descendants. As is the case with mammals, the avian HPT axis is involved in metabolism, growth, reproduction, and molting, with potential roles in bird migration.

Studies addressing the HPT axis' effect on metabolism in non-avian reptiles have shown that the oxygen consumption of reptile tissues increases in response to thyroid hormones at higher temperatures. Similarly to mammals, this axis appears to have effects on reptile growth and reproduction as seen in snakes, lizards, and turtles. The activity of the HPT axis also correlates with skin shedding, although it has been shown to have both stimulating and inhibitory effects on this process depending on the group (lizards and snakes, respectively) (Norris and Carr 2020).

The amphibian HPT axis is involved in reproduction, growth, and molting, but it is of special relevance in metamorphosis and the transition between the aquatic and terrestrial lifestyles. Thyroid hormones induce aquatic amphibian larvae to undergo metamorphosis into terrestrial adults. Unlike in amniotes, however, the hypothalamic stimulating factor of the HPT axis of amphibians is CRH rather than TRH, with important interlinks between the HPT and the hypothalamic–pituitary–adrenal (HPA) axes in relation to metamorphosis. Increased thyroid activity in larval amphibians arrests growth rather than stimulate it as is the case in amniotes, and this axis does not appear to play a role in growth in adult amphibians. Some amphibian skin adaptations to life in terrestrial environments are under direct control by thyroid hormones: increased thyroid activity promotes thickening and keratinization of the skin of anurans and salamanders, and molting during metamorphosis is also regulated by the HPT axis (Paul et al. 2022).

The teleostean HPT axis differs from that of other vertebrates in that it comes under inhibitory, rather than stimulatory control by hypothalamic factors: unlike in other groups, TRH, DA, and somatostatin prevent TSH release by the teleost anterior pituitary. Thyroid hormones have a greater degree of involvement in teleost physiology which, along with the prevalence of a diffuse thyroid gland, makes it difficult to establish direct and indirect effects of this axis. Similarly to birds, there is a positive correlation between thyroid activity and teleostean spawning and migratory behaviors (Campinho 2019).

The HPT axis of chondrichthyans has not been well studied, but evidence from sharks supports a role in reproductive cycles and nervous system development.

While the general function and organization of the thyroid follicles are conserved in all jawed vertebrates, the follicles of cyclostomes are devoid of colloid, and thus the thyroid hormone synthesis and storage take place intracellularly. Cyclostomes also lack thyrotropes and do not express TSH, and the function of the thyroid appears to be limited to reproduction (Hulbert 2000). Gonadotropes show cytological similarities to thyrotropes, and there is significant sequence conservation between the subunit genes of gonadotropins and TSH, which has led to the hypothesis that jawed vertebrate thyrotropes evolved from gonadotropes.

A primitive HPT axis is present in tunicates and lancelets, where a mucus-secreting pharyngeal gland called the endostyle has iodine-concentrating and thyroid hormone-producing capabilities. The endostyle is also present in the ammocete larvae of lampreys, where it reorganizes into the thyroid gland during metamorphosis. Multiple lines of evidence have identified the endostyle as being homologous to the vertebrate thyroid gland. The role of the endostyle and the HPT axis in protochordates appears to be related to development and metamorphosis (Heyland and Moroz 2005).

While all three main components of this axis (the hypothalamus, the pituitary gland, and the thyroid) first appeared during chordate evolution, as did endogenous synthesis of thyroid hormones, the presence of iodine-binding proteins dates back to the last common ancestor of all extant animals, with MIT and DIT being present in sponges and cnidarians, while thyroid hormones (T_3 and T_4) first evolved before the divergence of protostomes and deuterostomes roughly 550 million years ago. Thyroid hormones in protostomes are involved in embryogenesis, development, and metamorphosis (Morthorst et al. 2023).

4.5.3 The Hypothalamus–Pituitary System in Reproduction

The strength of the force of natural selection over reproduction is evidenced by the diverse and, in some cases, extreme strategies of reproduction seen in vertebrates. Some species, such as the house mouse, reach reproductive age 2 months after birth while others, like the Atlantic halibut, are only able to reproduce after 10 years of life. An axolotl can lay hundreds to thousands of eggs during spawning; others, such as the blue whale, give birth to a single calf as big as an elephant. A human is able to reproduce on more than one occasion, but salmon and the Australian so-called marsupial mice (*Antechinus* spp.) have a single event of reproduction and then die. Mating strategies can even include sequential hermaphroditism as seen in the blue-head wrasse, where a switch from female to male sex correlates with an increase in body size. However, the neuroendocrine elements of gonadal function and reproductive regulation are conserved across vertebrate species (Kanda 2019).

In tetrapods, gonadotropes are bi-hormonal pituitary cells that secrete LH and FSH, also called gonadotropins, in a pulsatile manner (Herbison 2016). FSH plays a crucial role during folliculogenesis in female individuals by acting upon the granulosa cells of the ovary. After follicle maturation, LH switches from a pulsatile secretion pattern to a hormone surge that promotes the release of steroid hormones (including estradiol and progesterone) and ovulation. In males, FSH and LH promote spermatogenesis, and LH also promotes the synthesis of testosterone in Leydig cells of the testis that acts as an inhibiting factor of further FSH and LH secretion. However, in teleost species, LH and FSH are produced in distinct subpopulations of gonadotropes that are in close proximity to each other, and LH and GnRH are important for ovulation but not folliculogenesis, as observed in knockout experiments of gene expression. In addition, the characterization of hagfish and lamprey gonadotropes revealed that, contrary to what happens in other vertebrates, there is a single gonadotropin that regulates reproduction (Fig. 4.1).

Independently of these differences, gonadotropes are regulated by a single molecule produced in the hypothalamus—GnRH. Mammalian GnRH was isolated and characterized in the early 1970s and was soon identified as a conserved decapeptide in many species of vertebrates, as well as in tunicates, where up to nine variants have been isolated. GnRH-like peptides with slightly longer amino acid sequences have been also identified in non-chordate invertebrates, termed corazonin (CRZ), adipokinetic hormone (AKH), and corazonin-related peptide (ACP). The variation of GnRH molecules across vertebrates due to gene duplication events has led researchers to distinguish three main groups by means of developmental origin of GnRH-expressing neurons, phylogenetic and functional analysis, or expression patterns in different areas of the brain: GnRH1, the first GnRH characterized in mammals, has been identified in all vertebrates except for cyclostomes and chondrichthyans; GnRH2 is widely expressed across vertebrate species; GnRH3 has an irregular pattern of expression in lampreys, the coelacanth, and some chondrichthyes and teleosts (Fig. 4.5). This pattern of GnRH expression has led to the idea that GnRH2 may be the ancestral form of GnRH, and that subsequent gene duplication events gave rise to the GnRH1 and GnRH3 peptides (Sower 2018).

The GnRH and CRZ receptors follow a similar pattern of diversification and expression to their modulator GnRH, with a remarkable reduction in GnRH forms that are coexpressed between tunicates and mammals. For example, the ancestor of mammals had two variants of this receptor but, in the majority of mammalian species, either one of the GnRH receptors is not functional as a consequence of a deletion in the amino acid sequence, or it has been lost altogether, indicating an increase in specialization of the hypothalamic–pituitary–gonadal (HPG) axis. The

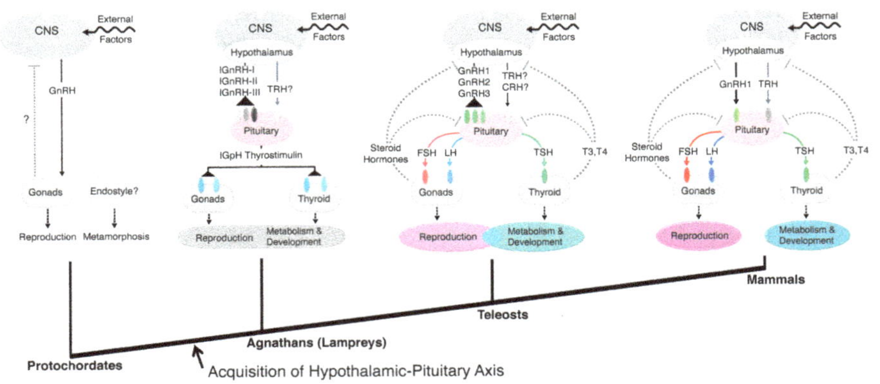

Fig. 4.5 Evolution of reproductive neuroendocrine regulation. The action of GnRH over reproduction has been reported in many lineages of animals. However, gonadotropin control of gonadal function is restricted to vertebrates where GnRH regulates the synthesis and secretion of pituitary hormones. GnRH also regulates the function of the thyroid gland in lower vertebrates, indicating its multifunctional role in neuroendocrine control of homeostasis. The evolution of the different axes (here represented by the gonadal and thyroid axes) has been the reduction of hypothalamic and pituitary regulators and specialization of hormones and receptors that regulate each axis, generating a high degree independence of biological processes. CRH, corticotropin-releasing hormone; CNS, central nervous system; FSH, follicle-stimulating hormone; LH, luteinizing hormone; GnRH, gonadotropin-releasing hormone; lGnRH, lamprey gonadotropin-releasing hormone; lGpH, lamprey glycoprotein hormone; lGpH-R, lamprey glycoprotein hormone receptor; T_3, triiodothyronine; T_4, thyroxine; TRH, thyrotropin-releasing hormone; TSH, thyroid-stimulating hormone; TSHR, thyroid-stimulating hormone receptor. Reprinted from Sower (2018), with permission from Elsevier

interaction of GnRH and its receptor in the membrane of vertebrate gonadotropes induces the mobilization of intracellular calcium that allows LH and FSH release, the expression of gonadotropin and GnRH receptor, and promotes cell-to-cell communication and vascular-gonadotrope contacts. In invertebrate species, which do not have a pituitary gland, the GnRH–GnRH receptor interaction also plays an important role in reproduction, development, and metamorphosis (Fig. 4.5). In tunicates, GnRH increases water flow through the oral and atrial siphon and stimulates the release of sperm and ova from the gonads to the water column where fertilization occurs. The GnRH-like peptides in mussels also directly activate gonadal organs, stimulate parapodium opening, inhibit feeding, and promote gripping of the substrate (Sakai et al. 2017). In addition, the mammalian GnRH receptor lacks the C-terminal tail and does not experience homologous desensitization where the receptor is unable to respond to GnRH and allows the cell machinery to recover from one event of gonadotropin secretion to another, but the biological significance of this property in reproduction (for example, enabling the LH surge necessary for ovulation) is controversial.

GnRH from the hypothalamus is released in a pulsatile manner rather than continuously, and is dependent on steroid regulation from the gonads, generating pulses of gonadotropin secretion. This pattern of hormone release was first described in the

1970s and 1980s in sheep, monkeys, and rats, with later studies confirming it as a general mode of secretion in the reproductive axis of vertebrates. Moreover, the loss of pulsatility, or rhythmicity, of GnRH and gonadotropins is strongly associated with infertility, hypogonadism, and polycystic ovary syndrome. These facts led researchers to explore and understand the mechanisms that generate the pulsatile secretion of gonadotropins. Firstly, advances in optogenetic methods have allowed the direct stimulation of GnRH neurons in mice with different frequencies, and 2 min of optogenetic stimulation at 10 Hz generated LH pulses as observed naturally, but lower frequencies or shorter windows of stimulation failed to stimulate LH secretion. It is noteworthy that the generation of LH pulses and surges in mice requires just a small fraction of the GnRH-producing neurons to be stimulated, 10% and 50%, respectively. This redundancy in the system may allow GnRH release to be maintained even after perturbations to the system (Herbison 2016).

Secondly, steroid hormones produced in the gonads are involved in the regulation of GnRH and gonadotropin release in a context-dependent manner. Estradiol, for example, regularly has a direct inhibitory role over gonadotrope and GnRH neuron activity in both males and females, but a switch to a stimulatory effect is observed during the LH surge in the estrous stage of the reproductive cycle of mice. Estrogens act in the pituitary through two receptor isoforms (ERα and ERβ). However, ERs are not expressed in GnRH neurons, indicating that GnRH release is indirectly regulated by a second interactor in the brain, which was later identified as a neuron population coexpressing Kiss, neurokinin B (NKB), and dynorphin (Dyn), because of which they are known as KNDy neurons (Herbison 2016; Han et al. 2018). KNDy neurons express estrogen receptors and are found in the periventricular and the arcuate nuclei of the hypothalamus in rodents, the preoptic area in primates, and the infundibular nucleus in humans (Sivalingam et al. 2022). KNDy neurons control GnRH release by the pulsatile secretion of Kiss, which acts throughout the Kiss receptor in GnRH neurons. Kiss of the arcuate nucleus is downregulated by high concentrations of estrogens, which has a consequent negative effect on GnRH release. On the other hand, Kiss from the paraventricular nucleus or the preoptic area is positively regulated by estrogens, indicating a more complex interaction in the hypothalamus of gonadotropin regulation that was previously thought.

In contrast to the detailed characterization of gonadotropin regulation in mammals, the pattern of secretion in other vertebrates needs further investigation, although it is reasonable to assume that the configuration of the gonadal axis may exist with some variations. The existence of Kiss and Kiss receptor orthologs in the genome of different groups of fishes has been identified: the Australian ghostshark, the spotted gar, the European eel, and the coelacanth. However, it appears that KNDy neurons are absent in non-mammalian vertebrates. In zebrafish and medaka, the expression of Kiss and NKB occurs in separate neurons, although GnRH neurons express the Kiss receptor, indicating a potential indirect regulation of GnRH release different from that observed in mammals. Birds exhibit expression of NKB and Dyn but not Kiss, and the negative feedback on GnRH release by estrogen appears to act in a different way (Sivalingam et al. 2022).

Other modulators of gonadotropin secretion have been reported in vertebrates, following a similar pattern of decrease in number between fishes and mammals (Kanda 2019). In teleosts, gonadotropes are directly innervated by neurons producing neurohormones other than GnRH, such as aminergic and GABAergic neurons that can inhibit or stimulate gonadotropin release, a type of regulation not seen in tetrapods. In particular, DA via its DA receptor has been identified as a potent inhibitory component of GnRH-responsiveness of the pituitary gland during the reproductive dormancy period in winter. Similarly, the pineal gland strongly influences puberty and spawning in fishes and amphibians through changes in melatonin concentration. Melatonin concentrations are correlated with the photoperiod, and removal of the pineal gland in frogs decreases melatonin concentrations and accelerates gonad development, but this factor has a decreased or null role in birds and mammals. It has also been reported that somatolactin stimulates steroidogenesis in salmon ovaries.

Finally, another aspect of the HPG axis evolution is the gradual specialization and separation from other neuroendocrine axes. As mentioned before, protochordates do not possess an HP system, but reproduction involves the action of GnRH. In lampreys, thyroid regulation is orchestrated by GnRH and the proto-glycotropes that release a gonadotropin-like hormone termed GpH that acts directly on thyroid gland cells expressing the gonadotropin receptor (Sower 2018). TSH and TRH are absent in cyclostomes and TSN does not have a stimulatory role over the thyroid gland, thus reproduction and metabolism are regulated through the same hypothalamic and pituitary factors, suggesting the HPG and HPT axes in vertebrates evolved from a single system. Moreover, TSN and GpH are both expressed in the same cells, reflecting the common origin of gonadotropes and thyrotropes in other vertebrates, as does the similarity in structure between GpHβ, lGpA2, and lGpB5 subunits in lampreys and LH, FSH, and TSH. Interestingly, the GnRH variants and their respective receptors are differentially regulated through the life cycle of lampreys, suggesting specific functions during development, metamorphosis, sexual maturation, and reproduction. The overlapping nature of the neuroendocrine axes in lampreys, with gradual specialization during vertebrate evolution, is consistent with the fact that hypothalamic and pituitary hormones from more basal vertebrates stimulate activity in tetrapods (Sower 2018). For example, LH from bullfrogs is able to stimulate the thyroid glands of reptiles and birds, while treatment with mammalian hormones tends to elicit no response from tissues in fish or amphibians.

In summary, the neuroendocrine control of vertebrate reproduction is dominated by GnRH at the hypothalamic level, gonadotropins from the pituitary gland, and steroid hormones from the gonads. In cyclostomes, there is a single gonadotropin hormone that controls the gonadal physiology and the thyroid gland. In the rest of vertebrates, there are two different gonadotropins—LH and FSH—that are specialized in different aspects of gonadal physiology. LH and FSH exist as separated subpopulations of gonadotropes in teleosts, but coexist in tetrapods. Gonadotropins of mammals are secreted in pulses and not continuously, and loss of pulsatility is correlated with reproductive dysfunction. However, the characterization of pulses in other vertebrates remains to be studied. In addition, the decrease in diversity of modulators of the gonadal axis in vertebrate evolution and its overlap with the thyroid axis in lampreys show that the evolutionary trend has been the restriction of modulation to fewer elements and the independence of this axis from other neuroendocrine processes.

4.6 Pulsatility and Plasticity of Cell–Cell Communication in Vertebrate Pituitary

Pituitary hormones are secreted in pulses, meaning that they follow a regular pattern of secretion with peaks of hormone concentrations in the blood (amplitude) occurring at intervals (frequency) of minutes, hours, days, and months. In mammals, the pulsatile pattern of pituitary secretion is a determinant of hormone release and has been described in every secretory cell type. For example, release of LH and FSH by gonadotropes in mice occurs around every 30 min and each hour, respectively. The loss of pulsatility and the continuous secretion, which maintains a constant hormone concentration in the body, results in target organ dysfunction and several pathologies (reviewed in Le Tissier et al. (2017)). Pulsatile secretion is the result of the interplay of cell–cell interaction and diverse elements of cell communication, which exhibit great plasticity.

4.6.1 Pulsatile Secretion

Pulsatile secretion has adaptive advantages over a continuous secretion, as it is more energy efficient and the information encoded in its amplitude and frequency generates a clear and precise code, similar to a Morse code, which when decoded by target organs elicits an appropriate response. Due to its quantity and its temporal delimitation, the signal is not lost as noise among other stimuli, and inter-peak periods allow the axis to recover from the stimulus (resensitization of the receptors, synthesis of new hormones, and resetting of the cell signaling machinery) and to be ready for a future regulatory pulse.

The existence of pulsatile pituitary secretion in vertebrates other than mammals has been poorly explored. This is mainly due to technical limitations when measuring hormones by radioimmunoassay or the enzyme-linked immunosorbent assay (ELISA), as these require large volumes and high frequency of blood sampling, both of which are not feasible in the majority of species. However, some evidence from in vitro and *ex vivo* hormone measurement experiments show a pulsatile pattern in species such as zebrafish (LH and FSH), goldfish, grass carp, and salmonids (GH), tilapia (ACTH and LH), frogs (LH), and chickens (GH, FSH, and LH). The development in recent years of an ultra-sensitive ELISA in mice, capable of measuring hormone concentrations in a few microliters of blood sample, makes it possible to obtain several samples from the same organism over a short period of time. The possibly universal nature of the pulsatile hormonal activity of the pituitary gland among vertebrates can thus be verified in the near future.

The pulsatile secretion of pituitary hormones has led to the search for the element or elements in the system that permit and control such a pattern of hormone release: the pulse generator. It was soon accepted that the hypothalamus must be the pulse generator because of the generalized control that the brain exerts on the body, as well as the pulsatile release of the hypothalamic–pituitary regulators themselves. The HPG axis was adopted as a model to study the nature of the pulse generator, where GnRH release generates a concordant secretion of LH pulses, and the modification of GnRH pulse frequency results in differential LH and FSH secretion, or loss of LH pulsatility. However, the model presents some complications and cannot be directly applied to other axes or species where pulse generation is composed of different elements, or directly associated to target organs (for an extended discussion, see Le Tissier et al. (2018)). The pituitary gland, on the other hand, possesses its own internal regulation of hormone secretion by means of cell network communication.

4.6.2 Cell–Cell Communication in the Pituitary Gland

Cell–cell interactions and communication are essential for pulsatile secretion and internal regulation, regardless of the zonal or scattered distribution of cell types in the vertebrate pituitary. The loss of the patchwork distribution of cells in the mammalian pituitary, together with histological studies in thin slices of tissue that lose the three-dimensional perspective of the gland, led to the idea that secretory cells are randomly distributed throughout the gland and that hormone secretion is the result of hypothalamic stimulation on isolated cells. However, the use of transgenic models and the advance of powerful techniques in microscopy allowed the study of the 3D structure and physiology of cells, demonstrating the network organization of every cell type in the pituitary from fish to mammals through distinct components of cell communication (Santiago-Andres et al. 2020) (Fig. 4.6).

Direct contact between cells via transmembrane channels or gap junctions is a powerful communication mechanism because the response of one cell can be transmitted to others using small molecules such as calcium, InsP3, and cAMP. The use of pharmacological blockers of gap junctions leads to the loss of cell population response synchronization and a decrease in secretion of LH and FSH in teleosts, and of almost all hormones in rodents. Interestingly, unlike in mammals, where LH and FSH are secreted by the same gonadotropes, teleostean FSH-cells do not express the GnRH receptor, but are able to respond to GnRH indirectly via LH-cells that are GnRH-activated and exhibit cell–cell coupling with FSH-cells. In addition, although the pituitary gland of teleosts shows neuroglandular regulation, not all the pituitary cells are in direct contact with neuronal terminals, but cell–cell coupling may propagate signals across the population, thus resulting in a synchronized response (Fig. 4.6).

Folliculostellate cells are a non-secretory population of the pituitary gland, arranged in a 3D network in all vertebrates. This population has the potential to regulate processes in the gland in several ways: it can respond to diverse stimuli

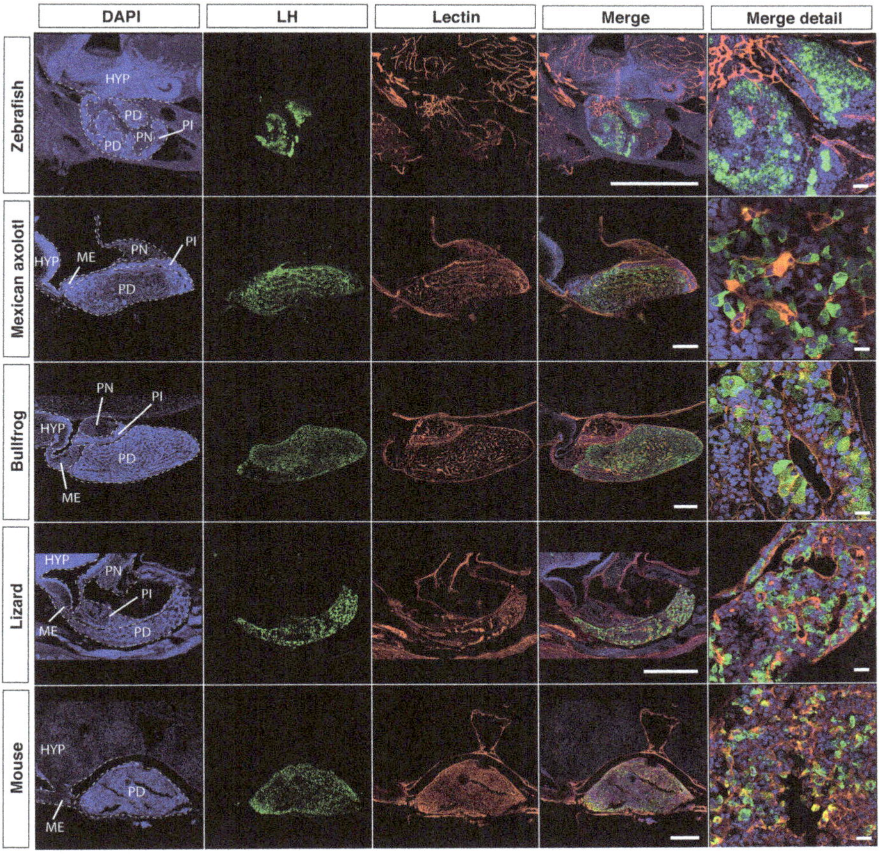

Fig. 4.6 Network organization of the gonadotrope cells in the pituitary of vertebrates. Confocal images of tissue slices showing sagittal views of the brain and pituitary where immunostaining was performed for LHβ (green), capillaries (red), and nucleus (blue) for the species described in Fig 4.4. First column shows the three main components of the pituitary gland, the hypothalamus and their connection through the median eminence in tetrapods. The second and third columns show the distribution of LH-expressing gonadotropes in the pituitary and the capillary distribution, respectively. The variation and complexity of communication modes in the hypothalamus–pituitary system and the network configuration of gonadotropes can be appreciated in the last columns. Note the cell–cell and cell–vasculature contacts in all species of vertebrates. HYP, hypothalamus; PD, pars distalis; PI, pars intermedia; PN, pars nervosa; ME, median eminence. Scale bars represent 400 μm in the first four columns and 20 μm for the last. Modified from Santiago-Andres et al. (2020) with permission

coming from the systemic circulation or from other pituitary cell types, and it secretes growth factors and cytokines that influence the activity of pituitary secretory cells. Additionally, it generates a modular communication circuit throughout the gland, enabling the connection and propagation of stimuli from distant cells, whether of the same type or of different types, via gap junctions. Furthermore, this population participates in the phagocytosis of secretory cells and contributes to the renewal of pituitary cells thanks to its multipotent nature. For these reasons, folliculostellate cells have been called the glial system of the anterior pituitary gland (Rizzoti et al. 2023). Furthermore, direct contact between GnRH terminals and folliculostellate cells has also been reported in the *pars tuberalis* of mammals. This provides an extra mode of communication between the hypothalamus and the pituitary gland, capable of propagating electrical signals across the gland quickly and efficiently.

Although the synchronous activity of cell types enables the correct functioning of the endocrine axes individually by cell–cell communication of the same cell type (homotypic networks), the whole system demands the coordination of endocrine function through communication between the different homotypic networks. The formation of heterotypic networks in the anterior pituitary gland is in part the result of paracrine communication, where the secretion of one cell type affects the activity of another. In mammals, hundreds of molecules have been identified as having a potential role in paracrine regulation, including the hormones secreted by the pituitary endocrine cells, and the receptors for these molecules have also been found in the gland.

Finally, the vascular system also plays an active role in cell signaling and the generation of pulses. It is not just important in delivery of regulatory factors from the hypothalamus across vertebrates, but many paracrine factors reach their target cells via the circulation, and the frequency and amplitude of hormone pulses are determined by the access of endocrine cells to the blood vessels (Fig. 4.6).

4.6.3 Plasticity and Communication

A major aspect of pituitary cell communication is its high plasticity, modifying the pituitary gland's response according to metabolic state, stress, reproductive cycles, growth, and aging. Examples of paracrine interaction between the cell types and plasticity in the pituitary can be found in Le Tissier et al. (2017) and Santiago-Andres et al. (2020). Even the classical plasticity in the number and size of anterior pituitary cells in response to physiological demand and system perturbations can be seen as a consequence of the plasticity of these modes of communication. For example, the increase in pituitary cell population after target organ removal is not only the consequence of cell proliferation, but also of transdifferentiation of other secretory cell types due to the spatial limitation of the pituitary gland size, the exceedance of which has been observed to have adverse effects, for instance, in cases of pituitary adenomas. In the same way, it has been demonstrated that the access of endocrine cells to blood vessels is subject to remodeling after changes in physiological demand.

The plasticity of the mechanisms of cell–cell communication in the pituitary gland has not been explored extensively in vertebrates other than mammals, but all the components of the interaction described are present in other groups. These components have a clear effect on pituitary homeostasis, but their contribution is expected to vary among species. For example, the diffusion of paracrine factors may be more important than vascular dynamics for communication in teleosts, and the transmission of signals for synchronized responses could depend more on gap junctions in homotypic networks than on the folliculostellate cell network. Another hypothesis to be tested is the number of elements involved in paracrine regulation in non-mammalian vertebrates, and whether the trend observed in the neuroendocrine axes to gain complexity and specificity of the elements within one axis, while gaining higher regulatory independence between them, applies to paracrine regulation of the gland as well.

4.7 Perspectives

The hypothalamus and pituitary gland of vertebrates form a complex system that regulates fundamental functions such as growth, survival, and reproduction, causing it to be strongly modulated by natural selection. The discovery of the neural regulation of the pituitary gland in the 1950s and 1960s led to the idea (almost a dogma) of a passive role for the pituitary gland in this system, which in turn biased future research in favor of a full regulatory role of the hypothalamus. Extensive research, however, especially that employing a comparative method while looking at the evolutionary history of organisms, has challenged this idea in the last few decades and re-shaped the paradigm on the role of the pituitary gland in vertebrate biology. There are reasons to believe that the pituitary gland is an evolutionary novelty acquired at, or shortly before, the origin of vertebrates, although some structures present in the closest relatives of this lineage have been proposed as homologous and potential precursors to the pituitary gland. Recent advances in molecular biology and an increase in the capacity to analyze whole genomes of many species have demonstrated that many genetic and molecular regulatory components of neuroendocrine regulation were present in the ancestors of vertebrates and protochordates, and even before the divergence of the chordates, as exemplified by the ability of GnRH to regulate reproduction in invertebrates, even in the absence of a pituitary gland. Genome duplication events, including at least two that are thought to have happened early in the vertebrate lineage history, are major occurrences in animal evolution, and are strongly associated with gene neofunctionalization and the emergence of new, previously inaccessible horizons in morphology, physiology, and ecological roles. In the next few years, the study of gene regulatory networks and evolutionary developmental biology will resolve the question of the origin of the pituitary gland using state-of-the-art research techniques that have thus far been restricted to few animal models.

The pituitary gland has enormous relevance in vertebrate evolution due to its function as an integrator of otherwise unconnected but fundamental processes in organisms. This can be observed in the gradual reduction of hypothalamic and peripheral factors that directly influence pituitary function and the formation of specific regulatory axes of biological processes in recent groups of vertebrates when compared to cyclostomes. The fine-tuning of pituitary function and the generation of hormone pulses also brought along a finer hypothalamic regulation that, over the course of vertebrate evolution, saw a change from a diffusion-based delivery of neurohormones to a neuroglandular and neurovascular communication in teleosts (the most diverse group of vertebrates) and tetrapods, respectively. However, the specific pressures that established the different types of communication in these two groups of vertebrates, and whether one is more efficient than the other in particular environments, are still unknown. The coexistence of the two modes of communication has been reported in some species and will be of help to answer this question. Additionally, there is a strong correlation between the origin of lungs and species that have a complex vascular system, and in which neurovascular communication predominates. Other changes in the pituitary configuration may also be related to main events in vertebrate evolution and diversification (Fig. 4.7).

In vertebrate evolutionary history, the rise in importance of pituitary regulation over multiple, distinct physiological processes happened in parallel with increasingly intricate mechanisms of cell–cell communication and plasticity that allowed it to generate hormone pulses in a cyclical manner (for example, the estrous cycle of reproduction) and sense the physiological state of different axes in order to maintain homeostasis, by means of resource management and compensation after system perturbations. These mechanisms include communication through gap junctions and formation of homotypic networks, interactions with folliculostellate cells, context-dependent reactions by the vasculature, and the formation of heterotypic networks with paracrine communication. The elements of cell communication in the pituitary have been identified in all species of vertebrates, from jawless fish to mammals. However, while their relative relevance in maintaining the highly synchronous hormone secretion has not yet been tested in the majority of species, the development of transgenic animals and powerful tools of microscopy analysis have started to yield some information about the complex regulation that occurs within and around the pituitary gland. Another yet unanswered question that may begin to be unraveled soon concerns the universality of pulsatile hormone secretion in vertebrates, which there is reason to believe may not be restricted to mammals. Finally, the generation of pulses cannot be explained by any single element of the system, but rather is the result of the concerted occurrence of multiple processes. Diderot expressed this idea with the sentence: "*Everything changes, all things pass, only the totality remains.*"

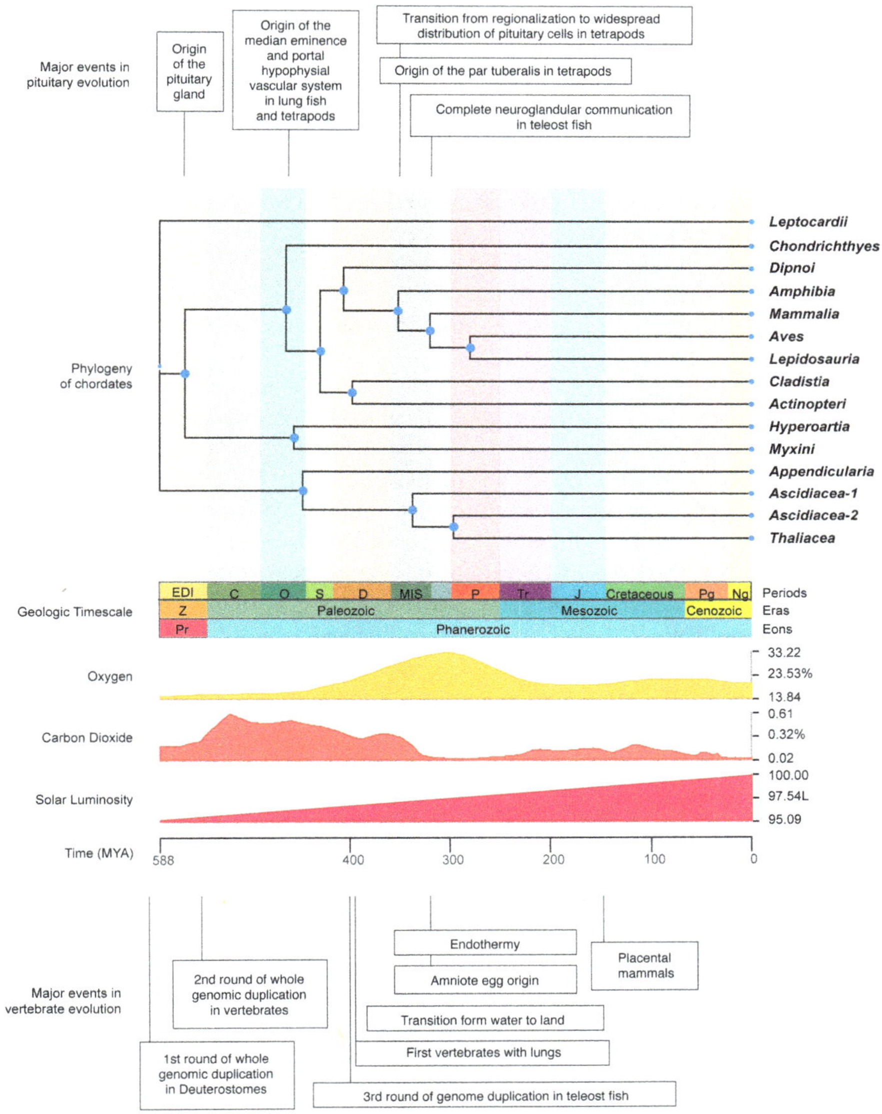

Fig. 4.7 Phylogenetic tree of the main groups of vertebrates and evolution. Main changes in pituitary and vertebrate evolution are shown in correlation with geologic time, and environmental changes in oxygen, carbon dioxide, and solar luminosity. This figure was extracted and edited from the TimeTree database (Kumar et al. 2022)

Key Literature

Campbell et al. (2004): Offers a mechanism for the origin of the neuroendocrine system in vertebrates by looking at the molecular elements that were already present in vertebrate related lineages (tunicates) with the potential to generate the signaling pathways known in vertebrates.

Sower (2018): Describes the pioneering work of Sower and collaborators in the discovery of the main features of lamprey pituitary and its relevance in the comprehension of pituitary evolution of vertebrates.

Trudeau and Somoza (2020): A recent description of the different modes of communication between the hypothalamus and pituitary of the main groups of living vertebrates and their physiological relevance.

Santiago-Andres et al. (2020): Summarizes the present evidence of the pituitary cell networks formation and the modes of cell–cell communication in all living vertebrates.

Kanda (2019): A comparison of the regulation of the hypothalamus–pituitary–gonadal axis across vertebrates emphasizing the conserved mechanisms of reproduction control in spite of differences in life histories.

Tessmar-Raible (2007): An example of deep homology and molecular tools in the investigation of neuroendocrine evolution comparing the function and gene expression of cells in the forebrain of annelids and zebrafish.

References

Acher R (1996) Molecular evolution of fish neurohypophysial hormones: neutral and selective evolutionary mechanisms. Gen Comp Endocrinol 102(2):157–172. Available at: https://doi.org/10.1006/gcen.1996.0057

Caldwell HK, Young WS (2006) Oxytocin and vasopressin: genetics and behavioral implications. In: Handbook of neurochemistry and molecular neurobiology. Springer, Boston, MA, pp 573–607. Available at: https://doi.org/10.1007/978-0-387-30381-9_25

Campbell RK, Satoh N, Degnan BM (2004) Piecing together evolution of the vertebrate endocrine system. Trends Genet TIG 20(8):359–366. Available at: https://doi.org/10.1016/j.tig.2004.06.005

Campinho MA (2019) Teleost metamorphosis: the role of thyroid hormone. Front Endocrinol 10:383. Available at: https://doi.org/10.3389/fendo.2019.00383

Chanet B, Meunier FJ (2014) The anatomy of the thyroid gland among "fishes": phylogenetic implications for the Vertebrata. Société Française d'Ichtyologie. Available at: https://doi.org/10.26028/CYBIUM/2014-382-002

Chowen JA, Frago LM, Argente J (2004) The regulation of GH secretion by sex steroids. Eur J Endocrinol 151 Suppl 3:U95–U100. Available at: https://doi.org/10.1530/eje.0.151u095

Dale HH (1906) On some physiological actions of ergot. J Physiol 34(3):163–206. Available at: https://doi.org/10.1113/jphysiol.1906.sp001148

Deyts C et al (2006) Morphological and gene expression similarities suggest that the ascidian neural gland may be osmoregulatory and homologous to vertebrate peri-ventricular organs. Eur J Neurosci 24(8):2299–2308. Available at: https://doi.org/10.1111/j.1460-9568.2006.05073.x

Donaldson ZR, Young LJ (2008) Oxytocin, vasopressin, and the neurogenetics of sociality. Science 322(5903):900–904. Available at: https://doi.org/10.1126/science.1158668

Du Vigneaud V (1956) Trail of sulfur research: from insulin to oxytocin. Science 123(3205):967–974. Available at: https://doi.org/10.1126/science.123.3205.967

Fekete C (2022) Pars tuberalis as a key regulator of neuroendocrine functions. Nat Rev Endocrinol 18(6):332–332. Available at: https://doi.org/10.1038/s41574-022-00667-6

Fontaine R et al (2022) Pituitary multi-hormone cells in mammals and fish: history, origin, and roles. Front Neuroendocrinol 67:101018. Available at: https://doi.org/10.1016/j.yfrne.2022.101018

Gahete MD et al (2009) Understanding the multifactorial control of growth hormone release by somatotropes: lessons from comparative endocrinology. Ann N Y Acad Sci 1163:137–153. Available at: https://doi.org/10.1111/j.1749-6632.2008.03660.x

Gerhart J, Kirschner M (2007) The theory of facilitated variation. Proc Natl Acad Sci U S A 104(Suppl 1):8582–8589. Available at: https://doi.org/10.1073/pnas.0701035104

Golan M et al (2015) Architecture of GnRH-gonadotrope-vasculature reveals a dual mode of gonadotropin regulation in fish. Endocrinology 156(11):4163–4173. Available at: https://doi.org/10.1210/en.2015-1150

Gorbman A, Nozaki M, Kubokawa K (1999) A brain-Hatschek's pit connection in amphioxus. Gen Comp Endocrinol 113(2):251–254. Available at: https://doi.org/10.1006/gcen.1998.7193

Gwee P-C et al (2008) Sequence and organization of coelacanth neurohypophysial hormone genes: evolutionary history of the vertebrate neurohypophysial hormone gene locus. BMC Evol Biol 8:93. Available at: https://doi.org/10.1186/1471-2148-8-93

Han SY et al (2018) Optical approaches for interrogating neural circuits controlling hormone secretion. Endocrinology 159(11):3822–3833. Available at: https://doi.org/10.1210/en.2018-00594

Herbison AE (2016) 'Control of puberty onset and fertility by gonadotropin-releasing hormone neurons', Nature reviews. Endocrinology 12(8):452–466. Available at: https://doi.org/10.1038/nrendo.2016.70

Heyland A, Moroz LL (2005) Cross-kingdom hormonal signaling: an insight from thyroid hormone functions in marine larvae. J Exp Biol 208(Pt 23):4355–4361. Available at: https://doi.org/10.1242/jeb.01877

Heyland A et al (2012) Distinct expression patterns of glycoprotein hormone subunits in the lophotrochozoan Aplysia: implications for the evolution of neuroendocrine systems in animals. Endocrinology 153(11):5440–5451. Available at: https://doi.org/10.1210/en.2012-1677

Holland LZ et al (2008) The amphioxus genome illuminates vertebrate origins and cephalochordate biology. Genome Res 18(7):1100–1111. Available at: https://doi.org/10.1101/gr.073676.107

Howell WH (1898) The physiological effects of extracts of the hypophysis cerebri and infundibular body. J Exp Med 3(2):245–258. Available at: https://doi.org/10.1084/jem.3.2.245

Hulbert AJ (2000) Thyroid hormones and their effects: a new perspective. Biol Rev Camb Philos Soc 75(4):519–631. Available at: https://doi.org/10.1017/s146479310000556x

Jönsson E, Björnsson BT (2002) Physiological functions of growth hormone in fish with special reference to its influence on behaviour. Fish Sci FS 68(sup1):742–748. Available at: https://doi.org/10.2331/fishsci.68.sup1_742

Kanda S (2019) Evolution of the regulatory mechanisms for the hypothalamic-pituitary-gonadal axis in vertebrates-hypothesis from a comparative view. Gen Comp Endocrinol 284:113075. Available at: https://doi.org/10.1016/j.ygcen.2018.11.014

Kaneko T (1996) Cell biology of somatolactin. Int Rev Cytol 169:1–24. Available at: https://doi.org/10.1016/S0074-7696(08)61983-X

Kano S (2010) Genomics and developmental approaches to an ascidian adenohypophysis primordium. Integr Comp Biol 50(1):35–52. Available at: https://doi.org/10.1093/icb/icq050

Kumar S et al (2022) TimeTree 5: an expanded resource for species divergence times. Mol Biol Evol 39(8) Available at: https://doi.org/10.1093/molbev/msac174

Lee LTO et al (2007) Discovery of growth hormone-releasing hormones and receptors in non-mammalian vertebrates. Proc Natl Acad Sci USA 104(7):2133–2138. Available at: https://doi.org/10.1073/pnas.0611008104

Le Tissier P et al (2017) An updated view of hypothalamic-vascular-pituitary unit function and plasticity. Nat Rev Endocrinol 13(5):257–267. Available at: https://doi.org/10.1038/nrendo.2016.193

Le Tissier P, Fiordelisio Coll T, Mollard P (2018) The processes of anterior pituitary hormone pulse generation. Endocrinology 159(10):3524–3535. Available at: https://doi.org/10.1210/en.2018-00508

Lévêque TF (1974) The endocrine hypothalamus: An historical review. Can J Neurol Sci 1(1):24–28. Available at: https://doi.org/10.1017/s0317167100019545

Li M et al (2017) Activities of amphioxus GH-like protein in osmoregulation: insight into origin of vertebrate GH family. Int J Endocrinol 2017:9538685. Available at: https://doi.org/10.1155/2017/9538685

Marquis TJ et al (2017) Comprehensive histological and immunological studies reveal a novel glycoprotein hormone and thyrostimulin expressing proto-glycotrope in the sea lamprey pituitary. Cell Tissue Res 367(2):311–338. Available at: https://doi.org/10.1007/s00441-016-2502-y

Morthorst JE et al (2023) Thyroid-like hormone signaling in invertebrates and its potential role in initial screening of thyroid hormone system disrupting chemicals. Integr Environ Assess Manag 19(1):63–82. Available at: https://doi.org/10.1002/ieam.4632

Norris DO, Carr JA (2020) Vertebrate endocrinology. Academic, London. Available at: https://play.google.com/store/books/details?id=L9PVDwAAQBAJ

Nozaki M (2013) Hypothalamic-pituitary-gonadal endocrine system in the hagfish. Front Endocrinol 4:200. Available at: https://doi.org/10.3389/fendo.2013.00200

Nozaki M, Gorbman A (1992) The question of functional homology of Hatschek's pit of amphioxus (Branchiostoma belcheri) and the vertebrate adenohypophysis. Zool Sci 9(2):387–395

Oliver G, Schäfer EA (1895) On the physiological action of extracts of pituitary body and certain other glandular organs: preliminary communication. J Physiol 18(3):277–279. Available at: https://doi.org/10.1113/jphysiol.1895.sp000565

Ott I, Scott JC (1910) The action of infundibulin upon the mammary secretion. Exp Biol Med 8(2):48–49. Available at: https://doi.org/10.3181/00379727-8-27

Pascual A, Aranda A (2013) Thyroid hormone receptors, cell growth and differentiation. Biochim Biophys Acta 1830(7):3908–3916. Available at: https://doi.org/10.1016/j.bbagen.2012.03.012

Paul B et al (2022) Thyroid and corticosteroid signaling in amphibian metamorphosis. Cells 11(10) Available at: https://doi.org/10.3390/cells11101595

Rizzoti K et al (2023) SOX9-positive pituitary stem cells differ according to their position in the gland and maintenance of their progeny depends on context. Sci Adv 9(40):eadf6911. Available at: https://doi.org/10.1126/sciadv.adf6911

Rousset B et al (2015) Chapter 2 Thyroid hormone synthesis and secretion. In: Feingold KR et al (eds) Endotext. MDText.com, South Dartmouth, MA. Available at: https://www.ncbi.nlm.nih.gov/pubmed/25905405

Sakai T et al (2017) Invertebrate gonadotropin-releasing hormone-related peptides and their receptors: an update. Front Endocrinol 8:217. Available at: https://doi.org/10.3389/fendo.2017.00217

Santiago-Andres Y, Golan M, Fiordelisio T (2020) Functional pituitary networks in vertebrates. Front Endocrinol 11:619352. Available at: https://doi.org/10.3389/fendo.2020.619352

Scharrer E, Scharrer B (1937) Über Drüsen-Nervenzellen und neurosekretorische Organe Bei Wirbellosen und Wirbeltieren. Biol Rev Camb Philos Soc 12(2):185–216. Available at: https://doi.org/10.1111/j.1469-185x.1937.tb01229.x

Sekiguchi T et al (2007) Further EST analysis of endocrine genes that are preferentially expressed in the neural complex of Ciona intestinalis: Receptor and enzyme genes associated with endocrine system in the neural complex. Gen Comp Endocrinol 150(2):233–245. Available at: https://doi.org/10.1016/j.ygcen.2006.08.010

Sivalingam M et al (2022) Conserved functions of hypothalamic kisspeptin in vertebrates. Gen Comp Endocrinol 317:113973. Available at: https://doi.org/10.1016/j.ygcen.2021.113973

Sower SA (2018) Landmark discoveries in elucidating the origins of the hypothalamic-pituitary system from the perspective of a basal vertebrate, sea lamprey. Gen Comp Endocrinol 264:3–15. Available at: https://doi.org/10.1016/j.ygcen.2017.10.016

Tessmar-Raible K et al (2007) Conserved sensory-neurosecretory cell types in annelid and fish forebrain: insights into hypothalamus evolution. Cell 129(7):1389–1400. Available at: https://doi.org/10.1016/j.cell.2007.04.041

Trudeau VL, Somoza G (2020) Multimodal hypothalamo-hypophysial communication in the vertebrates. Gen Comp Endocrinol 293:113475. Available at: https://doi.org/10.1016/j.ygcen.2020.113475
Uchida K et al (2010) Evolutionary origin of a functional gonadotropin in the pituitary of the most primitive vertebrate, hagfish. Proc Natl Acad Sci U S A 107(36):15832–15837. Available at: https://doi.org/10.1073/pnas.1002208107
Ukena K, Iwakoshi-Ukena E, Hikosaka A (2008) Unique form and osmoregulatory function of a neurohypophysial hormone in a urochordate. Endocrinology 149(10):5254–5261. Available at: https://doi.org/10.1210/en.2008-0607
Wagner GP (1989) The origin of morphological characters and the biological basis of homology. Evolution 43(6):1157–1171. Available at: https://doi.org/10.1111/j.1558-5646.1989.tb02566.x
Waxman DJ, O'Connor C (2006) Growth hormone regulation of sex-dependent liver gene expression. Mol Endocrinol 20(11):2613–2629. Available at: https://doi.org/10.1210/me.2006-0007
Weidner J et al (2020) Hormones as adaptive control systems in juvenile fish. Biol Open 9(2) Available at: https://doi.org/10.1242/bio.046144
Wittkowski W et al (1999) Cell and molecular biology of the pars tuberalis of the pituitary. Int Rev Cytol 185:157–194. Available at: https://doi.org/10.1016/s0074-7696(08)60151-5
Wright PA, Wood CM (2015) Regulation of ions, acid–base, and nitrogenous wastes in elasmobranchs. In: Shadwick RE, Farrell AP, Brauner CJ (eds) Fish physiology. Elsevier, Lieden, pp 279–345. Available at: https://doi.org/10.1016/b978-0-12-801286-4.00005-8
Zhang S, Ji X (2022) Hatschek's pit and origin of pituitary gland. Acta Oceanol Sin 41(12):1–6. Available at: https://doi.org/10.1007/s13131-022-2044-y

Further Reading

Establishment of Pituitary Networks During Development

Budry L et al (2011) Related pituitary cell lineages develop into interdigitated 3D cell networks. Proc Natl Acad Sci U S A 108(30):12515–12520 Available at: https://doi.org/10.1073/pnas.1105929108
Mollard P et al (2012) A tridimensional view of pituitary development and function. Trends Endocrinol Metab 23(6):261–269 Available at: https://doi.org/10.1016/j.tem.2012.02.004

Paracrine Communication in the Pituitary Gland

Denef C (2008) Paracrinicity: the story of 30 years of cellular pituitary crosstalk. J Neuroendocrinol 20(1):1–70 Available at: https://doi.org/10.1111/j.1365-2826.2007.01616.x

Folliculostellate Cell Network Functions

Le Tissier PR, Murray JF, Mollard P (2022) A new perspective on regulation of pituitary plasticity: the network of SOX2-positive cells may coordinate responses to challenge. Endocrinology 163(8) Available at: https://doi.org/10.1210/endocr/bqac089

Regulation of Dopamine-PRL Activity Mediated by PRL

Yip SH et al (2019) Elevated prolactin during pregnancy drives a phenotypic switch in mouse hypothalamic dopaminergic neurons. Cell Rep 26(7):1787–1799.e5 Available at: https://doi.org/10.1016/j.celrep.2019.01.067

A Comparative and Historical Discussion of GnRH Activity in Fishes and Mammals

Zohar Y et al (2022) A half century of fish gonadotropin-releasing hormones: Breaking paradigms. J Neuroendocrinol 34(5):e13069 Available at: https://doi.org/10.1111/jne.13069

Pulse Generation and Measurement

Campos P et al (2018) Pituitary cell and molecular. In: Encyclopedia of reproduction. Elsevier, pp 184–187. Available at: https://doi.org/10.1016/b978-0-12-801238-3.64577-4

Czieselsky K et al (2016) Pulse and surge profiles of luteinizing hormone secretion in the mouse. Endocrinology 157(12):4794–4802 Available at: https://doi.org/10.1210/en.2016-1351

Duittoz AH et al (2022) Development of the gonadotropin-releasing hormone system. J Neuroendocrinol 34(5):e13087 Available at: https://doi.org/10.1111/jne.13087

Hoa O et al (2019) Imaging and manipulating pituitary function in the awake mouse. Endocrinology 160(10):2271–2281 Available at: https://doi.org/10.1210/en.2019-00297

Hodson DJ et al (2012) Coordination of calcium signals by pituitary endocrine cells in situ. Cell Calcium 51(3-4):222–230 Available at: https://doi.org/10.1016/j.ceca.2011.11.007

Romanò N et al (2017) Multiple-scale neuroendocrine signals connect brain and pituitary hormone rhythms. Proc Natl Acad Sci U S A 114(9):2379–2382 Available at: https://doi.org/10.1073/pnas.1616864114

Part II

Evolution of the Hypothalamic-Pituitary-Endocrine Gland Axes in Vertebrates

Evolution of the GnRH Neuron

5

Allan E. Herbison

Abstract

The 8–11 amino acid peptides gonadotropin-releasing hormone (GnRH), adipokinetic hormone (AKH) and corazonin represent a superfamily that evolved in Bilateria nearly one billion years ago. While these peptides have been employed for many different types of intercellular signaling, a relatively consistent neuroendocrine-type function is evident. In invertebrates, AKH, and corazonin neurons are found to have prominent roles in coordinating energy metabolism responses. The classical hypophysiotropic GnRH neuron appears to become evident only during chordate evolution. The tunicate *Ciona intestinalis* has a neurogenic olfactory placode from which GnRH neurons are born that have roles in metamorphosis and adult fertility. Two rounds of whole-genome duplication in vertebrates and subsequent species-specific loss events have resulted in the variable existence of GnRH1, GnRH2, and GnRH3 neurons. The basal forebrain GnRH1 neurons always represent the hypophysiotropic neuroendocrine population, except in some fish where GnRH1 has been lost and is replaced by GnRH3. In rodents and primates, multiple extrahypothalamic neural populations coexpress low levels of GnRH1 that very likely have unrelated functions. The midbrain GnRH2 neurons are the most highly conserved population in vertebrates but appear to have little role in fertility control. Within mammals, the functional impact of subtle species variations in hypophysiotropic GnRH1 neuron distribution and morphology remain unknown. However, the mammalian requirement for pulsatile GnRH secretion, as well as a sophisticated mechanism for control-

A. E. Herbison (✉)
Department of Physiology, Development and Neuroscience, University of Cambridge, Cambridge, UK
e-mail: aeh36@cam.ac.uk

V. Grinevich, R. F. Oliveira (eds.), *Evolutionary and Comparative Neuroendocrinology*, Masterclass in Neuroendocrinology 17,
https://doi.org/10.1007/978-3-031-80209-6_5

ling the preovulatory surge, has likely resulted in the adoption of kisspeptin pulse and surge generators as core elements of the GnRH neuronal network.

Keywords

Evolution · GnRH · Corazonin · AKH · ACP · Kisspeptin · Invertebrate

5.1 Introduction

The gonadotropin-releasing hormone (GnRH) neuron is well established as the final output cell of the neural network controlling fertility in all vertebrates. While different species may vary the strategies they use to ultimately control gonadotropin secretion (Goodman 2015), the principle of a hypophysiotropic population of GnRH neurons remains constant. The relatively recent evolution of the vertebrate hypophysiotropic GnRH neuron system has occurred in association with the development of the hypothalamus and pituitary gland (Trudeau and Somoza 2020) and the remarkable migration of the GnRH neurons from the olfactory placode into the brain during embryonic development (Wray 2010).

The vertebrate decapeptide was initially named luteinizing hormone-releasing hormone (LHRH) (Schally et al. 1971) and luteinizing hormone-releasing hormone factor (Amoss et al. 1971) due to its first discovered role in stimulating the secretion of luteinizing hormone (LH) and follicle-stimulating hormone. This has since been revised to "GnRH." However, given the ubiquity of peptide signaling in cells, there is no particular reason why the functions of GnRH should be restricted to its first identified role. This is exemplified by evolution, where the wide use of GnRH-like signaling has gradually narrowed to the predominant but not exclusive role of GnRH in controlling fertility.

Several excellent reviews on the molecular phylogeny of GnRH receptors and their ligands have been published in recent years (Zandawala et al. 2018; Dufour et al. 2020). The aim of the present review is to examine the evolutionary development of the GnRH neuron itself and its control of gonadotropin secretion. A brief summary of the bilaterian evolution of GnRH-like signaling is provided before examining the more recent evolution of GnRH neurons in urochordates, the ancestral animals immediately preceding vertebrates, and the multiple aspects of vertebrate GnRH neurons themselves.

5.2 The GnRH-Adipokinetic Hormone-Corazonin Superfamily

A family of neuropeptides, including GnRH, AKH, corazonin, and AKH/corazonin-like peptide (ACP), has existed since the evolution of organisms with bilateral symmetry somewhere between 670 million and one billion years ago (Jekely 2013). Although substantial differences in amino acid sequence exist between these peptides, their receptors show some homology and have been used to interrogate the evolution of GnRH-like signaling. It is thought that a very early gene duplication resulted in GnRH and corazonin receptors and that a further duplication of the GnRH receptor in arthropods resulted in the AKH and ACP receptors (Sakai et al. 2017; Zandawala et al. 2018; Dufour et al. 2020) (Fig. 5.1). Thus, insects can have up to three GnRH-like signaling systems, whereas other clades have two. However, in many cases, GnRH-like receptors have then been lost in certain lineages or individual species; for example, corazonin receptors are absent in vertebrates and nematodes, while ACP disappeared in *Drosophila* within the arthropods (Zandawala et al. 2018) (Fig. 5.1).

The evolutionary path of GnRH receptor ligands is less clear and is still in the process of being established (Zandawala et al. 2018). Nevertheless, peptides in the GnRH family are all decapeptides with four residues perfectly conserved and share a common genetic organization (Fernald and White 1999). Alongside their existence in all vertebrates, the GnRH decapeptide is also now considered to exist in many other phyla (Zandawala et al. 2018; Dufour et al. 2020) (Fig. 5.1). The AKH and ACP ligands are only described in arthropods, whereas corazonin-like peptides

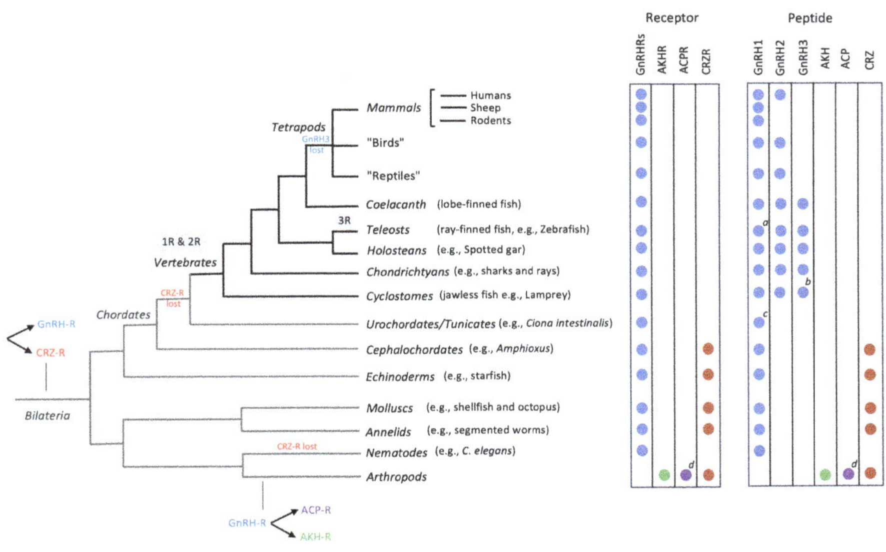

Fig. 5.1 A simplified diagram showing the known evolution of the GnRH family of neuropeptides and receptors. The phylogenetic tree is shown for Bilateria with the proposed key events indicated for GnRH signaling. An initial duplication resulted in GnRH and corazonin (CRZ) receptors with CRZ receptors then being lost soon after chordate evolution and also in nematodes. A subsequent duplication of the GnRH-R occurred in arthropods to generate adipokinetic hormone (AKH) receptors and AKH/corazonin-like peptide (ACP) receptors. The times of whole-genome duplication are indicated in blue boxes—two rounds at the beginning of vertebrate evolution and another specifically for teleosts. Note that the multiple GnRH receptors existing in different species are indicated as a single family (GnRHR). a Some teleost species have lost GnRH1; b the relationship of lamprey GnRH1-3 to the vertebrate GnRH1-3 remains unclear; c there are 11 tunicate GnRH peptides; d *Drosophila melanogaster* has selectively lost ACP signaling within the Arthropod family. (Adapted from (Zandawala, Tian et al. 2018; Dufour, Querat et al. 2020; Ogawa, Yamamoto et al. 2022)

are found in a wide range of invertebrates (Zandawala et al. 2018; Dufour et al. 2020) (Fig. 5.1). However, there are other ligands within the family that can fall between these broad classifications, with many, at least initially, termed "AKH-GnRH" peptides (Lindemans et al. 2009). Together with the receptors, this has resulted in a relatively consistent but nevertheless variable pattern of possible GnRH-like signaling across evolution (Fig. 5.1).

It is readily apparent that the GnRH-AKH-corazonin family of peptides has been used by different cells in different organisms to perform a very wide variety of functions across evolution (Sakai et al. 2020). While some of these roles have resembled endocrine or hormonal-type signaling, in particular for energy homeostasis, the GnRH superfamily of peptides does not appear to have been selected for a principal or uniform role in the control of reproduction until chordate evolution. Brief accounts of AKH, ACP, and corazonin signaling are provided below before progressing to GnRH.

5.2.1 AKH

The arthropod 8–10 amino acid peptide AKH is a neurohormone synthesized in the insect corpora cardiaca, an endocrine gland connected to the brain that is considered to be a very early forerunner of the pituitary gland (Stone et al. 1976). The principal role of AKH is in the control of energy metabolism. For example, in *Drosophila*, AKH is secreted at times of high energy expenditure and during stress to induce triglyceride and glycogen breakdown from the equivalent of the fly liver (Lee and Park 2004; Hughson 2021). Beyond the impact of AKH on energy status to modulate fertility, no direct role has been identified as yet for this neuropeptide in regulating reproduction (Ben-Menahem 2021).

5.2.2 ACP

This enigmatic GnRH-like signaling system was identified in insect species as having both a ligand and a receptor, intermediate in the sequence between the AKH and corazonin systems, and hence it was named AKH/corazonin-like peptide (Hansen et al. 2010). Notably, ACP does not exist outside the arthropod subphyla and even within it, it has been lost several times; for example, it is absent in *Drosophila*. To date, the function of ACP is unclear, with a single report indicating it has a role in regulating the lipid and carbohydrate supply in crickets (Fukumura et al. 2018).

5.2.3 Corazonin

Corazonin is an 11-amino acid peptide that was initially discovered in cockroaches for its cardiac acceleratory properties. While it is now apparent that corazonin signaling occurs in all Bilateria except for vertebrates, urochordates, and nematodes

(Zandawala et al. 2018), most functional studies to date have been undertaken in insects. Here, once again, a prominent role for corazonin in energy regulation has been detected, although this occurs alongside a wide spectrum of other functions, including some aspects of male-mating behavior in *Drosophila* (Ben-Menahem 2021; Khan et al. 2021) (see below).

5.3 GnRH-Like Neurons in Invertebrate Model Organisms

Given the importance of flies and worms as model organisms in biology, it is pertinent to consider the roles of GnRH-like peptides in these well-investigated species.

5.3.1 *Caenorhabditis elegans*

Nematodes are one of the few phyla outside vertebrates that have lost corazonin signaling and, thus, only have GnRH receptors (Zandawala et al. 2018) (Fig. 5.1). The genome of *C. elegans* encodes eight GnRH-like GPCRs (*gnrr-1* to *gnrr-8*), although the ligands of these receptors remain only partially characterized and are widely expressed in both larvae and adults (Van der Auwera et al. 2020). The *C. elegans* peptide most similar to the GnRH superfamily, originally named *Ce*-AKH-GnRH (now neuropeptide-like peptide 47 (NLP-47)), activates GNRR-1 expressed in oocytes and sperm, among other cells, and modulates egg-laying behavior in *C. elegans* (Lindemans et al. 2009). Indeed, *nlp-47* is one of the few NLPs to be conserved across nematodes, where it also has a uniform role, among others, in activating the ovijector reproductive muscle (McKay et al. 2022). However, the knockdown of either *nlp-47* or *gnrr1* in *C. elegans* only results in an approximate 30% reduction in egg laying (Lindemans et al. 2009).

The RPamide-family neuropeptides *nlp-2* and *nlp-22* also share some sequence similarity to GnRH/AKH and act on *gnrr-3* and *gnrr-6,* but do so to modulate sleep-wake transitions in *C. elegans* (Van der Auwera et al. 2020). Interestingly, *nlp-2* is expressed by a pair of pheromone-sensing olfactory neurons innervating motor circuit neurons expressing *gnrr-3* and *-6* to initiate wakefulness. Two other GnRH-like receptors, *gnrr-2* and *gnrr-6,* are thought to have roles in regulating protein homeostasis in *C. elegans.* Neurons expressing *gnrr-2* respond to early gonadal cues to promote reproductive maturity involving the redirection of energy reserves to the reproductive system and egg laying (Kishner et al. 2022).

5.3.2 *Drosophila melanogaster*

Drosophila do not have GnRH receptors but, instead, express both AKH and corazonin signaling pathways (Fig. 5.1). Their many roles in fruit flies have been reviewed recently (Ben-Menahem 2021). The octapeptide AKH is expressed in the neurons of the corpus cardiacum and is secreted primarily at times when high energy levels are required. The peptide acts somewhat like glucagon to promote the liberation of carbohydrates and triglycerides while also having cardio-stimulatory and orexigenic actions (Hughson 2021). Accordingly, the AKH receptor is expressed in multiple tissues, including the fat body and crop sack, which store carbohydrates and lipids. The deletion of AKH or the AKH receptor has not been found to have any impact on *Drosophila* fertility, and to date, this signaling system is not thought to have a role in regulating fruit fly fertility (Ben-Menahem 2021). However, AKH may modulate the mating behavior of starved male *Drosophila* (Lebreton et al. 2016).

A similar situation exists for corazonin signaling in *Drosophila,* in that it appears to have prominent roles in energy homeostasis and may also contribute to male-mating behavior. In adult *Drosophila,* there are six to eight corazonin-expressing neurons found in either hemisphere of the brain and four corazonin neurons in the abdominal ganglion of only males (Khan et al. 2021). Corazonin receptors are expressed widely but particularly in the fat body and salivary gland, and receptor deletions in these tissues suggest that corazonin acts primarily as a stress hormone to coordinate feeding behavior and carbohydrate metabolism (Kubrak et al. 2016). Related to fertility, the sexually dimorphic abdominal ganglion population of corazonin neurons has been found to control the transfer of sperm and seminal fluid during copulation as well as the duration of copulation in males (Tayler et al. 2012). Intriguingly, the reciprocally connected corazonin neurons in the male abdominal ganglion exhibit striking patterns of episodic activity through the dephosphorylation of calmodulin-dependent protein kinase II, which generates an approximately 6-minute timer gating the duration of copulation (Thornquist et al. 2020). However, the importance of the corazonin peptide signaling itself in these abdominal ganglion neurons remains unclear as corazonin-null flies remain fertile (Ben-Menahem 2021; Khan et al. 2021).

5.4 GnRH Neuron Evolution in Chordates

The chordate phylum includes all animals that have a notochord/vertebral column and dorsal nerve cord/brain at some point in their development. It consists of the vertebrates and their nearest relative, the urochordates, all descended from a common ancestor over 500 million years ago. The urochordates, also referred to as tunicates, exist as more than 3000 filter-feeding sac-like marine animals and are the closest invertebrates to the vertebrate phylum. This proximity has made them attractive organisms for examining evolutionary biology, with *Ciona intestinalis* in the Ascidian paraphylum receiving particular attention as a model invertebrate chordate (Sakai et al. 2020). This organism exists as mobile tadpoles before undergoing metamorphosis to become permanently attached to substrates on the ocean floor and taking up a soft tube-like structure resembling the intestine (Fig. 5.2).

Along with vertebrates, the urochordates lost corazonin signaling and only express GnRH (Fig. 5.1). Unusually, *C. intestinalis* has two GnRH genes, *Ci-gnrh-1* and *Ci-gnrh-2,* which each produce three tandem transcripts, which are translated into six distinct GnRH peptides—tGnRH3–8 (Adams et al. 2003; Sakai et al. 2020).

These *gnrh* genes are widely expressed in larvae but only present in neural tissue in adults, including the cerebral ganglion, dorsal strand plexus, and visceral nerves, which form the dorsal nervous system (Tsutsui et al. 1998; Terakado 2001; Kavanaugh et al. 2005). Immunohistochemistry in adult *C. intestinalis* has revealed the existence of primarily bipolar-shaped GnRH neurons occurring as long nose-to-tail trails within the dorsal strand plexus, which extends from the cerebral ganglion to the gonads (Tsutsui et al. 1998; Terakado 2001; Adams et al. 2003) (Fig. 5.2). Importantly, in the larval state, ciliated neurons expressing *Ci-gnrh-2* first appear in the proto-placodal ectoderm of *C. intestinalis,* considered to represent that earliest forerunner of the vertebrate neurogenic olfactory placode (Abitua et al. 2015). It is presently unknown whether GnRH neurons migrate into the dorsal strand plexus, but their bipolar morphology and nose-to-tail distribution are remarkably reminiscent of vertebrate GnRH neurons (Fig. 5.2).

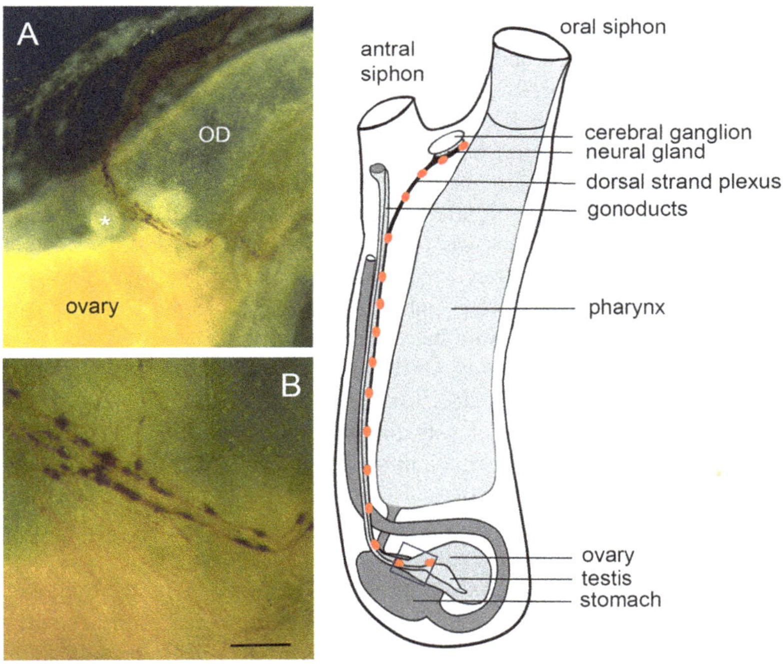

Fig. 5.2 The GnRH neuron system of the adult tunicate *Ciona intestinalis*. A schematic diagram is provided to the right showing the principal organs of the adult tunicate attached to the seafloor. The distribution of GnRH neurons running alongside the visceral nerves in the dorsal strand plexus running from the cerebral ganglion to the gonads is indicated by red dots. (**A**) Photomicrograph of GnRH-immunoreactive neurons taken at the approximate position indicated by the rectangle placed over the gonads in the schematic diagram. (**B**) Higher power view of GnRH neurons showing their bipolar morphology and nose-to-tail topography. OD, oviduct; the asterisk indicates an oocyte. The scale bar is 100 microns. (**A**) and (**B**) adapted from *Tsutsui, Yamamoto, Ito, and Oka. GnRH-immunoreactive neuronal system in the presumptive ancestral chordate Ciona intestinalis (Ascidian). Gen Comp Endocrinol 112, 426–432 (1998)*

Although GnRH and its receptors have many nonreproductive functions in the swimming larvae of *Ciona* (Okawa et al. 2020; Sakai et al. 2020), recent investigations have identified a puberty-like role with the $GABA_B$ receptor activation of GnRH neurons initiating their metamorphosis into sexually mature adults (Hozumi et al. 2020). Furthermore, it has been found that the administration of any of the tGnRH peptides results in body contraction and increased water flow, leading to the spawning of eggs and sperm (Terakado 2001; Adams et al. 2003). Although further studies are required to understand the mechanisms by which GnRH neurons can control gamete release, many of the hallmarks of the vertebrate hypophysiotropic GnRH neuron, such as their birth in a neurogenic placode, their potential migration, and their control of "puberty" and fertility, may exist in *Ciona intestinalis*.

5.5 GnRH Neurons in Vertebrate Evolution

Aided by two rounds of whole-genome duplication, vertebrates became endowed with an axial endoskeleton and a neural tube that developed into a brain. From a neuroendocrine perspective, this resulted in the evolution of a distinct hypothalamus and pituitary gland (Campbell et al. 2004; Trudeau and Somoza 2020). Two key events related to fertility control were the loss of corazonin signaling and quadruplication of *gnrh* so that, ultimately, vertebrates had three paralogous genes for GnRH (*gnrh1–3*), with *gnrh4* lost soon after the last duplication event (Zandawala et al. 2018; Dufour et al. 2020). This was then altered further by the loss of individual GnRH genes in specific vertebrate species, in particular *gnrh3* in all tetrapods and *gnrh2* in some mammalian species such as rodents and sheep (Fig. 5.1). The third round of genome-wide duplication that occurred only in teleosts did not, however, ultimately end up in a wider array GnRH gene expression in the ray-finned fish (Dufour et al. 2020; Ogawa et al. 2022). It is interesting to note, however, that hypophysiotropic GnRH neurons in teleosts directly innervate the pituitary gland rather than relying on intermediary median eminence and portal circulation (Trudeau and Somoza 2020; Ogawa et al. 2022).

In broad terms, *gnrh1* is expressed in the neurons of the hypothalamus, while *gnrh2* is found within the midbrain, and in various fish species, *gnrh3* is located in terminal nerve cells (Fig. 5.4). As such, the GnRH1 neurons of the hypothalamus almost always represent the hypophysiotropic population controlling gonadotropin secretion in vertebrates. The most common exceptions to this occur in the numerous "fish" subphyla. As recently detailed by Ogawa and colleagues (Ogawa et al. 2022), some fish species have a mixture of both GnRH1 and GnRH3 neurons that project in the direction of the pituitary gland. Further, the loss of *gnrh1* in certain teleosts, such as salmon and zebrafish, has resulted in GnRH3 becoming the hypophysiotropic hormone (Dufour et al. 2020; Ogawa et al. 2022).

While the GnRH neurons of vertebrates share many common features, there are several interesting differences that likely reflect evolutionary pressures within the phyla. A selection of key similarities and differences between vertebrates in the organization and function of GnRH neurons are considered below.

5.5.1 GnRH Neuron Migration in Vertebrates

A fundamental process in vertebrate reproductive biology is the migration of the hypophysiotropic GnRH neurons from the olfactory placode into the brain during embryonic development (Wray 2010). Precisely when this developed in chordate evolution remains unclear due to the report that the lamprey of the jawless cyclostome subphyla does not have detectable GnRH neurons in the olfactory placode (Tobet et al. 1996). However, as noted by Ogawa and colleagues (Ogawa et al. 2022), this remains inconclusive as no other jawless fish has been examined to date. Further, while speculative, the anatomy and topography of GnRH neurons in tunicates noted above (Fig. 5.2) raise the possibility that GnRH neuron migration may have its origins back further within the chordate phylum. Interestingly, in fish species, this olfactory migration extends to both GnRH1- and GnRH3-expressing neurons, regardless of whether they finally reside in the preoptic area/hypothalamus or are associated with the terminal nerve (Ogawa et al. 2022). At present, it is clear that all jawed vertebrates exhibit the remarkable migration of GnRH neurons from olfactory regions into the brain during development.

One subtle but substantial difference among vertebrates is the location at which individual GnRH neurons cease their migration along the migratory continuum, which extends from the olfactory bulbs (OBs) to the base of the hypothalamus. In some species, most GnRH neurons are located primarily within the OB and preoptic/anterior hypothalamic area, while in other species, individual GnRH neurons extend from the OB all the way through the preoptic area into the mediobasal hypothalamus (Barry 1979; Campbell et al. 2022). It is difficult to ascribe these different topographies to specific species or subphyla. For example, although most rodents, amphibians, and fish exhibit the "preoptic" pattern of GnRH neuron distribution, not all do so, and even very close subspecies of mole-rats, for instance, have both types of distribution (Coen et al. 2021; Ogawa et al. 2022; Hasunuma 2023). The "extended" GnRH neuron continuum has been reported in several species, including primates, dogs, bats, ferrets, and frogs, and in some cases, the majority of GnRH neurons can reside within the basal hypothalamus (Barry 1979; King and Anthony 1984; Campbell et al. 2022). The mechanisms responsible for individual GnRH neurons ceasing their migration along this migratory trajectory remain unknown. Notably, the locations of the hypophysiotropic GnRH neurons projecting to the median eminence are always found scattered throughout the GnRH neuron migratory pathway (Silverman et al. 1987; Goldsmith et al. 1990; Jansen et al. 1997). As such, the interesting but unexplained varying distributions of GnRH neuron cell bodies along the migratory continuum in vertebrates may have relatively little functional impact on the control of fertility.

5.5.2 GnRH1 Neuron Morphology

Likely due to their common axophilic pattern of migration into the brain, vertebrate GnRH1 neurons exhibit a highly conserved morphology. As identified initially in the seminal studies by Barry (1979), the great majority of GnRH1 neurons exhibit a bipolar morphology with two processes extending from each pole of a cell body that ranges from 10–40 microns in diameter. A minority of GnRH neurons exhibit more complex morphologies with three or more dendrites, although, curiously, multipolar GnRH1 neurons are much more common in sheep and springboks (Lehman et al. 1986; Robinson et al. 1997). Studies in mice and rats have shown that GnRH neuron dendritic remodeling can occur in the peripubertal period when the small subpopulation of GnRH1 neurons with complex morphologies are pruned to the bipolar shape (Cottrell et al. 2006; Ybarra et al. 2011). This process may be regulated in part by the fibroblast growth factor (Rochester et al. 2012). However, very little is known about why some GnRH1 neurons may change their dendritic morphology and what impact this has on function.

The recent use of genetic models in mice has overcome limitations around the dependence upon immunoreactive GnRH peptide location to provide a detailed view of GnRH1 neuron morphology (Campbell et al. 2022) (Fig. 5.3). These studies have shown that, typically, the two projections from GnRH neuron cell bodies are both dendrites. After the dorsally orientated dendrite undertakes a 180-degree hairpin loop, both dendrites project all the way to the ventral aspect of the arcuate nucleus before branching into multiple short axons that innervate vasculature within the median eminence (Herde et al. 2013; Herde and Herbison 2015). This very unusual morphology is accompanied by the location of the action potential

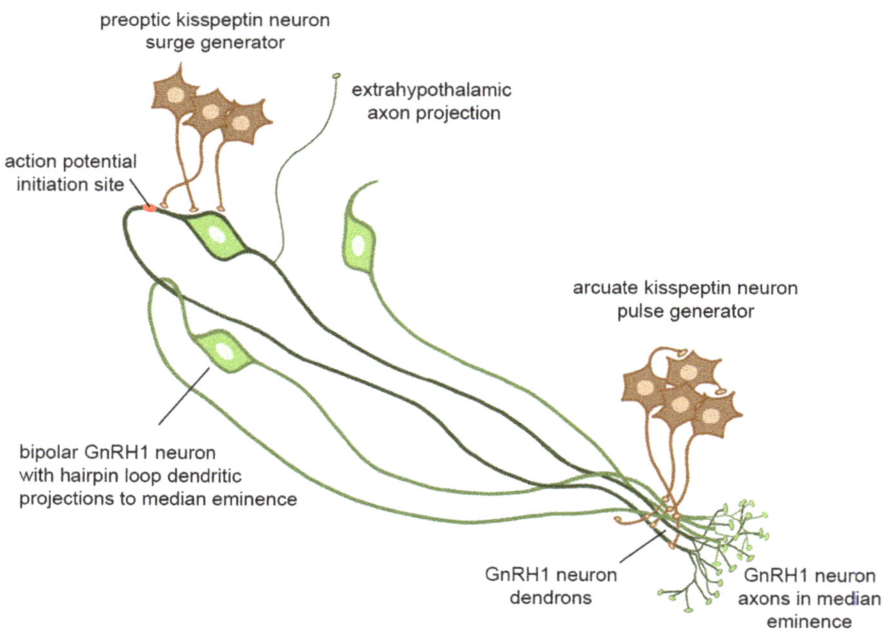

Fig. 5.3 Schematic diagram summarizing the features of the hypophysiotropic GnRH1 neuron in the mouse. See text for details

initiation site about 50 microns from the cell body in one of the dendrites and with synaptic innervation being concentrated on the primary proximal dendrite near the cell body as well as on the distal dendritic projections in the arcuate nucleus (Iremonger and Herbison 2012; Herde and Herbison 2015; Herbison 2021; Yeo et al. 2021). Since the distal projection conducts action potentials, like an axon, and receives substantial synaptic input, similar to a dendrite, it has been termed a "dendron" (Herde et al. 2013; Herbison 2021). There is abundant evidence that GnRH1 neurons also innervate multiple brain regions, and some GnRH neurons can be identified to elaborate what appears to be a typical axon from one of the proximal dendrites (Herde and Herbison 2015) (Fig. 5.3).

The degree to which the morphology of GnRH1 neurons in mice is shared by other mammalian or vertebrate species is unclear at this time. The generation of a *Gnrh1*-Cre rat has enabled the identification of a functional GnRH neuron dendron in this species, indicating that dendron exists in multiple rodent species (Yip et al. 2021). There are, however, only rare glimpses into the potential conserved nature of this morphology among other mammals. For example, immunohistochemical studies of GnRH neurons in humans and monkeys often show the same hairpin loops of the dorsal projection from the cell body (Barry 1979; Campbell et al. 2022). Also, an electron microscopic investigation in sheep reported synapses on GnRH neuron projections, thought likely to be dendrites, located in the ventrolateral arcuate nucleus, where the dendrons are found in rodents (Kuljis and Advis 1989). Although the GnRH1 neuron dendron can be identified by standard electron microscopy (Moore et al. 2018), genetic animal models make this much easier (Liu et al. 2021), and technological advances will likely be required to identify the dendron in other species.

5.5.3 GnRH2 Neurons

The GnRH2 paralog is considered the most ancient of the GnRH genes, with its sequence being conserved perfectly over 500 million years of evolution (Fernald and White 1999). Initially identified in chickens and referred to as chicken-II GnRH, the decapeptide is expressed widely in tetrapods and the various fish subphyla (Fig. 5.4) but has been lost through independent mutation events, deleting or preventing its expression in species such as rodents, sheep, and some primates (Stewart et al. 2009; Desaulniers et al. 2017). Mapping studies in vertebrates show a highly consistent expression of GnRH2 in a population of mesencephalic neurons, as well as in other brain regions in a species-specific manner (Rissman et al. 1995; Kasten et al. 1996; Urbanski et al. 1999; Millar et al. 2001; Ogawa et al. 2022; Hasunuma 2023). In contrast to GnRH1 neurons, midbrain GnRH2 neurons are born within the mesencephalon and project widely throughout the brain (Northcutt and Muske 1994; Xia et al. 2014; Ogawa et al. 2022). Expression in primates is perhaps the most extensive, with the localization of *Gnrh2* transcripts in the midbrain, hippocampus, and hypothalamus and near ubiquitous expression of the GnRH2 receptor

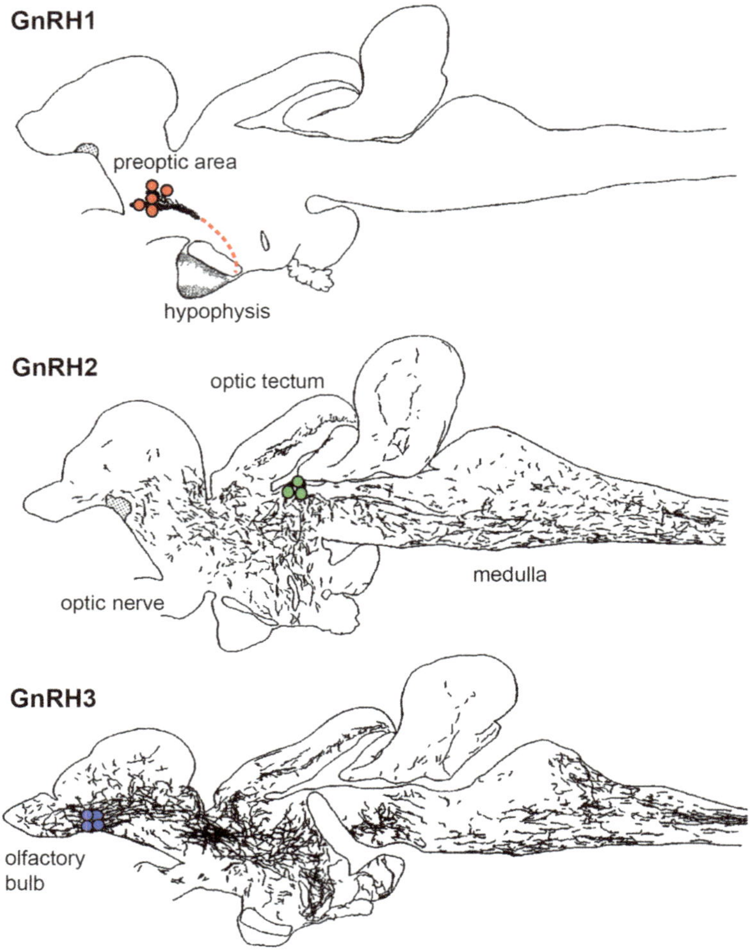

Fig. 5.4 Neuroanatomical distribution of GnRH1, GnRH2, and GnRH3 neurons in the teleost dwarf gourami. Adapted from *Karigo and Oka, Neurobiological study of fish brains gives insights into the nature of gonadotropin-releasing hormone 1–3 neurons. Front Endocrinol. 4:177 (2013)*

throughout the body (White et al. 1998; Urbanski et al. 1999; Millar et al. 2001; Desaulniers et al. 2017).

The initial excitement that GnRH2 might represent a parallel neuroendocrine regulator of fertility in primates has now waned (Desaulniers et al. 2017). Unfortunately, the loss of *Gnrh2* in the standard experimental model species has greatly hampered understanding of its function. Perhaps, commensurate with its broad expression, GnRH2 has been implicated in many different peripheral and central functions (Desaulniers et al. 2017; Umatani and Oka 2019; Ogawa et al. 2022). Given the roles of corazonin and AKH in regulating energy status and mating in prechordates, it is interesting that GnRH2 has also been proposed to link energy status with mating behavior and can control feeding in musk shrews (Temple et al.

2003; Kauffman and Rissman 2004). The administration of GnRH2 has also been observed to enhance courtship behavior in female sparrows (Maney et al. 1997) and marmoset proceptive sexual behavior (Barnett et al. 2006).

5.5.4 GnRH3 Neurons

The *gnrh3* gene is only expressed in fish subphyla, being a vertebrate GnRH paralog that was subsequently lost in all tetrapods (Fig. 5.1). As recently reviewed by Ogawa and colleagues (Ogawa et al. 2022), apart from certain teleosts that have lost GnRH1 and use GnRH3 as the hypophysiotropic hormone, GnRH3 operates as a neuromodulatory peptide within the brain. Nevertheless, it is interesting that GnRH3 neurons are thought to originate in olfactory regions and migrate into the brain using the same route as GnRH1 neurons, the principal difference being that GnRH3 neurons remain associated with the terminal nerve in the very rostral aspects of the GnRH neuron migratory continuum within the olfactory bulb (Ogawa et al. 2022) (Fig. 5.4). Interestingly, Fujimora and colleagues have very recently suggested that hypophysiotropic GnRH neurons in early teleosts may have initially co-expressed GnRH1 and GnRH3 (Fujimori et al. 2024).

Facilitated by genetic models, the most comprehensive morphological and electrophysiological assessment of GnRH3 neurons has been undertaken in medaka and dwarf gourami fish (Karigo and Oka 2013). The approximately 40 terminal nerve GnRH3 neurons found in each fish have complex dendritic morphologies, associate together in tight clusters, and also project very widely throughout the brain (Karigo and Oka 2013) (Fig. 5.4). Electrophysiological studies have demonstrated that these GnRH3 neurons exhibit a robust synchronous pacemaker pattern of activity that is very likely mediated by paracrine GnRH release as well as gap junctions (Abe and Oka 2000; Wayne et al. 2005; Haneda and Oka 2008). As such, it is thought that the terminal nerve GnRH3 neuron population provides a regular episodic pattern of neuromodulatory activity throughout the brain (Karigo and Oka 2013). The exact roles of this highly divergent GnRH3 output are unclear but may involve the facilitation of the motivation state related to nest building and sexual behavior in males (Yamamoto et al. 1997; Umatani et al. 2022), as well as sexual receptivity and mate preference in females (Okuyama et al. 2014). The behavior, mechanisms of firing, and functions of GnRH3 neurons are distinct from those of hypophysiotropic GnRH1 neurons (Ogawa et al. 2022).

5.5.5 Pulsatile GnRH Secretion in Mammals

All vertebrates need to release gametes to achieve sexual reproduction, and GnRH1 neurons have a central role in this process. Whether this involves the spawning of eggs and sperm or ovulation, the appropriately timed activation of GnRH1 neurons during the reproductive cycle to generate a "GnRH surge" is essential. An ultradian or pulsatile pattern of GnRH secretion is also found, but interestingly, this is thought to exist only in mammals (Kanda 2019; Ben-Menahem 2021). As noted by Kanda

(2019), the appearance of pulsatile gonadotropin secretion in evolution has occurred alongside the loss of the C-terminal tail of the GnRH receptor, making it resistant to desensitization, as well as an increased importance of kisspeptin in fertility control.

It now seems clear that a population of kisspeptin neurons in the arcuate nucleus represents the mammalian GnRH pulse generator (Herbison 2018; Plant 2019; Goodman et al. 2022). Studies in mice have demonstrated that synchronously firing kisspeptin neurons project to the clustered GnRH neuron distal dendrons just above the median eminence to evoke the discrete episodes of GnRH release that drive pulsatile LH secretion (Herbison 2018, 2021). This is a highly unusual mechanism of neuronal synchronization but provides an elegant solution to the problem of synchronizing the activity of a GnRH1 neuronal population that has its cell bodies scattered along the full extent of the migratory continuum. As pulsatile GnRH secretion probably only occurs in mammals, this mechanism of synchronization may not be shared among nonmammalian vertebrates. Although multiple kisspeptin paralogs are expressed in vertebrates (Dufour et al. 2020), the absence of kisspeptin in birds and its less critical role in fish species (Nakajo et al. 2018; Dufour et al. 2020; Ogawa et al. 2022) indicate that the kisspeptin pulse generator may indeed be specific to mammals.

As noted in the GnRH1 neuron morphology section, the degree to which the rodent pulse generator mechanism can be generalized to all mammals remains unclear. Indeed many mammals have GnRH1 neuron cell bodies residing near the arcuate kisspeptin neurons within the mediobasal hypothalamus (MBH). This raises the possibility that the arcuate kisspeptin neurons could innervate MBH GnRH1 neurons directly in a more conventional manner to initiate pulsatile GnRH secretion (Herbison 2021). For example, in sheep, the MBH population of GnRH1 neurons is preferentially activated at the time of an LH pulse, and arcuate kisspeptin neurons are known to project directly to MBH GnRH1 cell bodies in this species as well as in humans (Goodman et al. 2022). Hence, although the arcuate kisspeptin neurons are very likely the conserved GnRH pulse generator, how they bring about episodic GnRH secretion may vary slightly among mammalian species (Herbison 2021).

It is also worth considering whether species differences may exist in the surge behavior of GnRH1 neurons. In many spontaneously ovulating mammals, the preoptic population of kisspeptin neurons has a critical role in integrating circulating estradiol concentrations and, where appropriate, circadian inputs to drive the synchronous activation of GnRH neurons to trigger ovulation (Goodman et al. 2022; Kauffman 2022). It is interesting to note that while the initiating stimulus for ovulation is very different in induced ovulators, they are thought to ultimately activate preoptic kisspeptin neurons to evoke the GnRH surge (Inoue et al. 2011; El Allali et al. 2017). Hence, much like the "arcuate kisspeptin-GnRH1 neuron pulse module," there appears to be a common mammalian "preoptic kisspeptin-GnRH1 neuron surge module." Notably in rodents, these different modules are thought to operate at two separate compartments of the GnRH1 neuron: pulses at the distal dendron and surges at the GnRH1 neuron cell body (Herbison 2020) (Fig. 5.3). In addition, some mammalian species have co-opted special forms of fertility control, such as seasonal and pheromonal regulation, to suit their particular needs. This may result from species differences in the afferent control of the "kisspeptin-GnRH1 neuron pulse and surge modules" rather than fundamental differences in GnRH1 neurons themselves (Herbison 2016).

The absence of a synchronizing kisspeptin population for generating the GnRH surge in nonmammalian species suggests that GnRH1 neurons in these species use different mechanisms. In this light, it is very interesting to note that the hypophysiotropic GnRH1 neurons in cichlid fish utilize gap junctions to synchronize their behavior (Ma et al. 2015). It is also possible that a more prominent synchronizing role exists among the pituitary gonadotrophs in fish (Ogawa et al. 2022). It was also recently reported that estradiol may act directly on *esr2a*-expressing medaka GnRH1 neurons to increase their activity at the time of the surge (Ikegami et al. 2022). Functional gap junctions do not exist between GnRH1 neurons in mice (Campbell et al. 2011), and it is now well accepted that GnRH1 neurons do not express the relevant estrogen receptor in mammals (Herbison 2015). Thus, prior to the development of the synchronizing kisspeptin systems in mammals, it seems that GnRH1 neurons used other, more direct, mechanisms to control their firing to evoke ovulation or spawning.

5.5.6 Persistence of Extrahypothalamic GnRH1 Neurons in Primates

A substantial species difference exists between primates and other mammals in that large numbers of neurons expressing GnRH1 are found in extrahypothalamic brain regions in adult monkeys and humans (Rance et al. 1994; Campbell et al. 2022). Investigations in human embryos have identified that, in addition to the normal "ventral" migratory continuum into the hypothalamus, a second "dorsal" migratory route exists through which thousands of GnRH1 neurons populate extrahypothalamic brain regions (Casoni et al. 2016). This is highly reminiscent of the monkey, where "early" GnRH1 neurons populate multiple parts of the limbic and basal forebrain, while "late" migrating GnRH1 neurons represent the hypothalamic hypophysiotropic GnRH1 neurons (Quanbeck et al. 1997; Terasawa et al. 2001). In addition, it seems very likely that some populations of cholinergic neurons that originate and develop from within the brain go on to coexpress GnRH1 (Skrapits et al. 2021). These different GnRH1 neural populations have been referred to as "type 1," expressing high levels of *GNRH1* mRNA and representing the hypophysiotropic cell population; "type 2", expressing low levels of *GNRH1,* found mostly in the septum and bed nucleus of the stria terminalis (BNST); and "type 3" cells, with intermediate *GNRH1* mRNA levels, located primarily in the basal ganglia (Rance et al. 1994; Skrapits et al. 2021). Extrahypothalamic GnRH1 neurons originating from within the brain are also found in some regions of the mouse brain during postnatal development, but cease to express GnRH1 peptide as adults (Skynner et al. 1999).

The roles of GnRH1 coexpressed in extrahypothalamic neurons in primates are unknown (Skrapits et al. 2021). Given the wide and variable use of other "hypothalamic releasing factors" such as somatostatin and corticotropin-releasing factor as neuropeptides in the nervous system, there is unlikely to be any role for extrahypothalamic neurons coexpressing GnRH1 in fertility control. Interestingly, the cholinergic neurons of the human putamen coexpressing GNRH1 also have *GNRHR1* transcripts, suggesting an autocrine-type role for GnRH in these cells. Another interesting feature of types 2 and 3 neurons in adult primates, and in developing rodents, is that they synthesize the GnRH1 cleavage peptide $GnRH_{1-5}$ (Quanbeck et al. 1997; Skynner et al. 1999; Skrapits et al. 2021). The $GnRH_{1-5}$ peptide likely acts on GPR101 and has been found to have numerous actions in the brain and periphery, including some related to the control of reproductive behavior and hormone secretion (Cho-Clark et al. 2023). The observation that GnRH1 is coexpressed by extrahypothalamic neurons primarily in adult primates suggests roles for these neurons in higher cognitive functioning. Interestingly, there is gathering momentum for the potential therapeutic use of GnRH in neurodevelopmental cognitive disorders such as Down's syndrome and Alzheimer's disease (Prevot et al. 2023).

5.6 Summary

Gene duplication episodes in Bilateria and again in arthropods generated the GnRH-AKH-corazonin superfamily of peptides, which have now been employed for a billion years in different forms of intercellular communication. Whole-genome duplication events and the loss of specific genes in certain species have further resulted in a rich tapestry of GnRH-like signaling across evolution. The huge diversity of known roles involving the GnRH superfamily indicates that, like many peptide signaling systems, there was no common or overarching role early in evolution. Instead, different cells in different organisms used the GnRH family peptides for whatever purpose was advantageous. However, one common early feature would appear to be the use of the GnRH-AKH-corazonin superfamily in a neuroendocrine role, with the expression of the peptides in neurons that projected to or released their peptides for actions at distant organs. In the case of AKH and corazonin, a conserved role appears in the integration and control of energy metabolism. There is little evidence for these peptides being directly involved or being critical for reproduction in invertebrates. The naming of their mammalian homolog as LHRH has perhaps been overly burdensome for the course of AKH and corazonin exploration.

It seems that it is only from the time of urochordate evolution onward that the "GnRH neuron" becomes apparent in the control of fertility. This is first observed in the tunicates, where the many different GnRH peptides of *Ciona intestinalis* are detected to have a role in the surge-like release of sperm and eggs. Associated with this, there is evidence of the first olfactory placode from where GnRH neurons are born in the larval state and perhaps, even, the possibility of GnRH neuron migration down the dorsal strand plexus to directly control gonadal function.

The role of the hypophysiotropic GnRH neuron in the control of vertebrate fertility is undisputed. Nevertheless, the appearance of GnRH2 and GnRH3 following vertebrate whole-genome duplication has provided a more complex landscape for GnRH signaling in the brain. Due primarily to the lack of genetically manipulable animal models, the functions of GnRH2, the most conserved of the decapeptides, remain unclear. However, there are intriguing reminiscences to AKH and corazonin in some of its proposed roles in coordinating reproductive behavior and energy status. It is unlikely that GnRH2 has any direct role in fertility control, and, again, its name has probably been more of a hindrance than a help in deciphering the functions of this midbrain population of neurons. The GnRH3 of fish species is perhaps more enigmatic, given its similar placodal origins to GnRH1 neurons and the observation that GnRH3 neurons can assume the hypophysiotropic role of GnRH1 neurons in fish that have lost *Gnrh1*. In addition to the GnRH paralogs, it is also evident that many extrahypothalamic neuron populations coexpress GnRH1 and this can persist into adulthood in primate species. Again, as is often the case in development, this is likely to represent a cell population using whatever signaling program is available and advantageous rather than representing any unified role of extrahypothalamic GnRH in fertility control.

Finally, it is interesting to note that despite the huge commonalities between vertebrate GnRH1 neurons, there are subtle differences that may be insightful into their functioning. In particular, this includes the mammalian need for a pulse generator and the interesting way in which this has been achieved through synchronizing kisspeptin neurons that project to nearby GnRH neurons. Similarly, it includes the use of a second kisspeptin neuron population for surge induction in different types of ovulators. These kisspeptin neuron populations do not appear to be necessary in nonmammalian vertebrates, where, indeed, more direct mechanisms of GnRH1 neuron synchronization and control appear to exist. Presumably, the necessity for pulsatility and more complex control of the surge mechanism in mammals has required the establishment of a more sophisticated "neocortex-like" layer of kisspeptin neurons to control the GnRH1 neuron.

Acknowledgments The author thanks Prof. Yoshitaka Oka, University of Tokyo, and the members of the Herbison laboratory for commenting on an earlier version of the manuscript.

References

Abe H, Oka Y (2000) Modulation of pacemaker activity by salmon gonadotropin-releasing hormone (sGnRH) in terminal nerve (TN)-GnRH neurons. J Neurophysiol 83(5):3196–3200

Abitua PB, Gainous TB, Kaczmarczyk AN, Winchell CJ, Hudson C, Kamata K, Nakagawa M, Tsuda M, Kusakabe TG, Levine M (2015) The pre-vertebrate origins of neurogenic placodes. Nature 524(7566):462–465

Adams BA, Tello JA, Erchegyi J, Warby C, Hong DJ, Akinsanya KO, Mackie GO, Vale W, Rivier JE, Sherwood NM (2003) Six novel gonadotropin-releasing hormones are encoded as triplets on each of two genes in the protochordate, Ciona intestinalis. Endocrinology 144(5):1907–1919

Amoss M, Burges R, Blackwell R, Vale W, Fellows R, Guillemin R (1971) Purification, amino acid composition and N-terminus of the hypothalamic luteinizing hormone releasing hormone factor (LRF) of ovine origin. Biochem Biophys Res Commun 44:205–210

Barnett DK, Bunnell TM, Millar RP, Abbott DH (2006) Gonadotropin-releasing hormone II stimulates female sexual behavior in marmoset monkeys. Endocrinology 147(1):615–623
Barry J (1979) Immunohistochemistry of luteinizing hormone-releasing hormone-producing neurons of the vertebrates. Int Rev Cytol 60:179–221
Ben-Menahem D (2021) GnRH-related neurohormones in the fruit fly Drosophila melanogaster. Int J Mol Sci 22(9)
Campbell RK, Satoh N, Degnan BM (2004) Piecing together evolution of the vertebrate endocrine system. Trends Genet 20(8):359–366
Campbell RE, Ducret E, Porteous R, Liu X, Herde MK, Wellerhaus K, Sonntag S, Willecke K, Herbison AE (2011) Gap junctions between neuronal inputs but not gonadotropin-releasing hormone neurons control estrous cycles in the mouse. Endocrinology 152(6):2290–2301
Campbell RE, Coolen LM, Hoffman GE, Hrabovszky E (2022) Highlights of neuroanatomical discoveries of the mammalian gonadotropin-releasing hormone system. J Neuroendocrinol 34(5):e13115
Casoni F, Malone SA, Belle M, Luzzati F, Collier F, Allet C, Hrabovszky E, Rasika S, Prevot V, Chedotal A, Giacobini P (2016) Development of the neurons controlling fertility in humans: new insights from 3D imaging and transparent fetal brains. Development 143(21):3969–3981
Cho-Clark MJ, Watkins A, Wu TJ (2023) The role of GnRH metabolite, GnRH-(1-5), in endometrial cancer. Front Endocrinol (Lausanne) 14:1183278
Coen CW, Bennett NC, Holmes MM, Faulkes CG (2021) Neuropeptidergic and neuroendocrine systems underlying eusociality and the concomitant social regulation of reproduction in naked mole-rats: a comparative approach. Adv Exp Med Biol 1319:59–103
Cottrell EC, Campbell RE, Han SK, Herbison AE (2006) Postnatal remodeling of dendritic structure and spine density in gonadotropin-releasing hormone neurons. Endocrinology 147(8):3652–3661
Desaulniers AT, Cederberg RA, Lents CA, White BR (2017) Expression and role of gonadotropin-releasing hormone 2 and its receptor in mammals. Front Endocrinol (Lausanne) 8:269
Dufour S, Querat B, Tostivint H, Pasqualini C, Vaudry H, Rousseau K (2020) Origin and evolution of the neuroendocrine control of reproduction in vertebrates, with special focus on genome and gene duplications. Physiol Rev 100(2):869–943
El Allali K, El Bousmaki N, Ainani H, Simonneaux V (2017) Effect of the Camelid's seminal plasma ovulation-inducing factor/beta-NGF: a Kisspeptin target hypothesis. Front Vet Sci 4:99
Fernald RD, White RB (1999) Gonadotropin-releasing hormone genes: phylogeny, structure, and functions. Front Neuroendocrinol 20(3):224–240
Fujimori C, Sugimoto K, Ishida M, Yang C, Kayo D, Tomihara S, Sano K, Akazome Y, Oka Y, Kanda S (2024) Long-lasting redundant gnrh1/3 expression in GnRH neurons enabled apparent switching of paralog usage during evolution. iScience 27(3):109304
Fukumura K, Konuma T, Tsukamoto Y, Nagata S (2018) Adipokinetic hormone signaling determines dietary fatty acid preference through maintenance of hemolymph fatty acid composition in the cricket Gryllus bimaculatus. Sci Rep 8(1):4737
Goldsmith PC, Thind KK, Song T, Kim EJ, Boggan JE (1990) Location of the neuroendocrine gonadotropin-releasing hormone neurons in the monkey hypothalamus by retrograde tracing and immunostaining. J Neuroendocrinol 2:157–168
Goodman RL (2015) Neuroendocrine control of goandotropin secretion: comparative aspects. Knobil and Neill's physiology of reproduction. T. M. Plant and A. J Zeleznik 2:1537–1574
Goodman RL, Herbison AE, Lehman MN, Navarro VM (2022) Neuroendocrine control of gonadotropin-releasing hormone: pulsatile and surge modes of secretion. J Neuroendocrinol 34(5):e13094
Haneda K, Oka Y (2008) Coordinated synchronization in the electrically coupled network of terminal nerve gonadotropin-releasing hormone neurons as demonstrated by double patch-clamp study. Endocrinology 149(7):3540–3548
Hansen KK, Stafflinger E, Schneider M, Hauser F, Cazzamali G, Williamson M, Kollmann M, Schachtner J, Grimmelikhuijzen CJ (2010) Discovery of a novel insect neuropeptide signaling

system closely related to the insect adipokinetic hormone and corazonin hormonal systems. J Biol Chem 285(14):10736–10747

Hasunuma I (2023) Central regulation of reproduction in amphibians. J Exp Zool A Ecol Integr Physiol 341:219

Herbison AE (2015) Physiology of the adult GnRH neuronal network. Knobil and Neill's physiology of reproduction. T. M. Plant and A. J Zeleznik 1:399–467

Herbison AE (2016) Control of puberty onset and fertility by gonadotropin-releasing hormone neurons. Nat Rev Endocrinol 12(8):452–466

Herbison AE (2018) The gonadotropin-releasing hormone pulse generator. Endocrinology 159(11):3723–3736

Herbison AE (2020) A simple model of estrous cycle negative and positive feedback regulation of GnRH secretion. Front Neuroendocrinol 57:100837

Herbison AE (2021) The dendron and episodic neuropeptide release. J Neuroendocrinol e13024

Herde MK, Herbison AE (2015) Morphological characterization of the action potential initiation segment in GnRH neuron dendrites and axons of male mice. Endocrinology 156(11):4174–4186

Herde MK, Iremonger KJ, Constantin S, Herbison AE (2013) GnRH neurons elaborate a long-range projection with shared axonal and dendritic functions. J Neurosci 33(31):12689–12697

Hozumi A, Matsunobu S, Mita K, Treen N, Sugihara T, Horie T, Sakuma T, Yamamoto T, Shiraishi A, Hamada M, Satoh N, Sakurai K, Satake H, Sasakura Y (2020) GABA-induced GnRH release triggers chordate metamorphosis. Curr Biol 30(8):1555–1561 e1554

Hughson BN (2021) The glucagon-like adipokinetic hormone in Drosophila melanogaster - biosynthesis and secretion. Front Physiol 12:710652

Ikegami K, Kajihara S, Umatani C, Nakajo M, Kanda S, Oka Y (2022) Estrogen upregulates the firing activity of hypothalamic gonadotropin-releasing hormone (GnRH1) neurons in the evening in female medaka. J Neuroendocrinol 34(4):e13101

Inoue N, Sasagawa K, Ikai K, Sasaki Y, Tomikawa J, Oishi S, Fujii N, Uenoyama Y, Ohmori Y, Yamamoto N, Hondo E, Maeda K, Tsukamura H (2011) Kisspeptin neurons mediate reflex ovulation in the musk shrew (Suncus murinus). Proc Natl Acad Sci USA 108(42):17527–17532

Iremonger KJ, Herbison AE (2012) Initiation and propagation of action potentials in gonadotropin-releasing hormone neuron dendrites. J Neurosci 32(1):151–158

Jansen HT, Hileman SM, Lubbers LS, Kuehl DE, Jackson GL, Lehman MN (1997) Identification and distribution of neuroendocrine gonadotropin-releasing hormone neurons in the ewe. Biol Reprod 56:655–662

Jekely G (2013) Global view of the evolution and diversity of metazoan neuropeptide signaling. Proc Natl Acad Sci USA 110(21):8702–8707

Kanda S (2019) Evolution of the regulatory mechanisms for the hypothalamic-pituitary-gonadal axis in vertebrates-hypothesis from a comparative view. Gen Comp Endocrinol 284:113075

Karigo T, Oka Y (2013) Neurobiological study of fish brains gives insights into the nature of gonadotropin-releasing hormone 1-3 neurons. Front Endocrinol (Lausanne) 4:177

Kasten TL, White SA, Norton TT, Bond CT, Adelman JP, Fernald RD (1996) Characterization of two new preproGnRH mRNAs in the tree shrew: first direct evidence for mesencephalic GnRH gene expression in a placental mammal. Gen Comp Endocrinol 104(1):7–19

Kauffman AS (2022) Neuroendocrine mechanisms underlying estrogen positive feedback and the LH surge. Front Neurosci 16:953252

Kauffman AS, Rissman EF (2004) The evolutionarily conserved gonadotropin-releasing hormone II modifies food intake. Endocrinology 145(2):686–691

Kavanaugh SI, Root AR, Sower SA (2005) Distribution of gonadotropin-releasing hormone (GnRH) by in situ hybridization in the tunicate Ciona intestinalis. Gen Comp Endocrinol 141(1):76–83

Khan Z, Tondravi M, Oliver R, Vonhoff FJ (2021) Drosophila Corazonin neurons as a hub for regulating growth, stress responses, ethanol-related behaviors, copulation persistence and sexually dimorphic reward pathways. J Dev Biol 9(3)

King JC, Anthony EL (1984) LHRH neurons and their projections in humans and other mammals: species comparisons. Peptides 5(Suppl 1):195–207

Kishner M, Habaz L, Meshnik L, Meidan TD, Polonsky A, Ben-Zvi A (2022) Gonadotropin-releasing hormone-like receptor 2 inversely regulates somatic proteostasis and reproduction in Caenorhabditis elegans. Front Cell Dev Biol 10:951199

Kubrak OI, Lushchak OV, Zandawala M, Nassel DR (2016) Systemic corazonin signalling modulates stress responses and metabolism in Drosophila. Open Biol 6(11)

Kuljis RO, Advis JP (1989) Immunocytochemical and physiological evidence of a synapse between dopamine- and luteinizing hormone-releasing hormone-containing neurons in the ewe median eminence. Endocrinology 124:1579–1581

Lebreton S, Mansourian S, Bigarreau J, Dekker T (2016) The adipokinetic hormone receptor modulates sexual behavior, pheromone perception and pheromone production in a sex-specific and starvation-dependent manner in Drosophila melanogaster. Front Ecol Evol 3. https://doi.org/10.3389/fevo.2015.00151

Lee G, Park JH (2004) Hemolymph sugar homeostasis and starvation-induced hyperactivity affected by genetic manipulations of the adipokinetic hormone-encoding gene in Drosophila melanogaster. Genetics 167(1):311–323

Lehman MN, Robinson JE, Karsch FJ, Silverman AJ (1986) Immunocytochemical localization of luteinizing hormone-releasing hormone (LHRH) pathways in the sheep brain during anestrus and the mid-luteal phase of the estrous cycle. J Comp Neurol 244(1):19–35

Lindemans M, Liu F, Janssen T, Husson SJ, Mertens I, Gade G, Schoofs L (2009) Adipokinetic hormone signaling through the gonadotropin-releasing hormone receptor modulates egg-laying in Caenorhabditis elegans. Proc Natl Acad Sci USA 106(5):1642–1647

Liu X, Yeo SH, McQuillan HJ, Herde MK, Hessler S, Cheong I, Porteous R, Herbison AE (2021) Highly redundant neuropeptide volume co-transmission underlying episodic activation of the GnRH neuron dendron. elife 10

Ma Y, Juntti SA, Hu CK, Huguenard JR, Fernald RD (2015) Electrical synapses connect a network of gonadotropin releasing hormone neurons in a cichlid fish. Proc Natl Acad Sci USA 112(12):3805–3810

Maney DL, Richardson RD, Wingfield JC (1997) Central administration of chicken gonadotropin-releasing hormone-II enhances courtship behavior in a female sparrow. Horm Behav 32(1):11–18

McKay FM, McCoy CJ, Crooks B, Marks NJ, Maule AG, Atkinson LE, Mousley A (2022) In silico analyses of neuropeptide-like protein (NLP) profiles in parasitic nematodes. Int J Parasitol 52(1):77–85

Millar R, Lowe S, Conklin D, Pawson A, Maudsley S, Troskie B, Ott T, Millar M, Lincoln G, Sellar R, Faurholm B, Scobie G, Kuestner R, Terasawa E, Katz A (2001) A novel mammalian receptor for the evolutionarily conserved type II GnRH. Proc Natl Acad Sci USA 98(17):9636–9641

Moore AM, Prescott M, Czieselsky K, Desroziers E, Yip SH, Campbell RE, Herbison AE (2018) Synaptic innervation of the GnRH neuron distal dendron in female mice. Endocrinology 159(9):3200–3208

Nakajo M, Kanda S, Karigo T, Takahashi A, Akazome Y, Uenoyama Y, Kobayashi M, Oka Y (2018) Evolutionally conserved function of Kisspeptin neuronal system is nonreproductive regulation as revealed by nonmammalian study. Endocrinology 159(1):163–183

Northcutt RG, Muske LE (1994) Multiple embryonic origins of gonadotropin-releasing hormone (GnRH) immunoreactive neurons. Brain Res Dev Brain Res 78(2):279–290

Ogawa S, Yamamoto N, Hagio H, Oka Y, Parhar IS (2022) Multiple gonadotropin-releasing hormone systems in non-mammalian vertebrates: ontogeny, anatomy, and physiology. J Neuroendocrinol 34(5):e13068

Okawa N, Shimai K, Ohnishi K, Ohkura M, Nakai J, Horie T, Kuhara A, Kusakabe TG (2020) Cellular identity and Ca(2+) signaling activity of the non-reproductive GnRH system in the Ciona intestinalis type A (Ciona robusta) larva. Sci Rep 10(1):18590

Okuyama T, Yokoi S, Abe H, Isoe Y, Suehiro Y, Imada H, Tanaka M, Kawasaki T, Yuba S, Taniguchi Y, Kamei Y, Okubo K, Shimada A, Naruse K, Takeda H, Oka Y, Kubo T, Takeuchi H (2014) A neural mechanism underlying mating preferences for familiar individuals in medaka fish. Science 343(6166):91–94

Plant TM (2019) The neurobiological mechanism underlying hypothalamic GnRH pulse generation: the role of kisspeptin neurons in the arcuate nucleus. F1000Res 8

Prevot V, Tena-Sempere M, Pitteloud N (2023) New horizons: gonadotropin-releasing hormone and cognition. J Clin Endocrinol Metab 108(11):2747–2758

Quanbeck C, Sherwood NM, Millar RP, Terasawa E (1997) Two populations of luteinizing hormone-releasing hormone neurons in the forebrain of the rhesus macaque during embryonic development. J Comp Neurol 380(3):293–309

Rance NE, Young WS 3rd, McMullen NT (1994) Topography of neurons expressing luteinizing hormone-releasing hormone gene transcripts in the human hypothalamus and basal forebrain. J Comp Neurol 339(4):573–586

Rissman EF, Alones VE, Craig-Veit CB, Millam JR (1995) Distribution of chicken-II gonadotropin-releasing hormone in mammalian brain. J Comp Neurol 357(4):524–531

Robinson JE, Skinner DC, Skinner JD, Haupt MA (1997) Distribution and morphology of luteinising hormone-releasing hormone neurones in a species of wild antelope, the springbok (Antidorcas marsupialis). J Comp Neurol 389(3):444–452

Rochester JR, Chung WC, Hayes TB, Tsai PS (2012) Opposite-sex housing reactivates the declining GnRH system in aged transgenic male mice with FGF signaling deficiency. Am J Physiol Endocrinol Metab 303(12):E1428–E1439

Sakai T, Shiraishi A, Kawada T, Matsubara S, Aoyama M, Satake H (2017) Invertebrate gonadotropin-releasing hormone-related peptides and their receptors: an update. Front Endocrinol (Lausanne) 8:217

Sakai T, Yamamoto T, Matsubara S, Kawada T, Satake H (2020) Invertebrate gonadotropin-releasing hormone receptor signaling and its relevant biological actions. Int J Mol Sci 21(22)

Schally AV, Arimura A, Baba Y, Nair RM, Matsuo H, Redding TW, Debeljuk L (1971) Isolation and properties of the FSH and LH-releasing hormone. Biochem Biophys Res Commun 43(2):393–399

Silverman AJ, Jhamandas J, Renaud LP (1987) Localization of luteinizing hormone-releasing hormone (LHRH) neurons that project to the median eminence. J Neurosci 7(8):2312–2319

Skrapits K, Sarvari M, Farkas I, Gocz B, Takacs S, Rumpler E, Vaczi V, Vastagh C, Racz G, Matolcsy A, Solymosi N, Poliska S, Toth B, Erdelyi F, Szabo G, Culler MD, Allet C, Cotellessa L, Prevot V, Giacobini P, Hrabovszky E (2021) The cryptic gonadotropin-releasing hormone neuronal system of human basal ganglia. elife 10

Skynner MJ, Slater R, Sim JA, Allen ND, Herbison AE (1999) Promoter transgenics reveal multiple gonadotropin-releasing hormone-1-expressing cell populations of different embryological origin in mouse brain. J Neurosci 19(14):5955–5966

Stewart AJ, Katz AA, Millar RP, Morgan K (2009) Retention and silencing of prepro-GnRH-II and type II GnRH receptor genes in mammals. Neuroendocrinology 90(4):416–432

Stone JV, Mordue W, Batley KE, Morris HR (1976) Structure of locust adipokinetic hormone, a neurohormone that regulates lipid utilisation during flight. Nature 263(5574):207–211

Tayler TD, Pacheco DA, Hergarden AC, Murthy M, Anderson DJ (2012) A neuropeptide circuit that coordinates sperm transfer and copulation duration in Drosophila. Proc Natl Acad Sci USA 109(50):20697–20702

Temple JL, Millar RP, Rissman EF (2003) An evolutionarily conserved form of gonadotropin-releasing hormone coordinates energy and reproductive behavior. Endocrinology 144(1):13–19

Terakado K (2001) Induction of gamete release by gonadotropin-releasing hormone in a protochordate, Ciona intestinalis. Gen Comp Endocrinol 124(3):277–284

Terasawa E, Busser BW, Luchansky LL, Sherwood NM, Jennes L, Millar RP, Glucksman MJ, Roberts JL (2001) Presence of luteinizing hormone-releasing hormone fragments in the rhesus monkey forebrain. J Comp Neurol 439(4):491–504

Thornquist SC, Langer K, Zhang SX, Rogulja D, Crickmore MA (2020) CaMKII measures the passage of time to coordinate behavior and motivational state. Neuron 105(2):334–345 e339

Tobet SA, Chickering TW, Sower SA (1996) Relationship of gonadotropin-releasing hormone (GnRH) neurons to the olfactory system in developing lamprey (Petromyzon marinus). J Comp Neurol 376(1):97–111

Trudeau VL, Somoza GM (2020) Multimodal hypothalamo-hypophysial communication in the vertebrates. Gen Comp Endocrinol 293:113475

Tsutsui H, Yamamoto N, Ito H, Oka Y (1998) GnRH-immunoreactive neuronal system in the presumptive ancestral chordate, Ciona intestinalis (Ascidian). Gen Comp Endocrinol 112(3):426–432

Umatani C, Oka Y (2019) Multiple functions of non-hypophysiotropic gonadotropin releasing hormone neurons in vertebrates. Zool Lett 5:23

Umatani C, Yoshida N, Yamamoto E, Akazome Y, Mori Y, Kanda S, Okubo K, Oka Y (2022) Co-existing neuropeptide FF and gonadotropin-releasing hormone 3 coordinately modulate male sexual behavior. Endocrinology 163(2)

Urbanski HF, White RB, Fernald RD, Kohama SG, Garyfallou VT, Densmore VS (1999) Regional expression of mRNA encoding a second form of gonadotropin-releasing hormone in the macaque brain. Endocrinology 140(4):1945–1948

Van der Auwera P, Frooninckx L, Buscemi K, Vance RT, Watteyne J, Mirabeau O, Temmerman L, De Haes W, Fancsalszky L, Gottschalk A, Raizen DM, Nelson MD, Schoofs L, Beets I (2020) RPamide neuropeptides NLP-22 and NLP-2 act through GnRH-like receptors to promote sleep and wakefulness in C. elegans. Sci Rep 10(1):9929

Wayne NL, Kuwahara K, Aida K, Nagahama Y, Okubo K (2005) Whole-cell electrophysiology of gonadotropin-releasing hormone neurons that express green fluorescent protein in the terminal nerve of transgenic medaka (Oryzias latipes). Biol Reprod 73(6):1228–1234

White RB, Eisen JA, Kasten TL, Fernald RD (1998) Second gene for gonadotropin-releasing hormone in humans. Proc Natl Acad Sci USA 95(1):305–309

Wray S (2010) From nose to brain: development of gonadotrophin-releasing hormone-1 neurones. J Neuroendocrinol 22(7):743–753

Xia W, Smith O, Zmora N, Xu S, Zohar Y (2014) Comprehensive analysis of GnRH2 neuronal projections in zebrafish. Sci Rep 4:3676

Yamamoto N, Oka Y, Kawashima S (1997) Lesions of gonadotropin-releasing hormone-immunoreactive terminal nerve cells: effects on the reproductive behavior of male dwarf gouramis. Neuroendocrinology 65(6):403–412

Ybarra N, Hemond PJ, O'Boyle MP, Suter KJ (2011) Spatially selective, testosterone-independent remodeling of dendrites in gonadotropin-releasing hormone (GnRH) neurons prepubertally in male rats. Endocrinology 152(5):2011–2019

Yeo S-H, Herde MK, Herbison AE (2021) Morphological assessment of GABA and glutamate inputs to GnRH neurons in intact female mice using expansion microscopy. J Neuroendocrinol (in revision)

Yip SH, Campos P, Liu X, Porteous R, Herbison AE (2021) Innervation of GnRH neuron distal projections and activation by Kisspeptin in a new GnRH-Cre rat model. Endocrinology 162(1)

Zandawala M, Tian S, Elphick MR (2018) The evolution and nomenclature of GnRH-type and corazonin-type neuropeptide signaling systems. Gen Comp Endocrinol 264:64–77

Thyrotropin-Releasing Hormone (TRH)-Containing Neuronal System in the Brain: Organization, Function and Evolution

6

Praful S. Singru, Sumela Basu, Omprakash Singh, and Abhinav Srivastava

Abstract

Thyrotropin-releasing hormone (TRH) is a crucial regulator of the hypothalamus-pituitary-thyroid-axis and metabolism in mammals. TRH is highly conserved since it is detected in the genome/nervous system of phylogenetic clades across protostomes to deuterostomes. However, variations in the amino acid (aa) sequence, neuroanatomical distribution, and functional properties are quite common. While the 5–8 aa TRH-like peptides have been identified in protostomes, 4 aa TRH with pyroglutamate residue appears first in deuterostomes like echinoderms. However, its tripeptide amide nature was encountered first in *Branchiostoma* and vertebrates. In vertebrates, TRH is abundantly expressed in the brains of mammals and nonmammals. The organization of TRH-containing elements in the brain of cyclostomes through mammals shows profound evolutionary changes. Distinct hypophysiotropic TRHergic system in the brain of mammals and birds may have evolved from the cerebrospinal-fluid-contacting TRH neurons in the preoptic area/hypothalamus of lower vertebrates. The non-hypophysiotropic TRH system seems engaged in regulating the hypophysiotropic dopamine neurons in teleosts and mammals. Dense TRHergic neuronal groups were found in the brains of teleosts and mammals, followed by other vertebrates. TRH functions as a multifunctional hypophysiotropic factor and serves as an essential regulator of metamorphosis as well as metabolism in vertebrates. The central inhibitory action of neuropeptide Y (NPY) on the hypophy-

P. S. Singru (✉) · S. Basu · O. Singh · A. Srivastava
School of Biological Sciences, National Institute of Science Education and Research (NISER)-Bhubaneswar, Khurda, Odisha, India

Homi Bhabha National Institute (HBNI), Mumbai, India
e-mail: pssingru@niser.ac.in

V. Grinevich, R. F. Oliveira (eds.), *Evolutionary and Comparative Neuroendocrinology*, Masterclass in Neuroendocrinology 17,
https://doi.org/10.1007/978-3-031-80209-6_6

siotropic TRH neurons in mammals is conserved in birds. However, the NPY-TRH response to changing energy states is more rapid in birds. Its ancient origin, neurotransmitter-like characteristics, wide neuroanatomical presence, and conserved features suggest TRH to be an essential central regulator.

Keywords

Amphibians · Birds · Brain · Cyclostomes · Elasmobranch · Energy balance · Mammals · Neuroanatomy · Reptiles · Teleosts · TRH · TRH-dopamine interaction

6.1 Introduction

Pioneering studies, independently conducted in the laboratories of Drs. Andrew Schally and Roger Guillemin in 1969, demonstrated that hypothalamic extracts released different hormones from the pituitary. These were ultimately determined to be "releasing factors" or hormones. The releasing factor (RF) that stimulated the release of thyroid-stimulating hormone (TSH) was referred to as TSH-RF, TRF, or thyrotropin-releasing hormone (TRH). Later, the structure of TRH was elucidated and found to be a tripeptide amide (pyro-glutamine-histidine-proline-NH2) (Reichlin 1989). Early studies established that TRH drives the hypothalamus-pituitary-thyroid (HPT) axis, acts on the TSH cells in the pituitary gland, and regulates the synthesis and release of TSH (Harris et al. 1978).

In addition to its well-established role as a regulator of TSH and prolactin (PRL) secretion, TRH is widely employed by neural circuits in the brain, and its role in the regulation of a range of functions has been suggested. TRH regulates cognition and mood (Alvarez-Salas et al. 2022), physiological activities (respiratory frequency (Greer et al. 1996), cerebral vasodilatation (Koskinen and Bill 1984), thermoregulation (Lin et al. 1980)), and neuroendocrine processes (Fujihara and Shiino 1983). TRH produces distinct behavioral responses, including anticonvulsant effects in animal seizure models (Veronesi et al. 2007) and food reward (Puga et al. 2016). In addition, TRH regulates autonomic function (Riedel and Burke 1988).

TRH is a highly conserved neuropeptide. Using radioimmunoassay, the presence of immunoreactive TRH was demonstrated in the hypothalamic/brain tissue extracts from different vertebrates and amphioxus (Jackson and Reichlin 1974). Although the neuroanatomical distribution and functional significance of TRH have been extensively studied in mammals (Lechan et al. 1986; Segerson et al. 1987; Tsuruo et al. 1987; Merchenthaler et al. 1988; Fleming and Todd 1994; Guldenaar et al. 1996; Pu et al. 1996; Fliers et al. 1994; Mihály et al. 2000), relevant information in nonmammalian brains is limited. Using immunohistochemistry or in situ hybridization histochemistry, TRH/proTRH or proTRH mRNA expression was observed in the brain of cyclostome (sea lamprey, *Petromyzon marinus* (Del Carmen De Andrés et al. 2002)), an elasmobranch (dogfish, *Scyliorhinus canicula* (Teijido et al. 2002)), teleosts (catfish, *Clarias batrachus* (Singh et al. 2019); zebrafish, *Danio rerio* (Díaz et al. 2002); sea bass, *Dicentrarchus labrax* (Batten et al. 1990); and sockeye salmon, *Oncorhynchus nerka* (Ando et al. 1998)), amphibians [frogs, *Xenopus laevis* (Zoeller and Conway 1989; Bidaud et al. 2004), *Rana catesbeiana* (Seki et al. 1983; Mimnagh et al. 1987), and *Rana ridibunda* (Lamacz et al. 1989), toad, *Bufo arenarum* (Miranda and Affanni 2000); newt, *Pleurodeles waltl*, axolotl, *Ambystoma mexicanum*, and salamander, *Ambystoma tigrinum* (Domínguez et al. 2008)], reptiles [turtle, *Pseudemys scripta elegans* and gecko, *Gekko gecko* (López et al. 2008)], and birds (domestic mallard, *Anas platyrhynchos* (Józsa et al. 1988); pigeon, *Columba livia* (Péczely and Kiss 1988); chickens, *Gallus gallus* (Aoki et al. 2007); and zebra finch, *Taeniopygia guttata* (Singh et al. 2023)].

Herein, we review different facets, such as the nature of TRH and TRH-like peptides, the neuroanatomy of TRH-expressing elements, and functional attributes in different areas of the brain, and try to trace the evolutionary history of TRH-containing systems in the brain across different phylogenetic clades.

6.2 TRH Gene and Processing of PreproTRH

The structure of the TRH gene in mammals has been extensively studied and reviewed (Lee et al. 1988; Fekete and Lechan 2014). Briefly, the rodents and human genomes contain a single copy of the preproTRH gene. The rat TRH gene is composed of three exons and two introns (intron size: 750 and 450 bp) and encodes a 255 aa preproTRH (Gkonos et al. 1989). Of the three exons, while the first exon gives rise to the 5′ untranslated part of the mRNA, a signal peptide and a part of the amino terminus of the proTRH peptide are produced from the second exon. The remaining portion of the amino-terminal peptide, five copies of the TRH sequence separated by seven non-TRH peptide sequences, and the carboxyl-terminal peptide followed by the 3′ untranslated region of the mRNA are encoded by the third exon (Lee et al. 1988). A TATA box and regulatory elements reside in the promoter of the TRH gene. Site 4 in the human preproTRH gene is equipped with three negative thyroid hormone response elements (TRE) and functions as a cAMP response element (CRE), and in the rat preproTRH gene, this functions as TRE but not CRE. However, the presence of a CRE-2 site has been demonstrated 5′ to site 4, which serves as a primary CRE in the preproTRH gene. In addition, the proximal promoter region of the preproTRH gene contains the binding sites for other factors (Fekete and Lechan 2014).

The structure of the chicken preproTRH gene has also been described (Vandenborne et al. 2005). It contains two introns of 953 bp and 1203 bp. The deduced amino acid sequence of chicken preproTRH contains five exact copies of the TRH progenitor sequence and three copies of sequences resembling the progenitor sequence. Among the five TRH progenitor sequences, four are flanked by prohormone convertase cleavage sites (Lys-Arg or Arg-Arg), whereas the fifth one is flanked at the N-terminus by Lys-His, making it unrecognizable by the known prohormone convertases. Since a proteolytic enzyme that recognizes Lys-His as a cleavage site is unknown in chickens, it was proposed that the fifth TRH progenitor sequence cannot be proteolytically cleaved. Vandenborne et al. (2005) suggested that the gene coding the chicken TRH precursor has undergone frequent mutations during the diversion of birds from other vertebrates. The *T. guttata* proTRH showed 29.8% to 38.2% amino acid (aa) identity with other animals, but the putative progenitor TRH peptide sequence of this bird seems to be highly conserved (Singh et al. 2023).

The *trh* gene in the teleost *Monopterus albus* is 3327 bp in length, with three exons and two introns (Feng et al. 2022). The 270 aa preproTRH in this fish showed the presence of a 21 aa predicted signal peptide and eight copies of the TRH progenitor sequence (Feng et al. 2022). While the similarity in the composition of the *trh* gene in *M. albus* to that of medaka, chicken, rat, and human was observed, an analysis of the preproTRH sequences across vertebrates suggests the conserved nature of TRH (Aoki et al. 2007; Feng et al. 2022). The *trh* gene in *X. laevis* contains at least two exons and an intron (Bulant et al. 1992). Exon 2 encodes a 224 aa preproTRH, which contains seven copies of TRH progenitor sequences (Bulant et al. 1992). Interestingly, the TRH progenitor sequence was seen in the C-terminal hydrophilic region of preproTRH in teleosts and humans (Aoki et al. 2005).

Different animal species show variation in the number of TRH progenitor sequences in the proTRH. For example, the rat proTRH contains five (Gkonos et al. 1989), the human proTRH has six (Nillni 2010), proTRH in *Xenopus* harbors seven, the teleost proTRH contains six to eight (Aoki et al. 2005; Feng et al. 2022), and that in chicken and *T. guttata* showed only four TRH progenitor sequences (Aoki et al. 2007; Singh et al. 2023).

6.3 TRH-Like Peptides in the Nervous System of Invertebrates

Information on TRH-like peptides in invertebrates is confined to that in *Caenorhabditis elegans* and a few insects, and their functional significance is not well understood. TRH-1 activates the TRH-1 receptor in *C. elegans* as well as human TRH receptors (Van Sinay et al. 2017). The peptide is expressed in the pharyngeal motor neurons M4 and M5. While neuron M4 innervates pharyngeal muscle to control isthmus peristalsis, the role of M5 motor neurons is unclear. The TRH-1 neuropeptide seems to have no role in isthmus peristalsis in the worm, but it promotes growth, and its knockdown decreases body volume (Van Sinay et al. 2017). In the brain of the linden bug, *Pyrrhocoris apterus,* 4–6 EFLa immunoreactive neurons and their projections were seen in the lateral protocerebrum (Kotwica-Rolinska et al. 2020). Similarly, in the migratory locust, *Locusta migratoria*, EFLa neurons have been observed in the protocerebrum (Veenstra and Šimo 2020).

6.4 TRH-Containing Systems in the Brain of Vertebrates

With the exception of cyclostomes and reptiles, TRH-ir elements are extensively distributed in the brains of vertebrates, including elasmobranchs, teleosts, amphibians, birds, and mammals. The distribution of cells and fibers in each component of the brain is described below and presented in Table 6.1.

Table 6.1 TRH neuronal groups in the brain of vertebrates

		Fishes					
	Cyclotomata	Elasmobranchii	Teleostei	Amphibia	Reptilia	Aves	Mammalia
	Petromyzon marinus (Del Carmen De Andrés et al. 2002)	*Scyliorhinus canicula* (Teijido et al. 2002)	*Oncorhynchus nerka* (Ando et al. 1998), *Danio rerio* (Díaz et al. 2002), *Clarias batrachus* (Singh et al. 2019)	*Rana catesbeina* (Mimnagh et al. 1987; Seki et al. 1983), *Rana ridibunda* (Lamacz et al. 1989), *Bufo arenarum* (Miranda and Affanni 2000), *Xenopus laevis* (Bidaud et al. 2004), *Pleurodeles waltl*, *Ambystoma mexicanum*, *Ambystoma tigrinum* (Domínguez et al. 2008)	*Pseudemys scripta elegans* and *Gekko geko* (López et al. 2008)	*Anas platyrhynchos* (Józsa et al. 1988), *Columba livia* (Péczely and Kiss 1988), *Gallas gallas* (Aoki et al. 2007), *Taeniopygia guttata* (Singh et al. 2023)	Sprague-Dawley rats (Lechan et al. 1986; Segerson et al. 1987; Pu et al. 1996), Wistar-R-Amsterdam rats (Merchenthaler et al. 1988), albino Swiss rats (Fleming and Todd 1994), male albino rats (Tsuruo et al. 1987), humans (Fliers et al. 1994; Guldenaar et al. 1996; Mihály et al. 2000)
Telencephalon							
Olfactory bulb (OB)	No TRH neuons in OB	TRH neurons in the granule cell layer, Mitral cells TRH -ve, TRH terminals around mitral cells and dense TRH innervation in the glomerular layer (GL) of OB.	Granule cells and fibers in GCL and fibers in GL and MCL (Singh et al. 2019), cells in the inner cell layer (Ando et al. 1998) contain TRH	The GCL in the OB of *Xenopus* showed a high density of proTRH mRNA hybridization signal (Bidaud et al. 2004).	No TRH-ir in OB (Domínguez et al. 2008)	No TRH neurons in OB (Singh et al. 2023)	TRH neurons in the GL and EPL (Lechan et al. 1986; Segerson et al. 1987; Merchenthaler et al. 1988; Tsuruo et al. 1987)

(continued)

Table 6.1 (continued)

Telencephalic hemispheres	Striatum	Dorsal pallium, posterior	Ventral telencephalic subdivisions: central (Vc) and supracommissural (Vs) nuclei of ventral telencephalic area (Díaz et al. 2002; Singh et al. 2019), dorsal (Vd) and intermedialis (Vi) (Singh et al. 2019), ventral (Vv) (Ando et al. 1998; Díaz et al. 2002), and postcommissural (Vp) (Ando et al. 1998; Díaz et al. 2002; Singh et al. 2019)	Amygdala pars medialis (Apm) (Seki et al. 1983; Domínguez et al. 2008), amygdala (Mimnagh et al. 1987; Miranda and Affanni 2000)	Medial amygdala (López et al. 2008)	Diagonal band of Broca (Józsa et al. 1988)	Taenia tecta (Pu et al. 1996), diagonal band of Broca (Segerson et al. 1987; Merchenthaler et al. 1988; Pu et al. 1996)
		Lateral pallium, caudal level	Entopeduncular nucleus (EN) (Singh et al. 2019)	Diagonal band of Broca (Seki et al. 1983; Mimnagh et al. 1987; Miranda and Affanni 2000)			Medial septum (Lechan et al. 1986; Merchenthaler et al. 1988)
				Stiatum (Bidaud et al. 2004)			Septohypothalamic nucleus (Lechan et al. 1986; Merchenthaler et al. 1988; Tsuruo et al. 1987)
				Medial septum (Bidaud et al. 2004; Miranda and Affanni 2000)			Bed nucleus of stria terminalis (Merchenthaler et al. 1988; Pu et al. 1996)
				Nucleus accumbens (Acc) (Bidaud et al. 2004)			Anterior commissure nucleus (Segerson et al. 1987)

				Bed nucleus of the pallial commissure (BN) (Bidaud et al. 2004)			Nucleus accumbens (Lechan et al. 1986; Tsuruo et al. 1987)
							Caudate-putamen complex (Lechan et al. 1986; Segerson et al. 1987; Pu et al. 1996)
							Amygdaloid complex (Segerson et al. 1987; Lechan et al. 1986, Pu et al. 1996)
							Hippocampus (hilus of dentate gyrus, pyramidal cells in CA3 and CA4) (Lechan et al. 1986; Pu et al. 1996)
Diencephalon							
	Medial preoptic nucleus	Preoptic nucleus	Organum vasculosum of the lamina terminalis (OVLT) (Singh et al. 2019)	Preoptic area, anterior (POa) (Domínguez et al. 2008)/anterior preoptic nucleus (Poa) (Bidaud et al. 2004; Seki et al. 1983), pars anterior preoptic nucleus (aNPO) (Mimnagh et al. 1987), nucleus preopticus pars anterior (Miranda and Affanni 2000)	Paraventricular nucleus (PV) (López et al. 2008)	Nucleus praeopticus anterior, dorsal and ventral (Józsa et al. 1988)	Preoptic suprachiasmatic nucleus (Lechan et al. 1986; Segerson et al. 1987; Merchenthaler et al. 1988; Pu et al. 1996)
	Lateral preoptic nucleus	Caudal preoptic region, laterally and close to optic chiasm	Parvocellular preoptic nucleus (Díaz et al. 2002); nucleus preopticus periventricularis (NPP), anterior (NPPa) and posterior (NPPp) (Singh et al. 2019); nucleus preopticus parvocellularis anterioris (PPa) (Ando et al. 1998)	Preoptic area, posterior (POp) (Domínguez et al. 2008)	Periventricular hypothalamic nucleus (Ph) (López et al. 2008)	Periventricular preoptic nucleus (PPN) (Singh et al. 2023; Józsa et al. 1988)/nucleus preopticus periventricularis (Aoki et al. 2007)	SCN (Segerson et al. 1987; Fliers et al. 1994; Guldenaar et al. 1996)

(continued)

Table 6.1 continued

	Nucleus of the postoptic commissure (ncpo)	Thalamus ventromedialis	Magnocellular preoptic nucleus (Ando et al. 1998; Díaz et al. 2002)	Preoptic nucleus dorsal (Lamacz et al. 1989)		Medial preoptic nucleus (MPN) (Singh et al. 2023; Péczely and Kiss 1988; Aoki et al. 2007)	Periventricular nucleus (PeV) (Lechan et al. 1986; Tsuruo et al. 1987; Pu et al. 1996)
	Dorsal hypothalamic nucleus (dHy)	Periventricular cells in the medial, lateral, and intermediate regions of the hypothalamic lobes	Dorsal zone of periventricular hypothalamus (Hd) (Ando et al. 1998)	Anterobasal periventricular area (Lamacz et al. 1989)		Supraoptic nucleus (SON) (Péczely and Kiss 1988)	Paraventricular nucleus (PVN), anterior, medial, and periventricular parvocellular PVN (Segerson et al. 1987; Lechan et al. 1986; Tsuruo et al. 1987; Fliers et al. 1994; Pu et al. 1996)
	Ventral hypothalamic nucleus (vHy)	Nucleus sacci vasculosi	Nucleus postopticus lateralis (NPl) (Singh et al. 2019)	Magnocellular preoptic nucleus (Mg) (Bidaud et al. 2004; Miranda and Affanni 2000)		Ventral supraoptic nucleus (Sov) (Singh et al. 2023)	SON (Pu et al. 1996), dorsal to SON (Fliers et al. 1994)
				Suprachiasmatic nucleus (SC) (Bidaud et al. 2004; Domínguez et al. 2008)		Magnocellular preoptic nucleus (Péczely and Kiss 1988)	Perifornical area (Tsuruo et al. 1987; Guldenaar et al. 1996)
		Posterior tubercle nucleus		Ventral hypothalamus or ventral hypothalamic nucleus (VH) (Bidaud et al. 2004; Domínguez et al. 2008)		SCN (Péczely and Kiss 1988)	Medial preoptic nucleus (Lechan et al. 1986)
		Ventrocaudal region of the posterior hypothalamic lobe	Suprachiasmatic nucleus (SCN) (Díaz et al. 2002; Singh et al. 2019)	Infundibular nucleus (Lamacz et al. 1989; Mimnagh et al. 1987; Miranda and Affanni 2000; Seki et al. 1983)		Medial (mSCN) and visual (vSCN) SCN (Singh et al. 2023; Péczely and Kiss 1988)	Lateral preoptic nucleus (Lechan et al. 1986)
			Nucleus anterior hypothalami (NAH) (Singh et al. 2019)	Posterior tubercle (Bidaud et al. 2004)		Anterior hypothalamic nucleus (AH) (Singh et al. 2023)	Sexually dimorphic nucleus (SDN) (Segerson et al. 1987; Lechan et al. 1986; Merchenthaler et al. 1988; Fliers et al. 1994)

		Nucleus recessus posterioris (NRP) (Singh et al. 2019)	Nucleus of the periventricular organ (NPv) (Bidaud et al. 2004)		PVN (Singh et al. 2023; Józsa et al. 1988; Péczely and Kiss 1988; Aoki et al. 2007)	Anterior hypothalamus (AH) (Merchenthaler et al. 1988; Tsuruo et al. 1987; Guldenaar et al. 1996)
		Lateral hypothalamic nucleus (Díaz et al. 2002)	Thalamus (Domínguez et al. 2008)		Hypothalamic dorsomedial nucleus (DMN) (Singh et al. 2023)	Lateral hypothalamus (LH) (Lechan et al. 1986; Segerson et al. 1987; Merchenthaler et al. 1988; Guldenaar et al. 1996), basal lateral hypothalamus (Pu et al. 1996)
		Anterior tuberal nucleus (Díaz et al. 2002)			Lateral hypothalamic area (LH) (Singh et al. 2023; Józsa et al. 1988)/ regio lateralis hypothalami (Aoki et al. 2007)	Dorsomedial nucleus (DMN) (Segerson et al. 1987; Lechan et al. 1986; Merchenthaler et al. 1988; Tsuruo et al. 1987; Pu et al. 1996)
		Torus lateralis (Díaz et al. 2002)			Periventricular hypothalamic nucleus (Péczely and Kiss 1988; Aoki et al. 2007)	Posterior arcuate nucleus (ARC) (Lechan et al. 1986; Segerson et al. 1987; Merchenthaler et al. 1988; Pu et al. 1996)
		Diffuse nucleus of inferior hypothalamic lobes (Díaz et al. 2002)			Ventromedial hypothalamic nucleus (Peczely and Kiss 1988; Aoki et al. 2007)	Posterior hypothalamus (Merchenthaler et al. 1988)
		Habenula (Díaz et al. 2002)			Dorsomedial anterior thalamic nucleus (Péczely and Kiss 1988)	Ventromedial nucleus (Pu et al. 1996)
		Central posterior thalamic nucleus (Díaz et al. 2002)			Ovoidal nucleus (Peczely and Kiss 1988)	Premammillary nucleus (PM) (Merchenthaler et al. 1988)
		Lateral to periventricular nucleus of the posterior tuberculum (Díaz et al. 2002)			Infundibular nucleus (Péczely and Kiss 1988)	Reticular nucleus of the thalamus (Merchenthaler et al. 1988; Pu et al. 1996)
		Posterior tuberal nucleus (Díaz et al. 2002)				

(continued)

Table 6.1 continued

Mesencephalon							
			Rostrodorsal tegmentum (Díaz et al. 2002)	Optic tectum (Domínguez et al. 2008)	Mesencephalic tegmentum (Tegm) (López et al. 2008)	Central gray[a] (Józsa et al. 1988)	Periaquaductal grey (PAG) (Lechan et al. 1986; Segerson et al. 1987; Merchenthaler et al. 1988; Pu et al. 1996)
			Dorsal posterior tegmental nucleus (DPTN) (Singh et al. 2019)				Raphe magnus (Merchenthaler et al. 1988; Pu et al. 1996)
			Edinger-Westphal nucleus (Díaz et al. 2002)	Mesencephalic tectum (tect) (Bidaud et al. 2004)			
			Torus semicircularis (Ando et al. 1998; Díaz et al. 2002; Singh et al. 2019)				
			Stratum griesum centrale (SGC) (Singh et al. 2019)				
			Stratum album centrale (SAC) (Singh et al. 2019)				
			Nucleus of the lateral leminiscus (Díaz et al. 2002)				
			Interpeduncular nucleus (Díaz et al. 2002)				
			Superior reticular formation (Díaz et al. 2002)				

Rhombencephalon							
		Central gray	Nucleus isthmi (Díaz et al. 2002)	Nucleus reticularis medius (Rm) (Bidaud et al. 2004)	Descending vestibular nucleus (Veds) (López et al. 2008)	No TRH neurons (Singh et al. 2023)	Raphe pallidus (Merchenthaler et al. 1988)
		Superior raphe nucleus	Locus coeruleus (Díaz et al. 2002; Singh et al. 2019)				Raphe obscurus (Merchenthaler et al. 1988)
			Around the medial longidunal fasciculus (MLF) (Singh et al. 2019)				Raphe pyramidal/periolivar (Lechan et al. 1986)
			Facial lobe (Singh et al. 2019)				Pontine nuclei (Tsuruo et al. 1987)
			Glossopharyngeal and vagal lobes (Díaz et al. 2002)				Cochlear nucleus (Lechan et al. 1986)
			Vagal motor nucleus (Díaz et al. 2002) or motor nucleus of vagus nerve (Ando et al. 1998)				External cuneate nucleus (Lechan et al. 1986; Tsuruo et al. 1987)
			Nucleus ambiguus (Singh et al. 2019)				Lateral reticular nucleus (Lechan et al. 1986)
							Dorsal motor nucleus of vagus (Lechan et al. 1986; Segerson et al. 1987; Tsuruo et al. 1987)
Spinal cord							
		Trigeminal spinal nucleus					Superficial dorsal horn (Fleming and Todd 1994; Tsuruo et al. 1987)

[a] Only in one animal out of eight animals studied; -ve, negative

6.4.1 Telencephalon

6.4.1.1 Olfactory Bulb (OB)

Followed by the hypothalamus and spinal cord, the OB in rats is the third-highest TRH-containing CNS site (Kreider et al. 1981; Segerson et al. 1987). In rats, beaded TRH-ir axons were found in the olfactory nerve layer, whereas several small TRH-ir as well as pro-TRH mRNA-expressing perikarya were observed in the glomerular layer (GL) and larger neurons in the external plexiform layer (Lechan et al. 1986; Segerson et al. 1987; Tsuruo et al. 1987; Merchenthaler et al. 1988; Pu et al. 1996). In contrast, no TRH-ir was observed in the OB of *P. marinus* (Del Carmen De Andrés et al. 2002). In the OB of fishes (*S. canicula* and *C. batrachus*), TRH-ir was observed in the granule cells with a dense TRH innervation in the GL (Teijido et al. 2002; Singh et al. 2019). The mitral cells in the OB of *S. canicula* were TRH negative but were contacted by TRH-ir terminals (Teijido et al. 2002). Using in situ hybridization histochemistry, proTRH mRNA expression was observed in the cells of the internal cell layer in the OB of *O. nerka* (Ando et al. 1998) and the internal granular layer in the OB of the *X. laevis* (Bidaud et al. 2004). Species-specific differences were also observed since the OB of *R. catesbeiana* did not show TRH-ir (Mimnagh et al. 1987). While no TRH-ir neurons were found in the OB of *P. waltl*, *A. mexicanum*, and *A. tigrinum*, TRH-ir fibers were seen in the internal granular layer of *A. mexicanum* and *A. tigrinum* (Domínguez et al. 2008). Immunoreactive TRH was detected in the OB of *Gallas domesticus* (Jackson and Reichlin 1974), but isolated TRH-ir fibers were seen in the OB of *T. guttata* (Singh et al. 2023). Except for the cyclostomes, reptiles, and birds, the granule cells in the OB of other vertebrate species studied were TRH-ir. In addition, the TRHergic innervation of GL is seen in the OB of an elasmobranch, a teleost, and mammals. Thus, TRH would appear to be employed in the OB very early during vertebrate evolution; however, the functional significance has not been evaluated.

6.4.1.2 Striatum and Septum

In rats, TRH neurons were observed in the medial and cortical nuclei of the amygdala, the medial septum, the caudate-putamen complex, the septohypothalamic nucleus, the diagonal band of Broca, the bed nucleus of stria terminalis (BNST), the bed nucleus of the anterior commissure, the anterior commissural nucleus, and the nucleus accumbens (Lechan et al. 1986; Segerson et al. 1987; Tsuruo et al. 1987; Merchenthaler et al. 1988; Pu et al. 1996). Striatal TRH neurons project to the globus pallidus, pretectal area, and substantia nigra (SN), whereas septal TRH neurons project to TMN.

A relatively small number of TRH neurons were found in the striatum of *P. marinus* (Del Carmen De Andrés et al. 2002). In teleosts, TRH neurons and fibers were detected in different ventral telencephalic (V) parts, including the dorsal (Vd), central (Vc), ventral (Vv), supracommissural (Vs), postcommissural (Vp), and intermedialis (Vi) and the ventral and lateral parts of the ventral telencephalon (Singh et al. 2019). The Vd is homologous to the mammalian striatum, whereas the Vv and lateral part of V have been considered homologous to mammalian septum

(Yamamoto 2009). TRH cell bodies were also seen in the entopeduncular nucleus in the ventral telencephalon of *C. batrachus* (Singh et al. 2019), which project to the dorsomedial telencephalon (Turner et al. 2016) and may play a role in the regulation of energy balance. ProTRH mRNA expression was observed in the dorsal pallium, lateral septum, striatum, and BNST of *X. laevis* (Bidaud et al. 2004) and the amygdala of amphibians, *A. mexicanus*, *A. tigrinum*, and *R. catesbeina*, with fibers in the pallium, nucleus accumbens, striatum, and septum (Seki et al. 1983; Domínguez et al. 2008). TRH neurons were observed in the diagonal band of Broca in the amphibian brain (Seki et al. 1983; Mimnagh et al. 1987; Miranda and Affanni 2000). The pars medialis of the amygdala has been suggested to be partly equivalent to the BNST in mammals (Domínguez et al. 2008). Therefore, TRH neurons in the amygdala in the urodele amphibians and BNST of mammals may be a conserved feature. In reptiles, no TRH cell bodies but TRH fibers were found in the nucleus accumbens, caudal striatum, globus pallidus, medial septal nucleus, and diagonal band of Broca, but TRH neurons were seen in the medial amygdala (López et al. 2008). In *T. guttata*, TRH fibers were seen in the striatum and septum (Singh et al. 2023).

6.4.2 Preoptic Area (POA) and Hypothalamus

The details of the TRH-containing neuronal groups in the POA/hypothalamus of rodents and nonmammalian vertebrates are given in Table 6.1. The occurrence of TRH neurons in the POA and hypothalamus across all vertebrates has been suggested to be a conserved feature (López et al. 2008). In rodents, TRH neurons were found in the preoptic suprachiasmatic, suprachiasmatic (SCN), periventricular, medial, and lateral preoptic nuclei; perifornical area; sexually dimorphic nucleus; anterior and lateral hypothalamus; dorsomedial nucleus; posterior arcuate nucleus (Lechan et al. 1986; Segerson et al. 1987; Tsuruo et al. 1987); and posterior hypothalamic and premammillary nuclei (Merchenthaler et al. 1988). The projections of the nonhypophysiotropic TRH neurons in the tuberal lateral hypothalamus (TuLH), BNST, perifornical area, and anterior part of the paraventricular nucleus (PVN) are described in Sect. 6.7.1.

In the brain of lamprey, TRH neurons were observed in the medial and lateral preoptic nuclei along the ventricle, the nucleus of the postoptic commissure, and the dorsal and ventral hypothalamic nucleus (Del Carmen De Andrés et al. 2002). In teleosts, neurons in the organum vasculosum of the lamina terminalis (OVLT), the nucleus preopticus periventricularis anterior (NPPa) organized around the preoptic recess, the magnocelluar preoptic nucleus, the dorsal zone of the periventricular nucleus, the lateral part of the postoptic nucleus, the SCN, the nucleus anterior hypothalami, the nucleus recessus posterioris, and the tuberal and lateral hypothalamic nuclei contained TRH-ir (Singh et al. 2019) (Table 6.1). In *O. nerka*, the TRH mRNA-expressing neurons were found in the postcommissural part of the ventral telencephalon (Vp) and nucleus preopticus parvocellularis anterioris (PPa) (Ando et al. 1998). In *amphibians*, TRH neurons were seen in the anterior, posterior, and dorsal preoptic and magnocellular preoptic nuclei; anterobasal periventricular area; SCN; ventral hypothalamic nuclei; posterior tubercle; and nucleus of the PVO (Lamacz et al. 1989; Bidaud et al. 2004; Domínguez et al. 2008). In two species of reptiles, TRH neurons were located in the paraventricular (PV) and the hypothalamic periventricular (Ph) nuclei (López et al. 2008). In birds, the neurons in the anterior, dorsal, and ventral nucleus praeopticus, periventricular and medial preoptic, supraoptic, magnocellular preoptic, SCN, periventricular hypothalamic, anterior and lateral hypothalamic, hypothalamic paraventricular, dorsomedial, ventromedial, ovoid and infundibular nuclei were TRH-ir (Józsa et al. 1988; Péczely and Kiss 1988; Aoki et al. 2007; Singh et al. 2023).

In addition to the well-established role of the PVN TRH neurons as a major neuroendocrine regulator in mammals and birds (see Sect. 6.6), the TRHergic neuronal groups in the POA/hypothalamus of other vertebrates may also participate in neuroendocrine functioning. For example, the number of TRH mRNA-expressing neurons in the Vp and PPa of *O. nerka* showed seasonal variation (Ando et al. 1998). The number of TRH-expressing neurons in these regions was higher during the spawning season and lower prior to smolting (Ando et al. 1998). Similarly, the TRH neurons/fibers in NPPa in the POA of *C. batrachus* showed reproduction-phase-dependent changes (Singh et al. 2019). The TRH-ir in the NPPa of this fish increased from the preparatory and prespawning phases and reached a peak in the spawning phase (Singh et al. 2019).

6.4.3 Midbrain, Hindbrain, and Spinal Cord

The application of immunohistochemistry as well as in situ hybridization histochemistry revealed the presence of TRH/proTRH mRNA-expressing neurons in the periaqueductal gray (PAG) in the midbrain of rats (Lechan et al. 1986; Segerson et al. 1987; Pu et al. 1996). TRH neurons were also seen in the raphe, cochlear, external cuneate, lateral reticular, and pontine nuclei; the dorsal motor nucleus of the vagus; and the dorsal horn of the spinal cord in rats (Segerson et al. 1987; Tsuruo et al. 1987; Pu et al. 1996).

No TRH-ir cell bodies were localized in the midbrain and hindbrain of the lamprey, *P. marinus* (Del Carmen De Andrés et al. 2002). Similarly, no TRH-ir cell bodies were found in the mesencephalon of *S. canicula* (Teijido et al. 2002). However, two TRH neuronal groups were seen in the central gray and superior raphe nucleus in the rostral rhombencephalon and the trigeminal spinal nucleus in the spinal cord of *S. canicula* (Teijido et al. 2002). In teleosts, TRH-ir cell bodies were found in the rostrocauldal tegmentum, dorsal posterior tegmental nucleus, Edinger-Westphal nucleus, torus semicircularis, stratum griesum as well as album centrale, nucleus of the lateral lemniscus, interpeduncular nucleus, and superior reticular formation (Díaz et al. 2002; Singh et al. 2019). In the rhombencephalon of teleosts, TRH neurons were seen in the nucleus isthmi; in the locus coeruleus; around the medial longitudinal fasciculus; in the facial, glossopharyngeal, and vagal lobes; in the vagal motor nucleus; and in the nucleus ambiguus (Ando et al. 1998;

Díaz et al. 2002; Singh et al. 2019). No TRH-ir was seen in the nucleus raphe in teleosts (Singh et al. 2019). In amphibians, proTRH mRNA-expressing neurons were found in the mesencephalic tectum in the mesencephalon and nucleus reticularis medius in the rhombencephalon of *X. laevis* (Bidaud et al. 2004). TRH-ir neurons were found in the optic tectum in the mesencephalon of *A. mexicanum* and *A. tigrinum* (Domínguez et al. 2008). No TRH-ir cell bodies were observed in other mesencephalic and rhombencephalic regions of amphibians (Bidaud et al. 2004; Domínguez et al. 2008). TRH-ir neurons were localized to the mesencephalic tegmentum in *P. scripta elegans* and the descending vestibular nucleus in the rhombencephalon (López et al. 2008). No TRH-ir cell bodies were observed in the midbrain and hindbrain regions of *T. guttata* (Singh et al. 2023), but TRH neurons were seen in the central gray of *A. platyrhynchos* (Józsa et al. 1988). In the spinal cord, TRH neurons were found in the trigeminal spinal nucleus of *P. marinus* (Del Carmen De Andrés et al. 2002). No TRH cells were found in the spinal cord of other vertebrates.

6.4.4 Diversity in the Number of TRH Neuronal Groups across the CNS of Vertebrates

A great variation in the number of TRH-ir neuronal groups was observed in the brains of different vertebrates (Table 6.1). Six TRH-ir neuronal groups have been identified in the brain of *P. marinus*. In *S. canicula*, 13 different TRH neuronal groups were found in the brain. The majority of these groups were located in the telencephalon and ventral diencephalon, but only a few were present in the rostral rhombencephalon and spinal cord. In the brain of amphibians, TRH neurons were found in more than 20 different nuclei/regions. While only five TRH neuronal groups have been identified in the brains of two reptilian species, the brains of birds have so far been reported to contain 19 TRH neuronal groups. In the central nervous system of both teleosts and mammals, nearly 40 TRH-positive neuronal groups have been identified. These neuronal groups dominated in the OB, ventral telencephalon, POA, hypothalamus, midbrain, and hindbrain.

Localizing TRH-Containing Systems in the Brain: Opportunities and Challenges

Defining the anatomical framework for the distribution of TRH-containing elements in the brain is crucial to understanding the significance of the peptide. This also paves the way for establishing the association between the TRH-containing system and other neural elements in the brain. Significant information on the role of TRH in the brain has been generated by localizing TRH-producing structures. Dealing with the localization of TRH using immunohistochemistry offers advantages. TRH is cleaved from the proTRH peptide, and therefore antibodies raised against proTRH or TRH have been routinely used for the localization of TRH-expressing elements in the brain

(Lechan and Jackson 1982; Kádár et al. 2010; Singh et al. 2019). TRH is made of only three amino acids, and due to high specificity with the antisera at both the C- and N-terminals of the tripeptide, the cross-reactivity issue of the polyclonal anti-TRH antisera with other antigens has not been observed in the brain. Therefore, polyclonal anti-TRH antisera have been widely used for the localization of TRH-producing elements in the brain of different animals (Seki et al. 1983; Del Carmen De Andrés et al. 2002; Teijido et al. 2002; Singh et al. 2019, 2023). However, the immunohistochemistry protocol for the localization of TRH in the brain is technically challenging. While using the anti-proTRH antiserum, paraformaldehyde has been routinely used as a fixative (Lechan et al. 1986). Since TRH lacks the reactive amino acid essential for cross-linking by common fixatives such as formaldehyde, its visualization requires the use of a novel fixative. Acrolein reacts with the imidazole group of histidine, which is antigenically silent. Therefore, acrolein either alone or in combination with paraformaldehyde, administered by the transcardial route, has been widely used for the fixation of the brain during the application of the anti-TRH antiserum for immunolabeling (Lechan and Jackson 1982; Mimnagh et al. 1987); as well as several other publications following]. Intracerebroventricular colchicine pretreatment followed by transcardial perfusion with a paraformaldehyde solution containing acrolein produces bright labeling of TRH perikarya. In addition, a combination of fixatives, like paraformaldehyde-picric acid-glutaraldehyde (Tsuruo et al. 1987), Zamboni's fixative (Merchenthaler et al. 1988), or paraformaldehyde and glutaraldehyde (Lamacz et al. 1989; Del Carmen De Andrés et al. 2002), has also been used for the localization of TRH in the brain. In situ hybridization histochemistry has also been widely used for the localization of proTRH mRNA-expressing neuronal groups in the brain of different animals, including rodents (Segerson et al. 1987; Ando et al. 1998; Bidaud et al. 2004).

6.5 Evolution of TRH and Its Hypophysiotropic Activity

Analysis of TRH sequences in the genomes of different animal species across the phylogenetic clades shows the occurrence of TRH-like peptides in protostomes and deuterostomes. In protostomes, the TRH-like peptides contain 5–8 aa, whereas in deuterostomes, TRH has 3–4 aa (Van Sinay et al. 2017). FSEFLGamide has been identified as the ligand for TRH-type receptors in the annelid *Platynereis* (Bauknecht and Jékely 2015). TRH-like peptides have several similarities with the TRH neuropeptide family. TRH-like peptides in protostomes have a common core amino acid signature E-[L/F)-[L/F/V], whereas deuterostome TRH contains N-terminal pyroglutamate (Van Sinay et al. 2017). Both vertebrate preproTRH and the TRH precursor of protostomes contain multiple copies of TRH-like peptide sequences. The two TRH-like peptides in *C. elegans*, viz. (R)GRELF-NH2 and (R)ANELF-NH2, have been named TRH-1. TRH-1 may be comparable to the deuterostome TRH since both these peptides are capable of activating TRH receptors in *C. elegans*, *Platynereis*, and humans (Van Sinay et al. 2017). TRH-like peptides are named EFLamide in arthropods (Veenstra and Šimo 2020), EFLGamide in annelids (Van Sinay et al. 2017), and fulicins in a mollusk, *Achatina fulica* (Ohta et al. 1991; Van Sinay et al. 2017). EFLamide (EFLa) was originally isolated from the mite *Tetranychus urticae* (Veenstra et al. 2012).

The TRH neuropeptide signaling pathway is believed to be lost in a urochordate, *Ciona interestinalis* (Mirabeau and Joly 2013), but the TRH peptide comprising a 4 aa TRH peptide in echinoderms and 3 aa in chordates (*Branchiostoma* (pGlu-Ser-Pro-NH2) and vertebrates (pGlu-His-Pro-NH2)) has been identified. While all TRH-like and TRH peptide sequences have a C-terminal amide (-NH2) group, deuterostomian TRH show a common N-terminal pyroglutamate residue (Van Sinay et al. 2017). The glutamic acid residue in protostome TRH-like peptides was replaced by pyroglutamate in the deuterostome TRH.

Although TRH stimulates TSH secretion in mammals, the emergence of TRH as a regulator of TSH secretion in vertebrates is still less understood. In adult lampreys, TRH was detected in the brain but not TSH (Sage 1973). In teleosts, while TRH stimulates growth hormone (GH), melanocyte-stimulating hormone (MSH), adrenocorticotropic hormone, and PRL secretion (Galas et al. 2009), species-specific differences in the action of TRH on TSH secretion have been observed. In two teleosts (*Salmo gairdneri* and *Salvelinus alpinus*), TRH injection enhanced plasma T4 levels, and the action was probably mediated by thyrotrophs (Eales and Himick 1988). This was later demonstrated by experimental evidence showing the role of TRH in the regulation of the thyroid axis in teleosts. In bighead carp, *Aristichthys nobilis* (Chatterjee et al. 2001), and Japanese eel, *Anguilla japonica* (Han et al. 2004), TRH treatment enhanced, whereas T4 decreased TSH-beta mRNA expression in the pituitary cells. While TRH has limited TSH release activity, the tripeptide emerged as a crucial regulator of GH, MSH, and PRL in amphibians (Lamacz et al. 1989; Galas et al. 2009). TRH is reported to release TSH in birds (Scanes 1974; Galas et al. 2009), reptiles (Preece and Licht 1987; Chien et al. 2006), and mammals (Galas et al. 2009). Jackson and Reichlin (1974) suggested that thyroidal regulation by TRH might be a late evolutionary event. Although a range of diverse functions have been attributed to TRH, the role of the peptide as a molecule for TSH regulation across vertebrates emerges as the most conserved aspect of the peptide.

6.6 Emergence of the Hypophysiotropic TRH System in Vertebrates

The three anatomical modes for the hypothalamic regulation of the pituitary include (1) the portal system in the median eminence in tetrapods, (2) direct innervation of hypophysiotropic neurons in the pituitary in teleosts, and (3) diffusion in agnathans (Sower 2015). In addition to direct neuroendocrine axonal innervation on pituitary endocrine cells in teleosts, a recent study on *D. rerio* showed the existence of a neurovascular route (Golan et al. 2015). In addition to the innervation of the gonadotrophs, GnRH3 fibers were seen in close association with the permeable vasculature in the neurohypophysis and may allow the uptake of GnRH3 (Golan et al. 2015). In tetrapods, hypophysiotropic neurons in the hypothalamus secrete releasing hormones in the portal capillaries in the median eminence. These hormones reach the anterior pituitary to act on specific hormone-secreting cells, triggering their synthesis and release. Although studies have been conducted to determine the neuroanatomical location of hypophysiotropic neurons in the preoptic/hypothalamic of teleosts and tetrapods, the emergence of these neuroendocrine centres in the brain is still unclear. The hypophysiotropic TRH system in rodents, birds, reptiles, amphibians, and teleosts is described below. Since cerebrospinal fluid (CSF)-contacting TRH neurons, as well as neurosecretory neurons, have been suggested as a primitive character, we have presented the available information on CSF-contacting TRH neurons in the preoptic/hypothalamic regions of vertebrates. The idea that the hypophysiotropic TRHergic system in higher vertebrates originates from CSF-contacting TRH neurons is discussed.

6.6.1 The Hypophysiotropic TRH Neurons in the PVN of Mammals

Although neurons synthesizing TRH are widely distributed in the brains of mammals, those regulating the HPT axis reside in the PVN (Fekete and Lechan, 2007, 2014). Distinct differences in the organization of TRH neurons were observed in the PVN of rats, mice, and humans.

In the hypothalamus, the PVN is located on either side of the third ventricle and is composed of autonomic and neuroendocrine compartments. The neuroendocrine or neurosecretory part of the PVN consists of magnocellular and parvocellular neurons. Parvocellular neurons are further grouped into anterior, periventricular, medial, ventral, dorsal, and lateral parvocellular subdivisions (Swanson and Kuypers 1980). In rats, the TRH neurons are organized throughout the rostro-caudal extent of the PVN (Fekete and Lechan 2014). Using the retrograde neural tract tracing method, TRH neurons projecting to the median eminence were found in the anterior part of the periventricular nucleus and medial part of the PVN (Ishikawa et al. 1988). The hypophysiotropic TRH neurons reside in the periventricular and medial parvocellular subdivisions, and their axons project to the external zone of the median eminence (Fekete and Lechan 2014). We confirmed the distribution of TRH neurons in the rat PVN and dense TRH fibers in the external zone of the median eminence by using a specific sheep anti-TRH antiserum (the kind gift of Dr. Csaba Fekete, Department of Endocrine Neurobiology, Institute of Experimental Medicine, Budapest, Hungary) (Fig. 6.1A–G). From the axon terminals in the external zone of the median eminence, TRH is released into the portal vessels and reaches the pituitary TSH cells.

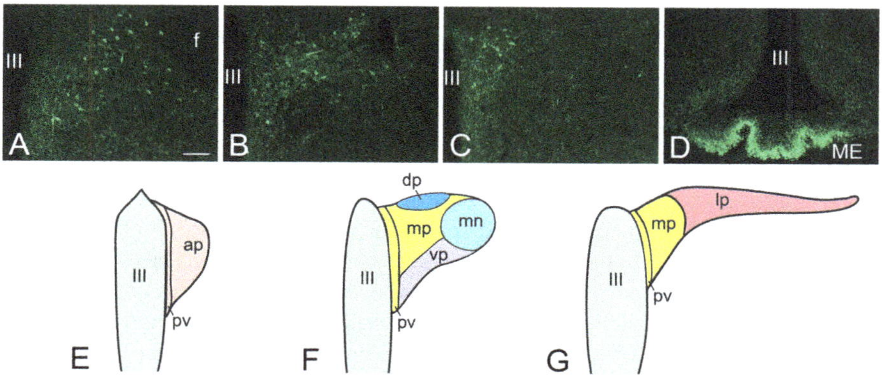

Fig. 6.1 Organization of the hypothalamic paraventricular nucleus (PVN) in rats showing the distribution of TRH neurons. Fluorescence photomicrographs of the coronal sections of the rat brain showing (**A**–**C**) TRH neurons throughout the rostro-caudal extent of the PVN and (**D**) TRH fibers in the external zone of the median eminence (ME). (**E**–**G**) Schematics showing the organization of the PVN and its subdivisions. The area shaded yellow in F and G harbors the hypophysiotropic TRH neurons. (**A**) and (**E**), anterior parvocellular subdivision of the PVN; (**B**) and (**F**), midlevel of the PVN; (**C**) and (**G**), caudal level of the PVN. ap, anterior parvocellular subdivision; dp, dorsal parvocellular subdivision; lp, lateral parvocellular subdivision; mn, magnocellular part; mp, medial parvocellular subdivision; pv, periventricular parvocellular subdivision; III, third ventricle. Scale bar = 100 μm

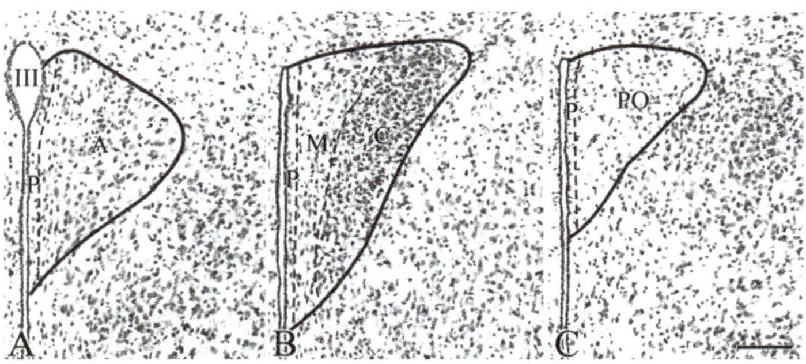

Fig. 6.2 Organization of the hypothalamic paraventricular nucleus (PVN) in mice. The Nissl-stained photomicrographs of the coronal sections of the hypothalamic paraventricular nucleus (PVN) at three different levels: (**A**) anterior, (**B**) mid, and (**C**) posterior. The different subdivisions of PVN are indicated with smaller labels: A, anterior part; C, compact part; M, midlevel medial part; P, periventricular zone; PO, posterior part. The hypophysiotropic TRH neurons reside in the compact part of the PVN. III, third ventricle. Scale bar = 100 μm. (Reproduced from Kádár et al. (2011) with permission)

Kádár et al. (2010) studied the cytoarchitecture of mouse PVN and found distinct differences in organization compared to that of rats. In mice, the PVN consists of the anterior, compact part; midlevel medial and posterior parts; and the periventricular zone (Kádár et al. 2010) (Fig. 6.2A–C). The compact zone contains equal proportions of parvocellular and magnocellular neurons, whereas the midlevel medial part is largely occupied by the parvocellular population. Although TRH neurons were observed throughout the anterior, mid, and posterior levels of the PVN, the neurons were more densely located in the anterior and mid-PVN levels. To identify the hypophysiotropic TRH neurons in the PVN of mice, Kádár et al. (2010) intravenously injected a retrograde neuronal tracer Fluoro-Gold (FG), and its retrograde accumulation in the PVN TRH neurons was studied. While the majority of the TRH neurons in the compact zone and only a few in the anterior part of the PVN were hypophysiotropic, none of the TRH neurons in the posterior part accumulated FG (Kádár et al. 2010). Unlike rats, the hypophysiotropic TRH neurons and the neurons synthesizing oxytocin and vasopressin (AVP) are intermingled within the compact zone in the mouse PVN (Kádár et al. 2010). These TRH neurons respond to hypothyroidism by enhancing proTRH mRNA in the medial, periventricular, and caudal levels of the PVN in rats (Lechan and Kakucska 1992). Unlike rats, except in the periventricular zone, the presence of TRH neurons was observed in all PVN compartments, with the largest number of these neurons residing in the compact part (Kádár et al. 2010). Further, the proTRH mRNA-expressing neurons responding to hypothyroidism reside only in the midlevel of the PVN in mice (Fekete and Lechan 2014).

In the human hypothalamus, while several parvocellular TRH neurons reside in the PVN, few magnocellular neurons expressing TRH were also observed (Fliers et al. 1994; Mihály et al. 2000). Although the parvocellular TRH neurons are organized throughout the rostro-caudal extent of the PVN, these were concentrated medially at the midlevel position (Guldenaar et al. 1996; Mihály et al. 2000). In addition, TRH neurons were also seen at dorsocaudal levels in the human PVN (Fliers et al. 1994, 2014). The TRH neurons in the PVN were round and small in size, with a bipolar and multipolar nature (Guldenaar et al. 1996; Mihály et al. 2000). The identity of the hypophysiotropic TRH neurons in the human PVN is still not known (Dudas and Merchenthaler 2020).

6.6.2 The Hypophysiotropic TRH Neurons in the PVN of Birds

The presence of TRH neurons has been observed in the PVN of different species of birds (*A. platyrhynchos* (Józsa et al. 1988), *C. livia* (Péczely and Kiss 1988), *T. guttata* (Singh et al. 2023)). The neuroanatomical organization of PVN TRH neurons and their fibers projecting to the median eminence of *T. guttata* is shown schematically in Fig. 6.3. The TRH neurons were clustered in the dorsal part of the PVN on either side of the ventricular wall (Fig. 6.3A–F, J–L). No TRH neurons were found in the magnocellular paraventricular part (Józsa et al. 1988). While the CRF and TRH neurons showed clear segregation in the PVN of rats, these two neuronal populations may potentially coexist in the PVN of *C. livia* (Péczely and Kiss 1988). In *T. guttata*, the PVN TRH neurons, which may serve as a hypophysiotropic TRHergic neuronal population, showed distinct axonal processes extending laterally and ventrally to enter the external zone of the median eminence (Fig. 6.3G–I, J–M) (Singh et al. 2023).

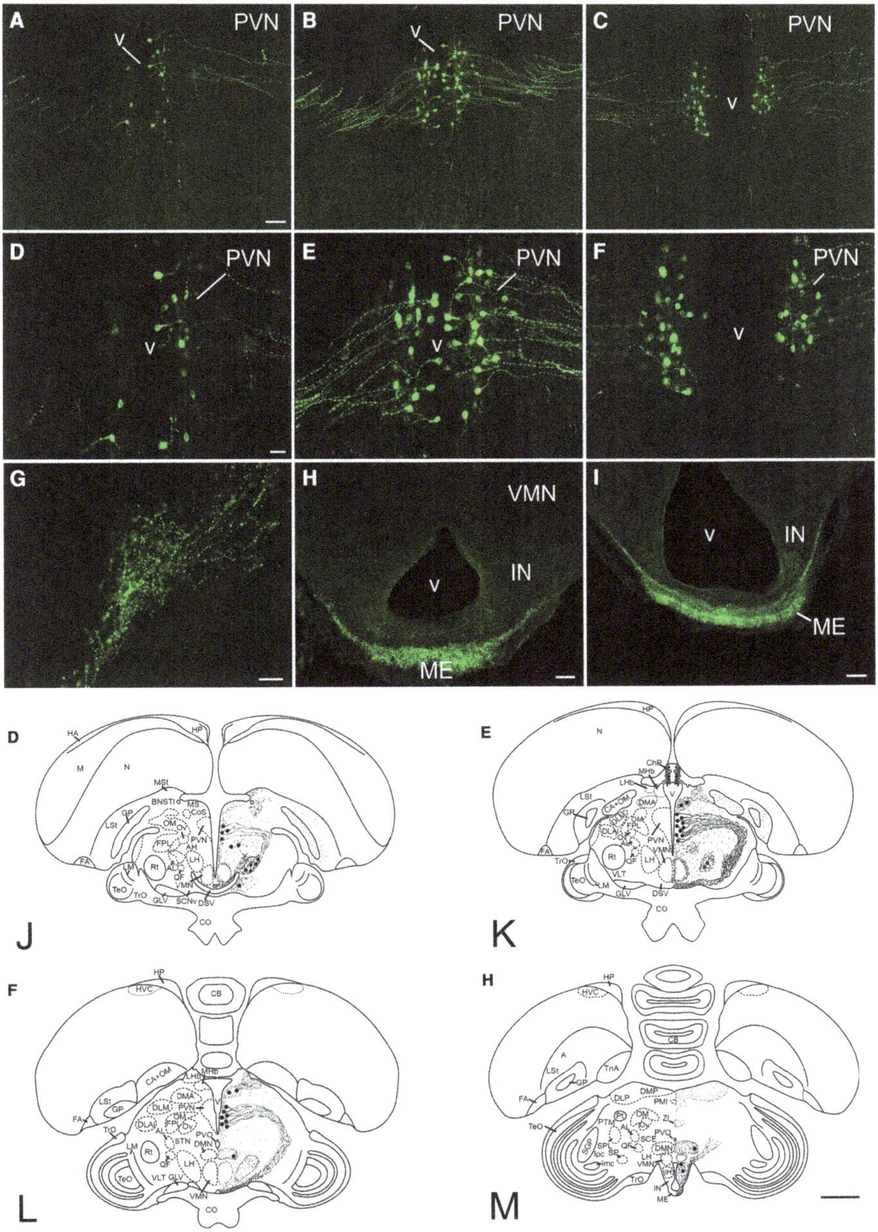

Fig. 6.3 The hypothalamic paraventricular nucleus (PVN) and its projection to the median eminence in *T. guttata*. Low (**A**–**C**) and high (**D**–**F**) magnification photomicrographs of the coronal sections of the brain of *T. guttata* showing the organization of TRH neurons (green) throughout the rostro-caudal extent of the PVN. (**G**–**I**) Fluorescence photomicrographs of the coronal hypothalamic sections showing TRH fibers/terminals in the median eminence (ME). Note that TRH fibers

6.6.3 The Hypophysiotropic TRH Neurons in the Brain of Teleosts, Amphibians, and Reptiles

Although the location of hypophysiotropic TRH neurons is well defined in birds and mammals, the relevant aspect in lower vertebrates is not well established. Species-specific differences in the presence of TRH in hypothalamic regions controlling pituitary function were observed in teleosts. The preoptic nucleus in fish is considered to be homologous to the PVN and supraoptic nucleus (SON) of mammals (Peter 1977). Distinct hypophysiotropic paraventricular (NPOpv) and supraoptic (NPOso) subdivisions of the nucleus preopticus (NPO) have been found in the brain of *C. batrachus* (Singh et al. 2012), but such an organization is not found in most fishes. Although considered homologous to the mammalian PVN/SON, no TRH neurons were found in the NPO of *C. batrachus* (Singh et al. 2019). TRH neurons were found in the magnocellular preoptic nucleus of a teleost, *Oncorhynchus tshawatscha* (Matz and Takahashi 1994). The presence of TRH at this location, with fibers extending to the pituitary, suggests that these TRH neurons may function as a hypothalamic releasing hormone (Matz and Takahashi 1994). In the hypothalamus of the frog *R. ridibunda*, while the vasotocin and mesotocin neurons were located dorsally, TRH neurons were seen in the ventral region, and a partial overlap was detected (Lamacz et al. 1989). In *X. laevis*, TRH mRNA-expressing neurons in the nucleus preopticus and infundibular nucleus may serve as a source of TRH innervation to the median eminence (Zoeller and Conway 1989). In a few reptilian species,

Fig. 6.3 (continued) originating from the PVN TRH neurons extend laterally and then descend ventrally into the external zone of the ME. (**J–M**) Schematics of the coronal sections through the rostro caudal extent of the hypothalamus of *T. guttata* showing the neuroanatomical areas on the left and the TRH immunoreactive cells (black circles) and fibers (dots and dashed lines) on the right side. The PVN TRH neurons with projections to the external zone of the ME are seen. A, acropallium; AH, anterior hypothalamus; AL, ansa lenticularis; BNSTl, lateral part of the BNST; CA, anterior commissure; CB, cerebellum; ChP, choroid plexus; CO, optic chiasm; CoS, commissural septal nucleus; DLA, nucleus dorsolateralis anterior thalami; DLM, medial nucleus of dorsolateral thalami; DLP, nucleus dorsolaterlis posterior thalami; DMA, nucleus dorsomedialis anterior thalami; DMN, hypothalamic dorsomedial nucleus; DMP, nucleus dorsomedialis posterior thalami; DSV, ventral supraoptic decussation; FA, frontal archo-pallial tract; FPL, fasciculus prosencephali lateralis; GLV, ventral part of the lateral geniculate nucleus; GP, globus pallidus; HA, apical part of the hyperpallium; HVC, higher vocal center; HP, hippocampal formation; IH, inferioris hypothalamic nucleus; Imc, magnocellular part of the isthmic nucleus; IN, infundibular nucleus; Ipc, parvocellular part of the isthmic nucleus; LH, lateral hypothalamic area; LHb, lateral habenular nucleus; LM, nucleus lentiformis mesencephalic; LSt, lateral striatum; M, mesopallium; MHb, medial habenular nucleus; MS, medial septum; MSt, medial striatum; N, nidopallium; OM, occipitomesencephalic track; Ov, nucleus ovoidalis; PMI, internal paramedian thalamic nucleus; Pt, pretectal nucleus; PTM, medial pretectal nucleus; PVO, paraventricular organ; QF, quintofrontal tract; Rt, nucleus rotundus; SCE, stratum cellular externum; SCNv, visual suprachiasmatic nucleus; SGP, stratum griseum periventriculare; SP, nucleus subpretectalis; SPL, lateral spiriform nucleus; STN, subthalamic nucleus; TeO, optic tectum; TnA, nucleus taenia of the amygdala; TrO, tractus opticus; V, ventricle; VLT, nucleus ventrolateralis thalami; VMN, ventromedial hypothalamic nucleus; ZI, zona incerta. Scale bar = 25 um in A–I and 1 mm in J–M. (Reproduced from Singh et al. 2023 with permission)

TRH neurons were found in the hypothalamic paraventricular and periventricular nuclei (López et al. 2008). Which TRH neuronal population contributes to the hypophysiotropic regulation in reptiles has remained unexplored.

While TRH fibers were observed in the pituitary of teleosts (Singh et al. 2019), TRH fibers/terminals have been reported in the external zone of the median eminence in amphibians (Domínguez et al. 2008), reptiles (López et al. 2008), birds (Józsa et al. 1988; Singh et al. 2023), and mammals (Merchenthaler et al. 1988).

6.6.4 The CSF-Contacting TRH Neurons in the POA/Hypothalamus and the Hypophysiotropic TRH System

The presence of CSF-contacting processes of neurosecretory neurons is the most primitive condition found in a lancelet, *Brachiostoma* (Obermüller-Wilén 1979). The cell bodies of these neurosecretory neurons were found in the ependymal cell layer, while their axons extend to the ventral surface of the brain (Obermüller-Wilén 1979). In ray-finned fishes, the dendrites of cells communicate with the ventricle (Knobloch and Grinevich 2014). The neurosecretory neurons in basal vertebrates are located in the preoptic nucleus, but in advanced vertebrates, this nucleus has become the paraventricular and supraoptic groups, along with accessory nuclei (Knobloch and Grinevich 2014).

Figure 6.4 shows a schematic of the occurrence of CSF-containing TRH neurons in the POA/hypothalamus of different vertebrate species. The presence of TRH neurons with distinct ventricular contacts is found in the POA/hypothalamus of cyclostomes through reptiles but not in birds and mammals. In *P. marinus*, CSF-contacting TRH neurons were found in the nucleus of the postoptic commissure (ncpo) and the dorsal (dHy) and ventral (vHy) hypothalamic nuclei (Fig. 6.4A) (Del Carmen De Andrés et al. 2002). From ncpo TRH neurons, a short apical process enters the postoptic recess, whereas a thin basal process travels laterally. The site where it terminates is not known (Del Carmen De Andrés et al. 2002). The occurrence of CSF-contacting TRH neurons has been suggested as a primitive feature (Del Carmen De Andrés et al. 2002). In *S. canicula*, CSF-contacting TRH neurons were not detected in the POA but were present in the dorsal hypothalamus and periventricular area characterized by ventricular dendrites (Fig. 6.4B) (Teijido et al. 2002). Similarly, no CSF-contacting TRH neurons were detected in the POA of teleosts, but TRH cells were found lining the wall of the lateral recess in the hypothalamus of *C. batrachus* (Fig. 6.4C) (Singh et al. 2019). In amphibians, TRH neurons in the anterior (POa) and posterior (POp) preoptic area and ventral hypothalamus showed CSF-contacting processes (Fig. 6.4D) (Domínguez et al. 2008). While one process of these neurons is embedded in the periventricular lining, the other extends toward the hypophysis. Although distinct hypophysiotropic TRH neurons have been described in the PVN of birds (Józsa et al. 1988; Péczely and Kiss 1988; Singh et al. 2023) (Fig. 6.3) and mammals, these neurons did not make CSF contacts (Fig. 6.4F, G). However, in two reptilian species (*P. scripta elegans* and *G. gekko*), the

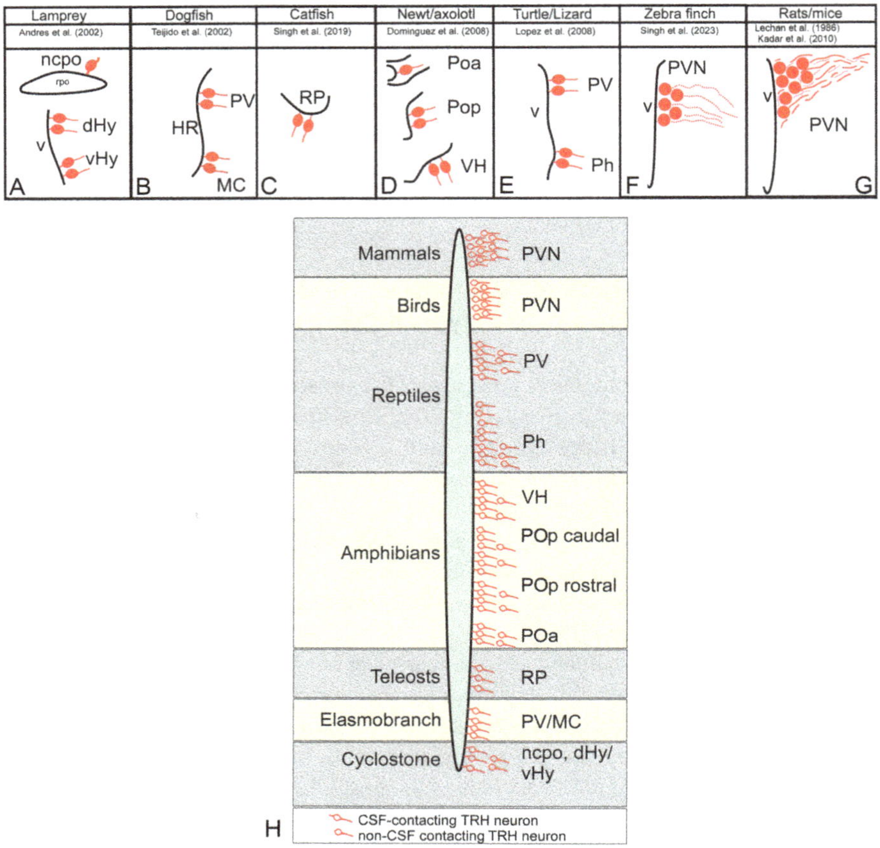

Fig. 6.4 Cerebrospinal fluid (CSF)-contacting TRH neurons in the preoptic area (POA)/hypothalamus of vertebrates and the emergence of the hypophysiotropic TRH system. Schematic of the POA/hypothalamic regions of different animals showing CSF-contacting TRH neurons (red) around the ventricle/recess in (**A**) lamprey, (**B**) dogfish, (**C**) catfish, (**D**) newt/axolotl and (**E**) turtle/lizard, and hypophysiotropic TRH neurons (red circles) and fibers (lines and dots) in the hypothalamic paraventricular nucleus (PVN) in (**F**) zebra finch and (**G**) rodents. (**H**) Schematic showing the TRH neurons with or without CSF contact lining the ventricle in the POA/hypothalamus of different vertebrates. Note the periventricular arrangement of cells. dHy, dorsal hypothalamic nucleus; HR, lateral hypothalamic recess; MC, migrating cells; ncpo, nucleus of the postoptic commissure; Ph, periventricular hypothalamic nucleus; POa, anterior preoptic area; POp, posterior preoptic area; PV, paraventricular nucleus; Pv, periventricular cells; PVN, hypothalamic paraventricular nucleus; RP, posterior recess; rpo, postoptic infundibular recess; V, ventricle; VH, ventral hypothalamus; vHy, ventral hypothalamic nucleus

CSF-contacting TRH neurons were found at two distinct locations, viz., the paraventricular (PV) and periventricular hypothalamic (Ph) nuclei (López et al. 2008) (Fig. 6.4E).

The CSF-contacting TRH neurons were arranged dorsoventrally along the ventricle at different levels across the phylogenetic orders (Fig. 6.4H). In addition to the

CSF-contacting neurons, the TRH neurons gradually appeared periventricularly adjacent to the ventricular lining in cyclostomes, amphibians, and reptiles (Fig. 6.4H). Periventricular TRH neurons have also been observed in the PVN of birds and mammals (Fig. 6.4H). However, in addition to the periventricular part, the hypophysiotropic TRH neurons also extend laterally to reside in the medial parvocellular part of the PVN in rodents (Fig. 6.4H). Del Carmen De Andrés et al. (2002) suggested that the preopticohypothalamic TRH neuronal groups have an ancient phylogenetic origin. The CSF-contacting TRH neurons in the POA/hypothalamus in lower vertebrates might have gradually transformed into hypophysiotropic neurons occupying the periventricular and medial portions of the PVN in higher vertebrates.

6.7 The Nonhypophysiotropic TRH System Indirectly Regulates Hypophysiotropic Function in Teleosts and Mammals

6.7.1 The Nonhypophysiotropic TRH Neurons in Rodents

In addition to the controller of the HPT axis, widely distributed TRH neurons in the brain may regulate a range of physiological and behavioral functions. However, little is known about the projections and significance of these nonhypophysiotropic TRH neurons in the brain.

Among the several nonhypophysiotropic TRH neuronal groups, the central projections of the anterior parvocellular subdivision of the PVN (PVNa), the tuberal lateral hypothalamus (TuLH), and perifornical/BNST TRH neurons have been investigated. The perifornical TRH neurons in rats innervate the lateral septum (Merchenthaler 1991). Rostrally in the PVN (Fig. 6.1A, E), a distinct group of nonhypophysiotropic TRH neurons are located in the PVNa. Using anterograde neuronal tracing, projections of the TRH neurons in the PVNa of rats were mapped (Wittmann et al. 2009b). The brain regions innervated by the PVNa TRH neurons are schematically shown in Fig. 6.5A. The PVNa TRH neurons innervate the anterior olfactory nucleus, dorsal endopiriform cortex, Islands of Caleja, major island, medial septal nucleus, horizontal and vertical limbs of the diagonal band, ventral pallidum, zona limitans, preoptic region, hypothalamus, thalamus, epithalamus, amygdala, hippocampus, substantia innominata, BNST, septum, midbrain, and hindbrain. The origin of TRH fiber innervation to the tuberomammillary nucleus (TMN) has been studied using retrograde neuronal tracing in rats (Sánchez-Jaramillo et al. 2022). The neurons in the septum, POA, BNST, perifornical area, anterior paraventricular nucleus, peduncular and tuberal lateral hypothalamus (TuLH), SCN, and medial amygdala project to the TMN (Sánchez-Jaramillo et al. 2022). Of these TMN projecting regions, only the TuLH TRH neurons contribute to the TRHergic innervation to TMN and may regulate energy balance by acting on histaminergic neurons (Sánchez-Jaramillo et al. 2022). The projections of the BNST/perifornical TRH neurons are shown schematically in Fig. 6.5B. The Urocortin 3

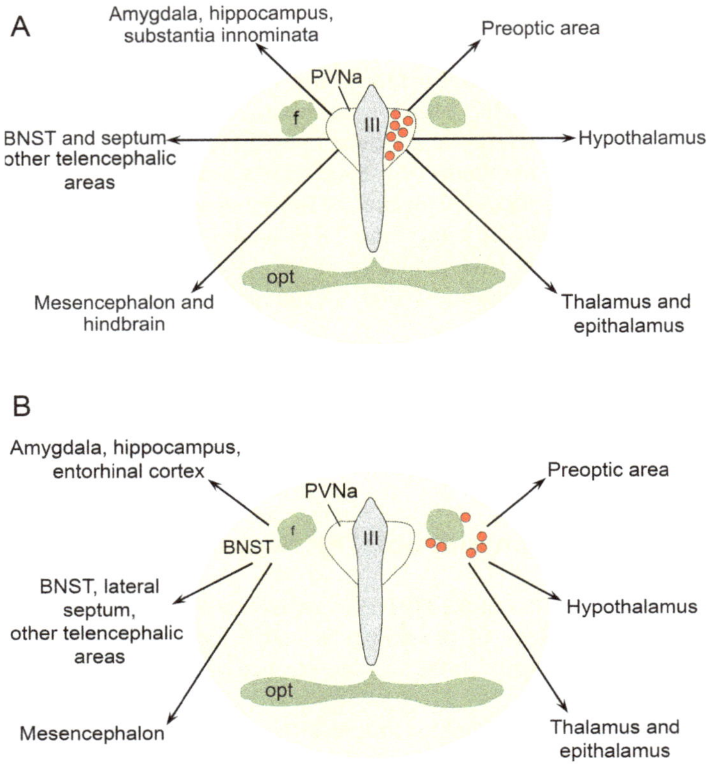

Fig. 6.5 Central projections of the nonhypophysiotropic TRH neurons in the (**A**) anterior parvocellular subdivision of the hypothalamic paraventricular nucleus (PVNa) and (**B**) perifornical area/bed nucleus of the stria terminalis (BNST). Schematics from an anti-TRH antiserum labelled coronal section through the hypothalamus of rat showing the location of the PVNa and perifornical area/BNST on the left and the presence of TRH neurons (red circles) on the right. While the projections of TRH neurons in the PVNa were studied using an anterograde neuronal tracing method (Wittmann et al. 2009b), those in the perifornical/BNST were analyzed using a Urocortin-3/TRH double-labeling approach (Wittmann et al. 2009a). The arrows indicate the projections of the TRH neurons to different brain regions. f, fornix; opt, optic tract; III, third ventricle

(UCN-3)-co-expressing TRH neurons in the BNST and perifornical area densely innervate the ventromedial and lateral septal nuclei, posterior division and interfascicular nucleus of the BNST, anterior hypothalamic area, preoptic nucleus, anterolateral part of the tuber cinereum area, ventral premammillary nucleus, ARC, medial amygdaloid nucleus, amygdalohippocampal area, ventral hippocampus and medial entorhinal cortex (Wittmann et al. 2009a). In addition, the TRH/UCN-3 projections were also seen in the forebrain areas associated with psychological stress and anxiety (Wittmann et al. 2009a). It has been suggested that the BNST/perifornical TRH neurons regulate feeding and behavior during stress (Wittmann et al. 2009a).

Exploring the Projections of Nonhypophysiotropic TRH Neurons in the Brain
Several TRH-producing neuronal groups are widely organized in the brain. However, the significance of many of these is still unclear. One of the challenges in establishing the significance of a neuronal group is to determine its connectivity in the brain. Neural tract tracing has emerged as a powerful tool for exploring the anatomy of neural circuits. The technique is based on the anterograde (toward the axon terminals) and retrograde (toward the cell body) transport systems engaged in the movement of material in neurons. The method for tracing tracts using an anterograde neuronal tracer, *Phaseolus vulgaris* leucoagglutinin (PHA-L), and the multidimensional approach has been described (Lanciego and Wouterlood 2020).

Using PHA-L, projections of the nonhypophysiotropic TRH neurons in the PVNa of rats were established (Wittmann et al. 2009b). PHA-L (2.5% in the phosphate buffer) is injected into the region of interest using iontophoresis. The neurons around the iontophoresis site take up PHA-L and show labeling of its cell body, dendrites, axons, and terminals. A precise and focal injection of the neuronal tracer in the area of interest in the brain can be achieved with stereotaxic surgery and iontophoresis. A glass micropipette with a tip size of 10–30 μm is pulled on a pipette puller (Lanciego and Wouterlood 2020; Wittmann et al. 2009b). The tip of the glass micropipette is loaded with the neuronal tracer and mounted on the holder of the stereotaxic instrument. Using the well-established stereotaxic atlas of the rat/mouse brain and following the surgical procedures, a small hole is drilled in the skull above the injection site, and the glass micropipette is then lowered into the brain to the region of interest. The neuronal tracer is then iontophoresed into the region of interest using ~6–10 μA positive DC current with 7 s on-off pulses (Lanciego and Wouterlood 2020; Wittmann et al. 2009b). Allowing enough time for the transport of the neuronal tracer, the brains are fixed using the transcardial perfusion method. The sets of coronal cryosections through the rostrocaudal extent of the brain are prepared. For the detection of the tracer, the sections are processed for immunohistochemistry or immunofluorescence using the antibodies raised against the neuronal tracer used. For example, Wittmann et al. (2009b) iontophoresed PHA-L in the PVNa and localized the injection spot using immunohistochemistry with anti-PHAL antibodies. A set of sections from the brain injection site were processed for double immunofluorescence using anti-PHAL and anti-TRH antisera. The PHA-L/TRH-labeled fibers/terminals in each section/region were analyzed. The anterograde neuronal projections of a neuronal group should be confirmed by employing retrograde neural tract tracing. For example, the retrograde neuronal tracer (cholera toxin-beta subunit (CtB) or FG) can be administered into the brain region that is found to be innervated by the axons of the neuronal group of interest. The retrograde transport and the accumulation of the tracer into the neurons of interest can be studied using double immunofluorescence. After identifying the projections of the PVNa TRH neurons in the rat brain (Fig. 6.5a), CtB was iontophoresed in selected nuclei to validate the origin of the fiber innervation from the PVNa TRH neurons (Wittmann et al. 2009b). Using a similar approach, recently, the projections of TuLH TRH neurons to TMN have also been established (Sánchez-Jaramillo et al. 2022).

The tip diameter of the glass micropipette, the precise stereotaxic coordinates, the duration of iontophoresis, and a power source that provides a constant current and maintains the set current and transport time are crucial factors in studies aimed at revealing neuronal connectivity. After the completion of iontophoresis, the glass micropipette should be kept in the same position for 10–15 min. The transport time for the tracer may vary. For example, to trace the projections of PVNa TRH neurons, postsurgical survival for 9–14 days was used (Wittmann et al. 2009b). Anti-PHAL or anti-CtB antibodies, raised in goats as well as rabbits, and a range of fluorescent dye-conjugated neuronal tracers are commercially available, making the multiplex immunofluorescence feasible. Researchers often encounter difficulty tracking the tip of the glass micropipette due to poor visibility. While loading the tip with a tracer or calculating the stereotaxic coordinates relative to bregma, the tip may accidentally hit the skull or other objects, bend, and break. To address this, Ter Horst et al. (1988) developed a device that brightly illuminates the tip of the glass micropipette with a light-emitting diode.

There is another method to identify the projections of a TRH or a neurochemical-containing neuronal group in the brain. Double immunofluorescence using an anti-TRH antiserum and an antiserum raised against a neurotransmitter/neuropeptide that is specifically expressed in the same group of the TRH neurons can be employed. If these two neurochemicals are highly colocalized in a group of neurons, the application of double immunofluorescence is useful in identifying the fibers or terminals of these neurons, distributed in various brain regions. Wittmann et al. (2009a) studied the projections of TRH neurons in the BNST and perifornical area. Urocortin-3 (Ucn-3)- and TRH-containing neurons reside in the BNST, perifornical area, dorsal supraoptic nucleus, and antero- and posterodorsal parts of the medial amygdaloid nucleus. Since almost all Ucn-3 neurons in the perifornical area and BNST coexpress TRH, the projections of these double-labeled neurons were analyzed in the rat brain (Wittmann et al. 2009a). Such an approach has been widely accepted for tracing the neuronal connectivity of a neuronal group. For example, the projections of ARC CART neurons to the preoptic area have been studied using a similar strategy (True et al. 2013). Since the ARC CART neurons coexpress α-MSH, the presence of CART as well as α-MSH double labeling in the fibers would indicate their origin from ARC.

6.7.2 Interaction Between Nonhypophysiotropic TRH and Hypophysiotropic DAergic Systems in the Brain of Rodents and Teleost

In addition to stimulating TSH synthesis and evoking its release from the pituitary, TRH stimulates PRL secretion in mammals. PRL secretion is controlled by DAergic neurons located in the hypothalamus. Three subpopulations of DA neurons, viz., tuberohypophyseal (THDA), tuberoinfundibular (TIDA), and periventricular hypophyseal (PHDA), contribute to the regulation of PRL secretion. The THDA and PHDA innervate the intermediate/neural lobe of the pituitary, transport DA to the anterior pituitary gland, and contribute to the basal regulation of PRL secretion (Grattan 2015). The TIDA neurons extend their axons to the external zone of the median eminence for the direct delivery of DA to the anterior pituitary and to inhibit PRL secretion (Grattan 2015). While the TIDA neurons reside in the dorsal part of the hypothalamic ARC (Fig. 6.6), the ARC is densely innervated by TRH fibers (Fig. 6.6Aa–Af, B) originating from the PVNa (Wittmann et al. 2009b). Although TRH has been found to act directly on lactotrophs, the peptide may interact with TIDA neurons, which in turn regulate PRL secretion (Fig. 6.6B) (Goldstein et al. 2007; Lyons et al. 2010). The application of double immunofluorescence showed TRH axons communicating with TIDA neurons, whereas TRH treatment changed the firing pattern of DA neurons (Lyons et al. 2010) and decreased tyrosine hydroxylase levels in the primary neuronal hypothalamic cultures (Goldstein et al. 2007). The change in the DA neuronal activity has been suggested as a mechanism underlying TRH-induced reduction in DA release, thereby facilitating PRL secretion (Lyons et al. 2010). The TRH-DA interaction is also evident in different neuronal groups, including the infundibular nucleus in the human hypothalamus (Dudas and Merchenthaler 2020).

The TRH-DA interaction in the brain seems to be conserved. In teleosts, the hypophysiotropic neurons directly enter the pituitary gland and control pituitary secretion. A group of DA neurons residing in the anteroventral POA of teleosts, identified as NPPa, is hypophysiotropic and plays an important role in the regulation of luteinizing hormone (LH) cells in the pituitary (Singh et al. 2019) (Fig. 6.7). While GnRH stimulates, DA inhibits LH secretion in teleosts. The application of a retrograde neuronal tracer in the pituitary labeled the hypophysiotropic DA neurons in the NPPa, but the TRH neurons in this neuronal group did not accumulate the tracer (Singh et al. 2019). The TRH neurons in the NPPa located lateral to the hypophysiotropic DA neurons densely innervate the NPPa DA neurons in *C. batrachus* (Singh et al. 2019) (Fig. 6.7Ag-p, B). The TRH seems to have an inhibitory effect on DA neurons in the NPPa of *C. batrachus* (Singh et al. 2019). TRH serves as a regulator of the hypophysiotropic DA neurons in teleosts as well as mammals, and the interaction between the two systems seems to be a highly conserved feature.

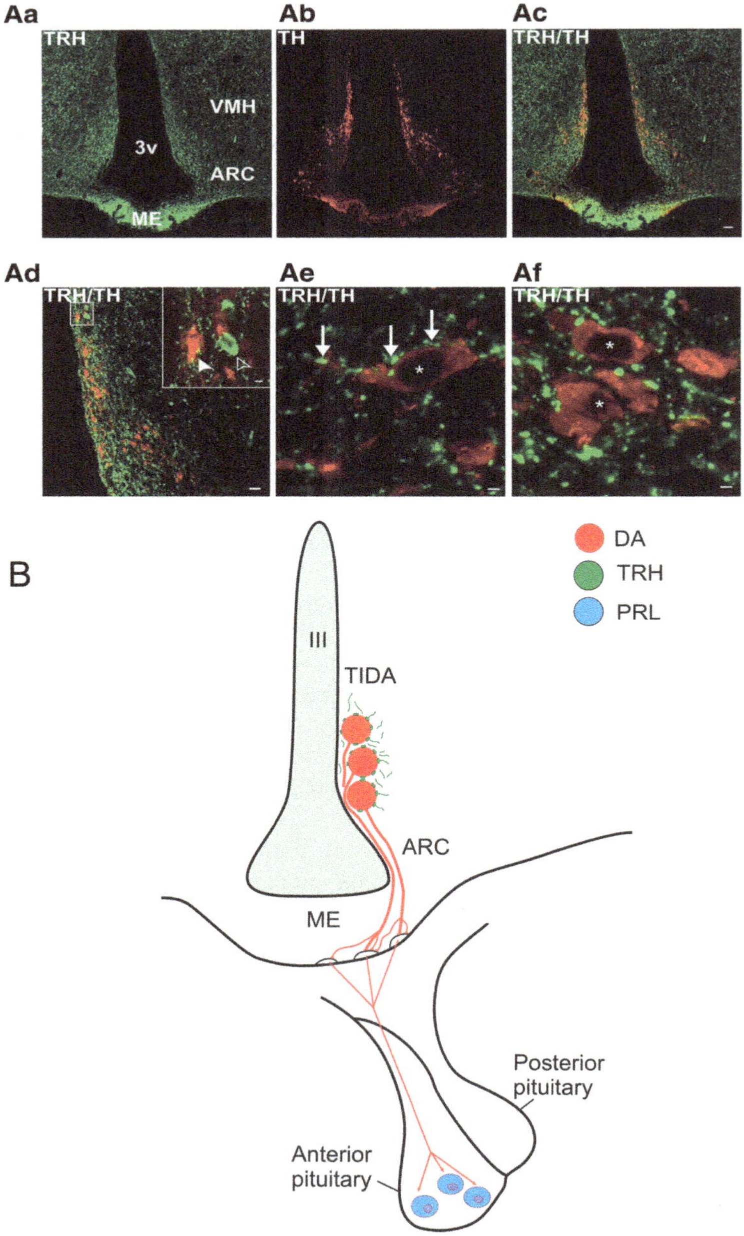

Fig. 6.6 Interaction between TRH- and dopamine (DA)-containing systems in the hypothalamus of rats. (**Aa–Af**) Fluorescence photomicrographs of the hypothalamic section at the level of the arcuate nucleus (ARC) showing the TRH fibers (green) innervating (arrows in **Ae**) tyrosine hydroxylase (TH)-positive tuberoinfundibular DA (TIDA) neurons (red). (**B**) Schematic showing TRH-TIDA interaction in the hypothalamus of mammals. The TIDA neurons involved in regulating prolactin (PRL) secretion reside in the dorsal part of the ARC. The TIDA neurons are hypophysiotropic and release DA in the portal vessels in the external zone of the median eminence (ME). The TIDA neurons receive dense TRHergic innervation (Lyons et al. 2010). 3v/III, third ventricle; VMH, ventromedial hypothalamic nucleus. Scale bar = 100 μm in **Aa–Ac**, 50 μm in **Ad**, 5 μm in **Ae** and **Af**, 10 μm in inset in **Ad**. (Reproduced "**Aa–Af**" with permission from Lyons et al. 2010)

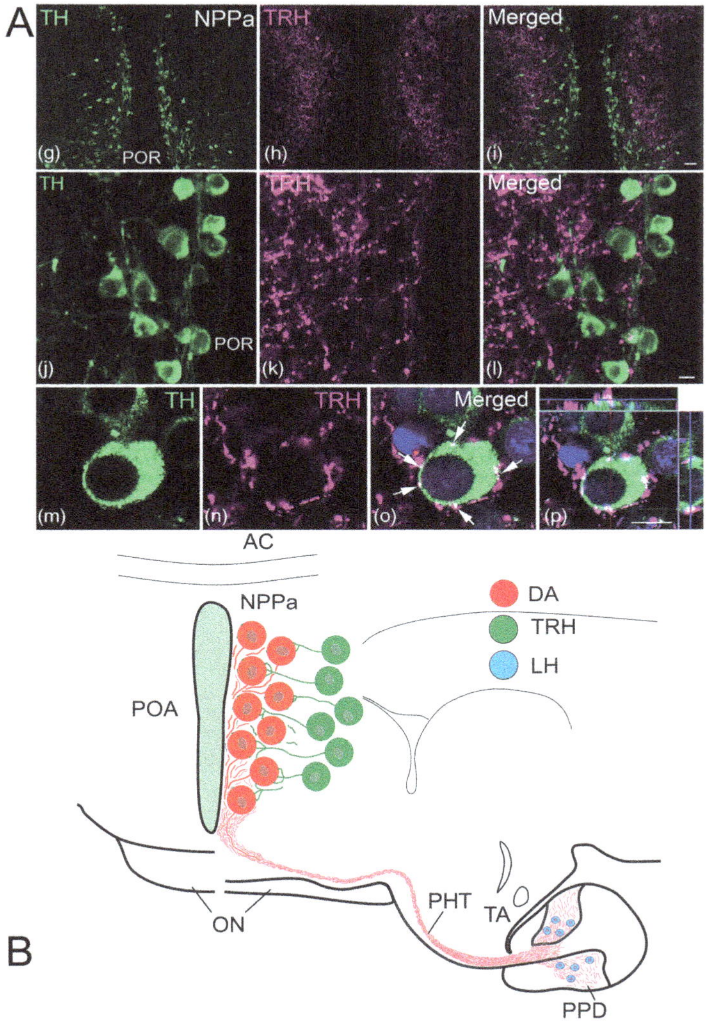

Fig. 6.7 Interaction between TRH- and dopamine (DA)-containing systems in the preoptic area (POA) of the catfish, *C. batrachus*. (**Ag–Ai**) Fluorescence and (**Am–Ap**) confocal microscope photomicrographs showing the association between tyrosine hydroxylase (TH) neurons (green) and TRH fibers (magenta) in the nucleus preopticus periventricularis (NPPa). (**Ag–Ai**) Low and (**Aj–Ap**) high magnification photomicrographs. (**Ap**) The orthogonal view of TH neurons in the NPPa innervated by TRH fibers. TH neurons contacted by TRH fibers are shown with arrows in Ao. (**B**) Schematic of the brain and pituitary gland of *C. batrachus* showing the interaction between TRH- and DA-containing systems in the POA. The DA-containing axons innervate luteinizing hormone (LH)-secreting cells (blue) in the proximal pars distalis (PPD) in the pituitary, and DA inhibits LH secretion. For representation, the neuroanatomical arrangement of the DA and TRH neurons in the POA of *C. batrachus* is shown in the schematic of the coronal section at the level of POA (left). The DAergic preoptico-hypophyseal tract (PHT) and its pituitary innervation are schematically shown as a sagittal view of the hypothalamus-pituitary complex (right). The hypophysiotropic DA neurons (red) and the nonhypophysiotropic TRH neurons (green) reside in the NPPa. The TRH neurons innervate the hypophysiotropic DA neurons in the NPPa. AC, anterior commissure; ON, optic nerve; POR, preoptic recess; TA, tuberal area. Scale bar = 25 μm in **g–i** and 5 μm in **j–p**. (Reproduced "**Ag–Ap**" with permission from Singh et al. 2019)

6.8 Evolution of TRH as a Regulator of Metamorphosis and Metabolism

Metamorphosis is an energy-intensive process, and therefore an enhanced metabolic rate is critical for restructuring and transforming the organism. Thyroid hormones play a crucial role in the initiation and maintenance of metamorphosis as well as in the control of metabolism. Thyroid hormone signaling controls metamorphosis in amphibians (Galas et al. 2009), urochordates, cephalochordates, and echinoderms (Heyland et al. 2004; Paris and Laudet 2008), suggesting a conserved role of thyroid hormones in the regulation of the metamorphosis of deuterostomes. However, in the metamorphosis of lampreys, the thyroid hormone T3 functions like a juvenile hormone (Youson 1997). The levels of T3 and T4 are reduced prior to metamorphosis in these animals (Youson 1997). Thyroid hormones play an important role in the metamorphosis of teleosts. The activity of the thyroid remains low during the premetamorphic stage but increases during the metamorphic climax (Campinho 2019). While a thyroid hormone surge plays a role in the metamorphosis of larval to juvenile transition in most teleosts, the second peak of thyroid hormones plays a role in smoltification in salmonids (Campinho 2019). T3 plays an important role in systemic metabolic regulation and controls metabolic adjustment during the onset of the metamorphic climax. The transition from prometamorphic to metamorphic stages is associated with profound alteration in metabolism (Zhu et al. 2020). Zhu et al. (2020) showed the metabolic reorganization of lipids, proteins, and carbohydrates in the liver and tail during the T3-driven metamorphic climax in the livers and tails of *Rana omeimontis* tadpoles.

In *X. laevis*, the brain immunoreactive TRH concentration was higher during the prometamorphosis and metamorphic climax (King and Millar, 1981). The higher TRH concentration during early prometamorphosis has been suggested to initiate thyroid hormone synthesis and metamorphosis (King and Millar, 1981). However, in *R. catesbeiana* the TRH content in the brain increased following metamorphosis and peaked during adulthood in *R. catesbeiana* (Prasad et al. 1982). In this frog, the expression of the TRH receptor TRHR3 was observed in the PRL cells of the anterior pituitary. The expression of TRHR3 mRNA in the anterior lobe of the pituitary and in PRL increased in tadpoles undergoing metamorphosis (Nakano et al. 2018). Nakano et al. (2018) suggested that the TRH-induced PRL release via TRHR3 is important in growth and metabolism in amphibians. While the thyroid hormone serves as a regulator of metamorphosis in basal vertebrates, the function might have been transformed to support metabolism and thermoregulation in birds and mammals.

6.9 The Role of TRH in the Central Regulation of Energy Balance: A Comparison Across Mammals and Birds

Hypophysiotropic TRH neurons in the PVN drive the HPT axis and metabolism in mammals. The hypothalamus of birds is also equipped with a well-defined hypophysiotropic PVN TRH neuronal system. In view of the highly energy-demanding physiological and behavioral adaptations in birds, it would be interesting to compare how the response of PVN TRH neurons differs in terms of the regulation of energy balance across birds and mammals.

In rodents, the biosynthesis of proTRH mRNA in the PVN is suppressed in response to fasting (Sánchez et al. 2008). Although the underlying central mechanism has been extensively studied, restoring the suppressed TRH levels to the ad libitum fed state after food intake is not well understood. Sánchez et al. (2008) demonstrated that only a small fraction (~10%) of the PVN TRH neurons were activated following early refeeding (Sánchez et al. 2008). The restoration of fasting-induced suppression of the PVN proTRH mRNA and serum TSH levels to the ad libitum fed state required a longer refeeding interval (Sánchez et al. 2008). While the hypophysiotropic TRH neurons in the PVN receive several complex neural inputs, the ARC serves as a crucial neural hub for the integration/processing of the peripheral signals. The ARC neurons issue efferent projections to the PVN TRH neurons (Fekete and Lechan 2014; Fliers et al. 2014), and these inputs modulate the response of the HPT axis to the changing energy states in the organism (Fig. 6.8). Two subsets of neurons residing in the medial and lateral parts of the ARC serve as crucial components of the neural circuitry regulating food intake and energy balance (Schwartz et al. 2000). The medial ARC contains neurons producing orexigenic neuropeptides (neuropeptide Y (NPY) and agouti-related protein (AGRP)) whereas the neurons producing anorectic neuropeptides (cocaine- and

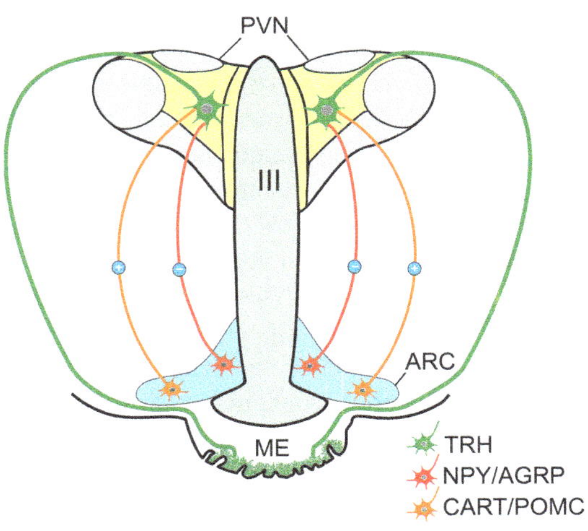

Fig. 6.8 Schematic showing the regulation of hypophysiotropic TRH neurons in the hypothalamic paraventricular nucleus (PVN) by the feeding-related neuropeptides of the hypothalamic arcuate nucleus (ARC) in mammals. The hypophysiotropic TRH neurons (green) in the medial parvocellular and periventricular subdivisions (shaded yellow) of the PVN project to the external zone of the median eminence (ME). TRH is released into the portal vessels in the ME and carried to the anterior pituitary for its action. The hypophysiotropic TRH neurons are innervated by two subsets of ARC neurons. The neurons producing neuropeptide Y (NPY) and agouti-related protein (AGRP) reside in the medial part of ARC, whereas cocaine- and amphetamine-regulated transcript (CART) and proopiomelanocortin (POMC)-synthesizing neurons are located in the lateral ARC. NPY/AGRP inhibit (−), whereas CART/POMC stimulate (+) proTRH mRNA expression in PVN neurons and regulate hypothalamus-pituitary thyroid axis. III, third ventricle

amphetamine-related transcript (CART) and proopiomelanocortin (POMC)) reside in the lateral ARC (Schwartz et al. 2000) (Fig. 6.8). While activation of NPY/AGRP neurons promotes feeding, enhanced CART/POMC signaling inhibits the feeding response. NPY and AGRP regulate the HPT axis and energy expenditure by their action on hypophysiotropic TRH neurons. The NPY and AGRP axons from the ARC densely innervate the PVN TRH neurons, and the intracerebroventricular administration of NPY or AGRP inhibits the hypophysiotropic TRH neurons. While the fasting-induced increase in the AGRP mRNA expression in the ARC was restored to baseline levels within 12 h of refeeding, the NPY mRNA levels in the ARC remained elevated throughout the 24 h refeeding (Sánchez et al. 2008). It has been proposed that the sustained elevation of the AGRP and NPY levels in the ARC during refeeding may not allow the restoration of TRH levels to normal fed levels. This may provide the animals time to replenish the energy stores exhausted during the fasting period (Sánchez et al. 2008).

Birds have emerged as an interesting model to study energy balance. They possess high daytime body temperatures and basal metabolic rates and employ a range of highly energy-demanding activities, including egg incubation, flight, and migration. As seen in mammals, the energy balance in birds is regulated by a comparable set of feeding-related neurons in the infundibular nucleus (IN, a homolog of mammalian ARC). The expression of NPY, POMC, AGRP, and CART and the coexpression of NPY/AGRP have been demonstrated in the IN of birds. These neuropeptides, when injected centrally in birds, produced effects on food intake comparable to those in mammals. NPY plays a crucial role in the central regulation of energy homeostasis in birds. The intracerebroventricular administration of NPY enhanced food intake in birds (Kuenzel et al. 1987). In addition, fasting enhanced NPY and lowered AGRP levels in the IN of Japanese quails (Boswell et al. 2002).

An extensive organization of the TRH neurons and fibers was observed throughout the rostro-caudal extent of the PVN in *T. guttata* and fiber innervation in the external zone of the median eminence (Singh et al. 2023) (Fig. 6.3). The PVN TRH neurons are densely innervated by the NPY-containing axons from IN (Figs. 6.9A–D and 6.10A, B). One of the interesting aspects of the central regulation of energy balance in birds is the response of TRH and NPY neurons in the PVN and IN, respectively, to the changes in energy states. In *T. guttata*, fasting robustly activates IN NPY neurons, whereas TRH-ir and proTRH mRNA levels in the PVN were reduced (Singh et al. 2016, 2023). Unlike in mammals, a 2–6 h refeeding interval restored the fasting-induced increase in NPY levels in IN and TRH levels in the PVN of *T. guttata* to the ad libitum fed state (Fig. 6.9E–M). Since these birds restored the fasting-induced loss in body weight much more rapidly than rodents, it was proposed that the rapid reduction in NPY levels may allow the birds to quickly restore depleted stores and regain weight after fasting (Singh et al. 2023). This rapid restoration of NPY levels and body weight to the prefasting stage may allow the hypophysiotropic TRH neurons and the physiologically challenging energy homeostasis mechanism in the birds essential for survival to quickly return to their normal functioning.

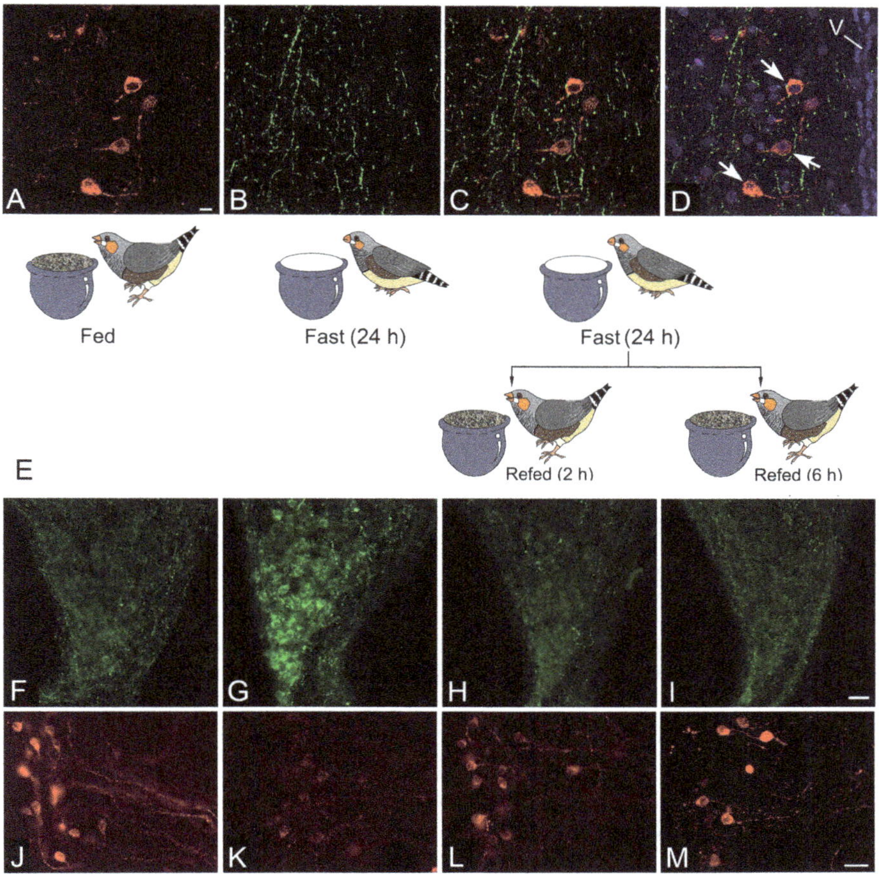

Fig. 6.9 Interaction between thyrotropin-releasing hormone (TRH) and neuropeptide Y (NPY) in the hypothalamic paraventricular nucleus (PVN) of *T. guttata* during dynamic changes in the energy states. (**A–D**) Photomicrographs showing the PVN TRH neurons (red) densely innervated by NPY axons (green, arrows in **D**). (**E**) Schematic of the experimental design. Birds were divided into three groups, viz., fed ad libitum, fasted, and fasted birds refed for either 2 or 6 h. Effect of fasting and refeeding on (**F–I**) NPY immunoreactivity in the infundibular nucleus (IN) and (**J–M**) TRH immunoreactivity in the PVN of *T. guttata*. (**F**, **J**) Fed, (**G**, **K**) fasted, and fasted birds refed either for (**H**, **L**) 2 or (**I**, **M**) 6 h. Fasting upregulates NPY-ir in the IN but suppresses TRH-ir in the PVN. Refeeding for 2–6 h restored TRH-ir to ad libitum fed state. Scale bar = 5 μm in **A–D** and 25 μm in **F–M**. (Reproduced with permission from Singh et al. 2023. Panels from three figures are used and the figure has been relabeled "**A–M**")

Perspective

In the course of evolution, TRH has undergone alterations in its amino acid sequence, the number of TRH progenitors in proTRH, neuronal groups and their distribution in the brain, and functions. The thyroid hormones have emerged as regulators of metamorphosis early in the evolution of the deuterostome and serve

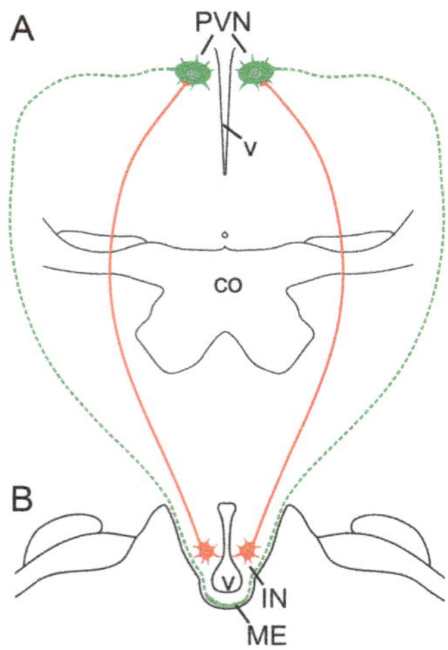

Fig. 6.10 Schematics of the coronal sections of the brain of *T. guttata* at the level of the (**A**) hypothalamic paraventricular (PVN) and (**B**) infundibular (IN) nuclei showing the interaction between thyrotropin-releasing hormone (TRH) (green) and neuropeptide Y (NPY) (red). The axons of the hypophysiotropic TRH neurons located in the PVN terminate in the external zone of the median eminence (ME). The IN is an avian homolog of the mammalian arcuate nucleus and contains NPY-producing neurons that respond to changes in energy states. The IN NPY neurons innervate PVN TRH neurons and regulate energy balance. CO, optic chiasm; V, ventricle

as a multifunctional hypophysiotropic regulator of a range of pituitary hormones in lower vertebrates. Although TRH seems to have established its control over TSH secretion early during vertebrate evolution, the emergence of TRH as a major regulator of the HPT axis and thermogenesis appears to be a recent evolutionary step in vertebrates. The hypothalamic origin of TRH innervation to the teleost pituitary and the median eminence of lower vertebrates needs to be identified. The presence of TRH in the POA/hypothalamus has been observed in the species across the phylogenetic clades and has been suggested as a conserved feature (López et al. 2008). The hypophysiotropic TRH system seems to have evolved from the CSF-contacting TRH neurons in the POA/hypothalamus of lower vertebrates, retaining the periventricular arrangement of cells in birds and mammals. Due to dramatic species-specific differences in the organization of the TRH-containing system in the brains of different vertebrates, additional studies exploring the neuroanatomy of TRH elements in the brain are required. The central inhibitory effect of NPY on the hypophysiotropic TRH neurons in *T. guttata* is comparable to that in rodents. However, even with the lowest number of TRH progenitor sequences in the proTRH among vertebrates, the PVN TRH neurons and TRH-NPY interaction in *T. guttata* were more rapidly responsive to the dynamic changes in energy states. The nonhypophysiotropic TRH system regulating hypophysiotropic DA neurons and pituitary function in teleosts appears conserved in mammals. In the human brain, TRH-containing terminals were seen on the DAergic neurons residing in the dorsal part of the PVN as well as in other hypothalamic DAergic neuronal groups. In addition, TRH also controls DA release in the nucleus accumbens and controls food reward in rodents. TRH may therefore serve as a crucial regulator of DAergic pathways in the brain, controlling a range of physiological activities and behavior. TRH-producing elements occur widely in the brains of vertebrates and are believed to perform crucial functions, yet the full spectrum of their functional attributes remains unexplored. Studies should focus on exploring the central projections and significance of non-hypophysiotropic/extrahypothalamic TRHergic neuronal groups in the brain. Jackson and Reichlin (1974) suggested that TRH displays features of a neurotransmitter. It has a low molecular weight, is soluble in water, and is quickly inactivated in blood/tissue. Given these characteristics, its high abundance in crucial brain regions, and the multiple copies produced from a proTRH peptide, TRH might have evolved as a more preferred neurochemical/neurotransmitter for neuronal communication.

Key References

1. Jackson IMD, Reichlin S (1974) Endocrinology 95, 854–862
[A comparative study on distribution of TRH in the hypothalamus and other brain regions of chordates].
2. Van Sinay E, Mirabeau O, Depuydt G, Van Hiel MB, Peymen K, Watteyne J, Zels S, Schoofs L, Beets I (2017) Proc. Natl. Acad. Sci. U S A.
[Analyzed the TRH aa sequences in different phylogenetic clades]
3. Singh O, Singh D, Mitra S, Kumar A, Lechan RM, Singru PS (2023) Endocrinology 164(3).
[The study demonstrates the importance of NPY-TRH interaction in the regulation of energy balance in a bird model]
4. Fekete C, Lechan RM (2007) Front Neuroendocrinol. 28(2–3):97–114. [Hypophysiotropic TRHergic system and its regulation].
5. Singh O, Pradhan DR, Nagalakashmi B, Kumar S, Mitra S, Sagarkar S, Sakharkar AJ, Lechan RM, Singru PS (2019) J. Comp. Neurol. 527(6):1070–1101.
[This study show a detail neuroanatomical distribution of TRH-containing neurons/fibers in the brain and interaction between TRH and dopaminergic system in the preoptic area of the catfish].

Acknowledgments We thank Professor Nishikant Subhedar, Indian Institute of Science Education and Research (IISER)-Pune, India for critical reading of this manuscript and suggestions. We are grateful to NISER-Bhubaneswar, Department of Atomic Energy (DAE, Grant # RIN4002), and the Science and Engineering Research Board (SERB), Department of Science and Technology (Grant # CRG/2021/007466), Govt. of India for support.

References

Alvarez-Salas E, García-Luna C, Soberanes-Chávez P, de Gortari P (2022) Role of the thyrotropin-releasing hormone of the limbic system in mood and eating regulation. J Integr Neurosci 21(2):47

Ando H, Ando J, Urano A (1998) Localization of mRNA encoding thyrotropin-releasing hormone precursor in the brain of sockeye salmon. Zool Sci 15:945–953

Aoki Y, Takahashi M, Masuda T, Tsukamoto T, Iigo M, Yanagisawa T (2005) Molecular cloning of prepro-thyrotropin-releasing hormone cDNAs from the common carp *Cyprinus carpio* and goldfish *Carassius auratus*. Gen Comp Endocrinol 141:84–92

Aoki Y, Ono H, Yasuo S, Masuda T, Yoshimura T, Ebihara S, Iigo M, Yanagisawa T (2007) Molecular evolution of prepro-thyrotropin-releasing hormone in the chicken (*Gallus gallus*) and its expression in the brain. Zool Sci 24:686–692

Batten TFC, Moons L, Cambre ML, Vandesande F, Seki T, Suzuki M (1990) Thyrotropin-releasing hormone-immunoreactive system in the brain and pituitary gland of the sea bass (*Dicentrarchus labrax*, Teleostei). Gen Comp Endocrinol 79:385–392

Bauknecht P, Jékely G (2015) Large-scale combinatorial deorphanization of *Platynereis* neuropeptide GPCRs. Cell Rep 12(4):684–693

Bidaud I, Galas L, Bulant M, Jenks BG, Ouwens DTWM, Jégou S, Ladram A, Roubos EW, Tonon M-C, Nicolas P, Vaudry H (2004) Distribution of the mRNAs encoding the thyrotropin-releasing hormone (TRH) precursor and three TRH receptors in the brain and pituitary of *Xenopus laevis*: effect of background color adaptation on TRH and TRH receptor gene expression. J Comp Neurol 477:11–28

Boswell T, Li Q, Takeuchi S (2002) Neurons expressing neuropeptide Y mRNA in the infundibular hypothalamus of Japanese quail are activated by fasting and co-express agouti-related protein mRNA. Brain Res Mol Brain Res 100:31–42

Bulant M, Richter K, Kuchler K, Kreil G (1992) A cDNA from brain of *Xenopus laevis* coding for a new precursor of thyrotropin-releasing hormone. FEBS Lett 296:292–296

Campinho MA (2019) Teleost metamorphosis: the role of thyroid hormone. Front Endocrinol (Lausanne) 10:383

Chatterjee A, Hsieh YL, Yu JYL (2001) Molecular cloning of cDNA encoding thyroid stimulating hormone β subunit of bighead carp *Aristichthys nobilis* and regulation of its gene expression. Mol Cell Endocrinol 174:1–9

Chien JT, Chowdhury I, Lin YS, Liao CF, Shen ST, Yu JYL (2006) Molecular cloning and sequence analysis of a cDNA encoding pituitary thyroid stimulating hormone β-subunit of the Chinese soft-shell turtle *Pelodiscus sinensis* and regulation of its gene expression. Gen Comp Endocrinol 146:74–82

Del Carmen De Andrés M, Anadón R, Manso MJ, González MJ (2002) Distribution of thyrotropin-releasing hormone immunoreactivity in the brain of larval and adult sea lampreys, *Petromyzon marinus* L. J Comp Neurol 453:323–335

Díaz ML, Becerra M, Manso MJ, Anadón R (2002) Distribution of thyrotropin-releasing hormone (TRH) immunoreactivity in the brain of the zebrafish (*Danio rerio*). J Comp Neurol 450:45–60

Domínguez L, López JM, González A (2008) Distribution of thyrotropin-releasing hormone (TRH) immunoreactivity in the brain of urodele amphibians. Brain Behav Evol 71:231–246

Dudas B, Merchenthaler I (2020) Thyrotropin-releasing hormone axonal varicosities appear to innervate dopaminergic neurons in the human hypothalamus. Brain Struct Funct 225(7):2193–2201

Eales JG, Himick BA (1988) The effects of TRH on plasma thyroid hormone levels of rainbow trout (*Salmo gairdneri*) and arctic charr (*Salvelinus alpinus*). Gen Comp Endocrinol 72:333–339

Fekete C, Lechan RM (2007) Negative feedback regulation of hypophysiotropic thyrotropin-releasing hormone (TRH) synthesizing neurons; role of neuronal afferents and type 2 deiodinase. Front Neuroendocrinol 28(2-3):97–114

Fekete C, Lechan RM (2014) Central regulation of hypothalamic-pituitary-thyroid axis under physiological and pathophysiological conditions. Endocr Rev 35:159–194

Feng K, Su J, Wu Z, Su S, Yao W (2022) Molecular cloning and expression analysis of thyrotropin-releasing hormone, and its possible role in gonadal differentiation in rice field eel *Monopterus albus*. Animals 12(13):1691

Fleming AA, Todd AJ (1994) Thyrotropin-releasing hormone- and GABA-like immunoreactivity coexist in neurons in the dorsal horn of the rat spinal cord. Brain Res 638:347–351

Fliers E, Noppen NW, Wiersinga WM, Visser TJ, Swaab DF (1994) Distribution of thyrotropin-releasing hormone (TRH)-containing cells and fibers in the human hypothalamus. J Comp Neurol 350(2):311–323

Fliers E, Boelen A, van Trotsenburg AS (2014) Central regulation of the hypothalamo-pituitary-thyroid (HPT) axis: focus on clinical aspects. Handb Clin Neurol 124:127–138

Fujihara N, Shiino M (1983) Stimulatory effect of thyrotrophin-releasing hormone (TRH) on the release of luteinizing hormone (LH) from rat anterior pituitary cells *in vitro*. Can J Physiol Pharmacol 61:186–189

Galas L, Raoult E, Tonon M-C, Okada R, Jenks BG, Castano JP, Kikuyama S, Malagon M, Roubos EW, Vaudry H (2009) TRH acts as a multifunctional hypophysiotropic factor in vertebrates. Gen Comp Endocrinol 164:40–50

Gkonos PJ, Tavianinif MA, Liu CC, Roost BA (1989) Thyrotropin-releasing hormone gene expression in normal thyroid parafollicular cells. Mol Endocrinol 3:2101–2109

Golan M, Zelinger E, Zohar Y, Levavi-Sivan B (2015) Architecture of GnRH-gonadotrope-vasculature reveals a dual mode of gonadotropin regulation in fish. Endocrinology 156:4163–4173

Goldstein J, Perello M, Nillni EA (2007) PreproThyrotropin-releasing hormone 178-199 affects tyrosine hydroxylase biosynthesis in hypothalamic neurons: a possible role for pituitary prolactin regulation. J Mol Neurosci 31(1):69–82

Grattan DR (2015) 60 years of neuroendocrinology: the hypothalamo-prolactin axis. J Endocrinol 226:T101–T122

Greer JJ, Carter JE, Allan DW (1996) Respiratory rhythm generation in a precocial rodent in vitro preparation. Respir Physiol 103:105–112

Guldenaar SE, Veldkamp B, Bakker O, Wiersinga WM, Swaab DF, Fliers E (1996) Thyrotropin-releasing hormone gene expression in the human hypothalamus. Brain Res 743(1–2):93–101

Han YS, Liao IC, Tzeng WN, Yu JYL (2004) Cloning of the cDNA for thyroid stimulating hormone β subunit and changes in activity of the pituitary-thyroid axis during silvering of the Japanese eel, *Anguilla japonica*. J Mol Endocrinol 32:179–194

Harris AR, Christianson D, Smith MS, Fang SL, Braverman LE, Vagenakis AG (1978) The physiological role of thyrotropin-releasing hormone in the regulation of thyroid-stimulating hormone and prolactin secretion in the rat. J Clin Invest 61:441–448

Heyland A, Reitzel AM, Hodin J (2004) Thyroid hormones determine developmental mode in sand dollars (Echinodermata: Echinoidea). Evol Dev 6:382–392

Ishikawa K, Taniguchi Y, Inoue K, Kurosumi K, Suzuki M (1988) Immunocytochemical delineation of thyrotrophic area: origin of thyrotropin-releasing hormone in the median eminence. Neuroendocrinology 47(5):384–388

Jackson IM, Reichlin S (1974) Thyrotropin-releasing hormone (TRH): distribution in hypothalamic and extrahypothalamic brain tissues of mammalian and submammalian chordates. Endocrinology 95:854–862.

Józsa R, Korf H-W, Csernns V, Mess B (1988) Thyrotropin-releasing hormone (TRH)-immunoreactive structures in the brain of the domestic mallard. Cell Tissue Res 251:441–449

Kádár A, Sánchez E, Wittmann G, Singru PS, Füzesi T, Marsili A, Larsen PR, Liposits Z, Lechan RM, Fekete C (2010) Distribution of hypophysiotropic thyrotropin-releasing hormone (TRH)-synthesizing neurons in the hypothalamic paraventricular nucleus of the mouse. J Comp Neurol 518:3948–3961

King JA, Millar RP (1981) TRH, GH-RIH, and LH-RH in metamorphosing *Xenopus laevis*. Gen Comp Endocrinol 44(1):20–27

Knobloch HS, Grinevich V (2014) Evolution of oxytocin pathways in the brain of vertebrates. Front Behav Neurosci 8:31

Koskinen L, Bill A (1984) Thyrotropin-releasing hormone (TRH) causes sympathetic activation and cerebral vasodilation in the rabbit. Acta Physiol Scand 122:127–136

Kotwica-Rolinska J, Krištofová L, Chvalová D, Pauchová L, Provazník J, Hejníková M, Sehadová H, Lichý M, Vaněčková H, Doležel D (2020) Functional analysis and localisation of a thyrotropin-releasing hormone-type neuropeptide (EFLa) in hemipteran insects. Insect Biochem Mol Biol 122:103376

Kreider MS, Winokur A, Krieger NR (1981) The olfactory bulb is rich in TRH immunoreactivity. Brain Res 217:69–77

Kuenzel WJ, Douglass LW, Davison BA (1987) Robust feeding following central administration of neuropeptide Y or peptide YY in chicks, *Gallus domesticus*. Peptides 8:823–828

Lamacz M, Hindelang C, Tonon MC, Vaudry H, Stoeckel ME (1989) Three distinct thyrotropin-releasing hormone-immunoreactive axonal systems project in the median eminence-pituitary complex of the frog *Rana ridibunda*. Immunocytochemical evidence for co-localization of thyrotropin-releasing hormone and mesotocin in fibers. Neuroscience 32:451–462

Lanciego JL, Wouterlood FG (2020) Neuroanatomical tract-tracing techniques that did go viral. Brain Struct Funct 225:1193–1224

Lechan RM, Jackson IMD (1982) Immunohistochemical localization of thyrotropin-releasing hormone in the rat hypothalamus and pituitary. Endocrinology 111:55–65

Lechan RM, Kakucska I (1992) Feedback regulation of thyrotropin-releasing hormone gene expression by thyroid hormone in the hypothalamic paraventricular nucleus. Ciba Found Symp 168:144–158

Lechan RM, Wu P, Jackson IMD (1986) Immunolocalization of the thyrotropin-releasing hormone prohormone in the rat central nervous system. Endocrinology 119:1210–1216

Lee SL, Stewart K, Goodman RH (1988) Structure of the gene encoding rat thyrotropin releasing hormone. J Biol Chem 263:16604–16609

Lin MT, Chandra A, Chern YF, Tsay BL (1980) Effects of thyrotrophic-releasing hormone (TRH) on thermoregulation in the rat. Experientia 36:1077–1078

López JM, Domínguez L, González A (2008) Immunohistochemical localization of thyrotropin-releasing hormone in the brain of reptiles. J Chem Neuroanat 36:251–263

Lyons DJ, Horjales-Araujo E, Broberger C (2010) Synchronized network oscillations in rat tuberoinfundibular dopamine neurons: switch to tonic discharge by thyrotropin-releasing hormone. Neuron 65:217–229

Matz SP, Takahashi TT (1994) Immunohistochemical localization of thyrotropin-releasing hormone in the brain of Chinook salmon (*Oncorhynchus tshawytscha*). J Comp Neurol 345:214–223

Merchenthaler I (1991) Co-localization of enkephalin and TRH in perifornical neurons of the rat hypothalamus that project to the lateral septum. Brain Res 544(1):177–180

Merchenthaler I, Csernus V, Csontos C, Petrusz P, Mess B (1988) New data on the immunocytochemical localization of thyrotropin-releasing hormone in the rat central nervous system. Am J Anat 181:359–376

Mihály E, Fekete C, Tatro JB, Liposits Z, Stopa Ed G, Lechan RM (2000) Hypophysiotropic thyrotropin-releasing hormone-synthesizing neurons in the human hypothalamus are innervated by neuropeptide Y, agouti-related protein, and α-melanocyte-stimulating hormone. J Clin Endocrinol Metab 85:2596–2603

Mimnagh KM, Bolaffi JL, Montgomery NM, Kaltenbach JC (1987) Thyrotropin-releasing hormone (TRH): immunohistochemical distribution in tadpole and frog brain. Gen Comp Endocrinol 66:394–404

Mirabeau O, Joly J-S (2013) Molecular evolution of peptidergic signaling systems in bilaterians. Proc Natl Acad Sci USA 110:E2028–E2037

Miranda LA, Affanni JM (2000) Thyrotropin-releasing hormone immunoreactivity in the brain and the pituitary during *Bufo arenarum* development. Int J Dev Neurosci 18:47–52

Nakano M, Hasunuma I, Minagawa A, Iwamuro S, Yamamoto K, Kikuyama S, Machida T, Kobayashi T (2018) Possible involvement of thyrotropin-releasing hormone receptor 3 in the release of prolactin in the metamorphosing bullfrog larvae. Gen Comp Endocrinol 267:36–44

Nillni EA (2010) Regulation of the hypothalamic thyrotropin releasing hormone (TRH) neuron by neuronal and peripheral inputs. Front Neuroendocrinol 31:134–156

Obermüller-Wilén H (1979) A neurosecretory system in the brain of the lancelet, *Branchiostoma lanceolatum*. Acta Zool 60:187–196

Ohta N, Kubota I, Takao T, Shimonishi Y, Yasuda-Kamatani Y, Minakata H, Nomoto K, Muneoka Y, Kobayashi M (1991) Fulicin, a novel neuropeptide containing a D-amino acid residue isolated from the ganglia of *Achatina fulica*. Biochem Biophys Res Commun 178:486–493

Paris M, Laudet V (2008) The history of a developmental stage: metamorphosis in chordates. Genesis 46:657–672

Péczely P, Kiss JZ (1988) Immunoreactivity to vasoactive intestinal polypeptide (VIP) and thyreotropin-releasing hormone (TRH) in hypothalamic neurons of the domesticated pigeon (*Columba livia*). Alterations following lactation and exposure to cold. Cell Tissue Res 251:485–494

Peter RE (1977) The preoptic nucleus in fishes: a comparative discussion of function-activity relationships. Am Zool 17:775–785

Prasad C, Wilber JF, Amborski RL (1982) Thyrotropin-releasing hormone: its distribution and metabolism during development in bullfrog (*Rana catesbeiana*). Dev Neurosci 5:293–297

Preece H, Licht P (1987) Effects of thyrotropin-releasing hormone in vitro on thyrotropin and prolactin release from the turtle pituitary. Gen Comp Endocrinol 67:247–255

Pu LP, Ma W, Barker JL, Loh P (1996) Differential coexpression of genes encoding prothyrotropin-releasing hormone (pro-TRH) and prohormone convertases (PC1 and PC2) in rat brain neurons: implications for differential processing of pro-TRH. Endocrinology 137:1233–1241

Puga L, Alcántara-Alonso V, Coffeen U, Jaimes O, de Gortari P (2016) TRH injected into the nucleus accumbens shell releases dopamine and reduces feeding motivation in rats. Behav Brain Res 306:128–136

Reichlin S (1989) TRH: historical aspects. Ann N Y Acad Sci 553:1–6

Riedel W, Burke SL (1988) Selective autonomic nervous control of thyroid hormone and calcitonin secretion during metabolic and cardiorespiratory activation by intracisternal thyrotropin-releasing hormone (TRH). J Auton Nerv Syst 24:157-173

Sage M (1973) The evolution of thyroidal function in fishes. Am Zool 13:899–905

Sánchez E, Singru PS, Acharya R, Bodria M, Fekete C, Zavacki AM, Bianco AC, Lechan RM (2008) Differential effects of refeeding on melanocortin-responsive neurons in the hypothalamic paraventricular nucleus. Endocrinology 149:4329–4335

Sánchez-Jaramillo E, Wittmann G, Menyhért J, Singru P, Gómez-González GB, Sánchez-Islas E, Yáñez-Recendis N, Pimentel-Cabrera JA, León-Olea M, Gereben B, Fekete C, Charli J-L, Lechan RM (2022) Origin of thyrotropin-releasing hormone neurons that innervate the tuberomammillary nuclei. Brain Struct Funct 227:2329–2347

Scanes CG (1974) Some in vitro effects of synthetic thyrotrophin releasing factor on the secretion of thyroid stimulating hormone from the anterior pituitary gland of the domestic fowl. Neuroendocrinology 15:1–9

Schwartz MW, Woods SC, Porte Jr D, Seeley RJ, Baskin DG (2000) Central nervous system control of food intake. Nature 404:661–671

Segerson TP, Hoefler H, Childers H, Wolfe HJ, Wu P, Jackson IMD, Lechan RM (1987) Localization of thyrotropin-releasing hormone prohormone messenger ribonucleic acid in rat brain by *in situ* hybridization. Endocrinology 121:98–107

Seki T, Nakai Y, Shioda S, Mitsuma T, Kikuyama S (1983) Distribution of immunoreactive thyrotropin-releasing hormone in the forebrain and hypophysis of the bullfrog, *Rana catesbeiana*. Cell Tissue Res 233:507–516

Singh U, Kumar S, Singru PS (2012) Interaction between dopamine- and isotocin-containing neurones in the preoptic area of the catfish, *Clarias batrachus*: role in the regulation of luteinising hormone cells. J Neuroendocrinol 24:1398–1411

Singh O, Kumar S, Singh U, Kumar V, Lechan RM, Singru PS (2016) Cocaine-and amphetamine-regulated transcript peptide (CART) in the brain of zebra finch, *Taeniopygia guttata*: organization, interaction with neuropeptide Y, and response to changes in energy status. J Comp Neurol 524:3014–3041

Singh O, Pradhan DR, Nagalakashmi B, Kumar S, Mitra S, Sagarkar S, Sakharkar AJ, Lechan RM, Singru PS (2019) Thyrotropin-releasing hormone (TRH) in the brain and pituitary of the teleost, *Clarias batrachus* and its role in regulation of hypophysiotropic dopamine neurons. J Comp Neurol 527:1070–1101

Singh O, Singh D, Mitra S, Kumar A, Lechan RM, Singru PS (2023) TRH and NPY interact to regulate dynamic changes in energy balance in the male zebra finch. Endocrinology 164(3)

Sower SA (2015) Breaking dogma on the hypothalamic-pituitary anatomical relations in vertebrates. Endocrinology156 (11): 3882–3884

Swanson LW, Kuypers HG (1980) The paraventricular nucleus of the hypothalamus: cytoarchitectonic subdivisions and organization of projections to the pituitary, dorsal vagal complex, and spinal cord as demonstrated by retrograde fluorescence double-labeling methods. J Comp Neurol 194:555–570

Teijido O, Manso MJ, Anadón R (2002) Distribution of thyrotropin-releasing hormone immunoreactivity in the brain of the dogfish *Scyliorhinus canicula*. J Comp Neurol 454:65–81

Ter Horst GJ, Mast JG, Vaartjes K (1988) A device for enhancing the visibility of the tip of a glass micropipette: application in neuroanatomy. Brain Res Bull 21(6):917–918

True C, Verma S, Grove KL, Smith MS (2013) Cocaine-and amphetamine-regulated transcript is a potent stimulator of GnRH and kisspeptin cells and may contribute to negative energy balance-induced reproductive inhibition in females. Endocrinology 154(8):2821–2832

Tsuruo Y, Hökfelt T, Visser T (1987) Thyrotropin releasing hormone (TRH)-immunoreactive cell groups in the rat central nervous system. Exp Brain Res 68:213–217

Turner KJ, Hawkins TA, Yáñez J, Anadón R, Wilson SW, Folgueira M (2016) Afferent connectivity of the zebrafish habenulae. Front Neural Circuits 10:30

Van Sinay E, Mirabeau O, Depuydt G, Van Hiel MB, Peymen K, Watteyne J, Zels S, Schoofs L, Beets I (2017) Evolutionarily conserved TRH neuropeptide pathway regulates growth in *Caenorhabditis elegans*. Proc Natl Acad Sci USA 114:E4065 LP-E4074

Vandenborne K, Roelens SA, Darras VM, Kuhn ER, Van der Geyten S (2005) Cloning and hypothalamic distribution of the chicken thyrotropin-releasing hormone precursor cDNA. J Endocrinol 186:387–396

Veenstra JA, Šimo L (2020) The TRH-ortholog EFLamide in the migratory locust. Insect Biochem Mol Biol 116:103281

Veenstra JA, Rombauts S, Grbić M (2012) *In silico* cloning of genes encoding neuropeptides, neurohormones and their putative G-protein coupled receptors in a spider mite. Insect Biochem Mol Biol 42:277–295

Veronesi MC, Kubek DJ, Kubek MJ (2007) Intranasal delivery of a thyrotropin-releasing hormone analog attenuates seizures in the amygdala-kindled rat. Epilepsia 48:2280–2286

Wittmann G, Füzesi T, Liposits Z, Lechan RM, Fekete C (2009a) Distribution and axonal projections of neurons co-expressing thyrotropin-releasing hormone and urocortin 3 in the rat brain. J Comp Neurol 517:825–840

Wittmann G, Füzesi T, Singru PS, Liposits Z, Lechan RM, Fekete C (2009b) Efferent projections of thyrotropin-releasing hormone-synthesizing neurons residing in the anterior parvocellular subdivision of the hypothalamic paraventricular nucleus. J Comp Neurol 515:313–330

Yamamoto N (2009) Studies on the teleost brain morphology in search of the origin of cognition. Jpn Psychol Res 51:154–167

Youson JH (1997) Is lamprey metamorphosis regulated by thyroid hormones? Am Zool 37:441–460

Zhu W, Chang L, Zhao T, Wang B, Jiang J (2020) Remarkable metabolic reorganization and altered metabolic requirements in frog metamorphic climax. Front Zool 17:30

Zoeller RT, Conway KM (1989) Neurons expressing thyrotropin-releasing hormone-like messenger ribonucleic acid are widely distributed in *Xenopus laevis* brain. Gen Comp Endocrinol 76:139–146

The Evolution of Hypothalamic-Pituitary-Thyroid Axis

7

Iván Lazcano, Patricia Joseph-Bravo, and Aurea Orozco

Abstract

The hypothalamic-pituitary-thyroid axis (HPT) is the main regulator of the synthesis and secretion of thyroid hormones. It is composed of the hypothalamus, the pituitary gland, and the thyroid gland, which communicate with each other through different messengers. A comparative analysis of vertebrates, including lampreys, fish, amphibians, reptiles, avians, and mammals, can recreate the evolution of this axis. The study of these species has facilitated a comprehensive understanding of the anatomy and functionality of the glands and molecules involved in the communication of the HPT axis during vertebrate evolution. Furthermore, scientists have been able to trace the rudiments of the glands in invertebrates, such as tunicates and amphioxus, and the evolution of ancestral

For simplicity, species-specific gene and protein nomenclature was not observed. For all species, gene symbols are italicized and first letter in capitals (e.g., *Trh*), and proteins are in capital letters and not italicized (e.g., TRH).

I. Lazcano
Instituto de Neurobiología, Universidad Nacional Autónoma de México (UNAM), Campus Juriquilla, Santiago de Querétaro, Mexico

P. Joseph-Bravo
Departamento de Genética del Desarrollo y Fisiología Molecular, Instituto de Biotecnología, Universidad Nacional Autónoma de México (UNAM), Chimalpa, Cuernavaca, Mexico

A. Orozco (✉)
Instituto de Neurobiología, Universidad Nacional Autónoma de México (UNAM), Campus Juriquilla, Santiago de Querétaro, Mexico

Escuela Nacional de Estudios Superiores, Unidad Juriquilla, Universidad Nacional Autónoma de México (UNAM), Campus Juriquilla, Santiago de Querétaro, Mexico
e-mail: aureao@unam.mx

V. Grinevich, R. F. Oliveira (eds.), *Evolutionary and Comparative Neuroendocrinology*, Masterclass in Neuroendocrinology 17,
https://doi.org/10.1007/978-3-031-80209-6_7

molecules in the vertebrate HPT axis. Analyzing changes in genes, molecules, proteins, and structures that comprise the HPT axis and its involvement in energy homeostasis and thermogenesis may provide insights into the evolutionary processes that have fine-tuned this neuroendocrine network.

Keywords

Thyroid hormones · Thyroid-stimulating hormone · Thyrotropin-releasing hormone · Vertebrate evolution · Hypothalamus · Pituitary · Thyroid gland · Energy homeostasis · Thermogenesis

7.1 Introduction

In mammals, the hypothalamic-pituitary-thyroid (HPT) axis maintains the metabolic rate, energy expenditure, and corporal temperature within a physiological range (Fekete and Lechan 2014; Joseph-Bravo et al. 2015a; Ortiga-Carvalho et al. 2016). It is composed of specialized neuroendocrine cells located in the paraventricular nucleus (PVN) of the hypothalamus, whose nerve endings secrete thyrotropin-releasing hormone (TRH) into the median eminence. Portal capillaries then transport this hormone to the pituitary, reaching the thyrotropes and stimulating the synthesis and release of thyroid-stimulating hormone (TSH, or thyrotropin) (Fekete and Lechan 2014; Joseph-Bravo et al. 2015a). TSH travels through the blood circulation to the thyroid gland, where it induces the synthesis and secretion of thyroid hormones (THs), the final effectors of this neuroendocrine network that, through efficient feedback processes, maintain the homeostasis of the axis (Chatzitomaris et al. 2017; Joseph-Bravo et al. 2021). Most research on the HPT axis has been done in mammals, but studies on its evolution are scarce in contrast to other endocrine axes, such as those of stress and reproduction (Denver 2009; Takahashi et al. 2016; Kanda 2019; Ortega et al. 2021). Studies in nonmammalian vertebrates have revealed conserved essential elements of the HPT axis, including peculiar anatomical, cellular, and molecular characteristics between classes. For example, some species do not have a median eminence or portal capillaries, but neuroendocrine cells directly innervate the pituitary gland. Only in mammals is TRH the TSH-releasing factor (TRF); most nonmammalian species use other peptides as TRFs (De Groef et al. 2006; Galas et al. 2009; Lazcano et al. 2021) (see Sect. 7.2). As vertebrates transitioned from ectotherms to endotherms and from aquatic to terrestrial environments, the HPT axis underwent significant adaptations, probably due to the demands of the new ecological niches. This highlights the importance of finely regulating this neuroendocrine network system and the synthesis and release of THs. In addition, the molecular components of the HPT axis in vertebrates, such as TRFs and their receptors, TSH, TSH receptors (TSH-R), and TH signaling components (TH carriers, transporters, deiodinase enzymes, and TH receptors (THRs)), have also adapted over time. Finally, these adaptations promoted the evolution of feedback and regulatory mechanisms involving organs and hormones in the HPT axis and stimulated the adjustment of TH synthesis and secretion in response to environmental cues. In this

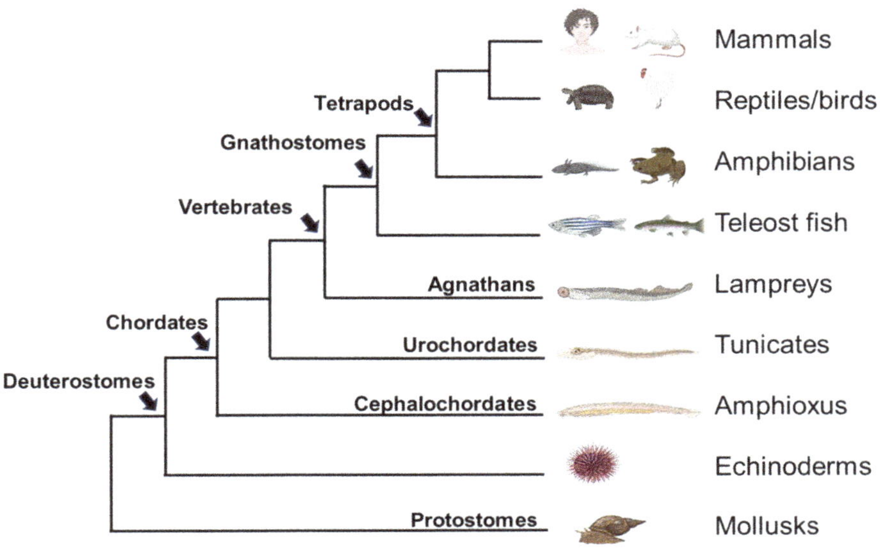

Fig. 7.1 Phylogenetic relationships between the species analyzed in this chapter. A simplified phylogenetic tree showing the evolutionary relationships between the invertebrates and vertebrates, where the evolution of the HPT axis has been studied. Some invertebrates are included in the phylogenetic tree. The origin of the most important groups is indicated (arrows). Figure created with BioRender.com

chapter, we explore the anatomical, cellular, and molecular components of the HPT axis and exemplify the diverse actions of this axis in different vertebrate groups. We also illustrate some of the evolutionary changes that have shaped this intricate endocrine network by analyzing invertebrates that are phylogenetically close to vertebrates (i.e., amphioxus and tunicates), basal vertebrates that lack jaws (i.e., lampreys), and jawed vertebrates (i.e., fish, amphibians, reptiles, birds and mammals) (Fig. 7.1).

7.2 Hypothalamus as the Central Regulator of the HPT Axis

7.2.1 Hypothalamic Hormones

The discovery that the hypothalamus contained releasing and inhibitory factors that control the secretion of pituitary hormones (Guillemin 2005) confirmed the hypothesis proposed by Harris in 1955 about the neural control of the pituitary gland (Raisman 1997). TRH was the first hypothalamic peptide to be characterized (reviewed in: Joseph-Bravo et al. (2015a)), followed by other releasing factors, including somatostatin, a peptide of 14 amino acids that inhibits the release of TSH, growth hormone, and corticotropin-releasing hormone (CRH). In mammals, CRH is released in response to stress and regulates the activity of the hypothalamus-pituitary-adrenal axis through adrenocorticotropin secretion from the pituitary (Vale et al. 1981; Hauger and Dautzenberg 2000; Guillemin 2005).

TRH is a tripeptide (pGlu-His-ProNH2) synthesized from a protein precursor gene that encodes several copies of the immediate precursor molecule Gln-His-Pro-Gly (prepro-TRH). During posttranslational modifications, specific enzymes convert Gln residues to pGlu, and terminal Gly is partially cleaved, leaving the amino group in proline as an amidated residue ProNH2 (reviewed in Joseph-Bravo et al. (2021)). The mature peptide and its mRNA have been found in the hypothalamic neuroendocrine cells of mammals (Lechan and Jackson 1982; Yamada et al. 1990), reptiles (López et al. 2008), birds (Aoki et al. 2007), amphibians (Mimnagh et al. 1987; Dominguez et al. 2008; López et al. 2008), and fish (Hamano et al. 1990; Matz and Takahashi 1994; Ando et al. 1998; Díaz et al. 2002), which is consistent with its role as a neurohormone. It has been recognized that TRH induces TSH synthesis and secretion in mammals. In many nonmammalian vertebrates, TRH does not stimulate TSH synthesis or release but does elicit the release of other pituitary hormones due to the presence of TRH receptors (TRH-Rs) in different pituitary cells. Diverse neuroendocrine effects of TRH in nonmammalian vertebrates have been documented (Galas et al. 2009). In vertebrates, the TRH gene contains several copies of the sequence that encodes prepro-TRH. These copies vary according to species; for example, there are six copies in humans (Yamada et al. 1990), five in rats and mice (Lechan et al. 1986; Satoh et al. 1992), four in chickens (Aoki et al. 2007), seven in frogs (Kuchler et al. 1990), and seven to eight copies in fish (Hamano et al. 1990; Matz and Takahashi 1994; Ando et al. 1998; Díaz et al. 2002). In all vertebrates (since agnathans), the structure of the mature TRH tripeptide is the same. TRH-like peptides have been reported in the whole brain of amphioxus and lamprey (Jackson and Reichlin 1974); however, these findings contrast with those from the amphioxus genome, where the *Trh* gene was not found (Holland et al. 2008). Other *Trh* orthologs have been found in various invertebrates, but they exhibit differences in their sequence (Lazcano et al. 2021). For example, in the sea urchin, researchers found a cDNA that encodes a protein containing ten copies of the peptide (Gln-Tyr-Pro-Gly-Gly). Through possible posttranslational modifications, this peptide could generate the sequence pGlu-Tyr-Pro-Gly-NH2, a peptide similar to TRH. Thus, it would seem that ancestral *Trh* genes contained a greater number of *TRH* copies and that prepro-TRH and the mature peptide have been conserved over the course of evolution, at least since the appearance of fish, suggesting TRH as part of vertebrate neuroendocrine innovation (Rowe and Elpick 2012).

CRH belongs to a family of neuropeptides, which includes urocortins, urotensin, and sauvagine. In mammals, CRH is a large peptide (41 amino acids) that is released in response to stress and controls the hypothalamic-pituitary-adrenal axis (Bale and Vale 2004; Aguilera and Liu 2012). Its protein precursor is synthesized in the neuronal cell body and compartmentalized into secretion granules, where they are processed into the final product upon their transport from the terminal. Both CRH and TRH peptides or their mRNAs have been localized in several brain regions; however, only the CRH or TRH present in the neurons of the PVN that project to the median eminence have a neuroendocrine or hypophysiotropic role in mammals

(Hauger and Dautzenberg 2000). Moreover, there is evidence indicating that CRH could be the main TSH inducer in fish, amphibians, reptiles, and birds (Denver 1988; Denver and Licht 1991; Geris et al. 1996; Larsen et al. 1998; De Groef et al. 2003; Okada et al. 2005), leading to the proposal that CRH is the main activator of the HPT axis in nonmammalian vertebrates. The presence of the *Crh* gene has been demonstrated in amphioxus and lampreys. It would be interesting to investigate whether CRH induces the synthesis and release of the TSH archetype thyrostimulin in these species (Sect. 7.3.2). Some vertebrate species, such as sharks, contain two genes that code for CRH (*Crh1* and *Crh2*). Although orthologs of the CRH gene have been found in several invertebrate species, only the vertebrate sequence contains features that confer the well-known biological activity in mammals. In humans, the N-terminal region of CRH activates the receptor, whereas the C-terminal region allows the hormone to specifically recognize it (On et al. 2019).

The invertebrate orthologs of *Trh* and *Crh* (Mirabeau and Joly 2013; Bauknecht and Jékely 2015; Van Sinay et al. 2017; On et al. 2019; Veenstra and Šimo 2020) suggested that they probably evolved into their current vertebrate homologs. At the neuroendocrine level, CRH acquired the ability to activate the HPT axis, a function later gained by TRH, the key HPT activator in mammals.

Finally, the glucagon-like peptide (GCGL), which is specific to nonmammalian vertebrates, has been found to regulate TSH secretion in chickens by acting through a newly discovered receptor, GCGL-R. In these birds, GCGL appears to have high specificity for inducing TSH secretion without affecting the secretion of other pituitary hormones, such as growth hormone, prolactin, luteinizing hormone, follicle-stimulating hormone, or adrenocorticotropic hormone (Huang et al. 2014).

7.2.2 Hypothalamus-Pituitary Interaction

At the morphological level, in all vertebrates, the hypothalamus communicates with the pituitary gland, but this interaction presents singularities between vertebrate classes (Fig. 7.2). Some rudiments of brain-pituitary interaction can be found in the invertebrate amphioxus, in which discrete contacts between the brain and Hatschek's pit (the homolog of the pituitary gland; see below) probably reflect the earliest neuroendocrine communication (Gorbman 1999). As described later, amphioxus releases thyrostimulin from Hatschek's pit, in turn stimulating the release of THs from the endostyle (see Sect. 7.4.2) (Wang et al. 2018; Zhang and Ji 2022). Whether thyrostimulin release is regulated by a thyrostimulin-releasing factor from the brain is still unknown. In lampreys and fish, at least two different types of communication have been proposed: (1) peptides are released into the median eminence and subsequently transported by a portal system, and (2) peptides are released directly to the anterior pituitary (Sower 2015; Golan et al. 2015). In terrestrial vertebrates, such as amphibians, reptiles, birds, and mammals, the TRF transport mechanism occurs through the portal system (Sower 2015). Vertebrate TRFs from the brain changed during evolution. While TRH is well established as the major mammalian TRF, CRH or GCGL can induce and regulate TSH synthesis and be release in fish,

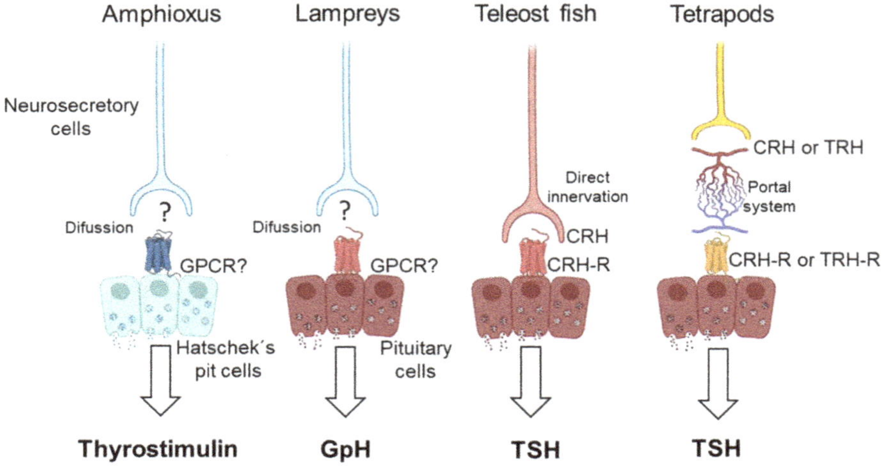

Fig. 7.2 Evolution of the communication between TRFs and the pituitary. A schematic representation of the adaptations in the HPT axis over time, where neuronal projections from the brain release a TRF that has changed throughout vertebrate evolution. A TRF-specific GPCR is expressed in thyrotropes and allows the synthesis and release of TSH (or an ancestral gene) after binding to its ligand. For simplicity, the bird TRF (GCGL) was not represented. Figure created with BioRender.com. Abbreviations: *CRH*, corticotropin-releasing hormone; *CRH-R*, corticotropin-releasing hormone receptor; *GPCR*, G-protein-coupled receptors; *GpH*, glycoprotein hormone; *TRH*, thyrotropin-releasing hormone; *TRH-R*, thyrotropin-releasing hormone receptor; *TSH*, thyroid-stimulating hormone

amphibians, reptiles, and birds (Fig. 7.2) (De Groef et al. 2006; Huang et al. 2014; Lazcano et al. 2021). The selective pressure that led TRFs to change during vertebrate evolution, without modifying the downstream hormones TSH and TH, is not clear.

7.2.3 TRF Regulation

Growing evidence indicates a coevolution between peptide ligands and their respective membrane receptors (Vaudry 2014). Additionally, other regulatory proteins that act in the extracellular space are thought to have coevolved and specifically inactivate certain peptides, including TRFs. This is the case for TRH and CRH, for which more complex coevolution of these ligands, receptors, and inactivation proteins/enzymes has been proposed (Lazcano et al. 2021). Most released peptides are degraded by peptidases with little substrate specificity. In both TRH and CRH peptides, the carboxy-terminal is amidated, making them resistant to proteolysis by general carboxypeptidases. However, TRH-degrading ectoenzyme (TRH-DE) is a specific protein that cleaves the modified amino-terminal of TRH (pyroGlu), removing its biological activity (Charli et al. 2020). This enzyme is anchored to the cell membrane, with its active site directed to the extracellular space (for a major review of peptidases, see Antczak et al. 2001). This ectoenzyme is present in tanycytes

(Box 7.1) and inactivates TRH released into the median eminence before it reaches the portal vessels (Sánchez et al. 2009). Invertebrate *Trh-DE* orthologs do not hydrolyze TRH, but fish, amphibians, reptiles, birds, and mammalian orthologs are known to specifically recognize and hydrolyze the TRH tripeptide due to the

Box 7.1 The Enigma of the Axolotl Neoteny

One of the most fascinating functions of the HPT axis is the induction of metamorphosis in amphibians. It is known that in frogs, there is activation of the HPT axis, which culminates in an increase in the plasma concentrations of THs, promoting the transformation of the tadpole into the classic frog phenotype that we know. However, some amphibians do not undergo metamorphosis. These neotenic species become sexually mature adults but retain a juvenile phenotype (Wakahara 1996). For a long time, researchers have tried to find out the origin of this adaptation, so far without a conclusive answer. The most studied neotenic amphibian is the *Ambystoma mexicanum*, or axolotl, which has been used as a study model in comparative endocrinology and regeneration (Crowner et al. 2019; Wells et al. 2021). Initial experiments showed that feeding axolotls with mammalian thyroid tissue induces the transformation of axolotls into salamanders, their terrestrial version (De Groef et al. 2018). Moreover, axolotl metamorphosis can be experimentally induced if exogenous TH is administered (Page and Randal Voss 2009; Lazcano et al. 2023a, b). Axolotls contain two genes for TRs that are THRA and THRB orthologs THRA and THRB orthologs, demonstrating that they are sensitive to THs and that their THRs are functional (Safi et al. 2004). The same metamorphic induction occurs if TSH of mammalian origin is administered, suggesting that the axis is functional at the level of TSH signaling and that their THS-R, located in the thyroid gland, is responsive to TSH (Taurog 1974). One possible explanation for the lack of natural metamorphosis is that the synthesis and secretion of axolotl TSH are not being efficiently stimulated by TRH or other TRFs. Indeed, the administration of TRH is not capable of inducing metamorphosis, and it is now known that this peptide is not the main activator of TSH synthesis and release. However, it induces the synthesis and release of other hormones, such as prolactin and growth hormone (Galas et al. 2009). In non-mammalian vertebrates, CRH has been attributed as the main activator of the HPT axis. Studies have demonstrated that CRH is present in the preoptic nucleus of amphibians whose nerve endings are located in the median eminence (De Groef et al. 2018). A hypothesis that could explain axolotl neoteny is a possible weak CRH signaling mediated by CRH-R2 in pituitary thyrotropes. However, CRH and CRH-R2 are relatively similar to other vertebrate orthologs and do not have mutations in their sequence that could explain the impaired signaling (De Groef et al. 2018). Nonetheless, the recent publication of the axolotl genome (Nowoshilow et al. 2018) provides an additional tool to unravel the mechanisms underlying the neoteny of this amphibian.

prevalence of particular amino acids in their sequences (Lazcano et al. 2021). Several studies have highlighted the importance of this enzyme in HPT axis control in mammals under conditions like cold exposure or fasting (Sánchez et al. 2009; Lazcano et al. 2015; Joseph-Bravo et al. 2021). In mammals and nonmammalian vertebrates, TRH regulates the synthesis and release of other hypophysiotropic hormones, including prolactin and growth hormone (Galas et al. 2009; Kanasaki et al. 2015). TRH-DE is also present in the pituitary (in lactotrophs but not in thyrotrophs), and its inhibition amplifies TRH's effects on prolactin release (Cruz et al. 2008). This suggests that TRH-DE could regulate the biological activity that TRH has in other pituitary cells in nonmammalian models, but this hypothesis needs to be explored.

Although CRH is not hydrolyzed by a specific peptidase, there is a CRH-binding protein (CRH-BP) that regulates its half-life in the extracellular space (Petraglia et al. 1993; Westphal 2006). *Crh-BP* orthologs have been found in several classes of invertebrates (Lazcano et al. 2021), although their biochemical and functional characterizations are better documented in vertebrates. Depending on vertebrate species, CRH-BP is a monomeric protein composed of 321–324 amino acids. All vertebrate proteins contain an N-linked glycosylation site and ten cysteine residues that form five disulfide bonds, which are important for CRH binding (Fischer et al. 1994; Seasholtz et al. 2002). CRH-BP is expressed in the brain and pituitary of mammals (Potter et al. 1991; Petraglia et al. 1993), birds (Wan et al. 2022), and amphibians (Valverde et al. 2001). CRH has ten times greater affinity for CRH-BP than for its membrane receptors (Sutton et al. 1995); it has thus been suggested that it sequesters CRH in the extracellular space, avoiding binding to its receptors in the pituitary or other tissues. Because CRH-BP regulates the actions of CRH, it has been proposed that it could regulate CRH bioactivity in nonmammalian vertebrates, as TRH-DE does for TRH.

7.3 Evolution of the Pituitary Gland and TSH as the Main Stimulator of the Thyroid Gland

Until the beginning of the twentieth century, the pituitary was considered the master endocrine gland because it maintained the homeostasis of several neuroendocrine axes in all vertebrates. The pituitary is a vertebrate-specific endocrine gland located downstream of the hypothalamus, present in fish and mammals. In marine invertebrates, specifically amphioxus and tunicates, Hatschek's pit region has been suggested as the pituitary homolog (for a detailed review of the biology of Hatschek's pit, see Zhang and Ji (2022)). Anatomically, the vertebrate pituitary is subdivided into three regions: the neurohypophysis, the intermediate lobe, and the adenohypophysis. The latter contains at least five different cell types. One of these cell types, thyrotropes, is responsible for synthesizing TSH (Santiago-Andres et al. 2021).

7.3.1 Vertebrate Thyrotropes Express Receptors for TRFs

Pituitary cells have evolved to recognize neurohormones through the use of G-protein-coupled receptors (GPCRs), the largest family of transmembrane receptors in eukaryotic cells (De Mendoza et al. 2014). TRFs stimulate thyrotropes by at least two GPCRs, the CRH-receptor (CRH-R) and TRH-R being the most studied receptors (Fig. 7.2). The evolution of CRH-R has been analyzed on multiple occasions using the available genomes of different groups of invertebrates and vertebrates; the presence of *CRH-R* has been demonstrated in amphioxus, tunicates, and lampreys and in the rest of the vertebrates (Cardoso et al. 2016; Cardoso et al. 2020, On et al. 2019). From fish to mammals, there are two genes referred to as type 1 and type 2 (*Crh-R1* and *Crh-R2*). In mammals, CRH is well recognized for its ability to stimulate the stress axis by inducing adrenocorticotropin synthesis and release from corticotropes (via CRH-R1); however, it induces TSH release via CRH-R2 in nonmammalian vertebrates (De Groef et al. 2003).

On the other hand, *TRH-R* orthologs have been documented in different animal species, including invertebrates, but interestingly, the vertebrate receptor contains certain amino acids that can recognize the vertebrate TRH sequence. Some *TRH-R* orthologs have been reported in invertebrates such as the *C. elegans* worm, which recognizes a TRH homolog that is different in sequence as compared to the vertebrate counterpart. The presence of TRH-R has not been reported in tunicates, amphioxus, or lampreys. Humans express only one *Trh-R* gene (Yamada et al. 1993), rodents express two, and fish, amphibians, and birds express three *TRH-R* genes (Harder et al. 2001; Li et al. 2020a; Mekuchi et al. 2011; Saito et al. 2011). In general, all vertebrate TRH-Rs contain the key sequences to recognize TRH (Fig. 7.3), so it could be deduced that these receptors are functional and probably have tissue-specific expression, as occurs in rodents, where TRH-R1 is pituitary-specific (Hinkle et al. 2012). Moreover, in mammals, TRH-Rs may be downregulated by ligand binding through internalization (Hinkle et al. 2012) and rapid mRNA degradation (Narayanan et al. 1992), but its regulation in nonmammalian vertebrates is unknown. Other TRH-Rs could mediate the additional different TRH effects, independent of the HPT axis.

7.3.2 TSH During Vertebrate Evolution

In most vertebrates (from teleost fish to mammals), TSH is a glycoprotein that belongs to the cystine knot-forming glycoprotein hormone family; it is composed of two subunits called the α- and β-chains. *Tsh* shares common ancestry with pituitary glycoproteins such as luteinizing hormone and follicle-stimulating hormone. A single gene encodes the α-chain in these three glycoproteins, but the β-chain, which confers function specificity to the hormone, is encoded in three different genes (Kawauchi and Sower 2006). These three genes result from genetic duplications of a single ancestral gene during vertebrate evolution (Farid and Szkudlinski 2004). Amphioxus do not express TSH but express thyrostimulin, a heterodimeric

```
Rattus norvegicus TRHR-1     ---------------------------------------MENET--VSEL----NQT---   12
Rattus norvegicus TRHR-2     -------------------------------------------------MDGPSNVSLI-   10
Chelonia mydas TRH-R1        ---------------------------------------MENSS----KFDSPTNISLA-   16
Gallus gallus TRHR-1         ---------------------------------------MENGT--GDEQ----NHTGL-   14
Danio rerio TRH-R1           ---------------------------------------MENGT--FSNG----TW----   11
Danio rerio TRH-R2           -------------------------------------MQENVSS----RMDTPTNISLV-   18
Danio rerio TRH-R3           MQDSDSLSVSKEEEERGRGLKLILLQIWRTKHLKLKIRQLEDQKTSLTSADPN-INSTII   59
Xenopus laevis TRH-R1        ---------------------------------------MDNVT--FAEL----NAT---   12
Xenopus laevis TRH-R2        ---------------------------------------MENSS----KFDSPTNISLA-   16
Xenopus laevis TRH-R3        ---------------------------------------MENQT--LSIGMPPVNATGMN   19

Rattus norvegicus TRHR-1     -ELPPQVAVALEYQVVTILLVVVICGLGIVGNIMVVLVVMRTKHMRTATNCYLVSLAVAD   71
Rattus norvegicus TRHR-2     --HGDTTLGLPEYKVVSVFLVLLVCTLGIVGNAMVILVVLTSRDMHTPTNCYLVSLALAD   68
Chelonia mydas TRH-R1        --SSDAVFDMIEYKAISIFLVLVICGVGIVGNIMVVLVVLTTREMKTPTNCYLVSLAVAD   74
Gallus gallus TRHR-1         -LLSSQEIVTAEYQVVTILLVLLICGLGIVGNIMVVLVVLRTKHMRTPTNCYLVSLAVAD   73
Danio rerio TRH-R1           ------TDNSVEYKVVSILLVILICGVGIVGNVMVILVVLTTKHMRTPTNCYLVSLAVAD   65
Danio rerio TRH-R2           -GSGDAVTESLEYKTVSVFLVLLVCGVGIVGNIMVVLVVLTTRHMRTPTNCYLVSLAVAD   77
Danio rerio TRH-R3           PTLSSKPENSLEVQLVTISLSLLICGVGIAGNIMVVLVVIQSKHMMTATNCYLVSLAVAD   119
Xenopus laevis TRH-R1        -ELQKREWHGLEYQVVTVFLVVVICGLGIVGNVMVVLVVLQTKHMRTPTNCYLVSLAIAD   71
Xenopus laevis TRH-R2        --NHDIVFSMNEYKAISVFLVLFICGIGIIGNIMVVLVVFTTRDMRTPTNCYLVSLAVAD   74
Xenopus laevis TRH-R3        ETVGNMPQNALEVQVVTISLVLLICGVGIAGNIMVVLVVLRTKHMMTPTNCYLVSLAIAD   79
                                        * : ::: * :.:* :** ** **:***: ::.* * *********:**

Rattus norvegicus TRHR-1     LMVLVAAGLPNITDSIYG-SWVYGYVGCLCITYLQYLGINASSCSITAFTIERYIAICHP   130
Rattus norvegicus TRHR-2     LLVLLAAGLPNVSDSLVG-HWIYGRAGCLGITYFQYLGINVSSFSILAFTVERYIAICHP   127
Chelonia mydas TRH-R1        LMVLVAAGLPNISDSLAG-TWIYGHTGCLGITYFQYLGINVSSCSITAFTVERYIAVCHP   133
Gallus gallus TRHR-1         LMVLVAAGLPNITESLYK-SWVYGYVGCLCITYLQYLGINASSFSITAFTIERYIAICHP   132
Danio rerio TRH-R1           LMVLTAAGLPNITEILFGGQWVYGYAGCLSITYFQYLGINASSCSITAFTIERYIAICHP   125
Danio rerio TRH-R2           LMVLVAAGLPNISESLMA-TWVYGHAGCLGITYFQYLGINASSCSITAFTVERYIAICHP   136
Danio rerio TRH-R3           LTVLMAAGLPNISEVT--SSWIYGYTGCVCITYLQYLGINVSSCSITAFTIERYIAICHS   177
Xenopus laevis TRH-R1        LIVLVAAGLPNITESVYG-SWVYGYIGCLCITYLQYLGINASSCSITAFTVERYLAICHP   130
Xenopus laevis TRH-R2        LMVLVAAGLPNVSDSLAG-TWIYGHAGCLGITYFQYLGINASSCSITAFTVERYIAICHP   133
Xenopus laevis TRH-R3        LIVLLAAGLPNISEVV--ASWVYGYVGCLCITYLQYLGINISACSITAFTVERYIAICHS   137
                             * ** ******:::      *:**  **: ***:****** *: ** ***:***:*:**

Rattus norvegicus TRHR-1     IKAQFLCTFSRAKKIIIFVWAFTSIYCMLWFFLLDLNISTYKDAIVISCGYKISRNYYSP   190
Rattus norvegicus TRHR-2     LRAQTVCTVARAKRIIAGIWGVTSLYCLLWFFLVDLNVR---DNQRLECGYKVPRGLYLP   184
Chelonia mydas TRH-R1        MRAHTMCTVSRAKRIIAFVWICTSIYCMLWFFLVDINVN---KSQHLECGYKVSRNLYLP   190
Gallus gallus TRHR-1         IKAQFLCTFSRAKKIIIFVWSFTSVYCMLWFFLLDLNIAVYKDTTVVSCGYKVSRSYYSP   192
Danio rerio TRH-R1           IKAQFMCTLSRAKKIIVLVWVLTSLYCVMWFYLLDTREKVYDNAVLVTCGYKVSRELYLP   185
Danio rerio TRH-R2           MRAQTVCTVSRAKRIIAVVWVFTCVYCMLWLFLVDIQVN---PDGRVQCGYRVSRDLYLP   193
Danio rerio TRH-R3           IKAKLICTVSRAKKIIACVWIFTAIYCIMWFFLVDTEETVYSNGVVVRCGYRVSRSLYMP   237
Xenopus laevis TRH-R1        IKAQFLCTISRAKKIIVFVWAFTSLYCLMWFFLLDLNTTIYKDATVVNCGYRVSRSYYSP   190
Xenopus laevis TRH-R2        MRAKTVCTVSRAKRIIAIVWIFTSIYCMLWFFLVDIHVN---KSQQVECGYKVSRNLYLP   190
Xenopus laevis TRH-R3        IKAQFICTVSRAKKIIAFVWVFTSMYCVMWFFLVDITEVKFADGVQVNCGYRVFRNLYTP   197
                             ::*: :**.:***:**  :*  *.:**::*::*:*           : ***::  * * *

Rattus norvegicus TRHR-1     IYLMDFGVFYVMPMILATVLYGFIARILFLNPIPSDPKENSKTWKNDSTHQNKNMN----   246
Rattus norvegicus TRHR-2     IYLLDFAVFFIGPLLVTLVLYGLIGRILFQSPLSQEA------WQKERQPH----GQSEA   234
Chelonia mydas TRH-R1        IYLLDFALFFVTPLVVATMLYGLIGRILFASPIPHQADSTAKPWRERSGREKNGVDTRES   250
Gallus gallus TRHR-1         IYMMDFGIFYVLPMVLATVLYGLIARILFLNPIPSDPKENSNTWRNDMAQQNKTVN----   248
Danio rerio TRH-R1           IYFTDYAVFYVIPLLLATVLYGLIARILFLNPLPSDPKESRRNWKKESSVHGN-------   238
Danio rerio TRH-R2           IYLIDFTIFYVIPLLLAIALYGLIARILYLNPLPHPPDSGTVIAPTLRRSCKE--PADGG   251
Danio rerio TRH-R3           IYFMDFTLFYVVPLLVALVLYGLIARILFRDPPSSNISNQRF-----TAQQGRRQNS---   289
Xenopus laevis TRH-R1        IYLIDFGIFYAVPMTLATVLYGLIARILFLNPIPSDPKENSKTWRNDVTHQNKAFN----   246
Xenopus laevis TRH-R2        IYLADFAIFYVTPLLVATILYGLIGRILFLSPIPQHPESTTERWREKSSKEKNESETEGN   250
Xenopus laevis TRH-R3        IYFLDFTIFYVIPLVLATVLYGLIARILFMNPLPSNPQDLSRMSSKH---YGKPYNS---   251
                             **: *: :*:  *: ::  ***:*.***: .*

Rattus norvegicus TRHR-1     LNTTNRCFNSTVSSRKQVTKMLAVVVILFALLWMPYRTLVVVNSFLSSPFQENWFLLFCR   306
Rattus norvegicus TRHR-2     APG---NCSRAKSSRKQATRMLAVVVLLFAVLWTPYRTLVLLNSFVAQPFLDPWVLLFCR   291
Chelonia mydas TRH-R1        KPSSRSRTRGALSSRKQVTKMLAVVVILFALLWMPYRTLVLVNSFMDNPFLNPWFVLFCR   310
Gallus gallus TRHR-1         SKMTNKSFNSTIASRRQVTKMLAVVVVLFAFLWMPYRTLVVVNSFLSSPFQENWFLLFCR   308
Danio rerio TRH-R1           --SRSSSNSTTVASRRQVTKMLAVVVILFALLWMPYRTLVVVNSFLKEAYLDTCFLLFCR   296
Danio rerio TRH-R2           KAIRQGRPKSALSSRKQVTKMLAVVVVLFALLWMPYRTLVLINSFVSTPYLDSWFILFCR   311
Danio rerio TRH-R3           LKVNC-KPNRAASSRKQVTQMLAVVVVVFALLWLPYRTLVVVNSLLDPPYLDTWFLLFSR   348
Xenopus laevis TRH-R1        SKMSSRCSNNTIASRRQVTKMLAVVVLLFAFLWMPYRTLVVVNSFLSHPYLQTWFVLFCR   306
Xenopus laevis TRH-R2        KPSNRLKTKGALSSRKQVTKMLAVVVVLFALLWMPYRTLVLVNSFMDKPYLDPWFLLFCR   310
Xenopus laevis TRH-R3        IKLSGKGNKNTANSRKQVTKMLAVVVILFALLWMPYRTLVVVNSFMDPPYLNVWFVLFYR   311
                                       :  **:*.*:******::**.** ******::**::   : :  .:** *
```

Fig. 7.3 Deduced amino acid sequence of TRH receptors (TRH-R) in different vertebrates. Yellow amino acids indicate important residues for TRH binding. Transmembrane domains (1–7) are indicated in seven different colors (green, red, dark blue, orange, purple, light blue, and ocher). (*) identical amino acids, (:) conserved residues, and (.) similar residues. The alignment was created using ClustalOmega. Accession numbers: *Rattus norvegicus* TRHR-1: NM_013047; *Rattus norvegicus* TRHR-2: NM_181364; *Chelonia mydas* TRH-R1: XM_007059298.3; *Gallus* TRHR-1: NM_204930; *Danio rerio* TRH-R1 XM_682154; *Danio rerio* TRH-R2: NM_001044346; *Danio rerio* TRH-R3 XM_688966; *Xenopus laevis* TRH-R1: AJ420782; *Xenopus laevis* TRH-R2: AJ420781; *Xenopus laevis* TRH-R3: AJ420780

```
Rattus norvegicus TRHR-1    ICIYLNSAINPVIYNLMSQKFRAAFRKLCNCKQKPTEK-AANYSVALNYSVIKES-----   360
Rattus norvegicus TRHR-2    TCVYTNSAVNPVVYSLMSQKFRAAFLKLCWCRAAGPQR-RAARVLTSNYSAAQETSEGTE   350
Chelonia mydas TRH-R1       LCVYTNSAINPVVYNLMSQKFRAAFKRLCKCGGEPL-P-RSLYMATASYSLAREPLPPTP   368
Gallus gallus TRHR-1        ICIYLNSAINPVIYNLMSQKFRAAFRKLCNCHLKRDKK-PANYSVALNYNVIKES-----   362
Danio rerio TRH-R1          LCIYMNSAINPIIYNAMSQKFRAAFHKLCHCRPQRSEK-PPAYSVALTYSVIKETSNGES   355
Danio rerio TRH-R2          TCIYANSAINPVVYNLMSQKFRSAFRGLYRCRREDLHQ-RTLSMLQSGYSLGRDPRICSN   370
Danio rerio TRH-R3          MCIYTNSAVNPIIYNAMSQKFRAAFKRLCDCKHRTDQGRAGTHTVTVYYGVIKELSSENS   408
Xenopus laevis TRH-R1       ICIYLNSAINPVIYNLMSQKFRAAFRKLCKCKKKRSEK-PTNYGVALNYSVIKESSNGGS   365
Xenopus laevis TRH-R2       ICVYANSAINPVIYNLMSQKFRTAFKRLCKCGHEETQR-RSMYMTTTSYSMVKDGVNTGG   369
Xenopus laevis TRH-R3       LCIYLNSAINPIIYNLMSQKFRAAFKNLRKCEQKRTEK-AAKYNIPVYYSVMKDSSH-ES   369
                             *:* ***:**::*. ******:**  *  *                *.  ::

Rattus norvegicus TRHR-1    -DRFSTE--------LDDITVTDTYVSTTKVSFD-----DTCLASEKNGPSSCTYGYSLT   406
Rattus norvegicus TRHR-2    KM----------------------------------------------------------   352
Chelonia mydas TRH-R1       QPQDHK--EGKQPPPM-SNMALPKQQQPPSRSKG-ENGDELYFSVV--------------   410
Gallus gallus TRHR-1        -DHFSSE--------IEDITVTNTYLSSAKTSIG-----DTCLSSEVRINA-C-------   400
Danio rerio TRH-R1          PDHYTTE--------LDDIRGAAEDLLPERKR--------LCFKEPAKDTLATA------   393
Danio rerio TRH-R2          GTRDTPKKVGSVTPEQKQNKGFENKVENTDRQISDGVENEINPSIKKNACANCDIDESKT   430
Danio rerio TRH-R3          MKKQ------------DNTNTDTFSPA---------------------------------   423
Xenopus laevis TRH-R1       PDHFSTE--------LEDITVTDTYLSTSKMSFD-----DTCLPT---------------   397
Xenopus laevis TRH-R2       DTKDK--------PTL-EKTNIIAK---QTNKAG-P-EGDLYYSVV--------------   401
Xenopus laevis TRH-R3       PDHDVTV--------QEDLNGF----PTKKVNFP-----QNCVDTTTTYSVA--------   404

Rattus norvegicus TRHR-1    AKQEKI------------------------------------------------------   412
Rattus norvegicus TRHR-2    ------------------------------------------------------------   352
Chelonia mydas TRH-R1       ------------------------------------------------------------   410
Gallus gallus TRHR-1        ------------------------------------------------------------   400
Danio rerio TRH-R1          ------------------------------------------------------------   393
Danio rerio TRH-R2          SEMPTVEDVDMSKINDNGKIMEESGNSAQDIIADMSNTQQTQRVSGEFTSTLGEAEQNDA   490
Danio rerio TRH-R3          ------------------------------------------------------------   423
Xenopus laevis TRH-R1       ------------------------------------------------------------   397
Xenopus laevis TRH-R2       ------------------------------------------------------------   401
Xenopus laevis TRH-R3       ------------------------------------------------------------   404
Rattus norvegicus TRHR-1    -------          412
Rattus norvegicus TRHR-2    -------          352
Chelonia mydas TRH-R1       -------          410
Gallus gallus TRHR-1        -------          400
Danio rerio TRH-R1          -------          393
Danio rerio TRH-R2          ELSHSVL          497
Danio rerio TRH-R3          -------          423
Xenopus laevis TRH-R1       -------          397
Xenopus laevis TRH-R2       -------          401
Xenopus laevis TRH-R3       -------          404
```

Fig. 7.3 (continued)

glycoprotein composed of GpA2 (Glycoprotein hormone-alpha 2) and GpB5 (Glycoprotein hormone-beta 5) subunits and predominantly expressed in Hatschek's pit. Thyrostimulin is the archetype of TSH signaling because it induces TH synthesis (Wang et al. 2018). Lampreys also express thyrostimulin and a second heterodimeric glycoprotein called GpH (glycoprotein hormone); this protein is made up of lGpA2 and lGpHβ subunits, representing the α- and β-subunits, respectively. Of note, GpH represents the common ancestor for *Tsh*, luteinizing hormone and follicle-stimulating hormone (Sower and Hausken 2017), suggesting that GpH overlaps thyroid and reproductive functions in lampreys. During vertebrate evolution, the pituitary gained the capability to synthesize different hormones to specifically activate the thyroid and reproductive axes. It is important to mention that TRFs and GPCRs that induce thyrostimulin secretion in amphioxus or GpH in lampreys have not been studied, so the roots of these interactions have yet to be explored.

Thyrostimulin binds to a specific receptor located in the amphioxus endostyle, whereas GpH binds to a specific receptor located in the endostyle/thyroid gland of pre- and postmetamorphic lamprey (see Sect. 7.4.2) (Sower and Hausken 2017; Wang et al. 2018). Specific duplications of the β-subunit have been reported in teleost and cartilaginous fish, but in the rest of the vertebrates, genetic duplications led to the divergence of the only vertebrate TSH-β subunit and TSH-R (Farid and Szkudlinski 2004; Dos Santos et al. 2011; Maugars et al. 2014). Evidence suggesting that TSH controls TH secretion in fish includes the effect of hypophysectomy on thyroid activity and in vitro studies showing that TSH and pituitary extracts stimulate thyroid function (Eales and Brown 1993; MacKenzie et al. 2009). In other vertebrates, TH secretion is regulated by TSH-TSH-R interactions in the pituitary gland (Holzer and Laudet 2013).

In summary, the pituitary gland evolved from a structure called Hatschek's pit, which is found in some marine invertebrates, such as amphioxus. The thyrotropes of the pituitary gland specialize in synthesizing TSH, the main stimulator of the vertebrate thyroid gland for TH synthesis and secretion. The evolution of TSH α- and β-subunits has been reconstructed due to the study of marine invertebrates (e.g., amphioxus) and ancestral vertebrates (e.g., lampreys).

7.4 The Evolution of the Thyroid Gland

7.4.1 Functional Evolution of Iodine-Containing Signaling Molecules

Iodine is an essential element present in many biological systems, ranging from bacteria to mammals (Crockford 2009). It is thought that the coupling of iodine to ancient biological systems could be due to its antioxidant capacity (for a complete review, see Venturi (2011)). Over time, iodine acquired new functions due to its incorporation into proteins and other chemical compounds. Because iodine is obtained from external sources, it is thought that the presence of these iodinated molecules is a sign of a nutrient-rich environment. In fact, it has been proposed that these iodinated molecules or iodinated messengers are employed to activate the transition or metamorphosis of some invertebrate marine species (Heyland and Moroz 2005). Mono- and diiodotyrosines and THs are examples of these iodinated messengers in the animal kingdom (Fig. 7.4) (Crockford 2009). Accordingly, although the thyroid gland is exclusive to the vertebrate lineage, iodinated messengers are not, and different invertebrate species synthesize these signaling molecules in pathways yet to be identified. In this context, endogenous and exogenous iodinated messengers regulate several aspects of development, growth, skeletogenesis, and metamorphosis in mollusks, echinoderms, tunicates, and amphioxus (Taylor and Heyland 2017).

7.4.2 The Dawn of the Thyroid Gland

One of the most important events in vertebrate endocrinology is the evolution of a gland capable of capturing, storing, and recycling iodine to optimize the synthesis of iodinated messengers, including THs. Indeed, the first step for TH synthesis is iodine dietary intake, followed by its active transport into the thyroid follicular cell, a single layer of thyrocytes encapsulating a colloid, the site where THs are synthesized (see below). Once inside the thyrocyte, iodine is oxidized by thyroid peroxidase and is incorporated into the benzene residue of tyrosines

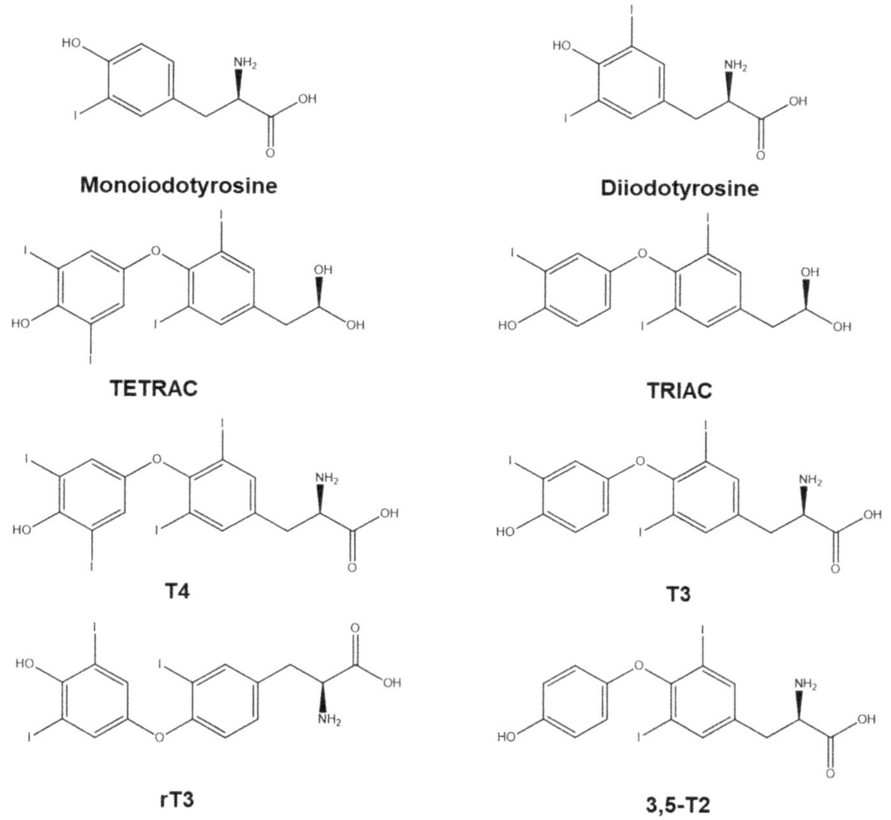

Fig. 7.4 Chemical structure of iodinated messengers. In invertebrates, some molecules, such as mono- and diiodotyrosine, can function as endocrine messengers, while TETRAC and TRIAC, secondary derivatives of TH, function as chemical messengers in early chordates. T4 and T3 are synthesized in the thyroid gland of vertebrates, while rT3 and 3,5-T2 are products of T4 and T3 deiodination. *Abbreviations*: *3,5-T2*, 3,5-diiodothyronine; *T3*, 3,5,3′-triiodothyronine; *T4*, thyroxine; *rT3*, reverse T3; *TETRAC*, 3,5,3′,5′-tetraiodothyroacetic acid; *TRIAC*, 3,5,3′-triiodothyroacetic acid

contained in the thyroglobulin protein, forming mono- and diiodotyrosines. To form TH, coupling reactions between the different mono- and diiodotyrosines need to occur. This interaction is favored by the three-dimensional structure of the protein; of the 132 tyrosine residues, only approximately five can undergo a coupling reaction (Carvalho and Dupuy 2017; Coscia et al. 2020). Once coupled, THs are then released by proteolysis and transported into the blood circulation (Carvalho and Dupuy 2017).

In some marine invertebrates (tunicates, amphioxus), the endostyle has been recognized as the rudimentary gland that is capable of storing iodine and synthesizing iodinated messengers, including THs (Taylor and Heyland 2017). The endostyle of the amphioxus is located in the pharyngeal region and functions for filter feeding (Kluge et al. 2005). The endostyle has been shown to express the molecular machinery for iodine metabolism and has been attributed as the site of TH production, probably using an ancestor of the thyroglobulin protein (Ogasawara 2000; Kluge et al. 2005; Paris et al. 2010; Belkadi et al. 2012); it is a rudimentary organ capable of storing iodine and synthesizing iodinated messengers, including THs. Over time, this organ evolved into the thyroid gland found in vertebrates. In support of this notion, the endostyle of the premetamorphic ammocoete larvae transforms into a thyroid gland during lamprey metamorphosis, confirming the homology and evolutionary transition of this endocrine structure (Kluge et al. 2005). In lampreys and fish, the thyroid gland is not encapsulated; instead, the thyroid follicles are scattered along the pharyngeal region, the heart, and the kidneys (Deal and Volkoff 2020). Thyroglobulin is found only in vertebrates, from lampreys to mammals, representing vertebrate innovation specialized in TH synthesis (Holzer et al. 2016). Other thyroid gland proteins, such as thyroid peroxidase, are not vertebrate- or tissue-specific proteins (Taurog 1999; Soleimani 2015). Of note, TSH regulates the expression of all these genes involved in TH synthesis (Jang et al. 2022). In conclusion, the study of marine invertebrates and basal vertebrate lamprey has demonstrated the functional homology between the endostyle and the thyroid gland. During vertebrate evolution, this specialized gland, capable of storing iodine and synthesizing THs, acquired the capability to be regulated by hormonal inputs from the pituitary.

7.5 TH Signaling During Vertebrate Evolution

The iodinated aromatic ring of THs confers lipophilic properties that, at physiological pH, have amphipathic characteristics (Tóth et al. 2012); thus, THs require different families of proteins for their transport, metabolism, and signaling. The evolution of TH carriers, transporters, deiodinases, and THRs reflects millions of years of specialization across the vertebrate lineage. These proteins are crucial for the correct and precise action of THs, regulating their cellular bioavailability and molecular effects in a tissue-specific manner. In this section, we describe TH signaling elements using a comparative approach, when possible.

7.5.1 TH Carriers

THs synthesized in the thyroid gland are secreted into the bloodstream, where they bind to plasmatic soluble proteins so that less than 1% of circulating TH is free. To date, albumin, transthyretin (TTR), and thyroxine-binding globulin (TBG) have been identified as TH carriers, and their dissociation constant, *Kd*, is variable (albumin: T4 (70 μM), T3 (1 μM); TTR: T4 (700 nM), T3 (140 nM); and TBG: T4 (0.1 nM), T3 (46 nM)) (Richardson 2007; McLean et al. 2017). Besides presenting low TH affinity, albumin is known to transport other molecules, making it highly unspecific as a TH carrier, in contrast to TBG, which has the highest affinity for both

THs, although this characteristic may prevent its release to the target cells (McLean et al. 2017). Because TBG is only expressed in mammals, TTR has been proposed as the main vertebrate TH carrier.

The *Ttr* gene is unique to vertebrates and has been identified in every representative class of lampreys (Manzon et al. 2007), fish (Santos et al. 2002), amphibians (Yamauchi et al. 1998; Prapunpoj et al. 2000), reptiles (Prapunpoj et al. 2002), birds (Duan et al. 1991), and several mammals and marsupials (Richardson 2007). TTR is a homotetrameric protein, mainly expressed in the liver and choroid plexus; the former provides most of the circulating TTR, whereas the latter is found in cerebrospinal fluid (CSF). TTR expression in the vertebrate central nervous system (CNS) is proposed as a recent innovation since it is not found in the brain of fish or amphibians, in contrast to reptiles, birds, and mammals, which do express TTR in the choroid plexus (Santos et al. 2002; Richardson 2007). Tissue-specific TTR expression is thought to depend on the developmental stage of vertebrates because increases in liver expression and serum concentrations have been observed to coincide with postembryonic peaks of TH concentrations (Yamauchi and Ishihara 2009). TTR has the highest affinity for T3 in fish, amphibians, birds, and reptiles, whereas in mammals, its affinity is higher for T4. The evolutionary significance of these differences is not entirely clear, but they have been suggested as an adaptation to the higher T4 concentrations found in mammalian serums as compared to those found in fish (Power et al. 2000), species in which the most abundant circulating hormone is T3 (Sefkow et al. 1996). In conclusion, at least three proteins synthesized in the liver have evolved to bind and efficiently distribute THs to peripheral tissues and the brain. The recent evolution of *Tbg* and the expression of TTR in the choroid plexus support adaptive evolutionary strategies for efficient TH distribution into the brain.

7.6 TH Transmembrane Intracellular Transport

Transmembrane intracellular transporters facilitate the exchange of THs between the extracellular space and the intracellular environment. The discovery and study of these transporters are relatively recent since it was previously thought that THs could travel passively through the plasma membrane due to their lipophilic nature. However, as mentioned, at physiological pH, THs are amphipathic and thus require active transport through these membrane proteins (Friesema et al. 1999). Importantly, it has been proposed that these transporters also allow TH passage from the bloodstream to the CSF and CNS, where some types of glial cells (tanycytes and astrocytes) also express these TH transmembrane transporters to exit locally produced T3 into the extracellular space, where it reaches other cell-like neurons to exert their biological actions (Emamnejad et al. 2022; Fekete and Lechan 2014; Russell 2018). Thus, transmembrane transporters seem to have a dual role in mobilizing THs to and from the intracellular space, at least in the brain.

Various families of TH intracellular transporters have been described: organic anion transporters (OATPs; SLC10, SLC17), monocarboxylate transporters (MCTs, also known as SLC16), and L-amino acid transporters (LATs); most of them are nonspecific for THs and can transport many other small molecules (Groeneweg et al. 2020). Neither MCT8, MCT10, OATP1C1, nor LAT have been reported in marine invertebrates or lampreys. It is possible that, in these species, other nonspecific transporters mediate TH cellular uptake. In other vertebrate classes, such as fish (Arjona et al. 2011), amphibians (Mughal et al. 2017), birds (Bourgeois et al. 2016), and mammals (Groeneweg et al. 2020), MCT8 is the most studied and structurally conserved. Paradoxically, mutations in the *Mct8* of frogs do not affect natural metamorphosis, suggesting that other TH transporters, which have yet to be identified, could be responsible for TH uptake during early development. The zebrafish (*Danio rerio*) has been a useful model to study MCT8 in major detail, even for Allan-Herndon-Dudley syndrome, since *Mct8* mutants resemble some characteristics of this human disease (Zada et al. 2017; Silva and Campinho 2023). In other nonmammalian vertebrates, the physiological roles of MCT10 and OATP1C1 have not been described in detail, but their expression has been reported in fish, amphibian, and avian species (Bourgeois et al. 2016; Mughal et al. 2017; Hernández-Linares et al. 2019; Admati et al. 2020). In zebrafish, *Oatp1c1* is expressed in astrocytes, neurons, and endothelial cells, resembling the mammalian-tissue-specific distribution. Mutations in this gene produce alterations in radial glial cells, neuronal axon size, and goiter (an increase in the size of the thyroid gland) (Admati et al. 2020).

Thus, TH distribution and uptake have evolved in favor of increasing distribution efficiency (in the blood and CSF) and also for tissue-specific uptake in peripheral tissues and the brain; therefore, vertebrates reflect the adaptability and complexity of these proteins. Some authors have suggested that in more recent vertebrates, TH action has favored the complexity of neuronal diversity and the evolution of new structures (e.g., the neocortex) and contributed to the acquired advance in cognitive capacities (Richardson 2007; Mantzouratou et al. 2021).

7.7 Deiodinases

Deiodinases are selenoproteins that catalyze the stereo-specific and sequential removal of iodine atoms from the prohormone T4, generating active and inactive isomers of both triiodo- and diiodothyronine (Fig. 7.5). Three distinct deiodinase isotypes catalyze this iodine removal: DIO1, DIO2, and DIO3, each with different catalytic properties and specific tissue and developmental expressions. Two deiodinases, DIO1 and DIO2, serve the activating or outer-ring deiodination (ORD) pathway by converting T4 into T3 and T3 into 3,5-diiodothyronine (3,5-T2), a TH that has recently been recognized as bioactive (Box 7.2). The inactivating or inner-ring deiodination (IRD) pathway is catalyzed primarily by DIO3, which converts T4 and T3 into inactive metabolites (reverse triiodothyronine (rT3) and 3, 3′-diiodothyronine, respectively). Thus, peripheral deiodinases tightly regulate, in an organ-specific manner, both circulating levels and the local intracellular concentrations of active and inactive TH (reviewed by Gereben et al. 2008).

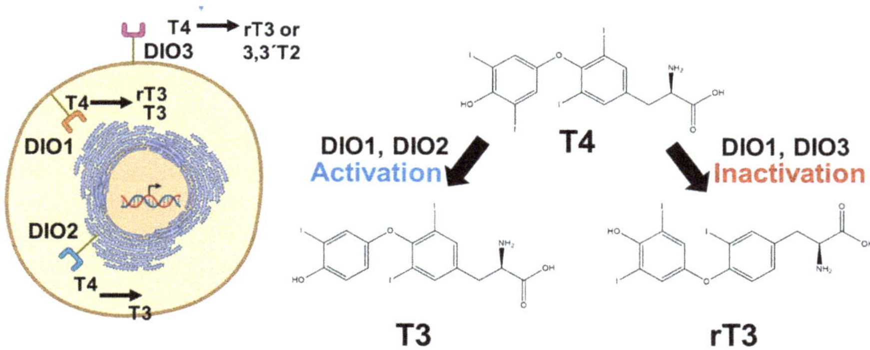

Fig. 7.5 Deiodinases are differentially localized at the subcellular level. DIO1 is a plasma membrane protein, whereas DIO2 is in the endoplasmic reticulum; both catalytic domains face the cytosol. DIO3 is anchored in the plasma membrane, with most of the molecule, including the active center, facing the extracellular space. D1 and DIO2 can generate bioactive T3 by removing the 5′ iodine from the outer ring, while DIO1 and DIO3 can generate inactive TH, such as rT3, where iodine 5 is removed from the inner ring. Figure created with BioRender.com. *Abbreviations: DIO1*, deiodinase 1; *DIO2*, deiodinase 2; *DIO3*, deiodinase 3; *T3*, 3,5,3′-triiodothyronine; *T4*, thyroxine; *rT3*, reverse T3

Box 7.2 Is 3,5-T2 Physiologically Relevant?

TH actions are mediated by the bioactive T3, which acts by binding to THRs. Since THs exert pleiotropic effects, interest has grown in identifying other possible TH-derived bioactive iodinated messengers that could explain this pleiotropy. Among these, 3,5-diiodothyronine (3,5-T2), an endogenous and naturally occurring metabolite of T3 outer-ring deiodination, has been shown to exert thyromimetic activity. In this context, 3,5-T2 has been extensively studied in terms of its extragenomic effects. Many studies in mammals (reviewed in: Sane et al. 2022) and a few in birds (Incerpi et al. 2002; Scapin et al. 2009; Sechman et al. 2011) have provided sound evidence that the effects of 3,5-T2 on mitochondrial respiration rate and lipid metabolism are protein synthesis independent and are mediated through nonreceptor mechanisms (reviewed in: Sane et al. 2022). These findings led scientists to propose 3,5-T2 as a possible drug to ameliorate obesity. However, it was later found that although chronic 3,5-T2 administration did directly stimulate energy expenditure and reduce body mass gain, it could also suppress TSH and, overall, impair thyroid function (Padron et al. 2014).

Although genomic 3,5-T2 actions have also been described, these have only lately begun to attract attention. In this context, it was shown that 3,5-T2 acts as an alternative ligand of the teleost and human isoforms of TRβ1. In vivo studies in fish have revealed that 3,5-T2 is physiologically relevant, suggesting that T3 could function as a prohormone and undergo a second outer-ring deiodination (via D2) to generate 3,5-T2. This additional TH signaling pathway would imply a further regulatory step that could partly explain the

(continued)

plurality in TH actions (Lazcano et al. 2019a, b; Orozco et al. 2014). The bioactivity of 3,5-T2 is not limited to teleosts or mammals; this TH has been shown to be bioactive in insects like *Locusta migratoria*, where it exerts an even more powerful effect than the juvenile hormone III in reducing the volume of the follicular cell (Davey 2000). The evolutionary analysis of the THR families has the potential to reveal protein sequence features that could determine protein structure and function. In this context, T3 and 3,5-T2 could both bind in the ligand pocket of the THR but remodel and stabilize the conformation of the LBD to generate distinct binding surfaces for coactivator proteins, which could regulate different genes, thereby producing new TR functions (Bridgham et al. 2010; Harms and Thornton 2010). The effects that 3,5-T2 have upon axolotl metamorphosis can illustrate this point. The Mexican axolotl can undergo metamorphosis by the exogenous administration of T4 and T3. It has recently been shown that 3,5-T2 treatment induces gill retardation (a first step of metamorphosis), but the withdrawal of this treatment elicits gill regeneration. These reversible effects are not observed when the axolotl is treated with T3. Differential effects on gill retraction after 3,5-T2 or T3 treatment are accompanied by the activation of different gene clusters (Lazcano et al. 2023a, b), suggesting an additional TH signaling pathway. It has yet to be explored whether 3,5-T2 has thyromimetic effects upon all TH target genes or on all vertebrates.

Despite their distinct functional roles in TH homeostasis, members of this enzyme family share a common structural organization, thus suggesting that they may have diverged from a common ancestral gene (Klootwijk et al. 2011). All deiodinases contain at their active or catalytic center the modified amino acid selenocysteine (SeCys), which is encoded by an in-frame UGA triplet that functions as a codon for the incorporation of the rare amino acid (Berry et al. 1991). Also, all deiodinases contain in their 3′ untranslated regions a selenocysteine insertion sequence (SECIS), which is a *cis*-acting signal required for the incorporation of SeCys into the protein during translation (Bianco et al. 2002; Orozco et al. 2012); their transcription also requires ancient and highly complex *trans*-acting machinery (Lu and Holmgren 2009; Palioura et al. 2009).

DIO1 is the only member of the family that catalyzes both the ORD and IRD of various TH derivatives. This complex dual catalytic activity suggests that DIO1 could contribute to the circulating T3 pool, favoring iodine recycling (Schneider et al. 2006; Gereben et al. 2008; Galton et al. 2009), which could be important during specific stages of high demand for TH, such as embryogenesis and metamorphosis in fish and amphibians, as demonstrated by its upregulation during these processes in several species (Shepherdley et al. 2002; Campinho et al. 2010; Itoh et al. 2010; Morvan-Dubois et al. 2013).

DIO2 exhibits a remarkable circadian rhythm entrained by the light/dark cycle, which has been documented in teleosts (García-G et al. 2004; Isorna et al. 2009), birds (Yoshimura et al. 2003), and mammals (Luna et al. 1995). This circadian rhythm is modulated in the latter by TSH released from the pars tuberalis in response to melatonin (Dardente and Migaud 2021). Furthermore, DIO2 expression in hypothalamic tanycytes is an important factor in regulating seasonal reproduction in mammals and birds (Williams and Bassett 2011; Ikegami and Yoshimura 2012).

DIO3 is the predominant deiodinase expressed during embryonic life. It controls TH homeostasis by protecting tissues from an excess of these hormones during species-specific ontogenetic programs. Thus, its hepatic expression is limited to embryonic life in most vertebrates, but during adulthood, the enzyme is also expressed in the liver of omnivorous species that devour whole prey. This finding, which is consistent with the protective role of DIO3 during embryogenesis, has led to the proposal that hepatic DIO3 helps prevent an inappropriate systemic overload of exogenous T3 after feeding (Martínez et al. 2008; Villalobos et al. 2010).

Depending on the organ and species, different hormonal, nutritional, and developmental signals, as well as physiological demands, modulate the expression and activity of deiodinases. TH availability is the most potent and well-studied regulator (reviewed in: Darras 2021). This regulation has been observed in all studied vertebrates. For example, during amphibian development, the premetamorphic surge of T3 rapidly stimulates DIO3 in frog tadpoles (reviewed in: Brown et al. 2005). Thus, the prevailing understanding is that the coordinated development in amphibia depends in part on the tissue-specific expression patterns of DIO2 and DIO3, safeguarding the required level of intracellular T3 in a given tissue at the appropriate stage of metamorphosis (Becker et al. 1997).

Deiodinases represent a sophisticated, tissue-specific on/off switch for regulating TH activity. The deiodination of tyrosine metabolites is an ancient feature of all chordates studied to date and, consequently, precedes the integration of the thyroid systems that characterize vertebrates. Deiodinases can be viewed as the fourth component of the HPT axis, and TH deiodination is the essential first step in the control mechanism of TH action (Darras 2021). From an evolutionary perspective, deiodinases may be considered pivotal players in the emergence and functional diversification of thyroid systems and their iodinated messengers. Deiodinase orthologs can be traced back to invertebrate chordates (for a review, see Kluge et al. 2005; Paris et al. 2008; Orozco et al. 2012). In the ascidian (tunicate) *Halocynthia roretzi*, at least one deiodinase homolog has been biochemically characterized. The enzyme presents a bona fide SECIS and conserves the signature string that characterizes vertebrate deiodinases and catalytically resembles DIO1. Three deiodinase homologs have been identified in the amphioxus; two have a SeCys at their catalytic site. The only one characterized has a Cys residue and specifically catalyzes the IRD of thyroacetic acid metabolites of T4 and T3 (Klootwijk et al. 2011). Furthermore, these acidic metabolites are the endogenous ligands of amphioxus THRs, and they control metamorphosis in this species, supporting the notion that even in invertebrate chordates, this distinct, inactivating deiodination pathway is physiologically relevant (Paris et al. 2008, 2010). Interestingly, these invertebrate chordate deiodinase homologs could have an even more ancient origin, and there is evidence that invertebrates, like mollusks (Wu et al. 2012) and echinoderms (Heyland et al. 2006), may express putative deiodinases. Thus, deiodinase functional diversity in living (extant) vertebrates seems to stem from a common ancestral molecular scaffold that could have already been present in the ancient Metazoan. Even though there is no information regarding the catalytic activities of the remaining two SeCys-containing putative amphioxus homologs, the singular substrate selectivity of Cys-containing putative deiodinase poses the intriguing question of whether ancestral deiodinases played a protective role as does the extant vertebrate IRD or whether it fulfilled multiple roles that were later divided among several enzymes. In this context, *Dio1* seems to be the oldest of the three deiodinases (Orozco et al. 2012). *Dio2* and *Dio3* seem to share an immediate common ancestor that could have resembled *Dio1* in that it contained both activating and inactivating activities; by duplication, the two resulting genes acquired specialized functions. Furthermore, *Dio2* appears to be the most recent deiodinase gene, a hypothesis that agrees with the notion that D*io*2 is the most specialized and finely regulated member of the family. It plays a key role in vertebrate neurogenesis, the mammalian homeostasis of the HPT axis, and tissue-specific regulation (Russo et al. 2021).

Natural variants of all three *Dios* have gained interest in the context of possible associations with altered metabolic function or diseases. One of the more relevant variants is D*io*2 Thr92Ala, which has decreased catalytic activity and is thus associated with impaired T3 formation (Köhrle and Frädrich 2021).

It has been proposed that Ala92 D*io*2 expression prevailed in ancient humanoids, who, as gatherers or hunters, had low carbohydrate intake. Modern humans (pleistocene) markedly increased carbohydrate ingestion, possibly requiring higher T3 production and iodine requirements (Ricci et al. 2020). This change in diet, as well as the evolutionary pressure to cope with nutritional iodine deficiency, might have favored the selection of the novel Thr92 D*io*2 variant for delivering more T3 (Köhrle and Frädrich 2021). These hypotheses highlight the importance of deiodinases in fine-tuning the cellular regulation of TH bioavailability—a remarkable event in evolution that might have been decisive for the explosive diversification of HPT regulatory strategies in vertebrates.

7.8 Thyroid Hormone Receptors

THs exert their actions mainly through genomic mechanisms (Vella and Hollenberg 2017), but nongenomic mechanisms have also been well documented (Davis et al. 2016, 2021). This latter mechanism is fundamental for the activity of iodinated messengers in marine invertebrates (Taylor et al. 2023). T3 is the bioactive hormone; its genomic effects are mediated by receptors that function as ligand-dependent transcription factors. In gnathostomes, or jawed vertebrates (fish, amphibians, reptiles, avians, and mammals), there are two genes: *Thra* and *Thrb*. The expression of these *Thr* genes can generate different isoforms of TRα and TRβ proteins through transcriptional start sites or alternative RNA splicing. It is thought that these genes

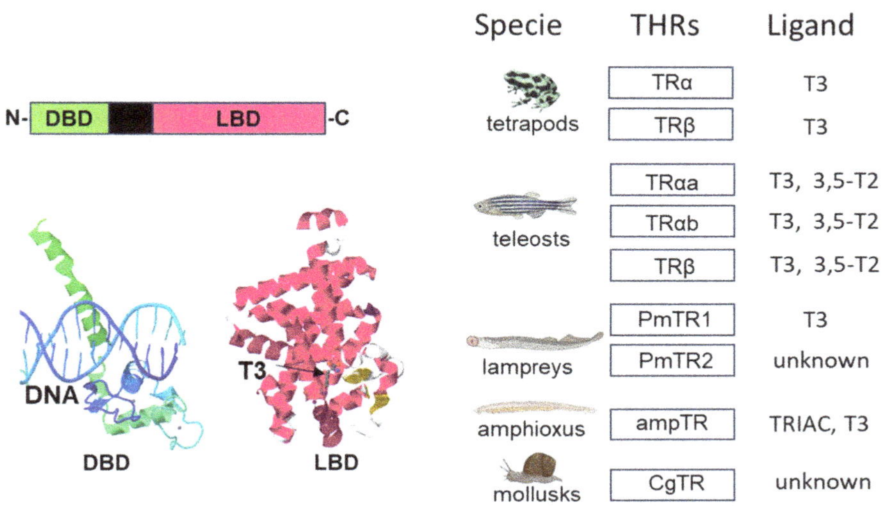

Fig. 7.6 THR structure and expression in metazoa. Left: THRs contain a DBD that allows TR/TRE interaction and an LBD that recognizes and binds T3. Right: schematic representation of THRs and their respective endogenous ligands. Figure created with BioRender.com and Protein Data Bank. PDB DOI: https://doi.org/10.2210/pdb3GWS/pdb and https://doi.org/10.2210/pdb2NLL/pdb. *Abbreviations*: *C*, C-terminal; *DBD*, DNA-binding domain; *LBD*, ligand-binding domain; *N*, N-terminal; *THR*, thyroid hormone receptor

originated from a duplication event that occurred during the early stages of vertebrate evolution (Kuraku et al. 2008; Holland and Ocampo Daza 2018). All functional vertebrate *Thr* homologs contain an amino-terminal domain (NTD), a DNA-binding domain (DBD), a hinge, and a ligand-binding domain (LBD) (Fig. 7.6). DBDs are highly conserved, whereas LBDs exhibit higher variability (Holzer and Laudet 2013; Vella and Hollenberg 2017). Interestingly, it has been shown that in all vertebrates, these receptors can be directly regulated by THs, mediated by their own receptors (Suzuki et al. 1994; Wong et al. 1998; Deveau et al. 2020). *THR* homologs are also present in a larger group of invertebrates (Fig. 7.6), suggesting that the genomic actions of iodinated messengers were specialized before the appearance of vertebrates. The molecular mechanisms of the TH-THR interaction and their target genes in invertebrate species are not fully elucidated, but, as for other TH signaling proteins, *Thr* homologs have been described in mollusks, echinoderms, and tunicates (Carosa et al. 1998; Wu et al. 2007; Paris et al. 2008; Taylor and Heyland 2017; Li et al. 2020b), although these receptors have not been shown to bind to T3, T4, or any other metabolite product of TH metabolism. Amphioxus expresses a *Thr* gene (*ampTR*), which encodes a nuclear receptor that preferentially binds TRIAC, a TR agonist that is a secondary product of TH metabolism (Paris et al. 2008). *ampTR* lacks the essential amino acids for TH binding present in vertebrate TRs, contains an Arg^{283} in the LDB, but favors binding to the OH of the carboxylic side chain of TRIAC (Paris et al. 2008, 2010). Lampreys, which represent the most ancient group of vertebrates, express two bona fide *TRs* (PmTR1 and PmTR2) (Manzon et al. 2007). Only PmTR1 is functional and can bind T3; therefore, it is thought that this receptor was further adapted to recognize the classical bioactive hormone of vertebrates with high affinity.

Of all vertebrates, teleost fish have the peculiarity of containing three or four functional THR genes. They contain the canonical *Thra* and *Thrb*, but some species have an additional gene, a product of teleost-specific whole-genome duplication (Glasauer and Neuhauss 2014). For example, eels, one of the earliest teleosts to appear on Earth, about 3.5 million years ago, contain two *Thra* and two *Thrb* genes (Kawakami et al. 2003). More recently, teleosts lost the duplication of the *Thrb* gene, keeping only two copies of *Thra* (*Thra*a and *Thrab*) and a single copy of *THRB*. Moreover, several teleost-specific isoforms are present in these species, which contain amino acid insertions (9 or 29) at the beginning of the LBD of *Thrb* (Marchand et al. 2001; Mendoza et al. 2013; Lazcano et al. 2019a, b). The functional meaning of these teleost isoforms has not been elucidated thus far, but experimental evidence shows the tissue- and time-specific expression of some isoforms, pointing to functional diversity (Lazcano and Orozco 2018; Lazcano et al. 2023a, b).

We can deduce that during vertebrate evolution, a protein from the nuclear receptor family was specialized in recognizing certain iodine messengers through the LBD. This protein, which was also bound to DNA via the DBD, decoded the binding of the ligand to its receptor by regulating DNA transcription, thereby allowing the direct regulation of gene expression through this type of receptor. This can be evidenced, for example, by the interaction between TRIAC and ampTR, where an iodinated messenger is recognized by a nuclear receptor encoded by a single gene. Given the significance of THs in processes such as metabolism, growth, and differentiation, the maintenance of a second gene that encodes another THR was favored, as is the case in more modern vertebrates. In this way, organisms could survive when one of these genes mutated and one of them compensated for the function of the other.

7.9 Acquisition of Negative Feedback During Vertebrate Evolution

One of the most fascinating adaptations of neuroendocrine axes is their ability to self-regulate. In humans, rats, and mice, T3 is responsible for HPT-negative feedback, inhibiting the transcription of *Trh* in the PVN and of TSH in thyrotropes (Fekete and Lechan 2014; Mendoza and Hollenberg 2017). T3 also diminishes the amount of TRH that reaches the pituitary by stimulating TRH-DE expression in tanycytes, causing increased TRH degradation at the median eminence (Sánchez et al. 2009). At the pituitary level, T3 lowers the expression of *Trh-R1* (Schomburg and Bauer 1995) and *Dio2* (Lechan and Fekete 2005; Fekete and Lechan 2014). TRH and TSH T3-negative regulation is mediated by TRβ2, although the mechanisms are still controversial, in contrast to those involved in T3-positive regulation (Joseph-Bravo et al. 2015b; Fekete and Lechan 2014; Mendoza and Hollenberg 2017).

The synthesis of THs, deiodinases, transporters, and receptors occur at different stages of brain development in rodents. TH feedback of TRH transcription appears much later since it requires tanycyte maturation around the third ventricle, where DIO2 converts T4 into T3. For the negative feedback to evolve to its present form in mammals, the neuroendocrine cells that participate in the activation of the HPT axis in the brain had to undergo multiple adjustments, including the selection of transporters in the plasma membranes of blood vessels and target cells required to access THs from the bloodstream and Thrs to decode the endocrine message. Tracing back to the lamprey, the most basal vertebrate, no information is yet available regarding the expression of any type of transporter that allows THs to be introduced into the CNS. However, studies have shown that the zebrafish blood-brain barrier (BBB) does express MCT8 (Zada et al. 2016, 2017). In fact, zebrafish with mutations in this transporter are known to manifest a series of symptoms related to Allan-Herndon-Dudley syndrome. This evidences MCT8-mediated TH transport from the bloodstream to the CNS in early vertebrates. However, despite evidence of TH-CNS transport, in vivo experiments in fish have shown that the administration of THs does not regulate TRH expression in the preoptic nucleus (the mammalian PVN homolog) (Geven et al. 2006; Iziga et al. 2010; Ogawa et al. 2013; Campinho et al. 2015). Therefore, the TRH neuron is not a substantial element of the HPT axis in teleosts or probably in other nonmammalian species. In fact, the appearance of a clear TH-negative feedback mechanism of the HPT axis traces back to chicken. Indeed, there is an increase in DIO2 expression in chick tanycytes after hatching, which coincides with the onset of T3 feedback regulation of the chicken HPT (Mohácsik et al. 2016). It would be desirable to further study the role of tanycytes in controlling TH transport and bioactivity, as well as other aspects that could regulate the HPT axis in other vertebrate models.

7.10 HPT Axis in Energy Homeostasis and Thermogenesis

7.10.1 Energy Homeostasis

Living organisms require energy for survival and have developed mechanisms that allow equilibrium within themselves and with the environment. Energy homeostasis involves a balance between energy intake and expenditure, which is attained in mammals through a concert of neuronal, endocrine, and metabolic signals decodified in the hypothalamus to modulate hunger and food intake as well as energy expenditure. Around 60–70% of total energy expenditure corresponds to biochemical reactions involved in the cellular functions of an organism at rest, and in thermoneutral conditions (basal/resting metabolic rate (BMR or RMR)), physical activity comprises nearly 20%, and the rest is due to diet- or cold-induced thermogenesis (Silva 2011; Tran et al. 2022). THs play an essential role in energy homeostasis as they regulate carbohydrate and lipid metabolism, mobilize energy reserves providing substrates to oxidizing tissues, induce the genesis of mitochondria, and regulate thermogenesis (Klieverik et al. 2009; Mullur et al. 2014; Bianco et al. 2019; Mendoza and Hollenberg 2017).

As mentioned (Sect. 7.2), the activity of the mammalian HPT axis in mammals is triggered by TRH synthesized and released from the hypophysiotropic neurons of the PVN (Lechan and Fekete 2005; Joseph-Bravo et al. 2015b; Costa-e-Sousa et al. 2023) and subject to multifactorial regulation (Joseph-Bravo et al. 2021). The TRH neuron has thus been proposed as a metabolic sensor (Hollenberg 2008) and an integrator of energy metabolism (Lechan and Fekete 2005) as its activity is inhibited by energy deficiency while stimulated by energy demands. TRH-hypophysiotropic neurons are tuned by T3 feedback and by the stimulatory activity of arcuate neurons, which synthetize the anorectic peptide POMC or the inhibitory orexigenic peptides NPY and Agrp (Fekete and Lechan 2014; Ortiga-Carvalho et al. 2016; Mendoza and Hollenberg 2017; Costa-e-Sousa et al. 2023); due to the loose BBB, these arcuate neurons sense metabolites circulating in the blood, such as insulin secreted by the pancreas in response to glucose stimulation and leptin secreted by adipose tissue after feeding (Vohra et al. 2022; Martelli and Brooks 2023). Situations of negative energy balance like fasting and food restriction inhibit TRH neurons and the HPT axis through the combined effects of arcuate afferents and circulating leptin and corticosterone, which modulate Dio2 activity in tanycytes and TRH-degrading enzymes, resulting in decreased synthesis and the release of TRH, TSH, and TH, as well as diminished RMR (Sánchez et al. 2009; Fekete and Lechan 2014; Lazcano et al. 2015; Joseph-Bravo et al. 2021). Chapter 9 describes the evolution of these mechanisms in more detail. In contrast, energy-demanding situations such as cold exposure or exercise stimulate TRH expression and the activity of the HPT axis (Uribe et al. 1993, 2014; Silva 2011; Jaimes-Hoy et al. 2021; Joseph-Bravo et al. 2015b).

7.10.2 Thermogenesis

Endothermy is the intrinsic mechanism that maintains a constant body temperature despite variations in the ambient temperature; the term homeotherm is assigned to species maintaining a core body temperature at a set point of around 37 °C in mammals and 42 °C in birds (Silva 2011); however, the term endotherm is more generally used. It is a highly regulated process that generates heat, not only by the chemical processes of adenosine triphosphate (ATP) synthesis and utilization, which comprise RMR (obligatory thermogenesis), but also by generating heat when induced by cold or the intake of hypercaloric food (adaptive or facultative thermogenesis). Endotherms respond acutely to cold exposure with shivering and with a longer metabolic response, called nonshivering thermogenesis (NST) (Silva 2011). Shivering is an acute repetitive muscle contraction-relaxation process, energetically costly, that produces heat from the energy of ATP hydrolysis by the myosin ATPase or by the sarcoendoplasmic reticulum Ca^{2+} ATPase (SERCA) (Bal and Periasamy 2020). NST is a long-lasting process that generates heat by two mechanisms: (a) in brown adipose tissue (BAT) uncoupling the electron transport on the inner mitochondrial membrane bypassing the synthesis of ATP and transforming the chemical energy into heat or (b) in skeletal muscle uncoupling Ca2+ mobilization by

SERCA1 in the contraction-relaxation events; the energy of ATP hydrolysis normally used for Ca^{2+} transport during muscle relaxation is instead released as heat; uncoupling proteins are required in both processes—uncoupling protein 1 (UCP1) in BAT and sarcolipin (SLN) in the muscle of mammals (Silva 2011; Nowack et al. 2017; Till et al. 2022). Additional means of thermoregulation include vasoconstriction, which diminishes heat loss and isolation systems such as subcutaneous lipid depots, fur in mammals, or feathers in birds. In situations of chronic cold exposure, subcutaneous white adipose tissue can acquire characteristics of BAT, increasing the number of mitochondria and UCP expression (Nguyen et al. 2023; Wu et al. 2023).

The concerted neuronal and neuroendocrine responses to cold have been well characterized in mammals, particularly in rodents (Fig. 7.7). The somatosensory neurons of the skin express ion channels (such as the transient receptor potential ion channel TRPM8), which sense an environmental temperature lower than that corresponding to thermoneutrality (28–30 °C in rodents and 28 °C in a naked man (Tran et al. 2022)), and convert thermal stimuli to electrical activity. The signal is propagated through multisynaptic contacts along the spinal cord, the brainstem, and the lateral parabrachial nucleus external part (LPBel) to reach the hypothalamic preoptic area, which then stimulates the neurons of the dorsomedial hypothalamic nucleus. From the latter nucleus, another multisynaptic is engaged, including the neurons of the Raphe nucleus and the premotor neurons of the spinal cord, which activate thermogenic organs such as skeletal muscles, causing shivering, or BAT, which is responsible for nonshivering thermogenesis (Morrison 2018; Kashio and Tominaga 2022; Nakamura and Morrison 2022) (Fig. 7.7a, b). In parallel, the PVN is stimulated by noradrenergic neurons from the locus coeruleus, which increase TRH synthesis and release, rapidly and transiently, stimulating TSH and T4 release (Fig. 7.7a) (Uribe et al. 1993: Joseph-Bravo et al. 2021). Stimulated adrenergic receptor Adrβ-3 in BAT increases cAMP-PKA activity, which activates the hormone-sensitive lipase, releasing free fatty acids, which are phosphorylated and integrated into the mitochondrial membrane, forming a channel with UCP1 (Nedergaard and Cannon 2018; Nguyen et al. 2023); PKA also activates Dio2, which transforms T4 to T3, stimulating the transcription of UCP-1 and the function of the adrenergic receptor, lipolysis, and many of the processes involved in BAT function (Fig. 7.7b) (Silva 2011; Yau and Yen 2020). NTS has also been recently demonstrated in the muscles of rodents and has been proposed to be used as an alternative or complementary system (Fig. 7.7c) (Bal and Periasamy 2020; Nowack et al. 2017). Additionally, cold-induced TSH may also act directly on BAT, increasing cAMP and Dio2 activity (Martinez-deMena et al. 2015). Other hypothalamic nuclei, such as the ventromedial nucleus (VMN), also participate in NTS, activating BAT in response to hypercaloric diets (diet-induced thermogenesis), cold exposure, or directly to other hormones such as leptin, estrogen, or TH (Contreras et al. 2017; Tran et al. 2022; Till et al. 2022).

The thermogenic role of BAT has been long recognized in rodents and human babies, and it was thought that in human adults, heat generation occurred only in muscles, but at the beginning of the twenty-first century, potent imaging techniques

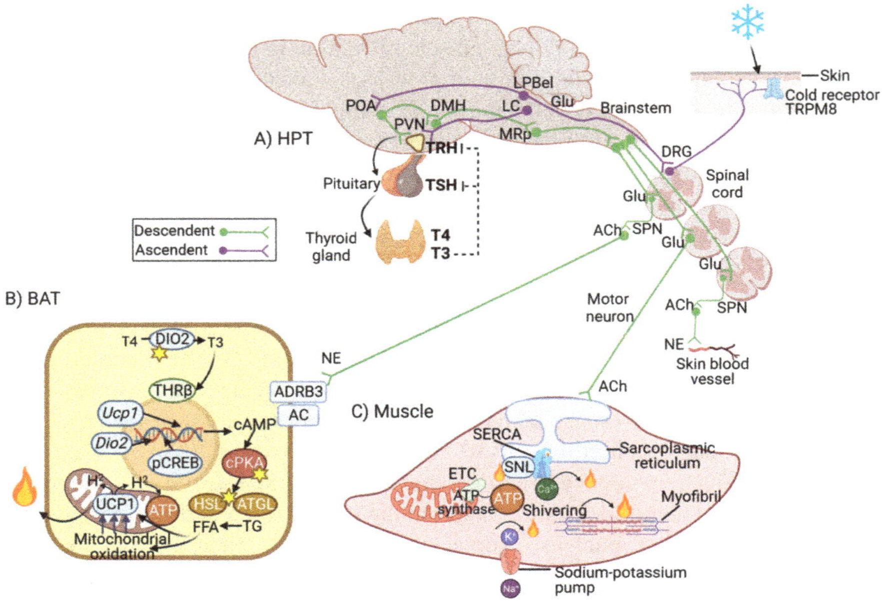

Fig. 7.7 Cold response in mammals. Cold sensed by skin receptors transmit signals by the ascending multisynaptic pathway through the spinal cord and nuclei in the brain stem and, finally, to the hypothalamus, activating the POA, which triggers the sympathetic response, causing shivering and nonshivering thermogenesis (muscle and BAT), and simultaneously, neurons from the brain stem activate the PVN, increasing the activity of the HPT axis (see text in Sect. 7.10.2). *Abbreviations*: *C*, adenylyl cyclase; *Ch*, acetylcholine; *ADRB3*, beta-3 adrenergic receptor; *ATGL*, adipose triglyceride lipase; *ATP*, adenosine triphosphate; *BAT*, brown adipose tissue; *Ca2+*, calcium ion; *cAMP*, cyclic adenosine monophosphate; *cPKA*, catalytic subunit of protein kinase A; *DIO2*, deiodinase 2; *Dio2*, deiodinase 2 gene; *DMH*, dorsomedial hypothalamic nucleus; *DRG*, dorsal root ganglio; *ETC*, electron transport chain; *FFA*, free fatty acids; *Glu*, glutamate; *H2*, dihydrogen cation; *HPT*, hypothalamus-pituitary-thyroid axis; *HSL*, hormone-sensitive lipase; *K+*, potassium ion; *LC*, locus coeruleus; *LPBel*, external lateral part of the lateral parabrachial nucleus; *MRp*, rostral; medullary raphe region; *NA*, noradrenaline; *Na+*, sodium ion; *NE*, norepinephrine; *pCREB*, phosphorylated cAMP response element-binding protein; *POA*, preoptic area; *PVN*; paraventricular hypothalamic nucleus; *SERCA*, sarco/endoplasmic reticulum; Ca2+ ATPase; *SNL*, sarcolipin; *SPN*; sympathetic preganglionic neurons; *T3*, triiodothyronine; *T4*, thyroxine; *TG*, triacylglycerol; *THRβ*, thyroid hormone receptor β; *TRH*, thyrotropin-releasing hormone; *TRPM8*, transient receptor 8; *TSH*, thyroid-stimulating hormone; *UCP1*, uncoupling protein 1; *Ucp1*, uncoupling protein 1 gene. Figure created with BioRender.com

allowed the identification of BAT in humans (Till et al. 2022). However, not all mammals have BAT, including rabbits, pigs, and wild boars; they produce NST solely by the muscle-SERCA system (Nowack et al. 2017; Bal and Periasamy 2020). Birds also generate heat by shivering and by NST in the muscle-SERCA system. These species lack BAT and UCP-1, and the responsible uncoupling protein remains under investigation. Avian SLN and SERCA present several amino acid variations; in the case of SLN, these substitutions are mainly located in the N and C ends, sites where the mammalian one interacts with SERCA. These differences cast doubt on this protein being responsible for uncoupling. Other uncoupling proteins, such as the ryanodin receptor or the adenine nucleotide translocator (ANT), have been proposed, but the mechanism involved is yet to be defined (Nowack et al. 2017).

Birds and mammals utilize fatty acids as an energy source for NST, while shivering is fueled by carbohydrates in mammals and lipids in birds. Both species share the regulatory mechanisms of thermogenesis involving norepinephrine, TH, and transcription factors such as PGC-1α (protein peroxisome-proliferator-activated receptor-γ coactivator-1α), which associates with the nuclear receptors PPARα and PPARγ, involved in many metabolic reactions related to thermogenesis and to lipid and ketone metabolism (Grigg et al. 2022; Wu et al. 2023).

Some animals adapt to extreme cold or food scarcity, entering torpor, which, together with hibernation, constitutes a strategy for survival and energy preservation (Legendre and Davesne 2020; Geiser 2020; Wu et al. 2023). When RMR falls in hibernated animals, metabolism is suppressed, and so is the core body temperature (attenuated at the level of sympathetic-motor neurons); some squirrels may achieve ambient temperatures and hibernate for 7–8 months using stored fat as the energy source. The low amount of glucose available switches metabolism to the utilization of fat-derived ketone bodies, limiting protein degradation and tissue damage (Wu et al. 2023). During hibernation, intermittent bouts of torpor and increased arousal and body temperature (by NST complemented with shivering) occur during hibernation. A prehibernation period precedes when animals eat more, increasing the size of white adipose tissue depots of BAT, and the thyroid gland, as well as the expression of Dio2 in BAT, TH-transport protein, and proteins involved in thyroid gland secretory activity. All of these changes favor more efficient thermogenesis in the torpor-arousal bouts of hibernation even though no clear differences are detected in the plasma concentrations of TH (Frare et al. 2021).

Other vertebrate lineages, such as reptiles, amphibians, and most fish, depend on the external environment as a source of heat (ectotherms); they present diverse strategies to preserve cellular function despite cold environments, some of which are similar to those in endotherms. An important one is the regulation of optimal membrane fluidity as low temperature alters their physical property, induces rigidity in the lipid bilayers, and affects their function. The mechanism is similar to that characterized in bacteria, where membrane sensors detect changes in membrane fluidity and signal the cell cytoplasm to change the composition of fatty acyl chains (Wu et al. 2023). Endotherms that hibernate or go into torpor also sense cold-induced changes in membrane fluidity using adiponectin membrane receptors without requiring adiponectin binding; cold induces noradrenergic activation, leading to fatty acid uptake in mitochondria and PPARα activation. Local thermogenesis may also occur with the energy from the by-products of metabolism (as described for obligatory thermogenesis) or of striated muscle activity obtained from movement; for example, swimming in fish, or movement of the muscle that surrounds the eye of certain fishes, keeps temperatures above that of the water. Another widely used mechanism is that of countercurrent, which takes place in vascular networks called *retia mirabilia* ("wonderful nets" by Galen), where arterial and venous vessels of opposite fluxes are tightly arranged by a protein, *apelin*; this close physical alignment allows countercurrent heat exchange from arterial to venous blood (Kahn 2015). Many examples illustrate this mechanism, for example, in sharks, where the heat from a warm tissue, like the muscle near the spinal cord, warms up the brain (Legendre and Davesne 2020).

Even when the thermogenic properties of the muscles of birds and mammals share many characteristics, doubt has been shed on the theory of the monophyletic evolution of these species (Nowack et al. 2017; Grigg et al. 2022). The expression and differences of specific molecules, like UCP1, SERCA, sarcolipin, and TPM8, relevant in thermogenesis or in such defined characteristics as aerobic capacity, higher RMR, NST, or adipose tissue browning, have been analyzed in attempts to understand the evolution of endothermy (Nowack et al. 2017; Legendre and Davesne 2020; Seebacher 2020; Grigg et al. 2022). It has been suggested that endothermy could have been achieved by recruiting additional mechanisms of thermogenesis to muscular work during locomotion, including specialized shivering thermogenesis, increases in mitochondrial density and membrane leakage, increases in sodium-potassium pump activity, and shifts in SERCA activity toward NST (Nowack et al. 2017). Another approach analyzes not the isolated features or the processes involved in thermogenesis but the regulatory mechanisms involved, such as the appearance of TH and their function as a driving force of the evolutionary process (Little and Seebacher 2013). It is beyond the aim of this section and the expertise of the writers to extend the discussion of the theories on the evolution of endothermy. We may consider the proposal to discard teleological and etiological arguments and look at endothermy as the evolutionary by-product of energy balance regulation rather than as an adaptation and interpret its evolution in the context of metabolic networks (Seebacher 2020).

7.10.3 Caveats

All the information discussed in this section corresponds to males since over 70% of research has been done on male rodents. Sex differences are now, at last, being characterized in humans and rodents. Energy expenditure, energy homeostasis, and thermogenesis strongly differ between sexes, with estrogens having a leading role and the sex dimorphic characteristics of the HPT axis (Parra-Montes De Oca et al. 2021; Lee et al. 2022; De Jesus and Henry 2023; Fernández-Peña et al. 2023).

7.11 Perspectives

The HPT axis is highly conserved across vertebrate species, showing its fundamental role in regulating vital functions such as growth, metamorphosis, metabolism, and energy homeostasis. Some adaptations in the structure of the HPT axis may have allowed organisms to adapt to different environmental conditions, such as variations in food availability, the aquatic or terrestrial environment, and temperature. For example, in endothermic organisms, the bird lineage loses the *Ucp1* gene, lacks BAT, and activates the HPT axis through CGCL and CRH, whereas in mammals, thermogenesis is activated via TRH, *Ucp1*, and BAT. Comparative studies in different invertebrate and vertebrate species have highlighted differences in HPT axis function, hormone levels, and receptor properties. The future study of this neuroendocrine network can help researchers understand how evolution has shaped the endocrine systems of various organisms and will allow the understanding of how neuroendocrine systems can adapt to environmental changes, which have been occurring for more than 500 million years. Analyzing changes in genes, molecules, and proteins that comprise the HPT axis can provide insights into the evolutionary processes that have promoted the fine-tuning of this neuroendocrine network.

There are still several aspects of the evolution of the thyroid axis that remain unexplored. One of them is the central activator of the HPT in some vertebrate groups. This area has never been explored in reptiles. Despite methodological limitations in the study of certain nonmammalian species, the secretion of a particular TRF can be activated (or inhibited) in genetic manipulation experiments using transgenic animals like zebrafish and *Xenopus laevis*. A deeper understanding of the HPT axis across vertebrates may lead us to reconsider current established schemes of thyroid physiology. Also, as previously mentioned, our current knowledge on the evolution of the HPT axis has been carried out in males, a point to ponder when analyzing the accumulated experimental evidence in this regard.

Research on the HPT axis in nonmammalian species has revealed how this neuroendocrine network has been evolving for about 500 million years (Fig. 7.8). What will be the molecular, cellular, and anatomical adaptations for the future?

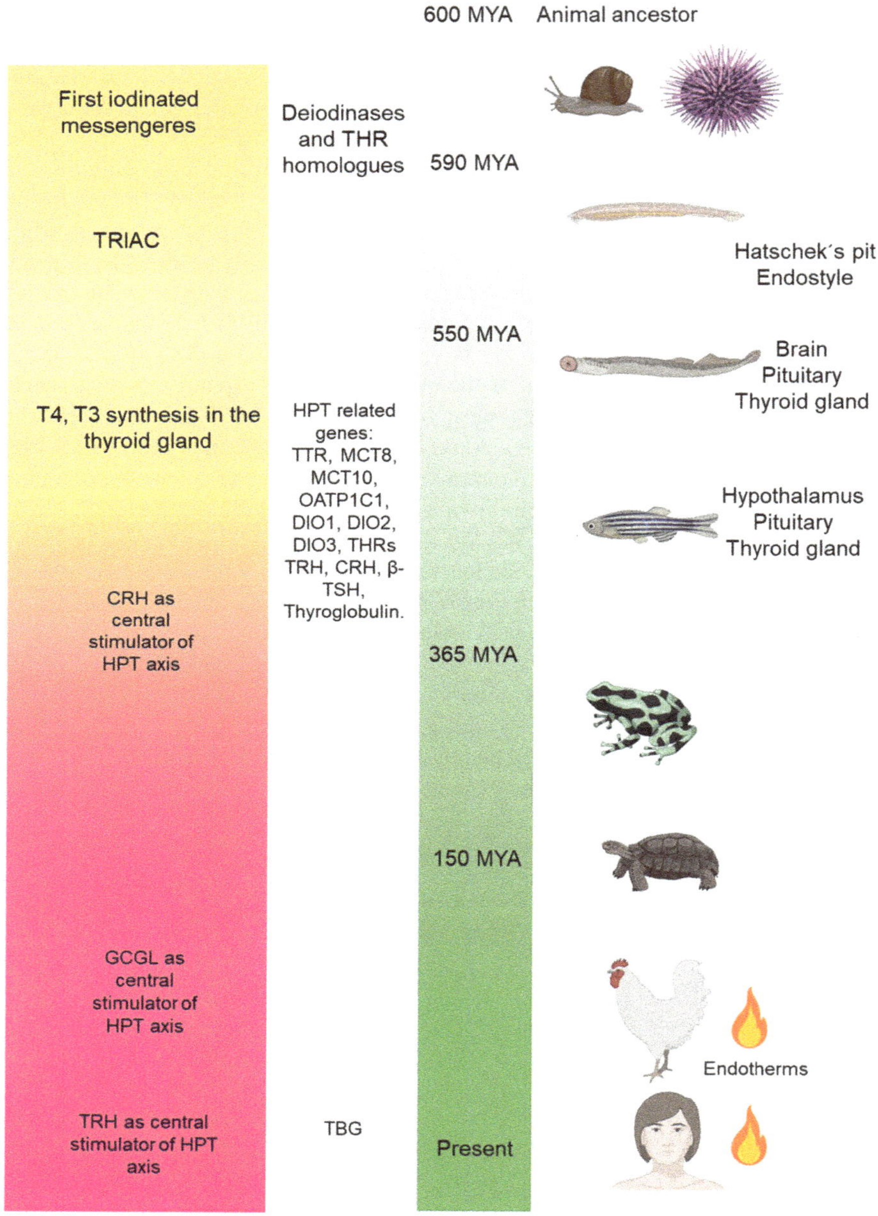

Fig. 7.8 Emergence of genes and structures relevant to the evolution of the HPT axis. The figure shows a timeline, represented in millions of years ago (MYA), showing the evolution of the thyroid gland, preceded by the existence of iodinated messenger molecules. It is considered that the thyroid gland appeared in most basal vertebrates, represented by the lamprey around 550 MYA. Homologous thyroid signaling genes can be found in some invertebrates, but it is in fish that they acquire specialized structures and functions to allow the HPT axis to function. The emergence of the pituitary and thyroid glands was preceded by Hatschek's pit and the endostyle, respectively,

Box 7.3 The Role of Tanycytes as Regulators of the HPT Axis

Communication between the hypothalamus and the pituitary is carried out by a microstructure located at the base of the brain: the median eminence. This structure is composed of axon terminals of neurosecretory cells. It receives blood flow through the portal system, contains ependymal cells, and allows the transport of chemical messengers from the brain to the periphery and vice versa. A specialized type of glial cells, denominated tanycytes, surround the walls of the floor of the third ventricle and are a fundamental part of the median eminence neuroarchitecture. There are at least four subtypes of tanycytes classified according to their anatomy, distribution, and the genes they express (α1, α2, β1, and β2) (Rodriguez et al. 2005). In particular, the cellular process of β2 tanycytes contacts the capillaries of the portal system and the nerve endings of neurosecretory cells, such as TRH neurons (Sánchez et al. 2009; Rodríguez-Rodríguez et al. 2019). In mammals, tanycytes express *DIO2*, the enzyme that produces T3 from T4 (Tu et al. 1997; Sánchez et al. 2009), as well as the MCT8 transporter involved in the negative feedback of the HPT axis, explaining the T4 entrance route to the hypothalamus to regulate the HPT axis at the central level (Heuer 2007). Tanycytes can also regulate the output of TRH neurons through a specific TRH-DE that degrades released TRH in the extracellular space of the median eminence. TRH-DE is activated by T3, thus regulating the amount of neuropeptide that can reach the anterior pituitary via the portal system (Sánchez et al. 2009; Lazcano et al. 2015). The presence of tanycytes in nonmammalian vertebrates has been poorly explored. Markers such as vimentin and glial fibrillary acidic protein (GFAP) have been used to identify this type of cell in different vertebrates, such as fish or amphibians (Kálmán et al. 1993; Lazzari et al. 1997; Lazzari and Franceschini 2004; Mohácsik et al. 2016; Ceriani and Whitlock 2021). Thus, the evolution of tanycytes and their control over the HPT axis in nonmammalian vertebrates is an area that should be explored.

Acknowledgments We thank Dr. Santiago Pech-Pool for his technical support and to, Angélica Gutiérrez-Mata for creating Fig. 7.8, and Jessica Gonzalez Norris for proofreading the manuscript. This work was supported by the following grants: UNAM PAPIIT IN204920, UNAM PAPIIT IA201122, UNAM PAPIIT IN206724, UNAM PAPIIT IN217422, and Ciencias de Frontera CONACyT 319880.

Fig. 7.8 (continued) which appeared in the first chordates. *Abbreviations*: *CRH*, corticotropin-releasing hormone; *DIO1*, deiodinase 1; *DIO2*, deiodinase 2; *DIO3*, deiodinase 3; *GCGL*, glucagon-like peptide; *GpH*, glycoprotein hormone; *HPT*, hypothalamus-pituitary-thyroid axis; *MCT8*, monocarboxylate transporter 8; *MCT10*, monocarboxylate transporter 10; *OATP1C1*, organic anion transporter 1C1; *T3*, 3,5,3′-triiodothyronine; *T4*, thyroxine; *TBG*, thyroxine-binding globulin; *THR*, thyroid hormone receptor; *TRIAC*, 3,5,3′-triiodothyroacetic acid; *TRH*, thyrotropin-releasing hormone; *TTR*, transthyretin; *TSH*, thyroid-stimulating hormone

Conflict of Interest The authors declare that they have nothing to disclose.

References

Admati I, Wasserman-Bartov T, Tovin A, Rozenblat R, Blitz E, Zada D, Lerer-Goldshtein T, Appelbaum L (2020) Neural alterations and hyperactivity of the hypothalamic–pituitary–thyroid axis in Oatp1c1 deficiency. Thyroid 30(1):161–174. https://doi.org/10.1089/thy.2019.0320

Aguilera G, Liu Y (2012) The molecular physiology of CRH neurons. Front Neuroendocrinol 33(1):67–84. https://doi.org/10.1016/j.yfrne.2011.08.002

Ando H, Ando J, Urano A (1998) Localization of mRNA encoding thyrotropin-releasing hormone precursor in the brain of Sockeye Salmon. Zool Sci 15(6):945–953. https://doi.org/10.2108/zsj.15.945

Antczak C, De Meester I, Bauvois B (2001) Ectopeptidases in pathophysiology. BioEssays 23(3):251–260. https://doi.org/10.1002/1521-1878(200103)23:3<251::AID-BIES1035>3.0.CO;2-O

Aoki Y, Ono H, Yasuo S, Masuda T, Yoshimura T, Ebihara S, Iigo M, Yanagisawa T (2007) Molecular evolution of prepro-thyrotropin-releasing hormone in the chicken (Gallus gallus) and its expression in the brain. Zool Sci 24(7):686–692. https://doi.org/10.2108/zsj.24.686

Arjona FJ, de Vrieze E, Visser TJ, Flik G, Klaren PHM (2011) Identification and functional characterization of zebrafish solute carrier Slc16a2 (Mct8) as a thyroid hormone membrane transporter. Endocrinology 152(12):5065–5073. https://doi.org/10.1210/en.2011-1166

Bal NC, Periasamy M (2020) Uncoupling of sarcoendoplasmic reticulum calcium ATPase pump activity by sarcolipin as the basis for muscle non-shivering thermogenesis. Philos Trans R Soc Lond B Biol Sci 375(1793):20190135. https://doi.org/10.1098/rstb.2019.0135

Bale TL, Vale WW (2004) CRF and CRF receptors: role in stress responsivity and other behaviors. Annu Rev Pharmacol Toxicol 44(1):525–557. https://doi.org/10.1146/annurev.pharmtox.44.101802.121410

Bauknecht P, Jékely G (2015) Large-scale combinatorial deorphanization of platynereis neuropeptide GPCRs. Cell Rep 12(4):684–693. https://doi.org/10.1016/j.celrep.2015.06.052

Becker KB, Stephens KC, Davey JC, Schneider MJ, Galton VA (1997) The type 2 and type 3 iodothyronine deiodinases play important roles in coordinating development in Rana catesbeiana tadpoles. Endocrinology 138(7):2989–2997. https://doi.org/10.1210/endo.138.7.5272

Belkadi A, Jacques C, Savagner F, Malthièry Y (2012) Phylogenetic analysis of the human thyroglobulin regions. Thyroid Res 5(1):3. https://doi.org/10.1186/1756-6614-5-3

Berry MJ, Banu L, Chen Y, Mandel SJ, David Kieffer J, Harney JW, Reed Larsen P (1991) Recognition of UGA as a Selenocysteine codon in type I deiodinase requires sequences in the 3′ untranslated region. Nature 353(6341):273–276. https://doi.org/10.1038/353273a0

Bianco AC, Salvatore D, Gereben B, Berry MJ, Reed Larsen P (2002) Biochemistry, cellular and molecular biology, and physiological roles of the iodothyronine selenodeiodinases. Endocr Rev 23(1):38–89. https://doi.org/10.1210/edrv.23.1.0455

Bianco AC, Dumitrescu A, Gereben B, Ribeiro MO, Fonseca TL, Fernandes GW, Bocco BMLC (2019) Paradigms of dynamic control of thyroid hormone signaling. Endocr Rev 40(4):1000–1047. https://doi.org/10.1210/er.2018-00275

Bourgeois NMA, Van Herck SLJ, Vancamp P, Delbaere J, Zevenbergen C, Kersseboom S, Darras VM, Visser TJ (2016) Characterization of chicken thyroid hormone transporters. Endocrinology 157(6):2560–2574. https://doi.org/10.1210/en.2015-2025

Bridgham JT, Eick GN, Larroux C, Deshpande K, Harms MJ, Gauthier MEA, Ortlund EA, Degnan BM, Thornton JW (2010) Protein evolution by molecular tinkering: diversification of the nuclear receptor superfamily from a ligand-dependent ancestor. PLoS Biol 8(10):e1000497. https://doi.org/10.1371/journal.pbio.1000497

Brown DD, Cai L, Das B, Marsh-Armstrong N, Schreiber AM, Juste R (2005) Thyroid hormone controls multiple independent programs required for limb development in *Xenopus Laevis* metamorphosis. Proc Natl Acad Sci 102(35):12455–12458. https://doi.org/10.1073/pnas.0505989102

Campinho MA, Galay-Burgos M, Sweeney GE, Power DM (2010) Coordination of deiodinase and thyroid hormone receptor expression during the larval to juvenile transition in sea bream (Sparus Aurata, Linnaeus). Gen Comp Endocrinol 165(2):181–194. https://doi.org/10.1016/j.ygcen.2009.06.020

Campinho MA, Silva N, Roman-Padilla J, Ponce M, Manchado M, Power DM (2015) Flatfish metamorphosis: a hypothalamic independent process? Mol Cell Endocrinol 404(March):16–25. https://doi.org/10.1016/j.mce.2014.12.025

Cardoso JCR, Bergqvist CA, Félix RC, Larhammar D (2016) Corticotropin-releasing hormone family evolution: five ancestral genes remain in some lineages. J Mol Endocrinol 57(1):73–86. https://doi.org/10.1530/JME-16-0051

Cardoso JCR, Bergqvist CA, Larhammar D (2020) Corticotropin-releasing hormone (CRH) gene family duplications in lampreys correlate with two early vertebrate genome doublings. Front Neurosci 14(July):672. https://doi.org/10.3389/fnins.2020.00672

Carosa E, Fanelli A, Ulisse S, Roberto Di Lauro J, Rall E, Jannini EA (1998) *Ciona intestinalis* nuclear receptor 1: a member of steroid/thyroid hormone receptor family. Proc Natl Acad Sci 95(19):11152–11157. https://doi.org/10.1073/pnas.95.19.11152

Carvalho DP, Dupuy C (2017) Thyroid hormone biosynthesis and release. Mol Cell Endocrinol 458(December):6–15. https://doi.org/10.1016/j.mce.2017.01.038

Ceriani R, Whitlock KE (2021) Gonadotropin releasing hormone (GnRH) triggers neurogenesis in the hypothalamus of adult zebrafish. Int J Mol Sci 22(11):5926. https://doi.org/10.3390/ijms22115926

Charli J-L (2020) Sexually dimorphic dynamics of thyroid axis activity during fasting in rats. Front Biosci 25(7):1305–1323. https://doi.org/10.2741/4857

Charli J-L, Rodríguez-Rodríguez A, Hernández-Ortega K, Cote-Vélez A, Uribe RM, Jaimes-Hoy L, Joseph-Bravo P (2020) The thyrotropin-releasing hormone-degrading ectoenzyme, a therapeutic target? Front Pharmacol 11(May):640. https://doi.org/10.3389/fphar.2020.00640

Chatzitomaris A, Hoermann R, Midgley JE, Hering S, Urban A, Dietrich B, Abood A, Klein HH, Dietrich JW (2017) Thyroid allostasis-adaptive responses of thyrotropic feedback control to conditions of strain, stress, and developmental programming. Front Endocrinol 8:163. https://doi.org/10.3389/fendo.2017.00163

Contreras C, Nogueiras R, Diéguez C, Rahmouni K, López M (2017) Traveling from the hypothalamus to the adipose tissue: the thermogenic pathway. Redox Biol 12(August):854–863. https://doi.org/10.1016/j.redox.2017.04.019

Coscia F, Taler-Verčič A, Chang VT, Sinn L, O'Reilly FJ, Izoré T, Renko M et al (2020) The structure of human thyroglobulin. Nature 578(7796):627–630. https://doi.org/10.1038/s41586-020-1995-4

Costa-e-Sousa RH, Rorato R, Hollenberg AN, Vella KR (2023) Regulation of thyroid hormone levels by hypothalamic thyrotropin-releasing hormone neurons. Thyroid 33(7):867–876. https://doi.org/10.1089/thy.2023.0173

Crockford SJ (2009) Evolutionary roots of iodine and thyroid hormones in cell-cell signaling. Integr Comp Biol 49(2):155–166. https://doi.org/10.1093/icb/icp053

Crowner A, Khatri S, Blichmann D, Randal Voss S (2019) Rediscovering the axolotl as a model for thyroid hormone dependent development. Front Endocrinol 10(April):237. https://doi.org/10.3389/fendo.2019.00237

Cruz R, Vargas MA, Uribe RM, Pascual I, Lazcano I, Yiotakis A, Matziari M, Joseph-Bravo P, Charli J-L (2008) Anterior pituitary Pyroglutamyl peptidase II activity controls TRH-induced prolactin release. Peptides 29(11):1953–1964. https://doi.org/10.1016/j.peptides.2008.07.011

Dardente H, Migaud M (2021) Thyroid hormone and hypothalamic stem cells in seasonal functions. Vitam Horm 116:91–131. https://doi.org/10.1016/bs.vh.2021.02.005

Darras VM (2021) Deiodinases: how nonmammalian research helped shape our present view. Endocrinology 162(6):bqab039. https://doi.org/10.1210/endocr/bqab039

Davey KG (2000) Do thyroid hormones function in insects? Insect Biochem Mol Biol 30(8–9):877–884. https://doi.org/10.1016/S0965-1748(00)00061-8

Davis PJ, Goglia F, Leonard JL (2016) Nongenomic actions of thyroid hormone. Nat Rev Endocrinol 12(2):111–121. https://doi.org/10.1038/nrendo.2015.205

Davis PJ, Mousa SA, Lin H-Y (2021) Nongenomic actions of thyroid hormone: the integrin component. Physiol Rev 101(1):319–352. https://doi.org/10.1152/physrev.00038.2019

De Groef B, Van der Geyten S, Darras VM, Kühn ER (2006) Role of corticotropin-releasing hormone as a thyrotropin-releasing factor in non-mammalian vertebrates. Gen Comp Endocrinol 146(1):62–68. https://doi.org/10.1016/j.ygcen.2005.10.014

De Groef B, Grommen SVH, Darras VM (2018) Forever young: endocrinology of paedomorphosis in the Mexican axolotl (Ambystoma mexicanum). Gen Comp Endocrinol 266(September):194–201. https://doi.org/10.1016/j.ygcen.2018.05.016

De Jesus AN, Henry BA (2023) The role of oestrogen in determining sexual dimorphism in energy balance. J Physiol 601(3):435–449. https://doi.org/10.1113/JP279501

Deal CK, Volkoff H (2020) The role of the thyroid axis in fish. Front Endocrinol 11(November):596585. https://doi.org/10.3389/fendo.2020.596585

Denver RJ (1988) Several hypothalamic peptides stimulate in vitro thyrotropin secretion by pituitaries of anuran amphibians. Gen Comp Endocrinol 72(3):383–393. https://doi.org/10.1016/0016-6480(88)90160-8

Denver R (2009) Stress hormones mediate environment-genotype interactions during amphibian development. Gen Comp Endocrinol 164(1):20–31. https://doi.org/10.1016/j.ygcen.2009.04.016

Denver RJ, Licht P (1991) Several hypothalamic peptides stimulate thyrotropin and growth hormone secretion by adult turtle pituitary glands. Comp Biochem Physiol A Physiol 100(3):603–606. https://doi.org/10.1016/0300-9629(91)90377-O

Deveau C, Jiao X, Suzuki SC, Krishnakumar A, Takeshi Yoshimatsu J, Hejtmancik F, Nelson RF (2020) Thyroid hormone receptor Beta mutations alter photoreceptor development and function in Danio Rerio (zebrafish). PLoS Genet 16(6):e1008869. https://doi.org/10.1371/journal.pgen.1008869

Díaz ML, Becerra M, Manso MJ, Anadón R (2002) Distribution of thyrotropin-releasing hormone (TRH) immunoreactivity in the brain of the zebrafish (*Danio Rerio*): TRH immunoreactivity in zebrafish brain. J Comp Neurol 450(1):45–60. https://doi.org/10.1002/cne.10300

Domínguez L, López JM, González A (2008) Distribution of thyrotropin-releasing hormone (TRH) immunoreactivity in the brain of Urodele amphibians. Brain Behav Evol 71(3):231–246. https://doi.org/10.1159/000122835

Dos Santos S, Mazan S, Venkatesh B, Cohen-Tannoudji J, Quérat B (2011) Emergence and evolution of the glycoprotein hormone and neurotrophin gene families in vertebrates. BMC Evol Biol 11(1):332. https://doi.org/10.1186/1471-2148-11-332

Duan W, Achen MG, Richardson SJ, Lawrence MC, Wettenhall REH, Jaworowski A, Schreiber G (1991) Isolation, characterization, cDNA cloning and gene expression of an avian transthyretin. Implications for the evolution of structure and function of transthyretin in vertebrates. Eur J Biochem 200(3):679–687. https://doi.org/10.1111/j.1432-1033.1991.tb16232.x

Eales JG, Brown SB (1993) Measurement and regulation of thyroidal status in teleost fish. Rev Fish Biol Fish 3(4):299–347. https://doi.org/10.1007/BF00043383

Emamnejad R, Dass M, Mahlis M, Bozkurt S, Ye S, Pagnin M, Theotokis P, Grigoriadis N, Petratos S (2022) Thyroid hormone-dependent oligodendroglial cell lineage genomic and non-genomic signaling through integrin receptors. Front Pharmacol 13(September):934971. https://doi.org/10.3389/fphar.2022.934971

Farid NR, Szkudlinski MW (2004) Minireview: structural and functional evolution of the thyrotropin receptor. Endocrinology 145(9):4048–4057. https://doi.org/10.1210/en.2004-0437

Fekete C, Lechan RM (2014) Central regulation of hypothalamic-pituitary-thyroid axis under physiological and pathophysiological conditions. Endocr Rev 35(2):159–194. https://doi.org/10.1210/er.2013-1087

Fernández-Peña C, Reimúndez A, Viana F, Arce VM, Señarís R (2023) Sex differences in thermoregulation in mammals: implications for energy homeostasis. Front Endocrinol 14(March):1093376. https://doi.org/10.3389/fendo.2023.1093376

Fischer WH, Behan DP, Park M, Potter E, Lowry PJ, Vale W (1994) Assignment of disulfide bonds in corticotropin-releasing factor-binding protein. J Biol Chem 269(6):4313–4316. https://doi.org/10.1016/S0021-9258(17)41780-7

Frare C, Williams CT, Drew KL (2021) Thermoregulation in hibernating mammals: the role of the "thyroid hormones system". Mol Cell Endocrinol 519(January):111054. https://doi.org/10.1016/j.mce.2020.111054

Friesema ECH, Docter R, Moerings EPCM, Stieger B, Hagenbuch B, Meier PJ, Krenning EP, Hennemann G, Visser TJ (1999) Identification of thyroid hormone transporters. Biochem Biophys Res Commun 254(2):497–501. https://doi.org/10.1006/bbrc.1998.9974

Galas L, Raoult E, Tonon M-C, Okada R, Jenks BG, Castaño JP, Kikuyama S, Malagon M, Roubos EW, Vaudry H (2009) TRH acts as a multifunctional hypophysiotropic factor in vertebrates. Gen Comp Endocrinol 164(1):40–50. https://doi.org/10.1016/j.ygcen.2009.05.003

Galton VA, Schneider MJ, Clark AS, Donald LS, Germain. (2009) Life without thyroxine to 3,5,3′-triiodothyronine conversion: studies in mice devoid of the 5′-deiodinases. Endocrinology 150(6):2957–2963. https://doi.org/10.1210/en.2008-1572

García-G C, Jeziorski MC, Valverde-R C, Orozco A (2004) Effects of iodothyronines on the hepatic outer-ring deiodinating pathway in killifish. Gen Comp Endocrinol 135(2):201–209. https://doi.org/10.1016/j.ygcen.2003.09.010

Geiser F (2020) Seasonal expression of avian and mammalian daily torpor and hibernation: not a simple summer-winter affair. Front Physiol 11(May):436. https://doi.org/10.3389/fphys.2020.00436

Gereben B, Zavacki AM, Ribich S, Kim BW, Huang SA, Simonides WS, Zeöld A, Bianco AC (2008) Cellular and molecular basis of deiodinase-regulated thyroid hormone signaling. Endocr Rev 29(7):898–938. https://doi.org/10.1210/er.2008-0019

Geris KL, Kotanen SP, Berghman LR, Kühn ER, Darras VM (1996) Evidence of a thyrotropin-releasing activity of ovine corticotropin-releasing factor in the domestic fowl (Gallus domesticus). Gen Comp Endocrinol 104(2):139–146. https://doi.org/10.1006/gcen.1996.0156

Geven EJW, Verkaar F, Flik G, Klaren PHM (2006) Experimental hyperthyroidism and central mediators of stress axis and thyroid axis activity in common carp (Cyprinus carpio L.). J Mol Endocrinol 37(3):443–452. https://doi.org/10.1677/jme.1.02144

Glasauer SMK, Neuhauss SCF (2014) Whole-genome duplication in teleost fishes and its evolutionary consequences. Mol Gen Genomics 289(6):1045–1060. https://doi.org/10.1007/s00438-014-0889-2

Golan M, Zelinger E, Zohar Y, Levavi-Sivan B (2015) Architecture of GnRH-gonadotrope-vasculature reveals a dual mode of gonadotropin regulation in fish. Endocrinology 156(11):4163–4173. https://doi.org/10.1210/en.2015-1150

Gorbman A (1999) Brain-Hatschek's pit relationships in amphioxus species. Acta Zool 80(4):301–305. https://doi.org/10.1046/j.1463-6395.1999.00027.x

Grigg G, Nowack J, Bicudo JEPW, Bal NC, Woodward HN, Seymour RS (2022) Whole-body endothermy: ancient, homologous and widespread among the ancestors of mammals, birds and crocodylians. Biol Rev 97(2):766–801. https://doi.org/10.1111/brv.12822

De Groef B, Goris N, Arckens L, Kühn ER, Darras VM (2003) Corticotropin-releasing hormone (CRH)-induced thyrotropin release is directly mediated through CRH receptor type 2 on thyrotropes. Endocrinology 144(12):5537–5544. https://doi.org/10.1210/en.2003-0526

Groeneweg S, van Geest FS, Peeters RP, Heuer H, Edward W, Visser. (2020) Thyroid hormone transporters. Endocr Rev 41(2):146–201. https://doi.org/10.1210/endrev/bnz008

Guillemin R (2005) Hypothalamic hormones a.k.a. hypothalamic releasing factors. J Endocrinol 184(1):11–28. https://doi.org/10.1677/joe.1.05883

Hamano K, Inoue K, Yanagisawa T (1990) Immunohistochemical localization of thyrotropin-releasing hormone in the brain of carp, Cyprinus carpio. Gen Comp Endocrinol 80(1):85–92. https://doi.org/10.1016/0016-6480(90)90151-B

Harder S, Dammann O, Buck F, Zwiers H, Lederis K, Richter D, Bruhn TO (2001) Cloning of two thyrotropin-releasing hormone receptor subtypes from a lower vertebrate (Catostomus commersoni): functional expression, gene structure, and evolution. Gen Comp Endocrinol 124(2):236–245. https://doi.org/10.1006/gcen.2001.7709

Harms MJ, Thornton JW (2010) Analyzing protein structure and function using ancestral gene reconstruction. Curr Opin Struct Biol 20(3):360–366. https://doi.org/10.1016/j.sbi.2010.03.005

Hauger RL, Dautzenberg FM (2000) Regulation of the stress response by corticotropin-releasing factor receptors. In: Michael Conn P, Freeman ME (eds) Neuroendocrinology in physiology and medicine. Humana Press, Totowa, NJ, pp 261–286. https://doi.org/10.1007/978-1-59259-707-9_15

Hernández-Linares Y, Olvera A, Villalobos P, Lozano-Flores C, Varela-Echavarría A, Luna M, Orozco A (2019) 3,5-T2 and 3,3′,5-T3 regulate cerebellar thyroid hormone signalling and myelin molecular dynamics in tilapia. Sci Rep 9(1):7359. https://doi.org/10.1038/s41598-019-43701-w

Heuer H (2007) The importance of thyroid hormone transporters for brain development and function. Best Pract Res Clin Endocrinol Metab 21(2):265–276. https://doi.org/10.1016/j.beem.2007.03.003

Heyland A, Moroz LL (2005) Cross-kingdom hormonal signaling: an insight from thyroid hormone functions in marine larvae. J Exp Biol 208(Pt 23):4355–4361. https://doi.org/10.1242/jeb.01877. PMID: 16339856

Heyland A, Reitzel AM, Price DA, Moroz LL (2006) Endogenous thyroid hormone synthesis in facultative planktotrophic larvae of the sand dollar Clypeaster rosaceus: implications for the evolutionary loss of larval feeding. Evol Dev 8(6):568–579. https://doi.org/10.1111/j.1525-142X.2006.00128.x

Hinkle PM, Gehret AU, Jones BW (2012) Desensitization, trafficking, and resensitization of the pituitary thyrotropin-releasing hormone receptor. Front Neurosci 6. https://doi.org/10.3389/fnins.2012.00180

Holland LZ, Ocampo Daza D (2018) A new look at an old question: when did the second whole genome duplication occur in vertebrate evolution? Genome Biol 19(1):209. https://doi.org/10.1186/s13059-018-1592-0

Holland LZ, Albalat R, Azumi K, Benito-Gutiérrez È, Blow MJ, Bronner-Fraser M, Brunet F et al (2008) The amphioxus genome illuminates vertebrate origins and cephalochordate biology. Genome Res 18(7):1100–1111. https://doi.org/10.1101/gr.073676.107

Hollenberg AN (2008) The role of the thyrotropin-releasing hormone (TRH) neuron as a metabolic sensor. Thyroid 18(2):131–139. https://doi.org/10.1089/thy.2007.0251

Holzer G, Laudet V (2013) Thyroid hormones and postembryonic development in amniotes. Curr Top Dev Biol 103:397–425. https://doi.org/10.1016/B978-0-12-385979-2.00014-9

Holzer G, Morishita Y, Fini J-B, Lorin T, Gillet B, Hughes S, Tohmé M et al (2016) Thyroglobulin represents a novel molecular architecture of vertebrates. J Biol Chem 291(32):16553–16566. https://doi.org/10.1074/jbc.M116.719047

Huang G, He C, Meng F, Li J, Zhang J, Wang Y (2014) Glucagon-like peptide (GCGL) is a novel potential TSH-releasing factor (TRF) in chickens: I) evidence for its potent and specific action on stimulating TSH mRNA expression and secretion in the pituitary. Endocrinology 155(11):4568–4580. https://doi.org/10.1210/en.2014-1331

Ikegami K, Yoshimura T (2012) Circadian clocks and the measurement of daylength in seasonal reproduction. Mol Cell Endocrinol 349(1):76–81. https://doi.org/10.1016/j.mce.2011.06.040

Incerpi S, De Vito P, Luly P, Spagnuolo S, Leoni S (2002) Short-term effects of thyroid hormones and 3,5-diiodothyronine on membrane transport systems in chick embryo hepatocytes. Endocrinology 143(5):1660–1668. https://doi.org/10.1210/endo.143.5.8767

Isorna E, Obregon MJ, Calvo RM, Vázquez R, Pendón C, Falcón J, Muñoz-Cueto JA (2009) Iodothyronine deiodinases and thyroid hormone receptors regulation during flatfish (*Solea senegalensis*) metamorphosis. J Exp Zool B Mol Dev Evol 312B(3):231–246. https://doi.org/10.1002/jez.b.21285

Itoh K, Watanabe K, Xiaoming W, Suzuki T (2010) Three members of the iodothyronine deiodinase family, *dio1*, *dio2* and *dio3*, are expressed in spatially and temporally specific patterns during metamorphosis of the flounder, *Paralichthys olivaceus*. Zool Sci 27(7):574–580. https://doi.org/10.2108/zsj.27.574

Iziga R, Ponce M, Infante C, Rebordinos L, Cañavate JP, Manchado M (2010) Molecular characterization and gene expression of thyrotropin-releasing hormone in Senegalese sole (Solea senegalensis). Comp Biochem Physiol B Biochem Mol Biol 157(2):167–174. https://doi.org/10.1016/j.cbpb.2010.05.013

Jackson IMD, Reichlin S (1974) Thyrotropin-releasing hormone (TRH): distribution in hypothalamic and extrahypothalamic brain tissues of mammalian and submammalian chordates. Endocrinology 95(3):854–862. https://doi.org/10.1210/endo-95-3-854

Jaimes-Hoy L, Pérez-Maldonado A, Bahena EN, De La Cruz N, Guarneros AR-R, Charli J-L, Soberón X, Joseph-Bravo P (2021) Sex dimorphic changes in *Trh* gene methylation and thyroid-axis response to energy demands in maternally separated rats. Endocrinology 162(8):bqab110. https://doi.org/10.1210/endocr/bqab110

Jang D, Eliseeva E, Klubo-Gwiezdzinska J, Neumann S, Gershengorn MC (2022) TSH stimulation of human thyroglobulin and thyroid peroxidase gene transcription is partially dependent on internalization. Cell Signal 90(February):110212. https://doi.org/10.1016/j.cellsig.2021.110212

Joseph-Bravo P, Jaimes-Hoy L, Uribe R-M, Charli J-L (2015a) 60 years of neuroendocrinology: TRH, the first hypophysiotropic releasing hormone isolated: control of the pituitary–thyroid axis. J Endocrinol 226(2):T85–T100. https://doi.org/10.1530/JOE-15-0124

Joseph-Bravo P, Jaimes-Hoy L, Charli J-L (2015b) Regulation of TRH neurons and energy homeostasis-related signals under stress. J Endocrinol 224(3):R139–R159. https://doi.org/10.1530/JOE-14-0593

Joseph-Bravo P, Jaimes-Hoy L, Rodríguez-Rodríguez A, De Oca MP-M, Uribe RM, Charli J-L (2021) Multifactorial regulation of the activity of hypophysiotropic thyrotropin-releasing hormone neurons. In: Grinevich V, Dobolyi Á (eds) Neuroanatomy of neuroendocrine systems, Masterclass in neuroendocrinology, vol 12. Springer, Cham, pp 251–294. https://doi.org/10.1007/978-3-030-86630-3_10

Kahn ML (2015) Vascular development for climate control. Dev Cell 33(3):241–242. https://doi.org/10.1016/j.devcel.2015.04.011

Kálmán M, Székely AD, Csillag A (1993) Distribution of glial fibrillary acidic protein-immunopositive structures in the brain of the domestic chicken (*Gallus domesticus*): GFAP in chicken brain. J Comp Neurol 330(2):221–237. https://doi.org/10.1002/cne.903300206

Kanasaki H, Oride A, Mijiddorj T, Kyo S (2015) Role of thyrotropin-releasing hormone in prolactin-producing cell models. Neuropeptides 54(December):73–77. https://doi.org/10.1016/j.npep.2015.08.001

Kanda S (2019) Evolution of the regulatory mechanisms for the hypothalamic-pituitary-gonadal axis in vertebrates–hypothesis from a comparative view. Gen Comp Endocrinol 284(December):113075. https://doi.org/10.1016/j.ygcen.2018.11.014

Kashio M, Tominaga M (2022) TRP channels in thermosensation. Curr Opin Neurobiol 75(August):102591. https://doi.org/10.1016/j.conb.2022.102591

Kawakami Y, Tanda M, Adachi S, Yamauchi K (2003) cDNA cloning of thyroid hormone receptor betas from the conger eel, Conger myriaster. Gen Comp Endocrinol 131(3):232–240. https://doi.org/10.1016/S0016-6480(02)00638-X

Kawauchi H, Sower SA (2006) The dawn and evolution of hormones in the adenohypophysis. Gen Comp Endocrinol 148(1):3–14. https://doi.org/10.1016/j.ygcen.2005.10.011

Klieverik LP, Coomans CP, Endert E, Sauerwein HP, Havekes LM, Voshol PJ, Rensen PCN, Romijn JA, Kalsbeek A, Fliers E (2009) Thyroid hormone effects on whole-body energy homeostasis and tissue-specific fatty acid uptake in vivo. Endocrinology 150(12):5639–5648. https://doi.org/10.1210/en.2009-0297

Klootwijk W, Friesema ECH, Visser TJ (2011) A nonselenoprotein from amphioxus deiodinates triac but not T3: is triac the primordial bioactive thyroid hormone? Endocrinology 152(8):3259–3267. https://doi.org/10.1210/en.2010-1408

Kluge B, Renault N, Rohr KB (2005) Anatomical and molecular reinvestigation of lamprey endostyle development provides new insight into thyroid gland evolution. Dev Genes Evol 215(1):32–40. https://doi.org/10.1007/s00427-004-0450-0

Köhrle J, Frädrich C (2021) Thyroid hormone system disrupting chemicals. Best Pract Res Clin Endocrinol Metab 35(5):101562. https://doi.org/10.1016/j.beem.2021.101562

Kuchler K, Richter K, Trnovsky J, Egger R, Kreil G (1990) Two precursors of thyrotropin-releasing hormone from skin of Xenopus laevis. Each contains seven copies of the end product. J Biol Chem 265(20):11731–11733

Kuraku S, Meyer A, Kuratani S (2008) Timing of genome duplications relative to the origin of the vertebrates: did cyclostomes diverge before or after? Mol Biol Evol 26(1):47–59. https://doi.org/10.1093/molbev/msn222

Larsen DA, Swanson P, Dickey JT, Rivier J, Dickhoff WW (1998) In vitro thyrotropin-releasing activity of corticotropin-releasing hormone-family peptides in coho salmon, Oncorhynchus kisutch. Gen Comp Endocrinol 109(2):276–285. https://doi.org/10.1006/gcen.1997.7031

Lazcano I, Orozco A (2018) Revisiting available knowledge on teleostean thyroid hormone receptors. Gen Comp Endocrinol 265(September):128–132. https://doi.org/10.1016/j.ygcen.2018.03.022

Lazcano I, Cabral A, Uribe RM, Jaimes-Hoy L, Perello M, Joseph-Bravo P, Sánchez-Jaramillo E, Charli J-L (2015) Fasting enhances pyroglutamyl peptidase II activity in tanycytes of the mediobasal hypothalamus of male adult rats. Endocrinology 156(7):2713–2723. https://doi.org/10.1210/en.2014-1885

Lazcano I, Hernández-Puga G, Robles JP, Orozco A (2019a) Alternative ligands for thyroid hormone receptors. Mol Cell Endocrinol 493(August):110448. https://doi.org/10.1016/j.mce.2019.05.007

Lazcano I, Rodríguez-Ortiz R, Villalobos P, Martínez-Torres A, Solís-Saínz JC, Orozco A (2019b) Knock-down of specific thyroid hormone receptor isoforms impairs body plan development in zebrafish. Front Endocrinol 10(March):156. https://doi.org/10.3389/fendo.2019.00156

Lazcano I, Rodríguez AR, Uribe RM, Orozco A, Joseph-Bravo P, Charli J-L (2021) Evolution of thyrotropin-releasing factor extracellular communication units. Gen Comp Endocrinol 305(May):113642. https://doi.org/10.1016/j.ygcen.2020.113642

Lazcano I, Olvera A, Pech-Pool SM, Sachs L, Buisine N, Orozco A (2023a) Differential effects of 3,5-T2 and T3 on the gill regeneration and metamorphosis of the Ambystoma Mexicanum (axolotl). Front Endocrinol 14(July):1208182. https://doi.org/10.3389/fendo.2023.1208182

Lazcano I, Pech-Pool SM, Olvera A, García-Martínez I, Palacios-Pérez S, Orozco A (2023b) The importance of thyroid hormone signaling during early development: lessons from the zebrafish model. Gen Comp Endocrinol 334(April):114225. https://doi.org/10.1016/j.ygcen.2023.114225

Lazzari M, Franceschini V (2004) Glial fibrillary acidic protein and vimentin immunoreactivity of astroglial cells in the central nervous system of the African lungfish, protopterus annectens (Dipnoi: Lepidosirenidae). J Morphol 262(3):741–749. https://doi.org/10.1002/jmor.10274

Lazzari M, Franceschini V, Ciani F (1997) Glial fibrillary acidic protein and vimentin in radial glia of Ambystoma mexicanum and Triturus carnifex: an immunocytochemical study. J Hirnforsch 38(2):187–194

Lechan RM, Fekete C (2005) Role of thyroid hormone deiodination in the hypothalamus. Thyroid 15(8):883–897. https://doi.org/10.1089/thy.2005.15.883

Lechan RM, Jackson IMD (1982) Immunohistochemical localization of thyrotropin-releasing hormone in the rat hypothalamus and pituitary. Endocrinology 111(1):55–65. https://doi.org/10.1210/endo-111-1-55

Lechan RM, Ping W, Jackson IMD, Wolf H, Cooperman S, Mandel G, Goodman RH (1986) Thyrotropin-releasing hormone precursor: characterization in rat brain. Science 231(4734):159–161. https://doi.org/10.1126/science.3079917

Lee JW, Profant M, Wang C (2022) Metabolic sex dimorphism of the brain at the gene, cell, and tissue level. J Immunol 208(2):212–220. https://doi.org/10.4049/jimmunol.2100853

Legendre LJ, Davesne D (2020) The evolution of mechanisms involved in vertebrate Endothermy. Philos Trans R Soc Lond B Biol Sci 375(1793):20190136. https://doi.org/10.1098/rstb.2019.0136

Li X, Li Z, Deng Y, Zhang J, Li J, Wang Y (2020a) Characterization of a novel thyrotropin-releasing hormone receptor, TRHR3, in chickens. Poult Sci 99(3):1643–1654. https://doi.org/10.1016/j.psj.2019.10.062

Li Y-F, Cheng Y-L, Chen K, Cheng Z-Y, Zhu X, Cardoso JCR, Liang X, Zhu Y-T, Power DM, Yang J-L (2020b) Thyroid hormone receptor: a new player in epinephrine-induced larval metamorphosis of the hard-shelled mussel. Gen Comp Endocrinol 287(February):113347. https://doi.org/10.1016/j.ygcen.2019.113347

Little AG, Seebacher F (2013) Thyroid hormone regulates cardiac performance during cold acclimation in zebrafish (*Danio Rerio*). J Exp Biol. https://doi.org/10.1242/jeb.096602

López JM, Domínguez L, González A (2008) Immunohistochemical localization of thyrotropin-releasing hormone in the brain of reptiles. J Chem Neuroanat 36(3–4):251–263. https://doi.org/10.1016/j.jchemneu.2008.06.006

Lu J, Holmgren A (2009) Selenoproteins. J Biol Chem 284(2):723–727. https://doi.org/10.1074/jbc.R800045200

Luna M, Guzmán G, Navarro L, Sánchez De La Peña S, Valverde-R C (1995) Circadian rhythm of type II 5′deiodinase activity in the rat hypothalamic-pituitary-adrenal axis. Endocrine 3(8):597–601. https://doi.org/10.1007/BF02953025

MacKenzie DS, Jones RA, Miller TC (2009) Thyrotropin in teleost fish. Gen Comp Endocrinol 161(1):83–89. https://doi.org/10.1016/j.ygcen.2008.12.010

Mantzouratou P, Lavecchia AM, Xinaris C (2021) Thyroid hormone signalling in human evolution and disease: a novel hypothesis. J Clin Med 11(1):43. https://doi.org/10.3390/jcm11010043

Manzon RG, Neuls TM, Manzon LA (2007) Molecular cloning, tissue distribution, and developmental expression of lamprey transthyretins. Gen Comp Endocrinol 151(1):55–65. https://doi.org/10.1016/j.ygcen.2006.11.017

Marchand O, Safi R, Escriva H, Van Rompaey E, Prunet P, Laudet V (2001) Molecular cloning and characterization of thyroid hormone receptors in teleost fish. J Mol Endocrinol 26(1):51–65. https://doi.org/10.1677/jme.0.0260051

Martelli D, Brooks VL (2023) Leptin increases: physiological roles in the control of sympathetic nerve activity, energy balance, and the hypothalamic–pituitary–thyroid axis. Int J Mol Sci 24(3):2684. https://doi.org/10.3390/ijms24032684

Martínez LM, Orozco A, Villalobos P, Valverde-R C (2008) Cloning and characterization of a type 3 iodothyronine deiodinase (D3) in the liver of the Chondrichtyan Chiloscyllium Punctatum. Gen Comp Endocrinol 156(3):464–469. https://doi.org/10.1016/j.ygcen.2008.02.012

Martinez-deMena R, Anedda A, Cadenas S, Obregon M-J (2015) TSH effects on thermogenesis in rat brown adipocytes. Mol Cell Endocrinol 404(March):151–158. https://doi.org/10.1016/j.mce.2015.01.028

Matz SP, Takahashi TT (1994) Immunohistochemical localization of thyrotropin-releasing hormone in the brain of Chinook Salmon (Oncorhynchus tshawytscha). J Comp Neurol 345(2):214–223. https://doi.org/10.1002/cne.903450205

Maugars G, Dufour S, Cohen-Tannoudji J, Quérat B (2014) Multiple thyrotropin β-subunit and thyrotropin receptor-related genes arose during vertebrate evolution. PLoS One 9(11):e111361. https://doi.org/10.1371/journal.pone.0111361

McLean TR, Rank MM, Smooker PM, Richardson SJ (2017) Evolution of thyroid hormone distributor proteins. Mol Cell Endocrinol 459(December):43–52. https://doi.org/10.1016/j.mce.2017.02.038

Mekuchi M, Saito Y, Aoki Y, Masuda T, Iigo M, Yanagisawa T (2011) Molecular cloning, gene structure, molecular evolution and expression analyses of thyrotropin-releasing hormone receptors from medaka (Oryzias latipes). Gen Comp Endocrinol 170(2):374–380. https://doi.org/10.1016/j.ygcen.2010.10.013

Mendoza A, Hollenberg AN (2017) New insights into thyroid hormone action. Pharmacol Ther 173(May):135–145. https://doi.org/10.1016/j.pharmthera.2017.02.012

Mendoza A, Navarrete-Ramírez P, Hernández-Puga G, Villalobos P, Holzer G, Renaud JP, Laudet V, Orozco A (2013) 3,5-T2 is an alternative ligand for the thyroid hormone receptor B1. Endocrinology 154(8):2948–2958. https://doi.org/10.1210/en.2013-1030

Mendoza D, Alex AS-P, Ruiz-Trillo I (2014) The evolution of the GPCR signaling system in eukaryotes: modularity, conservation, and the transition to metazoan multicellularity. Genome Biol Evol 6(3):606–619. https://doi.org/10.1093/gbe/evu038

Mimnagh KM, Bolaffi JL, Montgomery NM, Kaltenbach JC (1987) Thyrotropin-releasing hormone (TRH): immunohistochemical distribution in tadpole and frog brain. Gen Comp Endocrinol 66(3):394–404. https://doi.org/10.1016/0016-6480(87)90250-4

Mirabeau O, Joly J-S (2013) Molecular evolution of peptidergic signaling systems in bilaterians. Proc Natl Acad Sci U S A 110(22). https://doi.org/10.1073/pnas.1219956110

Mohácsik P, Füzesi T, Doleschall M, Szilvásy-Szabó A, Vancamp P, Hadadi É, Darras VM, Fekete C, Gereben B (2016) Increased thyroid hormone activation accompanies the formation of thyroid hormone-dependent negative feedback in developing chicken hypothalamus. Endocrinology 157(3):1211–1221. https://doi.org/10.1210/en.2015-1496

Morrison SF (2018) Efferent neural pathways for the control of brown adipose tissue thermogenesis and shivering. In: Handbook of clinical neurology, vol 156. Elsevier, pp 281–303. https://doi.org/10.1016/B978-0-444-63912-7.00017-5

Morvan-Dubois G, Fini JB, Demeneix BA (2013) Is thyroid hormone signaling relevant for vertebrate embryogenesis? Curr Top Dev Biol 103:365–396. https://doi.org/10.1016/B978-0-12-385979-2.00013-7

Mughal BB, Leemans M, Lima EC, Souza D, Le Mevel S, Spirhanzlova P, Visser TJ, Fini J-B, Demeneix BA (2017) Functional characterization of xenopus thyroid hormone transporters Mct8 and Oatp1c1. Endocrinology 158(8):2694–2705. https://doi.org/10.1210/en.2017-00108

Mullur R, Liu Y-Y, Brent GA (2014) Thyroid hormone regulation of metabolism. Physiol Rev 94(2):355–382. https://doi.org/10.1152/physrev.00030.2013

Nakamura K, Morrison SF (2022) Central sympathetic network for thermoregulatory responses to psychological stress. Auton Neurosci 237(January):102918. https://doi.org/10.1016/j.autneu.2021.102918

Narayanan CS, Fujimoto J, Geras-Raaka E, Gershengorn MC (1992) Regulation by thyrotropin-releasing hormone (TRH) of TRH receptor mRNA degradation in rat pituitary GH3 cells. J Biol Chem 267(24):17296–17303

Nedergaard J, Cannon B (2018) Brown adipose tissue as a heat-producing thermoeffector. In: Handbook of clinical neurology, vol 156. Elsevier, pp 137–152. https://doi.org/10.1016/B978-0-444-63912-7.00009-6

Nguyen VT-T, Vuong Van V, Van Pham P (2023) Brown adipocyte and Browning thermogenesis: metabolic crosstalk beyond mitochondrial limits and physiological impacts. Adipocytes 12(1):2237164. https://doi.org/10.1080/21623945.2023.2237164

Nowack J, Giroud S, Arnold W, Ruf T (2017) Muscle non-shivering thermogenesis and its role in the evolution of endothermy. Front Physiol 8(November):889. https://doi.org/10.3389/fphys.2017.00889

Nowoshilow S, Schloissnig S, Fei J-F, Dahl A, Pang AWC, Pippel M, Winkler S et al (2018) The axolotl genome and the evolution of key tissue formation regulators. Nature 554(7690):50–55. https://doi.org/10.1038/nature25458

Ogasawara M (2000) Overlapping expression of amphioxus homologs of the thyroid transcription factor-1 gene and thyroid peroxidase gene in the endostyle: insight into evolution of the thyroid gland. Dev Genes Evol 210(5):231. https://doi.org/10.1007/s004270050309

Ogawa S, Ng KW, Xue X, Ramadasan PN, Sivalingam M, Li S, Levavi-Sivan B, Lin H, Liu X, Parhar IS (2013) Thyroid hormone upregulates hypothalamic Kiss2 gene in the male Nile tilapia, Oreochromis niloticus. Front Endocrinol 4. https://doi.org/10.3389/fendo.2013.00184

Okada R, Ito Y, Kaneko M, Yamamoto K, Nicolas Chartrel J, Conlon M, Vaudry H, Kikuyama S (2005) Frog corticotropin-releasing hormone (CRH): isolation, molecular cloning, and biological activity. Ann N Y Acad Sci 1040(1):150–155. https://doi.org/10.1196/annals.1327.019

On JS, Wun AW, Arokiaraj R, Chow BKC (2019) Molecular evolution of CRH and CRHR subfamily before the evolutionary origin of vertebrate. Peptides 120(October):170087. https://doi.org/10.1016/j.peptides.2019.04.014. Epub 2019 Apr 28. PMID: 31042548

Orozco A, Valverde-R C, Olvera A, García-G C (2012) Iodothyronine deiodinases: a functional and evolutionary perspective. J Endocrinol 215(2):207–219. https://doi.org/10.1530/JOE-12-0258

Orozco A, Navarrete-Ramírez P, Olvera A, García-G C (2014) 3,5-diiodothyronine (T2) is on a role. A new hormone in search of recognition. Gen Comp Endocrinol 203(July):174–180. https://doi.org/10.1016/j.ygcen.2014.02.014

Ortega VA, Mercer EM, Giesbrecht GF, Arrieta M-C (2021) Evolutionary significance of the neuroendocrine stress axis on vertebrate immunity and the influence of the microbiome on early-life stress regulation and health outcomes. Front Microbiol 12(April):634539. https://doi.org/10.3389/fmicb.2021.634539

Ortiga-Carvalho TM, Chiamolera MI, Pazos-Moura CC, Wondisford FE (2016) Hypothalamus-pituitary-thyroid axis. In: Terjung R (ed) Comprehensive physiology, 1st edn. Wiley, pp 1387–1428. https://doi.org/10.1002/cphy.c150027

Padron AS, Neto RAL, Pantaleão TU, Santos MCDSD, Araujo RL, De Andrade BM, Da Silva M, Leandro JP, De Castro SW, Ferreira ACF, De Carvalho DP (2014) Administration of 3,5-diiodothyronine (3,5-T2) causes central hypothyroidism and stimulates thyroid-sensitive tissues. J Endocrinol 221(3):415–427. https://doi.org/10.1530/JOE-13-0502

Page RB, Randal Voss S (2009) Induction of Metamorphosis in Axolotls (*Ambystoma Mexicanum*). Cold Spring Harbor Protocols 2009(8):pdb.prot5268. https://doi.org/10.1101/pdb.prot5268

Palioura S, Lynn Sherrer R, Steitz TA, Söll D, Simonović M (2009) The human SepSecS-tRNA^{Sec} complex reveals the mechanism of Selenocysteine formation. Science 325(5938):321–325. https://doi.org/10.1126/science.1173755

Paris M, Escriva H, Schubert M, Brunet F, Brtko J, Ciesielski F, Roecklin D et al (2008) Amphioxus postembryonic development reveals the homology of chordate metamorphosis. Curr Biol 18(11):825–830. https://doi.org/10.1016/j.cub.2008.04.078

Paris M, Hillenweck A, Bertrand S, Delous G, Escriva H, Zalko D, Cravedi J-P, Laudet V (2010) Active metabolism of thyroid hormone during metamorphosis of amphioxus. Integr Comp Biol 50(1):63–74. https://doi.org/10.1093/icb/icq052

Parra-Montes De Oca MA, Sotelo-Rivera I, Gutiérrez-Mata A, Charli J-L, Joseph-Bravo P (2021) Sex dimorphic responses of the hypothalamus-pituitary-thyroid Axis to energy demands and stress. Front Endocrinol 12(October):746924. https://doi.org/10.3389/fendo.2021.746924

Petraglia F, Potter E, Cameron VA, Sutton S, Behan DP, Woods RJ, Sawchenko PE, Lowry PJ, Vale W (1993) Corticotropin-releasing factor-binding protein is produced by human placenta and intrauterine tissues. J Clin Endocrinol Metabol 77(4):919–924. https://doi.org/10.1210/jcem.77.4.8408466

Potter E, Behan DP, Fischer WH, Linton EA, Lowry PJ, Vale WW (1991) Cloning and characterization of the cDNAs for human and rat Corticotropin releasing factor-binding proteins. Nature 349(6308):423–426. https://doi.org/10.1038/349423a0

Power DM, Elias NP, Richardson SJ, Mendes J, Soares CM, Santos CRA (2000) Evolution of the thyroid hormone-binding protein, transthyretin. Gen Comp Endocrinol 119(3):241–255. https://doi.org/10.1006/gcen.2000.7520

Prapunpoj P, Yamauchi K, Nishiyama N, Richardson SJ, Schreiber G (2000) Evolution of structure, ontogeny of gene expression, and function of *Xenopus Laevis* transthyretin. Am J Phys Regul Integr Comp Phys 279(6):R2026–R2041. https://doi.org/10.1152/ajpregu.2000.279.6.R2026

Prapunpoj P, Richardson SJ, Schreiber G (2002) Crocodile transthyretin: structure, function, and evolution. Am J Phys Regul Integr Comp Phys 283(4):R885–R896. https://doi.org/10.1152/ajpregu.00042.2002

Raisman G (1997) An urge to explain the incomprehensible: Geoffrey Harris and the discovery of the neural control of the pituitary gland. Annu Rev Neurosci 20(1):533–566. https://doi.org/10.1146/annurev.neuro.20.1.533

Ricci C, Kakularam KR, Marzocchi C, Capecchi G, Riolo G, Boschin F, Kuhn H, Castagna MG, Cantara S (2020) Thr92Ala polymorphism in the type 2 deiodinase gene: an evolutionary perspective. J Endocrinol Investig 43(12):1749–1757. https://doi.org/10.1007/s40618-020-01287-5

Richardson SJ (2007) Cell and molecular biology of transthyretin and thyroid hormones. Int Rev Cytol 258:137–193. https://doi.org/10.1016/S0074-7696(07)58003-4

Rodriguez E, Blazquez J, Pastor F, Pelaez B, Pena P, Peruzzo B, Amat P (2005) Hypothalamic tanycytes: a key component of brain–endocrine interaction. Int Rev Cytol 247:89–164. https://doi.org/10.1016/S0074-7696(05)47003-5

Rodríguez-Rodríguez A, Lazcano I, Sánchez-Jaramillo E, Uribe RM, Jaimes-Hoy L, Joseph-Bravo P, Charli J-L (2019) Tanycytes and the control of thyrotropin-releasing hormone flux into portal capillaries. Front Endocrinol 10(June):401. https://doi.org/10.3389/fendo.2019.00401

Rowe ML, Elphick MR (2012) The neuropeptide transcriptome of a model echinoderm, the sea urchin Strongylocentrotus purpuratus. Gen Comp Endocrinol 179(3):331–344. https://doi.org/10.1016/j.ygcen.2012.09.009. PMID: 23026496.

Russell JA (2018) Fifty years of advances in neuroendocrinology. Brain Neurosci Adv 2(January):239821281881201. https://doi.org/10.1177/2398212818812014

Russo SC, Salas-Lucia F, Bianco AC (2021) Deiodinases and the metabolic code for thyroid hormone action. Endocrinology 162(8):bqab059. https://doi.org/10.1210/endocr/bqab059

Safi R, Bertrand S, Marchand O, Duffraisse M, De Luze A, Vanacker J-M, Maraninchi M, Margotat A, Demeneix B, Laudet V (2004) The axolotl (*Ambystoma Mexicanum*), a neotenic amphibian, expresses functional thyroid hormone receptors. Endocrinology 145(2):760–772. https://doi.org/10.1210/en.2003-0913

Saito Y, Mekuchi M, Kobayashi N, Kimura M, Aoki Y, Masuda T, Azuma T, Fukami M, Iigo M, Yanagisawa T (2011) Molecular cloning, molecular evolution and gene expression of cDNAs encoding thyrotropin-releasing hormone receptor subtypes in a teleost, the sockeye Salmon (Oncorhynchus nerka). Gen Comp Endocrinol 174(2):80–88. https://doi.org/10.1016/j.ygcen.2011.07.011

Sánchez E, Vargas MA, Singru PS, Pascual I, Romero F, Fekete C, Charli J-L, Lechan RM (2009) Tanycyte pyroglutamyl peptidase II contributes to regulation of the hypothalamic-pituitary-thyroid axis through glial-axonal associations in the median eminence. Endocrinology 150(5):2283–2291. https://doi.org/10.1210/en.2008-1643

Sane R, Wirth EK, Köhrle J (2022) 3,5-T2-an endogenous thyroid hormone metabolite as promising lead substance in anti-steatotic drug development? Meta 12(7):582. https://doi.org/10.3390/metabo12070582

Santiago-Andres Y, Golan M, Fiordelisio T (2021) Functional pituitary networks in vertebrates. Front Endocrinol 11(January):619352. https://doi.org/10.3389/fendo.2020.619352

Santos CRA, Anjos L, Power DM (2002) Transthyretin in fish: state of the art. Clin Chem Lab Med 40(12). https://doi.org/10.1515/CCLM.2002.215

Satoh T, Yamada M, Monden T, Iizuka M, Mori M (1992) Cloning of the mouse hypothalamic preprothyrotropin-releasing hormone (TRH) cDNA and tissue distribution of its mRNA. Mol Brain Res 14(1–2):131–135. https://doi.org/10.1016/0169-328X(92)90020-C

Scapin S, Leoni S, Spagnuolo S, Fiore AM, Incerpi S (2009) Short-term effects of thyroid hormones on Na^+-K^+-ATPase activity of Chick embryo hepatocytes during development: focus on signal transduction. Am J Phys Cell Phys 296(1):C4–C12. https://doi.org/10.1152/ajpcell.90604.2007

Schomburg L, Bauer K (1995) Thyroid hormones rapidly and stringently regulate the messenger RNA levels of the thyrotropin-releasing hormone (TRH) receptor and the TRH-degrading ectoenzyme. Endocrinology 136(8):3480–3485. https://doi.org/10.1210/endo.136.8.7628384. PMID: 7628384

Schneider MJ, Fiering SN, Thai B, Wu S-y, St Germain E, Parlow AF, St. Germain DL, Galton VA (2006) Targeted disruption of the type 1 Selenodeiodinase gene (Dio1) results in marked changes in thyroid hormone economy in mice. Endocrinology 147(1):580–589. https://doi.org/10.1210/en.2005-0739

Seasholtz A, Valverde R, Denver R (2002) Corticotropin-releasing hormone-binding protein: biochemistry and function from fishes to mammals. J Endocrinol 175(1):89–97. https://doi.org/10.1677/joe.0.1750089

Sechman A, Pawlowska K, Hrabia A (2011) Effect of 3,3′,5-triiodothyronine and 3,5-diiodothyronine on progesterone production, cAMP synthesis, and mRNA expression of STAR, CYP11A1, and HSD3B genes in granulosa layer of chicken preovulatory follicles. Domest Anim Endocrinol 41(3):137–149. https://doi.org/10.1016/j.domaniend.2011.05.007

Seebacher F (2020) Is Endothermy an evolutionary by-product? Trends Ecol Evol 35(6):503–511. https://doi.org/10.1016/j.tree.2020.02.006

Sefkow AJ, DiStefano JJ, Iii BA, Himick SB, Brown, and J.G. Eales. (1996) Kinetic analysis of thyroid hormone secretion and interconversion in the 5-day-fasted rainbow trout, Oncorhynchus mykiss. Gen Comp Endocrinol 101(2):123–138. https://doi.org/10.1006/gcen.1996.0014

Shepherdley CA, Daniels CB, Orgeig S, Richardson SJ, Evans BK, Darras VM (2002) Glucocorticoids, thyroid hormones, and iodothyronine deiodinases in embryonic saltwater crocodiles. Am J Phys Regul Integr Comp Phys 283(5):R1155–R1163. https://doi.org/10.1152/ajpregu.00015.2002

Silva JE (2011) Physiological importance and control of non-shivering facultative thermogenesis. Front Biosci S3(1):352–371. https://doi.org/10.2741/s156

Silva N, Campinho MA (2023) In a zebrafish biomedical model of human Allan-Herndon-Dudley syndrome impaired MTH signaling leads to decreased neural cell diversity. Front Endocrinol 14(May):1157685. https://doi.org/10.3389/fendo.2023.1157685

Soleimani M (2015) The multiple roles of pendrin in the kidney. Nephrol Dial Transplant 30(8):1257–1266. https://doi.org/10.1093/ndt/gfu307

Sower SA (2015) Breaking dogma on the hypothalamic-pituitary anatomical relations in vertebrates. Endocrinology 156(11):3882–3884. https://doi.org/10.1210/en.2015-1778

Sower SA, Hausken KN (2017) A lamprey view on the origins of neuroendocrine regulation of the thyroid axis. Mol Cell Endocrinol 459(December):21–27. https://doi.org/10.1016/j.mce.2017.04.012

Sutton SW, Behan DP, Lahrichi SL, Kaiser R, Corrigan A, Lowry P, Potter E, Perrin MH, Rivier J, Vale WW (1995) Ligand requirements of the human corticotropin-releasing factor-binding protein. Endocrinology 136(3):1097–1102. https://doi.org/10.1210/endo.136.3.7867564

Suzuki S, Miyamoto T, Opsahl A, Sakurai A, DeGroot LJ (1994) Two thyroid hormone response elements are present in the promoter of human thyroid hormone receptor Beta 1. Mol Endocrinol 8(3):305–314. https://doi.org/10.1210/mend.8.3.8015548

Takahashi A, Kanda S, Abe T, Oka Y (2016) Evolution of the hypothalamic-pituitary-gonadal axis regulation in vertebrates revealed by knockout Medaka. Endocrinology 157(10):3994–4002. https://doi.org/10.1210/en.2016-1356

Taurog A (1974) Effect of TSH and long-acting thyroid stimulator on thyroid 131I-metabolism and metamorphosis of the Mexican axolotl (Ambystoma mexicanum). Gen Comp Endocrinol 24(3):257–266. https://doi.org/10.1016/0016-6480(74)90180-4

Taurog A (1999) Molecular evolution of thyroid peroxidase. Biochimie 81(5):557–562. https://doi.org/10.1016/S0300-9084(99)80110-2

Taylor E, Heyland A (2017) Evolution of thyroid hormone signaling in animals: non-genomic and genomic modes of action. Mol Cell Endocrinol 459(December):14–20. https://doi.org/10.1016/j.mce.2017.05.019

Taylor E, Wynen H, Heyland A (2023) Thyroid hormone membrane receptor binding and transcriptional regulation in the sea urchin Strongylocentrotus purpuratus. Front Endocrinol (Lausanne) 26(14):1195733. https://doi.org/10.3389/fendo.2023.1195733. PMID: 37305042; PMCID: PMC10250714

Till A, Fries C, Fenske WK (2022) Brain-to-BAT - and Back?: crosstalk between the central nervous system and thermogenic adipose tissue in development and therapy of obesity. Brain Sci 12(12):1646. https://doi.org/10.3390/brainsci12121646

Tóth G, Hosztafi S, Kovács Z, Noszál B (2012) The site-specific basicity of thyroid hormones and their precursors as regulators of their biological functions. J Pharm Biomed Anal 61(March):156–164. https://doi.org/10.1016/j.jpba.2011.11.020

Tran LT, Park S, Kim SK, Lee JS, Kim KW, Kwon O (2022) Hypothalamic control of energy expenditure and thermogenesis. Exp Mol Med 54(4):358–369. https://doi.org/10.1038/s12276-022-00741-z

Tu HM, Kim S-W, Salvatore D, Bartha T, Gabor Legradi P, Larsen R, Lechan RM (1997) Regional distribution of type 2 thyroxine deiodinase messenger ribonucleic acid in rat hypothalamus and pituitary and its regulation by thyroid hormone. Endocrinology 138(8):3359–3368. https://doi.org/10.1210/endo.138.8.5318

Uribe RM, Redondo JL, Charli JL, Joseph-Bravo P (1993) Suckling and cold stress rapidly and transiently increase TRH mRNA in the paraventricular nucleus. Neuroendocrinology 58(1):140–145. https://doi.org/10.1159/000126523

Uribe RM, Jaimes-Hoy L, Ramírez-Martínez C, García-Vázquez A, Romero F, Cisneros M, Cote-Vélez A, Charli J-L, Joseph-Bravo P (2014) Voluntary exercise adapts the hypothalamus-pituitary-thyroid axis in male rats. Endocrinology 155(5):2020–2030. https://doi.org/10.1210/en.2013-1724

Vale W, Spiess J, Rivier C, Rivier J (1981) Characterization of a 41-residue ovine hypothalamic peptide that stimulates secretion of corticotropin and β-endorphin. Science 213(4514):1394–1397. https://doi.org/10.1126/science.6267699

Valverde RA, Seasholtz AF, Cortright DN, Denver RJ (2001) Biochemical characterization and expression analysis of the Xenopus laevis corticotropin-releasing hormone binding protein. Mol Cell Endocrinol 173(1–2):29–40. https://doi.org/10.1016/S0303-7207(00)00437-8

Van Sinay E, Mirabeau O, Depuydt G, Van Hiel MB, Peymen K, Watteyne J, Zels S, Schoofs L, Beets I (2017) Evolutionarily conserved TRH neuropeptide pathway regulates growth in *Caenorhabditis elegans*. Proc Natl Acad Sci 114(20). https://doi.org/10.1073/pnas.1617392114

Vaudry H (2014) Molecular evolution of GPCRS: what we know and what the future holds. J Mol Endocrinol 52(3):E1–E2. https://doi.org/10.1530/JME-14-0103

Veenstra JA, Šimo L (2020) The TRH-Ortholog EFLamide in the migratory locust. Insect Biochem Mol Biol 116(January):103281. https://doi.org/10.1016/j.ibmb.2019.103281

Vella KR, Hollenberg AN (2017) The actions of thyroid hormone signaling in the nucleus. Mol Cell Endocrinol 458(December):127–135. https://doi.org/10.1016/j.mce.2017.03.001

Venturi S (2011) Evolutionary significance of iodine. Curr Chem Biol 5(3):155–162. https://doi.org/10.2174/2212796811105030155

Villalobos P, Orozco A, Valverde-R C (2010) Molecular cloning and characterization of a type 3 iodothyronine deiodinase in the pine snake Pituophis deppei. Gen Comp Endocrinol 169(2):167–173. https://doi.org/10.1016/j.ygcen.2010.08.001

Vohra MS, Benchoula K, Serpell CJ, Hwa WE (2022) AgRP/NPY and POMC neurons in the arcuate nucleus and their potential role in treatment of obesity. Eur J Pharmacol 915(January):174611. https://doi.org/10.1016/j.ejphar.2021.174611

Wakahara M (1996) Heterochrony and neotenic salamanders: possible clues for understanding the animal development and evolution. Zool Sci 13(6):765–776. https://doi.org/10.2108/zsj.13.765

Wan Y, Zhang Z, Lin D, Wang X, Huang T, Jiancheng S, Zhang J, Li J, Wang Y (2022) Characterization of CRH-binding protein (CRHBP) in chickens: molecular cloning, tissue distribution and investigation of its role as a negative feedback regulator within the hypothalamus–pituitary–adrenal axis. Genes 13(10):1680. https://doi.org/10.3390/genes13101680

Wang P, Liu S, Yang Q, Liu Z, Zhang S (2018) Functional characterization of Thyrostimulin in amphioxus suggests an ancestral origin of the TH signaling pathway. Endocrinology 159(10):3536–3548. https://doi.org/10.1210/en.2018-00550

Wells KM, Kelley K, Baumel M, Vieira WA, McCusker CD (2021) Neural control of growth and size in the axolotl limb regenerate. eLife 10(November):e68584. https://doi.org/10.7554/eLife.68584

Westphal NJ (2006) CRH-BP: the regulation and function of a phylogenetically conserved binding protein. Front Biosci 11(1):1878. https://doi.org/10.2741/1931

Williams GR, Duncan Bassett JH (2011) Local control of thyroid hormone action: role of type 2 deiodinase. J Endocrinol 209(3):261–272. https://doi.org/10.1530/JOE-10-0448

Wong J, Liang VC-T, Sachs LM, Shi Y-B (1998) Transcription from the thyroid hormone-dependent promoter of the Xenopus laevis thyroid hormone receptor βA gene requires a novel upstream element and the initiator, but not a TATA box. J Biol Chem 273(23):14186–14193. https://doi.org/10.1074/jbc.273.23.14186

Wu W, Niles EG, LoVerde PT (2007) Thyroid hormone receptor orthologues from invertebrate species with emphasis on Schistosoma mansoni. BMC Evol Biol 7(1):150. https://doi.org/10.1186/1471-2148-7-150

Wu T, Shi X, Zhou Z, Wang L, Wang M, Wang L, Huang M, Yang C, Song L (2012) An iodothyronine deiodinase from Chlamys farreri and its induced mRNA expression after LPS stimulation. Fish Shellfish Immunol 33(2):286–293. https://doi.org/10.1016/j.fsi.2012.05.011

Wu G, Baumeister R, Heimbucher T (2023) Molecular mechanisms of lipid-based metabolic adaptation strategies in response to cold. Cells 12(10):1353. https://doi.org/10.3390/cells12101353

Yamada M, Radovick S, Wondisford FE, Nakayama Y, Weintraub BD, Wilber JF (1990) Sequences of interest: cloning and structure of human genomic DNA and hypothalamic cDNA encoding human preproThyrotropin-releasing hormone. Mol Endocrinol 4(4):551–556. https://doi.org/10.1210/mend-4-4-551

Yamada M, Monden T, Satoh T, Satoh N, Murakami M, Iriuchijima T, Kakegawa T, Mori M (1993) Pituitary adenomas of patients with acromegaly express thyrotropin-releasing hormone receptor messenger RNA: cloning and functional expression of the human thyrotropin-releasing hormone receptor gene. Biochem Biophys Res Commun 195(2):737–745. https://doi.org/10.1006/bbrc.1993.2107

Yamauchi K, Ishihara A (2009) Evolutionary changes to transthyretin: developmentally regulated and tissue-specific gene expression: transthyretin gene expression during development. FEBS J 276(19):5357–5366. https://doi.org/10.1111/j.1742-4658.2009.07245.x

Yamauchi K, Takeuchi H-a, Overall M, Dziadek M, Munro SLA, Schreiber G (1998) Structural characteristics of bullfrog (Rana Catesbeiana) transthyretin and its cDNA. Comparison of its pattern of expression during metamorphosis with that of lipocalin. Eur J Biochem 256(2):287–296. https://doi.org/10.1046/j.1432-1327.1998.2560287.x

Yau WW, Yen PM (2020) Thermogenesis in adipose tissue activated by thyroid hormone. Int J Mol Sci 21(8):3020. https://doi.org/10.3390/ijms21083020

Yoshimura T, Yasuo S, Watanabe M, Iigo M, Yamamura T, Hirunagi K, Ebihara S (2003) Light-induced hormone conversion of T4 to T3 regulates photoperiodic response of gonads in birds. Nature 426(6963):178–181. https://doi.org/10.1038/nature02117

Zada D, Tovin A, Lerer-Goldshtein T, Appelbaum L (2016) Pharmacological and BBB-targeted genetic therapies for thyroid hormone-dependent hypomyelination. Dis Models Mech. https://doi.org/10.1242/dmm.027227

Zada D, Blitz E, Appelbaum L (2017) Zebrafish—an emerging model to explore thyroid hormone transporters and psychomotor retardation. Mol Cell Endocrinol 459(December):53–58. https://doi.org/10.1016/j.mce.2017.03.004

Zhang S, Ji X (2022) Hatschek's pit and origin of pituitary gland. Acta Oceanol Sin 41(12):1–6. https://doi.org/10.1007/s13131-022-2044-y

Key References

These papers review the regulation of HPT axis in mammals

Fekete C, Lechan RM (2014) Central regulation of hypothalamic-pituitary-thyroid axis under physiological and pathophysiological conditions. Endocr Rev 35(2):159–194

Joseph-Bravo P, Jaimes-Hoy L, Uribe R-M, Charli J-L (2015) 60 YEARS OF NEUROENDOCRINOLOGY: TRH, the first hypophysiotropic releasing hormone isolated: control of the pituitary–thyroid axis. J Endocrinol 226(2):T85–T100

These papers review the effects of iodinate messengers in invertebrates

Crockford SJ (2009) Evolutionary roots of iodine and thyroid hormones in cell-cell signaling. Integr Comp Biol 49(2):155–166

Taylor E, Heyland A (2017) Evolution of thyroid hormone signaling in animals: non-genomic and genomic modes of action. Mol Cell Endocrinol 459(December):14–20

Comprehensive reviews about vertebrate TRFs and their implications in HPT axis regulation

Galas L, Raoult E, Tonon M-C, Okada R, Jenks BG, Castaño JP, Kikuyama S, Malagon M, Roubos EW, Vaudry H (2009) TRH acts as a multifunctional hypophysiotropic factor in vertebrates. Gen Comp Endocrinol 164(1):40–50

Lazcano I, Rodríguez AR, Uribe RM, Orozco A, Joseph-Bravo P, Charli J-L (2021) Evolution of thyrotropin-releasing factor extracellular communication units. Gen Comp Endocrinol 305(May):113642

These papers review the evolution and functional networks in vertebrate pituitary

Kawauchi H, Sower SA (2006) The dawn and evolution of hormones in the adenohypophysis. Gen Comp Endocrinol 148(1):3–14

Santiago-Andres Y, Golan M, Fiordelisio T (2021) Functional pituitary networks in vertebrates. Front Endocrinol 11(January):619352

Zhang S, Ji X (2022) Hatschek's pit and origin of pituitary gland. Acta Oceanol Sin 41(12):1–6

Comprehensive reviews about the evolution of TH signaling

Holzer G, Roux N, Laudet V (2017) Evolution of ligands, receptors and metabolizing enzymes of thyroid signaling. Mol Cell Endocrinol 459(December):5–13

Form and Function of Growth Hormone Releasing Hormone in Vertebrates

8

Emilio J. Vélez and Suraj Unniappan

Abstract

Growth of an organism is determined by many factors, including the hormonal milieu. Among the hormones, growth hormone releasing hormone (GHRH), produced by the brain, and growth hormone (GH), secreted from the pituitary, are the most widely studied major regulators of growth in vertebrates. GHRH is the most important brain-derived positive regulator of GH secretion from the pituitary. While its primary and earliest identified role is its stimulatory action on pituitary somatotrophs, emerging research has identified several other effects on tissues including nervous, respiratory, cardiovascular and gonadal tissues, and in regulating metabolic, aging and sleep. Thus, in addition to the stimulation of growth processes mediated by GH, GHRH has direct or indirect regulatory effects on various other cells. These actions are mediated by a G-protein-coupled receptor (GPCR) named GHRH-R and associated cell signaling pathways. GHRH-R is widely presented in the above-mentioned target tissues where GHRH acts. The overarching goal of this chapter is to provide a state-of-the-art comparative review of GHRH biology in vertebrates. Tissue-specific localization, and pituitary and extra-pituitary functions, and mechanism of action of GHRH and GHRH-R are discussed in-depth, while clinical aspects of GHRH are

E. J. Vélez
Université de Pau et des Pays de l'Adour, E2S UPPA, INRAE, UMR1419 Nutrition Métabolisme et Aquaculture, Saint-Pée-sur-Nivelle, France
e-mail: emilio-jose.velez-velazquez@inrae.fr

S. Unniappan (✉)
Laboratory of Integrative Neuroendocrinology, Department of Veterinary Biomedical Sciences, Western College of Veterinary Medicine, University of Saskatchewan, Saskatoon, SK, Canada
e-mail: suraj.unniappan@usask.ca

V. Grinevich, R. F. Oliveira (eds.), *Evolutionary and Comparative Neuroendocrinology*, Masterclass in Neuroendocrinology 17,
https://doi.org/10.1007/978-3-031-80209-6_8

not included. Evolutionary perspectives of GHRH and GHRH-R form and function are also included.

Keywords

Growth hormone · Growth · Vertebrates · Brain · Pituitary · Receptor · Evolution

8.1 Introduction

Growth hormone releasing hormone (GHRH) is a brain hormone that stimulates the release of growth hormone (GH) from the anterior pituitary gland (adenohypophysis) of mammals. It was discovered more than 40 years ago by two independent research groups (Thorner et al. 1982; Guillemin et al. 1982; Bloch et al. 1983). Parker and Sherwood (1990) purified GHRH from the hypothalamic extracts of two strains of salmon (*Oncorhynchus keta* and *Oncorhynchus kisutch*) and identified the GHRH cDNA from catfish (*Clarias macrocephalus*) (McRory et al. 1995). Later, GHRH-like immunoreactivity was identified in the brain of zebra fish (Castro et al. 2009). Meanwhile, GHRH cDNAs from zebra fish, goldfish and African clawed fish, and zebrafish and goldfish GHRH receptor (GHRH-R) cDNAs were reported by Leo et al. (2007). Thus, GHRH research in non-mammals progressed in parallel to mammalian GHRH work. While structurally unrelated to GHRH, ghrelin (Kojima et al. 1999) emerged as the second most potent regulator of GH synthesis and secretion, and this discovery was impactful to the GHRH research field. GHRH research spanning four decades resulted in very impactful discoveries on both pituitary and extra-pituitary roles for GHRH in vertebrates. While preparing this chapter in the summer of 2023, a PubMed search for "GHRH" yielded 3719 results. This chapter discusses the evolutionary perspectives of GHRH and GHRH-R and their biological functions in vertebrates using selected relevant literature from this search. Some milestones in GHRH research presented here are summarized in Fig. 8.1.

8.2 GHRH: Gene to Peptide

In mammals, GHRH and pituitary adenylate cyclase activating polypeptide (PACAP) are present in two distinct genes (Moretti et al. 2002). Meanwhile, GHRH and PACAP are present in the same gene in non-mammals and invertebrates (McRory and Sherwood 1997). Gene and exon duplications during evolution led to GHRH/PACAP1 and VIP/PACAP2 genes in invertebrates including tunicates and non-mammals. The GHRH gene in mammals contains five exons and seven introns (Fig. 8.2), although an exception to this, where an untranslated sixth exon was reported, is in gonads and hypothalamus (Sherwood et al. 2000). The transcription of the GHRH gene results in a mRNA that codes for a precursor peptide containing a signal peptide, GHRH and a C-peptide. Upon processing by furin at the signal

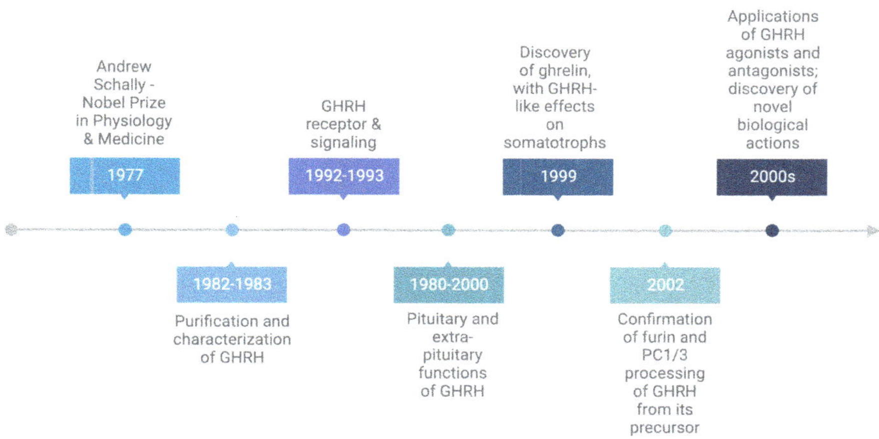

Fig. 8.1 Timeline of milestones in the GHRH research field (BioRender.com)

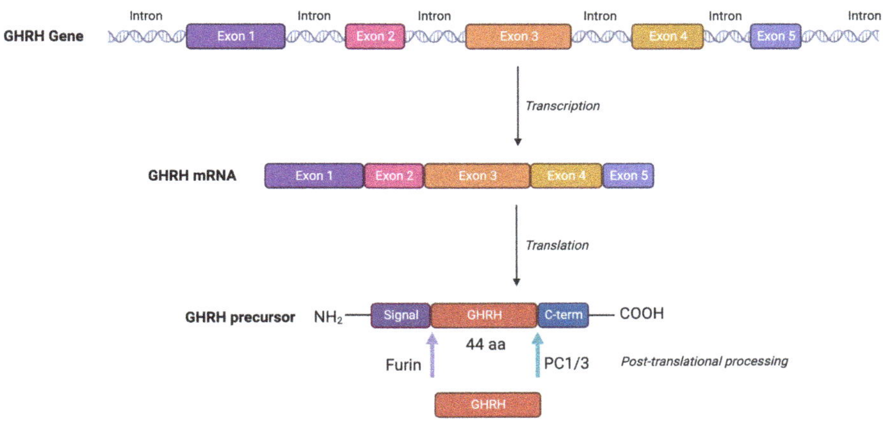

Fig. 8.2 Mouse GHRH gene intron:exon organization and the processing of GHRH precursor to form the 44 amino acid mature GHRH by the action of furin and prohormone convertase 1/3 (PC1/3) is shown here (BioRender.com)

peptide-GHRH junction and prohormone convertase 1 at the GHRH-C-peptide region, the mature, bioactive GHRH is formed (Zhu et al. 2002; Dey et al. 2004) (Fig. 8.2).

The length of mature GHRH in mammals in general is 44 amino acids, while in non-mammals it is a shorter peptide (*see protein sequence GenBank IDs listed under* Fig. 8.3a *for mature peptide lengths across species*). Dey et al. (2004) determined that two enzymes are important for mature GHRH processing from its precursor. In the N-terminal of pro-GHRH, furin had the most potency in cleaving an intermediate peptide 12–84 amino acids long. Further processing of this form by PC1/3 led to the formation of GHRH 1–44, the mature peptide. The same authors

A

Species	% Identity GHRH
Human	100
Chimpanzee	100
Lion	90.9
Narwhal	88.6
Cow	88.6
Japanese medaka	70.6
European seabass	70.6
Goldfish	67.7
Northern pike	66.7
Rainbow trout	65.0
Electric eel	64.7
Mouse	59.5
Brown snake	55.9
Painted turtle	54.6
Crocodile	54.6
Elephant shark	54.6
Atlantic cod	54.1
Ostrich	50.0
Wall lizard	50.0
Zebra finch	47.7

B

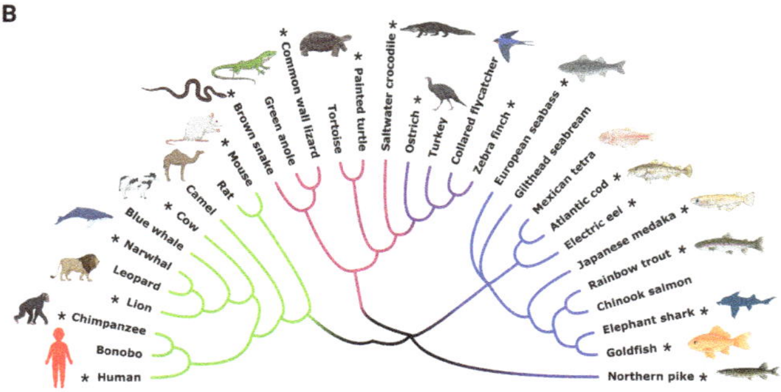

Fig. 8.3 (**a**) **Sequence analysis comparison of GHRHs**. Percentages of amino acid sequence identity between human GHRH and those from the species marked with asterisks in the (**b**) **phylogenetic analysis of GHRH peptides**. Sequences used for the analysis were retrieved from Ensembl Release 110 (Martin et al. 2023); African ostrich (*Struthio camelus camelus*): ENSSCUT00000010349. Arabian camel (*Camelus dromedarius*): ENSCDRT00005030030. Atlantic cod (*Gadus morhua*): ENSGMOT00000034254. Australian saltwater crocodile (*Crocodylus porosus*): ENSCPRT00005020264. Blue whale (*Balaenoptera musculus*): ENSBMST00010018610. Bonobo (*Pan paniscus*): ENSPPAT00000047251. Chimpanzee (*Pan troglodytes*): ENSPTRT00000025066. Chinook salmon (*Oncorhynchus tshawytscha*): ENSOTST00005027958. Collared flycatcher (*Ficedula albicollis*): ENSFALT00000038033. Common wall lizard (*Podarcis muralis*): ENSPMRT00000035237. Cow (*Bos taurus*): ENSBTAT00000016891. Eastern brown snake (*Pseudonaja textilis*): ENSPTXT00000010966. Electric eel (*Electrophorus electricus*): ENSEEET00000025688. Elephant shark (*Callorhinchus milii*): ENSCMIT00000037404. European sea bass (*Dicentrarchus labrax*): ENSDLAT00005080858. Gilthead seabream (*Sparus aurata*): ENSSAUT00010072881. Goldfish (*Carassius auratus*): ENSCART00000072462. Goode's thornscrub tortoise (*Gopherus evgoodei*): ENSGEVT00005013877. Green anole (*Anolis carolinensis*): ENSACAT00000011827. Human (*Homo sapiens*): ENST00000709405. Japanese medaka (*Oryzias latipes*): ENSORLT00000045469. Leopard (*Panthera pardus*): ENSPPRT00000004867. Lion (*Panthera leo*): ENSPLOT00000003024. Mexican tetra (*Astyanax mexicanus*): ENSAMXT00000003134. Mouse (*Mus musculus*): ENSMUST00000029172. Narwhal (*Monodon monoceros*): ENSMMNT00015004499. Northern pike (*Esox lucius*): ENSELUT00000037438. Norway rat (*Rattus norvegicus*): ENSRNOT00055046288. Painted turtle (*Chrysemys picta bellii*): ENSCPBT00000029600. Rainbow trout (*Oncorhynchus mykiss*): ENSOMYT00000049395. Turkey (*Meleagris gallopavo*): ENSMGAT00000002444. Zebra finch (*Taeniopygia guttata*): ENSTGUT00000005121. Multiple sequence analysis was performed with Clustal Omega (Madeira et al. 2022), and the phylogenetic and molecular evolutionary analyses were conducted using MEGA version 11 (Tamura et al. 2021) and iTOL version 6.8.1 (Letunic and Bork 2021)

found that GHRH was abundant in the Golgi apparatus and secretory vesicles of neuroendocrine cells.

Compared to PACAP, GHRH has evolved significantly, which is reflected in its less conserved sequence within mammals versus non-mammals and invertebrates. For example, the GHRHs of Japanese medaka and European sea bass have higher percentage of identity with the human GHRH than murine GHRH (Fig. 8.3a). The phylogenetic analysis (Fig. 8.3b) shows that GHRHs cluster according to their class for fish, reptiles, birds, and mammals, with a slight exception in the case of rodents, as previously reported (Lee et al. 2007). Despite this, the main functions of GHRH are relatively well conserved across mammals and non-mammals, as will be discussed in the following sections.

8.3 Pituitary Functions of GHRH

8.3.1 Regulation of GH Synthesis and Secretion

Growth hormone (GH) is synthesized, stored, and secreted by the somatotrophs residing in the anterior pituitary (adenohypophysis). GH plays a central role in stimulating hepatic insulin-like growth factors (IGFs), which are involved in vital processes such as protein synthesis, cell proliferation, and differentiation (Duan et al. 2010; Vélez et al. 2017). Together, GHRH, GH, and IGFs constitute the endocrine hypothalamus-pituitary-liver (GHRH-GH-IGF) axis, serving as an integrative hub that regulates various metabolic and physiological processes, including somatic growth and tissue development (Vélez and Unniappan 2021) (Fig. 8.4).

The production and release of GH are predominantly regulated by two hypothalamic factors, namely GHRH (stimulator) and somatostatin (inhibitor), through specific binding to G-protein-coupled receptors (GPCRs) (Martari and Salvatori 2009). Indeed, the main function of GHRH is the regulation of GH at the pituitary level and therefore of the GH/IGF axis (Fig. 8.4). Accordingly, it is worth focusing in the following sections on the regulation of GHRH itself, its receptors, and its mechanism of action.

GHRH is a GH-stimulating peptide across vertebrates (Lee et al. 2007; Xu 2016; Ávila-Mendoza et al. 2018). The regulation of GHRH secretion is conserved from fish to mammals and is modulated by the GH/IGF axis (Fig. 8.4), establishing negative feedback loops to fine-tune hormone levels [reviewed in Vélez and Unniappan (2021)]. GH can act in an autocrine or paracrine manner within the somatotrophs (ultra-short-loop feedback) to regulate its own synthesis and release while also influencing GHRH release by the hypothalamus (short-loop feedback) (Wong et al. 2006; Canosa et al. 2007; Gahete et al. 2016). Similarly, IGF-1 can directly inhibit GH secretion by acting on the pituitary and indirectly by regulating GHRH release (long-loop feedback) (Berelowitz et al. 1981). GHRH, in turn, can induce an ultra-short-loop feedback in the hypothalamus (Lasko et al. 2001). While long-loop feedback mechanisms appear to be well conserved across vertebrates, information on ultra-short-loops in nonmammalian vertebrates remains limited (Vélez and Unniappan 2021). Furthermore, the GH/IGF axis also regulates the secretion of the GH-inhibitor somatostatin (Lu et al. 2019). Somatostatin signaling can counteract the cascades activated by GHRH (discussed in detail below) and can directly or indirectly inhibit GHRH neurons, thereby modulating GHRH secretion (MacGregor and Leng 2005; Rubinek and Modan-Moses 2016; Osterstock et al. 2016).

Besides the feedback regulator discussed above, additional factors including neurotransmitters, peptides, and hormones have been involved directly or indirectly

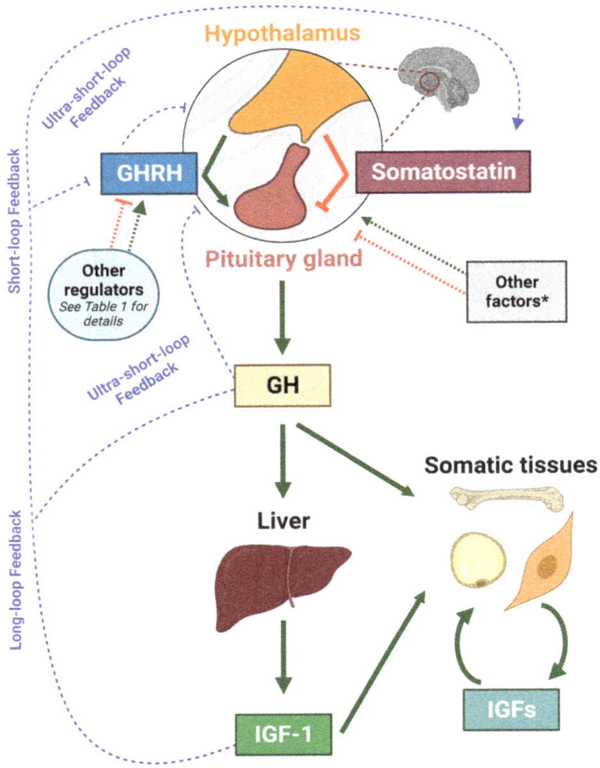

Fig. 8.4 **Schematic representation of the hypothalamus (GHRH/somatostatin)—pituitary (GH)—liver (IGF) endocrine axis**, the main regulator of somatic growth along vertebrates. The synthesis and secretion of GH by pituitary somatotrophs are primarily regulated by two key hypothalamic factors: the stimulator GHRH and the inhibitor somatostatin. GH, predominantly acting in the liver, stimulates the secretion of IGF-1. IGF-1, in turn, exerts its effects through autocrine, paracrine, and endocrine pathways in various somatic tissues, influencing processes such as protein synthesis, cell proliferation, and metabolism. Feedback mechanisms play a crucial role in regulating GH and IGF-1 levels. Both GH and IGF-1 can control their own levels through long-loop and short-loop feedback mechanisms. Additionally, GH, GHRH, and somatostatin can regulate their levels through an ultra-short-loop feedback mechanism. In addition, other factors can modulate the axis indirectly through the regulation of GHRH actions (please see Table 8.1), or by influencing directly pituitary GH (BioRender.com)

in the regulation of GHRH (for some examples see Table 8.1). The effects of some of these regulators in Table 8.1 have not been studied in all groups of vertebrates. It has been reported that some of those factors share a common role in different groups (e.g., thyrotropin releasing hormone and thyroid hormones), while others seem to exert the opposite effect (e.g., serotonin and norepinephrine).

Table 8.1 List of GHRH regulators and their roles in various vertebrates

Other GHRH regulators	Groups	Role (↑/NC/↓)	References
Dopamine	Mammals	↑	Vance et al. (1987)
	Birds	NotDet	–
	Reptiles	↑	Denver and Licht (1990)
	Fish	↑	Melamed et al. (1996)
Serotonin	Mammals	↑	Devlin et al. (2001), Gahete et al. (2009)
	Birds	↓	Hall et al. (1984)
	Reptiles	↓	Denver and Licht (1990)
	Fish	↓	Somoza and Peter (1991)
Norepinephrine	Mammals	↓	Soyoola et al. (1994)
	Birds	↓	Donoghue et al. (1990)
	Reptiles	↑	Denver and Licht (1990)
	Fish	↓	Lee et al. (2000)
Intermedin/Adrenomedullin-2	Mammals	↓	Taylor et al. (2006)
	Birds, reptiles, and fish	NotDet	–
Glucocorticoids (hydrocortisone, corticosterone, dexamethasone)	Mammals	↑/NC	Lam et al. (1996), Korytko and Cuttler (1997), Miller and Mayo (1997), Luque et al. (2004)
	Birds	↓	Porter et al. (2006)
	Reptiles	NotDet	–
	Fish	↓	Sadoul and Vijayan (2016)
Estrogens (estradiol)	Mammals	↓	Lam et al. (1996), Hudmon et al. (2009)
	Birds and reptiles	NotDet	–
	Fish	↓	Zhang et al. (2022)
Thyrotropin releasing hormone (TRH), and thyroid hormones (T3 and T4)	Mammals	↑	Miki et al. (1995), Korytko and Cuttler (1997)
	Birds	↑	Porter et al. (2006)
	Reptiles	↑	Denver and Licht (1990)
	Fish	↑	Melamed et al. 1995)
Androgens (testosterone)	Mammals	↑	Zeitler et al. (1990)
	Birds and reptiles	NotDet	–
	Fish	↓	Zhang et al. (2022)

Legends: ↑ = Stimulation/NC = No Change/↓ = Inhibition/NotDet = Not Determined

8.3.2 GHRH-Receptors: Evolution and Anatomical Localization

GHRH receptors (GHRH-Rs) belong to the class B/secretin family of GPCRs (Pioszak and Hay 2020), characterized by their distinctive seven-transmembrane (7-TM) helical structure (Fig. 8.5a). Additionally, this family of receptors are known for possessing six conserved cysteine residues in the extracellular amino-terminus

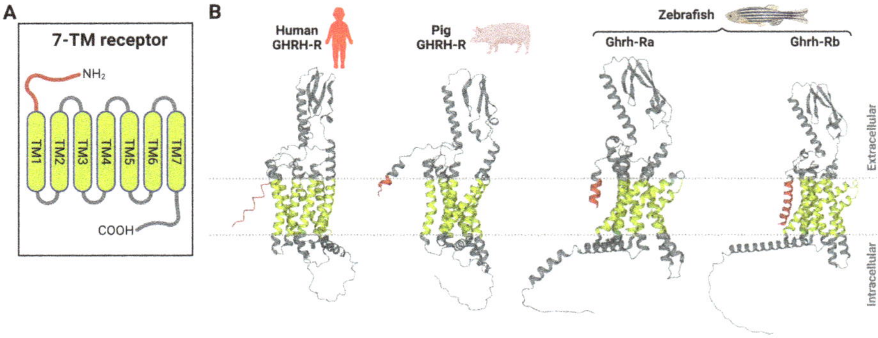

Fig. 8.5 (**a**) **Simplified representation of the GHRH 7TM receptor structure**. (**b**) **Molecular structure** of the human and pig GHRH-Rs, as well as of the two receptors of the zebra fish (Ghrh-Ra and Ghrh-Rb), retrieved from Transmembrane Protein Structure Database (TmAlphaFold database) (Dobson et al. 2023); Human (*Homo sapiens*) GHRH-R: Q02643. Pig (*Sus scrofa*) GHRH-R: P34999. Zebra fish (*Danio rerio*) Ghrh-Ra and Ghrh-Rb: A0A0R4IKR9 and E7FF79, respectively (BioRender.com)

and several phosphorylation sites within the C-intracellular domain (Mayo 1992; Mayo et al. 2000; Cunha and Mayo 2002; Malagón et al. 2006). Remarkably, this structural configuration remains highly conserved across the GHRH-Rs of vertebrates, including fish (Fig. 8.5b).

While the receptor length remains relatively the same in most mammals (around 423 amino acids), it exhibits more variability in non-mammals. For instance, the GHRH receptor in the electric eel is comprised of 403 residues, whereas it extends to 458 amino acids in the goldfish. When we undertake a comparative analysis with the human GHRH-R sequence, we discover an identity surpassing 80% among mammals (Fig. 8.6a). In contrast, conservation drops to approximately 50% when comparing fish GHRH-Rs and falls even further in birds and reptiles. This trend aligns with the evolutionary history of the GHRH receptor in vertebrates (Fig. 8.6b).

The *GHRH-R* gene, responsible for encoding this receptor, contains putative binding sites for several transcription factors including cAMP response elements (CRE), an estrogen-responsive element (ERE) and the pituitary-specific transcription factor Pit-1 (Petersenn and Schulte 2000). This last factor is indispensable for the expression of *GHRH-R* (Gaylinn 2002) and plays a pivotal role in regulating the GH gene, but also that of prolactin (PRL) and thyroid-stimulating hormone (TSH) beta-subunit (González-Parra et al. 1996). Consequently, the activity of Pit-1 can be associated with the anatomical localization of the GHRH receptor.

In mammals, GHRH-Rs are predominantly expressed in the anterior pituitary (mainly in somatotrophs). Nevertheless, GHRH-R is also reported in brain regions, the gonads, placenta, and the kidney (Cunha and Mayo 2002). Data on GHRH-R gene and protein expression in multiple species are available in the Expression Atlas (Moreno et al. 2022) and they suggest that receptor localization is primarily confined to the brain and/or the pituitary in other vertebrates as well. This localization of GHRH-Rs provides valuable insights into potential biological outcomes of GHRH signaling in these tissues.

A

Species	% Identity GHRH-R
Human	100
Chimpanzee	97.6
Narwhal	86.5
Lion	85.8
Cow	83.7
Mouse	81.5
Painted turtle	60.4
Ostrich	59.8
Crocodile	59.1
Zebra finch	58.8
Wall lizard	55.7
Atlantic cod	52.3
Rainbow trout	52.2
Electric eel	51.3
Japanese medaka	50.7
European seabass	50.4
Goldfish	50.0
Northern pike	50.0
Brown snake	46.9
Elephant shark	41.7

B

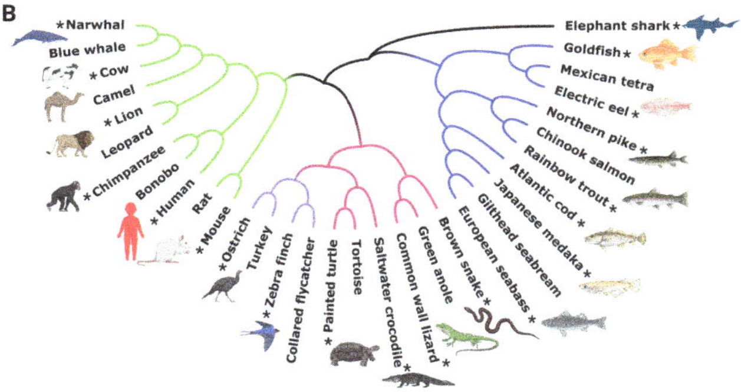

Fig. 8.6 (a) **Sequence analysis comparison of GHRH-Rs**. Percentages of amino acid sequence identity between the human GHRH-R and those from the species marked with asterisks in the (**b**) **phylogenetic analysis of GHRH-Rs**. Sequences used for the analysis were retrieved from Ensembl Release 110 (Martin et al. 2023); African Ostrich (*Struthio camelus camelus*): ENSSCUP00000004364. Arabian camel (*Camelus dromedarius*): ENSCDRP00005003980. Atlantic cod (*Gadus morhua*): ENSGMOP00000019484. Australian saltwater crocodile (*Crocodylus porosus*): ENSCPRP00005002794. Blue whale (*Balaenoptera musculus*): ENSBMSP00010016771. Bonobo (*Pan paniscus*): ENSPPAP00000012743. Chimpanzee (*Pan troglodytes*): ENSPTRP00000073434. Chinook salmon (*Oncorhynchus tshawytscha*): ENSOTSP00005017794. Collared flycatcher (*Ficedula albicollis*): ENSFALP00000000883. Common wall lizard (*Podarcis muralis*): ENSPMRP00000015235. Cow (*Bos taurus*): ENSBTAP00000026967. Eastern brown snake (*Pseudonaja textilis*): ENSPTXP00000014142. Electric eel (*Electrophorus electricus*): ENSEEEP00000002144. Elephant shark (*Callorhinchus milii*): ENSCMIT00000007094. European sea bass (*Dicentrarchus labrax*): ENSDLAP00005006964. Gilthead seabream (*Sparus aurata*): ENSSAUP00010022798. Goldfish (*Carassius auratus*): ENSCART00000095399. Goode's thornscrub tortoise (*Gopherus evgoodei*): ENSGEVP00005022188. Green anole (*Anolis carolinensis*): ENSACAP00000037827. Human (*Homo sapiens*): ENSP00000320180. Japanese medaka (*Oryzias latipes*): ENSORLP00000020808. Leopard (*Panthera pardus*): ENSPPRP00000002356. Lion (*Panthera leo*): ENSPLOP00000013368. Mexican tetra (*Astyanax mexicanus*): ENSAMXP00000000898. Mouse (*Mus musculus*): ENSMUSP00000068120. Narwhal (*Monodon monoceros*): ENSMMNP00015010025. Northern pike (*Esox lucius*): ENSELUP00000057331. Norway rat (*Rattus norvegicus*): ENSRNOP00000016087. Painted turtle (*Chrysemys picta bellii*): ENSCPBP00000005597. Rainbow trout (*Oncorhynchus mykiss*): ENSOMYP00000047403. Turkey (*Meleagris gallopavo*): ENSMGAP00000001280. Zebra finch (*Taeniopygia guttata*): ENSTGUP00000000347. Multiple sequence analysis was performed with Clustal Omega (Madeira et al. 2022), and the phylogenetic and molecular evolutionary analyses were conducted using MEGA version 11 (Tamura et al. 2021) and iTOL version 6.8.1 (Letunic and Bork 2021)

8.3.3 Mechanism of Action

When GHRH is secreted from the hypothalamus and it binds to GHRH-Rs on somatotrophs, it initiates a cascade of events that primarily regulate the synthesis and release of GH (Fig. 8.7). The GHRH regulation of GH implicates two main second messengers: cyclic AMP (cAMP) and Ca^{2+}-mediated signaling pathways (Cunha and Mayo 2002; Malagón et al. 2006).

At the mechanistic level, ligand-receptor binding triggers the association of a heterotrimeric G protein (composed of α, β, and γ subunits) with the receptor's cytosolic domain (Lu et al. 2019). This association leads to the binding of G protein

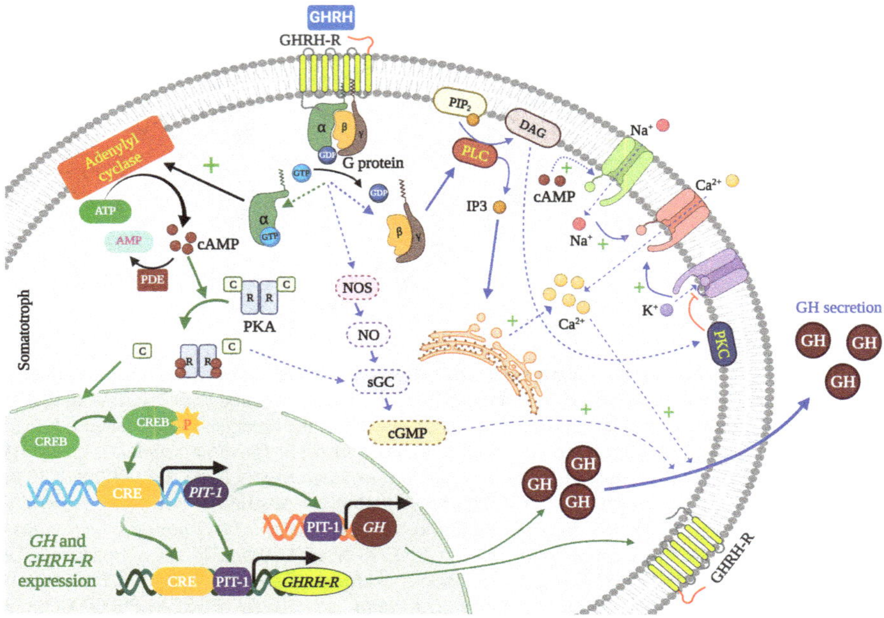

Fig. 8.7 Schematic representation of the signaling cascade initiated by GHRH in somatotrophs and resulting in the synthesis of GH and GHRH-R (green arrows), and the release of GH (blue arrows). Briefly, the control of GH synthesis is initiated once GHRH binds to GHRH-Rs, activating the adenylyl cyclase (AC) enzyme and leading to increased cAMP levels, which in turn activates protein kinase A (PKA). Phosphorylated CREB, a PKA target, stimulates the expression of *PIT-1*, a transcription factor that boosts GH and GHRH-R synthesis. On the other side, the activation of AC by GHRH can also stimulate protein lipase C (PLC), resulting in increased IP3 levels and activation of protein kinase C (PKC). In turn, cAMP-mediated activation of Na^+ channels, the blocking of K^+ channels by PKC, and the IP3-mediated mobilization of calcium (Ca^{2+}) from the endoplasmic reticulum, together cause membrane depolarization facilitating GH secretion. Other signaling pathways can contribute to such regulation [*please see below*] (BioRender.com)

to guanosine triphosphate (GTP), resulting in the dissociation of the complex, including the decoupling of Gα- and Gβγ-subunits (Oldham and Hamm 2008). Free Gα-subunits activate the adenylate cyclase (AC) enzyme, which converts adenosine triphosphate (ATP) into cAMP (Hernández-Ramírez et al. 2018). cAMP then binds to the regulatory subunits of the tetrameric protein kinase A (PKA), causing their dissociation and activation (Schernthaner-Reiter et al. 2020). These catalytic subunits act as serine-threonine kinases, phosphorylating various substrates, including the cAMP-responsive element-binding protein (CREB) (Shabb 2001; Peverelli et al. 2014). Phosphorylated CREB can bind to CRE-binding sites in the promotors of both the *PIT-1* and *GHRH-R* genes (Bertherat 1997; Peverelli et al. 2014). PIT-1 regulates the expression of the *GH* gene (Cunha et al.), as well as reinforces the expression of the *GHRH-R* (Gaylinn 2002). This transcriptional regulation appears to be conserved across vertebrates (Argenton et al. 1993). On the other hand, increased cAMP levels will activate Na^+ channels, leading to membrane depolarization (Tsaneva-Atanasova et al. 2007). This membrane depolarization, in turn, activates Ca^{2+} channels, facilitating calcium influx and enabling GH exocytosis and release (Wong et al. 2006; Canosa et al. 2007; Hernández-Ramírez et al. 2018). Additionally, a parallel pathway involves the activation of the protein lipase C (PLC) by the Gβγ-subunit released during the initial steps of the GHRH signaling cascade (Oldham and Hamm 2008). Activated PLC increases inositol triphosphate (IP3) levels, triggering Ca^{2+} release from the endoplasmic reticulum (Steyn 2018), and activates protein kinase C (PKC) to block membrane hyperpolarization by K^+ channels (Hernández-Ramírez et al. 2018), further supporting membrane depolarization and facilitating GH exocytosis.

GH secretion can also be stimulated through a cyclic GMP (cGMP)-dependent mechanism associated with nitric oxide (NO) levels (Hartt et al. 1995; Wong et al. 2006). While this mechanism appears cAMP-independent, interactions between cAMP's downstream PKA and the cGMP-synthesizing enzyme, the soluble guanylyl cyclase (sGC), could be present (Malagón et al. 2006). Notably, these signaling pathways, converging to regulate GH secretion by modulating Ca^{2+} influx, are remarkably well conserved among vertebrates, including fish (Vélez and Unniappan 2021).

8.3.4 Other Effects of GHRH on Pituitary Somatotrophs

While most of the research on pituitary actions of GHRH focused on how it influences GH synthesis and secretion, there is significant evidence on additional roles for GHRH on somatotrophs. GHRH stimulates somatotroph proliferation in mice (Mayo et al. 1988; Pombo et al. 2000; Zeitler and Siriwardana 2000). The primary cell signaling pathway that mediates this action is the mitogen-activated protein (MAP) kinase pathway (Cunha and Mayo 2002). It was found that MAP kinase-mediated pathways are also involved in the proliferative actions of GHRH on CHO cells (Pombo et al. 2000) and rat somatotrophs GH4 (Zeitler and Siriwardana 2000). In a GH3 somatotroph cell line, GHRH-induced c-myc and cyclin D1 (Solloso et al. 2008), two positive regulators of cell cycle. Pit-1 was required for this effect of GHRH on GH3 cells. The critical nature of GHRH on somatotroph proliferation is evident from the reduced anterior pituitary somatotroph population in GHRH knockout mice (Alba and Salvatori 2004; Alba et al. 2005). This deficit in the somatotroph population (pituitary hypoplasia) was partially reversed when a long acting GHRH analog JI-38 was administered to GHRH-KO mice (Alba et al. 2005). Taken together, these complementary studies using multiple approaches clearly point to a critical role for GHRH on adenohypophyseal somatotroph proliferation. To the best of our knowledge, information regarding other functions of GHRH in somatotrophs in non-mammals is scarce or even absent. Notwithstanding, a study in largemouth bass fish showed that the impairment of *ghrh* gene affects cellular proliferation at embryonic stages and compromises their viability (Ma et al. 2014), suggesting that the somatotroph proliferative role of GHRH is conserved beyond mammals.

8.4 Extra-Pituitary Functions

The main extra-pituitary functions of GHRH reported in nonclinical research are on cardiovascular, respiratory, and reproductive systems, endocrine pancreas, aging, feeding and metabolism, and sleep (Fig. 8.8). Although these results were primarily from research carried out using cell lines or nonhuman/primate mammalian models, we will address the corresponding literature on nonmammalian vertebrates where available.

8.4.1 Cardiovascular System

GHRH is known to elicit various cellular and systemic effects to regulate cardiovascular physiology in model organisms. Endothelium is a simple squamous epithelium that lines the lumen of all vascular tissues. GHRH modulates the permeability of this epithelium. Both GHRH and its agonist MR-409 decreased transepithelial resistance in bovine pulmonary artery endothelial cells (BPEAC cells). In contrast, a GHRH antagonist MIA-602 caused the opposite effect (Uddin et al. 2019). Meanwhile, in vitro application of a GHRH antagonist protected BPEAC cell barrier functions and attenuated hydrochloric acid-induced paracellular permeability

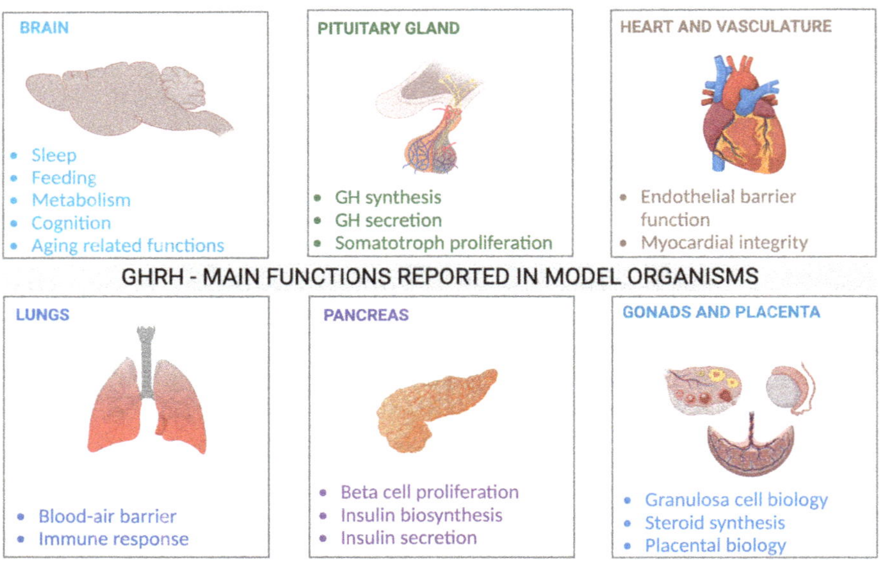

Fig. 8.8 Summary of tissue-specific biological actions of GHRH in vertebrates. Most of the functions listed here were characterized primarily using rodent models or cell lines. As discussed in the text, some of these effects were also found in some non-mammals based on the references cited. Due to the species-specific differences seen, only primary modulatory effects are listed. Please refer to the text for specific details on these effects, and consult the corresponding primary literature listed for additional details (BioRender.com)

(Barabutis et al. 2023). Further, the same researchers found that GHRH-R antagonist protection of endothelial barrier function in response to hydrogen peroxide is mediated by the maintenance of p53 phosphorylation and the activation of cofilin, an actin-binding protein with important roles in cellular junctional protein abundance. As in the previously discussed study, GHRH-antagonist JV-1-36 also prevented hyperpermeability induced by hydrogen peroxide (Akhter et al. 2023). These results suggest that GHRH-R antagonists can protect against oxidative stress in endothelial cells. GHRH-R activation by an agonist (JI-38) helped recovery from myocardial infarction in Fisher 344 rats (Kanashiro-Takeuchi et al. 2012). JI-38 caused significant cell mitosis, reduced infarct size, and enhanced cardiac performance (Kanashiro-Takeuchi et al. 2012). Similar lines of cardioprotective effects of GHRH were reported in other studies as well (Granata et al. 2009; Penna et al. 2013; Gesmundo et al. 2017; Rieger et al. 2021; Dulce et al. 2023). GHRH decreased infarct size and reversed contractile dysfunction due to ischemia reperfusion injury in male rats. This effect was blocked when GHRH was co-administered with JV-1-36, a GHRH antagonist. Using the same antagonist and GHRH, Granata et al. (2009) obtained very similar results in rat heart cells in vitro and in vivo. Rieger et al. (2021) found that MR-409, a GHRH agonist elicits similar positive effects on cardiac tissues in female Yorkshire pigs modeled for heart failure with preserved ejection fraction. The main beneficial effects included improved diastolic function, and an enhancement of cardiomyocyte calcium amplitude. Meanwhile, cardiac inflammatory markers and pro-BNP were significantly lower in the MR-409-treated pigs compared to placebo controls. Based on these studies, GHRH is considered a potent cardioprotector and regulator of heart physiology in mammals.

Information on the effects of GHRH on the cardiovascular system of non-mammals is scarce. However, it has been reported that during embryogenesis of the orange-spotted grouper (fish) the highest expression of the GHRH receptors takes place during the formation of the heart (Qian et al. 2012), suggesting a potential cardiovascular role of GHRH in fish species as well. Other evidence for a cardiovascular function of GHRH in non-mammals is unavailable.

8.4.2 Respiratory System

Many studies of GHRH effects on lungs focused on endothelial cells and barrier formation and this aspect is discussed in the previous section. JI-34, a GHRH agonist, caused a significant reduction in pulmonary capillary leak in lipopolysaccharide (LPS)-treated mice in vivo (Czikora et al. 2014). In the same study, in vitro treatment of MVEC cells resulted in cadherin expression maintenance and attenuation of LPS-induced barrier defects. When lung (NCI-H720) and bronchial (NCI-H727) cell lines were treated with the GHRH-R antagonists MZ-4-71 and MZ-5-16, cell proliferation was significantly decreased (Kidd et al. 2008). Subcutaneous injection of GHRH-R antagonist MIA-602 significantly reduced lung inflammation and fibrosis caused by an intraperitoneal injection of bleomycin (Zhang et al. 2019). GHRH knockout mice were more prone to *Streptococus pneumoniae* infection compared to wild-type littermates (Farhat et al. 2018). In the GHRH knockout mice, *S. pneumoniae* infection resulted in relatively more bacterial load and significantly more bacteriemia. The histopathology of lung tissue from infected mice showed infiltration by macrophages and neutrophils. A significant increase in pro-inflammatory cytokine mRNAs in the lungs was also detected in GHRH knockout mice. All these results help to conclude that GHRH is a pro-respiratory peptide that positively modulates lung physiology in mammals.

In teleost fish, the respiratory system is involved in other functions, including the excretion of ammonia, acid-base regulation, and osmoregulation. Seawater acclimation, also known as smoltification, represents a critical phase in the development and maturity of some salmonid species that during their development migrate from rivers to seawaters. GHRH is expressed in fish gills (Nam et al. 2011). GHRH-GH modulation of branchial epithelial growth factors regulates the proliferation and differentiation of the chloride cells involved in the active secretion of ions (Sakamoto and McCormick 2006), a key step in smoltification. In agreement with this, a recent study reported increased expression of *ghrh* in response to 7-days of seawater-challenge in the salmonid rainbow trout (Xiang et al. 2022). These outcomes suggest that similarly to what has been observed in mammals, the respiratory system is another target of GHRH actions in teleosts as well. However, the GHRH effects on fish gill tissue are primarily on ionoregulatory processes.

8.4.3 Reproduction

Reproductive regulatory tissues, especially the brain-pituitary-gonadal (HPG) axis of both male and female mammals, express GHRH and GHRH-R mRNA and protein. Whether GHRH directly regulates hypothalamic gonadotropin-releasing hormone (GnRH) is currently unknown. While pituitary somatotrophs are the primary target cells of GHRH in the anterior pituitary, a subpopulation of cells that express both GH and gonadotropins (luteinizing hormone/LH and follicle stimulating hormone/FSH) are proposed and are called somatogonadotrophs (Childs 2000). When bio-GHRH, a biotinylated analog of GHRH was injected into rats, it was found to bind to gonadotropin producing cells in both males and females, with significantly more gonadotrophs staining for GHRH (Childs et al. 1999). Whether GHRH could regulate these cells is unclear. Immunoneutralization of GHRH resulted in delayed sexual development in rats (Arsenijevic et al. 1989), which suggests a potential role for GHRH in male reproductive development. Ovarian and testicular cells express GHRH mRNA and GHRH is known to have local roles in gonads (Moretti et al. 2002). In line with this, GHRH stimulated stem cell factor mRNA abundance in rat Sertoli cell culture through a pKA-cAMP signaling pathway (Steinmetz et al. 2000). However, GHRH neutralization using antibodies did not affect spermatogenesis in male rats (Awoniyi et al. 1999). GHRH stimulated basal progesterone secretion from rat granulosa cells cultured in vitro, while GHRH reduced progesterone secretion in the presence of LH (Kotani et al. 1998). A direct effect of GHRH on sex steroid hormones was not found in female rats in vivo or granulosa cells cultured in vitro (Hugues et al. 1996). Meanwhile, IGF-1 significantly stimulated steroidogenesis in both conditions, suggesting a role for IGF-1 in mediating any effect for GHRH on gonadal steroids. Endogenous GHRH in rat ovary and in granulosa was reported (Hugues et al. 1996), suggesting a local action for this peptide within the ovary. For example, GHRH stimulated maturation of follicle and cumulus-enclosed oocytes from pregnant mare gonadotropin-treated rats and in follicle-enclosed oocytes from immature rats (Apa et al. 1995). This effect disappeared when GH was immunoneutralized, pointing to a critical role for GH in enabling GHRH effects. Rat placenta was also found to express GHRH (Bagnato et al. 1992) with expression detected as early as gestational day 7, gradually and significantly increasing on days 14, 17, and 20. In agreement with this, GHRH mRNA was detected in rat placenta, and the peptide of expected size was measured in placental perfusate (Margioris et al. 1990). The placental function of GHRH remains unclear, but it could have cell proliferative or endothelial roles found in other tissues.

Beyond mammals, there is evidence that GHRH has direct and indirect effects on the reproductive system in some other vertebrate groups. For example, GHRH induces GH release and increases both cell proliferation and metabolic activity in chicken testicular cell cultures (Martínez-Moreno et al. 2014). In reptiles, a more recent study reported that *ghrh* is expressed both in the testes and the ovaries of the green iguana (Ávila-Mendoza et al. 2018), suggesting a direct role for GHRH in the reproduction of reptiles, which should be further studied. In contrast, research in tilapia fish has demonstrated that exogenous administration of heterologous (i.e., mammalian) GHRH increased estradiol levels in females, and increased testosterone in both male and female fish. These effects were not observed in hypophysectomized animals (Kelly and Kohler 2007), implicating pituitary-derived gonadotropins or even GH in mediating such effects of GHRH on gonadal steroids. Nevertheless, it cannot be ruled out that a homologous GHRH could exert direct modulatory effects on fish reproduction.

8.4.4 Endocrine Pancreas

Research over the past decade has made significant progress on understanding GHRH and its roles on the endocrine pancreas. Most of these studies focused on GHRH agonist effects on islet beta cells and glucoregulation. MR-409, a GHRH-R agonist has been shown to cause pancreatic islet beta cell proliferation (Rodrigues-Dos-Santos et al. 2021; Louzada et al. 2023). More recent research has shown that MR-409 decreases beta cell death and enhances insulin secretion from murine and human islets in vitro (Louzada et al. 2023). Mechanistically, it was found that Akt-IRS2-PKA pathway known to have a significant role in beta cell proliferation and growth was activated in a beta cell line in vitro. In agreement with this, the same study, authors found that when MR-409 is administered in vivo into mice with a beta cell toxin streptozotocin-induced type 1 diabetes resulted in improved glucose handling and pancreatic islet beta cell mass (Louzada et al. 2023). MR-409 attenuated oxidative stress and enhanced survival in INS-1E cells under ER stress in vitro

(Rodrigues-Dos-Santos et al. 2021). MR-409 increased cAMP content and phosphorylation of CREB in INS-1 cells in vitro in another study (Zhang et al. 2015). GHRH-knockout mice had smaller islets and with a significantly lower number of beta cells (Lasher and Sun 2023). Administration of glucagon did not cause any change in blood glucose levels, suggesting an absence of gluconeogenic effects of glucagon in GHRH knockout mice. In support of this, the abundance of glucagon receptors in hepatocytes was found reduced in the GHRH-knockout mice. Zhang et al. (2015) reported that treatment of rat islets with MR-409 increased cell proliferation, islet mass and insulin mRNA abundance in vitro. In the same study, subcutaneous administration of MR-409 significantly reduced the adverse effects of streptozotocin on the pancreatic islet cells of NOD/SCID mice. Pretreatment of pancreatic islets with MR-409 before transplantation caused positive effects and these included an increase in serum insulin and IGF-1 in diabetic mice. Ludwig et al. (2012) found that GHRH agonist JI-36 pretreatment caused a similar beneficial effect in transplanted bioartificial pancreas in rats, in which improved insulin secretion and glucose tolerance were detected. The same agonist (JI-36) was found to reduce INS-1 cell apoptosis, and this effect was reversed by the co-application of a GHRH antagonist MIA-602 (Ludwig et al. 2010). JI-36 also elicited positive effects on glucose homeostasis after islet cell transplantation in NOD-SCID mice. Overall, GHRH agonists appear to have profound protective effects on islet beta cells and to promote insulin synthesis and secretion in rodents.

Information on the role of GHRH in the endocrine regulation of glucose homeostasis in other groups of vertebrates is scarce. Interestingly, it has been reported that while in homeotherms (i.e., mammals and birds) the main regulator of glucose uptake is insulin, in the muscle of poikilotherms (i.e., reptiles and fish) this function is mainly played by IGF-1 [reviewed in Vélez et al. (2017)]. In this regard, different fish skeletal muscles show a surprising abundance of IGF-receptors compared to the number of insulin receptors (Párrizas et al. 1994, 1995; Moon et al. 1996; Castillo et al. 2002, 2004). Hence, IGF-1 stimulates glucose uptake with greater potency than insulin does in rainbow trout (Castillo et al. 2004) and gilthead sea bream (Montserrat et al. 2007, 2012) muscle cells. Thus, IGF-1 could be considered to be the main regulator of glucose homeostasis in fish, and likely, in reptiles. Considering that one of the main downstream effectors of the GHRH actions is IGF-1 (Fig. 8.4), similar to what has been reported in mammals (although indirectly), GHRH regulates glucose metabolism in poikilotherms.

8.4.5 Aging

The endocrine milieu is tightly linked to the age of the animal, and it is possible that both GH and GHRH milieus are implicated in age-related changes in physiology. For example, endogenous GHRH levels are dependent on development stage and age of rodents and contributes to age-related changes in GH in rats. The number of GHRH immunoreactive neurons in the hypothalamic arcuate nucleus significantly decrease with age (Kuwahara et al. 2004). GHRH supplementation using a plasmid construct helped improve skin structure and prevented osteoporosis in old mice (Ye et al. 2021). A significant increase in trabecular bone and markers of bone health were detected in the GHRH-treated mice. This indicates a positive role for GHRH on bone and integument. The GHRH agonist MR-409 discussed under the cardiovascular system was effective in improving cardiac functions in old mice (Xiang et al. 2021). Daily subcutaneous injection of MR-409 for 6 months starting at 15 months of age caused a reversal of many aging effects and improved overall cardiac function and cellular energy production, and reduced senescence markers. However, GHRH-R knockout (Zhang et al. 2020) and GHRH-R-GH-R double knockout mice had extended lifespan (Icyuz et al. 2021). A similar increase in longevity was also found in GHRH knockout mice (Sun et al. 2013). The increase in lifespan seen in these knockout mice are likely due to compensatory changes in the endocrine and metabolic milieu, including insulin levels and insulin action. In line with this, treatment (daily subcutaneous injection) of 10-month-old mice for several months with a GHRH-R antagonist MZ-5-156 caused an increase in life expectancy by 8 weeks (Banks et al. 2010). In the early stages of this treatment, an improvement in cognitive functions was also found. GHRH causes an increase in long-term memory in male Wistar rats (Schneider-Rivas et al. 1995). The contrasting effects seen in pharmacological studies versus alterations in endogenous GHRH warrant further consideration. To our knowledge, the role of GHRH in aging has not been studied in non-mammals. It has been reported that the levels of *GHRH* decrease with aging in broiler chickens (Piórkowska et al. 2018). It is possible that GHRH plays age-related roles in other groups of vertebrates to enable physiolgical homeostasis.

8.4.6 Feeding and Metabolism

It is logical to expect a metabolic role for GHRH, a growth regulator, and there are several lines of evidence to support this notion. Recinella et al. (2021) subcutaneously injected MIA-690 (GHRH-R antagonist) or MR-409 (GHRH-R agonist) and measured food intake and metabolism. The former increased feeding and body weight, while the latter had no effect on either of these parameters. Hypothalamic AgRP and norepinephrine mRNA abundance was also increased by MIA-690. This increase in AgRP is possibly one of the mediators of increased food intake in MIA-690 chronically treated mice. In agreement with this, the disruption of GHRH-R resulted in increased feeding and circulating levels of the peripherally originated orexigen ghrelin and increased hypothalamic neuropeptide Y (NPY) and AgRP mRNA abundance in mice (Recinella et al. 2017). Intracerebroventricular or subcutaneous injection of GHRH inhibited food intake in LOU/C/Jall (LOU) rat strain rats and selectively inhibited lipid intake in both cases (Veyrat-Durebex et al. 2001). While most of these studies show an appetite inhibitory role for GHRH, in hamsters opposite effects were reported (Vaccarino et al. 1995). Injection of GHRH into the suprachiasmatic nucleus led to an increase in feeding in hamsters. Hoffman et al. (2020) found that brown adipose tissue (BAT) metabolism and amino acid metabolism-related genes were significantly increased in GHRH knockout mice.

Leone et al. (2015) reported increased locomotor activity and an increase in body temperature. These two changes were attributed as the reasons for increased feeding seen in these mice. BAT weight and BAT UCP-1 mRNA abundance were also significantly elevated in GHRH knockout mice. More recently, a role for GHRH on lipid metabolism in type 1 diabetes in rats was determined (Romero et al. 2016). Rats with streptozotocin-induced type 1 diabetes were treated with GHRH antagonist MIA-602. In diabetic rats, small intestinal GHRH receptor mRNA abundance was increased. MIA-602 treatment decreased intestinal GHRH-R mRNA abundance, lipemia and ApoB-48 levels in diabetic rats. GHRH-induced GLP-1 signaling impairment was detected in diabetic rats and this is possibly a reason for the dyslipidemia also seen in diabetic rats. Another model of GHRH knockout mice generated using CRISPR/Cas9 exhibited significant decrease in body weight, energy expenditure, respiratory exchange ratio, and lean mass. Meanwhile, their fat mass was significantly elevated (Icyuz et al. 2020).

Among non-mammals, metabolic roles of GHRH were explored in birds and fish. The administration of homologous GHRH decreased food intake in chicks (Tachibana et al. 2015). In contrast, injection with both low and high doses of heterologous (i.e., human) GHRH enhanced food intake in rainbow trout (Shepherd et al. 2007). While research methods vary between these studies, it appears that species specificity exists in GHRH regulation of metabolism. GHRH and metabolism in nontraditional models is a relatively unexplored area of research that requires further attention.

8.4.7 Sleep

Sleep is another process promoted by GHRH. While most of the data on this topic are from human studies, some studies in model organisms are available. Both GHRH knockout mice and GHRH-R knockout *lit/lit* mice exhibit non-rapid eye movement (REM) sleep disturbances (Liao et al. 2015). GHRH infusion in mice resulted in an increase in sleep, while the administration of a GHRH-R antagonist yielded the opposite effect (Liao et al. 2015). GHRH administration influenced non-REM sleep in many species (Liao et al. 2015). Microinjection of GHRH into the preoptic area induced non-REM sleep duration and intensity, while the injection of a GHRH-R antagonist [(*N*-Ac-Tyr^1,d-Arg^2)-GRF(1–29)-NH] into the same brain region caused the opposite effect in rats (Zhang et al. 1999). Normalization of hypothalamic GHRH content in a rat model of upper airway obstruction decreased the wake duration (Tarasiuk et al. 2011). In spontaneously dwarf rats that have reduced hypothalamic GHRH content, intracerebroventricular administration of GHRH caused an increase in non-REM sleep (Peterfi et al. 2006). Endogenous GHRH expression in the hypothalamus is also influenced by sleep. For example, an 8-hour sleep deprivation resulted in a significant reduction in GHRH-R mRNA and GHRH-binding sites in the brains of rats (Gardi et al. 2002). Similarly, 4- or 8-hour sleep deprivation significantly reduced hypothalamic GHRH contents measured by a radioimmunoassay in rats (Gardi et al. 1999). Dwarf rats that exhibit compromised

GHRH signaling in the brain and pituitary have decreased non-REM sleep (Obál et al. 2001). Together, these data provide a significant role for GHRH in sleep regulation in rodents. No data are available on this topic from non-mammals.

8.5 Perspectives

After four decades of research since its discovery in early 1980s, GHRH is now considered to be a multifunctional hormone with tissue-specific roles on several physiological processes in vertebrates. These include the biology of somatotrophs and endothelial cells, sleep, cardiovascular and respiratory functions, islet cell biology, feeding and metabolism and reproduction. The advancement in design, synthesis, and validation of numerous GHRH agonists and GHRH-R antagonists helped the evolution of GHRH from a GH/somatotroph regulator to a major extra-pituitary regulator. The recent discovery of additional tissue-specific effects of GHRH highlights the significance of this peptide. The evolutionarily conserved form and function of GHRH across mammals and non-mammals further emphasize its importance. The fundamental understanding of GHRH biology using model organisms certainly contributed to developing GHRH as a therapeutic target in human health and diseases. Whether GHRH has applications in non-mammals warrants further consideration. For example, a GH transgenic fish with accelerated growth is commercially available for consumption. Can GHRH also be used for such practical applications in non-humans, including species of veterinary importance? Future research is strongly expected to determine new roles for GHRH, further understanding of its already characterized functions, and GHRH applications in both mammals and non-mammals.

Acknowledgments Research in the laboratory of Suraj Unniappan is supported by funding from the Natural Sciences and Engineering Council of Canada Discovery Grant and the University of Saskatchewan Centennial Enhancement Chair Program. EJV is supported by the European Union's Horizon 2020 research and innovation program under the Marie Sokolowski-Curie grant agreement No 101030643 (Project MEAL fish). Figures provided in this chapter were created using BioRender.com tools (license purchased).

Key Literature

Guillemin et al. (1982)—*This article reported the discovery of GHRH from pancreatic tumor.*

Thorner et al. (1982)—*A second article published in the same year that reported the same identification using independent approaches.*

Bloch et al. (1983)—*This provided the first report on "GHRH" and its immunoreactive neurons in the brain of rodents.*

Mayo et al. (1988)—*The first study that showed that GHRH overexpression in mice led to somatotroph proliferation.*

Mayo (1992)—*Discovery of GHRH receptor and cell signaling was reported in this.*

Godfrey (1993)—*Identification of signature sequences in GHRH-R that are required for mediating GHRH functions were reported here.*

Zhu et al. (2002)—*This research identified furin and PC1/3 as the major enzymes involved in the processing of GHRH from its precursor protein.*

Alba and Salvatori (2004)—*The first report on GHRH-deficient mice.*

Lee et al. (2007)—*This article describes the identification of several GHRH and GHRH-R sequences from several non-mammals.*

References

Akhter MS, Kubra K-T, Barabutis N (2023) Protective effects of GHRH antagonists against hydrogen peroxide-induced lung endothelial barrier disruption. Endocrine 79(3):587–592. https://doi.org/10.1007/s12020-022-03226-1

Alba M, Salvatori R (2004) A mouse with targeted ablation of the growth hormone-releasing hormone gene: a new model of isolated growth hormone deficiency. Endocrinology 145(9):4134–4143. https://doi.org/10.1210/en.2004-0119

Alba M, Schally AV, Salvatori R (2005) Partial reversibility of growth hormone (GH) deficiency in the GH-releasing hormone (GHRH) knockout mouse by postnatal treatment with a GHRH analog. Endocrinology 146(3):1506–1513. https://doi.org/10.1210/en.2004-1044

Apa R, Lanzone A, Miceli F, Mastrandrea M, Macchione E, Caruso A, Mancuso S, Canipari R (1995) Growth hormone-releasing factor stimulates meiotic maturation in follicle- and cumulus-enclosed rat oocyte. Mol Cell Endocrinol 112(2):195–201. https://doi.org/10.1016/0303-7207(95)03599-3

Argenton F, Vianello S, Bernardini S, Jacquemin P, Martial J, Belayew A, Colombo L, Bortolussi M (1993) The transcriptional regulation of the growth hormone gene is conserved in vertebrate evolution. Biochem Biophys Res Commun 192(3):1360–1366. https://doi.org/10.1006/bbrc.1993.1566

Arsenijevic Y, Wehrenberg WB, Conz A, Eshkol A, Sizonenko PC, Aubert ML (1989) Growth hormone (GH) deprivation induced by passive immunization against rat GH-releasing factor delays sexual maturation in the male rat. Endocrinology 124(6):3050–3059. https://doi.org/10.1210/endo-124-6-3050

Ávila-Mendoza J, Pérez-Rueda E, Urban-Sosa V, Carranza M, Martínez-Moreno CG, Luna M, Arámburo C (2018) Characterization and distribution of GHRH, PACAP, TRH, SST and IGF1 mRNAs in the green iguana. Gen Comp Endocrinol 255:90–101. https://doi.org/10.1016/j.ygcen.2017.09.027

Awoniyi CA, Veeramachaneni DN, Roberts D, Tucker KE, Chandrashekar V, Schlaff WD (1999) Suppression of growth hormone does not affect ongoing spermatogenesis in rats. J Androl 20(1):102–108

Bagnato A, Moretti C, Ohnishi J, Frajese G, Catt KJ (1992) Expression of the growth hormone-releasing hormone gene and its peptide product in the rat ovary. Endocrinology 130(3):1097–1102. https://doi.org/10.1210/endo.130.3.1537276

Banks WA, Morley JE, Farr SA, Price TO, Ercal N, Vidaurre I, Schally AV (2010) Effects of a growth hormone-releasing hormone antagonist on telomerase activity, oxidative stress, longevity, and aging in mice. Proc Natl Acad Sci USA 107(51):22272–22277. https://doi.org/10.1073/pnas.1016369107

Barabutis N, Kubra K-T, Akhter MS (2023) Growth hormone-releasing hormone antagonists protect against hydrochloric acid-induced endothelial injury in vitro. Environ Toxicol Pharmacol 99:104113. https://doi.org/10.1016/j.etap.2023.104113

Berelowitz M, Szabo M, Frohman LA, Firestone S, Chu L, Hintz RL (1981) Somatomedin-C mediates growth hormone negative feedback by effects on both the hypothalamus and the pituitary. Science 212(4500):1279–1281. https://doi.org/10.1126/science.6262917

Bertherat J (1997) Nuclear effects of the cAMP pathway activation in somatotrophs. Horm Res 47(4–6):245–250. https://doi.org/10.1159/000185471

Bloch B, Brazeau P, Ling N, Bohlen P, Esch F, Wehrenberg WB, Benoit R, Bloom F, Guillemin R (1983) Immunohistochemical detection of growth hormone-releasing factor in brain. Nature 301(5901):607–608. https://doi.org/10.1038/301607a0

Canosa LF, Chang JP, Peter RE (2007) Neuroendocrine control of growth hormone in fish. Gen Comp Endocrinol 151(1):1–26. https://doi.org/10.1016/j.ygcen.2006.12.010

Castillo J, Le Bail P-Y, Paboeuf G, Navarro I, Weil C, Fauconneau B, Gutiérrez J (2002) IGF-I binding in primary culture of muscle cells of rainbow trout: changes during in vitro development. Am J Physiol Regul Integr Comp Physiol 283(3):R647–R652. https://doi.org/10.1152/ajpregu.00121.2002

Castillo J, Codina M, Martínez ML, Navarro I, Gutiérrez J (2004) Metabolic and mitogenic effects of IGF-I and insulin on muscle cells of rainbow trout. Am J Physiol Regul Integr Comp Physiol 286(5):R935–R941. https://doi.org/10.1152/ajpregu.00459.2003

Castro A, Becerra M, Manso MJ, Tello J, Sherwood NM, Anadón R (2009) Distribution of growth hormone-releasing hormone-like peptide: immunoreactivity in the central nervous system of the adult zebrafish (Danio rerio). J Comp Neurol 513(6):685–701. https://doi.org/10.1002/cne.21977. PMID: 19235874

Childs GV (2000) Growth hormone cells as co-gonadotropes: partners in the regulation of the reproductive system. Trends Endocrinol Metab 11(5):168–175. https://doi.org/10.1016/s1043-2760(00)00252-6

Childs GV, Unabia G, Miller BT, Collins TJ (1999) Differential expression of gonadotropin and prolactin antigens by GHRH target cells from male and female rats. J Endocrinol 162(2):177–187. https://doi.org/10.1677/joe.0.1620177

Cunha SR, Mayo KE (2002) Ghrelin and growth hormone (GH) secretagogues potentiate GH-releasing hormone (GHRH)-induced cyclic adenosine 3′,5′-monophosphate production in cells expressing transfected GHRH and GH secretagogue receptors. Endocrinology 143(12):4570–4582. https://doi.org/10.1210/en.2002-220670

Czikora I, Sridhar S, Gorshkov B, Alieva IB, Kasa A, Gonzales J, Potapenko O, Umapathy NS, Pillich H, Rick FG, Block NL, Verin AD, Chakraborty T, Matthay MA, Schally AV, Lucas R (2014) Protective effect of growth hormone-releasing hormone agonist in bacterial toxin-induced pulmonary barrier dysfunction. Front Physiol 5:259. https://doi.org/10.3389/fphys.2014.00259

Denver RJ, Licht P (1990) Modulation of neuropeptide-stimulated pituitary hormone secretion in hatchling turtles. Gen Comp Endocrinol 77(1):107–115. https://doi.org/10.1016/0016-6480(90)90211-4

Devlin RH, Biagi CA, Yesaki TY, Smailus DE, Byatt JC (2001) Growth of domesticated transgenic fish. Nature 409(6822):781–782. https://doi.org/10.1038/35057314

Dey A, Norrbom C, Zhu X, Stein J, Zhang C, Ueda K, Steiner DF (2004) Furin and prohormone convertase 1/3 are major convertases in the processing of mouse pro-growth hormone-releasing hormone. Endocrinology 145(4):1961–1971. https://doi.org/10.1210/en.2003-1472

Dobson L, Szekeres LI, Gerdán C, Langó T, Zeke A, Tusnády GE (2023) TmAlphaFold database: membrane localization and evaluation of AlphaFold2 predicted alpha-helical transmembrane protein structures. Nucleic Acids Res 51(D1):D517–D522. https://doi.org/10.1093/nar/gkac928

Donoghue DJ, Perez FM, Diamante BS, Malamed S, Scanes CG (1990) Influence of catecholamines, prostaglandins and thyroid hormones on growth hormone secretion by chicken pituitary cells in vitro. Domest Anim Endocrinol 7(1):35–42. https://doi.org/10.1016/0739-7240(90)90052-2

Duan C, Ren H, Gao S (2010) Insulin-like growth factors (IGFs), IGF receptors, and IGF-binding proteins: roles in skeletal muscle growth and differentiation. Gen Comp Endocrinol 167(3):344–351. https://doi.org/10.1016/j.ygcen.2010.04.009

Dulce RA, Kanashiro-Takeuchi RM, Takeuchi LM, Salerno AG, Wanschel ACBA, Kulandavelu S, Balkan W, Zuttion MSSR, Cai R, Schally AV, Hare JM (2023) Synthetic growth hormone-releasing hormone agonist ameliorates the myocardial pathophysiology characteristic of heart failure with preserved ejection fraction. Cardiovasc Res 118(18):3586–3601. https://doi.org/10.1093/cvr/cvac098

Farhat K, Bodart G, Charlet-Renard C, Desmet CJ, Moutschen M, Beguin Y, Baron F, Melin P, Quatresooz P, Parent A-S, Desmecht D, Sirard J-C, Salvatori R, Martens H, Geenen VG (2018) Growth hormone (GH) deficient mice with GHRH gene ablation are severely deficient in vaccine and immune responses against Streptococcus pneumoniae. Front Immunol 9:2175. https://doi.org/10.3389/fimmu.2018.02175

Gahete MD, Durán-Prado M, Luque RM, Martínez-Fuentes AJ, Quintero A, Gutiérrez-Pascual E, Córdoba-Chacón J, Malagón MM, Gracia-Navarro F, Castaño JP (2009) Understanding the multifactorial control of growth hormone release by somatotropes: lessons from comparative endocrinology. Ann N Y Acad Sci 1163:137–153. https://doi.org/10.1111/j.1749-6632.2008.03660.x

Gahete MD, Luque RM, Castaño JP (2016) Models of GH deficiency in animal studies. Best Pract Res Clin Endocrinol Metab 30(6):693–704. https://doi.org/10.1016/j.beem.2016.11.001

Gardi J, Obál F, Fang J, Zhang J, Krueger JM (1999) Diurnal variations and sleep deprivation-induced changes in rat hypothalamic GHRH and somatostatin contents. Am J Phys 277(5):R1339–R1344. https://doi.org/10.1152/ajpregu.1999.277.5.R1339

Gardi J, Taishi P, Speth R, Obal F, Krueger JM (2002) Sleep loss alters hypothalamic growth hormone-releasing hormone receptors in rats. Neurosci Lett 329(1):69–72. https://doi.org/10.1016/s0304-3940(02)00587-6

Gaylinn BD (2002) Growth hormone releasing hormone receptor. Recept Channels 8(3–4):155–162

Gesmundo I, Miragoli M, Carullo P, Trovato L, Larcher V, Di Pasquale E, Brancaccio M, Mazzola M, Villanova T, Sorge M, Taliano M, Gallo MP, Alloatti G, Penna C, Hare JM, Ghigo E, Schally AV, Condorelli G, Granata R (2017) Growth hormone-releasing hormone attenuates cardiac hypertrophy and improves heart function in pressure overload-induced heart failure. Proc Natl Acad Sci USA 114(45):12033–12038. https://doi.org/10.1073/pnas.1712612114

Godfrey P, Rahal JO, Beamer WG, Copeland NG, Jenkins NA, Mayo KE (1993) GHRH receptor of little mice contains a missense mutation in the extracellular domain that disrupts receptor function. Nat Genet 4(3):227–232. https://doi.org/10.1038/ng0793-227. PMID: 8395283

González-Parra S, Chowen JA, Garciá Segura LM, Argente J (1996) Ontogeny of pituitary transcription factor-1 (Pit-1), growth hormone (GH) and prolactin (PRL) mRNA levels in male and female rats and the differential expression of Pit-1 in lactotrophs and somatotrophs. J Neuroendocrinol 8(3):211–225. https://doi.org/10.1046/j.1365-2826.1996.04526.x

Granata R, Trovato L, Gallo MP, Destefanis S, Settanni F, Scarlatti F, Brero A, Ramella R, Volante M, Isgaard J, Levi R, Papotti M, Alloatti G, Ghigo E (2009) Growth hormone-releasing hormone promotes survival of cardiac myocytes in vitro and protects against ischaemia-reperfusion injury in rat heart. Cardiovasc Res 83(2):303–312. https://doi.org/10.1093/cvr/cvp090

Guillemin R, Brazeau P, Böhlen P, Esch F, Ling N, Wehrenberg WB (1982) Growth hormone-releasing factor from a human pancreatic tumor that caused acromegaly. Science 218(4572):585–587. https://doi.org/10.1126/science.6812220

Hall TR, Harvey S, Chadwick A (1984) Serotonin and acetylcholine affect the release of prolactin and growth hormone from pituitary glands of domestic fowl in vitro in the presence of hypothalamic tissue. Acta Endocrinol 105(4):455–462. https://doi.org/10.1530/acta.0.1050455

Hartt DJ, Ogiwara T, Ho AK, Chik CL (1995) Cyclic GMP stimulates growth hormone release in rat anterior pituitary cells. Biochem Biophys Res Commun 214(3):918–926. https://doi.org/10.1006/bbrc.1995.2374

Hernández-Ramírez LC, Trivellin G, Stratakis CA (2018) Cyclic 3′,5′-adenosine monophosphate (cAMP) signaling in the anterior pituitary gland in health and disease. Mol Cell Endocrinol 463:72–86. https://doi.org/10.1016/j.mce.2017.08.006

Hoffman JM, Poonawalla A, Icyuz M, Swindell WR, Wilson L, Barnes S, Sun LY (2020) Transcriptomic and metabolomic profiling of long-lived growth hormone releasing hormone knock-out mice: evidence for altered mitochondrial function and amino acid metabolism. Aging (Albany NY) 12(4):3473–3485. https://doi.org/10.18632/aging.102822

Hudmon A, Davenport G, Coleman ES, Sartin JL (2009) Low doses of estradiol partly inhibit release of GH in sheep without affecting basal levels. Domest Anim Endocrinol 37(3):181–187. https://doi.org/10.1016/j.domaniend.2009.05.004

Hugues JN, Miro F, Smyth CD, Hillier SG (1996) Effects of growth hormone releasing hormone on rat ovarian steroidogenesis. Hum Reprod 11(1):50–54. https://doi.org/10.1093/oxfordjournals.humrep.a019033

Icyuz M, Fitch M, Zhang F, Challa A, Sun LY (2020) Physiological and metabolic features of mice with CRISPR/Cas9-mediated loss-of-function in growth hormone-releasing hormone. Aging (Albany NY) 12(10):9761–9780. https://doi.org/10.18632/aging.103242

Icyuz M, Zhang F, Fitch MP, Joyner MR, Challa AK, Sun LY (2021) Physiological and metabolic characteristics of novel double-mutant female mice with targeted disruption of both growth hormone-releasing hormone and growth hormone receptor. Aging Cell 20(4):e13339. https://doi.org/10.1111/acel.13339

Kanashiro-Takeuchi RM, Takeuchi LM, Rick FG, Dulce R, Treuer AV, Florea V, Rodrigues CO, Paulino EC, Hatzistergos KE, Selem SM, Gonzalez DR, Block NL, Schally AV, Hare JM (2012) Activation of growth hormone releasing hormone (GHRH) receptor stimulates cardiac reverse remodeling after myocardial infarction (MI). Proc Natl Acad Sci USA 109(2):559–563. https://doi.org/10.1073/pnas.1119203109

Kelly AM, Kohler CC (2007) Mammalian growth hormone-releasing hormone indirectly stimulates sex steroid production in tilapia. J World Aquacult Soc 31(3):416–425. https://doi.org/10.1111/j.1749-7345.2000.tb00891.x

Kidd M, Schally AV, Pfragner R, Malfertheiner MV, Modlin IM (2008) Inhibition of proliferation of small intestinal and bronchopulmonary neuroendocrine cell lines by using peptide analogs targeting receptors. Cancer 112(6):1404–1414. https://doi.org/10.1002/cncr.23303

Kojima M, Hosoda H, Date Y, Nakazato M, Matsuo H, Kangawa K (1999) Ghrelin is a growth-hormone-releasing acylated peptide from stomach. Nature 402(6762):656–660. https://doi.org/10.1038/45230. PMID: 10604470

Korytko AI, Cuttler L (1997) Thyroid hormone and glucocorticoid regulation of pituitary growth hormone-releasing hormone receptor gene expression. J Endocrinol 152(2):R13–R17. https://doi.org/10.1677/joe.0.152r013

Kotani E, Usuki S, Kubo T (1998) Effect of growth hormone releasing hormone on luteinizing hormone stimulated progestin biosynthesis in cultured rat ovarian granulosa cells. Gynecol Endocrinol 12(5):307–313. https://doi.org/10.3109/09513599809012831

Kuwahara S, Sari DK, Tsukamoto Y, Tanaka S, Sasaki F (2004) Age-related changes in growth hormone (GH) cells in the pituitary gland of male mice are mediated by GH-releasing hormone but not by somatostatin in the hypothalamus. Brain Res 998(2):164–173. https://doi.org/10.1016/j.brainres.2003.10.060

Lam KS, Lee MF, Tam SP, Srivastava G (1996) Gene expression of the receptor for growth-hormone-releasing hormone is physiologically regulated by glucocorticoids and estrogen. Neuroendocrinology 63(6):475–480. https://doi.org/10.1159/000127075

Lasher AT, Sun LY (2023) Distinct physiological characteristics and altered glucagon signaling in GHRH knockout mice: implications for longevity. Aging Cell 22(12):e13985. https://doi.org/10.1111/acel.13985

Lasko CM, Korytko AI, Wehrenberg WB, Cuttler L (2001) Differential GH-releasing hormone regulation of GHRH receptor mRNA expression in the rat pituitary. Am J Physiol Endocrinol Metab 280(4):E626–E631. https://doi.org/10.1152/ajpendo.2001.280.4.E626

Lee EK, Chan VC, Chang JP, Yunker WK, Wong AO (2000) Norepinephrine regulation of growth hormone release from goldfish pituitary cells. I. Involvement of alpha2 adrenoreceptor and interactions with dopamine and salmon gonadotropin-releasing hormone. J Neuroendocrinol 12(4):311–322. https://doi.org/10.1046/j.1365-2826.2000.00455.x

Lee LTO, Siu FKY, Tam JKV, Lau ITY, Wong AOL, Lin MCM, Vaudry H, Chow BKC (2007) Discovery of growth hormone-releasing hormones and receptors in nonmammalian vertebrates. Proc Natl Acad Sci USA 104(7):2133–2138. https://doi.org/10.1073/pnas.0611008104

Leone S, Chiavaroli A, Shohreh R, Ferrante C, Ricciuti A, Manippa F, Recinella L, Di Nisio C, Orlando G, Salvatori R, Vacca M, Brunetti L (2015) Increased locomotor and thermogenic activity in mice with targeted ablation of the GHRH gene. Growth Hormon IGF Res 25(2):80–84. https://doi.org/10.1016/j.ghir.2014.12.007

Letunic I, Bork P (2021) Interactive Tree Of Life (iTOL) v5: an online tool for phylogenetic tree display and annotation. Nucleic Acids Res 49(W1):W293–W296. https://doi.org/10.1093/nar/gkab301

Liao F, Zhang TJ, Mahan TE, Jiang H, Holtzman DM (2015) Effects of growth hormone-releasing hormone on sleep and brain interstitial fluid amyloid-β in an APP transgenic mouse model. Brain Behav Immun 47:163–171. https://doi.org/10.1016/j.bbi.2014.09.005

Louzada RA, Blandino-Rosano M, Flores S, Lubaczeuski C, Cui T, Sha W, Cai R, Schally AV, Bernal-Mizrachi E (2023) GHRH agonist MR-409 protects β-cells from streptozotocin-induced diabetes. Proc Natl Acad Sci USA 120(25):e2209810120. https://doi.org/10.1073/pnas.2209810120

Lu M, Flanagan JU, Langley RJ, Hay MP, Perry JK (2019) Targeting growth hormone function: strategies and therapeutic applications. Signal Transduct Target Ther 4:3. https://doi.org/10.1038/s41392-019-0036-y

Ludwig B, Ziegler CG, Schally AV, Richter C, Steffen A, Jabs N, Funk RH, Brendel MD, Block NL, Ehrhart-Bornstein M, Bornstein SR (2010) Agonist of growth hormone-releasing hormone as a potential effector for survival and proliferation of pancreatic islets. Proc Natl Acad Sci USA 107(28):12623–12628. https://doi.org/10.1073/pnas.1005098107

Ludwig B, Rotem A, Schmid J, Weir GC, Colton CK, Brendel MD, Neufeld T, Block NL, Yavriyants K, Steffen A, Ludwig S, Chavakis T, Reichel A, Azarov D, Zimermann B, Maimon S, Balyura M, Rozenshtein T, Shabtay N, Vardi P, Bloch K, de Vos P, Schally AV, Bornstein SR, Barkai U (2012) Improvement of islet function in a bioartificial pancreas by enhanced oxygen supply and growth hormone releasing hormone agonist. Proc Natl Acad Sci USA 109(13):5022–5027. https://doi.org/10.1073/pnas.1201868109

Luque RM, Kineman RD, Park S, Peng X-D, Gracia-Navarro F, Castaño JP, Malagon MM (2004) Homologous and heterologous regulation of pituitary receptors for ghrelin and growth hormone-releasing hormone. Endocrinology 145(7):3182–3189. https://doi.org/10.1210/en.2003-1626

Ma DM, Han LQ, Bai JJ, Li SJ, Fan JJ, Yu LY, Quan YC (2014) A 66-bp deletion in growth hormone releasing hormone gene 5′-flanking region with largemouth bass recessive embryonic lethal. Anim Genet 45(3):421–426. https://doi.org/10.1111/age.12143

MacGregor DJ, Leng G (2005) Modelling the hypothalamic control of growth hormone secretion. J Neuroendocrinol 17(12):788–803. https://doi.org/10.1111/j.1365-2826.2005.01370.x

Madeira F, Pearce M, Tivey ARN, Basutkar P, Lee J, Edbali O, Madhusoodanan N, Kolesnikov A, Lopez R (2022) Search and sequence analysis tools services from EMBL-EBI in 2022. Nucleic Acids Res 50(W1):W276–W279. https://doi.org/10.1093/nar/gkac240

Malagón MM, Vázquez-martínez R, Martínez-fuentes AJ, Gracia-navarro F, Castaño JP (2006) CHAPTER 93 - growth hormone-releasing hormone. In: Kastin AJ (ed) Handbook of biologically active peptides. Academic Press, Burlington, pp 663–671

Margioris AN, Brockmann G, Bohler HC, Grino M, Vamvakopoulos N, Chrousos GP (1990) Expression and localization of growth hormone-releasing hormone messenger ribonucleic acid in rat placenta: in vitro secretion and regulation of its peptide product. Endocrinology 126(1):151–158. https://doi.org/10.1210/endo-126-1-151

Martari M, Salvatori R (2009) Diseases associated with growth hormone-releasing hormone receptor (GHRHR) mutations. Prog Mol Biol Transl Sci 88:57–84. https://doi.org/10.1016/S1877-1173(09)88003-4

Martin FJ, Amode MR, Aneja A, Austine-Orimoloye O, Azov AG, Barnes I, Becker A, Bennett R, Berry A, Bhai J, Bhurji SK, Bignell A, Boddu S, Branco Lins PR, Brooks L, Ramaraju SB, Charkhchi M, Cockburn A, Da Rin FL, Davidson C, Dodiya K, Donaldson S, El Houdaigui B, El Naboulsi T, Fatima R, Giron CG, Genez T, Ghattaoraya GS, Martinez JG, Guijarro C, Hardy M, Hollis Z, Hourlier T, Hunt T, Kay M, Kaykala V, Le T, Lemos D, Marques-Coelho D, Marugán JC, Merino GA, Mirabueno LP, Mushtaq A, Hossain SN, Ogeh DN, Sakthivel MP,

Parker A, Perry M, Piližota I, Prosovetskaia I, Pérez-Silva JG, Salam AIA, Saraiva-Agostinho N, Schuilenburg H, Sheppard D, Sinha S, Sipos B, Stark W, Steed E, Sukumaran R, Sumathipala D, Suner M-M, Surapaneni L, Sutinen K, Szpak M, Tricomi FF, Urbina-Gómez D, Veidenberg A, Walsh TA, Walts B, Wass E, Willhoft N, Allen J, Alvarez-Jarreta J, Chakiachvili M, Flint B, Giorgetti S, Haggerty L, Ilsley GR, Loveland JE, Moore B, Mudge JM, Tate J, Thybert D, Trevanion SJ, Winterbottom A, Frankish A, Hunt SE, Ruffier M, Cunningham F, Dyer S, Finn RD, Howe KL, Harrison PW, Yates AD, Flicek P (2023) Ensembl 2023. Nucleic Acids Res 51(D1):D933–D941. https://doi.org/10.1093/nar/gkac958

Martínez-Moreno CG, López-Marín LM, Carranza M, Giterman D, Harvey S, Arámburo C, Luna M (2014) Growth hormone (GH) and GH-releasing hormone (GHRH): co-localization and action in the chicken testis. Gen Comp Endocrinol 199:38–45. https://doi.org/10.1016/j.ygcen.2014.01.013

Mayo KE (1992) Molecular cloning and expression of a pituitary-specific receptor for growth hormone-releasing hormone. Mol Endocrinol 6(10):1734–1744. https://doi.org/10.1210/mend.6.10.1333056

Mayo KE, Hammer RE, Swanson LW, Brinster RL, Rosenfeld MG, Evans RM (1988) Dramatic pituitary hyperplasia in transgenic mice expressing a human growth hormone-releasing factor gene. Mol Endocrinol 2(7):606–612. https://doi.org/10.1210/mend-2-7-606

Mayo KE, Miller T, DeAlmeida V, Godfrey P, Zheng J, Cunha SR (2000) Regulation of the pituitary somatotroph cell by GHRH and its receptor. Recent Prog Horm Res 55:237–266. discussion 266–267

McRory JE, Parker DB, Ngamvongchon S, Sherwood NM (1995) Sequence and expression of cDNA for pituitary adenylate cyclase activating polypeptide (PACAP) and growth hormone-releasing hormone (GHRH)-like peptide in catfish. Mol Cell Endocrinol 108:169–177

McRory J, Sherwood NM (1997) Two protochordate genes encode pituitary adenylate cyclase-activating polypeptide and related family members. Endocrinology 138(6):2380–2390. https://doi.org/10.1210/endo.138.6.5167

Melamed P, Eliahu N, Levavi-Sivan B, Ofir M, Farchi-Pisanty O, Rentier-Delrue F, Smal J, Yaron Z, Naor Z (1995) Hypothalamic and thyroidal regulation of growth hormone in tilapia. Gen Comp Endocrinol 97(1):13–30. https://doi.org/10.1006/gcen.1995.1002

Melamed P, Gur G, Elizur A, Rosenfeld H, Sivan B, Rentier-Delrue F, Yaron Z (1996) Differential effects of gonadotropin-releasing hormone, dopamine and somatostatin and their second messengers on the mRNA levels of gonadotropin II beta subunit and growth hormone in the teleost fish, tilapia. Neuroendocrinology 64(4):320–328. https://doi.org/10.1159/000127135

Miki N, Ono M, Murata Y, Ohsaki E, Tamitsu K, Ri T, Demura H, Yamada M (1995) Thyroid hormone regulation of gene expression of the pituitary growth hormone-releasing factor receptor. Biochem Biophys Res Commun 217(3):1087–1093. https://doi.org/10.1006/bbrc.1995.2880

Miller TL, Mayo KE (1997) Glucocorticoids regulate pituitary growth hormone-releasing hormone receptor messenger ribonucleic acid expression. Endocrinology 138(6):2458–2465. https://doi.org/10.1210/endo.138.6.5184

Montserrat N, Sánchez-Gurmaches J, García de la Serrana D, Navarro MI, Gutiérrez J (2007) IGF-I binding and receptor signal transduction in primary cell culture of muscle cells of gilthead sea bream: changes throughout in vitro development. Cell Tissue Res 330(3):503–513. https://doi.org/10.1007/s00441-007-0507-2

Montserrat N, Capilla E, Navarro I, Gutiérrez J (2012) Metabolic effects of insulin and IGFs on gilthead sea bream (Sparus aurata) muscle cells. Front Endocrinol (Lausanne) 3:55. https://doi.org/10.3389/fendo.2012.00055

Moon TW, Castejón C, Baños N, Maestro MA, Plisetskaya EM, Gutiérrez J, Navarro I (1996) Insulin and IGF-I binding in isolated trout cardiomyocytes. Gen Comp Endocrinol 103(3):264–272. https://doi.org/10.1006/gcen.1996.0121

Moreno P, Fexova S, George N, Manning JR, Miao Z, Mohammed S, Muñoz-Pomer A, Fullgrabe A, Bi Y, Bush N, Iqbal H, Kumbham U, Solovyev A, Zhao L, Prakash A, García-Seisdedos D, Kundu DJ, Wang S, Walzer M, Clarke L, Osumi-Sutherland D, Tello-Ruiz MK, Kumari S, Ware D, Eliasova J, Arends MJ, Nawijn MC, Meyer K, Burdett T, Marioni J, Teichmann

S, Vizcaíno JA, Brazma A, Papatheodorou I (2022) Expression atlas update: gene and protein expression in multiple species. Nucleic Acids Res 50(D1):D129–D140. https://doi.org/10.1093/nar/gkab1030

Moretti C, Mencacci C, Frajese GV, Cerilli M, Frajese G (2002) Growth hormone-releasing hormone and pituitary adenylate cyclase-activating polypeptide in the reproductive system. Trends Endocrinol Metab 13(10):428–435. https://doi.org/10.1016/s1043-2760(02)00632-x

Nam B-H, Moon J-Y, Kim Y-O, Kong HJ, Kim W-J, Kim K-K, Lee S-J (2011) Molecular and functional analyses of growth hormone-releasing hormone (GHRH) from olive flounder (Paralichthys olivaceus). Comp Biochem Physiol B Biochem Mol Biol 159(2):84–91. https://doi.org/10.1016/j.cbpb.2011.02.006

Obál F, Fang J, Taishi P, Kacsóh B, Gardi J, Krueger JM (2001) Deficiency of growth hormone-releasing hormone signaling is associated with sleep alterations in the dwarf rat. J Neurosci 21(8):2912–2918. https://doi.org/10.1523/JNEUROSCI.21-08-02912.2001

Oldham WM, Hamm HE (2008) Heterotrimeric G protein activation by G-protein-coupled receptors. Nat Rev Mol Cell Biol 9(1):60–71. https://doi.org/10.1038/nrm2299

Osterstock G, Mitutsova V, Barre A, Granier M, Fontanaud P, Chazalon M, Carmignac D, Robinson ICAF, Low MJ, Plesnila N, Hodson DJ, Mollard P, Méry P-F (2016) Somatostatin triggers rhythmic electrical firing in hypothalamic GHRH neurons. Sci Rep 6:24394. https://doi.org/10.1038/srep24394

Parker DB, Sherwood NM (1990) Evidence of a growth hormone-releasing hormone-like molecule in salmon brain, Oncorhynchus keta and O. Kisutch. Gen Comp Endocrinol 79(1):95–102. https://doi.org/10.1016/0016-6480(90)90092-z. PMID: 2113021

Párrizas M, Planas J, Plisetskaya EM, Gutiérrez J (1994) Insulin binding and receptor tyrosine kinase activity in skeletal muscle of carnivorous and omnivorous fish. Am J Phys 266(6 Pt 2):R1944–R1950. https://doi.org/10.1152/ajpregu.1994.266.6.R1944

Párrizas M, Plisetskaya EM, Planas J, Gutiérrez J (1995) Abundant insulin-like growth factor-1 (IGF-1) receptor binding in fish skeletal muscle. Gen Comp Endocrinol 98(1):16–25. https://doi.org/10.1006/gcen.1995.1040

Penna C, Settanni F, Tullio F, Trovato L, Pagliaro P, Alloatti G, Ghigo E, Granata R (2013) GH-releasing hormone induces cardioprotection in isolated male rat heart via activation of RISK and SAFE pathways. Endocrinology 154(4):1624–1635. https://doi.org/10.1210/en.2012-2064

Peterfi Z, Obal F, Taishi P, Gardi J, Kacsoh B, Unterman T, Krueger JM (2006) Sleep in spontaneous dwarf rats. Brain Res 1108(1):133–146. https://doi.org/10.1016/j.brainres.2006.06.016

Petersenn S, Schulte HM (2000) Structure and function of the growth-hormone-releasing hormone receptor. Vitam Horm 59:35–69. https://doi.org/10.1016/s0083-6729(00)59003-7

Peverelli E, Mantovani G, Lania AG, Spada A (2014) cAMP in the pituitary: an old messenger for multiple signals. J Mol Endocrinol 52(1):R67–R77. https://doi.org/10.1530/JME-13-0172

Piórkowska K, Żukowski K, Połtowicz K, Nowak J, Wojtysiak D, Derebecka N, Wesoły J, Ropka-Molik K (2018) Transcriptomic changes in broiler chicken hypothalamus during growth and development. Int J Genomics 2018:e6049469. https://doi.org/10.1155/2018/6049469

Pioszak AA, Hay DL (2020) Chapter five - RAMPs as allosteric modulators of the calcitonin and calcitonin-like class B G protein-coupled receptors. In: Langmead CJ (ed) Advances in pharmacology. Academic Press, pp 115–141

Pombo CM, Zalvide J, Gaylinn BD, Diéguez C (2000) Growth hormone-releasing hormone stimulates mitogen-activated protein kinase. Endocrinology 141(6):2113–2119. https://doi.org/10.1210/endo.141.6.7513

Porter TE, Ellestad LE, Fay A, Stewart JL, Bossis I (2006) Identification of the chicken growth hormone-releasing hormone receptor (GHRH-R) mRNA and gene: regulation of anterior pituitary GHRH-R mRNA levels by homologous and heterologous hormones. Endocrinology 147(5):2535–2543. https://doi.org/10.1210/en.2005-1534

Qian Y, Yan A, Lin H, Li W (2012) Molecular characterization of the GHRH/GHRH-R and its effect on GH synthesis and release in orange-spotted grouper (Epinephelus coioides). Comp Biochem Physiol B Biochem Mol Biol 163(2):229–237. https://doi.org/10.1016/j.cbpb.2012.06.004

Recinella L, Leone S, Ferrante C, Chiavaroli A, Shohreh R, Di Nisio C, Vacca M, Orlando G, Salvatori R, Brunetti L (2017) Effects of growth hormone-releasing hormone gene targeted ablation on ghrelin-induced feeding. Growth Hormon IGF Res 37:40–46. https://doi.org/10.1016/j.ghir.2017.10.007

Recinella L, Chiavaroli A, Orlando G, Ferrante C, Gesmundo I, Granata R, Cai R, Sha W, Schally AV, Brunetti L, Leone S (2021) Growth hormone-releasing hormone antagonistic analog MIA-690 stimulates food intake in mice. Peptides 142:170582. https://doi.org/10.1016/j.peptides.2021.170582

Rieger AC, Bagno LL, Salerno A, Florea V, Rodriguez J, Rosado M, Turner D, Dulce RA, Takeuchi LM, Kanashiro-Takeuchi RM, Buchwald P, Wanschel ACBA, Balkan W, Schulman IH, Schally AV, Hare JM (2021) Growth hormone-releasing hormone agonists ameliorate chronic kidney disease-induced heart failure with preserved ejection fraction. Proc Natl Acad Sci USA 118(4):e2019835118. https://doi.org/10.1073/pnas.2019835118

Rodrigues-Dos-Santos K, Soares GM, Guimarães DSPSF, Araújo TR, Vettorazzi JF, Zangerolamo L, Marconato-Júnior E, Cai R, Sha W, Schally AV, Boschero AC, Barbosa HCL (2021) Effects of growth hormone-releasing hormone agonistic analog MR-409 on insulin-secreting cells under cyclopiazonic acid-induced endoplasmic reticulum stress. Mol Cell Endocrinol 535:111379. https://doi.org/10.1016/j.mce.2021.111379

Romero MJ, Lucas R, Dou H, Sridhar S, Czikora I, Mosieri EM, Rick FG, Block NL, Sridhar S, Fulton D, Weintraub NL, Bagi Z, Schally AV (2016) Role of growth hormone-releasing hormone in dyslipidemia associated with experimental type 1 diabetes. Proc Natl Acad Sci USA 113(7):1895–1900. https://doi.org/10.1073/pnas.1525520113

Rubinek T, Modan-Moses D (2016) Klotho and the growth hormone/insulin-like growth factor 1 axis: novel insights into complex interactions. Vitam Horm 101:85–118. https://doi.org/10.1016/bs.vh.2016.02.009

Sadoul B, Vijayan MM (2016) 5 - Stress and growth. In: Schreck CB, Tort L, Farrell AP, Brauner CJ (eds) Fish physiology. Academic Press, pp 167–205

Sakamoto T, McCormick SD (2006) Prolactin and growth hormone in fish osmoregulation. Gen Comp Endocrinol 147(1):24–30. https://doi.org/10.1016/j.ygcen.2005.10.008

Schernthaner-Reiter MH, Trivellin G, Stratakis CA (2020) Chaperones, somatotroph tumors and the cyclic AMP (cAMP)-dependent protein kinase (PKA) pathway. Mol Cell Endocrinol 499:110607. https://doi.org/10.1016/j.mce.2019.110607

Schneider-Rivas S, Rivas-Arancibia S, Vázquez-Pereyra F, Vázquez-Sandoval R, Borgonio-Pérez G (1995) Modulation of long-term memory and extinction responses induced by growth hormone (GH) and growth hormone releasing hormone (GHRH) in rats. Life Sci 56(22):PL433–PL441. https://doi.org/10.1016/0024-3205(95)00171-2

Shabb JB (2001) Physiological substrates of cAMP-dependent protein kinase. Chem Rev 101(8):2381–2411. https://doi.org/10.1021/cr000236l

Shepherd BS, Johnson JK, Silverstein JT, Parhar IS, Vijayan MM, McGuire A, Weber GM (2007) Endocrine and orexigenic actions of growth hormone secretagogues in rainbow trout (Oncorhynchus mykiss). Comp Biochem Physiol A Mol Integr Physiol 146(3):390–399. https://doi.org/10.1016/j.cbpa.2006.11.004

Sherwood NM, Krueckl SL, McRory JE (2000) The origin and function of the pituitary adenylate cyclase-activating polypeptide (PACAP)/glucagon superfamily. Endocr Rev 21(6):619–670. https://doi.org/10.1210/edrv.21.6.0414

Solloso A, Barreiro L, Seoane R, Nogueira E, Cañibano C, Alvarez CV, Zalvide J, Diéguez C, Pombo CM (2008) GHRH proliferative action on somatotrophs is cell-type specific and dependent on Pit-1/GHF-1 expression. J Cell Physiol 215(1):140–150. https://doi.org/10.1002/jcp.21295

Somoza GM, Peter RE (1991) Effects of serotonin on gonadotropin and growth hormone release from in vitro perifused goldfish pituitary fragments. Gen Comp Endocrinol 82(1):103–110. https://doi.org/10.1016/0016-6480(91)90301-l

Soyoola EO, Burgess MF, Bird RC, Kemppainen RJ, Williams JC, Sartin JL (1994) Neurotransmitter receptor agonists regulate growth hormone gene expression in cultured ovine pituitary cells. Proc Soc Exp Biol Med 207(1):26–33. https://doi.org/10.3181/00379727-207-43786

Steinmetz R, Lazzaro N, Rothrock JK, Pescovitz OH (2000) Effects of growth hormone-releasing hormone-related peptide on stem cell factor expression in cultured rat Sertoli cells. Endocrine 12(3):323–327. https://doi.org/10.1385/ENDO:12:3:323

Steyn FJ (2018) Growth hormone releasing hormone (GHRH). In: Choi S (ed) Encyclopedia of signaling molecules. Springer, Cham, pp 2271–2275

Sun LY, Spong A, Swindell WR, Fang Y, Hill C, Huber JA, Boehm JD, Westbrook R, Salvatori R, Bartke A (2013) Growth hormone-releasing hormone disruption extends lifespan and regulates response to caloric restriction in mice. eLife 2:e01098. https://doi.org/10.7554/eLife.01098

Tachibana T, Sugimoto I, Ogino M, Sakirul Islam Khan M, Masuda K, Ukena K, Wang Y (2015) Central administration of chicken growth hormone-releasing hormone decreases food intake in chicks. Physiol Behav 139:195–201. https://doi.org/10.1016/j.physbeh.2014.11.043

Tamura K, Stecher G, Kumar S (2021) MEGA11: molecular evolutionary genetics analysis version 11. Mol Biol Evol 38(7):3022–3027. https://doi.org/10.1093/molbev/msab120

Tarasiuk A, Berdugo-Boura N, Troib A, Segev Y (2011) Role of growth hormone-releasing hormone in sleep and growth impairments induced by upper airway obstruction in rats. Eur Respir J 38(4):870–877. https://doi.org/10.1183/09031936.00197610

Taylor MM, Bagley SL, Samson WK (2006) Intermedin/Adrenomedullin-2 inhibits growth hormone release from cultured, primary anterior pituitary cells. Endocrinology 147(2):859–864. https://doi.org/10.1210/en.2005-0949

Thorner MO, Perryman RL, Cronin MJ, Rogol AD, Draznin M, Johanson A, Vale W, Horvath E, Kovacs K (1982) Somatotroph hyperplasia. Successful treatment of acromegaly by removal of a pancreatic islet tumor secreting a growth hormone-releasing factor. J Clin Invest 70(5):965–977. https://doi.org/10.1172/jci110708

Tsaneva-Atanasova K, Sherman A, van Goor F, Stojilkovic SS (2007) Mechanism of spontaneous and receptor-controlled electrical activity in pituitary somatotrophs: experiments and theory. J Neurophysiol 98(1):131–144. https://doi.org/10.1152/jn.00872.2006

Uddin MA, Akhter MS, Singh SS, Kubra K-T, Schally AV, Jois S, Barabutis N (2019) GHRH antagonists support lung endothelial barrier function. Tissue Barriers 7(4):1669989. https://doi.org/10.1080/21688370.2019.1669989

Vaccarino FJ, Sovran P, Baird JP, Ralph MR (1995) Growth hormone-releasing hormone mediates feeding-specific feedback to the suprachiasmatic circadian clock. Peptides 16(4):595–598. https://doi.org/10.1016/0196-9781(95)00018-f

Vance ML, Kaiser DL, Frohman LA, Rivier J, Vale WW, Thorner MO (1987) Role of dopamine in the regulation of growth hormone secretion: dopamine and bromocriptine augment growth hormone (GH)-releasing hormone-stimulated GH secretion in normal man. J Clin Endocrinol Metab 64(6):1136–1141. https://doi.org/10.1210/jcem-64-6-1136

Vélez EJ, Unniappan S (2021) A comparative update on the neuroendocrine regulation of growth hormone in vertebrates. Front Endocrinol (Lausanne) 11:614981. https://doi.org/10.3389/fendo.2020.614981

Vélez EJ, Lutfi E, Azizi S, Perelló M, Salmerón C, Riera-Codina M, Ibarz A, Fernández-Borràs J, Blasco J, Capilla E (2017) Understanding fish muscle growth regulation to optimize aquaculture production. Aquaculture 467:28–40

Veyrat-Durebex C, Gaudreau P, Boghossian S, Alliot J (2001) Effects of peripheral and central administration of GHRH on feeding in aging LOU rats. Peptides 22(12):2119–2126. https://doi.org/10.1016/s0196-9781(01)00560-5

Wong AOL, Zhou H, Jiang Y, Ko WKW (2006) Feedback regulation of growth hormone synthesis and secretion in fish and the emerging concept of intrapituitary feedback loop. Comp Biochem Physiol A Mol Integr Physiol 144(3):284–305. https://doi.org/10.1016/j.cbpa.2005.11.021

Xiang P, Jing W, Lin Y, Liu Q, Shen J, Hu X, Chen J, Cai R, Hare JM, Zhu W, Schally AV, Yu H (2021) Improvement of cardiac and systemic function in old mice by agonist of growth hormone-releasing hormone. J Cell Physiol 236(12):8197–8207. https://doi.org/10.1002/jcp.30490

Xiang K, Yang Q, Liu M, Yang X, Li J, Hou Z, Wen H (2022) Crosstalk between growth and osmoregulation of GHRH-SST-GH-IGF axis in triploid rainbow trout (Oncorhynchus mykiss). Int J Mol Sci 23(15):8691. https://doi.org/10.3390/ijms23158691

Xu Z (2016) Subchapter 18B - growth hormone-releasing hormone. In: Takei Y, Ando H, Tsutsui K (eds) Handbook of hormones. Academic Press, San Diego, p 144-e18B-4

Ye R, Wang H-L, Zeng D-W, Chen T, Sun J-J, Xi Q-Y, Zhang Y-L (2021) GHRH expression plasmid improves osteoporosis and skin damage in aged mice. Growth Hormon IGF Res 60–61:101429. https://doi.org/10.1016/j.ghir.2021.101429

Zeitler P, Argente J, Chowen-Breed JA, Clifton DK, Steiner RA (1990) Growth hormone-releasing hormone messenger ribonucleic acid in the hypothalamus of the adult male rat is increased by testosterone. Endocrinology 127(3):1362–1368. https://doi.org/10.1210/endo-127-3-1362. PMID: 2117526

Zeitler P, Siriwardana G (2000) Stimulation of mitogen-activated protein kinase pathway in rat somatotrophs by growth hormone-releasing hormone. Endocrine 12(3):257–264. https://doi.org/10.1385/ENDO:12:3:257

Zhang J, Obál F, Zheng T, Fang J, Taishi P, Krueger JM (1999) Intrapreoptic microinjection of GHRH or its antagonist alters sleep in rats. J Neurosci 19(6):2187–2194. https://doi.org/10.1523/JNEUROSCI.19-06-02187.1999

Zhang X, Cui T, He J, Wang H, Cai R, Popovics P, Vidaurre I, Sha W, Schmid J, Ludwig B, Block NL, Bornstein SR, Schally AV (2015) Beneficial effects of growth hormone-releasing hormone agonists on rat INS-1 cells and on streptozotocin-induced NOD/SCID mice. Proc Natl Acad Sci USA 112(44):13651–13656. https://doi.org/10.1073/pnas.1518540112

Zhang C, Cai R, Lazerson A, Delcroix G, Wangpaichitr M, Mirsaeidi M, Griswold AJ, Schally AV, Jackson RM (2019) Growth hormone-releasing hormone receptor antagonist modulates lung inflammation and fibrosis due to bleomycin. Lung 197(5):541–549. https://doi.org/10.1007/s00408-019-00257-w

Zhang F, Icyuz M, Liu Z, Fitch M, Sun LY (2020) Insulin sensitivity in long-lived growth hormone-releasing hormone knockout mice. Aging (Albany NY) 12(18):18033–18051. https://doi.org/10.18632/aging.103588

Zhang M, Yang Q, Shi R, Wang J, Zhang Z, Yang Y, Li W, Chen S, Wang N (2022) Effects of long-term sex steroid hormones (estradiol and testosterone)–supplemented feeds on the growth performance of Chinese tongue sole (Cynoglossus semilaevis). Fish Physiol Biochem 48(5):1365–1375. https://doi.org/10.1007/s10695-022-01125-w

Zhu X, Zhou A, Dey A, Norrbom C, Carroll R, Zhang C, Laurent V, Lindberg I, Ugleholdt R, Holst JJ, Steiner DF (2002) Disruption of PC1/3 expression in mice causes dwarfism and multiple neuroendocrine peptide processing defects. Proc Natl Acad Sci USA 99(16):10293–10298. https://doi.org/10.1073/pnas.162352599. Epub 2002 Jul 26. PMID: 12145326; PMCID: PMC124907

Evolution of the Melanocortin System

9

Esther Leal, Borja Muriach,
and José Miguel Cerdá-Reverter

Abstract

The melanocortin system is probably the most complex neuroendocrine system known in vertebrates. Not only does it integrate different agonists encoded by a common precursor, the so-called proopiomelanocortin (Pomc), or non-Pomc agonists such as β-defensin, endogenous antagonists such as Asip (agouti-signalling protein) and Agrp (agouti-related protein) as well as five different receptors (Mcr), but also some accessory proteins that promote the functional expression or modify the pharmacological profile of Mcrs. The different duplications of the genome (R1, R2 and R3) together with some co-evolutionary and tandem duplication events have broadened the complexity of the system. According to this complexity, not only are evolutionary patterns controversial, but several hypotheses have attempted to explain the evolutionary relationships among members of the melanocortin system. In this review, we summarise the different evolutionary hypotheses given for the various melanocortin partners.

Keywords

Proopiomelanocortin (Pomc) Agouti-related protein (Agrp) · Agouti-signalling protein (Asip) · Melanocortin receptor (Mcr) · Melanocortin receptor accessory protein (Mrap)

E. Leal · B. Muriach · J. M. Cerdá-Reverter (✉)
Fish NeuroBehaviour Lab, Department of Fish Physiology and Biotechnology, Instituto de Acuicultura de Torre de la Sal, Consejo Superior de Investigaciones Científicas (IATS-CSIC), Ribera de Cabanes, Castellón, Spain
e-mail: jm.cerda.reverter@csic.es

V. Grinevich, R. F. Oliveira (eds.), *Evolutionary and Comparative Neuroendocrinology*, Masterclass in Neuroendocrinology 17,
https://doi.org/10.1007/978-3-031-80209-6_9

9.1 Introduction

The melanocortin system is perhaps the most complex endocrine system in vertebrates. The dominant player is proopiomelanocortin (POMC), a multiplex precursor that encodes several bioactive peptides (Nakanishi et al. 1979). These peptides are basically the melanocyte-stimulating hormones (MSHs), the adrenocorticotropic hormone (ACTH) and the opioid peptide β-endorphin, localised at the C-terminal end of the precursor. POMC structure involves different domains, each containing a core melanocortin sequence (HFRW) which identifies the different melanocyte-stimulating hormones (MSHs) of the precursor (Navarro et al. 2016). The N-terminal region/domain contains the pro-γ-MSH domain, the central domain includes α-MSH as the N-terminal sequence of the ACTH and the C-terminal domain incorporates the β-MSH but also contains an opioid core sequence (YGGF) of the β-endorphin (Fig. 9.1).

Not only is POMC one of the most abundant transcripts in the pituitary gland, but it is also expressed in two discrete groups of neurons in the mammalian encephalon, the arcuate nucleus of the hypothalamus and the nucleus of the solitary tract in the medulla (Bagnol et al. 1999). The post-transcriptional processing determines which peptides are released from the precursor, depending on the battery of prohormone convertases present in the tissue. For instance, in the corticotrophs of the anterior pituitary, proprotein convertase subtilisin/kexin type 1 (Pcsk1) activity releases

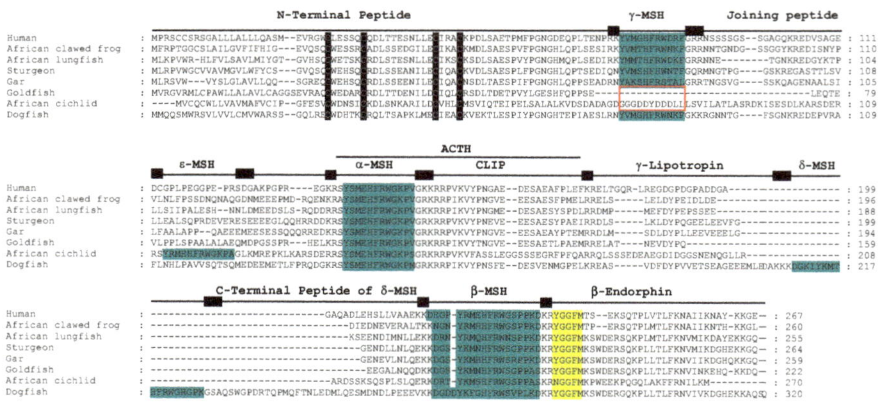

Fig. 9.1 Alignment of proopiomelanocortin (POMC) sequences from human, African clawed frog form B *(Xenopus laevis)* (Deen et al. 1991), African lungfish (*Protopterus annectens*) (Amemiya et al. 1999a), white sturgeon form B (*Acipenser transmontanus* (Amemiya et al. 1997), gar (*Lepisosteus osseus*) (Dores et al. 1997), goldfish (*Carassius auratus*) (Cerda-Reverter et al. 2003), African cichlid fish (*Haplochromis burtoni*) (Harris et al. 2013) and dogfish (*Squalus acanthias*) (Amemiya et al. 1999b). White letters on a black background indicate fully conserved cysteine residues in all POMC sequences. Green and yellow boxes show MSH peptides and endorphin core, respectively. The uncoloured boxed area demarcates the γ-MSH segment in species lacking this peptide. Black boxes indicate endoproteolytic cleavage sites. Adapted from Cerda-Reverter et al. (2003)

pro-γ-MSH, ACTH and β-lipotropin, whereas cleavage by Pcsk1 and Pcsk2 in the melanotrophs of the *pars intermedia* produces α-MSH and β-endorphin (Castro and Morrison 1997).

Melanocortin peptides signal through a set of five G-coupled receptors which are positively linked to the production of cAMP. Melanocortin receptors (Mcrs) mediate a wide range of critical physiological functions including the regulation of energy balance, skin/fur pigmentation and stress response. The signalling and binding profile of these receptors is fine-tuned by the melanocortin receptor accessory proteins (Mraps), a family of transmembrane proteins that physically interact with the receptors to promote their trafficking to the membrane and/or regulate sensitivity to the melanocortin agonist (Cerdá-Reverter et al. 2011).

The melanocortin system is the only endocrine system regulated by the binding of endogenous antagonists, agouti-signalling protein (Asip) and agouti-related protein (Agrp). Asip is a competitive antagonist of α-MSH at Mc1r, to modulate pigment synthesis and the fate of pigment cell precursors. Agrp binds Mc3r and Mc4r to regulate energy balance (Cerdá-Reverter et al. 2011). To further complicate matters, some of both Mc1r and Mc4r are constitutively activated, which means that these receptors are capable of signalling in the absence of agonists, and the binding of Asip and Agrp to Mc1r and Mc4r respectively can block such constitutive activity, in the same way as inverse agonists can (Cone 2006). This is the complexity that the evolutionary process has come up against, yet this intricacy also allows for extremely fine regulation of the system which, in turn, gives an idea of the importance of the melanocortin system in the survival of vertebrates.

In addition, some non-Pomc melanocortin agonists are responsible for different pigmentary patterns in mammals. For instance, β-defensin 103 (CBD103) has been shown to have a high affinity for Mc1r (Candille et al. 2007). Furthermore, mutations in mahogany (attractin, *Atrn*) and mahoganoid (mahogunin ring finger 1, *Mgrn1*) rescue the pigmentation abnormalities and obese phenotype of agouti mice, acting downstream of Asip and upstream or at the level of Mc1r/Mc4r. Both Atrn and Mgrn1 regulate Mc4r degradation in the lysosome, and mutations in these genes promote receptor trafficking back to the membrane (Overton and Leibel 2011).

The *phylum* Chordata can be split into three main branches, which have evolved over the last 500 million years: the protochordates, or animals without an internal skeleton, the agnathans, or bilaterians with internal reinforce but without a hinged jaw and the gnathostomes, or bilateral organisms with an internal cartilaginous or bony support and hinged jaws. The melanocortin system is present only in agnathans and gnathostomes, but no traces of melanocortins have been found in earlier evolutionary stages.

9.2 POMC

9.2.1 The Opioid Family

POMC belongs to the opioid/orphanin gene family, in which the peptide precursors have either an opioid core sequence (YGGF) or orphanin/nociceptin (FGGF) core sequences (Dores et al. 2002). Proenkephalin (PENK), prodynorphin (PDYN), and nociceptin/proorphanin (PNOC), in addition to POMC, are included in this family (Danielson and Dores 1999). Apart from the core motifs, the members differ considerably among each other and between species, yet they all have six cysteine residues at the N-terminal region of the molecule (PENK, PDYN and PNOC), although only four are found in POMC. In addition, every family gene has a single intron just after the region encoding the signal peptide (Larhammar et al. 2015). Such common structural features suggest a common ancestral opioid gene that probably emerged early during the chordate evolution. No related genes have been found in the genome of cephalochordates (amphioxus and tunicates). Subsequently, the opioid/orphanin family expanded in parallel with the duplication rounds of the vertebrate genome (1R and 2R), and the teleost specific gene duplication (TSGD or R3) (Sundström et al. 2008).

In addition to gene homology and structure analyses, synteny studies have shown that all four opioid genes are located on three chromosomes. In the human genome, PDYN and POMC are located on chromosomes 20 and 2, respectively (Fig. 9.2), while chromosome 8 carries both the PENK and PNOC genes. In the chicken genome, the PDYN gene is also located on chromosome 20, PENK is located on chromosome 2, but PNOC is linked to POMC on chromosome 3. PNOC is therefore linked to POMC on chromosome 3 and not to PENK, as is the case on human chromosome 8. POMC and PNOC share the same chromosome in the opossum (*Monodelphis domestica*) as well as in all teleost fish species studied (Sundström et al. 2010; Larhammar et al. 2015). In the soft-shell turtle (*Pelodiscus sinensis*), both genes are only 1 Mb apart and share the same chromosome in spotted gar (*Lepisosteus oculatus*), a basal non-teleost ray-finned fish. This species is exceptional in studies of vertebrate genome evolution since it has fewer chromosomal rearrangements than the teleost fish and many mammals, including humans. The close localisation of POMC and PNOC in most species suggests that the origin was the result of local gene duplication followed by fusion/integration of an ACTH/MSH gene. The alternative grouping of the opioid gene family in the human genome could be the result of a gene translocation event that took place in the antecessor of placental mammals. In summary, the expansion of the opioid peptide system seems to be the result of two complete genome duplications (1R and 2R) and a single local duplication. There are several possible scenarios that differ in the timing of the local duplication which generated POMC and PNOC that could have taken place before 1R (scenario 1), after 1R but before 2R (scenario 2), or after 2R (scenario 3) (Fig. 9.2) (Navarro et al. 2016).

Agnathans, such as lampreys, also have melanocortin peptides, but these are encoded by two separate genes, proopiocortin (POC) and proopiomelanotropin

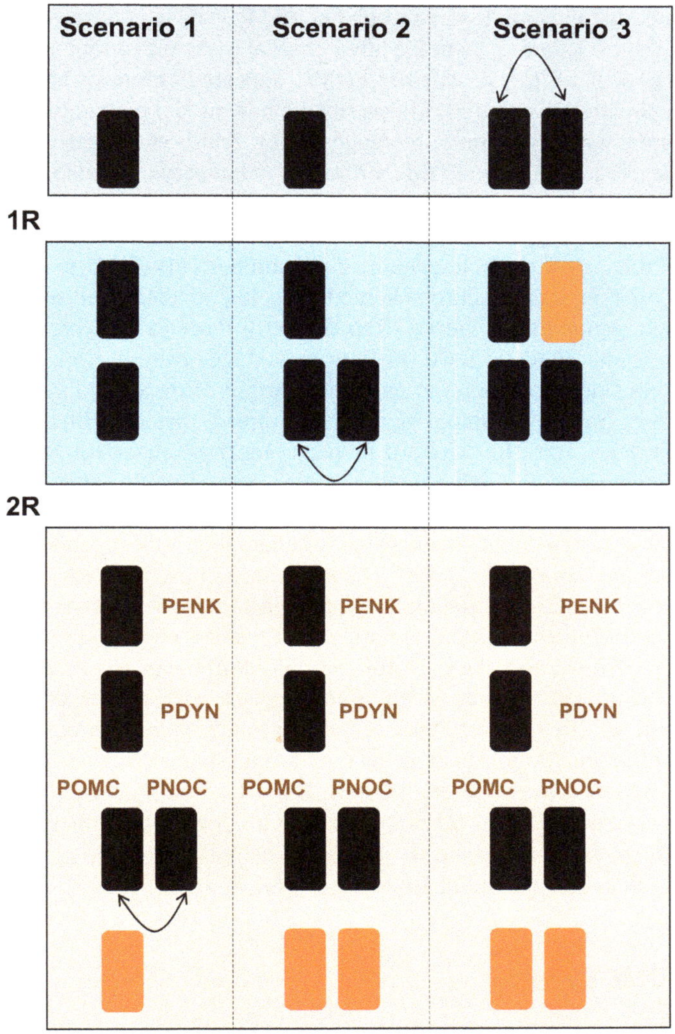

Fig. 9.2 Evolutionary history for the opioid peptide and nociceptin/orphanin family by a combination of whole-genome and local duplications. Three different scenarios have been proposed to explain the timing of the duplication events that have taken place in the common ancestor leading to POMC and PNOC. Local duplication, indicated by arrows, generating both POMC and PNOC could have taken place after the 2R (scenario 1), after the 1R (scenario 2) or prior to the 1R (scenario 3). POMC, proopiomelanocortin; PENK, proenkephalin; PDYN, prodynorphin; PNOC, nociceptin/proorphanin. The figure is an adapted version of the figure published in Sundström et al. (2010) and Navarro et al. (2016)

(POM). POC encodes ACTH, a β-MSH-related peptide and a β-endorphin sequence, whereas POM contains a MSH-B (related to β-MSH), a MSH-A (related to α-MSH) and a β-endorphin sequence (Takahashi et al. 1995a, b). Since lampreys have POMC sequences, the duplication leading to PNOC and POMC may have occurred before the split of the cyclostomes from the other chordates. As the timing of the 2R is still unclear, it is difficult to say whether POMC appeared before or after the second genome duplication. Dores (2013) described a new model that admits lampreys as 2R organisms, which was initially suggested by Smith et al. (2013). This model supports the existence of two POMC orthologs in lampreys, but also anticipates the existence of uncharacterised PENK and PDYN genes in lamprey. The genome sequencing project suggests that genome-wide patterns of conserved synteny could have arisen through two whole-genome duplication events (2R hypothesis) accompanied by large numbers of chromosome losses, but an alternative scenario involving one whole-genome duplication (1R) preceded by three distinct chromosome-scale duplication events requires fewer evolutionary stages (Smith et al. 2018). In any case, such considerations point to an ancient origin of the opioid system in vertebrates, with every human opioid gene having already been established more than 500 million years ago. But why not perhaps older? Searches for opioid genes in invertebrate genomes, including deuterostomes, have failed to identify opioid precursors or receptors (reviewed in Larhammar et al. 2015).

It is widely accepted that teleost fish have undergone a third genome duplication (3R) (Meyer and Van de Peer 2005). The gene copies have been neo-functionalised (by the new copy), sub-functionalised (both copies share the function of the original copy) or pseudo-functionalised (the sequence of the new copy degenerates and loses its function). This is also the case for the opioid/orphanin gene family, as Pomc duplicates have been reported in all teleost species with sequenced genome (see Sundström et al. 2010 for references) and sub-functionalisation has been demonstrated in Tetraodon (de Souza et al. 2005). In some species, such as barfin flounder (*Verasper moseri*) (Takahashi et al. 2005), Burton's mouthbrooder (*Haplochromis burtoni*) or medaka (*Oryzias latipes*) (Harris et al. 2013), three distinct genes have been found, yet these appear to be the result of lineage-specific gene duplication events other than genome duplication (Sundström et al. 2010).

9.2.2 POMC Bauplan

Obviously, POMC is not a conventional opioid precursor since it is the only precursor in the family that contains melanocortin sequences in its primary structure. The presence of these "foreign" sequences makes POMC the most complex precursor in the opioid family. Such complexity probably arose when a DNA segment encoding an MSH core sequence was inserted into the endorphin gene, since both the N-terminal region and carboxyterminal peptide maintain structural/sequence identities with the other three opioid prepropeptides. Therefore, not only does POMC retain four out of the six cysteine residues which characterise opioid precursors, but also a C-terminal endorphin core (YGGF). Alternatively, a cumulative process of successive mutations in the ancestral gene could lead to multiple melanocortin cores/sequences as a result of unequal cross-over events (Dores and Baron 2011).

One of the most intriguing questions in the POMC evolution is when this "alien" fragment encoding the melanocortin was potentially inserted into the endorphin gene. Sundström et al. (2010) provided evidence that this segment was introduced following the two vertebrate tetraploidisations (R1 and R2), when a distinct prepro-endorphin gene arose. However, it cannot be excluded that a gene encoding ACTH/melanocortin arose independently much earlier, as genes encoding melanocortin receptors have also been suggested to have arisen because of tetraploidisation from a single gene (Cortés et al. 2014).

When analysing the primary structure of POMC throughout the evolutionary process, it can be observed that the structural plan varies between species, and these modifications mainly concern the number of MSH core sequences or domains. Every tetrapod species has three MSH core sequences (γ-MSH, α-MSH and β-MSH), suggesting an intragenic duplication of the ancestral MSH gene. Lobe-finned or sarcopterygian fish have an identical structural plan with the same MSH domains, yet elasmobranch fish have an additional MSH domain called δ-MSH. Some cichlids and pomacentrids have an alternative domain called ε-MSH, which may have arisen from a duplication of the ACTH domain (see Navarro et al. 2016; Rocha et al. 2019 for reviews). The presence of the same structural plan in tetrapods and lobe-finned fish suggests that the ancestral gnathostome had a three-domain structure that diverged in cartilaginous, ray and lobe-finned fish. Today, three different plans meet in time. The ancestral organisation (γ-MSH, α-MSH and β-MSH) has been preserved in lobe-finned fish (lungfishes and coelacanths) and subsequently in tetrapod species. The second organisation plan is unique to cartilaginous fish, which have added a fourth MSH peptide (δ-MSH). This fourth domain is located between the ACTH and β-MSH domains and is thought to have evolved from a duplication of the C-terminal β-MSH/β-endorphin segment. The presence of this structure in elasmobranch and holocephalan species suggests that the duplication occurred early after the divergence of the cartilaginous lineage. The third organisational plan is found in ray-finned fish (Actinopterygii), which currently include Polypteriformes (e.g. bichirs), Acipenseriformes (e.g. sturgeons and paddlefish), Lepisosteiformes (e.g. spotted gar), Amiiformes (bowfin fishes) and teleosts (Fig. 9.3). Most ray-finned fish exhibit changes in the γ-MSH domain and the zenith is reached in teleost fish, where complete deletion of the domain is observed. The loss of the γ-MSH domain was most likely progressive throughout the evolutionary process (Dores and Baron 2011). Indeed, basal Actinopterygii still retain the ancestral building instruction, i.e. γ-MSH, α-MSH/ACTH and β-MSH/β-endorphin, but substitutions in the dibasic pairs flanking the MSH sequence make peptide processing and release unlikely. These N- and C-terminal cleavage motifs are clearly absent from the precursors of Polypteriformes and Acipenseriformes, once more making peptide processing unfeasible. Moreover, in some species, such as sturgeon, Pomc has mutations in the core sequence HFRW of γ-MSH, which is critical for binding to melanocortin receptors. All this progressive degeneration of the γ-MSH domain in basal ray-finned fish eventually led to the complete deletion of this domain, even affecting the spacer region between γ-MSH and α-MSH/ACTH in

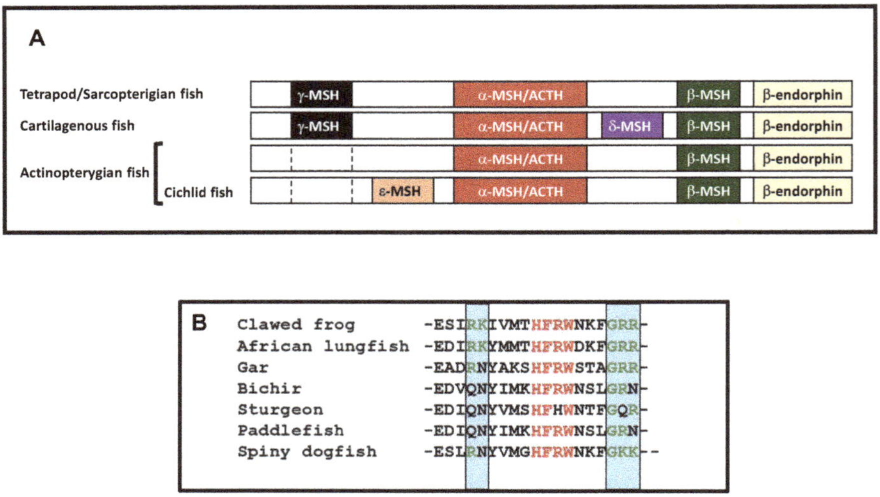

Fig. 9.3 (**a**) Structural plans of POMC precursor in main vertebrate groups. Both tetrapod and lobe-finned fish share the same POMC building plan. Cartilaginous fish exhibit an additional MSH peptide called δ-MSH whereas Actinopterygian fish lack γ-MSH peptide. Some teleost fish have an additional ε-MSH peptide as a putative tandem duplication of the α-MSH-ACTH segment. (**b**) Comparison of γ-MSH sequences showing the potential degenerative evolutionary process leading to the loss of γ-MSH peptide in teleost fish. The γ-MSH sequence and the flanked proposed endo-proteolytic cleavage sites for tetrapod, lobe-finned fish (lungfish), cartilaginous (spiny dogfish) and four ray-finned fish are presented. The N-terminal cleavage site (**a**) and the C-terminal cleavage and α-amidation site (**b**) are boxed. An arginine (R)-lysine (K) dibasic pair flanks tetrapod and lobe-finned fish γ-MSH peptide. An arginine (R) residue is still present in gar but lysine is replaced by an asparagine (N) residue, a common substitution in many fish species including elasmobranchs. It is uncertain whether this dibasic pair (RN) can promote γ-MSH cleavage, yet this motif is completely absent in Polypteriformes and Acipenseriformes precursors, which makes γ-MSH processing arguable. In addition, the C-terminal cleavage motif (RR or KK) is also absent in bichir, sturgeon and paddlefish. Moreover, sturgeon precursor carries a substitution in the MSH core sequence compromising the activity of the peptide. Adapted from Navarro et al. (2016)

teleost fish (Figs. 9.1 and 9.3). However, the massive deletion in the N-terminal region of the teleost precursors does not affect peptide processing in the pituitary.

9.3 Melanocortin Receptors

The presence of melanocortin and opioid peptides in the POMC precursors implies that it has co-evolved with the constraints imposed by different receptor families, i.e. melanocortin and opioid receptors. Melanocortin receptors are members of the rhodopsin family of seven-pass transmembrane G protein-coupled receptors. The genome of bony gnathostome has five paralogue receptors (MC1R-MC5R), which show different degrees of selectivity for melanocortin peptides. MC2R is the only receptor that exclusively binds ACTH, and it is not activated by the different MSHs, although it requires interaction with MRAP1 for its functional expression (Agulleiro et al. 2010). The remaining MCRs bind MSH peptides with different affinities but can also be activated by ACTH. MC1R seems to bind α-MSH more efficiently, whereas MC4R and MC3R seem to be more efficient in binding β-MSH and γ-MSH, respectively (Schiöth et al. 2005). Such receptors also have different expression domains according to their specific functions. MC1R is expressed mainly in the skin, to control melanogenesis and thus skin/fur colour. MC2R is expressed mainly in the adrenal cortex and binds pituitary ACTH to stimulate glucocorticoid secretion during stress response. MC3R and MC4R are expressed mainly in the encephalon, to regulate energy balance and thus body weight.

Like POMC, the origin of melanocortin receptors appears to be rooted in the early divergence of the chordate phylum, as their genes have been annotated in the genomes of hagfish, lamprey, elasmobranch, teleost and tetrapod species, yet they are absent from the genomes of echinoderms, cephalocordates and urochordates as well as protostomes (Västermark and Schiöth 2011). However, the number of receptors, binding profiles and functional expression vary among fish species. All five receptors have been found in elasmobranch genomes but no complete set has been characterised in a single species. MC1R–MC3R have been characterised in elephant shark (*Callorhynchus milii*, Västermark and Schiöth 2011) whereas MC3–MC5R have been cloned in the spiny dogfish (*Scualus acanthias*, Ringholm et al. 2003; Klovins et al. 2004). Our recent exploration of the complete sequenced genome of the elephant shark (http://www.ensembl.org/index.html) shows the presence of MC4R and MC5R, suggesting that the full set is present in elasmobranch species (unpublished data). The number of receptors also differs in teleost fish. Zebrafish have six MCRs, with two copies of melanocortin MC5R, whereas pufferfish have only four, with no melanocortin MC3R and only one copy of melanocortin MC5R (Logan et al. 2003).

The evolutionary origin of the melanocortin receptor family is also linked to the rounds of genome duplication, as is the opioid gene family, but the origin of MC5R is controversial. According to the 2R script, the ancestral gnathostome genome should have four melanocortin receptor paralogs, so it is plausible that the fifth receptor arose from a local duplication event. Synteny studies in the genomes of different species have revealed that MC1R, MC2R, MC3R and MC4R are localised in different chromosomes, whereas MC5R is placed in the same chromosome as MC2R but in the opposite direction. This suggests that MC1R, MC3R and MC4R could be the result of two rounds of duplication (R) whereas MC2R and MC5R arose from the local duplication event of a common antecessor (MC2R/MC5R). As elasmobranch fish have both MC2R and MC5R, the local duplication event could take place in the ancestral gnathostome. In support of this idea, it has been demonstrated that melanocortin receptors, with the exception of a spliced variant of MC1R, lack introns. However, Fugu MC2R and MC5R have one and three introns, respectively. The position of the MC2R intron in the DRY motif is conserved with the position of one of the introns in MC5R, suggesting a close phylogenetic relationship between the two receptors. Following this scheme, Baron et al. (2009) proposed an evolutionary tree in which the oldest known versions of MCRs are found in lampreys, i.e. MCRa and MCRb (Haitina et al. 2007). Such receptors represent two branches of the ancestral MCR duplication that may have occurred during 1R. MCRa then duplicated into MC1R and the ancestral MC2R/MC5R, whereas MCRb split into MC3R and MC4R (Fig. 9.4a). However, Västermark and Schiöth (2011) challenged this hypothesis by using the latest phylogenetic approaches to show that MC4R and MC5R are the most closely related, while MC2R and MC5R are the most divergent paralogues. Thus, the phylogenetic analysis cluster together MC4R and MC5R in addition to MC3R and lamprey MCRb. On the other hand, MC1R and MC2R cluster together with lamprey MCRa as a basal sequence. These phylogenetic relationships have led the authors to reconsider the hypothesis of MCR family evolution, as shown in Fig. 9.4b. In this scheme, the sequences of the cyclostome/gnathostome ancestor gave rise to the two main evolutionary branches of MCRs after 1R. MCRa was then duplicated into MC1R and MC2R, while MCRb split into MC3R and the ancestral version of MC4R/MC5R, which later split into MC4R and MC5R. Therefore, this hypothesis supports the idea that MC5R arose because of a local duplication of MC4R, probably in the ancestral gnathostome, and subsequently the MC5R locus was moved to the chromosome carrying the MC2R locus (Fig. 9.4b).

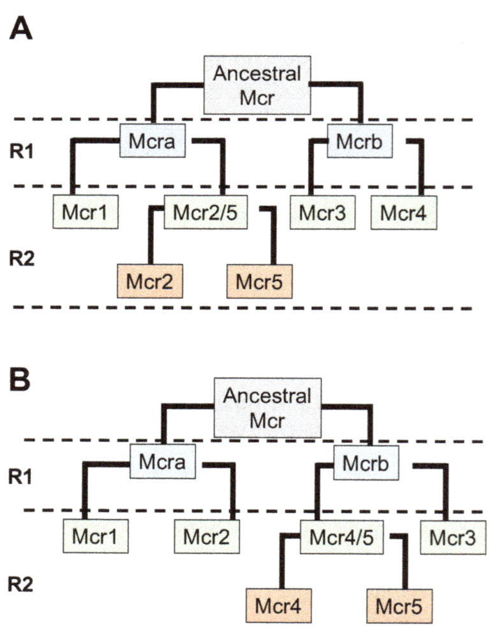

Fig. 9.4 Evolutionary schemes showing alternative hypotheses for the evolution of melanocortin receptors. (**a**) Evolutionary hypothesis and nomenclature proposed by Baron et al. (2009). (**b**) Evolutionary hypothesis and nomenclature proposed by Västermark and Schiöth (2011). See text for details

9.4 Melanocortin Receptor Accessory Proteins (MRAPs)

Early pharmacological experiments with Mcrs showed that Mc2r never reaches functional expression when expressed in conventional cell lines such as HEK293 or CHO (Schimmer et al. 1995). However, when the receptor is expressed in cell lines derived from the adrenal gland, ACTH can promote receptor signalling (Noon et al. 2002). The reason for this dysfunction is that the receptor cannot reach the plasma membrane when expressed in normal cell lines, suggesting that the receptor needs a "transporter or accessory protein" to deliver it to the membrane. A candidate gene, highly expressed in the adrenal gland, was identified by homozygosity mapping, using a single nucleotide polymorphism (SNP) microarray in individuals showing familial glucocorticoid deficiency (FGD) with no MC2R mutations (Metherell et al. 2005). This rare autosomal disorder results in resistance to ACTH, i.e. high plasma levels of pituitary hormones but no adrenal glucocorticoids, thus untreated patients can die from hypoglycaemia or recurrent infections. In fact, this gene had previously been cloned and shown to be involved in the differentiation process of 3T3-L1 cells, from fibroblasts to adipocytes, and named adipose tissue-specific low molecular weight protein (Xu et al. 2002). This protein was subsequently renamed melanocortin 2 receptor accessory protein (Mrap). Identity screening of the human genome using Mrap revealed another gene (*C6orf117*) encoding a small, single-transmembrane domain protein that is 27% identical to human *Mrap*. This suggested the presence of an additional *Mrap* in the human genome (Metherell et al. 2005), which was eventually identified and named *Mrap2* (Chan et al. 2009; Sebag and Hinkle 2009). This has caused some confusion in the nomenclature, as most authors continue to refer to the first characterised protein as Mrap rather than Mrap1.

Our preliminary experiments expressing piscine *mc2r* in HEK-293 cells failed to generate an ACTH-stimulated cAMP response. However, when co-transfected with human *Mrap1*, fish *mc2r* was able to mediate ACTH-induced cAMP production, suggesting the presence of a conserved Mc2r trafficking system in non-mammalian vertebrates. Taking advantage of the genomic facilities available for zebrafish, we identified three distinct fish accessory proteins, Mrap1, Mrap2a and Mrap2b, by identity screening. We also found Mrap orthologs in most fish genomes, but only zebrafish had an *mrap2* paralogue (Agulleiro et al. 2010; Cerdá-Reverter et al. 2011; Cortés et al. 2014). In fact, *mrap* genes have been described in most vertebrate species, including sea lamprey (*Petromizum marinus*), which belongs to agnatha, one of the most ancient lineages of vertebrates (Västermark and Schiöth 2011). In a major study, Valsalan et al. (2013) examined more than 40 genomes to identify *mrap1* and *mrap2* genes, and used various molecular markers, synteny studies and *dN/dS* values to analyse the overall evolution of the *mrap* gene family. Only one ambiguous homologue of an Mrap2-like protein was identified in Agnata (diverged from other vertebrates about 530 Mya). The elephant shark (460 Mya) has both genes, i.e. *mrap1* and *mrap2* (Dores 2016), suggesting that both paralogs may have been present in the genome of the ancestral gnathostome and that the duplication event producing separate genes occurred before the divergence of the elephant shark from other jawed vertebrates. The presence of an *mrap1* ortholog in the genome of a cartilaginous fish further suggest that the *mrap1* and the *mrap2* were the result of the 2R genome duplication event (Dores and Garcia 2015; Dores 2016).

Lamprey Mrap was found to share sequence features with Mrap2, confirming previous suggestions (Västermark and Schiöth 2011) that Mrap2 is likely an ancient branch in the evolutionary history of the Mrap family, but the existence of additional lamprey *mrap* orthologue cannot be completely ruled out (Dores 2016). As previously shown, the functional expression of MC2R requires the physical interaction with Mrap1 to reach plasma membrane and bind ACTH, yet studies in cartilaginous fish have demonstrated that the type 2 receptor can signal in the absence of any Mrap (Reinick et al. 2012; Takahashi et al. 2016). This question opens new evolutionary incognita on the physiological roles and evolutionary perspectives of the *mrap* gene family that only further research on the melanocortin system can solve.

An agreeable hypothesis by Robert Dores' group (Dores and Garcia 2015; Dores 2016; Dores and Chapa 2021) proposed an interactive coevolution between Mc2r and Mrap systems. They proposed that the Mc2r-independence of Mrap system is an ancestral trait, i.e. an ancient feature which all gnathostome and elephant shark have retained However, at some evolutionary point, potentially at the time of ancestral bony fish, Mc2r was unable to reach the plasma membrane, thus losing its functionality, yet the Mrap system "rescued" the functional expression of the receptor by trafficking it from the endoplasmic reticulum to the membrane and thus restoring the hypothalamic-pituitary-adrenal/interrenal axis in terms of ACTH response and glucocorticoids synthesis.

9.5 Endogenous Antagonist

Atypically, melanocortin receptor signalling is also modulated by an endogenous antagonist. It is the only endocrine system known to exhibit such fine regulation by endogenous antagonists. Agouti signalling and agouti-related protein are structurally related proteins that share a cysteine-rich C-terminal domain consisting of ten spatially conserved cysteine residues that form five disulphide bridges essential for their structural stability and biological function (Cortés et al. 2014). This cysteine pattern is similar to that of conotoxins and agatoxins, suggesting that agouti-like proteins adopt an inhibitory cysteine knot (ICK) fold. The N-terminal region of Asip is filled by a basic domain with a high proportion of arginine and lysine residues. This leads to a proline-rich region immediately preceding the cysteine-rich C-terminal domain. The integrity of this basic domain is also essential for the full activity of the Asip protein (Miltenberger et al. 1999, 2002). Agrp sequences lack both the basic and proline-rich domains, but have a processing site before the cysteine domain where the propeptide is cleaved (Creemers et al. 2006) (Fig. 9.5).

Tetrapod species have a single copy of the *asip* and *agrp* genes, but the teleost genome retains four paralogues, originally named *asip1*, *asip2*, *agrp1* and *agrp2* (Kurokawa et al. 2006). Kurokawa and colleagues suggested that *asip2* and *agrp2* are ohnolog genes (duplicated genes resulting from genome duplication) of *asip1* and *agrp1*, respectively, generated during teleost-specific genome duplication (3R).

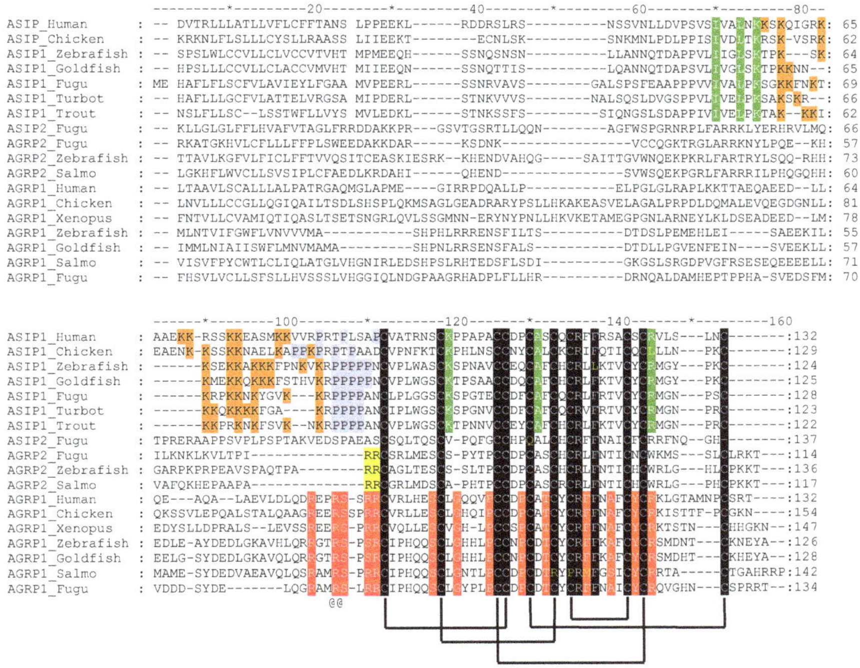

Fig. 9.5 Alignment of agouti-signalling protein (Asip) and agouti-related protein Agrp) amino acid sequences. Dashes were introduced to improve alignment. Black boxes show amino acid residues conserved in all sequences. Green boxes show residues conserved only in Asip1 sequences. Red boxes indicate residues conserved only in Agrp1 sequences. Orange boxes indicate basic residues before proline domains in Asip1 sequences. Purple boxes indicate proline residues before cysteine domains in Asip1 sequences. Lines joining cysteine residues indicate putative disulphide bonds formed in the cysteine domain. Alignment positions 106–107 marked with @ are conserved for Agrp1 post-transcriptional processing. Yellow boxes indicate dibasic pairs conserved in Agrp2 sequences. Yellow letters indicate exceptions in the Agrp1 sequence in salmonids (positions 133, 135 and 137) and Asip2 sequences in Fugu (position 130)

Both Agrp1 and Asip1 share a cystine knot with the structure C-x(6)-C-x(6)-C, whereas Agrp2 and Asip2 exhibit a shortened version of the cystine knot C-x(6)-C-x(5)-C and were originally named A2 peptides (Klovins and Schiöth 2005). Posterior synteny studies challenged Kurokawa's hypothesis (Braasch and Postlethwait 2011), since the *agrp2* chromosomal region does not share a synteny relationship with fish *agrp1* or with tetrapod *agrp*. The *agrp2* and *asip2* regions are syntenic with a region of human chromosome 8, which, shares paralogism with the *asip* region on chromosome 20. Braasch/Postlethwait's model maintains that the endogenous antagonist precursor was duplicated twice during early genome duplication rounds (R1, R2). Both *agrp2* and *asip2* were lost in the tetrapod genome but *asip2* was retained in the teleost genome. After the TSGD (R3), the additional copy of *agrp* gene was also lost from the teleost genome yet both copies of the ASIP2 gene were retained. These copies are named *asip2* and *agrp2* (Kurokawa et al. 2006) but the new model suggests calling them *asip*2a and *asip*2b, respectively (Braasch and Postlethwait 2011). The reconstruction of the phylogeny by Schiöth and colleagues (2011) using an elephant shark sequence seems to support Kurokawa's hypothesis. When AGRP is used to root the phylogenetic analysis, the results support Braasch and Postlehwait's hypothesis, yet when the tree is rooted by the shark sequence, the AGRP2 and ASIP2 clusters group with the AGRP cluster, supporting Kurokawa's nomenclature. The potential evolutionary hypothesis is summarised in Fig. 9.6.

Structural and functional data may shed some light on the evolutionary controversy. Agrp peptides are processed by a well conserved dibasic motif preceding the cysteine domain (see Fig. 9.5). This dibasic pair is also conserved in Agrp2 sequences, suggesting that Agrp2 peptides are also processed to release the active

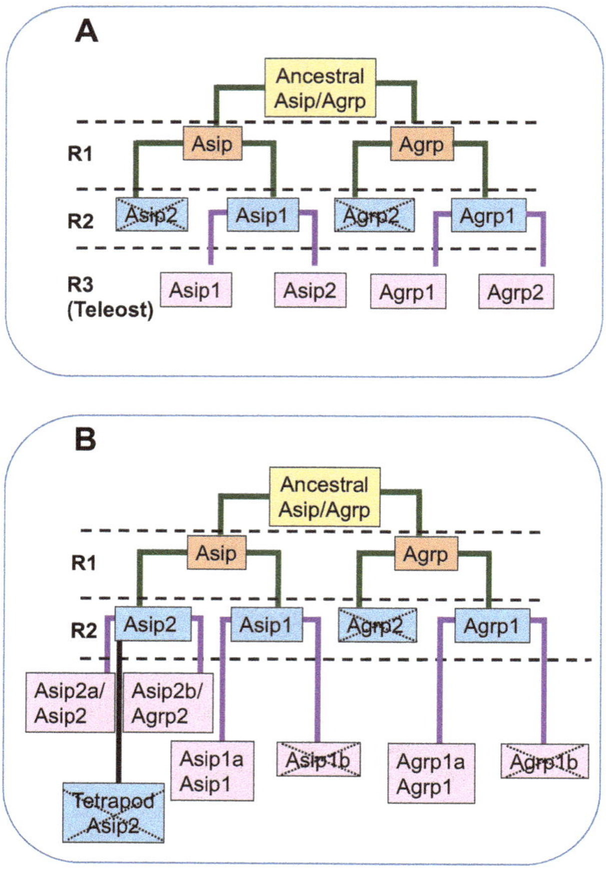

Fig. 9.6 Evolutionary schemes showing alternative hypotheses for the evolution of melanocortin endogenous antagonists. (**a**) Evolutionary hypothesis and nomenclature supported by Kurokawa et al. (2006). (**b**) Evolutionary hypothesis and nomenclature proposed by Braasch and Postlethwait (2011). See text for details. Dashed lines indicate the third round of genome duplication in teleost fish

C-terminal peptide. This motif is not present in Asip1 and Asip2 peptides. In addition, all Agrp1 peptides have a C-terminal extension after the last cysteine residue, which is absent in Asip1 and Asip2 sequences but present in Agrp2 peptides. This short tail does not appear to confer any binding property to the molecule since the last 12 amino acids of the human Agrp peptide can be removed without affecting Mc3r or Mc4r binding (Jackson et al. 2002). Therefore, the function of this conserved short C-terminal extension in all Agrp1 and Agrp2 sequences remains unknown. It is known that agouti peptides can interact with molecules other than melanocortin receptors (He et al. 2001), suggesting, at this point, an undiscovered intermolecular interaction mediated by segments outside the melanocortin receptor.

Asip1 sequences also show a proline-rich region just before the cysteine domain and a basic region in the N-terminal part of the peptide. Proline and basic regions seem to be characteristic of Asip1 peptides, since no other peptides of the family show such features. In summary, both evolutionary scenarios are theoretically possible, but Braasch's hypothesis is less parsimonious and less structurally supported than Västermark's proposal.

A number of invertebrate agouti-like sequences from the phylum Arthropoda and Ascomycota have also been annotated (Västermark et al. 2012). Some of the arthropod sequences retain the C-x(6)-C-C structure of the agouti-like sequences, are similar in length to the agouti peptides (130 amino acids), and have similar exon-intron structures to the agouti-like sequences, supporting a common origin. It is unlikely that these peptides act by binding to the MCRs, as the first evidence for these receptors was found in agnatha (Haitina et al. 2007). Phylogenetic analyses show no relationship between the arthropod sequences or any branch of the agouti-like sequences (Agrp, Asip or A2), suggesting that these sequences arose by horizontal gene transfer, but this remains to be demonstrated (Västermark et al. 2012).

9.6 Conclusions

The melanocortin system is a complex vertebrate hormone system which most likely evolved before the appearance of jawless vertebrates, more than 500 M years ago. The system diversified during the two rounds of genome duplication. Several evolutionary processes are represented within this family, including internal tandem duplications within the *Pomc* gene to generate up to four different Msh sequence cores, or local duplication to generate two different receptors, *mc2r*/*mc5r* and *mc4r*/*mc5r*, according to the Dores and Schiöth hypotheses, respectively. New intron acquisition occurred by reverse splicing and, indeed, gene loss processes after genome duplication. The evolution of the family is also linked to putative co-evolutionary processes, agreeably represented by the absence of *mc3r* and the major receptor ligand γ-Msh in teleost fish and the possible rescue of Mc2r function by the functional divergence of the *mrap2* copy, now named *mrap1*. Recently, we have shown that the Mc4r/Mrap2 interaction confers Acth sensitivity to the receptor. We also believe that these interactive processes have evolutionary significance, as the receptor/accessory protein interaction could provide new mechanisms for GPCRs to modify their binding spectrum and also lead to the sensitivity of new target tissues to the peptide hormones.

Acknowledgements This study was financed by funds from MCIN/AEI/10.13039/501100011033 (PID2022-136288OB-C33 to J.M.C.-R.) and, as appropriate, by "ERDF A way of making Europe", by the "European Union" or by the "European Union Next Generation EU/PRTR".

Conflicts of Interest The authors declare that the research was conducted without any commercial or financial relationships that could be construed as a potential conflict of interest.

References

Agulleiro MJ, Roy S, Sánchez E, Puchol S, Gallo-Payet N, Cerdá-Reverter JM (2010) Role of melanocortin receptor accessory proteins in the function of zebrafish melanocortin receptor type 2. Mol Cell Endocrinol 320:145–152

Amemiya Y, Takahashi A, Dores RM, Kawauchi H (1997) Sturgeon proopiomelanocortin has a remnant of gammamelanotropin. Biochem Biophys Res Commun 230(2):452–456. https://doi.org/10.1006/bbrc.1996.5987

Amemiya Y, Takahashi A, Meguro H, Kawauchi H (1999a) Molecular cloning of lungfish proopiomelanocortin cDNA. Gen Comp Endocrinol 115(3):415–421. https://doi.org/10.1006/gcen.1999.7327

Amemiya Y, Takahashi A, Suzuki N, Sasayama Y, Kawauchi H (1999b) A newly characterized melanotropin in proopiomelanocortin in pituitaries of an elasmobranch, Squalus acanthias. Gen Comp Endocrinol 114(3):387–395. https://doi.org/10.1006/gcen.1999.7256

Bagnol D, Lu XY, Kaelin CB, Day HE, Ollmann M, Gantz I, Akil H, Barsh GS (1999) Watson SJ Anatomy of an endogenous antagonist: relationship between Agouti-related protein and proopiomelanocortin in brain. J Neurosci 15;19(18):RC26. https://doi.org/10.1523/JNEUROSCI.19-18-j0004.1999

Baron A, Veo K, Angleson J, Dores RM (2009) Modeling the evolution of the MC2R and MC5R genes: studies on the cartilaginous fish, *Heterondotus francisci*. Gen Comp Endocrinol 161:13–19

Braasch I, Postlethwait JH (2011) The teleost agouti-related protein 2 gene is an ohnolog gone missing from the tetrapod genome. Proc Natl Acad Sci 108:E47–E48

Candille SI, Kaelin CB, Cattanach BM, Yu B, Thompson DA, Nix MA, Kerns JA, Schmutz SM, Millhauser GL, Barsh GS (2007) A β-defensin mutation causes black coat color in domestic dogs. Science 318:1418–1423

Castro MG, Morrison E (1997) Post-translational processing of proopiomelanocortin in the pituitary and the brain. Crit Rev Neurobiol 11:35–57

Cerdá-Reverter JM, Schiöth HB, Peter RE (2003) The central melanocortin system regulates food intake in goldfish. Regul Pept 115(2):101–113. https://doi.org/10.1016/s0167-0115(03)00144-7

Cerdá-Reverter JM, Agulleiro MJ, Sánchez E, Guillot R (2011) Fish melanocortin system. Eur J Pharmacol 660:53–60

Chan LF, Webb TR, Chung TT, Meimaridou E, Cooray SN, Guasti L, Chapple JP, Egertová M, Elphick MR, Cheetham ME, Metherell LA, Clark AJ (2009) MRAP and MRAP2 are bidirectional regulators of the melanocortin receptor family. Proc Natl Acad Sci 106:6146–6151

Cone RD (2006) Studies on the physiological functions of the melanocortin system. Endocr Rev 27:736–749

Cortés R, Navarro S, Agulleiro MJ, Guillot R, García-Herranz V, Sánchez E, Cerdá-Reverter JM (2014) Evolution of the melanocortin system. Gen Comp Endocrinol 209:3–10

Creemers JW, Pritchard LE, Gyte A, Le Rouzic P, Meulemans S, Wardlaw SL, Zhu X, Steiner DF, Davies N, Armstrong D, Lawrence CB, Luckman SM, Schmitz CA, Davies RA, Brennand JC, White A (2006) Agouti-related protein is posttranslationally cleaved by proprotein convertase 1 to generate agouti-related protein (AGRP)83-132: interaction between AGRP83-132 and melanocortin receptors cannot be influenced by syndecan-3. Endocrinology 147:1621–1631

Danielson PB, Dores RM (1999) Molecular evolution of the opioid/orphanin gene family. Gen Comp Endocrinol 113:169–186

Deen PM, Terwel D, Bussemakers MJ, Roubos EW, Martens GJ (1991) Martens GJ Structural analysis of the entire proopiomelanocortin gene of Xenopus laevis. Eur J Biochem 201(1):129–137. https://doi.org/10.1111/j.1432-1033.1991.tb16265.x

de Souza FS, Bumaschny VF, Low MJ, Rubinstein M (2005) Subfunctionalization of expression and peptide domains following the ancient duplication of the proopiomelanocortin gene in teleost fishes. Mol Biol Evol 22:2417–2427

Dores RM (2013) Observations on the evolution of the melanocortin receptor gene family: distinctive features of the melanocortin-2 receptor. Front Neurosci 7:28

Dores RM (2016) Hypothesis and theory: revisiting views on the co-evolution of the melanocortin receptors and the accessory proteins, MRAP1 and MRAP2. Front Endocrinol 28(7):79

Dores RM, Baron AJ (2011) Evolution of POMC: origin, phylogeny, posttranslational processing, and the melanocortins. Ann N Y Acad Sci 1220:34–48

Dores RM, Chapa E (2021) Hypothesis and theory: evaluating the co-evolution of the melanocortin-2 receptor and the accessory protein MRAP1. Front Endocrinol 12:747843

Dores RM, Garcia Y (2015) Views on the co-evolution of the melanocortin-2 receptor, MRAPs, and the hypothalamus/pituitary/adrenal-interrenal axis. Mol Cell Endocrinol 408:12–22

Dores RM, Smith TR, Rubin DA, Danielson P, Marra LE, Youson JH (1997) Deciphering posttranslational processing events in the pituitary of a neopterygian fish: cloning of a gar proopiomelanocortin cDNA. Gen Comp Endocrinol 107(3):401–413. https://doi.org/10.1006/gcen.1997.6947

Dores RM, Lecaude S, Bauer D, Danielson PB (2002) Analyzing the evolution of the opioid/orphanin gene family. Mass Spectrom Rev 21:220–243

Haitina T, Klovins J, Takahashi A, Löwgren M, Ringholm A, Enberg J, Kawauchi H, Larson ET, Fredriksson R, Schiöth HB (2007) Functional characterization of two melanocortin (MC) receptors in lamprey showing orthology to the MC1 and MC4 receptor subtypes. BMC Evol Biol 7:101

Harris RM, Dijkstra PD, Hofmann HA (2013) Complex structural and regulatory evolution of the pro-opiomelanocortin gene family. Gen Comp Endocrinol 195:107–115

He L, Gunn TM, Bouley DM, Lu X-Y, Watson SJ, Schlossman SF, Duke-Cohan JS, Barsh GS (2001) A biochemical function for attractin in agouti-induced pigmentation and obesity. Nat Genet 27:40–47

Jackson PJ, McNulty JC, Yang Y-K, Thompson DA, Chai B (2002) Design, pharmacology, and NMR structure of a minimized cysteine knot with agouti-related protein activity. Biochemistry 41:7565–7572

Klovins J, Schiöth HB (2005) Agouti-related proteins (AGRPs) and agouti-signaling peptide (ASIP) in fish and chicken. Ann N Y Acad Sci 1040:363–367

Klovins J, Haitina T, Ringholm A, Löwgren M, Fridmanis D, Slaidina M, Stier S, Schiöth HB (2004) Cloning of two melanocortin (MC) receptors in spiny dogfish: MC3 receptor in cartilaginous fish shows high affinity to ACTH-derived peptides while it has lower preference to gamma-MSH. Eur J Biochem 271:4320–4331

Kurokawa T, Murashita K, Uji S (2006) Characterization and tissue distribution of multiple agouti-family genes in pufferfish, *Takifugu rubripes*. Peptides 27:3165–3175

Larhammar D, Bergqvist C, Sundström G (2015) Ancestral vertebrate complexity of the opioid system. Vitam Horm 97:95–122

Logan DW, Bryson-Richardson RJ, Pagán KE, Taylor MS, Currie PD, Jackson IJ (2003) The structure and evolution of the melanocortin and MCH receptors in fish and mammals. Genomics 81:184–191

Metherell LA, Chapple JP, Cooray S, David A, Becker C, Rüschendorf F, Naville D, Begeot M, Khoo B, Nürnberg P, Huebner A, Cheetham ME, Clark AJ (2005) Mutations in MRAP, encoding a new interacting partner of the ACTH receptor, cause familial glucocorticoid deficiency type 2. Nat Genet 37:166–170

Meyer A, Van de Peer Y (2005) From 2R to 3R: evidence for a fish-specific genome duplication (FSGD). Bioessays 27(9):937–945. https://doi.org/10.1002/bies.20293

Miltenberger RJ, Mynatt RL, Bruce BD, Wilkinson WO, Woychik RP, Michaud EJ (1999) An agouti mutation lacking the basic domain induces yellow pigmentation but not obesity in transgenic mice. Proc Natl Acad Sci USA 96:8579–8584

Miltenberger RJ, Wakamatsu K, Ito S, Woychik RP, Russell LB, Michaud EJ (2002) Molecular and phenotypic analysis of 25 recessive, homozygous-viable alleles at the mouse agouti locus. Genetics 160(2):659–674. https://doi.org/10.1093/genetics/160.2.659

Nakanishi S, Inoue A, Kita T, Nakamura M, Chang ACY, Cohen SN, Numa S (1979) Nucleotide sequence of cloned cDNA for bovine corticotropin β-lipotropin precursor. Nature 288:610–613

Navarro S, Soletto L, Puchol S, Rotllant J, Soengas JL, Cerdá-Reverter JM (2016) 60 YEARS OF POMC: POMC: an evolutionary perspective. J Mol Endocrinol 56:113–118

Noon LA, Franklin JM, King PJ, Goulding NJ, Hunyady L, Clark AJ (2002) Failed export of the adrenocorticotrophin receptor from the endoplasmic reticulum in non-adrenal cells: evidence in support of a requirement for a specific adrenal accessory factor. J Endocrinol 174:17–25

Overton JD, Leibel RL (2011) Mahoganoid and mahogany mutations rectify the obesity of the yellow mouse by effects on endosomal traffic of MC4R protein. J Biol Chem 286:18914–18929

Reinick CL, Liang L, Angleson JK, Dores RM (2012) Identification of an MRAP-independent melanocortin-2 receptor: functional expression of the cartilaginous fish, Callorhinchus milii, melanocortin-2 receptor in CHO cells. Endocrinology 153:4757–4765

Ringholm A, Klovins J, Fredriksson R, Poliakova N, Larson ET, Kukkonen JP, Larhammar D, Schiöth HB (2003) Presence of melanocortin (MC4) receptor in spiny dogfish suggests an ancient vertebrate origin of central melanocortin system. Eur J Biochem 270:213–221

Rocha A, Godino-Gimeno A, Cerdá-Reverter JM (2019) Evolution of proopiomelanocortin. Vitam Horm 111:1–16

Schimmer BP, Kwan WK, Tsao J, Qiu R (1995) Adrenocorticotropin-resistant mutants of the Y1 adrenal cell line fail to express the adrenocorticotropin receptor. J Cell Physiol 163:164–171

Schiöth HB, Haitina T, Ling MK, Ringholm A, Fredriksson R, Cerdá-Reverter JM, Klovins J (2005) Evolutionary conservation of the structural, pharmacological, and genomic characteristics of the melanocortin receptor subtypes. Peptides 26:1886–1900

Sebag JA, Hinkle PM (2009) Regions of melanocortin 2 (MC2) receptor accessory protein necessary for dual topology and MC2 receptor trafficking and signaling. J Biol Chem 284:610–618

Smith JJ, Kuraku S, Holt C, Sauka-Spengler T, Jiang N, Campbell MS, Yandell MD, Manousaki T, Meyer A, Bloom OE et al (2013) Sequencing of the sea lamprey (*Petromyzon marinus*) genome provides insights into vertebrate evolution. Nat Genet 45:415–421

Smith JJ, Timoshevskaya N, Ye C, Holt C, Keinath MC, Parker HJ, Cook ME, Hess JE, Narum SR, Lamanna F, Kaessmann H, Timoshevskiy VA, Waterbury CKM, Saraceno C, Wiedemann LM, Robb SMC, Baker C, Eichler EE, Hockman D, Sauka-Spengler T, Yandell M, Krumlauf R, Elgar G, Amemiya CT (2018) The sea lamprey germline genome provides insights into programmed genome rearrangement and vertebrate evolution. Nat Genet 50:270–277

Sundström G, Larsson TA, Larhammar D (2008) Phylogenetic and chromosomal analyses of multiple gene families syntenic with vertebrate Hox clusters. BMC Evol Biol 8:254

Sundström G, Dreborg S, Larhammar D (2010) Concomitant duplications of opioid peptide and receptor genes before the origin of jawed vertebrates. PLoS One 6:e10512

Takahashi A, Amemiya Y, Nozaki M, Sower SA, Joss J, Gorbman A, Kawauchi H (1995a) Isolation and characterization of melanotropins from lamprey pituitary glands. Int J Pep Res 46:197–204

Takahashi A, Amemiya Y, Sarashi M, Sower SA, Kawauchi H (1995b) Melanotropin and corticotropin are encoded on two distinct genes in the lamprey, the earliest evolved extant vertebrate. Biochem Biophys Res Comm 213:490–498

Takahashi A, Amano M, Itoh T, Yasuda A, Yamanome T, Amemiya Y, Sasaki K, Sakai M, Yamamori K, Kawauchi H (2005) Nucleotide sequence and expression of three subtypes of proopiomelanocortin mRNA in barfin flounder. Gen Comp Endocrinol 141:291–303

Takahashi A, Davis P, Reinick C, Mizusawa K, Sakamoto T, Dores RM (2016) Characterization of melanocortin receptors from stingray Dasyatis akajei, a cartilaginous fish. Gen Comp Endocrinol 232:115–124

Valsalan R, Krishnan A, Almén MS, Fredriksson R, Schiöth HB (2013) Early vertebrate origin of melanocortin 2 receptor accessory proteins (MRAPs). Gen Comp Endocrinol 188:123–132

Västermark Å, Schiöth HB (2011) The early origin of melanocortin receptors, agouti-related peptide, agouti signalling peptide, and melanocortin receptor accessory proteins, with emphasis on pufferfishes, elephant shark, lampreys, and amphioxus. Eur J Pharmacol 660:61–69

Västermark Å, Krishnan A, Houle ME, Fredriksson R, Cerdá-Reverter JM, Schiöth HB (2012) Identification of distant Agouti-like sequences and reevaluation of the evolutionary history of the Agouti-related peptide (AgRP). PLoS One 7:e40982

Xu A, Choi KL, Wang Y, Permana PA, Xu LY, Bogardus C, Cooper GJ (2002) Identification of novel putative membrane proteins selectively expressed during adipose conversion of 3T3-L1 cells. Biochem Biophys Res Commun 293:1161–1167

The Corticotropin-Releasing Hormone: Evolutionary Origins and Functional Trajectories of an Ancient Peptide Ligand-Receptor System

10

Jan M. Deussing

Abstract

The corticotropin-releasing hormone (CRH) system constitutes an evolutionarily ancient and conserved peptidergic network, the origins of which can be traced back to early bilaterian evolution, before the divergence of protostomes and deuterostomes. The extant CRH system-related ligands, cognate receptors and the binding protein descend from ancestral genes that were subjected to lineage-specific gene duplication or deletion events during evolution. In vertebrates, the present diversity of ligands and receptors is a result of two (2R) or even three (3R) rounds of whole-genome duplication (WGD). The early appearance of the CRH system during evolution and its conservation suggest its original involvement in fundamental processes that are essential for survival, including the maintenance of metabolic and physiological homoeostasis. CRH in vertebrates is well known as a secretagogue controlling the hypothalamic-pituitary-adrenal/interrenal (HPA/I) axis. The diuretic hormone 44 (DH44), which is the CRH homolog in arthropods, exerts control over multiple physiological processes ranging from nutrient sensing to reproductive behaviour. The bandwidth of functions assigned to this neuropeptide reflects different modes of organization that have adopted the CRH system as a central component of the neuroendocrine stress response system, integrating sensory information to convey physiological responses to external and internal alterations. In most vertebrates, CRH is not only regulating the HPA/I axis but simultaneously exercises control over the hypothalamic-pituitary-thyroid (HPT)

J. M. Deussing (✉)
Max Planck Institute of Psychiatry, Molecular Neurogenetics, Munich, Germany
e-mail: deussing@psych.mpg.de

V. Grinevich, R. F. Oliveira (eds.), *Evolutionary and Comparative Neuroendocrinology*, Masterclass in Neuroendocrinology 17,
https://doi.org/10.1007/978-3-031-80209-6_10

axis. The observation that this function is lost in mammals and is restricted to the control of the HPA/I axis exemplifies how this system has been subjected to a constant process of adaptation during evolution. In this respect, the investigation of the CRH system in various taxa of animal evolution beyond humans and rodents has the potential to provide novel insights into our comprehension of this central neuroendocrine stress response system.

Keywords

Corticotropin-releasing hormone · Corticotropin-releasing factor · Urocortin · Hypothalamic-pituitary-adrenal/interrenal axis · HPA/I axis · Evolution · Hypothalamic-pituitary-thyroid axis · HPT axis

List of Abbreviations

AC	Anterior cell
ACTH	Adrenocorticotropic hormone
CA	Corpora allata
CC	Corpora cardiaca
CRH	Corticotropin-releasing hormone
CRHR	Corticotropin-releasing hormone receptor
CRHBP	Corticotropin-releasing hormone binding protein
CRF	Corticotropin-releasing factor
DH44	Diuretic hormone 44
DH44R	Diuretic hormone 44 receptor
DN	Dorsal neuron
GAC	Gnathostome ancestral chromosome
HP	Hypothalamic pituitary
HPA/I	Hypothalamic-pituitary-adrenal/interrenal
HPT	Hypothalamic-pituitary-thyroid
HOX	Homeobox
LNC	Lateral neurosecretory cell
MNC	Median neurosecretory cell
MT	Malphigian tubule
NPO	Neurosecretory preoptic area
PI	Pars intercerebralis
SEZ	Subesophageal zone
sNLv	Small lateral ventral neuron
SVG	Sauvagine
TRF	Thyrotropin-releasing factor
TSH	Thyroid-stimulating hormone

UI	Urotensin I
UCN	Urocortin
VNC	Ventral nerve cord
WGD	Whole-genome duplication

10.1 Introduction

The neuropeptide CRH is recognized as the canonical activator of the HPA/I axis. Upon physical or psychological stress, CRH is released from parvocellular neurons of the paraventricular nucleus of the hypothalamus to the pituitary portal vasculature to activate its high-affinity CRH receptor type 1 (CRHR1) on anterior pituitary corticotropes, thereby inducing the release of adrenocorticotropic hormone (ACTH) into the general circulation. ACTH, in turn, activates the synthesis and release of glucocorticoids from the adrenal cortex—cortisol in humans and corticosterone in rodents—which act on a wide variety of target tissues to shape an adaptive stress response. Eventually, glucocorticoids impinge on this stress hormone cascade by exerting a negative feedback control over the HPA axis. Besides orchestrating the neuroendocrine stress response, CRH acts as a neuromodulator, directly affecting the autonomic and behavioural stress responses (Deussing and Chen 2018). Evidence from patients and rodent models suggests that a dysregulated CRH system contributes to a maladaptive stress response, posing a major risk factor for stress-related disorders. Along these lines, small molecule CRHR1-antagonists have been tested in clinical trials, but lacked significant efficacy in any of the tested indications, stalling further development of the CRH system as a potential target for therapeutic intervention (Spierling and Zorrilla 2017).

Research on the CRH system has largely focussed on placental mammals, including humans and particularly mice and rats as the dominating model organisms in biomedical research. However, the first member of the CRH peptide family was isolated from the skin of the South American tree frog *Phyllomedusa sauvagii* (Montecucchi et al. 1980; Montecucchi and Henschen 1981). This peptide, designated as sauvagine (SVG), was shown to stimulate the release of ACTH and β-endorphin from the mammalian anterior pituitary (Montecucchi and Henschen 1981). In the same year, the first mammalian CRH was isolated from sheep and characterized as a 41-amino acid peptide that stimulates the release of ACTH in vitro and in vivo (Vale et al. 1981).

Identification of the Corticotropin-Releasing Factor/Hormone

In the 1930s already, one of the pioneers of stress research, Hans Seyle, had postulated a factor capable of releasing ACTH from the pituitary to activate the synthesis and secretion of adrenal corticosteroids. Based on earlier work by Ernst and Berta Scharrer, Wolfgang Bargmann and others, Geoffrey Harris coined the concept of neurohumoral control of anterior pituitary secretion and demonstrated that the hypothalamic-hypophysial portal vasculature plays an essential role (Fink 2015). Experimental evidence for the existence of a factor capable of stimulating the release of ACTH from the anterior pituitary had already been provided in 1955 (Guillemin and Rosenberg 1955; Saffran and Schally 1955a, b). In fact, this was the first hypothalamic factor to be detected, which was mainly owing to the circumstance that out of all hormones secreted by the anterior pituitary, only for ACTH had a reliable method of detection already been established. Andrew Schally and Murray Saffran were the first to call the ACTH-release stimulating substance corticotropin-releasing factor (CRF) and proposed to use the term "factor" for the unknown hypothalamic substances yet to be isolated, which were to be renamed as "hormones" when they were definitely identified and characterized. The first experimental demonstration of its existence led to intense research efforts spearheaded by the teams of Andrew Schally and Roger Guillemin to identify CRF (Klavdieva 1996). Both groups isolated the first hypothalamic releasing and inhibiting factors, a scientific breakthrough for which both of them were awarded with the Nobel Prize in Medicine or Physiology in 1977. However, the biochemical nature of CRF remained mysterious even after decades of intense research efforts. It was not until 1981 that Wylie W. Vale, a former postdoc of Roger Guillemin, and his team at the Salk Institute finally succeeded in isolating CRH from 490,000 ovine hypothalamic fragments that had previously been used by Roger Guillemin's lab for the isolation of the gonadotropin-releasing hormone (Vale et al. 1981).

Shortly after the discovery of ovine CRH, urotensin I (UI), another CRH-like peptide, was isolated from the urophysis of the common carp *Cyprinus carpio* and the white sucker *Catostomus commersonii* (Ichikawa et al. 1982; Lederis et al. 1982). The isolation of the diuretic hormone (DH, later designated as DH44) as a first CRH-like peptide from the tobacco hornworm *Manduca sexta* further proved the presence of CRH-like peptides in various animal phyla, dating the family's origin to early in evolution, even before divergence of protostomes and deuterostomes (Kataoka et al. 1989). The ortholog of SVG and UI was finally isolated from the rat and termed urocortin (UCN, today also designated as UCN1), indicating its homology to UI and CRH (Vaughan et al. 1995). The subsequent discovery of urocortin 2 (UCN2, also known as stresscopin-related peptide) and urocortin 3 (UCN3, also known as stresscopin) by in-silico cloning completed the family of CRH-like peptides in mammals (Hsu and Hsueh 2001; Lewis et al. 2001; Reyes et al. 2001).

In parallel with the discovery of novel CRH-like peptide ligands, the quest for cognate receptors commenced. However, before the characterization of membrane bound receptors, another molecule with high affinity for CRH was identified. The CRH binding protein (CRHBP) turned out as a secreted high-affinity glycoprotein (Behan et al. 1989; Potter et al. 1991). The CRHBP serves to regulate the availability of CRH and UCN1 but not of UCN2 and UCN3. Subsequently, two mammalian CRHRs were isolated from human, mouse and rat, showing 70% sequence similarity. CRHR1 binds CRH and UCN1 (Chen et al. 1993; Perrin et al. 1993), while CRHR2 is able to bind all four ligands of the CRH peptide family (Kishimoto et al. 1995; Perrin et al. 1995; Stenzel et al. 1995). In the meantime, and in agreement with the identification of CRH-like peptides, CRHRs and CRHBP have also been isolated and characterized from a broad range of non-mammalian species, supporting the relevance of the CRH system as an evolutionarily ancient stress response system.

This chapter will shed light on the intricate evolutionary origins of the different elements of the CRH system. It will focus on the physiological implications of this peptidergic system beyond its well-known function in regulating the neuroendocrine stress response in humans and rodents. Instead, the chapter will investigate the CRH-controlled neuroendocrine systems in two non-mammalian model organisms—zebrafish and drosophila—to illustrate how the CRH system evolved in different phyla to integrate sensory information and shape an adaptive response to environmental demands. As a final aspect, the chapter will touch upon the hypophysiotropic properties of CRH with regards to its capacity to simultaneously stimulate the HPA/I and hypothalamic-pituitary-thyroid (HPT) axes—a quality of CRH that has been largely underappreciated while focussing on the mammalian perspective.

10.2 The Evolutionary Origins of the CRH System: Ligands, Receptors and Binding Protein

During evolution, genes have regularly been subjected to duplications, either caused by tandem duplication or by replication of larger chromosomal regions which might have involved the entire genome. Today, the hypothesis that the vertebrate genome has undergone two rounds of WGD (2R hypothesis) is widely accepted (Dehal and Boore 2005). Teleost fish have even undergone a third tetraploidization, which has fertilized their evolution into the species richest vertebrate clade (Glasauer and Neuhauss 2014). Moreover, gene duplication can result from transposition of DNA sequences or retro-transposition of RNA transcripts. These processes are well-known evolutionary forces providing a source for genetic novelty (Hsueh and Feng 2020). Generally, duplications produce genetic redundancy, enabling one copy of the gene to be relieved of selective pressure. In a subsequent process of neofunctionalization, mutations accumulating in the coding or regulatory regions have the potential to assign a novel function or expression pattern to the gene copy. Alternatively, both copies can be equally prone to accumulate mutations, which can even be detrimental as long as the two copies are able to maintain gene function by complementing each other. In this process of subfunctionalization, the task of the original gene is distributed among the two copies possessing non-redundant functions without developing novel properties (Hsueh and Feng 2020; Mirabeau and Joly 2013). In the case of peptide ligands and their cognate receptors, additional mechanisms of co-evolution have to be postulated as ligand-receptor pairs usually act in concert to elicit physiological responses. Hence, each pair has to evolve in a coordinated fashion by gaining new functions or expression patterns without losing the partner (Hsueh and Feng 2020). Nevertheless, the origin of these functional pairs still remains largely mysterious, as there is no definitive answer to whether: (i) peptides evolved as ligands for pre-existing receptors, (ii) receptor subtypes evolved as targets for extant ligands or (iii) both came into existence at the same time (Krieger 1983). Compared to other peptidergic ligand-receptor pairs, the CRH system shows a remarkable degree of conservation across all bilaterian phylogenetic groups analysed to date (Mirabeau and Joly 2013). Extensive comparison of genome fragments with conserved synteny allowed extrapolation of the configuration of early gnathostome ancestral chromosomes (GAC). This genome comparison revealed that most peptide and receptor genes are clustered in three GAC linkage groups, suggesting that the ancestral genes of the CRH system and various other peptide subfamilies emerged through local tandem gene duplications before the two rounds of WGD occurring in vertebrates (Hwang et al. 2013).

10.2.1 CRH-Like Peptides

The increasing accessibility of genome sequences of numerous species across the animal kingdom has allowed the disentangling of the evolution of the CRH system with high resolution (Cardoso et al. 2016; Endsin et al. 2017; Lovejoy 2009; On et al. 2019). The diuretic hormone 44 (DH44) has been identified as a CRH peptide homolog in numerous arthropods (Audsley et al. 1995; Kataoka et al. 1989), dating the origins of the CRH peptide family back to early metazoan evolution prior to the divergence of protostomes and deuterostomes. CRH homologs have also been identified in different branches of invertebrate deuterostomes, such as the urochordates (tunicates) *Ciona intestinalis* and *Ciona savignyi*, the cephalochordate (lancelet) *Branchiostoma floridae*, the echinoderms *Asterias rubens* and *Strongylocentrotus purpuratus* as well as the hemichordate *Saccoglossus kowalevskii* (Elphick et al. 2018; Mirabeau and Joly 2013). In the course of vertebrate evolution, the CRH system diversified significantly, which matches well with the 2R hypothesis of WGD predicting the existence of four CRH-like genes in extant vertebrates (Lovejoy and Balment 1999). The model, favoured for a long time, suggested that in a first tetraploidization event during early vertebrate evolution (1R), a prototypic CRH would have given rise to the precursors of the CRH/UCN1 subfamily as well as to the UCN2/UCN3 subfamily. The second tetraploidization (2R) would then have produced the four CRH-related peptides that were eventually identified in rodents, humans and other placental mammals as CRH, UCN1, UCN2 and UCN3. However, recent analyses of the ever-growing number of genome sequences from different species have provided evidence for a different evolutionary path of the CRH peptide family. A first hint was the identification of two CRH-like peptides in the holocephalan genome of *Callorhinchus milii* (elephant shark, a cartilaginous fish) which share only 50% sequence homology among each other (Nock et al. 2011). The elephant shark represents an early branch of the vertebrate lineage with a comparably slow evolution. Thus, CRH-like peptides identified in this lineage may reflect the earliest gnathostome representatives of the family. The authors had speculated that this CRH paralog originated from an early lineage-specific gene duplication event in the evolution of holocephali, however, it turned out that CRH2 is also present in the genomes of many other vertebrates, including marsupials, monotremes, lizards, turtles, birds and several fishes, but not in placental mammals and teleost fishes (Grone and Maruska 2015b). Comparison of peptide sequences from various vertebrate clades and particularly detailed analyses of synteny conservation of neighbouring gene families in respective chromosomal regions has shed additional light on the evolution of the CRH peptide family (Cardoso et al. 2016). The genes of the CRH family are located in two separate paralogons: CRH, CRH2 and UCN1 are found in the same paralogon as the opioid peptides and receptors, whereas UCN2 and UCN3 are located in a different paralogon harbouring the visual opsin and oxytocin/vasopressin receptor genes (Lagman et al. 2013; Larhammar et al. 2015). The gene topology of the CRH family is very similar to those of other gene families in the two paralogons, supporting their generally shared evolutionary history which has its origin in the 2R events. These analyses suggest that the vertebrate progenitor possessed already two ancestral CRH peptide genes that served as founders of both CRH subfamilies. Two rounds of WGDs in combination with secondary losses gave rise to a triplet comprising CRH, CRH2 and UCN1 as well as a pair consisting of UCN2 and UCN3 (Fig. 10.1). This latest model is supported by the observation that all five CRH-related peptides are present not only in the elephant shark but have also been identified in *Latimeria chalumnae* (West Indian Ocean coelacanth, a lobe-finned fish) and *Lepisosteus oculatus* (spotted gar, a basal ray-finned fish), which also represent early-branching vertebrate lineages with slowly evolving genomes (Cardoso et al. 2016). Additional support for this hypothesis of CRH family evolution emerged from the identification of five CRH-related genes in non-jawed vertebrates, the cyclostomes (lampreys) *Petromyozon marinus* and *Lethenteron camtschaticum*, which likely represent orthologs of the gnathostome genes. The CRH genes in lampreys also share conserved synteny with respective neighbouring genes in gnathostomes, suggesting that the two whole-genome duplication events occurred before the divergence of cyclostomes and gnathostomes (Cardoso et al. 2020). Taken together, these findings support the evolvement of five CRH-like peptides early during vertebrate evolution, of which CRH2 has been independently lost in placental mammals and teleost fish (Fig. 10.1).

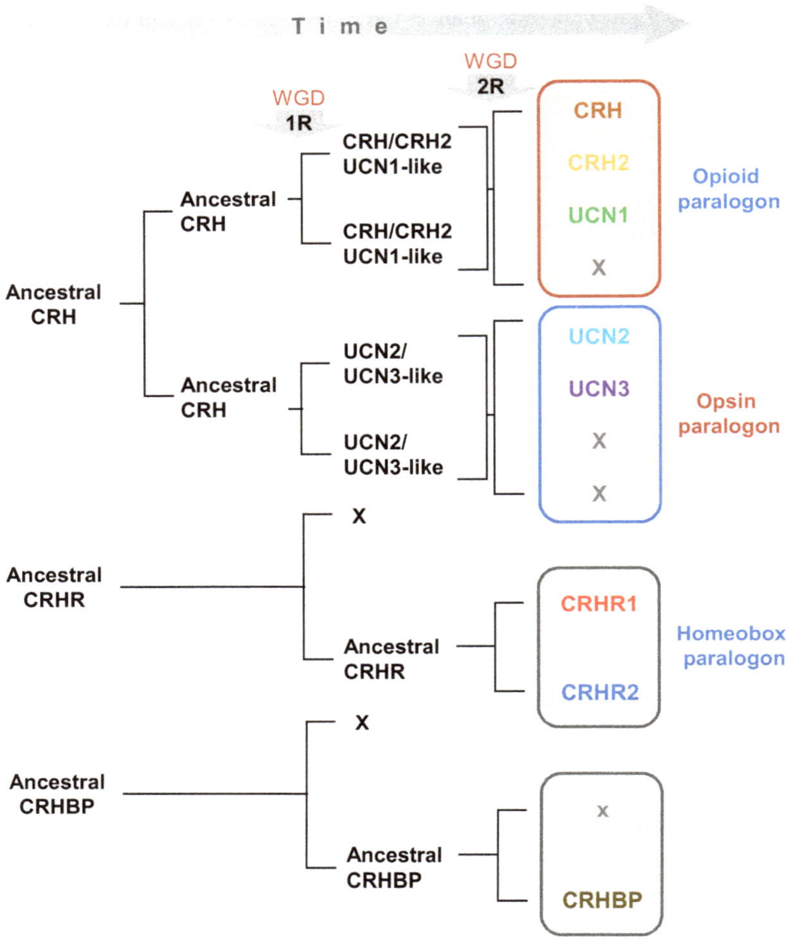

Fig. 10.1 **Evolution of CRH-like peptides, receptors and the binding protein in vertebrates.** Extant CRH system components are derived from ancestral genes that have been subjected to two rounds of whole-genome duplication (WGD) early in vertebrate evolution. The depicted scenario is based on the model proposed by Cardoso et al. (2016). For more details, see main text

10.2.2 CRH Receptors

CRH receptors belong to the family B1 of secretin-like receptors. The members of the B1 family evolved from a common ancestral gene precursor and represent a prime example of tandem gene duplication events occurring with increasing organismal complexity during metazoan evolution (Cardoso et al. 2006; Yun et al. 2015). Coinciding with the appearance of their ligands, the origins of CRH receptors can be traced back to the beginning of bilateral animal evolution before separation of protostomes and deuterostomes 500 million years ago (Cardoso et al. 2014; Chang and Hsu 2004; Lovejoy et al. 2014; On et al. 2019). The deuterostome CRH receptor and the protostome DH44 receptor are orthologs which evolved from a bilaterian ancestral gene. Homologs of the arthropod DH44R gene have been identified in molluscs, annelids and platyhelminths. In general, only a single CRH-like receptor is present in arthropods and protochordates, while two receptors are found in chordates. In protostomes, lineage-specific duplications but also losses have occurred which are likely owing to their comparably high evolutionary rate (Wyder et al. 2007). *Drosophila melanogaster*, for example, harbours two DH44 receptors: DH44R1 (CG8422) and DH44R2 (CG23370). These DH44Rs share a high degree of homology, supporting the assumption that they duplicated in diptera independently of the two vertebrate receptors (Jagge and Pietrantonio 2008). Interestingly, CRH/DH44 receptors have been entirely lost in nematodes including the model organism *Caenorhabditis elegans*. CRH-like receptors are also present in representatives of early diverging chordate lineages including echinoderms, hemichordates, cephalochordates and tunicates. As other members of the B1 family, CRH receptors have been located in the homeobox (HOX) paralogon. Reconstruction of early GACs revealed that CRHR1 and CRHR2 are distributed on GAC E2 and E1, which contain HOXB and HOXA, respectively. The absence of CRHR forms on GAC E0 and E3 suggests that these CRHR forms were likely lost prior to the divergence of teleost fishes and land vertebrates (Hwang et al. 2013). There is no clear evidence of CRHRs in basal metazoans such as placozoa, porifera, ctenophora or cnidaria (Lovejoy et al. 2014). In the course of vertebrate radiation, CRHR1 and CRHR2 have been retained in all major lineages. However, it remains unclear whether the first round of duplication resulted in an ancestral CRHR1- and an ancestral CRHR2-like gene which then underwent another round of duplication in which the second copy of the respective ancestor got lost. Alternatively, only a single CRHR1/CRHR2 copy survived R1 and in R2 gave rise to CRHR1 and CRHR2 (On et al. 2019). The fact that no pseudogenes have been identified to date suggests that the expected paralogs from the second round of WGD may have vanished rather early in chordate evolution (Fig. 10.1). The teleost genome benefitted from another tretraploidization event (3R) which resulted in a CRHR1a and CRHR1b, while CRHR2 duplicates seemingly have been lost early in the teleost lineage. To date, there is only a single exception from the default vertebrate configuration encompassing two CRHRs—the catfish *Ameiurus nebulosus*. Catfish are diploid and underwent another lineage specific genome expansion most likely resulting in three CRHRs (Arai et al. 2001).

10.2.3 CRH Binding Protein

The CRHBP does not bear any sequence similarity to other proteins and apparently constitutes an autonomous protein family. While gene or genome duplication events have resulted in multiple CRH peptide ligands and receptors, only a single CRHBP has been identified and isolated from numerous invertebrate and vertebrate species (Behan et al. 1993; Huising and Flik 2005; Lee et al. 2008; Liu et al. 2011). This suggests a somewhat similar situation as for the CRHRs, where one lineage was lost after the first genome duplication and then split into two forms as a result of the second event. However, one of these genes was subsequently eliminated, leading to only a single CRHBP gene (Fig. 10.1). Alternatively, CRHBP may have been excluded from these duplication events (Endsin et al. 2017). To date, neither of the above-mentioned hypotheses can be confirmed.

10.3 Evolutionary Adaptations of the Neuroendocrine CRH System

The hypothalamic-pituitary (HP) system is specific to vertebrates and plays a central role in development as well as in physiological and behavioural adaptation to environmental alterations. It can be considered as an evolutionary innovation enabling the neuroendocrine control over numerous complex functions. The first HP system most likely emerged in early vertebrates. Extant agnathans, such as lampreys, reflect an evolutionary intermediate stage of HP system development, eventually leading to the emergence of the highly specialized endocrine axes present in jawed vertebrates (Denver 2009; Sower 2018). In mammals, the role of CRH in HPA axis activity and regulation has been extensively studied since its discovery in 1981 (Deussing and Chen 2018; Herman et al. 2016). However, the CRH system in non-mammalian model organisms in the contexts of HPA axis regulation and beyond has just begun to emerge.

10.3.1 Genome Duplication and Subfunctionalization: Insights from *Danio rerio*

Central application of CRH in zebrafish induces behaviours which are reminiscent of classical stress responses including suppression of feeding, increased locomotor activity and anxiogenic behaviours (Matsuda 2013). Teleost fishes, similar to placental mammals, independently lost the fifth ancestral CRH-like peptide—CRH2. However, since teleosts were subjected to a third round of WGD prior to their ecological and evolutionary radiation, they often possess duplicates of their mammalian homologs. This is also the case for CRH for which two orthologs have been identified and designated as CRHa and CRHb (Grone and Maruska 2015b). Salmoniformes, which experienced another round of WGD, even possess two copies of CRHa (CRHa1, CRHa2) and CRHb (CRHb1, CRHb2) (Grone and Maruska 2015a) (Fig. 10.2a). Initially, CRHb was believed to represent the only CRH ortholog in zebrafish as previous studies had suggested that all duplications of HPI axis-related genes had been lost in zebrafish evolution in the last 33 million years (Alsop and Vijayan 2009; Chandrasekar et al. 2007). Accordingly, most expression and functional data is related to CRHb. In fish, stress-induced HPI axis activation promotes secretion of CRH from neurons of the neurosecretory preoptic area (NPO), which is the fish equivalent of the PVN (Herget et al. 2014). Similarly to the mammalian PVN, NPO CRHb^{+} neurons often co-express AVP and project to the rostral

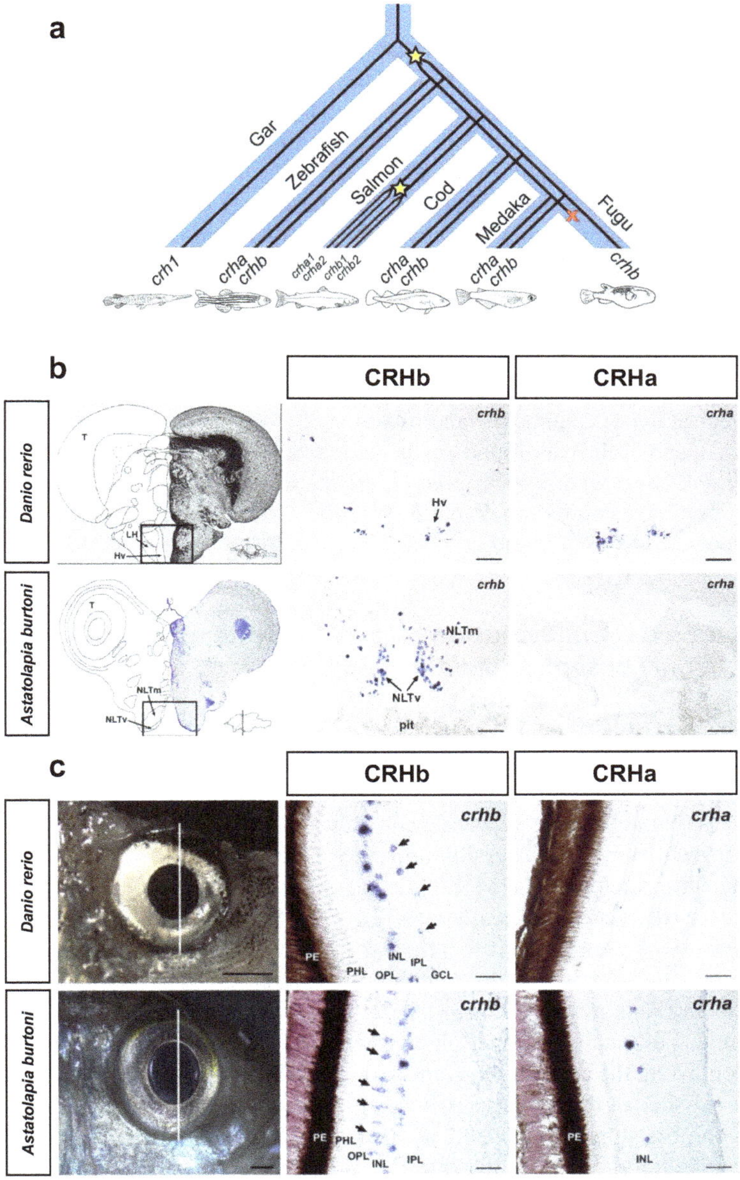

Fig. 10.2 **Phylogenetic origins and expression patterns of CRHa and CRHb in zebrafish.** (**a**) History of CRH gene duplications and losses in teleost fishes proposing a duplication of CRH in early teleosts. CRH is indicated as crh1 to discriminate it from CRH2, which has been lost in teleosts. While salmoniformes experienced an additional WGD, resulting in a duplication of CRHa and CRHb, CRHa has seemingly been lost in the fugu lineage. Gene duplication events are indicated by yellow stars and gene losses by a red X. (**b**, **c**) Comparative analysis of CRHa and CRHb mRNA expression by in situ hybridization in the brain (**b**) and retina (**c**) of the zebrafish *Danio rerio* and the African cichlid *Astatotilapia burtoni*. CRHb is expressed in the hypothalamus and retina of both species. In contrast, CRHa shows a species-specific expression pattern in both tissues. Depicted panels are from Grone and Maruska (2015a) with permission from the publisher

pars distalis of the anterior pituitary to stimulate ACTH release which, in turn, reaches the interrenal gland—the counterpart of the adrenal gland—to initiate the synthesis and release of cortisol from steroidogenic interrenal cells (Eachus et al. 2022). The fish HPI axis is already operational in early life around hatching, suggesting that it plays an import role in adjusting the physiology to the dynamically changing environment (Flik et al. 2006). CRHb neurons in the NPO seem to be individually specialized with regards to the stimuli they respond to, e.g. heat, salinity or acidity. Moreover, they promote rather rapid locomotor responses to various stressors (Vom Berg-Maurer et al. 2016). Mature CRHa and CRHb share a high degree of homology and are likely to be subjected to a process of subfunctionalization, which might entail differences in spatial and temporal gene expression, also in dependence of the environmental conditions. Along these lines and in contrast to CRHb, the expression of CRHa in zebrafish is restricted to the lateral tuberal nucleus in the ventral hypothalamus (Chandrasekar et al. 2007; Grone and Maruska 2015a). The fact that CRHa was found in the retina but not in the brain of the cichlid *Astatotilapia burtoni* might suggest a model in which duplicated neuropeptide genes may evolve expression domains that reflect the specific ecological and behavioural specializations of different species (Fig. 10.2b, c) (Grone and Maruska 2015a).

10.3.2 Convergent Evolution of the CRH System: Insights from *Drosophila melanogaster*

The utilization of hormonal axes to regulate fundamental physiological processes is an evolutionarily old concept. Thus, it is not surprising that some basic functional and structural counterparts of the HP system can also be found in arthropods including insects (Nassel and Zandawala 2020; Scharrer 1987). The central hormonal system in insects comprises lateral neurosecretory cells (LNCs) in the pars lateralis, median neurosecretory cells (MNCs) in the pars intercerebralis (PI) and neurosecretory cells in the subesophageal zone (SEZ) (de Velasco et al. 2007). In addition, neurosecretory systems are present in the ventral nerve cord (VNC) and in the intestine, in the latter represented by enteroendocrine cells. LNCs and MNCs project to the periphery and are in contact with the circulation or with other secretory cells producing additional hormones. In *Drosophila*, the axons of MNCs and LNCs spread out along the aorta, foregut and crop duct (Nassel and Zandawala 2019). The main target sites of these central neurosecretory cells are the corpora cardiaca (CC) and the corpora allata (CA), which are in direct contact with the anterior blood vessel (the insect aorta) and, together with the prothoracic gland, constitute the ring-gland. Despite basic similarities, there are also fundamental differences between the HP- and MNC/LNC-systems. For instance in mammals, the majority of hormones are produced by endocrine cells of the anterior pituitary and released into the general circulation via a vascular network where they target peripheral glands which, in turn, produce secondary hormones (Herman et al. 2016). In insects, only adipokinetic hormone and limostatin are produced by endocrine cells of the CC and secreted to the anterior blood vessel and the peripheral production of hormones is limited to the CA, PG and gonads. In addition, the LNCs and MNCs release many peptide hormones into the circulation via neurohaemal areas, whereas in mammals, this is the case only for vasopressin and oxytocin. Another characteristic of insects is the presence of a large number of neurosecretory cells in the abdominal ganglia. In mammals, a respective counterpart associated with the spinal cord is missing. Interestingly, fishes possess with Dahlgren cells a neurosecretory system in the posterior spinal cord that sends axons to the neurohaemal area of the urophysis. This vascularized gland-like structure produces, besides UI, CRH and other neuropeptides, and is involved in osmoregulation, reproduction and blood circulation (McCrohan et al. 2007; Rajamannar et al. 2021).

From the many peptide hormones identified in the mammalian HP system, CRH/DH44 is among the few that were also found in the LNC/MNC system of insects (Nassel and Zandawala 2019). Until today, DH44 has been identified in numerous insect species, including *Drosophila melanogaster*, *Bombyx mori*, *Anopheles gambiae*, *Apis mellifera* and *Tribolium castaneum* (Nassel and Winther 2010). The 44-amino acid peptide DH44, first isolated from the hornworm *Manduca sexta*, was demonstrated to stimulate fluid secretion in Malpighian tubules (MTs), suggesting a key role in the production of urine and osmoregulation (Kataoka et al. 1989). DH44-expressing (DH44^{+}) neurons have been identified in the PI and in the VNC. From the two DH44Rs, DH44R1 is expressed mainly in the brain, while DH44R2 is expressed in the midgut and in principal cells of MTs. DH44 released from neurons of the VNC acts on DH44R2 on MT principal cells to stimulate body fluid secretion (Cabrero et al. 2002; Kataoka et al. 1989). Conversely, the inhibition of DH44R2 is able to suppress body fluid secretion and, thus, can serve as a mechanism to increase stress tolerance under conditions of desiccation and starvation (Cannell et al. 2016; Zandawala et al. 2018). In the adult *Drosophila* brain, DH44 is expressed by three pairs of MNCs of the PI region (Fig. 10.3a–c). These neurons co-express *Drosophila* insulin-like peptide 2 and share the morphology of insulin producing cells (Dus et al. 2015). Efferents of these DH44^{+} PI neurons innervate the crop (the fly stomach), while their dendrites are localized in the SEZ. The ingestion of nutritive sugar activates DH44^{+} PI neurons and promotes DH44 release, characterizing them as sensors for nutritive sugars, particularly under conditions of starvation. Flies in which DH44^{+} neurons were specifically inhibited failed to select nutritive sugars. Similar observations were made upon inactivation of the peptide DH44 itself or its receptors. On the contrary, activation of DH44R1 neurons induced feeding-related behaviours to facilitate ingestion and digestion of nutritive food and activated gut motility and excretion (defecation) (Dus et al. 2015) (Fig. 10.3d). These DH44^{+} neurons are under tight control of two peripheral satiety signals originating from (i) crop stretch, which is mediated by the mechanosensory channel piezo after food ingestion, and by (ii) VNC neurosecretory neurons expressing the peptide hugin, which are activated by increased glucose levels (Oh et al. 2021). Besides sensing sugar, DH44^{+} PI neurons have also been described as key sensors to detect amino acids and to enhance food intake via activation of respective amino transporters (Chen and Dahanukar 2018; Yang et al. 2018).

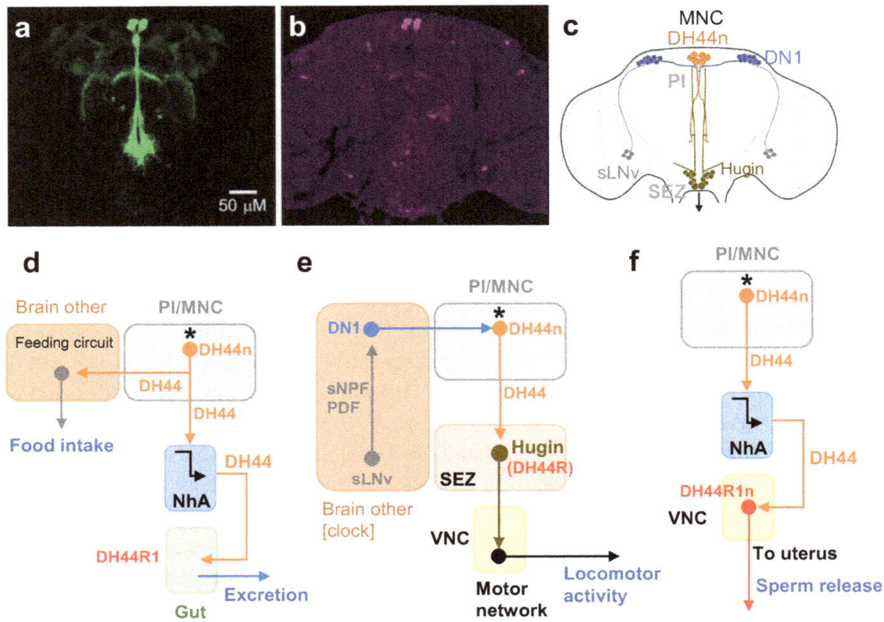

Fig. 10.3 Expression and neurosecretory function of DH44-controlled circuits in Drosophila. (**a**, **b**) DH44 is expressed in three pairs of median neurosecretory cells (MNCs) in the pars intercerebralis (PI) of adult Drosophila. DH44-GAL4-driven GFP (**a**) labels the same neurons as a DH44-specific antibody (**b**). (**c**) Schematic illustration of the Drosophila central nervous system depicting the localization of DH44-producing median neurosecretory cells (DH44n) in the PI. (**d**–**f**) Schemes depicting hormonal axes involving DH44 producing MNCs in adult fly. (**d**) Nutrient-sensing DH44 neurons (marked by an asterisk) induce feeding and defecation (excretion). Hormonally released DH44 activates DH44R1-expressing muscles in the intestine to induce excretion. In addition, DH44 also acts on circuits in the brain regulating feeding. (**e**) Illustration of DH44-expressing cells in relation to hugin-producing cells, and clock neurons including dorsal cluster 1 neurons (DN1) and small lateral ventral neurons (sLNv). The scheme depicts interactions between sLNv, D1 and DH44 neurons. The sLNv neurons act via sNPF or PDF on DN1 cells to activate DH44 neurons that, in turn, act on a subset of DH44R1-expressing hugin cells. Finally, hugin cells activate motor neuron circuits in the ventral nerve cord (VNC). This pathway regulates locomotor activity and also modulates feeding rhythms. (**f**) In female flies, the MNCs release DH44 that acts on DH44R1-expressing efferent neurons (DH44R1n) that innervate the uterus and thereby induces sperm release from spermatheca. Panels **a** and **b** are from Zandawala et al. (2018) with permission from the publisher. Panels **c**–**f** were modified according to Nassel and Zandwala (2020)

Molecular profiling identified DH44 expressed by PI neurons as a circadian output effector that is required for normal rest-activity rhythms. Along these lines, the selective activation or ablation of DH44^{+} PI cells was shown to cause pronounced signs of arrhythmicity (Cavanaugh et al. 2014). In addition, these DH44^{+} PI neurons, via dorsal neurons (DNs), receive the circadian signal generated by small lateral ventral clock neurons (sNLv), releasing peptides sNPF and PDF. DH44^{+} neurons convey the signal via a subset of DH44R1^{+} SEZ neurons which co-express the neuropeptide hugin to the VNC and thereby directly control circadian locomotor activity (King et al. 2017) (Fig. 10.3e). As DH44 is involved in conveying circadian rhythms, it is not completely unexpected that it also plays a role in regulating sleep-wake patterns. It was established that DH44$^+$ PI neurons promote wakefulness when thermal-sensing anterior cells (AC) detect an ambient temperature shift by secreting acetylcholine to a subset of DN neurons (DN1p) which, in turn, inhibit the activity of DH44$^+$ PI neurons. Hence, this AC-DN1p-DH44$^+$ PI circuit is able to integrate thermosensory inputs to promote wakefulness (Jin et al. 2021). The repertoire of stimuli integrated by DH44$^+$ PI neurons has recently been further expanded by the observation that tracheal dendritic neurons responding to altered CO_2-levels also activate DH44-mediated signalling in a larval stage (Huckesfeld et al. 2021). Finally, DH44$^+$ PI neurons have been identified as components of a neural circuit in female flies, which is relevant for post-copulatory sexual selection by regulating whether to retain or eject sperm from the uterus (Lee et al. 2015) (Fig. 10.3f). In male flies, the general downregulation of DH44 or DH44R1 has been associated with increased aggression (Kim et al. 2018). Taken together, the extensive exploration of the function of the DH44/DH44R system in *Drosophila* in the last decade has revealed how this peptidergic system has developed independently to serve as a versatile integrator of sensory information regulating a broad range of processes, including feeding and excretion, locomotor activity and circadian rhythms as well as reproductive and aggression behaviour, to maintain physiological and behavioural homoeostasis in insects.

10.4 CRH as a Common Regulator of the HPA and HPT Axes in Non-Mammalian Vertebrates

CRH was identified as the long sought-after factor capable of stimulating the release of ACTH from pituitary corticotropes (Vale et al. 1981). A dysregulation of the HPA axis has been observed in various disorders (Agorastos and Chrousos 2022) and thus has stimulated the extensive investigation of the CRH system's function in the context of HPA axis activity and regulation, with a particular focus on mammals (Deussing and Chen 2018). In mammals, the hypophysiotropic action of CRH is exclusively restricted to the HPA axis. However, in the search for peptides with thyrotropin-stimulating hormone (TSH, thyrotropin)-releasing activity, Denver and colleagues revealed that CRH and related peptides can additionally act as potent TSH secretagogues in non-mammalian vertebrates such as amphibians and reptiles. In cultivated pituitaries of non-mammalian vertebrates, exogenous CRH, UI or SVG proved themselves as thyrotropin-releasing factors (TRFs) (Denver 1988; Denver and Licht 1989). CRH turned out to be even more potent than TRH in stimulating TSH release from cultivated frog pituitaries (Denver 1988; Ito et al. 2004; Okada et al. 2004). The TRF property of CRH was also confirmed by the observation that CRH treatment reduced the TSH immunoreactivity of pituitaries from adult and larval amphibians (Malagon et al. 1991; Miranda et al. 2000). A similar effect of CRH on the HPT axis was observed in birds (De Groef et al. 2005; Watanabe et al. 2018). Chicken embryos injected with CRH showed, besides a rise in corticosterone, a marked increase in thyroid hormones (Meeuwis et al. 1989). Comparable observations have been made throughout the entire life cycle of the chicken (Geris et al. 2003). In fishes such as the coho salmon *Oncorhynchus kisutch*, CRH but also SVG and UI were able to dose-dependently increase TSH secretion, whereas the CRHR inhibitor α-helical CRH(9–41) blocked the stimulatory effect of exogenous CRH (Larsen et al. 1998). Taken together, these findings qualify CRH as a common regulator of the HPT and the HPT/I axis in non-mammalian vertebrates (De Groef et al. 2006) (Fig. 10.4).

How is the dual hypophysiotropic action of CRH achieved? The biological activity of CRH is mediated by its two cognate receptors, CRHR1 and CRHR2. Detailed analyses of CRHR expression in the chicken pituitary revealed a cell-type specific expression of CRHRs. CRHR1 is confined to corticotropes, whereas thyrotropes express only CRHR2 (De Groef et al. 2003a, b). The central role of CRHR2 was also functionally proven by using the CRHR2-specific agonist UCN3, which, similar to CRH, also stimulated the secretion of TSH. Conversely, the specific CRHR2 antagonist anti-SVG-30 efficiently blocked the stimulating effect of CRH (De Groef et al. 2003b). The specific role of CRHR2 in conveying the CRH effect with respect to TSH/thyrotropin release was confirmed independently in amphibians by similar functional experiments in the bullfrog *Rana catesbeiana* (Okada et al. 2007). Thyroid hormones also exert a negative feedback on CRH-mediated TSH secretion, which might be related to a downregulation of CRHR2 expression that is unaffected by corticosteroids (De Groef et al. 2005; Geris et al. 1999).

The recent identification of CRH2 in chicken discredited previous reports suggesting a selective loss of the fifth ancient member of the CRH family (Bu et al. 2019; Cardoso et al. 2016; Grone and Maruska 2015b). Chicken CRH2 is highly expressed in the hypothalamus and is about 15-fold more potent in activating CRHR2 compared to CRHR1. Interestingly, chicken CRH2 is also far less effective than CRH in stimulating ACTH secretion from cultivated pituitaries. However, CRH2 was proven to be an efficient inducer of TSHβ expression. These findings suggest that CRH2 in addition to CRH and TRH is able to stimulate/regulate the HPT axis. A recent model proposes that CRH activates ACTH secretion primarily via CRHR1, while CRH2 controls the expression and release of TSH via CRHR2 (Bu et al. 2019). To what extent this model can be generalized to other vertebrate lineages harbouring CRH2 is an intriguing question, mandating further investigations (Fig. 10.4).

In vertebrates, thyroid hormones play an important role in life-stage transitions that are required to adapt to habitat changes during the life cycle connected to birth, hatching or post-embryonic developmental events (Paris and Laudet 2008; Rousseau et al. 2021). Thyroid hormones, together with corticosteroids, regulate metabolic and developmental processes related to life stage transitions, such as amphibian metamorphosis, smoltification in salmonids and hatching in precocial birds (Watanabe et al. 2016). In non-mammalian vertebrates, CRH has been demonstrated to directly affect amphibian metamorphosis (Denver 1999). Thus CRH, as a common regulator of the HPT and HPA/I axis, might provide the means to synergistically orchestrate the coordination of these complex transition processes by thyroid hormones and corticosteroids (Watanabe et al. 2016).

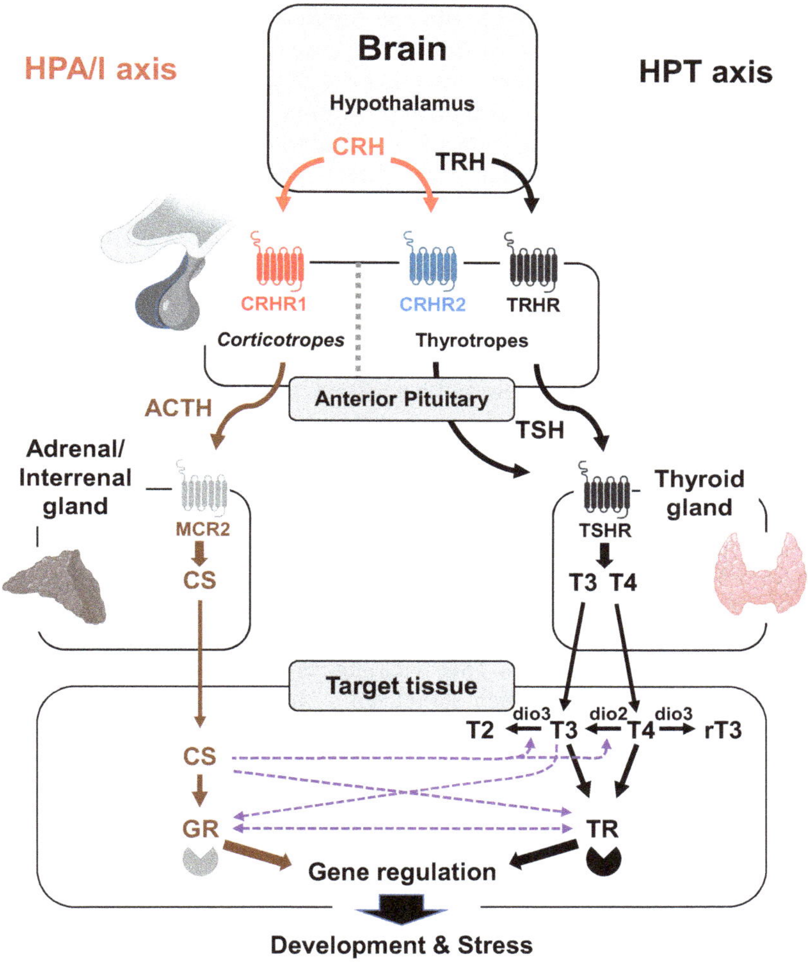

Fig. 10.4 Interactions between HPA/I and HPT axes during vertebrate developmental transitions. The figure summarizes current knowledge on the interactions between the HPA/I and HPT axes during developmental transitions in non-mammalian vertebrates involved in amphibian and teleost metamorphoses and sauropsid egg hatching. Purple arrows indicate regulatory interactions controlled by corticosteroids and thyroid hormones detailed in Paul et al. (2022). Abbreviations: *ACTH*, adrenocorticotropic hormone; *CRH*, corticotropin-releasing hormone; *CRHR*, corticotropin-releasing hormone receptor; *CS*, corticosteroids; *dio*, deiodinase; *GR*, glucocorticoid receptor; *MC2R*, melanocortin receptor 2; *T4*, thyroxine; *T3*, triiodothyronine; *rT3*, reverse triiodothyronine; *T2*, diiodothyronine; *TR*, thyroid hormone receptor; *TRH*, thyrotropin-releasing hormone; *TRHR*, thyrotropin-releasing hormone receptor; *TSH*, thyrotropin; *TSHR*, thyrotropin receptor

Whether the capacity of CRH to stimulate thyroid hormone secretion was present from the beginning or rather developed during vertebrate evolution is currently unclear. On the other end, there is clear evidence that CRH has lost its thyrotropic activity in the mammalian lineage (Watanabe et al. 2016). Experimentally, CRH is not able to stimulate TSH release from the human, rat or mouse pituitary (Holsboer et al. 1986; Lesch et al. 1989; Vale et al. 1981; Villalobos et al. 2004). This raises the question why and how CRH has lost its thyrotropic activity. As detailed above, the capacity of CRH to stimulate TSH release in non-mammalian vertebrates is mediated by CRHR2, expressed by thyrotropes of the anterior pituitary. In all mammalian species investigated so far, CRHR expression in the pituitary is dominated by CRHR1, which is almost exclusively expressed by corticotropes but not in thyrotropes. In the rat pituitary, CRHR2 expression is rather low and has largely been confined to gonadotropes (Kageyama et al. 2003; Watanabe et al. 2016). Similarly, in mice, CRHR2 has been exclusively localized to gonadotropes using CRHR2-Cre-based reporter mice (Raftogianni et al. 2018). These observations indicate that the expression of CRHR2 in anterior pituitary thyrotropes has been lost in the mammalian lineage. At the moment, it can only be speculated whether this is a direct consequence of mutations/modifications acquired in regulatory elements of the CRHR2 gene or is related to other presently unknown causes (De Groef et al. 2019).

The selective pressure that has led to the loss of CRHR2 expression in thyrotropes and of the CRH system's TSH-releasing activity is currently unknown, but the suspicion arises that this might be related to the establishment of intrauterine development and birth (Watanabe et al. 2016). Birth in mammals, similar to metamorphosis in anurans, is related to a transition from an aquatic (amniotic fluid) to a terrestrial habitat, and thus requires significant adaptations, e.g. preparation for pulmonary breathing (Ren and Flamant 2023; Sachs and Buchholz 2017). Similarly to that in non-mammalian vertebrates, the foetal and postnatal development of humans and rodents is also shaped by thyroid hormones and corticosteroids, although the activity of CRH is restricted to regulation of the HPA axis only. In this context, it is noteworthy that the human placenta starts to expresses CRH early in gestation and, due to a positive feed-forward mechanism, this expression increases exponentially towards the end of pregnancy. Therefore, it has been suggested that CRH plays a role in the physiology of pregnancy and as part of a placental clock in the timing of birth (Herrera et al. 2022; Kalantaridou et al. 2010). Accordingly, placental CRH has also been implicated in pregnancy-related complications such as preterm labour and pre-eclampsia. However, the fact that placental expression of CRH has exclusively evolved in anthropoid primates by co-option of a retroviral enhancer element (Dunn-Fletcher et al. 2018) suggests that this adaptation might have evolved independent of the lost control over the HPT axis (Power and Schulkin 2006). To what extent CRH is crucial for intrauterine development of the primate embryo or to what extent it has been substituted in other species, e.g. rodents, by other CRH-like peptides, is currently unclear.

Perspectives

The evolutionary genesis in animals of the CRH system, composed of its five peptide ligands, two receptors and a single binding protein, has been fully disentangled with the help of the ever-increasing accessibility of whole-genome sequences from numerous species across the animal kingdom. The phylogenetic analyses confirmed its ancient origins dating back to early bilaterians and the remarkable conservation of its components throughout evolution. The identification of CRH2 as the fifth ancient family member, which has been lost in placental mammals and teleosts, was a major leap forward in the comprehension of the systems' evolution. Previous attempts to trace the origins of CRH-like peptides might have been biased by the four CRH family members identified in mammals that perfectly matched the hypothesized two rounds of WGD. The identification of pseudogenes reflecting lineage-specific inactivation of genes which evolved from previous duplication events would help to further strengthen this model. Generally, phylogeny-driven approaches are able to provide important insights into the evolution of peptidergic systems but are naturally restricted by the availability of genome sequences. In particular, genome sequences from early stages of metazoan evolution would allow to further substantiate the current evolutionary model and the origins of the CRH system. Despite a better understanding of the evolutionary legacy of CRH-like peptides, the selective forces that led, for example, to the loss of CRH2 in placental mammals and teleosts remain entirely unclear. While CRH2 has been independently lost during teleost evolution, the lineage-specific tetraploidization produced two copies of CRH. In the zebrafish *D. rerio*, CRHb has conserved the original function of CRH as the canonical regulator of the HPI axis orchestrating the neuroendocrine stress response. The functional relevance of CRHa has not been revealed yet, but it has adopted a deviating, highly restricted and species-specific expression pattern, which might reflect the plasticity of the ongoing process of subfunctionalization.

The early evolutionary origin of the CRH system is evidenced by its presence in protostomes and deuterostomes. The identification of the CRH homolog DH44 in insects may indicate the original function of the CRH system, which could be related to osmoregulation and metabolic homoeostasis. Moreover, recent findings from the fruit fly impressively demonstrate how this ancient peptidergic system has evolved into a complex regulatory network that integrates environmental information to convey adaptive responses, including behavioural changes. Similar to the control of the HPA/I axis in vertebrates, neurosecretory DH44^{+} neurons of the PI exert central control over these responses, representing a fascinating example of convergent evolution. It is exciting to discern how the DH44 system in Drosophila has reached an enormous degree of complexity and specialization, while, at the same time, life without CRH is also possible, as can be seen in the nematode *C. elegans*. The absence of CRH from *C. elegans* raises the question of how an otherwise highly conserved and seemingly indispensable peptidergic system has become obsolete or has been successfully substituted.

The development of the HP unit can be traced back to early vertebrate evolution, with the CRH system being an integral part of this evolutionary innovation. In non-mammalian vertebrates, CRH has been demonstrated to possess a potent thyrotropin-releasing activity besides its well-known function as a secretagogue in the context of the HPA axis. In particular, the synergistic developmental effects of thyroid hormones and corticosteroids in the life cycle of many organisms are likely to have evolutionarily favoured a simultaneous coordination of HPA/I and HPT axes by the CRH system. To date, it is unclear which selective pressure in mammals entailed the loss of this dual hypophysiotropic activity and the connection to intrauterine development and birth remains largely speculative.

Taken together, these findings underscore the benefit and importance of a more holistic view of peptidergic systems. Tracing the evolutionary origins provides insights into the original physiological role of a given system. In case of the CRH system, we have seen how it has evolved independently into a central component of respective neurosecretory systems in vertebrates and arthropods. The loss of its capacity to stimulate the HPT axis in mammals reflects its specialization during vertebrate evolution. On the other hand, gene duplication events provide novel opportunities for sub- or neofunctionalization. Extending the perspective from humans and mammalian model organisms to the whole spectrum of metazoan organisms will clearly help to better understand this ancient peptidergic system. To what extent uncommon model organisms can be integrated into research approaches is currently not clear, and is a herculean challenge, as the experimental tool boxes available for long-established model organisms are largely missing. However, gene editing via CRISPR/Cas technology might provide necessary remedies to overcome this obstacle. Moreover, the conservation of the CRH system and its presence in classical model organisms such as zebrafish and fruit fly provide interesting alternatives to complement studies based on mouse models. These non-mammalian model organisms, whose central nervous system is more easily accessible in larval and adult stages than in mice, provide not only the genetic means to address individual CRH system components but also the opportunity to precisely manipulate and image CRH-related circuits in vivo.

Acknowledgements and Funding This work was supported by Max Planck Society and the Gunter Sachs Donation. I thank Elfi Fesl for proofreading and editing of the manuscript.

References

Agorastos A, Chrousos GP (2022) The neuroendocrinology of stress: the stress-related continuum of chronic disease development. Mol Psychiatry 27(1):502–513

Alsop D, Vijayan MM (2009) Molecular programming of the corticosteroid stress axis during zebrafish development. Comp Biochem Physiol A Mol Integr Physiol 153(1):49–54

Arai M, Assil IQ, Abou-Samra AB (2001) Characterization of three corticotropin-releasing factor receptors in catfish: a novel third receptor is predominantly expressed in pituitary and urophysis. Endocrinology 142(1):446–454

Audsley N, Kay I, Hayes TK, Coast GM (1995) Cross reactivity studies of CRF-related peptides on insect Malpighian tubules. Comp Biochem Physiol A Physiol 110(1):87–93

Behan DP, Linton EA, Lowry PJ (1989) Isolation of the human plasma corticotrophin-releasing factor-binding protein. J Endocrinol 122(1):23–31

Behan DP, Potter E, Lewis KA, Jenkins NA, Copeland N, Lowry PJ, Vale WW (1993) Cloning and structure of the human corticotrophin releasing factor-binding protein gene (CRHBP). Genomics 16(1):63–68
Bu G, Fan J, Yang M, Lv C, Lin Y, Li J, Meng F, Du X, Zeng X, Zhang J, Li J, Wang Y (2019) Identification of a novel functional corticotropin-releasing hormone (CRH2) in chickens and its roles in stimulating pituitary TSHbeta expression and ACTH secretion. Front Endocrinol (Lausanne) 10:595
Cabrero P, Radford JC, Broderick KE, Costes L, Veenstra JA, Spana EP, Davies SA, Dow JA (2002) The Dh gene of Drosophila melanogaster encodes a diuretic peptide that acts through cyclic AMP. J Exp Biol 205(Pt 24):3799–3807
Cannell E, Dornan AJ, Halberg KA, Terhzaz S, Dow JAT, Davies SA (2016) The corticotropin-releasing factor-like diuretic hormone 44 (DH44) and kinin neuropeptides modulate desiccation and starvation tolerance in Drosophila melanogaster. Peptides 80:96–107
Cardoso JC, Pinto VC, Vieira FA, Clark MS, Power DM (2006) Evolution of secretin family GPCR members in the metazoa. BMC Evol Biol 6:108
Cardoso JC, Felix RC, Bergqvist CA, Larhammar D (2014) New insights into the evolution of vertebrate CRH (corticotropin-releasing hormone) and invertebrate DH44 (diuretic hormone 44) receptors in metazoans. Gen Comp Endocrinol 209:162–170
Cardoso JC, Bergqvist CA, Felix RC, Larhammar D (2016) Corticotropin-releasing hormone family evolution: five ancestral genes remain in some lineages. J Mol Endocrinol 57(1):73–86
Cardoso JCR, Bergqvist CA, Larhammar D (2020) Corticotropin-releasing hormone (CRH) gene family duplications in lampreys correlate with two early vertebrate genome doublings. Front Neurosci 14:672
Cavanaugh DJ, Geratowski JD, Wooltorton JR, Spaethling JM, Hector CE, Zheng X, Johnson EC, Eberwine JH, Sehgal A (2014) Identification of a circadian output circuit for rest: activity rhythms in drosophila. Cell 157(3):689–701
Chandrasekar G, Lauter G, Hauptmann G (2007) Distribution of corticotropin-releasing hormone in the developing zebrafish brain. J Comp Neurol 505(4):337–351
Chang CL, Hsu SY (2004) Ancient evolution of stress-regulating peptides in vertebrates. Peptides 25(10):1681–1688
Chen YD, Dahanukar A (2018) DH44 neurons: gut-brain amino acid sensors. Cell Res 28(11):1048–1049
Chen R, Lewis KA, Perrin MH, Vale WW (1993) Expression cloning of a human corticotropin-releasing-factor receptor. Proc Natl Acad Sci USA 90(19):8967–8971
De Groef B, Geris KL, Manzano J, Bernal J, Millar RP, Abou-Samra AB, Porter TE, Iwasawa A, Kuhn ER, Darras VM (2003a) Involvement of thyrotropin-releasing hormone receptor, somatostatin receptor subtype 2 and corticotropin-releasing hormone receptor type 1 in the control of chicken thyrotropin secretion. Mol Cell Endocrinol 203(1–2):33–39
De Groef B, Goris N, Arckens L, Kuhn ER, Darras VM (2003b) Corticotropin-releasing hormone (CRH)-induced thyrotropin release is directly mediated through CRH receptor type 2 on thyrotropes. Endocrinology 144(12):5537–5544
De Groef B, Vandenborne K, Van As P, Darras VM, Kuhn ER, Decuypere E, Geris KL (2005) Hypothalamic control of the thyroidal axis in the chicken: over the boundaries of the classical hormonal axes. Domest Anim Endocrinol 29(1):104–110
De Groef B, Van der Geyten S, Darras VM, Kuhn ER (2006) Role of corticotropin-releasing hormone as a thyrotropin-releasing factor in non-mammalian vertebrates. Gen Comp Endocrinol 146(1):62–68
De Groef B, Wawrzyczek SK, Watanabe Y, Noy EB, Reser DH, Grommen SVH (2019) Evolutionary origin of the type 2 corticotropin-releasing hormone receptor gamma splice variant. Genes Cells 24(4):318–323
de Velasco B, Erclik T, Shy D, Sclafani J, Lipshitz H, McInnes R, Hartenstein V (2007) Specification and development of the pars intercerebralis and pars lateralis, neuroendocrine command centers in the drosophila brain. Dev Biol 302(1):309–323

Dehal P, Boore JL (2005) Two rounds of whole genome duplication in the ancestral vertebrate. PLoS Biol 3(10):e314
Denver RJ (1988) Several hypothalamic peptides stimulate in vitro thyrotropin secretion by pituitaries of anuran amphibians. Gen Comp Endocrinol 72(3):383–393
Denver RJ (1999) Evolution of the corticotropin-releasing hormone signaling system and its role in stress-induced phenotypic plasticity. Ann N Y Acad Sci 897:46–53
Denver RJ (2009) Structural and functional evolution of vertebrate neuroendocrine stress systems. Ann N Y Acad Sci 1163:1–16
Denver RJ, Licht P (1989) Neuropeptides influencing in vitro pituitary hormone secretion in hatchling turtles. J Exp Zool 251(3):306–315
Deussing JM, Chen A (2018) The Corticotropin-releasing factor family: physiology of the stress response. Physiol Rev 98(4):2225–2286
Dus M, Lai JS, Gunapala KM, Min S, Tayler TD, Hergarden AC, Geraud E, Joseph CM, Suh GS (2015) Nutrient sensor in the brain directs the action of the brain-gut axis in drosophila. Neuron 87(1):139–151
Eachus H, Ryu S, Placzek M, Wood J (2022) Zebrafish as a model to investigate the CRH axis and interactions with DISC1. Curr Opin Endocr Metab Res 26:100383
Elphick MR, Mirabeau O, Larhammar D (2018) Evolution of neuropeptide signalling systems. J Exp Biol 221(Pt 3)
Endsin MJ, Michalec O, Manzon LA, Lovejoy DA, Manzon RG (2017) CRH peptide evolution occurred in three phases: evidence from characterizing sea lamprey CRH system members. Gen Comp Endocrinol 240:162–173
Fink G (2015) 60 YEARS OF NEUROENDOCRINOLOGY: MEMOIR: Harris' neuroendocrine revolution: of portal vessels and self-priming. J Endocrinol 226(2):T13–T24
Flik G, Klaren PH, Van den Burg EH, Metz JR, Huising MO (2006) CRF and stress in fish. Gen Comp Endocrinol 146(1):36–44
Geris KL, Laheye A, Berghman LR, Kuhn ER, Darras VM (1999) Adrenal inhibition of corticotropin-releasing hormone-induced thyrotropin release: a comparative study in pre- and posthatch chicks. J Exp Zool 284(7):776–782
Geris KL, De Groef B, Kuhn ER, Darras VM (2003) In vitro study of corticotropin-releasing hormone-induced thyrotropin release: ontogeny and inhibition by somatostatin. Gen Comp Endocrinol 132(2):272–277
Glasauer SM, Neuhauss SC (2014) Whole-genome duplication in teleost fishes and its evolutionary consequences. Mol Gen Genomics 289(6):1045–1060
Grone BP, Maruska KP (2015a) Divergent evolution of two corticotropin-releasing hormone (CRH) genes in teleost fishes. Front Neurosci 9:365
Grone BP, Maruska KP (2015b) A second corticotropin-releasing hormone gene (CRH2) is conserved across vertebrate classes and expressed in the hindbrain of a basal neopterygian fish, the spotted gar (Lepisosteus oculatus). J Comp Neurol 523(7):1125–1143
Guillemin R, Rosenberg B (1955) Humoral hypothalamic control of anterior pituitary: a study with combined tissue cultures. Endocrinology 57(5):599–607
Herget U, Wolf A, Wullimann MF, Ryu S (2014) Molecular neuroanatomy and chemoarchitecture of the neurosecretory preoptic-hypothalamic area in zebrafish larvae. J Comp Neurol 522(7):1542–1564
Herman JP, McKlveen JM, Ghosal S, Kopp B, Wulsin A, Makinson R, Scheimann J, Myers B (2016) Regulation of the hypothalamic-pituitary-adrenocortical stress response. Compr Physiol 6(2):603–621
Herrera CL, Maiti K, Smith R (2022) Preterm birth and corticotrophin-releasing hormone as a placental clock. Endocrinology 164(2)
Holsboer F, Gerken A, von Bardeleben U, Grimm W, Beyer H, Muller OA, Stalla GK (1986) Human corticotropin-releasing hormone in depression--correlation with thyrotropin secretion following thyrotropin-releasing hormone. Biol Psychiatry 21(7):601–611
Hsueh AJW, Feng Y (2020) Discovery of polypeptide ligand-receptor pairs based on their co-evolution. FASEB J 34(7):8824–8832

Huckesfeld S, Schlegel P, Miroschnikow A, Schoofs A, Zinke I, Haubrich AN, Schneider-Mizell CM, Truman JW, Fetter RD, Cardona A, Pankratz MJ (2021) Unveiling the sensory and interneuronal pathways of the neuroendocrine connectome in drosophila. Elife 10

Huising MO, Flik G (2005) The remarkable conservation of corticotropin-releasing hormone (CRH)-binding protein in the honeybee (Apis mellifera) dates the CRH system to a common ancestor of insects and vertebrates. Endocrinology 146(5):2165–2170

Hwang JI, Moon MJ, Park S, Kim DK, Cho EB, Ha N, Son GH, Kim K, Vaudry H, Seong JY (2013) Expansion of secretin-like G protein-coupled receptors and their peptide ligands via local duplications before and after two rounds of whole-genome duplication. Mol Biol Evol 30(5):1119–1130

Ichikawa T, McMaster D, Lederis K, Kobayashi H (1982) Isolation and amino acid sequence of urotensin I, a vasoactive and ACTH-releasing neuropeptide, from the carp (Cyprinus carpio) urophysis. Peptides 3(5):859–867

Ito Y, Okada R, Mochida H, Hayashi H, Yamamoto K, Kikuyama S (2004) Molecular cloning of bullfrog corticotropin-releasing factor (CRF): effect of homologous CRF on the release of TSH from pituitary cells in vitro. Gen Comp Endocrinol 138(3):218–227

Jagge CL, Pietrantonio PV (2008) Diuretic hormone 44 receptor in Malpighian tubules of the mosquito Aedes aegypti: evidence for transcriptional regulation paralleling urination. Insect Mol Biol 17(4):413–426

Jin X, Tian Y, Zhang ZC, Gu P, Liu C, Han J (2021) A subset of DN1p neurons integrates thermosensory inputs to promote wakefulness via CNMa signaling. Curr Biol 31(10):2075–2087.e6

Kageyama K, Li C, Vale WW (2003) Corticotropin-releasing factor receptor type 2 messenger ribonucleic acid in rat pituitary: localization and regulation by immune challenge, restraint stress, and glucocorticoids. Endocrinology 144(4):1524–1532

Kalantaridou SN, Zoumakis E, Makrigiannakis A, Lavasidis LG, Vrekoussis T, Chrousos GP (2010) Corticotropin-releasing hormone, stress and human reproduction: an update. J Reprod Immunol 85(1):33–39

Kataoka H, Troetschler RG, Li JP, Kramer SJ, Carney RL, Schooley DA (1989) Isolation and identification of a diuretic hormone from the tobacco hornworm, Manduca sexta. Proc Natl Acad Sci USA 86(8):2976–2980

Kim YK, Saver M, Simon J, Kent CF, Shao L, Eddison M, Agrawal P, Texada M, Truman JW, Heberlein U (2018) Repetitive aggressive encounters generate a long-lasting internal state in Drosophila melanogaster males. Proc Natl Acad Sci USA 115(5):1099–1104

King AN, Barber AF, Smith AE, Dreyer AP, Sitaraman D, Nitabach MN, Cavanaugh DJ, Sehgal A (2017) A peptidergic circuit links the circadian clock to locomotor activity. Curr Biol 27(13):1915–1927 e5

Kishimoto T, Pearse RV 2nd, Lin CR, Rosenfeld MG (1995) A sauvagine/corticotropin-releasing factor receptor expressed in heart and skeletal muscle. Proc Natl Acad Sci USA 92(4):1108–1112

Klavdieva MM (1996) The history of neuropeptides II. Front Neuroendocrinol 17(1):126–153

Krieger DT (1983) Brain peptides: what, where, and why? Science 222(4627):975–985

Lagman D, Ocampo Daza D, Widmark J, Abalo XM, Sundstrom G, Larhammar D (2013) The vertebrate ancestral repertoire of visual opsins, transducin alpha subunits and oxytocin/vasopressin receptors was established by duplication of their shared genomic region in the two rounds of early vertebrate genome duplications. BMC Evol Biol 13:238

Larhammar D, Bergqvist C, Sundstrom G (2015) Ancestral vertebrate complexity of the opioid system. Vitam Horm 97:95–122

Larsen DA, Swanson P, Dickey JT, Rivier J, Dickhoff WW (1998) In vitro thyrotropin-releasing activity of corticotropin-releasing hormone-family peptides in coho salmon, Oncorhynchus kisutch. Gen Comp Endocrinol 109(2):276–285

Lederis K, Letter A, McMaster D, Moore G, Schlesinger D (1982) Complete amino acid sequence of urotensin I, a hypotensive and corticotropin-releasing neuropeptide from Catostomus. Science 218(4568):162–165

Lee KW, Rhee JS, Raisuddin S, Gi Park H, Lee JS (2008) A corticotropin-releasing hormone binding protein (CRH-BP) gene from the intertidal copepod, Tigriopus japonicus. Gen Comp Endocrinol 158(1):54–60

Lee KM, Daubnerova I, Isaac RE, Zhang C, Choi S, Chung J, Kim YJ (2015) A neuronal pathway that controls sperm ejection and storage in female drosophila. Curr Biol 25(6):790–797

Lesch KP, Muller U, Rupprecht R, Kruse K, Schulte HM (1989) Endocrine responses to growth hormone-releasing hormone, thyrotropin-releasing hormone and corticotropin-releasing hormone in depression. Acta Psychiatr Scand 79(6):597–602

Liu L, Yu X, Meng F, Guo X, Xu B (2011) Identification and characterization of a novel corticotropin-releasing hormone-binding protein (CRH-BP) gene from Chinese honeybee (Apis cerana cerana). Arch Insect Biochem Physiol 78(3):161–175

Lovejoy DA (2009) Structural evolution of urotensin-I: reflections of life before corticotropin releasing factor. Gen Comp Endocrinol 164(1):15–19

Lovejoy DA, Balment RJ (1999) Evolution and physiology of the corticotropin-releasing factor (CRF) family of neuropeptides in vertebrates. Gen Comp Endocrinol 115(1):1–22

Lovejoy DA, Chang BS, Lovejoy NR, del Castillo J (2014) Molecular evolution of GPCRs: CRH/CRH receptors. J Mol Endocrinol 52(3):T43–T60

Malagon MM, Ruiz-Navarro A, Torronteras R, Gracia-Navarro F (1991) Effects of ovine CRF on amphibian pituitary ACTH and TSH cells in vivo: a quantitative ultrastructural study. Gen Comp Endocrinol 83(3):487–497

Matsuda K (2013) Regulation of feeding behavior and psychomotor activity by corticotropin-releasing hormone (CRH) in fish. Front Neurosci 7:91

McCrohan CR, Lu W, Brierley MJ, Dow L, Balment RJ (2007) Fish caudal neurosecretory system: a model for the study of neuroendocrine secretion. Gen Comp Endocrinol 153(1–3):243–250

Meeuwis R, Michielsen R, Decuypere E, Kuhn ER (1989) Thyrotropic activity of the ovine corticotropin-releasing factor in the chick embryo. Gen Comp Endocrinol 76(3):357–363

Mirabeau O, Joly JS (2013) Molecular evolution of peptidergic signaling systems in bilaterians. Proc Natl Acad Sci USA 110(22):E2028–E2037

Miranda LA, Affanni JM, Paz DA (2000) Corticotropin-releasing factor accelerates metamorphosis in Bufo arenarum: effect on pituitary ACTH and TSH cells. J Exp Zool 286(5):473–480

Montecucchi PC, Henschen A (1981) Amino acid composition and sequence analysis of sauvagine, a new active peptide from the skin of Phyllomedusa sauvagei. Int J Pept Protein Res 18(2):113–120

Montecucchi PC, Anastasi A, de Castiglione R, Erspamer V (1980) Isolation and amino acid composition of sauvagine. An active polypeptide from methanol extracts of the skin of the South American frog Phyllomedusa sauvagei. Int J Pept Protein Res 16(3):191–199

Nassel DR, Winther AM (2010) Drosophila neuropeptides in regulation of physiology and behavior. Prog Neurobiol 92(1):42–104

Nassel DR, Zandawala M (2019) Recent advances in neuropeptide signaling in drosophila, from genes to physiology and behavior. Prog Neurobiol 179:101607

Nassel DR, Zandawala M (2020) Hormonal axes in drosophila: regulation of hormone release and multiplicity of actions. Cell Tissue Res 382(2):233–266

Nock TG, Chand D, Lovejoy DA (2011) Identification of members of the gonadotropin-releasing hormone (GnRH), corticotropin-releasing factor (CRF) families in the genome of the holocephalan, Callorhinchus milii (elephant shark). Gen Comp Endocrinol 171(2):237–244

Oh Y, Lai JS, Min S, Huang HW, Liberles SD, Ryoo HD, Suh GSB (2021) Periphery signals generated by Piezo-mediated stomach stretch and Neuromedin-mediated glucose load regulate the drosophila brain nutrient sensor. Neuron 109(12):1979–1995.e6

Okada R, Yamamoto K, Koda A, Ito Y, Hayashi H, Tanaka S, Hanaoka Y, Kikuyama S (2004) Development of radioimmunoassay for bullfrog thyroid-stimulating hormone (TSH): effects of hypothalamic releasing hormones on the release of TSH from the pituitary in vitro. Gen Comp Endocrinol 135(1):42–50

Okada R, Miller MF, Yamamoto K, De Groef B, Denver RJ, Kikuyama S (2007) Involvement of the corticotropin-releasing factor (CRF) type 2 receptor in CRF-induced thyrotropin release by the amphibian pituitary gland. Gen Comp Endocrinol 150(3):437–444

On JSW, Arokiaraj AWR, Chow BKC (2019) Molecular evolution of CRH and CRHR subfamily before the evolutionary origin of vertebrate. Peptides 120:170087

Paris M, Laudet V (2008) The history of a developmental stage: metamorphosis in chordates. Genesis 46(11):657–672

Paul B, Sterner ZR, Buchholz DR, Shi YB, Sachs LM (2022) Thyroid and corticosteroid signaling in amphibian metamorphosis. Cells 11(10)

Perrin MH, Donaldson CJ, Chen R, Lewis KA, Vale WW (1993) Cloning and functional expression of a rat brain corticotropin releasing factor (CRF) receptor. Endocrinology 133(6):3058–3061

Perrin M, Donaldson C, Chen R, Blount A, Berggren T, Bilezikjian L, Sawchenko P, Vale W (1995) Identification of a second corticotropin-releasing factor receptor gene and characterization of a cDNA expressed in heart. Proc Natl Acad Sci USA 92(7):2969–2973

Potter E, Behan DP, Fischer WH, Linton EA, Lowry PJ, Vale WW (1991) Cloning and characterization of the cDNAs for human and rat corticotropin releasing factor-binding proteins. Nature 349(6308):423–426

Power ML, Schulkin J (2006) Functions of corticotropin-releasing hormone in anthropoid primates: from brain to placenta. Am J Hum Biol 18(4):431–447

Raftogianni A, Roth LC, Garcia-Gonzalez D, Bus T, Kuhne C, Monyer H, Spergel DJ, Deussing JM, Grinevich V (2018) Deciphering the contributions of CRH receptors in the brain and pituitary to stress-induced inhibition of the reproductive axis. Front Mol Neurosci 11:305

Rajamannar P, Arokiadhas I, Levkowitz G, Biran J (2021) The neurohypophysis and urophysis: ancient piscine neurovascular interfaces. In: Grinevich V, Dobolyi Á (eds) Neuroanatomy of neuroendocrine systems. Springer, Cham, pp 95–118

Ren J, Flamant F (2023) Thyroid hormone as a temporal switch in mouse development. Eur Thyroid J 12(2)

Rousseau K, Dufour S, Sachs LM (2021) Interdependence of thyroid and corticosteroid signaling in vertebrate developmental transitions. Front Ecol Evol 9

Sachs LM, Buchholz DR (2017) Frogs model man: in vivo thyroid hormone signaling during development. Genesis 55(1–2)

Saffran M, Schally AV (1955a) In vitro bioassay of corticotropin: modification and statistical treatment. Endocrinology 56(5):523–532

Saffran M, Schally AV (1955b) The release of corticotrophin by anterior pituitary tissue in vitro. Can J Biochem Physiol 33(3):408–415

Scharrer B (1987) Insects as models in neuroendocrine research. Annu Rev Entomol 32:1–16

Sower SA (2018) Landmark discoveries in elucidating the origins of the hypothalamic-pituitary system from the perspective of a basal vertebrate, sea lamprey. Gen Comp Endocrinol 264:3–15

Spierling SR, Zorrilla EP (2017) Don't stress about CRF: assessing the translational failures of CRF(1) antagonists. Psychopharmacology 234(9–10):1467–1481

Stenzel P, Kesterson R, Yeung W, Cone RD, Rittenberg MB, Stenzel-Poore MP (1995) Identification of a novel murine receptor for corticotropin-releasing hormone expressed in the heart. Mol Endocrinol 9(5):637–645

Vale W, Spiess J, Rivier C, Rivier J (1981) Characterization of a 41-residue ovine hypothalamic peptide that stimulates secretion of corticotropin and beta-endorphin. Science 213(4514):1394–1397

Vaughan J, Donaldson C, Bittencourt J, Perrin MH, Lewis K, Sutton S, Chan R, Turnbull AV, Lovejoy D, Rivier C et al (1995) Urocortin, a mammalian neuropeptide related to fish urotensin I and to corticotropin-releasing factor. Nature 378(6554):287–292

Villalobos C, Nunez L, Garcia-Sancho J (2004) Anterior pituitary thyrotropes are multifunctional cells. Am J Physiol Endocrinol Metab 287(6):E1166–E1170

Vom Berg-Maurer CM, Trivedi CA, Bollmann JH, De Marco RJ, Ryu S (2016) The severity of acute stress is represented by increased synchronous activity and recruitment of hypothalamic CRH neurons. J Neurosci 36(11):3350–3362

Watanabe Y, Grommen SVH, De Groef B (2016) Corticotropin-releasing hormone: mediator of vertebrate life stage transitions? Gen Comp Endocrinol 228:60–68

Watanabe Y, Grommen SVH, De Groef B (2018) Thyrotropic activity of corticotropin-releasing hormone in an altricial bird species, the zebra finch (Taeniopygia guttata). Gen Comp Endocrinol 258:99–108

Wyder S, Kriventseva EV, Schroder R, Kadowaki T, Zdobnov EM (2007) Quantification of ortholog losses in insects and vertebrates. Genome Biol 8(11):R242

Yang Z, Huang R, Fu X, Wang G, Qi W, Mao D, Shi Z, Shen WL, Wang L (2018) A post-ingestive amino acid sensor promotes food consumption in drosophila. Cell Res 28(10):1013–1025

Yun S, Furlong M, Sim M, Cho M, Park S, Cho EB, Reyes-Alcaraz A, Hwang JI, Kim J, Seong JY (2015) Prevertebrate local gene duplication facilitated expansion of the neuropeptide GPCR superfamily. Mol Biol Evol 32(11):2803–2817

Zandawala M, Marley R, Davies SA, Nassel DR (2018) Characterization of a set of abdominal neuroendocrine cells that regulate stress physiology using colocalized diuretic peptides in drosophila. Cell Mol Life Sci 75(6):1099–1115

Key Literature

Cardoso JC, Bergqvist CA, Felix RC, Larhammar D (2016) Corticotropin-releasing hormone family evolution: five ancestral genes remain in some lineages. J Mol Endocrinol 57(1):73–86 *Extensive phylogenetic analysis establishing the model of five ancestral CRH-like peptides*

Cavanaugh DJ, Geratowski JD, Wooltorton JR, Spaethling JM, Hector CE, Zheng X, Johnson EC, Eberwine JH, Sehgal A (2014) Identification of a circadian output circuit for rest:activity rhythms in Drosophila. Cell 157(3):689–701 *Among the first studies addressing the pysiological role of DH44-related circuits*

De Groef B, Goris N, Arckens L, Kuhn ER, Darras VM (2003b) Corticotropin-releasing hormone (CRH)-induced thyrotropin release is directly mediated through CRH receptor type 2 on thyrotropes. Endocrinology 144(12):5537–5544 *First evidence that the CRH-mediated HPT axis activation is depending on CRHR2*

Deussing JM, Chen A (2018) The corticotropin-releasing factor family: physiology of the stress response. Physiol Rev 98(4):2225–2286 *Comprehensive review of the CRH family of neuropeptides and their cognate receptors*

Dunn-Fletcher CE, Muglia LM, Pavlicev M, Wolf G, Sun MA, Hu YC, Huffman E, Tumukuntala S, Thiele K, Mukherjee A, Zoubovsky S (2018) Anthropoid primate–specific retroviral element THE1B controls expression of CRH in placenta and alters gestation length PLOS. Biology 16(9):e2006337-10.1371/journal.pbio.2006337

Elphick MR, Mirabeau O, Larhammar D (2018) Evolution of neuropeptide signalling systems. J Exp Biol 221(Pt 3) *Extensive study on the evolution of neuropeptides and their receptors.*

Grone BP, Maruska KP (2015b) A second corticotropin-releasing hormone gene (CRH2) is conserved across vertebrate classes and expressed in the hindbrain of a basal neopterygian fish, the spotted gar (Lepisosteus oculatus). J Comp Neurol 523(7):1125–1143 *First description of CRH2 initating the current evolutionary model*

Herman JP, McKlveen JM, Ghosal S, Kopp B, Wulsin A, Makinson R, Scheimann J, Myers B (2016) Regulation of the hypothalamic-pituitary-adrenocortical stress response. Compr Physiol 6(2):603–621 *Detailed review on mechanisms and circuits regulating HPA axis activity*

Hsu SY, Hsueh AJW (2001) Human stresscopin and stresscopin-related peptide are selective ligands for the type 2 corticotropin-releasing hormone receptor. Nat Med 7(5):605–611. https://doi.org/10.1038/87936

Lewis K, Li C, Perrin MH, Blount A, Kunitake K, Donaldson C, Vaughan J, Reyes TM, Gulyas J, Fischer W, Bilezikjian L, Rivier J, Sawchenko PE, Vale WW (2001) Identification of urocortin. III an additional member of the corticotropin-releasing factor (CRF) family with high affinity for the CRF2 receptor Proceedings of the National Academy of Sciences 98(13):7570–7575. https://doi.org/10.1073/pnas.121165198

Nassel DR, Winther AM (2010) Drosophila neuropeptides in regulation of physiology and behavior. Prog Neurobiol 92(1):42–104 *Excellent and detailed overview on neuropeptide systems in Drosophila*

Reyes TM, Lewis K, Perrin MH, Kunitake KS, Vaughan J, Arias CA, Hogenesch JB, Gulyas J, Rivier J, Vale WW, Sawchenko PE (2001) Urocortin. II: A member of the corticotropin-releasing factor (CRF) neuropeptide family that is selectively bound by type 2 CRF receptors Proceedings of the National Academy of Sciences 98(5):2843–2848. https://doi.org/10.1073/pnas.051626398

Rousseau K, Dufour S, Sachs LM (2021) Interdependence of thyroid and corticosteroid signaling in vertebrate developmental transitions. Front Ecol Evol 9 *A broad overview on the interactions of HPA/I and HPT axes*

Vale W, Spiess J, Rivier C, Rivier J (1981) Characterization of a 41-residue ovine hypothalamic peptide that stimulates secretion of corticotropin and beta-endorphin. Science 213(4514):1394–1397 *First isolation and characterization of CRH*

Co-evolution of PACAP, VIP and Related Peptides and Their Receptors as Neuroendocrine Signalling Dyads in Vertebrates

11

Perspectives for Mammalian Physiology, Pharmacology and Medicine

Limei Zhang, Vito Hernandez, and Lee E. Eiden

Abstract

VIP and PACAP, originating from a shared primordial peptide-encoding gene in vertebrates, have an intertwined evolutionary history through two whole-genome duplications inherited by mammals. Based on their binding affinities they also share three structurally and evolutionarily related receptors. This review provides a brief history of VIP and PACAP, the peptides co-encoded with PACAP and VIP within their prohormone precursor proteins, and the PACAP, VIP, PACAP-related peptide (PRP) and VIP-related peptide (VRP/PHM/PHI) receptors. These are embedded within the growth hormone-releasing hormone/secretin division of the secretin superfamily of peptides and receptors which also includes the endocrine peptide-receptor dyads comprising the CRH/urocortin 1–3, calcitonin/CGRP/adrenomedullin (ADM)/amylin (IAPP), parathyroid hormone/TIP39 and GIP/glucagon/GLP-1/glucagon-related peptide subfamilies.

L. Zhang · V. Hernandez
Department of Physiology, School of Medicine, National Autonomous University of Mexico, Mexico City, Mexico

Section on Molecular Neuroscience, Laboratory of Cellular and Molecular Regulation, NIMH-IRP, Bethesda, MA, USA

L. E. Eiden (✉)
Section on Molecular Neuroscience, Laboratory of Cellular and Molecular Regulation, NIMH-IRP, Bethesda, MA, USA
e-mail: eidenl@mail.nih.gov

V. Grinevich, R. F. Oliveira (eds.), *Evolutionary and Comparative Neuroendocrinology*, Masterclass in Neuroendocrinology 17,
https://doi.org/10.1007/978-3-031-80209-6_11

Our review emphasizes four main points about the PACAP/VIP peptide signalling system. First, while PACAP and VIP and their receptors display evolutionary interdependence, evolutionary pressures on ligands and receptors are not perfectly aligned. Second, evolution of peptide and receptor structure and function is also linked to the evolution of broader cellular features such as prohormone production, processing, vesicular packaging and G-protein-coupling mechanisms. Therefore, functional deductions cannot rely solely on paralogous lineages or amino acid sequences. Third, the evolution of neuropeptides and their expanding functions, from hormone to paracrine factor to neurotransmitter, involves intricate control of peptide and receptor gene expression by cell-specific transcription factors. Fourth, understanding PACAP and VIP evolution offers insights into conservation of function, particularly within mammals, which can inform translational research for peptide-based therapeutics in humans. These might also be provisionally applied to other neuropeptides, including glucagon, GIP, secretin, CRH, GRP, calcitonin, CGRP, TIP39, adrenomedullin and amylin, that interact with family B1G-protein-coupled receptors.

Keywords

Pituitary adenylate cyclase-activating polypeptide · Vasoactive intestinal polypeptide · PACAP-related peptide · PHI/PHM · Family B1 peptide receptors · Vertebrate neuropeptide evolution

List of Abbreviations

ACTH	Adrenocorticotropic hormone
ADM	Adrenomedullin
cAMP	Cyclic adenosine monophosphate
CCK	Cholecystokinin
CGRP	Calcitonin gene-related peptide
CRH	Corticotropin-releasing hormone
FSH	Follicle-stimulating hormone
GH	Growth hormone
GHRH	Growth hormone-releasing hormone
GIP	Gastric inhibitory peptide
GLP	Glucagon-like peptide
GPCR	G-Protein-coupled receptor
IAPP	Islet amyloid polypeptide (amylin)
LDCV	Large dense-core vesicle
LHRH	Luteinizing hormone-releasing hormone
LH	Luteinizing hormone
PACAP	Pituitary adenylate cyclase-activating polypeptide
PAC1	PACAP receptor type 1

PHI/PHM	Peptide histidine isoleucine (PHI) or Peptide histidine methionine, human counterpart of PHI (PHM); associated with preproVIP and referred to here as VRP
PRL	Prolactin
PRP	PACAP-related peptide
SCT	Secretin
TIP39	Tuberoinfundibular peptide of 39 residues
TRH	Thyrotropin (thyroid hormone stimulating hormone)-releasing hormone
VIP	Vasoactive intestinal polypeptide
VPAC1	VIP/PACAP receptor type 1
VPAC2	VIP/PACAP receptor type 2
VRP	VIP-related peptide (*aka* PHI/PHM depending upon species)

11.1 Introduction

The neuropeptide vasoactive intestinal polypeptide (VIP) was isolated from the pig intestine and characterized by Sami Said in the laboratory of Victor Mutt in 1970 (Said and Mutt 1970). Said, initially searching for a lung vasodilator peptide because such a factor was postulated on physiological grounds, used a vasodilator bioassay to examine fractions remaining on methylcellulose columns after purification of gut peptides, in a general separation scheme for peptides from pig intestine developed in the Mutt laboratory (Said 2007). Since, like secretin (SCT), VIP was isolated from gut, a potential incretin role for VIP was considered. However, the high concentrations of VIP in the gut were later appreciated to be due to its presence in neurons, rather than the enterochromaffin cells that harbour secretin, glucagon-like peptide-1 (GLP-1), gastrin, and other gut peptides like motilin, pancreastatin and cholecystokinin (CCK), also discovered in the Mutt laboratory. In addition to its presence in lung and gut, VIP was found in brain and in immunocytes, and thus its neurotransmitter and potential immunokine-like actions also became appreciated (Said and Rosenberg 1976; Voice et al. 2002).

Pituitary adenylate cyclase-activating polypeptide (PACAP) was discovered almost 20 years later by a very similar logic to that employed in the discovery of VIP by Said and Mutt. Arimura and colleagues examined side fractions from the peptide extraction of sheep hypothalamus used to obtain hypophysiotropic hormones, employing a bioassay involving cyclic adenosine monophosphate (cAMP) generation in perfused anterior pituitary gland, and based, in practice as well as principle, on the cAMP-elevating effects of luteinizing hormone-releasing hormone (LHRH) and other hormone-releasing hormones (Arimura 1992). Systematic bioassay led to the discovery of the pituitary adenylate cyclase-activating polypeptides, PACAP38, and later PACAP27 (Kimura et al. 1990; Miyata et al. 1989). Their sequencing allowed a comparison with the previously discovered VIP, and the sequence similarity strongly implied an evolutionary relationship between these peptides. Similarly to VIP, PACAP was also detected in the immune system and in brain neurons (Arimura 1992). Despite its high concentration in hypothalamus, PACAP's hypophysiotropic role in mammals is less prominent than the earlier discovered peptide hormones, corticotropin-releasing hormone (CRH), LHRH, thyrotropin-releasing hormone (TRH), growth hormone-releasing hormone (GHRH) and somatostatin ((Rawlings and Hezareh 1996), but see Koves et al. (2020)). Thus, while there is evidence for the involvement of both peptides in hypophysiotropic function in some mammalian species (Palkovits et al. 1995; Vijayan et al. 1979), neither PACAP nor VIP appear to play a major role in hypophysiotropic regulation in most species. However, like VIP, PACAP was soon found to have multiple functions outside of the tissue, in this case hypothalamus, from which it was first isolated.

Initial ideas about the functions of VIP and PACAP in mammals, by the early yet prescient assessments of Said and Arimura, (Said and Rosenberg 1976; Arimura 1992), were guided by their great similarity of structure, and an implication of hormonal function based on their site of initial identification. In fact, the similarity of structure and the later similarity of gene structure upon cloning of VIP and PACAP (see Sherwood et al. (2000)) validated the assignment of VIP and PACAP to the so-called 'family B1' peptide group. It has since been confirmed that VIP and PACAP are indeed highly evolutionarily related, and that their initial functions in ancestral vertebrates were most likely as hypophysiotropic hormones (Sherwood et al. 2000). The same is true of the cognate receptors for VIP and PACAP, and related peptides, which comprise a small and unique GPCR family (see Sect. 11.3).

The evolution of VIP and PACAP and their receptors within the vertebrate lineage represents a progression from a presumed neuroendocrine function toward highly diverse functions, unique to each peptide, within neuronal, neuroendocrine and immune cells. This parallels the evolutionary path of other members of the B1 peptide/peptide receptor family (see Sect. 11.3). Some of these have either predominantly retained hypophysiotropic (CRH, GHRH) or peripheral gut hormonal roles (secretin, GLP-1, glucagon, calcitonin and gastric inhibitory peptide (GIP)). Like PACAP and VIP, all of the above-named family B1 peptides, except for GIP, do have documented neurotransmitter roles. CGRP is especially noteworthy as a neurotransmitter/neuromodulator in both central and peripheral nervous systems.

Here, we describe PACAP and VIP expression in various mammalian tissues and organs and provide a retrospective of how the peptides came to exist in mammals via genome, gene and exon duplication in the vertebrate lineage, beginning more than 600 million years ago. This provides insights into how, genetically, and why, physiologically, VIP and PACAP and related peptides might have acquired their distinct functions, and lost others. A brief comparison of vertebrates with peptide-expressing invertebrates such as flies, worms and crustaceans (see Sect. 11.8) provides additional evolutionary perspective on the neuropeptide-GPCR genomic niche, as exemplified by PACAP and VIP.

11.2 From Whole-Genome Duplications to Point Mutations: Drivers for Evolution of Diversity Among the Family B1 Ligand-Receptor Dyads

Gene duplication is an essential source of raw material for the evolution of organismic novelty and diversity (Lynch and Conery 2000). The postulation of early and sequential vertebral whole-genome duplication (the 1R + 2R hypothesis) (Cardoso et al. 2020; Dehal and Boore 2005; Hwang et al. 2013; Kuzmin et al. 2020, 2022; Lamb 2021; On et al. 2015) provides a useful frame of reference to assess the competing processes of elimination of gene duplicates, and stabilization of duplications through neofunctionalization. The 1R and 2R events likely occurred at the base of the vertebrates, beginning 500–600 million years ago (Dehal and Boore 2005). The first is thought to have occurred soon after the separation of chordates from non-chordates, and the second soon after chordate differentiation into the urochordate and vertebrate lineages. The "1R + 2R" hypothesis predicts that a given gene is an original member within one of 17 ancestral vertebrate chromosomal regions, inherited through successive whole-genome duplications as so-called paralogons (Lamb 2021). Gene members of each paralogon are therefore potentially represented, with accumulated alterations, four times within a vertebral genome. Thus, a given peptide- or receptor-encoding gene might have as many as three paralogs corresponding to its appearance as a singleton within the primordial paralogon.

If the genomes of modern-day vertebrates indeed reflect these processes, successive whole-genome duplications would result in the retention of all four potential copies of a given pre-duplication gene assuming sufficient diversification among them; the retention of only one copy because of elimination of all duplications; or the retention of two or three copies if diversification has not been successful for all four copies. Further sub-lineages, to the level of speciation, exhibit 'post-whole genome' chromosomal, gene and exon duplications, gene and exon losses and rearrangements and alternative exon splicing. Mutations in the coding regions of genes, and in gene promoters, exert meaningful variation in the inheritable functions of peptide-receptor interactions that regulate myriad species-specific physiological tasks and lifestyles. Thus, the two major whole-genome duplications may be considered to represent the genetic inheritance of modern vertebrates such as mammals, and the structural matrix for the additional genetic alterations that define differences between closely-related species, including individual gene duplication, exon duplication and elimination, exon cassette insertion and point mutations affecting both promoters and coding regions of the gene. All of the above, singly and in combination, represent drivers for diversity of ligand-receptor patterning and function in each organism.

11.3 Vertebrate Evolution of VIP, VRP, PACAP, PRP and Their Receptors: A Structural Overview

Using the 1R + 2R construct to position the relationships between extant prohormone-containing PACAP- and VIP-related sequences, Cardoso et al. hypothesized that an ancestral PACAP-like peptide-encoding gene underwent duplication of the exon containing the peptide sequence prior to the 1R whole-genome duplication, resulting in two sequences, then duplicated again at 2R, resulting in a total of four prohormone-containing genes, giving rise to the sequences of these peptides found in modern vertebrate lineages (Fig. 11.1; (Cardoso et al. 2007, 2020)). PACAP and VIP retained their primordial 'co-peptides', i.e. PRP and VRP, while this exon was subsequently lost from GHRH and SCRT.

Although the cognate receptors for the secretin sub-family peptides have been produced within the same vertebrate evolutionary matrix as the peptides themselves, receptor evolution is somewhat more complex, answering to functional constraints on preservation of coupling to receptor guanine nucleotide exchange function and arrestin interaction. Notably, the designation of family B1 peptides and receptors, as such, is based on an initial classification of GPCRs themselves into five families, designated A-E by Fredriksson and colleagues (Fredriksson et al. 2003), with a further division of family B into the peptide (B1) and adhesion (B2) receptor families (Nordstrom et al. 2009). We retain the family B1 designation here (see Fig. 11.2) rather than the sometimes-used 'secretin superfamily', because group E (Fig. 11.2, and see Hwang et al. (2013)) within family B1 is also often called the 'secretin subfamily', albeit GHRH subfamily might be more appropriate. It is noteworthy that while the genomes of non-chordate animals contain putative coding information for PACAP-like peptides, there is no evidence for the corresponding receptors,

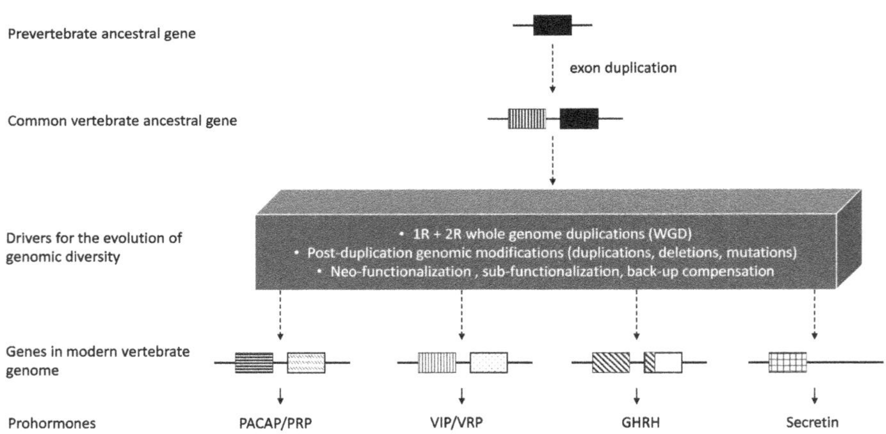

Fig. 11.1 Schematic illustration of the evolution of the vertebrate secretin peptide subfamily. The six peptides that constitute the endogenous ligands of the secretin subfamily of family B1 peptide ligands/G-protein-coupled receptors are hypothesized to derive from a pre-vertebrate ancestral gene duplication, followed by two successive whole-genome duplication (1R + 2R) events occurring at the base of vertebrate evolution, followed by a combination of neofunctionalization, exon expansion and exon deletion within the two peptide-encoding cassette of each of the four genes. Modified from: Mustafa and Eiden (2006)

and thus, the R1 + R2 hypothesis is restricted to the evolution of the PACAP/VIP peptides and their receptors within the vertebrates (Cardoso et al. 2020).

The sequences of the six peptides PACAP, PRP, VIP, VRP, GHRH and SCT are shown in Fig. 11.3. PACAP is the most conserved of these peptides. Especially noteworthy is the conservation of the PACAP28-38 sequence, despite the fact that a lack of conservation of this sequence might be expected, on the grounds that the other peptides all feature sequences of less than 30 amino acids that bind with high affinity to their respective, evolutionarily highly-related receptors, and that PACAP27 binds to the PAC1 receptor with an affinity equivalent to that of PACAP38. Nevertheless, there is a structural basis for distinct attachment sites on the PAC1 receptor for amino acids beyond 27 (Piper et al. 2022).

Corresponding analysis of the evolution of the receptors for PACAP, VIP, PRP, GHRH and secretin reveal a similar process with respect to the effects of 1R + 2R whole-genome duplication, but with considerably more complexity. This reflects the fact that the overlap among PACAP, VIP, PRP and VRP and their receptors is complexly entangled with the overlap between all of these and other ligand-receptor dyads within the family B1 ligand-receptor family (Cardoso et al. 2007). Thus, family B1 appears to have evolved via a combination of local and whole-genome duplication, with some gene (local) duplications occurring before and some after the 1R + 2R whole-genome duplications of the ancestral vertebrate genome (Hwang

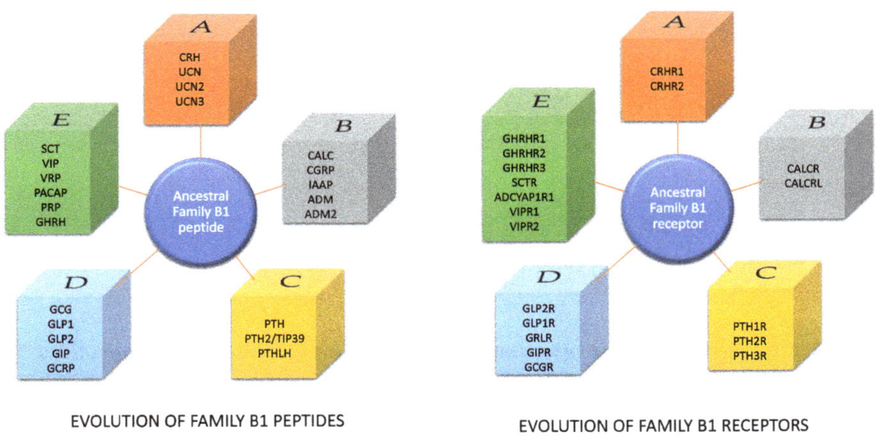

Fig. 11.2 Phylogeny of the family B1 peptide and receptor families. The phylogenetic relationships among the family B peptide and receptor families are illustrated after Hwang et al. (2013), and represent a reconstruction from three early gnathostome ancestral chromosome (GAC) linkage groups. The diagram calls attention to the division of family B peptides into five groups, commonly called group A (CRH and urocortins; group B (calcitonin, CGRP, adrenomedullin and amylin/IAPP), group C (PTH, PTH-like hormone and PTH2/TIP39), group D (glucagon, glucagon-like peptide and gastric inhibitory peptide) and group E (GHRH, secretin, PACAP and PRP, and VIP and VRP), with corresponding phylogenetic relationships among their cognate G-protein-coupled receptors. Phylogenetic relationships are inferred by a combination of synteny-based genomic analysis and sequence-based translated protein analysis

PACAP
HSDGIFTDSY SRYRKQMAVK KYLAAVLGKR YKQRVKNK Human
HSDGIFTDSY SRYRKQMAVK KYLAAVLGKR YKQRVKNK Mouse
HIDGIFTDSY SRYRKQMAVK KYLAAVLGKR YKQRVKNK Chicken
HSDGIFTDSY SRYRKQMAVK KYLAAVLGKR YRQRYRNK Salmon

PRP
DVAHGILNEA YRKVLDQLSA GKHLQSLVAR GVGGSLGGGA GDDAEPLS Human
DVAHEILDKA YRKVLDQLSA RKYLQSVVAR GAGDEPRRHA VDDPAPLT Mouse
-HADGIFSKA YRKLLGQLSA RNYLHSLMAK RV----GSGL GDEAEPLS Chicken
-HADGMFNKA YRKALGQLSA RKYLHSLMAK RVGG--GSTM EDDSEPLS Salmon

VIP
HSDAVFTDNY TRLRKQMAVK KYLNSILN Human
HSDAVFTDNY TRLRKQMAVK KYLNSILN Mouse
HSDAVFTDNY SRFRKQMAVK KYLNSVLT Chicken
HSGAVFSDNY SHYGKQMAIK KYLKSALSN Salmon

VRP
HADGVFTSDF SKLLGQLSAK KYLESLM Human
HADGVFTSDF SKLLGQLSAK KYLESLI Mouse
HADGIFTSVY SHLLAKLAVK RYLHSLI Chicken
HADGLFTSGY SKLLGQLSAK DYLESLIA Salmon

GHRH
YADAIFTNSY RKVLGQLSAR KLLQDIMSRQ QGESNQERGA RARL Human
HVDAIFTTNY RKLLSQLYAR KVIQDIMNKQ GERIQEQRAR LS Mouse
HADGIFSKAY RKLIGQLSAR NYLRSLMAKR VGGASSGLGD EAEPLS Chicken
HADGMFNKAY RKALGQLSAR KYLHSLMAKR VGGGSTMEDD SEPLS Salmon

SCT
HSDGTFTSEL SRLREGARLQ RLLQGLV Human
HSDGMFTSEL SRLQDSARLQ RLLQGLV Mouse
HSDGLFTSEY SKMRGNAQVQ KFIQNLM Chicken

Fig. 11.3 **Sequences across representative vertebrates for secretin-subfamily peptides**. The sequence similarities between the six peptides of the secretin subfamily are depicted. Sequence conservation is greatest for PACAP and least for GHRH. Note the SCT gene has been lost in fish

et al. 2013). Figures 11.1 and 11.2 illustrate this progression from the hypothetical vertebral ancestor through land animals, but a 3R event in teleost fish makes a more comprehensive structural narrative difficult.

11.4 Ligand-Receptor Co-Evolution in Family B1

It is intuitively attractive to suppose that ligands and receptors co-evolve, yet non-vertebrate chordates, for example, possess a PACAP gene but not genes for its recognized receptors in vertebrates (Cardoso et al. 2007). Likewise, the gene encoding a putative PRP receptor is thought to have been lost in the vertebrate chordata subphylum (Chow et al. 2003), albeit PRP persists with remarkable sequence conservation throughout mammalian species.

As illustrated in the cases of lipophilic hormone receptors and CRH binding protein, it appears that receptors find their ligands after first becoming established as proteins due to other attributes (Duessing et al., this volume; (Bridgham et al. 2006)). While it must be true that at some level ligands and receptors stabilize each other genetically to create functionally durable ligand-receptor dyads, co-evolution represents an integration of the separate evolutionary pathways for each that remains mysterious at the level of actual molecular convergence or mutual stabilization. This is in part because there is promiscuity or overlap between receptors for these ligands. In fact, there is evidence that PACAP engages other receptors, besides the canonical ones, during development, and that binding of PACAP to these receptors could underlie developmentally related functions of the peptide ((Bakalar et al. 2022; Bakalar et al. 2023) and references therein).

The process of receptor deorphanization has also radically changed our perception of the ligand-receptor co-evolution process (Foster et al. 2019; Yosten et al. 2020). Systematic receptor deorphanization reveals that there is considerable pharmacological cross-talk, or specificity overlap, among peptides and their receptors (Foster et al. 2019). Neofunctionalization for a peptide ligand-receptor dyad during vertebrate evolution presumably depends on the actual opportunity for functional engagement between corresponding peptide-secreting, and peptide receptor-bearing cells, whether neuronal, inflammatory, immune, endocrine or neuroendocrine. This in turn is a complex function of the 'broadcast radius' or effective range of the peptide after its secretion/release (Chini et al. 2017); the location of the receptor; the coupling of the receptor to downstream signalling components within the target cell; and the duration of the signal relative to downstream metabotropic functions and their temporal range. Progress in these areas has sharpened our perspectives on chemoanatomical considerations of potential peptide action and action radius, in both sessile and non-sessile tissues. The advent of in vivo sensors, to detect when and where peptides are actually secreted, physiologically, such as the one developed for PACAP based on the PAC1 receptor (Cola et al. 2024), will no doubt aid in further understanding of how peptide-receptor interactions are stabilized within physiological circuits, especially in complex tissues such as the brain.

An interesting example of peptide-receptor co-evolution is the requirement for C-terminal amidation, a consequence of prohormone processing of family B1 peptides. For several of these, C-terminal amidation is in fact a requirement for activity at the cognate receptor but, notably, not for PACAP (Emery et al. 2016). Yet another aspect of the evolution of family B1 ligand-receptor dyads was uncovered in the discovery of RAMPs, or receptor activity-modifying proteins (Lorenzen et al. 2019;

Morfis et al. 2003). These single-pass transmembrane proteins bind to specific receptors within family B1, most notably the receptors CALCRL and CALCR (Fig. 11.2), which recognize calcitonin, CGRP, amylin or adrenomedullin, depending upon the co-expression of these receptors with RAMP1, 2 or 3 (Christopoulos et al. 2003; Morfis et al. 2003). More recently, analysis of receptor-RAMP interactions in vitro has led to the suggestion that RAMPs also play a role in ligand recognition by PACAP and VIP receptors, although this has yet to be substantiated experimentally (Lorenzen et al. 2019).

A further connection between stabilization and neofunctionalization in the mutual evolution of family B1 peptides and receptors is the overlap between PACAP and VIP recognition of the three receptors PAC1, VPAC1 and VPAC2. The relative potencies are generally considered to be PACAP > VIP at PAC1, and PACAP = VIP at VPAC1 and VPAC2 (Harmar et al. 2012). May and colleagues documented a lack of compensation in the distinct functional deficits of PACAP and VIP knockout mice, expected if the two peptides indeed extensively share all three receptors physiologically (Girard et al. 2006). Within at least one system, the suprachiasmatic nucleus, VIPR knockout recapitulates the VIP knockout phenotype, and PAC1 knockout the PACAP knockout phenotype (Aton et al. 2005). This suggests that while pharmacological experiments indicate the potential for receptor sharing by PACAP and VIP, some degree of specialization of PACAP for PAC1 occurs. Establishing whether characteristics of ligand-receptor interaction predicted in vitro do apply to the actual engagement of PACAP versus VIP to the VPAC1 and VPAC2 receptors in vivo is important. This does not mean that the pharmacology of PACAP/VIP receptors cannot be exploited therapeutically using either ligand, but only that 'pharmacology does not necessarily recapitulate phylogeny', or even physiology, due to circumstances that might prevent endogenous ligand engagement with receptor, including the availability of sufficient concentration of ligand for efficient receptor activation in vivo. Despite the observed preference of PACAP for PAC1 in the brain, divergence in PACAP and PAC1 knockout phenotypes, in the periphery, suggests the use of VPAC receptors by PACAP there (Rasbach et al. 2018; Splitthoff et al. 2020). Family B1 receptor sensors, including the PAC1 sensor mentioned (Cola et al. 2024), in genetically tractable models such as mice and fish, are now available to address these questions more concretely.

The affinities of VRP (PHI/M) and PRP for PAC1, VPAC1 and VPAC2 establish their potential role in ligand-receptor co-evolution in the PACAP/VIP family (Alexander et al. 2016). Pharmacologically, there is no receptor for PRP in mammals, and indeed it has been proposed that the PRP receptor has been lost during evolution, existing in fish, but not in tetrapods (Tam et al. 2007). Despite this, it is notable that the structure of PRP is highly conserved across mammalian species, which is not expected for an amino acid sequence without a function requiring a high-affinity interaction with a receptor. In fact, it has been noted that changes in this sequence within the human PACAP prohormone compared to other primates are higher than expected, potentially a clue to an evolutionary pressure related to function for this peptide (Wang et al. 2005). Obviously, naming a receptor after a ligand does not necessarily make it so. Attempts to deorphanize peptide receptors reveals overlap and a range of affinities among peptides and receptors suggestive of 'opportunistic binding', so that unambiguous assignment of ligand to receptor reflecting in vivo physiology is not always possible, even in simple organisms (Beets et al. 2023). The LHRH-LHRHR dyad, extending beyond the chordate lineage to the ancestors of the earliest metazoans, yeast, is instructive. Here, the yeast alpha-mating factor (αMF) is not only structurally similar to the hypophysiotropic hormone LHRH, but LHRH and αMF can interact with each other's receptors (Loumaye et al. 1982). This suggests a potential evolutionary mechanism, with αMF acting as a 'primordial LHRH' with a general role in initiating mating itself, evolving to a more specific function in more complex organisms like mammals to facilitate hormone cascades that, along with behaviour, makes reproduction possible. LHRH is found, in the frog, both in the hypothalamus and in other endocrine and neuronal sites, suggesting a pathway for neofunctionalization taken by this peptide in some vertebrates (amphibians) but not others (Eiden et al. 1982; Jones et al. 1984). The cross-family and even cross-phylum ligand-receptor pharmacology conducted for LHRH, would be perhaps worthwhile to extend to PACAP and VIP, and especially to VRP and PRP, and their receptors (see Sect. 11.8).

11.5 Evolution of PACAP and VIP Function During Vertebrate Tissue Diversification

In this section, we consider the *functional* evolution of signalling by PACAP and VIP and their receptors in late vertebrates, in parallel with their *genetic* evolution described in the preceding sections. The secretin subfamily developed functionally from acting as neuroendocrine factors in early vertebrates, and as still seen in modern teleosts, to neurotransmitter-like functions in mammals. We would like to place the evolution of peptides and their receptors into the larger picture of the evolution of whole organ and regulatory systems, in which each ligand-receptor dyad contributes to expanded functions for homeostatic regulation, cytoprotection, inflammation, and prey, predator and social behaviours. Unfortunately, an extensive knowledge of the evolution of organ systems and their interplay, i.e., the evolution of physiology, is not available to us from prehistoric genetic or fossil records. Nevertheless, we can look at the evolution of the organ systems and subsystems in which VIP and PACAP signalling exists and examine constraints on the basic properties of intercellular communication—the essential metazoan role of peptide ligands and their receptors—in the immune, endocrine, neuroendocrine, neuronal and inflammatory systems in which PACAP and VIP function.

The evolution of the gut, the brain and the complex endocrine and neuroendocrine systems that bind them together constitute a basic narrative of animal evolution. This is reflected in the branch points that signal the emergence of a feeding tube (deuterostomy); an organized network of neuronal ganglia (cephalization); a cellular system of immunity and inflammation; and a system of locomotion uniting strategies for food- and water-seeking, toxin-escaping and predation with decision-making. This section highlights the evolution of PACAP/VIP from an ostensibly hypophysiotropic role, largely supplanted by the related peptide GHRH, to a more pleiotropic role in increasingly complex vertebrates, i.e. mammals, across a range of tissue types. A full explication of the gene regulatory networks for neural crest development, intimately involved in the formation of endocrine tissues and a jaw-like structure, for the mesoderm-derived gut and for other within-vertebrate innovations and specializations, does not yet exist. Thus, an accounting of how members of the secretin superfamily, including VIP and PACAP, came to colonize tissue compartments functionally and differentially as they evolved and specialized cannot be systematic. Accordingly, the sections below are essentially anecdotal; emphasizing how the PACAP/VIP group distinguishes itself from other family B peptide-receptor dyads within group E and from groups A–D (see Fig. 11.2), and how these distinctions among evolved function might relate to physiology, and potentially pharmacology and therapeutics, associated with each group.

PACAP and VIP as Hypothalamic Hypophysiotropic Hormones PACAP and VIP are members of the secretin subfamily of family B1 peptides and their receptors (subfamily E in Fig. 11.2). Secretin is a peptide of the gut with perhaps more in common, in mammalian physiology, with group D, which includes the incretins GLP-1 and GIP, as well as glucagon, although a central nervous system role for secretin has recently been uncovered (Zhang et al. 2024). Subfamily E might be better termed the 'GHRH' subfamily, as the ancestral gene is likely to have been functionally most important as a releasing factor expressed in primordial hypothalamus, and controlling growth hormone release during and after development in proto-chordate hypothalamo-pituitary anlagen such as CNS-Hatschek's Pit of the cephalochordate amphioxus (McRory and Sherwood 1997; On et al. 2022). PACAP is a secretagogue for GH, PRL, ACTH and LH in various ex vivo and in vitro pituitary preparations ((Sherwood et al. 2000) and references therein; (Koves et al. 2020)). PACAP elevates prolactin, GH and LH blood levels in the ewe, as expected from the presence of PAC1 receptors on anterior pituitary cells (Sawangjaroen and Curlewis 1994), but is without effect, at physiological doses, on hormone release from perfused sheep pituitary ex vivo (Sawangjaroen et al. 1997; Sawangjaroen and Curlewis 1994). PACAP deletion from the genome of mice does not appear to result in major endocrine abnormalities, as would be expected if it were of major physiological importance as a hypothalamic releasing factor (Sherwood et al. 2007). An evolutionary perspective for PACAP's hypophysiotropic function is that a prominent GH-releasing activity in teleosts, and perhaps other non-tetrapod vertebrates, has been supplanted by GHRH in mammals (Sherwood et al. 2000). A similar evolutionary fate, loss of GHRH function and re-purposing for specialized function (neofunctionalization) within other tissue compartments, can be envisaged for VIP.

PACAP in the Retina PACAP has been a prominent neurotransmitter within the retinal system since intrinsically photosensitive retinal ganglion cells, ipRGCs, developed in mammals. These cells, which express the 'intrinsic photopigment' melanopsin (see Panda et al. (2005) and references therein) also express PACAP

(Hannibal et al. 2002). The role of PACAP in the ipRGC illustrates how neuropeptide co-transmission can increase the adaptability of basic physiological control mechanisms, such as circadian rhythmicity. Light modulates a circadian rhythm within the suprachiasmatic nucleus that maintains organismic sleep and metabolic rhythmicity. PACAP expression in these cells also waxes and wanes in a circadian fashion and appears to affect late night phase advance, but not early night phase delay (Lindberg et al. 2019). How this effect has contributed to mammalian survival is unknown. It is of interest that PACAPergic retinohypothalamic innervation of the SCN is complemented by local VIP expression. There is an emerging consensus that VIP acts through VPAC receptors to regulate rhythmic entrainment within the SCN, while PACAP acts through PAC1 receptors in phase entrainment (see Sect. 11.4). This function of PACAP may be evolutionarily connected to its role in regulating pupillary reflex, via a link between retinohypothalamic and parasympathetic circuits (Engelund et al. 2012; Keenan et al. 2016).

PACAP and VIP in the Autonomic Nervous System A major physiological role has emerged for PACAP as a sympathetic neurotransmitter involved in stress transduction (Guerineau 2019; Rudecki and Gray 2016; Stroth et al. 2011, 2013) and as a sensory neurotransmitter in neurons mediating light, pain, auditory, olfactory and taste perception (Drescher et al. 2006; Hannibal et al. 2002; Hegg et al. 2003; Moller et al. 1993; Wang et al. 2018; Zhang et al. 2021). Axelrod and colleagues first postulated that a non-cholinergic transmitter might function as a stress transducer at the adrenomedullary synapse (Thoenen et al. 1969). Wakade and colleagues initially proposed VIP for this role (Malhotra et al. 1989). However, the discovery of PACAP, and its potent catecholamine secretagogue effect in vivo (Edwards and Jones 1994), shifted focus to it, and PACAP was shown to be present in the cholinergic innervation of the adrenal medulla, and to be required to sustain the massive catecholamine release required for survival during prolonged hypoglycaemia (Hamelink et al. 2002). PACAP acts preganglionically in the sympathetic nervous system during cold stress responses (Gray et al. 2002), with its signalling properties at this site documented experimentally in cultured sympathetic neurons (Braas and May 1996), and in the parasympathetic system (Tompkins and Parsons 2008).

PACAP and VIP in Central Nervous System Function PACAP is important for stress responding both via the sympathetic nervous system, the hypothalamo-pituitary adrenal (HPA) axis, and within the brain (Stroth et al. 2011). VIP is found mainly in inhibitory neurons in the cerebral cortex, and is a marker for a large class of GABAergic interneurons in both rodent and human cortex (Smith et al. 2019) (Allen Brain Atlas Transcriptomics at <https://portal.brain-map.org/overview>). Both peptides have been mapped to diverse neuronal projections, circuits and nuclei throughout the mammalian brain (Hannibal 2002; Loren et al. 1979; Sims et al. 1980). PACAP is expressed in GABAergic neurons in rostral (olfactory) brain and in cerebellum, and mostly in glutamatergic neurons throughout the rest of the brain (Zhang et al. 2021). Involvement of both peptides in regulation of astroglial function has been championed by Magistretti and colleagues, especially in cortical tissue (Magistretti et al. 1998). The role of PACAP as a neuroprotectant has been observed in the contexts both of ischaemia and stroke, and inflammation following nerve injury (Armstrong et al. 2008; Reglödi et al. 2000).

PACAP and Sensory Nervous System PACAP is ubiquitous in sensorimotor-associated neuronal projections (Zhang et al. 2021). PACAP is strongly induced in dorsal root ganglion cells under conditions of chronic pain (Dickinson and Fleetwood-Walker 1999; Moller et al. 1993). Blockade of PACAP action via intrathecal administration of antagonist peptide has been reported to diminish allodynia associated with chronic pain (Takasaki et al. 2019). These actions of PACAP are reported to be mediated via PAC1 receptors located on astrocytes rather than other neurons in the spinal cord (Ohnou et al. 2016). It is not yet known whether this function of PACAP has evolved specifically in mammals, if it is conserved among all mammals including primates, and if it is widespread throughout the brain.

PACAP and Neuroprotection PACAP is pharmacologically neuroprotective in stroke and ischaemia (Reglödi et al. 2000), and deletion of endogenous PACAP exacerbates stroke damage (Chen et al. 2006; Nonaka et al. 2020; Ohtaki et al. 2008), demonstrating neuroprotection as an endogenous function of the peptide in brain and other tissues. Understanding how this function for the peptide might have evolved within mammals, or earlier in evolution, awaits further characterization of the cellular and molecular mechanisms through which neuroprotection occurs (Ohtaki et al. 2006).

PACAP and VIP in Immune/Inflammatory Function There is not yet a comprehensive immunophysiology for VIP and PACAP, including how and when they are expressed in lymphocytes and macrophages; what mechanisms control their secretion; and where the receptors with which PACAP and VIP interact are located (see Vaudry et al. (2009)). Waschek and colleagues have identified a potential role for both PACAP and VIP in a broad range of anti-inflammatory actions leading to neuroprotection after axonal injury, involving VIP and PACAP mRNA induction in injured neurons following axotomy (Armstrong et al. 2006, 2008). VIP and PACAP are implicated in a host of other anti-inflammatory processes, including septic shock (Delgado et al. 2003). However without definitive in situ localization and targeted genetic manipulation studies, it is not yet clear if VIP and PACAP have entered a distinct tissue compartment during their evolution to more pleiotropic function, or whether they are released mainly by neurons under conditions of injury, and mediate anti-inflammatory effects important in neuronal autoimmunity and other processes indirectly (Waschek 2013).

VIP and PACAP in Development VIP was identified as a neuronal survival factor first in spinal cord cultures (Brenneman and Eiden 1986), and then in whole-embryo developmental studies (Gressens et al. 1993). Neurodevelopmental roles for both VIP and PACAP may be related to their neuroprotective effects (Brenneman 2007). However, the receptors through which these important effects of VIP and PACAP are exerted are not yet characterized, limiting our understanding of how they might have arisen evolutionarily. At least some of PACAP's developmental effects could involve non-family B1 receptors for which PACAP has reasonable affinity and efficacy (Foster et al. 2019). At least one of these, the receptor for the opiate peptide BAM22P, is expressed in adult mice exclusively in sensory neurons (Lembo et al. 2002). The use of alternative receptors outside family B1 by either PACAP or VIP suggests pathways for evolutionary diversity more complicated than the co-evolution of family B1 peptides and their cognate receptors.

Evolution of various aspects of peptide production, packaging, secretion, proteolytic breakdown and receptor signalling are all considerations in how VIP and PACAP and related peptides have acquired functions within various tissue compartments as hormones, paracrine factors, neurotransmitters and inflammatory actors during mammalian evolution. Clearly, there is much additional information to be learned about where, how and when VIP and PACAP are released in endocrine, neuroendocrine, immune and neuronal environments, before these evolutionary questions can be properly addressed. Kuzmin and colleagues have posed the question about duplicated genes in general, whether evolutionary trajectories of duplicated genes are dictated by combined "functional and structural entanglement factors" (Kuzmin et al. 2020, 2022) such as those listed above for VIP and PACAP first messenger function.

11.6 Diversity and Evolution of Signalling Function

The divergent evolution of the molecular features of the VIP/PACAP ligand-receptor interaction within the B1 lineage is critical to understanding the evolution of the family and in practical terms, assessing PACAP/VIP-based drug targets for migraine, depression and other disorders in animal models. One reason why speculation about neofunctionalization in the secretin family is risky is that organism-specific endocrinology, neurochemistry, immunology and tissue compartmentation of these peptides is not equally well-characterized in representative organisms within various vertebrate clades, where structure and function might be compared. However, some comparisons between VIP and PACAP, and other family B1 ligand-receptor dyads are worthwhile as we consider the signalling properties of PACAP and VIP receptors.

The known VIP/PACAP receptors VPAC1, VPAC2 and PAC1 are present within all the major vertebrate families, or at least in which commonly used experimental animal models exist. All three are Gs-protein-coupled receptors both according to empirical investigation and to structural predictions (e.g. PredictProtein, http://ppopen.rostlab.org). However, soon after the discovery of PACAP, its major receptor was cloned and characterized functionally (Fig. 11.4), from both rodent and human, and found to couple to Gq as well (Pisegna et al. 1996; Pisegna and Wank 1993, 1996; Spengler et al. 1993). This coupling had two unique features. First, it was associated only with an exon insertion, the so-called 'hop' cassette, and second, it conferred Gq-coupling for activation by PACAP-38, but not PACAP-27, at least for rodent PAC1 (Spengler et al. 1993). Thus, its evolutionary significance, in terms

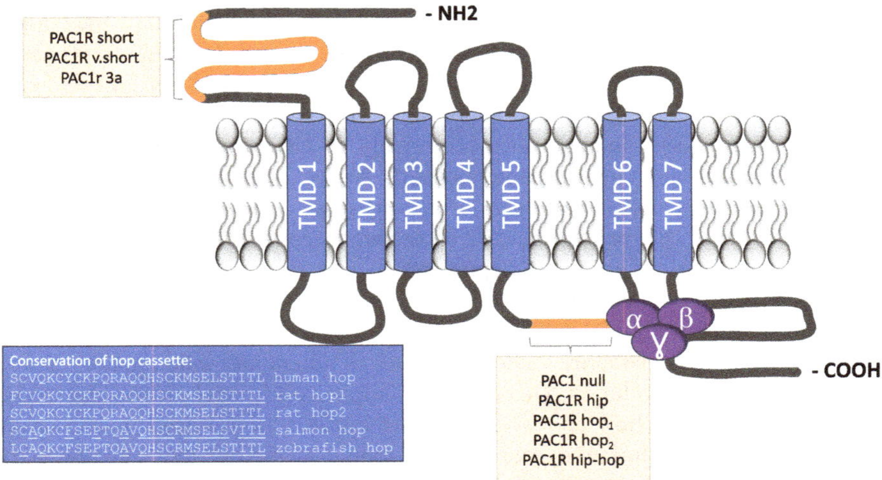

Fig. 11.4 **Alternative exon insertions in PAC1**. The PAC1 receptor is depicted with N-terminal and hip, hop, and hiphop insertions in IC3

of neofunctionalization, is twofold: a new unique role for PACAP38 compared to PACAP27, and a new and unique role for the hop exon variant of PAC1. Figure 11.3 depicts the internally evolved diversity of the PAC1 receptor, including the intracellular loop 3 insertions that are reported to impart coupling to Gq and Gs. Coupling to Gq may be related to PACAP's secretagogue function through its ability to link IP3 elevation to calcium mobilization, associated with the hop variant of PAC1 (Mustafa et al. 2007). The hop cassette is also associated with specific stress-related behaviours in zebrafish (Biran et al. 2020; Woods et al. 2014). However, PACAP-induced secretion of LDCV contents from chromaffin cells appears to be both PLC- and cAMP-dependent, consistent with coupling to both Gs and Gq (Morales et al. 2023). Expansion of signalling coupling from exclusively Gs, to also include Gq, may represent a key evolved function for PAC1 among the B1 Gs-coupled receptors, especially if this is related to a secretagogue function for PACAP unrelated to its ability to cause secretion via cAMP elevation, as per its original name.

Most of the family B1 ligand-receptor dyads can function similarly to PACAP and PAC1 for signalling in a neuroendocrine cell line, and this includes linkage through the guanine nucleotide exchange factor RapGEF2, to activation of the MAP kinase ERK (Xu et al. 2021). Signalling by PACAP in brain, however, lacks connection to the RapGEF2 cAMP→ERK signalling pathway, first discovered and characterized in neuroendocrine cell lines (Emery and Eiden 2012), even though connection to the pathway does exist for other Gs-coupled GPCRs in rodent brain (Jiang et al. 2017). May et al. have reported that PACAP signalling appears to rely upon endosomally transported PAC1 in a brainstem-to-amygdala pathway critically involved in adaptation to chronic pain (Missig et al. 2014). Thus, signalling modalities themselves have evolved, or at least diversified, to restrict PACAP (and VIP) function in certain cell types, perhaps due to the evolution of mechanisms for partitioning receptors into subcellular compartments in which signalling partnerships are promoted or restricted, such as at axosomatic compared to axodendritic synapses in CNS neurons.

The tendency for bricolage (Duboule and Wilkins 1998), or the evolutionary borrowing of pre-existing mechanisms for new physiological purposes, informs understanding of neofunctionalization of PACAP, VIP and their receptors in terms of tissue distribution and function. This may be especially important in the nervous system, where unique properties of G-protein coupling and intracellular signalling, at the molecular level, may endow some neuropeptides with distinct functions within complex circuits mediating stress, pain, threat and other inputs requiring complex integration prior to behavioural response.

11.7 Evolution of Neuropeptide Promoters and Transcription Factors

The evolution of PACAP and VIP from a GHRH-like hypophysiotropic signalling function to other hormonal and neurotransmitter functions throughout the brain, and in other tissues, implies evolution not only of peptide and receptor function, but of the tissue specificities of the promoters for the genes, encoding both peptides and receptors, that allows their expression in new tissue types. The discovery of the POU domain (Pit-1) elements of the growth hormone and prolactin genes (Herr et al. 1988; Nelson et al. 1988) heralded the arrival of a mechanistic explanation for tissue- and cell-specific expression of hormones and other gene-encoded first messengers by transcription factors with restricted tissue expression. Neuropeptide promoters are excellent markers for nervous system neofunctionalization, as well as powerful tools to gain genetic access to populations of neurons with discrete functions in the brain. Neuropeptide promoter Cre-driver mice have been used to demonstrate that hypothalamic galaninergic neurons drive parental behaviour (Kohl et al. 2018; Wu et al. 2014), calcitonin/CGRP neurons of parabrachial nucleus of brain stem drive threat-related behaviours (Campos et al. 2018; Palmiter 2018), PACAPergic neurons of MPOA drive torpor regulation (Hrvatin et al. 2020), VIP interneurons of cortex control integration of sensorimotor determinants of cortical output (Lee et al. 2019), and PACAP neurons of frontal cortex and brainstem drive endocrine and behavioural responses to stress, respectively (Jiang et al. 2023). It is worth noting that these tools have been underutilized to explore neuropeptide function itself. Furthermore, because it is possible to introduce Cre recombinase (so-called Cre-knock-in) into the 3′ untranslated exon of various neuropeptide genes, instead of building exogenous promoter-driven Cre transgenes, there has been very little study of the cis-active elements within neuropeptide and neuropeptide receptor genes that regulate their tissue-specific expression. With little attention paid to promoter-specific, lineage-determining transcription factors that have driven neuropeptide diversification from hormonal, to neuronal, to immune expression (and subsequent function), there is a corresponding lack of knowledge about how mutations in promoter regions might have driven neofunctionalization of neuropeptides and neuropeptide receptors by the evolution of their expression in new types of cells and tissues. In summary, a paucity of information exists to guide a systematic understanding of promoter evolution as a driver for neuropeptide neofunctionalization via colonization of new tissue and organ compartments.

Some information does exist about the promoter organization of the VIP and PACAP genes and the elements that might control tissue-specific expression. Studies of the VIP gene indicate two levels of regulation. A so-called tissue-specifier element (TSE) that controls the expression of VIP within some and not other cell types is located about 5 kb upstream of the transcription initiation of the prohormone gene. A cAMP-response element (CRE), located near the site of transcription initiation, is responsible for the cAMP-dependent increase in the rate of transcription above that of basal transcription in permissive cells (Hahm and Eiden 1998). Expression of an alternate transcript for the VIP gene in the anterior pituitary is controlled by glucocorticoids (Chew et al. 1997), and the promoter also harbours a cytokine-response element controlling regulation by ciliary neurotrophic factor (CNTF) (Symes et al. 1994). Although less is known about the PACAP promoter, it has been reported to harbour cis-active elements that confer control by the hypothalamic transcription factor TTF, which interestingly may allow circadian control of PACAP expression (Kim et al. 2002). Clues from transcription regulation of other neuropeptide genes in neuronal and neuroendocrine as well as immune contexts might shed some light on the convergence of genetic and functional neuropeptide evolution (Burbach 2002; Goodman 1990).

11.8 PACAP/VIP Peptides in Non-Chordates

This chapter is focused on the vertebral branch of the phylum Chordata. Here, an 'ancestral vertebrate' can be invoked to trace the evolution of PACAP/VIP peptides and their cognate receptors. In vertebrate tetrapods, these have evolved both structurally and functionally in organisms with increasingly diverse and complex tissue organization. PACAP-like peptides also exist in other vertebrate classes, and in non-chordates such as Arthropoda, and are of particular interest in animals that are frequently used as model organisms to study features of peptides in neuronal, endocrine and immune tissues.

Arthropoda and Nematoda contain much-studied model organisms, including *Drosophila*, *Cancer borealis* and *C. elegans* (see Langenhan et al. 2015; Macosko et al. 2009; Nusbaum et al. 2017). Information about peptide-GPCR signalling applicable to PACAP/VIP function in mammals and throughout the animal kingdom has been gleaned from the study of neuropeptide families and functions in these phyla. The *C. elegans* genome, for example, contains three gene families that encode insulin-like, FMRFamide-like (flp) (Li et al. 1999) and neuropeptide-like peptides, with the latter expressed mainly in neurons (Nathoo et al. 2001). While no flp or nlp peptides have obvious lineage connection to family B1, Janssen and colleagues have documented that *C. elegans* PDH (pigment-dispersing hormone) sequences have strong-to-moderate sequence homology to insect PDF neuropeptides and to crustacean PDHs, and that while most nematode GPCRs are within family A, there are several related to family B (calcitonin) (Janssen et al. 2010).

How does parallel pre-vertebrate and non-vertebrate evolution inform our understanding of family B1 peptide-receptor signalling in mammals? Accounts of how peptides from distant vertebrate classes and non-vertebral phyla can mimic the actions of family B peptides in mammals begin with exendin, the Gila monster-derived GLP-1 agonist, on which a battery of modern GLP-1 agonist drugs are based (Holst 2008). It extends to the interaction between sandflies and their mammalian hosts, in which the vasodilator actions of maxadilan (CDATCQFRKAIDDCQKQAHHSNVLQTSVQTTATFTSMDTSQLPGNS VFKECMKQKKEFKAGK), which has little sequence similarity with PACAP, allows the parasite to feast on blood from dilated vessels within mammalian skin (Lerner et al. 2007). Whether maxadilan-PAC1 interaction is a conserved feature of a peptide-receptor dyad in the sandfly, or simply represents parallel evolution or accident is unknown. Nevertheless, cryo-EM analysis of the docking sites of PACAP and maxadilan on PAC1 (Lu et al. 2022) illustrates a general feature of the peptide-GPCR docking site relevant to peptide-receptor co-evolution, which is that the binding pocket for peptides on their receptors is generally larger and less compact than that of smaller non-peptide GPCR ligands. This presumably provides greater opportunity for multiple binding modes, greater structural flexibility and, potentially, genetic stabilization of multiple and overlapping ligand-peptide interactions during species evolution (Foster et al. 2019).

Amnesiac is a neuropeptide discovered in Drosophila in one of the genetic screens for which the fly became famous as a research subject (Feany and Quinn 1995). Like maxadilan, amnesiac has little similarity to PACAP, but does have adjacent sequences with some similarity to PRP (underlined, with identical or similar amino acids in bold) and PACAP (italics, with identical or similar amino acids in bold): "MLWRCTAYYCFTLFFLLFRASALRRRVVSGSKGSAALALCRQFE **QLSA**SRR***EE***_CRTTQLRYH_***Y***_H_***R***_NGAQS_***RSL***_C_***AAVL***_CC_***KR***SYIPRPNFS CFSLVFPVGQRFAAARTRFGPTLVASWPLCNDSETKVLTKWPSCSLIGRRSV PRGQPKFSRENPRALSPSLLGEMR". Despite the weak homologies to both PRP and PACAP, it was later reported that the action of an endogenous peptide in fly released by high-frequency stimulation of motor neuron input to the fly larval wall, and causing contraction, was mimicked by mammalian PACAP (Zhong and Pena 1995). However, the PACAP-like function of an unknown peptide at the neuromuscular junction has not been carefully phenocopied by *amnesiac* mutations in the fly, so that in this case, the parallels between chordate and non-chordate peptide systems remain to be fully explored. In any case, the ability of PACAP to activate a Ras-dependent signalling pathway in the fly leading to ERK activation presaged the discovery that PACAP itself, in common with other family B1 peptides acting at their cognate receptors, can activate a cAMP-dependent pathway through the guanine nucleotide exchanger RapGEF2 (see Sect. 11.6).

Crustaceans have provided a particularly good model for how neuropeptide co-transmission can modulate programmed electrical response profiles in downstream neurons. Studying stomatogastric ganglion, Marder and colleagues have determined that microcircuits responsible for specific behaviours and responses in the crab (Nusbaum et al. 2017) gain flexibility from small amine transmitter (SMAT) and neuropeptide co-transmission. The evolutionary significance for peptide-SMAT co-transmission in mammals is just now being explored (Ilanges et al. 2022).

11.9 Summary

Gene duplication followed by neofunctionalization, among both the secretin superfamily peptides/prohormones and their receptors, defines the present status of neuropeptide-receptor function in mammals. PACAP and VIP with their associated peptides PRP and VRP, as well as GHRH and secretin itself, are members of a sub family within the secretin superfamily, along with their receptors. GHRH and its receptors, and secretin and the secretin receptor constitute well-defined ligand-receptor dyads with distinct functions in the hypothalamus and in the endocrine gut, respectively. PACAP and VIP appear to have undergone more extensive neofunctionalization. This applies to receptor utilization, which is strongly overlapping not only among the three canonical and evolutionarily related receptors PAC1, VPAC1 and VPAC2, but, in the case of PACAP, to less related receptors such as the BAM-22P receptor and MRGLR, both outside the secretin receptor (family B1) superfamily of GPCRs.

VIP and PACAP have adopted distinct non-endocrine functions during functional diversification in mammals, including roles as first messengers of the nervous system (neurotransmitters) and in inflammatory and immune signalling (as neurokines?). The actions of neuropeptide neurotransmitters in general, and PACAP and VIP in particular, remain to be fully defined at the electrophysiological and transmitter level in most physiological instances (Eiden et al. 2022), although mechanisms by which PACAP can affect postsynaptic neurons and hormone-receptive endocrine cells in addition to cAMP elevation are potentially legion (Boucher et al. 2021; May et al. 2021).

Although it is risky to view vertebrate genetic diversity from the perspective of an evolutionary snapshot, it appears that VIP and PACAP expression and function are very much in flux across mammals from an evolutionary point of view. Thus, VIP and PACAP have been proposed as retaining hypophysiotropic function in some but not all mammals. VIP seems to be established as a transmitter in a prominent population of GABAergic (inhibitory) interneurons in cortex from rodent to human but its expression is more variegated across species in other parts of the central and peripheral nervous systems. More granular mammalian comparisons in VIP and PACAP expression throughout peripheral and central nervous systems may indicate both conservation and specialization of function in various primate and rodent species. Dramatic mammalian interspecies differences in the expression of other peptides, like galanin and enkephalin, are reported in the median eminence and portal circulation, in anterior pituitary and in the peripheral nervous system and adrenal medulla (Elde et al. 1980; Schultzberg et al. 1978). The relevance of these patterns of species-specific expression for species-specific function, however, is currently lacking.

Can conservation of chemo-anatomy across species also be a clue to conservation of function? Does the relative conservation of sequence of PRP, for example (Fig. 11.3), imply function in the modern mammal, and at which receptor(s)? Neuropeptides essentially compete, evolutionarily, to develop specialist lineages to take on specific adaptive tasks within broader physiological mechanisms of homoeostasis and allostasis, among mammals, fish, birds, and non-vertebrates including nematodes and crustaceans. Understanding the specific advantages of expression, mode of release and receptor signalling characteristic of each peptide-receptor dyad may highlight peptide-specific points of intervention in therapeutics for human disorders created by the evolutionary honing of adaptive mechanisms. This might be particularly true for peptide-specific (or ‘peptide-coded’) behaviours that have become maladaptive, even to the point of no longer promoting survival, in certain modern environments (Dore et al. 2013; Hammack and May 2015; Kormos and Gaszner 2013; Levinstein et al. 2022; Li et al. 2023; Pinhasov et al. 2011; Rajbhandari et al. 2023; Slabe et al. 2023).

Glutamate and GABA, the primordial information-conveying ligands of the nervous system, acquired necessary diversity by spawning a host of receptors with nuanced properties for, among other things, building experience-dependent signalling gain and intensity into glutamatergic excitatory and GABAergic inhibitory/excitatory synaptic properties. Regulatory peptides, on the other hand, have acquired diversity at synapses by adding to GPCR metabotropic signal transduction ‘subgenres’ of function predicated upon signalling radius, time-dependent information transfer for secretagogue, gene induction and cellular excitability functions, and regional specialization for coordination of basic homoeostasis to allostatic responses to threat, pain, hunger, thirst and sleep. These are all integrated with the primordial metabolic drives providing energy for action, including reproduction. Whether insights gained through this growing body of knowledge can be harnessed for therapeutics remains to be seen. This does appear to be the case for the GLP-1 family, in treatment of metabolic disorders involving both peripheral and central sites of action (Graaf et al. 2016), and perhaps will also be the case for VIP and PACAP, and related peptides (Jaggi et al. 2008; Liao et al. 2019; Lu et al. 2022; O’Donnell et al. 1994). In attempting to find functional significance in evolutionary relationship among peptides and their receptors, it is worth remembering that organisms are not finished evolutionary products, making reverse-engineering a risky past-time. However, while the presence of receptors for a peptide in a given tissue may not always imply physiological function, it does provide pharmacological and therapeutic opportunity.

Acknowledgements and Funding Funding was received from the National Autonomous University of Mexico (UNAM), under Grant Agreements UNAM-PAPIIT-IG200121 and from Mexican Nacional Council for Humanity, Science and Technology (CONAHCYT), under Grant Agreement CF-2023-243. Fellowships from the UNAM-DGAPA-PASPA (LZ, VSH) and CONAHCYT (LZ), and a Fulbright-García Robles Fellowship (VSH) for a sabbatical research year in the Section on Molecular Neuroscience of the NIMH-IRP, NIH, USA are gratefully acknowledged. LEE was supported in this work by NIMH-IRP Project MH002386.

References

Alexander SPH, Davenport AP, Kelly E, Marrion N, Peters JA et al (2016) The concise guide to pharmacology 2015/16: G protein-coupled receptors. Br J Pharmacol 172(24):5744–5869

Arimura A (1992) Pituitary adenylate cyclase-activating polypeptide (PACAP): discovery and current status of research. Regul Peptides 37:287–303

Armstrong BD, Abad C, Chhith S, Cheung-Lau G, Hajji OE, Coute AC, Ngo DH, Waschek JA (2006) Impairment of axotomy-induced pituitary adenylyl cyclase-activating peptide gene expression in T helper 2 lymphocyte-deficient mice. Neuroreport 17(3):309–312

Armstrong BD, Abad C, Chhith S, Cheung-Lau G, Hajji OE, Nobuta H, Waschek JA (2008) Impaired nerve regeneration and enhanced neuroinflammatory response in mice lacking pituitary adenylyl cyclase activating peptide. Neuroscience 151(1):63–73

Aton SJ, Colwell CS, Harmar AJ, Waschek J, Herzog ED (2005) Vasoactive intestinal polypeptide mediates circadian rhythmicity and synchrony in mammalian clock neurons. Nat Neurosci 8:476–483

Bakalar D, Sweat S, Drossel G, Jiang SZ, Samal BS, Stroth N, Xu W, Zhang L, Zhang H, Eiden LE (2022) Relationships between constitutive and acute gene regulation, and physiological and behavioral responses, mediated by the neuropeptide PACAP. Psychoneuroendocrinology 135:1054478

Bakalar D, Gavrilova O, Jiang SZ, Zhang HY, Roy S, Williams SK, Liu N, Wisser S, Usdin TB, Eiden LE (2023) Constitutive and conditional deletion reveals distinct phenotypes driven by developmental versus neurotransmitter actions of the neuropeptide PACAP. J Neuroendocrinol 35(11):e13286

Beets I, Zels S, Vandewyer E, Demeulemeester J, Caers J, Baytemur E, Courtney A, Golinelli L, Hasakiogullari I, Schafer WR, Vertes PE, Mirabeau O, Schoofs L (2023) System-wide mapping of peptide-GPCR interactions in C. elegans. Cell Rep 42(9):113058

Biran J, Gliksberg M, Shirat I, Swaminathan A, Levitas-Djerbi T, Appelbaum L, Levkowitz G (2020) Splice-specific deficiency of the PTSD-associated gene PAC1 leads to a paradoxical age-dependent stress behavior. Sci Rep 10(1):9559

Boucher MN, May V, Braas KM, Hammack SE (2021) PACAP orchestration of stress-related responses in neural circuits. Peptides 142:170554

Braas KM, May V (1996) Pituitary adenylate cyclase-activating polypeptides, PACAP-38 and PACAP-27, regulation of sympathetic neuron catecholamine, and neuropeptide Y expression through activation of type I PACAP/VIP receptor isoforms. Ann N Y Acad Sci 805:204–216

Brenneman DE (2007) Neuroprotection: a comparative view of vasoactive intestinal peptide and pituitary adenylate cyclase-activating polypeptide. Peptides 28(9):1720–1726

Brenneman DE, Eiden LE (1986) Vasoactive intestinal peptide and electrical activity influence neuronal survival. Proc Natl Acad Sci USA 83:1159–1162

Bridgham JT, Carroll SM, Thornton JW (2006) Evolution of hormone-receptor complexity by molecular exploitation. Science 312(5770):97–101

Burbach JP (2002) Regulation of gene promoters of hypothalamic peptides. Front Neuroendocrinol 23(4):342–369

Campos CA, Bowen AJ, Roman CW, Palmiter RD (2018) Encoding of danger by parabrachial CGRP neurons. Nature 555(7698):617–622

Cardoso JC, Vieira FA, Gomes AS, Power DM (2007) PACAP, VIP and their receptors in the metazoa: insights about the origin and evolution of the ligand-receptor pair. Peptides 28(9):1902–1919

Cardoso JCR, Garcia MG, Power DM (2020) Tracing the origins of the pituitary adenylate-cyclase activating polypeptide (PACAP). Front Neurosci 14:366

Chen Y, Samal B, Hamelink CR, Xiang CC, Chen Y, Chen M, Vaudry D, Brownstein MJ, Hallenbeck JM, Eiden LE (2006) Neuroprotection by endogenous and exogenous PACAP following stroke. Regul Pept 137(1-2):4–19

Chew LJ, Burke ZD, Morgan H, Gozes I, Murphy D, Carter DA (1997) Transcription of the vasoactive intestinal peptide gene in response to glucocorticoids: differential regulation of alternative transcripts is modulated by a labile protein in rat anterior pituitary. Mol Cell Endocrinol 130(1–2):83–91

Chini B, Verhage M, Grinevich V (2017) The action radius of oxytocin release in the mammalian CNS: from single vesicles to behavior. Trends Pharmacol Sci 38(11):982–991

Chow BK-C, Pang RTK, Ng SS-M (2003) Molecular evolution of PACAP precursor and PACAP receptors. In: Vaudry H, Arimura A (eds) Pituitary adenylate cyclase-activating polypeptide. Kluwer Academic Publishers, Norwell, MA, pp 25–47

Christopoulos A, Christopoulos G, Morfis M, Udawela M, Laburthe M, Couvineau A, Kuwasako K, Tilakaratne N, Sexton PM (2003) Novel receptor partners and function of receptor activity-modifying proteins. J Biol Chem 278(5):3293–3297

Cola RB, N'itethammer SN, Rajamannar P, Gresch A, Bhat MA, Assoumou K, Willians ET, Hauck P, Hartrampf N, Benke D, Stoeber M, L'evkowitz G, Melzer S, Patriarchi T (2024) Probing PAC1 receptor activation across species with an engineered sensor. bioRxiv

Dehal P, Boore JL (2005) Two rounds of whole genome duplication in the ancestral vertebrate. PLoS Biol 3(10):e314

Delgado M, Abad C, Martinez C, Juarranz MG, Leceta J, Ganea D, Gomariz RP (2003) PACAP in immunity and inflammation. Ann N Y Acad Sci 992:141–157

Dickinson T, Fleetwood-Walker SM (1999) VIP and PACAP: very important in pain? Trends Pharmacol Sci 20:324–329

Dore R, Iemolo A, Smith KL, Wang X, Cottone P, Sabino V (2013) CRF mediates the anxiogenic and anti-rewarding, but not the anorectic effects of PACAP. Neuropsychopharmacology 38(11):2160–2169

Drescher MJ, Drescher DG, Khan KM, Hatfield JS, Ramakrishnan NA, Abu-Hamdan MD, Lemonnier LA (2006) Pituitary adenylyl cyclase-activating polypeptide (PACAP) and its receptor (PAC1-R) are positioned to modulate afferent signaling in the cochlea. Neuroscience 142(1):139–164

Duboule D, Wilkins AS (1998) The evolution of 'bricolage'. Trends Genet 14(2):54–59

Edwards AV, Jones CT (1994) Adrenal responses to the peptide PACAP in conscious functionally hypophysectomized calves. Am J Phys 266:E870–E876

Eiden LE, Loumaye E, Sherwood N, Eskay RL (1982) Two chemically and immunologically distinct forms of luteinizing hormone-releasing hormone are differentially expressed in frog neural tissues. Peptides 3(3):323–327

Eiden LE, Hernandez VS, Jiang SZ, Zhang L (2022) Neuropeptides and small-molecule amine transmitters: cooperative signaling in the nervous system. Cell Mol Life Sci 79(9):492

Elde R, Haber S, Ho R, Holets V, de Lanerolle N, Maley B, Micevych P, Seybold V (1980) Interspecies conservation and variation in peptidergic neurons. Peptides 1:21–26

Emery AC, Eiden LE (2012) Signaling through the neuropeptide GPCR PAC1 induces neuritogenesis via a single linear cAMP- and ERK-dependent pathway using a novel cAMP sensor. FASEB J 26:3199–3211

Emery AC, Alvarez RA, Abboud P, Xu W, Westover CD, Eiden MV, Eiden LE (2016) C-terminal amidation of PACAP-38 and PACAP-27 is dispensable for biological activity at the PAC1 receptor. Peptides 79:39–48

Engelund A, Fahrenkrug J, Harrison A, Luuk H, Hannibal J (2012) Altered pupillary light reflex in PACAP receptor 1-deficient mice. Brain Res 1453:17–25

Feany MB, Quinn WG (1995) A neuropeptide gene defined by the drosophila memory mutant amnesiac. Science 268:869–873

Foster SR, Hauser AS, Vedel L, Strachan RT, Huang XP, Gavin AC, Shah SD, Nayak AP, Haugaard-Kedstrom LM, Penn RB, Roth BL, Brauner-Osborne H, Gloriam DE (2019) Discovery of human signaling systems: pairing peptides to G protein-coupled receptors. Cell 179(4):895–908 e21

Fredriksson R, Lagerstrom MC, Lundin LG, Schioth HB (2003) The G-protein-coupled receptors in the human genome form five main families. Phylogenetic analysis, paralogon groups, and fingerprints. Mol Pharmacol 63(6):1256–1272

Girard BA, Lelievre V, Braas KM, Razinia T, Vizzard MA, Ioffe Y, El Meskini R, Ronnett GV, Waschek JA, May V (2006) Noncompensation in peptide/receptor gene expression and distinct behavioral phenotypes in VIP- and PACAP-deficient mice. J Neurochem 99(2):499–513

Goodman RH (1990) Regulation of neuropeptide gene expression. Annu Rev Neurosci 13:111–127

Graaf C, Donnelly D, Wootten D, Lau J, Sexton PM, Miller LJ, Ahn JM, Liao J, Fletcher MM, Yang D, Brown AJ, Zhou C, Deng J, Wang MW (2016) Glucagon-like Peptide-1 and its class B G protein-coupled receptors: a long march to therapeutic successes. Pharmacol Rev 68(4):954–1013

Gray SL, Yamaguchi N, Vencova P, Sherwood NM (2002) Temperature-sensitive phenotype in mice lacking pituitary adenylate cyclase-activating polypeptide. Endocrinology 143(10):3946–3954

Gressens P, Hill JM, Gozes I, Fridkin M, Brenneman DE (1993) Growth factor function of vasoactive intestinal peptide in whole cultured mouse embryos. Nature 362:155–158

Guerineau NC (2019) Cholinergic and peptidergic neurotransmission in the adrenal medulla: a dynamic control of stimulus-secretion coupling. IUBMB Life 72(4):553–567

Hahm SH, Eiden LE (1998) Cis-regulatory elements controlling basal and inducible VIP gene transcription. Ann N Y Acad Sci 865:10–26

Hamelink C, Tjurmina O, Damadzic R, Young WS, Weihe E, Lee H-W, Eiden LE (2002) Pituitary adenylate cyclase activating polypeptide is a sympathoadrenal neurotransmitter involved in catecholamine regulation and glucohomeostasis. Proc Natl Acad Sci USA 99:461–466

Hammack SE, May V (2015) Pituitary adenylate cyclase activating polypeptide in stress-related disorders: data convergence from animal and human studies. Biol Psychiatry 78:167–177

Hannibal J (2002) Pituitary adenylate cyclase-activating peptide in the rat central nervous system: an immunohistochemical and in situ hybridization study. J Comp Neurol 453(4):389–417

Hannibal J, Hindersson P, Knudsen SM, Georg B, Fahrenkrug J (2002) The photopigment melanopsin is exclusively present in pituitary adenylate cyclase-activating polypeptide-containing retinal ganglion cells of the retinohypothalamic tract. J Neurosci 22:RC191

Harmar AJ, Fahrenkrug J, Gozes I, Laburthe M, May V, Pisegna JR, Vaudry D, Vaudry H, Waschek JA, Said SI (2012) Pharmacology and functions of receptors for vasoactive intestinal peptide and pituitary adenylate cyclase-activating polypeptide: IUPHAR review. Br J Pharmacol 166(1):4–17

Hegg CC, Au E, Roskams AJ, Lucero MT (2003) PACAP is present in the olfactory system and evokes calcium transients in olfactory receptor neurons. J Neurophysiol 90(4):2711–2719

Herr W, Sturm RA, Clerc RG, Corcoran LM, Baltiore D, Sharp PA, Ingraham HA, Rosenfeld MG, Finney M, Ruvkun G, Horvitz HR (1988) The POU domain: A large conserved region in the mammalian Pit-1, Oct-1,Oct-2 and Caenorrhabditis elegans Unc-86 gene products. Genes Dev 2:1513–1516

Holst JJ (2008) Pharmacology of GLP-1-based therapies. Br J Diabetes Vasc Dis 8(Suppl 2):S8–S18

Hrvatin S, Sun S, Wilcox OF, Yao H, Lavin-Peter AJ, Cicconet M, Assad EG, Palmer ME, Aronson S, Banks AS, Griffith EC, Greenberg ME (2020) Neurons that regulate mouse torpor. Nature 583(7814):115–121

Hwang JI, Moon MJ, Park S, Kim DK, Cho EB, Ha N, Son GH, Kim K, Vaudry H, Seong JY (2013) Expansion of secretin-like G protein-coupled receptors and their peptide ligands via local duplications before and after two rounds of whole-genome duplication. Mol Biol Evol 30(5):1119–1130

Ilanges A, Shiao R, Shaked J, Luo JD, Yu X, Friedman JM (2022) Brainstem ADCYAP1(+) neurons control multiple aspects of sickness behaviour. Nature 609(7928):761–771

Jaggi M, Prasad S, Singh AT, Praveen R, Dutt S, Mathur A, Sharma R, Gupta N, Ahuja R, Mukherjee R, Burman AC (2008) Anticancer activity of a peptide combination in gastrointestinal cancers targeting multiple neuropeptide receptors. Investig New Drugs 26(6):489–504

Janssen T, Lindemans M, Meelkop E, Temmerman L, Schoofs L (2010) Coevolution of neuropeptidergic signaling systems: from worm to man. Ann N Y Acad Sci 1200:1–14

Jiang SZ, Xu W, Emery AC, Gerfen CR, Eiden MV, Eiden LE (2017) NCS-Rapgef2, the protein product of the neuronal Rapgef2 gene, is a specific activator of D1 dopamine receptor-dependent ERK phosphorylation in mouse brain. eNeuro 4(5)

Jiang SZ, Zhang HY, Eiden LE (2023) PACAP controls endocrine and Behavioral stress responses via separate brain circuits. Biol Psychiatry Glob Open Sci 3(4):673–685

Jones SW, Adams PR, Brownstein MJ, Rivier JE (1984) Teleost luteinizing hormone-releasing hormone: action on bullfrog sympathetic ganglia is consistent with role as neurotransmitter. J Neurosci 4(2):420–429

Keenan WT, Rupp AC, Ross RA, Somasundaram P, Hiriyanna S, Wu Z, Badea TC, Robinson PR, Lowell BB, Hattar SS (2016) A visual circuit uses complementary mechanisms to support transient and sustained pupil constriction. Elife 5

Kim MS, Hur MK, Son YJ, Park JI, Chun SY, D'Elia AV, Damante G, Cho S, Kim K, Lee BJ (2002) Regulation of pituitary adenylate cyclase-activating polypeptide gene transcription by TTF-1, a homeodomain-containing transcription factor. J Biol Chem 277(39):36863–36871

Kimura C, Ohkubo S, Ogi K, Hosoya M, Itoh Y, Onda H, Miyata A, Jiang L, Dahl RR, Stibbs HH et al (1990) A novel peptide which stimulates adenylate cyclase: molecular cloning and characterization of the ovine and human cDNAs. Biochem Biophys Res Commun 166(1):81–89

Kohl J, Babayan BM, Rubinstein ND, Autry AE, Marin-Rodriguez B, Kapoor V, Miyamishi K, Zweifel LS, Luo L, Uchida N, Dulac C (2018) Functional circuit architecture underlying parental behaviour. Nature 556(7701):326–331

Kormos V, Gaszner B (2013) Role of neuropeptides in anxiety, stress, and depression: from animals to humans. Neuropeptides 47(6):401–419

Koves K, Szabo E, Kantor O, Heinzlmann A, Szabo F, Csaki A (2020) Current state of understanding of the role of PACAP in the hypothalamo-hypophyseal gonadotropin functions of mammals. Front Endocrinol (Lausanne) 11:88

Kuzmin E, VanderSluis B, Nguyen Ba AN, Wang W, Koch EN, Usaj M, Khmelinskii A, Usaj MM, van Leeuwen J, Kraus O, Tresenrider A, Pryszlak M, Hu MC, Varriano B, Costanzo M, Knop M, Moses A, Myers CL, Andrews BJ, Boone C (2020) Exploring whole-genome duplicate gene retention with complex genetic interaction analysis. Science 368(6498)

Kuzmin E, Taylor JS, Boone C (2022) Retention of duplicated genes in evolution. Trends Genet 38(1):59–72

Lamb TD (2021) Analysis of paralogons, origin of the vertebrate karyotype, and ancient chromosomes retained in extant species. Genome Biol Evol 13(4)

Langenhan T, Barr MM, Bruchas MR, Ewer J, Griffith LC, Maiellaro I, Taghert PH, White BH, Monk KR (2015) Model organisms in G protein-coupled receptor research. Mol Pharmacol 88(3):596–603

Lee AT, Cunniff MM, See JZ, Wilke SA, Luongo FJ, Ellwood IT, Ponnavolu S, Sohal VS (2019) VIP interneurons contribute to avoidance behavior by regulating information flow across hippocampal-prefrontal networks. Neuron 102(6):1223–1234.e4

Lembo PM, Grazzini E, Groblewski T, O'Donnell D, Roy MO, Zhang J, Hoffert C, Cao J, Schmidt R, Pelletier M, Labarre M, Gosselin M, Fortin Y, Banville D, Shen SH, Strom P, Payza K, Dray A, Walker P, Ahmad S (2002) Proenkephalin A gene products activate a new family of sensory neuron-- specific GPCRs. Nat Neurosci 5(3):201–209

Lerner EA, Iuga AO, Reddy VB (2007) Maxadilan, a PAC1 receptor agonist from sand flies. Peptides 28(9):1651–1654

Levinstein MR, Bergkamp DJ, Lewis ZK, Tsobanoudis A, Hashikawa K, Stuber GD, Neumaier JF (2022) PACAP-expressing neurons in the lateral habenula diminish negative emotional valence. Genes Brain Behav 21:e12801

Li C, Nelson LS, Kim K, Nathoo A, Hart AC (1999) Neuropeptide gene families in the nematode Caenorhabditis elegans. Ann N Y Acad Sci 897:239–252

Li Y, Andero R, Luchkina NV, Suh J, Ross RA, Lowell BB, Carlezon WA, Ressler KJ, Bolshakov VY (2023) PACAP-mediated gating of anxiety-controlling circuits. bioRxiv

Liao C, de Molliens MP, Schneebeli ST, Brewer M, Song G, Chatenet D, Braas KM, May V, Li J (2019) Targeting the PAC1 receptor for neurological and metabolic disorders. Curr Top Med Chem

Lindberg PT, Mitchell JW, Burgoon PW, Beaule C, Weihe E, Schafer MK, Eiden LE, Jiang SZ, Gillette MU (2019) Pituitary adenylate cyclase-activating peptide (PACAP)-glutamate co-transmission drives circadian phase-advancing responses to intrinsically photosensitive retinal ganglion cell projections by suprachiasmatic nucleus. Front Neurosci 13:1281

Loren I, Emson PC, Fahrenkrug J, Bjorklund A, Alumets J, Hakanson R, Sundler F (1979) Distribution of vasoactive intestinal polypeptide in the rat and mouse brain. Neuroscience 4(12):1953–1976

Lorenzen E, Dodig-Crnkovic T, Kotliar IB, Pin E, Ceraudo E, Vaughan RD, Uhlen M, Huber T, Schwenk JM, Sakmar TP (2019) Multiplexed analysis of the secretin-like GPCR-RAMP interactome. Sci Adv 5(9):eaaw2778

Loumaye E, Thorner J, Catt KJ (1982) Yeast mating pheromone activates mammalian gonadotrophs: evolutionary conservation of a reproductive hormone? Science 218(4579):1323–1325

Lu J, Piper SJ, Zhao P, Miller LJ, Wootten D, Sexton PM (2022) Targeting VIP and PACAP receptor signaling: new insights into designing drugs for the PACAP subfamily of receptors. Int J Mol Sci 23(15)

Lynch M, Conery JS (2000) The evolutionary fate and consequences of duplicate genes. Science 290(5494):1151–1155

Macosko EZ, Pokala N, Feinberg EH, Chalasani SH, Butcher RA, Clardy J, Bargmann CI (2009) A hub-and-spoke circuit drives pheromone attraction and social behaviour in C. elegans. Nature 458(7242):1171–1175

Magistretti PJ, Cardinaux JR, Martin JL (1998) VIP and PACAP in the CNS: regulators of glial energy metabolism and modulators of glutamatergic signaling. Ann N Y Acad Sci 865:213–225

Malhotra RK, Blank M, Wakade TD, Wakade AR (1989) Vasoactive intestinal polypeptide (VIP) serves as another neurotransmitter in the rat adrenal medulla. Regul Peptides 26:168–168

May V, Johnson GC, Hammack SE, Braas KM, Parsons RL (2021) PAC1 receptor internalization and endosomal MEK/ERK activation is essential for PACAP-mediated neuronal excitability. J Mol Neurosci

McRory J, Sherwood NM (1997) Two protochordate genes encode pituitary adenylate cyclase-activating polypeptide and related family members. Endocrinology 138:2380–2390

Missig G, Roman CW, Vizzard MA, Braas KM, Hammack SE, May V (2014) Parabrachial nucleus (PBn) pituitary adenylate cyclase activating polypeptide (PACAP) signaling in the amygdala: implication for the sensory and behavioral effects of pain. Neuropharmacology 86:38–48

Miyata A, Arimura A, Dahl RR, Minamino N, Uehara A, Jiang L, Culler MD, Coy DH (1989) Isolation of a novel 38 residue-hypothalamic polypeptide which stimulates adenylate cyclase in pituitary cells. Biochem Biophys Res Commun 164:567–574

Moller K, Zhang Y-Z, Hakanson R, Luts A, Sjölund B, Uddman R, Sundler F (1993) Pituitary adenylate cyclase activating peptide is a sensory neuropeptide: immunocytochemical and immunochemical evidence. Neuroscience 57:725–732

Morales A, Mohan R, Chen X, Coffman BL, Bendahmane M, Watch L, West JL, Bakshi S, Traynor JR, Giovannucci DR, Kammermeier PJ, Axelrod D, Currie KPM, Smrcka AV, Anantharam A (2023) PACAP and acetylcholine cause distinct Ca2+ signals and secretory responses in chromaffin cells. J Gen Physiol 155(2)

Morfis M, Christopoulos A, Sexton PM (2003) RAMPs: 5 years on, where to now? Trends Pharmacol Sci 24(11):596–601

Mustafa T, Eiden LE (2006) Secretin superfamily: PACAP, VIP, and related neuropeptides. In: Lajtha A, Lim R (eds) Handbook of neurochemistry and molecular neurobiology. Springer, Boston, MA. https://doi.org/10.1007/978-0-387-30381-9_21

Mustafa T, Grimaldi M, Eiden LE (2007) The hop cassette of the PAC1 receptor confers coupling to Ca2+ elevation required for pituitary adenylate cyclase-activating polypeptide-evoked neurosecretion. J Biol Chem 282(11):8079–8091

Nathoo AN, Moeller RA, Westlund BA, Hart AC (2001) Identification of neuropeptide-like protein gene families in Caenorhabditis elegans and other species. Proc Natl Acad Sci USA 98(24):14000–14005

Nelson C, Albert VR, Elsholtz HP, Lu LI-W, Rosenfeld MG (1988) Activation of cell-specific expression of rat growth hormone and prolactin genes by a common transcription factor. Science 239:1400–1405

Nonaka N, Banks WA, Shioda S (2020) Pituitary adenylate cyclase-activating polypeptide: protective effects in stroke and dementia. Peptides 130:170332

Nordstrom KJ, Lagerstrom MC, Waller LM, Fredriksson R, Schioth HB (2009) The secretin GPCRs descended from the family of adhesion GPCRs. Mol Biol Evol 26(1):71–84

Nusbaum MP, Blitz DM, Marder E (2017) Functional consequences of neuropeptide and small-molecule co-transmission. Nat Rev Neurosci 18(7):389–403

O'Donnell M, Garippa RJ, Rinaldi N, Selig WM, Simko B, Renzetti L, Tannu SA, Wasserman MA, Welton A, Bolin DR (1994) Ro 25-1553: a novel, long-acting vasoactive intestinal peptide agonist. Part I: in vitro and in vivo bronchodilator studies. J Pharmacol Exp Ther 270(3):1282–1288

Ohnou T, Yokai M, Kurihara T, Hasegawa-Moriyama M, Shimizu T, Inoue K, Kambe Y, Kanmura Y, Miyata A (2016) Pituitary adenylate cyclase-activating polypeptide type 1 receptor signaling evokes long-lasting nociceptive behaviors through the activation of spinal astrocytes in mice. J Pharmacol Sci 130(4):194–203

Ohtaki H, Nakamachi T, Dohi K, Aizawa Y, Takaki A, Hodoyama K, Yofu S, Hashimoto H, Shintani N, Baba A, Kopf M, Iwakura Y, Matsuda K, Arimura A, Shioda S (2006) Pituitary adenylate cyclase-activating polypeptide (PACAP) decreases ischemic neuronal cell death in association with IL-6. Proc Natl Acad Sci USA 103(19):7488–7493

Ohtaki H, Nakamachi T, Dohi K, Shioda S (2008) Role of PACAP in ischemic neural death. J Mol Neurosci 36(1-3):16–25

On JS, Chow BK, Lee LT (2015) Evolution of parathyroid hormone receptor family and their ligands in vertebrate. Front Endocrinol (Lausanne) 6:28

On JSW, Su L, Shen H, Arokiaraj AWR, Cardoso JCR, Li G, Chow BKC (2022) PACAP/GCGa is an important modulator of the amphioxus CNS-Hatschek's pit axis, the homolog of the vertebrate hypothalamic-pituitary axis in the basal chordates. Front Endocrinol (Lausanne) 13:850040

Palkovits M, Somogyvari-Vigh A, Arimura A (1995) Concentrations of pituitary adenylate cyclase activating polypeptide (PACAP) in human brain nuclei. Brain Res 699(1):116–120

Palmiter RD (2018) The parabrachial nucleus: CGRP neurons function as a general alarm. Trends Neurosci 41(5):280–293

Panda S, Nayak SK, Campo B, Walker JR, Hogenesch JB, Jegla T (2005) Illumination of the melanopsin signaling pathway. Science 307(5709):600–604

Pinhasov A, Nesher E, Gross M, Turgeman G, Kreinin A, Yadid G (2011) The role of the PACAP signaling system in depression. Curr Pharm Des 17(10):990–1001

Piper SJ, Deganutti G, Lu J, Zhao P, Liang YL, Lu Y, Fletcher MM, Hossain MA, Christopoulos A, Reynolds CA, Danev R, Sexton PM, Wootten D (2022) Understanding VPAC receptor family peptide binding and selectivity. Nat Commun 13(1):7013

Pisegna JR, Wank SA (1993) Molecular cloning and functional expression of the pituitary adenylate cyclase-activating polypeptide type I receptor. Proc Natl Acad Sci U S A 90(13):6345–6349

Pisegna JR, Wank SA (1996) Cloning and characterization of the signal transduction of four splice variants of the human pituitary adenylate cyclase activating polypeptide receptor. Evidence for dual coupling to adenylate cyclase and phospholipase C. J Biol Chem 271(29):17267–17274

Pisegna JR, Moody TW, Wank SA (1996) Differential signaling and immediate-early gene activation by four splice variants of the human pituitary adenylate cyclase-activating polypeptide receptor (hPACAP-R). Ann N Y Acad Sci 805:54–64. discussion 64-6

Rajbhandari AK, Barson JR, Gilmartin MR, Hammack SE, Chen BK (2023) The functional heterogeneity of PACAP: stress, learning, and pathology. Neurobiol Learn Mem 203:107792

Rasbach E, Splitthoff P, Bonaterra GA, Schwarz A, Mey L, Schwarzbach H, Eiden LE, Weihe E, Kinscherf R (2018) PACAP deficiency aggravates atherosclerosis in ApoE deficient mice. Immunobiology

Rawlings SR, Hezareh M (1996) Pituitary adenylate cyclase activating polypeptide (PACAP) and PACAP/vasoactive intestinal polypeptide receptors: action on the anterior pituitary gland. Endocr Rev 17:4–29

Reglödi D, Somogyvari-Vigh A, Vigh S, Kozicz T, Arimura A (2000) Delayed systemic administration of PACAP38 is neuroprotective in transient middle cerebral artery occlusion in the rat. Stroke 31:1411–1417

Rudecki AP, Gray SL (2016) PACAP in the defense of energy homeostasis. Trends Endocrinol Metab

Said SI (2007) The discovery of VIP: initially looked for in the lung, isolated from intestine, and identified as a neuropeptide. Peptides 28(9):1620–1621

Said SI, Mutt V (1970) Polypeptide with broad biological activity: isolation from small intestine. Science 169(951):1217–1218

Said SI, Rosenberg RN (1976) Vasoactive intestinal polypeptide: abundant immunoreactivity in neural cell lines and normal nervous tissue. Science 192(4242):907–908

Sawangjaroen K, Curlewis JD (1994) Effects of pituitary adenylate cyclase-activating polypeptide (PACAP) and vasoactive intestinal polypeptide (VIP) on prolactin, luteinizing hormone and growth hormone secretion in the ewe. J Neuroendocrinol 6(5):549–555

Sawangjaroen K, Anderson ST, Curlewis JD (1997) Effects of pituitary adenylate cyclase-activating polypeptide (PACAP) and vasoactive intestinal polypeptide (VIP) on hormone secretion from sheep pituitary cells in vitro. J Neuroendocrinol 9(4):279–286

Schultzberg M, Lundberg JM, Hokfelt T, Terenius L, Brandt J, Elde RP, Goldstein M (1978) Enkephalin-like immunoreactivity in gland cells and nerve terminals of the adrenal medulla. Neuroscience 3:1169–1186

Sherwood NM, Krueckl SL, McRory JE (2000) The origin and function of the pituitary adenylate cyclase-activating polypeptide (PACAP)/glucagon superfamily. Endocr Rev 21:619–670

Sherwood NM, Adams BA, Isaac ER, Wu S, Fradinger EA (2007) Knocked down and out: PACAP in development, reproduction and feeding. Peptides 28(9):1680–1687

Sims KB, Hoffman DL, Said SI, Zimmerman EA (1980) Vasoactive intestinal polypeptide (VIP) in mouse and rat brain: an immunocytochemical study. Brain Res 186(1):165–183

Slabe Z, Balesar RA, Verwer RWH, Van Heerikhuize JJ, Pechler GA, Zorovic M, Hoogendijk WJG, Swaab DF (2023) Alterations in pituitary adenylate cyclase-activating polypeptide in major depressive disorder, bipolar disorder, and comorbid depression in Alzheimer's disease in the human hypothalamus and prefrontal cortex. Psychol Med 53(16):7537–7549

Smith SJ, Sumbul U, Graybuck LT, Collman F, Seshamani S, Gala R, Gliko O, Elabbady L, Miller JA, Bakken TE, Rossier J, Yao Z, Lein E, Zeng H, Tasic B, Hawrylycz M (2019) Single-cell transcriptomic evidence for dense intracortical neuropeptide networks. Elife 8

Spengler D, Bweber C, Pantaloni C, Hlsboer F, Bockaert J, Seeburg PH, Journot L (1993) Differential signal transduction by five splice variants of the PACAP receptor. Nature 365:170–175

Splitthoff P, Rasbach E, Neudert P, Bonaterra GA, Schwarz A, Mey L, Schwarzbach H, Eiden LE, Weihe E, Kinscherf R (2020) PAC1 deficiency attenuates progression of atherosclerosis in ApoE deficient mice under cholesterol-enriched diet. Immunobiology 225(3):151930

Stroth N, Holighaus Y, Ait-Ali D, Eiden LE (2011) PACAP: a master regulator of neuroendocrine stress circuits and the cellular stress response. Ann N Y Acad Sci 1220(1):49–59

Stroth N, Kuri BA, Mustafa T, Chan SA, Smith CB, Eiden LE (2013) PACAP controls adrenomedullary catecholamine secretion and expression of catecholamine biosynthetic enzymes at high splanchnic nerve firing rates characteristic of stress transduction in male mice. Endocrinology 154(1):330–339

Symes A, Lewis S, Corpus L, Rajan P, Hyman SE, Fink JS (1994) STAT proteins participate in the regulation of the vasoactive intestinal peptide gene by the ciliary neurotrophic factor family of cytokines. Mol Endocrinol 8:1750–1763

Takasaki I, Nakamura K, Shimodaira A, Watanabe A, Du Nguyen H, Okada T, Toyooka N, Miyata A, Kurihara T (2019) The novel small-molecule antagonist of PAC1 receptor attenuates formalin-induced inflammatory pain behaviors in mice. J Pharmacol Sci 139(2):129–132

Tam JK, Lee LT, Chow BK (2007) PACAP-related peptide (PRP)--molecular evolution and potential functions. Peptides 28(9):1920–1929

Thoenen H, Mueller RA, Axelrod J (1969) Trans-synaptic induction of adrenal tyrosine hydroxylase. J Pharm Exp Ther 169:249–254

Tompkins JD, Parsons RL (2008) Identification of intracellular signaling cascades mediating the PACAP-induced increase in Guinea pig cardiac neuron excitability. J Mol Neurosci 36(1-3):292–298

Vaudry D, Falluel-Morel A, Bourgault S, Basille M, Burel D, Wurtz O, Fournier A, Chow BK, Hashimoto H, Galas L, Vaudry H (2009) Pituitary adenylate cyclase-activating polypeptide and its receptors: 20 years after the discovery. Pharmacol Rev 61(3):283–357

Vijayan E, Samson WK, Said SI, McCann SM (1979) Vasoactive intestinal peptide: evidence for a hypothalamic site of action to release growth hormone, luteinizing hormone, and prolactin in conscious ovariectomized rats. Endocrinology 104(1):53–57

Voice JK, Dorsam G, Chan RC, Grinninger C, Kong Y, Goetzl EJ (2002) Immunoeffector and immunoregulatory activities of vasoactive intestinal peptide. Regul Pept 109(1-3):199–208

Wang YQ, Qian YP, Yang S, Shi H, Liao CH, Zheng HK, Wang J, Lin AA, Cavalli-Sforza LL, Underhill PA, Chakraborty R, Jin L, Su B (2005) Accelerated evolution of the pituitary adenylate cyclase-activating polypeptide precursor gene during human origin. Genetics 170(2):801–806

Wang L, Gillis-Smith S, Peng Y, Zhang J, Chen X, Salzman CD, Ryba NJP, Zuker CS (2018) The coding of valence and identity in the mammalian taste system. Nature 558(7708):127–131

Waschek JA (2013) VIP and PACAP: neuropeptide modulators of CNS inflammation, injury, and repair. Br J Pharmacol 169(3):512–523

Woods IG, Schoppik D, Shi VJ, Zimmerman S, Coleman HA, Greenwood J, Soucy ER, Schier AF (2014) Neuropeptidergic signaling partitions arousal behaviors in zebrafish. J Neurosci 34(9):3142–3160

Wu Z, Autry AE, Bergan JF, Watabe-Uchida M, Dulac CG (2014) Galanin neurons in the medial preoptic area govern parental behaviour. Nature 509(7500):325–330

Xu W, Dahlke S, Emery A, Sung M, Chepurny O, Holz GG, Eiden LE (2021) Cyclic AMP-dependent activation of ERK via GLP-1 receptor signaling requires the neuroendocrine cell-specific guanine nucleotide exchanger NCS-RapGEF2. J Neuroendocrinol 33:e12974

Yosten GL, Harada CM, Haddock C, Giancotti LA, Kolar GR, Patel R, Guo C, Chen Z, Zhang J, Doyle TM, Dickenson AH, Samson WK, Salvemini D (2020) GPR160 de-orphanization reveals critical roles in neuropathic pain in rodents. J Clin Invest 130(5):2587–2592

Zhang L, Hernandez VS, Gerfen CR, Jiang SZ, Zavala L, Barrio RA, Eiden LE (2021) Behavioral role of PACAP reflects its selective distribution in glutamatergic and GABAergic neuronal subpopulations. Elife 10

Zhang F, Qiao W, Wei JA, Tao Z, Chen C, Wu Y, Lin M, Ng KMC, Zhang L, Yeung KW, Chow BKC (2024) Secretin-dependent signals in the ventromedial hypothalamus regulate energy metabolism and bone homeostasis in mice. Nat Commun 15(1):1030

Zhong Y, Pena LA (1995) A novel synaptic transmission mediated by a PACAP-like neuropeptide in drosophila. Neuron 14(3):527–536

Relevant Literature

Arimura A (1992) Pituitary adenylate cyclase activating polypeptide (PACAP): discovery and current status of research. Regul Peptides 37:287–303 A trenchant summary of PACAP by its discoverer. A revealing look at how the PACAP field developed

Cardoso JCR, Garcia MG, Power DM (2020) Tracing the origins of the pituitary adenylate-cyclase activating polypeptide (PACAP). Front Neurosci 14:366 A cogent summary of attempts by endocrinologists and evolutionary biologists to put PACAP and VIP in the context of vertebrate evolution from the point of view of a pre-vertebral ancestor gene

Emery AC, Eiden LE (2012) Signaling through the neuropeptide GPCR PAC1 induces neuritogenesis via a single linear cAMP- and ERK-dependent pathway using a novel cAMP sensor. FASEB J 26:3199–3211

Fredriksson R, Lagerstrom MC, Lundin LG, Schioth HB (2003) The G-protein-coupled receptors in the human genome form five main families. Phylogenetic analysis, paralogon groups, and fingerprints. Mol Pharmacol 63(6):1256–1272 Organization of the GPCRs into five main families, one of which is the current family B1 (or secretin family), composed exclusively of peptide-liganded receptors

Hamelink C, Tjurmina O, Damadzic R, Young WS, Weihe E, Lee H-W, Eiden LE (2002) Pituitary adenylate cyclase-activating polypeptide is a sympathoadrenal neurotransmitter involved in catecholamine regulation and glucohomeostasis. Proc Natl Acad Sci USA 99:461–466 With this and the later report of Stroth et al., the identity of PACAP as a sympathetic neurotransmitter is put on a firm footing

Hannibal J, Hindersson P, Knudsen SM, Georg B, Fahrenkrug J (2002) The photopigment melanopsin is exclusively present in pituitary adenylate cyclase-activating polypeptide-containing retinal ganglion cells of the retinohypothalamic tract. J Neurosci 22:RC191 A fascinating and far-reaching discovery-intrinsically photosensitive retinal ganglionic cells that evolved in the mammalian lineage is equipped with the neuropeptide PACAP as well as the SMAT glutamate

Hwang JI, Moon MJ, Park S, Kim DK, Cho EB, Ha N, Son GH, Kim K, Vaudry H, Seong JY (2013) Expansion of secretin-like G protein-coupled receptors and their peptide ligands via local duplications before and after two rounds of whole-genome duplication. Mol Biol Evol 30:1119–1130 The report cogently summarizes the phylogenetic evidence, based upon sequence and paralogon analysis, for how the secretin sub-family of the family B peptide ligand-receptor dyads evolved in vertebrates

Jiang SZ, Zhang HY, Eiden LE (2023) PACAP controls endocrine and behavioral stress responses via separate brain circuits. Biol Psychiatry Glob Open Sci 3:673–685 Demonstration of a neurotransmitter role for PACAP in two separate brain circuits related to the stress response

Missig G, Roman CW, Vizzard MA, Braas KM, Hammack SE, May V (2014) Parabrachial nucleus (PBn) pituitary adenylate cyclase activating polypeptide (PACAP) signaling in the amygdala: implication for the sensory and behavioral effects of pain. Neuropharmacology 86:38–48 Elucidation of two novel modes of signaling via the PACAP receptor PAC1 in addition to the classical coupling via Gs to cAMP-dependent activation of protein kinase A

Miyata A, Arimura A, Dahl RR, Minamino N, Uehara A, Jiang L, Culler MD, Coy DH (1989) Isolation of a novel 38 residue-hypothalamic polypeptide which stimulates adenylate cyclase in pituitary cells. Biochem Biophys Res Commun 164:567–574 The report for the discovery of PACAP

Pisegna JR, Wank SA (1996) Cloning and characterization of the signal transduction of four splice variants of the human pituitary adenylate cyclase activating polypeptide receptor: evidence for dual coupling to adenylate cyclase and phospholipase C. J Biol Chem 271:17267–17274 These reports established the structure of PAC1 and its evolution as a multi-functional and multiply-coupled GPCR through exon cassette insertion

Reglödi D, Somogyvari-Vigh A, Vigh S, Kozicz T, Arimura A (2000) Delayed systemic administration of PACAP38 is neuroprotective in transient middle cerebral artery occlusion in the rat. Stroke 31:1411–1417 First report to document a role for PACAP in neuroprotection in the context of stroke-presaged much additional work on the neuroprotective role of PACAP in stroke, ischemia, and toxic insult to the brain and several other organs

Said SI, Rosenberg RN (1976) Vasoactive intestinal polypeptide: abundant immunoreactivity in neural cell lines and normal nervous tissue. Science 192:907–908 Describes the discovery—function, sequence and significance--of VIP

Sherwood NM, Krueckl SL, McRory JE (2000) The origin and function of the pituitary adenylate cyclase-activating polypeptide (PACAP)/glucagon superfamily. Endocrine Rev 21:619–670 After the discovery of PACAP just eleven years earlier, this is a comprehensive account of the origin of PACAP and VIP within a larger family and from a GHRH basic functional unit

Spengler D, Bweber C, Pantaloni C, Holsboer F, Bockaert J, Seeburg PH, Journot L (1993) Differential signal transduction by five splice variants of the PACAP receptor. Nature 365:170–175

Stroth N, Kuri BA, Mustafa T, Chan SA, Smith CB, Eiden LE (2013) PACAP controls adrenomedullary catecholamine secretion and expression of catecholamine biosynthetic enzymes at high splanchnic nerve firing rates characteristic of stress transduction in male mice. Endocrinology 154:330–339 Building on previous work of Wakade and colleagues, and others, this report provides a comprehensive data set that makes the case for PACAP as a bone fide neurotransmitter of physiological importance in the peripheral nervous system

Zhang L, Hernandez VS, Gerfen CR, Jiang SZ, Zavala L, Barrio RA, Eiden LE (2021) Behavioral role of PACAP signaling reflects its selective distribution in glutamatergic and GABAergic neuronal subpopulations. Elife 10:e61718 Comprehensive documentation of excitatory and inhibitory co-trasmitters of PACAPergic neurons of the rodent CNS

Part III

Evolution, Comparative Physiology, and Behavioral Effects of Neurohypophyseal Peptides in Vertebrates

12 Evolution of Electrophysiological Properties of Magnocellular Neurons

Gareth Leng

Abstract

As summarized with prescient insight by Berta Scharrer, classical neurosecretory neurons "differ from more conventional neural elements by specializing in the manufacture of chemical mediators to a degree comparable to that of gland cells and by engaging in modes of information transfer that markedly digress from standard synaptic transmission." Their products are peptides, and she noted that these cells "seem to reach back to the very beginnings of the evolvement of a neural apparatus, when all of the endocrine functions had to be carried out by pluripotential neurons.... Gene duplication and enzymatic cleavage can be presumed to have been involved in the development, from an ancestral universal precursor, of peptidic entities with increasingly diversified functional properties" (Scharrer, Gen Comp Endocrinol 34:50–62, 1978). In all metazoans, multifunctional neurosecretory neurons release peptides that can act as both endocrine signals (neurohormones) and as locally acting paracrine/autocrine signals to other neurons. In lower invertebrates, such as annelids, at least half of the neurons in the cerebral ganglion are capable of both hormonal and local signaling. In this chapter, I discuss aspects of the evolution of the mammalian hypothalamo-neurohypophysial neurons from their invertebrate origins.

Keywords

Vasotocin · Isotocin · Mesotocin · Osmoregulation · Phasic firing

G. Leng (✉)
Centre for Discovery Brain Research, The University of Edinburgh, Edinburgh, UK
e-mail: Gareth.leng@ed.ac.uk

V. Grinevich, R. F. Oliveira (eds.), *Evolutionary and Comparative Neuroendocrinology*, Masterclass in Neuroendocrinology 17,
https://doi.org/10.1007/978-3-031-80209-6_12

12.1 Introduction

The hypothalamus and its hormones have their evolutionary roots in *Urbilateria*, wormlike marine organisms that are the last common ancestors of vertebrates, flies and worms. In *Urbilateria*, it appears (from comparative genomics) that the peptide-secreting cells were multisensory and multifunctional, with properties of both neurons and endocrine cells; they responded to cues from the ancient marine environment—including light, salinity, temperature and oxygen levels, and communicated with other cells by neurotransmitters and endocrine/paracrine factors to regulate their metabolism and reproduction. One of the peptides made by these cells was an ancestor of vasopressin and oxytocin (Scharrer 1978).

The first electrophysiological studies of hypothalamic neurosecretory cells were performed in teleosts (goldfish (Kandel 1964) and tilapia (Yagi and Bern 1965)), which offered easy experimental access to these systems. By contrast, to record from neurosecretory cells of the mammalian hypothalamus seemed to be a futile endeavor, given the scarcity of these neurons, located deep in the brain amongst many other cell types. But a breakthrough came when Kinji Yagi introduced the technique of antidromic identification, which enabled magnocellular neurons to be identified by virtue of their projections to the neurohypophysis (Yagi et al. 1966). This began three decades of intensive electrophysiological investigation of the hypothalamo-neurohypophysial system (HNS). With the introduction of ventral surgical approaches, which made the rat supraoptic nucleus (a hypothalamic region that contains only magnocellular neurons) directly accessible, these neurons became probably the most fully characterized of any in the CNS. It became possible to study their behavior during neuroendocrine reflexes in vivo, and to relate this to their secretory activity and to their membrane properties as elucidated in in vitro preparations.

Here we consider the evolutionary history of magnocellular neurons, and consider the electrophysiological features that they have come to be known by. All vertebrates possess an HNS comprising neurosecretory neurons that synthesize either an oxytocin-like peptide or a vasopressin-like peptide, which they secrete from axonal nerve endings at a specialized neurohemal organ—the neurohypophysis. Thus, the story of the evolution of these systems takes us back to before the origins of the vertebrates.

12.2 Evolutionary Origins of Separate Neurohypophysial Peptides

Most animals belong to one of two evolutionary lineages: Protostomia (most invertebrates, including insects); and Deuterostomia (some invertebrates and all vertebrates). Vasopressin- and oxytocin-like peptides have been identified in both lineages, indicating that they originated before the split of Proto- and Deuterostomia, 640–760 million years ago.

Most invertebrates have only one such peptide; inotocin is most commonly present in insects, nematocin in nematodes, conopressin in gastropods, cephalotocin in cephalopods, and annetocin in annelids. The earliest vertebrate fossils, from about 530 million years ago, are of *agnatha*—jawless fish. Hagfish and lamprey, the only extant representatives of *agnatha*, have only vasotocin, and, in brain regions that correspond to the hypothalamus, vasotocin is expressed in neurons that project to the neural lobe of the pituitary and/or to a neurohemal region homologous to the median eminence of mammals (Nozaki and Gorbman 1983; Goossens et al. 1977). Jawed descendants of *agnatha* appeared about 440 million years ago. So far as is known, all modern jawed fish, like all extant vertebrates other than lamprey and hagfish, have both vasotocin and an oxytocin-like peptide. Thus, it appears that the vasotocin gene was duplicated in an ancestor of all extant jawed fish.

Jawed fish comprise cartilaginous fish and bony fish. All cartilaginous fish have vasotocin, but different species express many different oxytocin-like hormones (and sometimes two oxytocin-like hormones, reflecting a further gene duplication). It has been suggested that this diversity reflects neutral evolution of the oxytocin-like peptide gene due to the relative absence of selection pressure immediately following gene duplication.

Bony fish evolved into ray-finned fish (most modern fishes), and lobe-finned fish, living examples of which are coelacanths and lungfish. Ray-finned fish arose about 400 million years ago and now comprise ~30,000 known species; these have massively divergent morphologies and live in a wide range of environments, including both freshwater and seawater, engaging a range of different osmoregulatory mechanisms. However, all express vasotocin and (usually) isotocin, an oxytocin homolog (Stafflinger et al. 2008). Like mammalian magnocellular neurons, some of the vasotocin and isotocin neurons project to the neurohypophysis from whence they secrete their product as a circulating hormone, and some project centrally to interact with other neuronal populations.

Tetrapods (amphibians, reptiles, birds, and mammals) all evolved from lobe-finned fish. Coelacanths, lungfish, amphibians, reptiles, and birds all express vasotocin and an oxytocin homolog, usually mesotocin.

12.3 Evolutionary Origins of the Separate Magnocellular Neurons

When a peptide gene is duplicated, one copy is free to mutate, and subsequent mutations in the coding region may lead to a new peptide, while mutations in the regulatory region may lead to expression at a different site or in response to different stimuli. For two descendants of a peptide to have differentiated functions, their secretion must be regulated independently; this generally requires that they be housed in different cells. However, the ‘new’ oxytocin-like peptides did not end up in preexisting neurons but in a new population of magnocellular neurons. It seems that duplication of the vasotocin gene was followed by a mutation that led to the differentiation of two populations of magnocellular neurons in which the two copies were differentially expressed.

It has been proposed that every cell type has its own molecular "password," comprising a few genes that determine its identity. In zebra fish embryos, differential expression of the transcription factors, *nk2.1* and *pax6*, divides the developing forebrain into two. The *nk2.1+* region gives rise to the hypothalamus and preoptic area, where a subset of vasotocin cells is determined by the expression of the transcription factors *rx3* and orthopedia (*otp*), and a microRNA, *miR-7*. These cells are specified by the password (*miR-7+*, *nk2.1+*, *rx+*, *otp+*)—and the same password defines the homologous cell type in the annelid *Platynereis dumerilii* (Tessmar-Raible et al. 2007).

Most of the vasotocin cells in zebra fish do not project to the neurohypophysis. However, in the dorsal preoptic area, vasotocin cells are mingled with isotocin cells, and these do project to the neurohypophysis. In this region, two transcriptional regulators, *orthopedia b* and *simple-minded 1*, are required for expression of vasotocin and isotocin.

But do similar cells in the same brain region have different passwords? In the mammalian HNS oxytocin and vasopressin are expressed in separate neurons that are located very close to each other and which have many properties in common, as detailed below. In the rat supraoptic nucleus, apparently *all* neurons contain either oxytocin or vasopressin, and while oxytocin cells are predominantly in the dorsal part and vasopressin cells in the ventral part, they are extensively intermingled.

Given that ray-finned fish diverged from the mammalian lineage more than 400 million years ago, distinctive commonalities between fish and mammalian neurohypophysial systems are likely to have very ancient origins. To test whether the genes that specify the magnocellular cell identities are conserved between rayed fish and mammals, Venkatesh et al. (1997) produced transgenic rats by inserting segments of puffer fish DNA that included the isotocin and vasotocin genes and their regulatory regions. In the brains of these rats, isotocin was expressed only in oxytocin cells, so it seems that the fish isotocin gene recognizes the mammalian oxytocin cell, which must therefore have the same password as fish isotocin cells. Vasotocin was not expressed in the hypothalamus, so it seemed that the construct lacked the appropriate elements for targeting vasotocin expression, but experiments with a slightly different sequence produced, in transgenic mice, expression of isotocin mRNA in magnocellular oxytocin cells and vasotocin mRNA in magnocellular vasopressin cells (Gilligan et al. 2003).

In the transgenic rats, expression of both oxytocin mRNA and isotocin mRNA was increased when rats were denied drinking water for 3 days; but in the mice, vasotocin mRNA expression was unchanged when drinking water was replaced by 2% saline for a week. In rats, such salt loading produces a raised plasma osmotic pressure, the activity of magnocellular neurons is elevated, and within 3 days, the pituitary content of both oxytocin and vasopressin is depleted by more than 50%. The expression of oxytocin and vasopressin mRNAs increases markedly over this time, probably as a response to intracellular signals associated with the increased electrical activity. Thus, it appears that the vasotocin transgene construct lacked some elements that are necessary for such activity-dependent changes in expression. Nevertheless, these studies imply that magnocellular oxytocin and vasopressin cells have separate molecular passwords which arose early in vertebrate evolution. They also show that duplication of a gene per se does not lead to separate populations of cells: isotocin and oxytocin were co-expressed in magnocellular neurons, as were vasopressin and vasotocin.

12.3.1 Intermediate Phenotypes

So far as is known, in all vertebrates, oxytocin-like peptides and vasopressin-like peptides are located in separate cells. Like most bold statements in biology, this is not 'exactly' true. In rodents, magnocellular neurons generally express either abundant vasopressin mRNA and traces of oxytocin mRNA, or abundant oxytocin mRNA and traces of vasopressin mRNA, while a few contain similar amounts of each (Glasgow et al. 1999).

After a tandem gene duplication, both genes are expected to be expressed in the same cells with a doubling of the cellular gene product. However, studies in *Drosophila* suggest that tandem gene duplication often results in overproduction of the gene product beyond the expected doubling, hence "some aspect of the duplicated sequence appears to generate a synergistic effect on expression whenever two identical genes are adjacent to each other" (Loehlin and Carroll 2016). As overproduction of a secreted product is likely to be maladaptive, natural selection may favor mechanisms to suppress expression from an adjacent identical gene, and indeed it appears that the transcription of oxytocin and vasopressin is controlled by distinct promoters that inhibit each other (Gainer 2012). Given such mutual inhibition, three outcomes are possible for any particular cell: a stable state in which vasopressin expression dominates, a stable state where oxytocin expression dominates, and an unstable state where both are expressed at intermediate amounts.

During sustained challenge, more magnocellular neurons express both peptides at a similar level, so is this biochemical plasticity adaptive—does the proportion of cells that secrete vasopressin or oxytocin vary to better meet a sustained demand? In rodents, oxytocin and vasopressin are both needed in response to salt intake or dehydration: oxytocin for natriuresis and vasopressin for antidiuresis. In these circumstances it does not seem useful to increase the population of one cell type at the expense of the other. On the other hand, lactation requires oxytocin but not vasopressin; however, to convert vasopressin cells into functional oxytocin cells would require not just a biochemical switch, but also a rewiring of neuronal connections to allow them to be activated by suckling.

The trace levels of vasopressin in oxytocin cells or oxytocin in vasopressin cells may be of no physiological significance: biology is always a bit messy. Magnocellular neurons make massive amounts of their principal products—enough to deliver biologically effective concentrations in the systemic circulation. In these cells, many other peptides (including cholecystokinin, met-enkephalin, nesfatin, dynorphin, apelin, galanin, corticotrophin-releasing factor, and neuropeptide Y) are expressed at low levels. Some, like dynorphin, have important roles, but why so many are expressed at low amounts is not clear. Perhaps the cost of eliminating such expression is greater than that of tolerating some biologically irrelevant expression. And the 'messiness' of biological systems makes them more evolvable: if natural selection efficiently eliminated everything useless, where would be the raw material to fuel a change of function when a new ecological niche emerged?

12.3.2 Intermediate Phenotypes of Magnocellular Neurons

Some magnocellular neurons have an intermediate *functional* phenotype; they have electrophysiological properties characteristic of vasopressin cells, yet also respond to suckling with the milk-ejection bursts that characterize oxytocin cells. We do not know if these intermediates are also the biochemical intermediates. Is there perhaps some cellular mechanism that differentiates magnocellular neurons not only by whether they make mainly oxytocin or vasopressin, but also by the differential expression of all the *other* genes that contribute to their identities? For example: oxytocin cells have oxytocin receptors, while vasopressin cells have vasopressin receptors. They have different complements of coexisting peptides: cholecystokinin is predominantly co-expressed with oxytocin, while dynorphin is predominantly co-expressed with vasopressin. Different complements of membrane ion channels give rise to different electrophysiological characteristics and differences in stimulus-secretion coupling.

During development, cell identity emerges not in a flash, but in a flow. As newborn cells migrate, the cells that they encounter exchange messages and both are changed by that exchange. Differential expression of one gene leads to differential expression of others, including genes involved in chemotaxis that attract particular synaptic inputs—and differential innervation leads to further changes in gene expression, and so on. The secretory phenotype of magnocellular neurons can influence the pattern of their afferent innervation: in the rat, noradrenergic fibers from the ventrolateral medulla densely innervate magnocellular vasopressin cells; however, in the Brattleboro rat, which is unable to secrete vasopressin, vasopressin cells receive a reduced noradrenergic input while oxytocin cells are hyperinnervated, suggesting that "the target neuron peptide (vasopressin) is necessary for the maintenance of normal noradrenergic innervation patterns" (Sladek and Johnson 2013). In the olfactory bulb, oxytocin signaling can influence morphological and synaptic development, and it has a continuing role in the functional plasticity of this site in adults (Pekarek et al. 2022). And in birds, neonatal exposure of the brain to vasotocin or a vasotocin antagonist can produce lifelong changes in affiliative behavior (Baran 2017).

The processes that determine cell identity are complex, yet deliver functionally coherent outcomes—recognizable cellular characteristics and broadly appropriate physiological behaviors. So, how does evolution refine the many different processes involved and adapt them to the differing needs of different species?

12.4 Similarities (and Differences) Between Oxytocin Neurons and Vasopressin Neurons

Magnocellular vasotocin neurons diverged into two sets of neurons at or soon after the duplication of the vasotocin gene in an early bony fish, so commonalities between these two sets are likely to reflect features present in a common ancestor.

In vertebrates, homologs of oxytocin and vasopressin are expressed in distinct populations of magnocellular neurons, located very close to each other. In mammals, the largest aggregations are in the paraventricular and supraoptic nuclei, together containing about 60% of all magnocellular neurons. In the rat supraoptic nucleus apparently *all* neurons contain either oxytocin or vasopressin, and while oxytocin cells predominate in the dorsal part and vasopressin cells in the ventral part, they are extensively intermingled. The neurons look similar: each has two or three large, unbranching dendrites that contain many large dense-cored vesicles (LDCVs); in the supraoptic nucleus most of the dendrites course ventrally to form a mat at the ventral surface of the brain, where they are entwined with the processes of specialized astrocytes. Normally, these processes engulf each dendrite, isolating it from its neighbors, but in lactation the glial processes retract from the oxytocin dendrites, and reorganize to 'bundle' them into assemblies of about eight to ten dendrites. Within each bundle the dendrites are in direct apposition, but the bundles, being wrapped by glial cell processes, are isolated from each other. A similar reorganization during dehydration affects both oxytocin and vasopressin neurons. As this intimate and dynamic association between specialized astrocytes and neurons is unusual, its presence in both oxytocin and vasopressin cells suggests that it was also a feature of the ancestral vasotocin cells.

The close apposition of dendrites within bundles facilitates cross-talk between them, mediated by intercellular messengers. These messengers include: nitric oxide, produced in response to raised intracellular [Ca^{2+}]; endocannabinoids; adenosine; the contents of small synaptic-like vesicles that are likely to contain glutamate; and the contents of the LDCVs that contain oxytocin or vasopressin (and coexisting peptides). These messengers have many cellular targets: endocannabinoids, target afferent synapses that express the CB1 receptor to exert a retrograde inhibition of afferent input. Nitric oxide has autocrine inhibitory actions but also stimulates GABA release from local afferent neurons. The messengers also target endothelial cells to regulate blood flow, and the glial cells to influence their morphological reorganization, and possibly to facilitate the elimination of secreted products from the extracellular fluid. These effects are exerted locally to the source neuron—but the peptides that are released have potentially a much larger sphere of action (Leng and Ludwig 2008; Ludwig and Leng 2006; Chini et al. 2017).

As noted above, the dendrites house LDCVs that contain oxytocin or vasopressin. Each LDCV contains ~80,000 molecules of the precursor proteins from which oxytocin and vasopressin are cleaved. As at the nerve terminals in the neurohypophysis, LDCVs are released by Ca^{2+}-dependent exocytosis, but in dendrites this is not the Ca^{2+} entry that accompanies action potentials, which is the regulatory mode of secretion from the neurohypophysis. In dendrites, the LDCVs are trapped in a scaffold of actin well away from the sites of Ca^{2+} entry at the plasma membrane. However, factors that trigger mobilization of intracellular Ca^{2+}stores can result in exocytosis of some LDCVs in the absence of action potentials. Moreover, these factors can produce a disassembly of the actin scaffold and a 'margination' of the LDCVs to regions bordering the dendritic plasma membrane (Tobin et al. 2012). This 'priming' brings LDVCs into the range of influence of voltage-gated Ca^{2+} channels, making electrical activity a regulator of dendritic peptide release. Thus, central release and peripheral secretion can occur synchronously or independently, depending on the afferent signal and physiological state (Ludwig and Leng 2006).

So it seems that the ancient ancestors of oxytocin and vasopressin cells had properties of both neurons and endocrine cells. They generated spikes in response to afferent inputs, and released vasotocin by calcium-dependent exocytosis from LDCVs to exert distant hormonal actions, but also from dendrites to exert local paracrine and/or autocrine actions, and they also communicated with other neurons by the release of glutamate at conventional synapses.

In many species, the peptide has important influences on reproductive behavior. For example, in the pond snail *Lymnaea stagnalis* conopressin is present in most neurons of the anterior lobe of the right cerebral ganglion, most of which project to the penial complex via the penis nerve, and stimulation of these neurons induces penile eversion. Once everted, the penial complex probes for the female gonopore, intromission occurs, and sperm is transferred down the vas deferens and penis into the vagina. Conopressin-containing nerve fibers also innervate the vas deferens, and injections of conopressin induce contractions of the vas deferens, suggesting that conopressin is also involved in ejaculation (Lockard et al. 2017). In the leech *Hirudo verbena* conopressin can induce elements of reproductive behavior, including a stereotyped twisting that is involved in aligning the gonopores (the genital pores through which eggs and sperm can be exchanged) in preparation for copulation (Wagenaar et al. 2010). And in the nematode *C. elegans* deletion of either nematocin or the two receptors at which it acts results in marked defects in mate searching, mate recognition, and mating itself (Beets et al. 2013).

But while central projections of neurons that make oxytocin- and vasopressin-like peptides are abundant in many species, whether LDCVs are released at their synapses is contentious (Leng and Ludwig 2008; Chini et al. 2017); in mammals, the magnocellular neurons are glutamatergic, hence the synapses, which contain many classical small synaptic vesicles but few LDCVs, may be sites of conventional transmitter release only. Indeed, in *C. elegans* most if not all of the behavioral effects of endogenous nematocin reflect actions at neurons distant from the sites of release. A complete wiring diagram is available for the nervous system of *C. elegans*, and this indicates that "the vast majority of nematocin receptor-expressing cells do not have direct synaptic connections to a peptide-expressing cell, and therefore that nematocin must communicate extrasynaptically with its target cells" (Lockard et al. 2017).

12.5 Electrophysiological Phenotypes

In rodents, magnocellular neurons have some distinctive electrophysiological properties (Fig. 12.1). They are not spontaneously active, but in response to synaptic excitation they display prolonged action potentials (spikes), reflecting an unusually large voltage-gated Ca^{2+} entry—a feature recognized in neurosecretory cells in the goldfish by Kandel (1964). But the most distinctive feature of the magnocellular neurons is their bursting behavior. In rats, vasopressin cells are activated by raised extracellular [Na^+], and during continued systemic dehydration most adopt a 'phasic' firing pattern of activity, comprising bursts of 4–10 spikes/s that last for 20–80 s or more, separated by silent intervals that last 15–40 s (Sabatier et al. 2004). However, the phasic firing is not synchronized; even adjacent vasopressin cells show apparently uncorrelated spike activity, and burst cycle times vary markedly between cells. By contrast, while oxytocin cells are also activated by raised extracellular [Na^+], they do not show any bursting activity; however, in response to suckling in lactating animals they display intense 'milk-ejection bursts' associated with milk let-down, and during parturition they display similar bursts associated with the delivery of the young. These bursts occur quasi-synchronously amongst the oxytocin cells, and, compared to the bursts in vasopressin cells, they are much shorter (~2 s), more intense (with instantaneous firing rates of up to 100 spikes/s), and relatively infrequent (typically at intervals of ~10 min during continuous suckling (Belin et al. 1984)).

12.5.1 Mechanisms Underlying Burst Firing

Bursting is frequently observed in neurosecretory cells of many different types in diverse species. However, bursting can (and does) arise in many different ways.

Bursting can arise either as a result of external 'forcing' or from an interplay of intrinsic factors with opposing effects on neuronal excitability and which operate on different timescales. One well-studied example is the R15 neuron of the mollusk, *Aplysia californica* (Fig. 12.2). This is a large, multifunctional peptidergic neuron, thought to play a role in egg-laying, cardiovascular regulation and water balance; in response to changes in the salinity of the seawater it secretes a peptide that stimulates water retention (Bablanian and Treistman 1983). It displays "parabolic bursting," so-called because during a burst, the intervals between successive spikes first progressively shorten and then progressively lengthen until the burst stops (Ghosh et al. 2009). This pattern arises because of an underlying rhythmic variation in the membrane potential (Stinnakre and Tauc 1973). At the start of a burst, spikes activate an inward Ca^{2+} current; this current is slowly inactivating and hence the depolarizing effect of successive spikes summates. However, the Ca^{2+} entry activates a slower, hyperpolarizing K^+ current. As a result, the excitability of the cell declines, slowing the firing rate and reducing the activity-dependent excitation. This ultimately stops the burst, and a new burst will not start until that hyperpolarization has decayed (Gorman and Hermann 1982). A similar mechanism may underlie bursts in

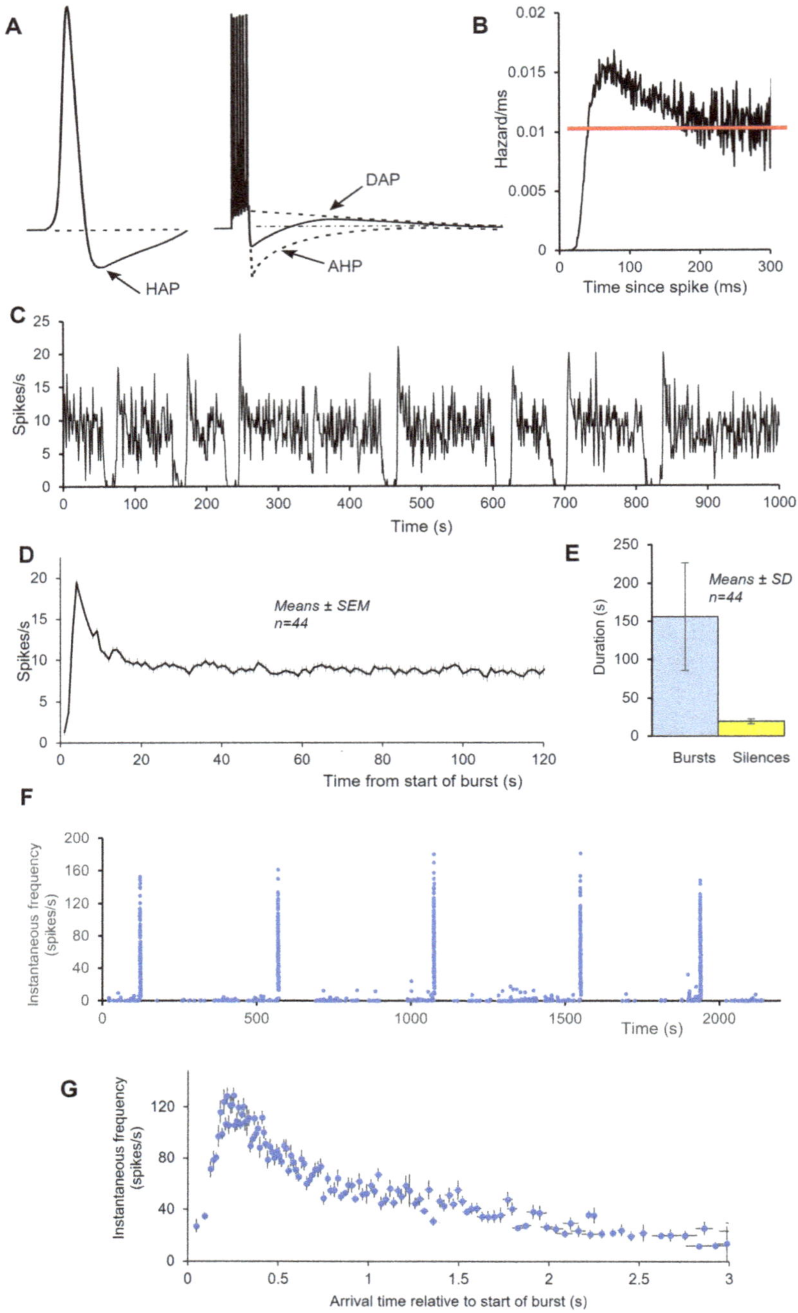

Fig. 12.1 Bursting in magnocellular neurons in the rat. (**a**) Activity-dependent conductances in magnocellular vasopressin cells. Every spike produces a large Ca^{2+} entry which activates a Ca^{2+}-dependent K^{+} conductance which hyperpolarizes the cell strongly; this HAP decays away over about 50 ms. The Ca^{2+}entry also activates two smaller, long-lasting conductances that result in a

mammalian GnRH neurons (Moran et al. 2016), and in Dahlgren cells in teleosts (see below).

The phasic activity of vasopressin cells does not reflect parabolic bursting, though there are some similarities (MacGregor and Leng 2012). After each spike in a vasopressin neuron, Ca^{2+}-dependent K^+ channels open resulting in a hyperpolarizing afterpotential (HAP) that imposes a relative refractory period that extends for 40–60 ms. This is superimposed upon a depolarizing afterpotential (DAP) which outlasts the HAP; thus, every spike is followed by a period of hypoexcitability (identified with the HAP) followed by a period of hyperexcitability (identified with the DAP). The DAP involves several conductances: there is evidence that Ca^{2+} modulation of a resting K^+ current, a nonspecific cation channel and regulation of intracellular Ca^{2+} buffering by calbindin may all be implicated. Because the DAP is very long-lasting (>500 ms) it can summate in response to successive spikes—hence it produces a prolonged depolarization associated with a prolonged burst. The spike activity activates a slow hyperpolarizing current that opposes the DAP, and this produces a prolonged afterhyperpolarization (AHP) after the end of a phasic burst.

Fig. 12.1 (continued) DAP and an AHP. (**b**) As a result of the superposition of a HAP and a DAP, spikes are followed by a period of hypo-excitability and then by a period of hyper-excitability. The graph is a hazard function constructed in 1-ms bins from 62,743 spikes recorded over 7700 s from a vasopressin cell. This function calculates the probability of a spike occurring in each time bin after a spike, given that a spike has not occurred in the intervening time; it thus plots the evolution of neuronal excitability after a spike (Wagenaar et al. 2010). The red line indicates the mean excitability of the cell at >200 ms after a spike, when the influence of the preceding spike upon excitability is minimal. The hypo-excitability attributable to the HAP is visible in the period 0–40 ms when the hazard function is below the red line; hyper-excitability (when the fast DAP is dominant) is visible as the period 40–200 ms when the hazard function is above the red line. (**c**) The first 1000-s of recording of the vasopressin cell analyzed in (**b**), plotted as spikes/s in 1-s bins. This cell fires 'phasically,' in long bursts separated by periods with few or no spikes ('silences'); it is firing at the high end of the range of firing rates observed in phasic vasopressin cells, at an overall mean rate of 8.1 spikes/s. (**d**) Mean burst profile of the 44 recorded bursts recorded in this cell (mean ± SEM spikes per 1-s bin relative to the first bin with >3 spikes after a silent period). Bursts reach a maximum firing rate within 4 s of the burst onset, and reach a plateau of constant firing rate within about 20 s. The plateau is maintained at a constant level until the burst ends. (**e**) Mean burst and silence durations of the 44 bursts; the error bars show the standard deviation, not the standard error, to show the variability of burst and silence duration in the same cell. Burst durations varied between 49 s and 305 s; silence durations between 12 s and 25 s. (**f**) Extract of 2300 s of spike activity recorded from an oxytocin cell in an anesthetized lactating rat during a period of suckling. The overall mean firing rate is 0.45 spikes/s, typical of oxytocin cells recorded in these circumstances. The arrival time of each spike is indicated by a blue dot, and is plotted at a height proportional to the inverse of the preceding interspike interval—thus a spike occurring 20 ms after the previous spike has an 'instantaneous frequency' of 1/0.02 = 50 spikes/s. In vasopressin cells—and in oxytocin cells between milk-ejection bursts—almost no interspike intervals are this short—but in a milk-ejection burst, most are even shorter. The bursts are intense, very brief (2.4–3.4 s), and are followed by many seconds of quiescence. (**g**) This plot shows the mean profile of 13 successive milk-ejection bursts recorded in the oxytocin cell shown in (**f**) (Brown et al. 2004). The bursts had between 105 and 125 spikes. For each burst, the spikes were numbered from first to last; the graph plots, for each spike number, the mean ± SEM arrival time against the mean ± SEM instantaneous frequency: thus, for example the peak of the burst occurred at the 15th spike of the burst, with a mean instantaneous frequency of 129 ± 5.9 spikes/s, and it occurred 252 ± 16 ms after the first spike. About 90% of the spikes occur within 2 s of the burst onset

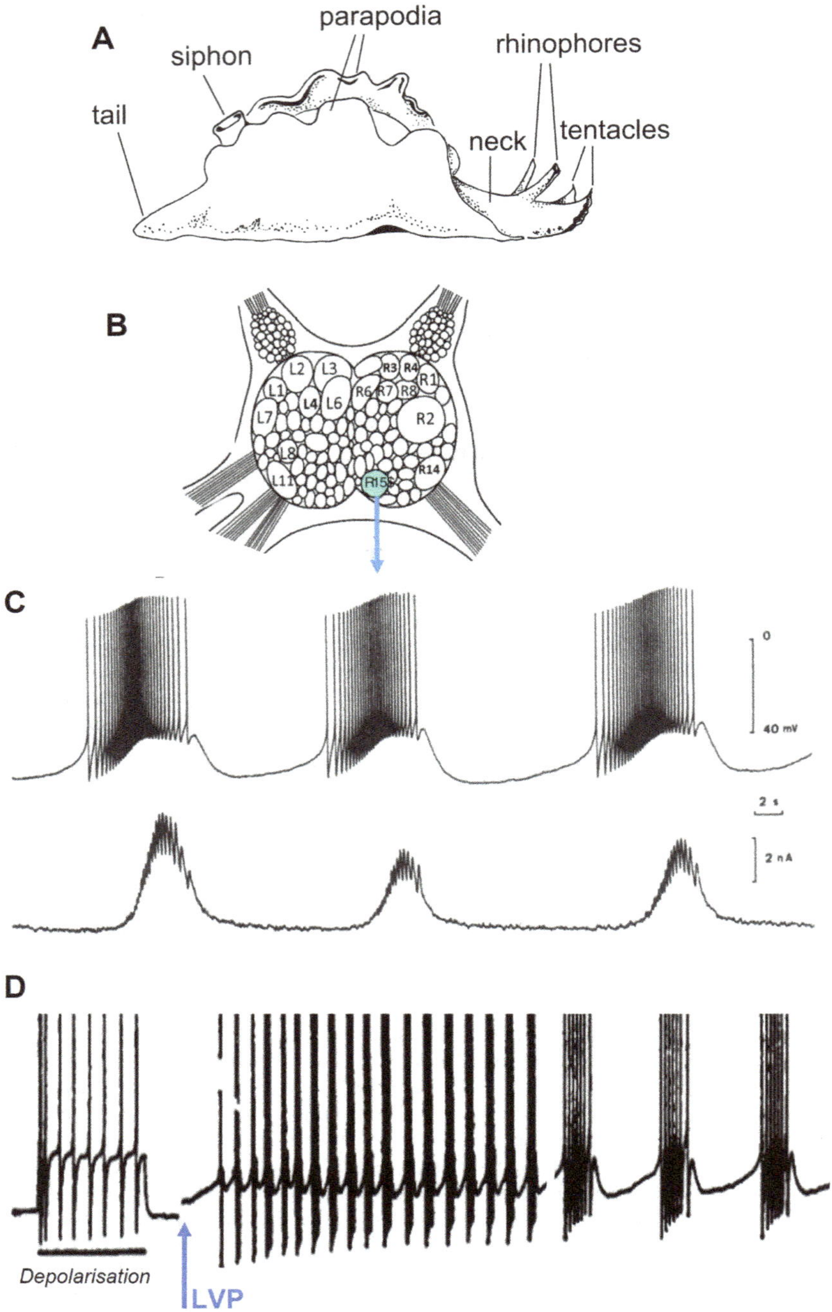

Fig. 12.2 Bursting in Aplysia R15. (**a**) Sketch of *Aplysia californica*, a marine mollusk with a relatively simple nervous system (from Vorster et al. (2014)). (**b**) Schematic of the abdominal nerve ganglion, indicating the position of neuron R15 (modified from Akhmedov et al. (2013)). (**c**)

These mechanisms are thus similar in some respects to those in Aplysia R15 neurons, but there are important differences. The bursts in R15 are spontaneously generated, in that they persist in the absence of synaptic activity. By contrast, bursts in vasopressin cells require synaptic activity, without which the neurons are silent. Moreover, the bursts do not decay gradually to silence as do bursts in R15, but rather, within a few seconds of the burst onset, they reach a relatively stable plateau of spiking activity. As spike-dependent depolarization balances the spike-dependent hyperpolarization in this condition, such bursts can continue more or less indefinitely unless some other event terminates them (MacGregor and Leng 2012). A burst may be terminated by a strong transient synaptic inhibition, but also can be terminated by a synaptic excitation when the resulting increase in spike activity produces a large enough hyperpolarization. Thus vasopressin cells are 'bistable oscillators' (Sabatier and Leng 2007): a phasically firing cell has two stable states—a silent state and an active state, and alternation between these states occurs as a result of external perturbations. Importantly, one common perturbation that can end a phasic burst is the activity-dependent dendritic release of dynorphin, and phasic firing can be eliminated by blocking the actions of dynorphin by a kappa-opioid antagonist (Brown et al. 2004).

12.5.2 Bursting in Oxytocin Neurons

The mechanisms underlying bursting are different again in oxytocin cells: they have a HAP and an AHP but only a small, fast DAP. Like vasopressin neurons, oxytocin neurons are activated by increases in extracellular [Na^+], but they do not display bursts. However, in a lactating rat, suckling triggers dendritic oxytocin release (Moos et al. 1989), probably by activating a peptidergic input that evokes intracellular Ca^{2+} release. The dendrites of oxytocin cells express oxytocin receptors, and oxytocin acting at these can trigger further Ca^{2+} mobilization, causing priming of the vesicle stores, making oxytocin available for activity-dependent release. As oxytocin also depolarizes oxytocin cells, once the vesicle stores are primed, electrical activity triggers dendritic release that triggers further activity, producing an explosive burst that is terminated by the AHP and by depletion of the store of vesicles available for immediate release (Rossoni et al. 2008).

Fig. 12.2 (continued) Intracellular recording (upper trace) from an R15 neuron injected with aequorin, to enable intracellular [Ca^{2+}] to be monitored simultaneously (lower trace). Spontaneous bursts of spikes are shown, superimposed upon a rhythmically oscillating membrane potential; the bursts are associated with rhythmic Ca^{2+} entry and are parabolic on form—the interspike intervals progressively decline at the start of the burst and then progressively increase again until the burst stops (modified from Stinnakre and Tauc (1973)). (**d**) Intracellular recording from an R15 neuron that was initially silent; the spikes have been truncated in this figure. When exposed to a constant depolarization (left trace) the neuron fired action potentials but generated no bursts. However, when exposed to lysine vasopressin (LVP), it generated repeated bursts of spikes, shown at an expanded time scale on the right (modified from Barker and Gainer (1974))

An autoexcitatory action is also apparent in the snail *Lymnaea stagnalis*. As mentioned, in this species conopressin is synthesized in neurons that control penile eversion and ejaculation. These behaviors require a short episode of large hormone secretion, and this is supported by local autoexcitatory actions of conopressin on putative conopressin neurons. These involve induction of both a high-voltage activated pacemaker current and a low-voltage activated pacemaker current, and conotoxin also broadens action potential duration, with the anticipated consequence of facilitating secretion (van Soest and Kits 1998).

12.5.3 Autoregulatory Actions of Vasopressin

While dendritic dynorphin release controls the firing pattern of vasopressin cells, dendritic vasopressin release also has a role. Exogenous vasopressin inhibits vasopressin cells when they are active, and can activate them when they are silent, but the absence of vasopressin does not appear to impede the expression of phasic firing per se. Thus dendritic vasopressin release has a paracrine effect—its level in the extracellular fluid reflects the overall level of activity in the population, and its effects bring the activity of individual vasopressin cells towards the average (Gouzenes et al. 1998). Because each vasopressin cell has a finite store of vesicles that can be replenished only slowly, the store in a hyperactive cell will be swiftly depleted—so it may be important that physiological demand for vasopressin is shared equitably amongst the vasopressin cells.

The earliest demonstration that neurohypophysial peptides might have a specific effect on bursting behavior came from studies of invertebrate neurosecretory cells. In 1974, Barker and Gainer reported that exogenously applied vasopressin could modulate the bursting of R15 but had no effect on other identified neurons in this species—and that both vasopressin and oxytocin could enhance similar bursting activity in a neurosecretory cell (cell 11) in the land snail *Otala lactea*. The authors noted that "The results obtained with this invertebrate system may thus serve as a model to describe a new form of cellular communication in the nervous system whereby peptides may modulate electrical activity" (Barker and Gainer 1974), a prescient prediction amply fulfilled in subsequent decades. Interestingly, the effects reported by Barker and Gainer—and similar effects of vasopressin in other invertebrate preparations—revealed excitatory actions of vasopressin, rather than the inhibitory actions that are dominant in rat vasopressin cells.

To summarize, many different types of neurosecretory cells in diverse species possess an intrinsic 'tool-kit' capable of generating bursting patterns of electrical activity. This includes a positive-feedback mechanism in the form of a slow, activity-dependent depolarizing current; a HAP, which prevents positive feedback from producing excessive spiking; and a slow activity-dependent hyperpolarization which opposes the slow depolarization and which decays more slowly, resulting in a post-burst hyperpolarization that causes a silent period between bursts. Whether these mechanisms produce bursts at all, and if so what kind of bursts, depends on the densities of the many ion channels involved, and the mechanisms of buffering changes in intracellular Ca^{2+}.

12.6 Adaptive Significance of Burst Firing

Vasopressin cells are not identical clones but are heterogeneous in their electrophysiological properties, and these properties vary according to physiological state. Although many fire phasically, each does so through a slightly different assemblage of the necessary elements. Different cells receive different inputs from chance variation, resulting in different levels of synaptic input, and this will produce different levels of spontaneous spike activity. However, the electrophysiological properties are adaptable. Completely silent neurons are useless, so prolonged electrical quiescence results in upregulation of intrinsic depolarizing mechanisms; conversely overactivity is wasteful, and results in upregulation of intrinsic hyperpolarizing mechanisms. Such adaptability is also present in the adult, and during chronic stimulation, adaptations to the changed input state results in an increased prevalence of phasic firing. But does phasic patterning in itself have any adaptive significance?

Asynchronous phasic firing is also observed in type 1 Dahlgren cells of the flounder, large neuroendocrine neurons that are located at the caudal end of the spinal cord in fish. These cells innervate the urophysis—a neurohemal gland peculiar to fish—and they produce two peptide hormones, urotensins I and II, which, in teleosts, have cardiovascular actions and regulate water and electrolyte balance by actions on ion transport at the bladder, intestine and gills. These cells discharge spikes in long bursts, even longer (at ~120 s) than most bursts in vasopressin cells, separated by yet longer silent periods (~200 s), and with a lower intraburst frequency (<5 spikes/s). Unlike vasopressin cells, Dahlgren cells are spontaneously active; spikes are followed by a large transient hyperpolarization (analogous to the HAP in vasopressin cells), followed by a fast DAP which arises as a hyperpolarization-induced rebound depolarization, via L-type Ca^{2+} channels. This DAP is relatively brief (~100 ms), but is large enough to regenerate spiking; the initial burst of spikes evokes a slow afterdepolarization (ADP) that then sustains the burst, and this also appears to depend on L-type Ca^{2+} channels. These channels were first characterized by their slow activation and inactivation kinetics, and their requirement of prolonged depolarization for activation. However, L-type Ca^{2+} channels containing the a1D subunit activate at more hyperpolarized potentials, so Dahlgren cells may express two subtypes of the L-type channel, one with the a1D subunit (underlying the DAP) and one without it (underlying the ADP). The Dahlgren cells do not display an AHP like that in vasopressin cells, but Ca^{2+}-dependent K^+ channels are involved in terminating bursts, through a role in frequency adaptation of spiking. Thus, Dahlgren cells show spontaneous parabolic bursting, albeit with much slower dynamics than in R15 cells (Brierley et al. 2004).

So, vasopressin cells in mammals and Dahlgren cells in teleosts are both involved in regulating antidiuresis, plasma volume and electrolyte balance but through secretion of different peptide hormones, and both display long asynchronous bursts of spikes that arise through different mechanisms. In neither case can the bursting be explained as a mechanism to generate rhythmic secretion, as the asynchronicity ensures that no such rhythm can exist for the hormone levels in the circulation.

12.6.1 Efficiency of Burst Patterns for Stimulus-Secretion Coupling

In 1972, Hal Gainer, reflecting on the bursting of neurosecretory cells in Aplysia and *Otala lactea*, found it "tempting to speculate that the burst pattern would be a particularly efficacious pattern for a secretion process" (Gainer 1972). This idea was quickly followed up by many studies of stimulus-secretion coupling. In these experiments, oxytocin and vasopressin secretion was evoked by electrical stimulation of the isolated rat neurohypophysis, applied in different patterns to mimic the patterns of spike activity recorded from oxytocin and vasopressin cells. The outcomes supported Gainer's speculation: oxytocin secretion was strongly facilitated when stimuli were applied in brief bursts that mimicked milk-ejection bursts. For vasopressin, a different pattern was most effective—20-s bursts of stimulation at 13 Hz separated by silent intervals of 30 s, a pattern similar to that observed in vasopressin cells in dehydrated rats (Bicknell 1988).

It was thus clear that the efficacy of a pattern of stimulation depended on the membrane properties of the nerve terminals—and these were different between oxytocin cells and vasopressin cells. In other words, the firing patterns of the two neuronal types were each efficient because of the particular characteristics of their respective nerve terminals—so did the electrophysiological phenotype evolve to fit the terminal properties or vice versa? Or does bursting activity have some other functional benefits?

Some such benefits are obvious for milk-ejection bursts. Each burst delivers a large abrupt pulse of oxytocin into the circulation, reaching concentrations high enough to trigger milk let-down from the mammary gland despite its relative insensitive to oxytocin. The brevity of the signal minimizes receptor desensitization, the long gap between pulses allows full recovery of sensitivity, and the requirement for a high concentration of oxytocin minimizes incidental milk let-down in the absence of suckling. Similar considerations apply to other endocrine systems, and many hormones—including the anterior pituitary hormones—are secreted in intermittent pulses rather than continuously. As secretion from the anterior pituitary is governed by hypothalamic releasing factors, release of a pulse from the pituitary typically requires release of a pulse of the releasing factor—and this is delivered by a burst of electrical activity that occurs synchronously amongst the releasing-factor neurons, achieved by a combination of intrinsic burst-generating properties and a mechanism for conditional intercommunication that provides mutual excitation (Leng and Brown 1997).

As similar constraints are likely to apply to any peptide hormone released into the systemic circulation in any species, it is perhaps unsurprising that neurosecretory cells commonly exhibit burst-generating properties and are receptive to local feedback from their hormonal products. Interestingly, the vasopressin cells are exceptions in this, in that interactions between them are exclusively inhibitory, and vasopressin is not released in a pulsatile pattern. To try to understand this, it is perhaps worth thinking carefully about osmoregulation.

12.7 Regulating Water and Electrolyte Balance

In mammals the importance of vasopressin in maintaining water and electrolyte balance is undeniable: to survive without vasopressin, the Brattleboro rat must drink about its own body weight of water each day, producing copious amounts of highly dilute urine. Vasopressin has two actions that are important in this: by its actions at V2 receptors in the kidney it promotes water retention through antidiuresis); and through vasoconstrictor actions at V1 receptors on the peripheral vasculature it can regulate the plasma volume to compensate for body water loss. Vasopressin is only one of many factors regulating vascular tone, but antidiuresis is uniquely identified with vasopressin.

Whereas mammals are constantly exposed to the threat of dehydration due to their high respiratory water loss, fish are vulnerable because their extracellular fluid is in close contact with water of varying salinity across the respiratory epithelia of the gills. Marine teleosts, freshwater teleosts, and marine elasmobranchs all use different strategies for osmoregulation. While the kidneys play a role, extrarenal mechanisms are important: at the gills (where uptake and extrusion of ions occur against the concentration gradients imposed by a freshwater or seawater environment); at the gut (particularly for scavenging water for fish that live in seawater); at the rectal gland (for chloride and sodium excretion in elasmobranchs); and at the urinary bladder. Teleost fishes maintain their plasma osmolarity at about one-third that of seawater. In fresh water, this involves excreting copious dilute urine while actively importing ions across the gills. In seawater, teleosts counteract the passive gain of ions and loss of water by drinking seawater, absorbing water and salts across the gut, and exporting excess Na^+ across the gills and divalent ions through the kidney (Takei et al. 2014). Many hormones are involved, including natriuretic factors; urotensins, which are specific to fish; and some mammalian hormones that have no analogous role in mammals. Thus, in teleosts, cortisol has been viewed as 'the' seawater-adapting hormone, and prolactin as 'the' fresh-water adapting hormone, while thyroid hormones, growth hormone and insulin-like growth factor also have a role in salinity acclimation (Seale et al. 2021; Seale and Breves 2022). So where does vasotocin fit in this? In teleosts, vasotocin is not an antidiuretic hormone—it has a diuretic action.

The electrophysiological behavior of isotocin and vasotocin cells has been studied in the rainbow trout, using confocal Ca^{2+} imaging in an isolated preparation (Saito and Urano 2001). Both cell types displayed repeated episodes of increased intracellular $[Ca^{2+}]$. Vasotocin cells displayed regular pulses with an average duration of ~50 s and an average cycle time of ~180 s; isotocin cells had an average pulse duration of ~20 s and an average cycle time of ~320 s. The pulses were associated with bursts of action potentials, apparently resulting from spike-triggered Ca^{2+} entry, as in other neurosecretory cells. The bursts of spikes had a rapid onset followed by a plateau, but ended with an oscillating activity not seen in parabolic bursters or in vasopressin cells (but seen in some phasic cells in the rat suprachiasmatic nucleus (Tsuji et al. 2016)). The Ca^{2+} pulses were highly correlated between pairs of vasotocin cells, and so were the pulses in isotocin cells, but there was no correlation between the activities of isotocin cells and vasotocin cells. At least some of the isotocin cells seem be electrically coupled, but the vasotocin cells were not; these cells had processes that are directly apposed, so may intercommunicate by dendritic peptide release or via conventional synapses (Saito et al. 2004). Both the vasotocin cells and the isotocin cells responded to lowering extracellular [Na^+] with an *increase* in bursting activity, as expected for cells that release a diuretic hormone, but the opposite response to that of vasopressin cells to this stimulus.

12.7.1 Role of Vasotocin in Osmoregulation in Jawless Fish

But teleosts are not part of the mammalian lineage. Perhaps we should look most carefully at the role of vasotocin in fish that are—in lamprey and hagfish, the two surviving representatives of jawless fish; and in coelacanth and lungfish, surviving representatives of lobe-finned fish.

Today, hagfishes live in deep ocean waters, and unlike all other bony fish, both the osmolarity and salinity of their body fluids is similar to that of the surrounding environment; this absence of osmoregulation may be one of many degenerate characteristics (including degenerate eyes) reflecting their long persistence in a very stable habitat.

Osmoregulation is very different in lampreys. All modern lampreys display a long-lived freshwater larval period, followed by metamorphosis into a sexually immature juvenile, and then into mature adult which differ dramatically from the larva. However, this life cycle evolved in the Triassic, about 250 million years ago, so modern lampreys combine ancient and evolved features. While some lampreys live in marine waters or estuaries for part of their life, all live and breed for long periods in fresh water, and lampreys captured in freshwater can regulate their water and electrolyte balance successfully when transferred to 50% seawater. To do so, they drink large amounts of seawater, and the excess Na^+, K^+, and Cl^- is mainly absorbed by the intestine. However, Ca^{2+} and Mg^{2+} are not readily absorbed by the intestine; some is excreted in the urine, but the urine is never hyperosmotic to the plasma, and much of the excess is excreted at the gills (Pickering and Morris 1970). Lampreys express vasotocin mRNA in the hypothalamus, and mRNAs for the oxytocin receptor, vasopressin 1a and V1b receptors and a subtype of V2 receptors are all expressed in the forebrain (Mayasich and Clarke 2016). As in ray-finned fish, vasotocin in lamprey is diuretic; at the kidney, vasotocin produces a vasoconstriction that induces both diuresis and an increased glomerular filtration rate.

In cartilaginous fishes, the plasma osmolarity is close to (or higher than) that of seawater, but the plasma [Na^+] is only about a third of that of sea water: this arises mainly because of a high plasma urea content—a feature that enables efficient water extraction. By contrast, bony fish, with one exception, have a plasma osmolarity only about one-third that of seawater. The exception is the coelacanth, in which the plasma osmolarity is close to that of the sea; it has low plasma [Na^+] like other bony fishes, but high levels of urea like cartilaginous fishes (Pang et al. 1977).

The Australian lungfish *Neoceratodus forsteri* was amongst many diverse species studied by Bill Sawyer, a pioneer of comparative physiology whose bioassays became critically important to the development of specific agonists and antagonists to oxytocin and vasopressin (Manning et al. 1992). He showed that the lungfish responded to intravenous vasotocin with increased dorsal aortic blood pressure, increased urine flow, and tubular rejection of Na^+—responses like those induced by infusions of isosmolar or hyposmolar NaCl solutions. This suggested that vasotocin is secreted in response to volume expansion in lungfishes, that it acts on the kidney as a diuretic hormone (Sawyer 1970), and that this effect is linked to its vasopressor activity. Thus, in 1977, Sawyer proposed that "*the most primitive neurohypophyseal functions were related to cardiovascular regulation and that the neurohypophysis acquired its osmoregulatory functions later in vertebrate evolution*" (Sawyer 1977).

12.7.2 Osmoregulation in Mammals

Thus, there is an abrupt break in the chain of functional evolution; in lungfish, as in lamprey, vasotocin cells are activated by a fall in osmotic pressure to produce a diuresis via effects at V1 receptors in the vasculature of the kidney; in mammals, vasopressin cells are activated by a *rise* in osmotic pressure to produce *anti*diuresis via effects at V2 receptors on the distal convoluted tubule and collecting ducts of the kidney. Just as it has been proposed that in ray-finned fish, oxytocin was freed from selection pressure and enabled to evolve variety of homologs, perhaps the conquest of the terrestrial environment freed vasopressin from its role in volume regulation as exercised in fish by evoking diuresis, to acquire a new role in tetrapods as an antidiuretic hormone. In man, vasopressin has natriuretic effects in some experimental conditions, but, in the physiological range of plasma vasopressin concentrations vasopressin has antinatriuretic effects (Bankir 2001).

For mammals, antidiuresis, when required, demands constant attention for prolonged periods—indeed, until water becomes available to drink. However inefficient continuous activity at vasopressin receptors might be, it may yet be more effective than intermittent activity.

Continuous exposure to a hormone will only activate (and subsequently desensitize) all of the available receptors if present at a saturating concentration. In man, vasopressin normally circulates at a concentration of about ~1 pg/ml, and maximal urine concentration is achieved at a concentration that is just tenfold higher, at ~10 pg/ml (10^{-12} M). The circulating concentration is linearly proportional to plasma [Na^+], and maximal antidiuresis is achieved with about a 5% increase in this. As the V2 receptors in the human kidney have an affinity for vasopressin of 1.2×10^{-9} M; half-maximal receptor binding of vasopressin is achieved at a concentration two orders of magnitude higher than the concentration needed for maximal antidiuresis. Thus, the system controlling antidiuresis in mammals appears to have evolved to become highly sensitive to low circulating concentrations of vasopressin, and to small changes in these concentrations. This is likely to be important to allow the finite store of vasopressin to meet what may be a very prolonged demand.

For the same reasons it may be important that the rate of delivery of vasopressin in given circumstances is stable. Strangely, the phasic firing of vasopressin cells in rats is particularly well suited for this. Both the firing rate of oxytocin cells and the intraburst firing rate of vasopressin cells are surprisingly stable given that spikes are triggered by random fluctuations in synaptic input. This is in part because the long HAP 'spaces' successive spikes, but also because of the negative feedbacks of spike activity upon excitability (via the AHP) and on glutamatergic afferents (via cannabinoids), and activity-dependent activation of inhibitory inputs (via nitric oxide). These slow feedbacks mean that the variance of firing rate when measured over several seconds is much less than predicted from the variance over short periods (Maicas Royo et al. 2016)—the cells 'defend' their firing rate by responding to perturbations with a counterbalancing response. This is true of both oxytocin and vasopressin cells—but the bistability of phasic firing cells introduces another element. In conditions of chronically raised plasma [Na^+], most vasopressin cells fire phasically, and because this activity is asynchronous, vasopressin is secreted from the neurohypophysis in a continuous stream. In these same conditions, oxytocin cells are also activated, firing more slowly than the vasopressin cells fire in bursts, but firing continuously, with no silent intervals. In these circumstances, a transient increase in synaptic excitation will evoke an increase in oxytocin cell activity and cause a subsequent perturbation in oxytocin secretion, but the same is not true for the vasopressin system. A transient increase in synaptic excitation to vasopressin cells will affect them in a way that depends on their firing state at that time. If a cell is perturbed at the beginning of a silent period, when it is strongly hyperpolarized by the AHP, the perturbation will be ineffective; towards the end of a silent period, the perturbation will merely advance the next burst by a few seconds. Early in a burst, the perturbation will have little effect because the cell is firing at close to its physiological maximum. In the late stages, an excitatory perturbation will produce a small increase in firing rate but then stop the burst. Thus, overall, the perturbation will have little effect on vasopressin secretion. Asynchronous bistability means that the vasopressin cell population acts as a low-pass filter—the population is sensitive to small sustained changes in input level, but damps the effects of transient changes (MacGregor and Leng 2012).

12.8 Conclusions

In widely different species, neurosecretory cells that make peptides related to oxytocin and vasopressin display electrophysiological bursting behavior; however, the bursting activity arises from different cellular mechanisms in different cells, and is organized in different ways to fulfil very different functions. Cells of the HNS in fish and mammals are both osmosensitive, but the former are inhibited by raise extracellular [Na^+] while the latter are activated by it; vasotocin in fish is a diuretic hormone, vasopressin in tetrapods is an antidiuretic hormone. These features do not indicate functional continuity through evolution, rather an abrupt break with the emergence of tetrapods. If there is true functional continuity from invertebrates to the mammalian HNS, it is more apparent in reproduction—particularly in mating, where continuity might be seen between the functions of the conopressin cells of land snails and the oxytocin cells of mammals.

References

Akhmedov K, Rizzo V, Kadakkuzha BM, Carter CJ, Magoski NS, Capo TR, Puthanveettil SV (2013) Decreased response to acetylcholine during aging of aplysia neuron R15. PLoS One 8:e84793

Bablanian GM, Treistman SN (1983) Seawater osmolarity influences bursting pacemaker activity in intact Aplysia californica. Brain Res 271:342–345

Bankir L (2001) Antidiuretic action of vasopressin: quantitative aspects and interaction between V1a and V2 receptor-mediated effects. Cardiovasc Res 51:372–390

Baran NM (2017) Sensitive periods, vasotocin-family peptides, and the evolution and development of social behavior. Front Endocrinol (Lausanne) 8:189

Barker JL, Gainer H (1974) Peptide regulation of bursting pacemaker activity in a molluscan neurosecretory cell. Science 184:1371–1373

Beets I, Temmerman L, Janssen T, Schoofs L (2013) Ancient neuromodulation by vasopressin/oxytocin-related peptides. Worm 2:e24246

Belin V, Moos F, Richard P (1984) Synchronization of oxytocin cells in the hypothalamic paraventricular and supraoptic nuclei in suckled rats: direct proof with paired extracellular recordings. Exp Brain Res 57:201–203

Bicknell RJ (1988) Optimizing release from peptide hormone secretory nerve terminals. J Exp Biol 139:51–65

Brierley MJ, Bauer CS, Lu W, Riccardi D, Balment RJ, McCrohan CR (2004) Voltage- and Ca2+-dependent burst generation in neuroendocrine Dahlgren cells in the teleost Platichthys flesus. J Neuroendocrinol 16:832–841

Brown CH, Ludwig M, Leng G (2004) Temporal dissociation of the feedback effects of dendritically co-released peptides on rhythmogenesis in vasopressin cells. Neuroscience 124:105–111

Chini B, Verhage M, Grinevich V (2017) The action radius of oxytocin release in the mammalian CNS: from single vesicles to behavior. Trends Pharmacol Sci 38:982–991

Gainer H (1972) Electrophysiological behavior of an endogenously active neurosecretory cell. Brain Res 39:403–418

Gainer H (2012) Cell-type specific expression of oxytocin and vasopressin genes: an experimental odyssey. J Neuroendocrinol 24:528–538

Ghosh A, Roy D, Jirsa VK (2009) Simple model for bursting dynamics of neurons. Phys Rev E Stat Nonlinear Soft Matter Phys 80:041930

Gilligan P, Brenner S, Venkatesh B (2003) Neurone-specific expression and regulation of the pufferfish isotocin and vasotocin genes in transgenic mice. J Neuroendocrinol 15:1027–1036

Glasgow E, Kusano K, Chin H, Mezey E, Young WS 3rd, Gainer H (1999) Single cell reverse transcription-polymerase chain reaction analysis of rat supraoptic magnocellular neurons: neuropeptide phenotypes and high voltage-gated calcium channel subtypes. Endocrinology 140:5391–5401

Goossens N, Dierickx K, Vandesande F (1977) Immunocytochemical demonstration of the-hypothalamo-hypophysial vasotocinergic system of Lampetra fluviatilis. Cell Tissue Res 177:317–323

Gorman AL, Hermann A (1982) Quantitative differences in the currents of bursting and beating molluscan pace-maker neurones. J Physiol 333:681–699

Gouzenes L, Desarmenien MG, Hussy N, Richard P, Moos FC (1998) Vasopressin regularizes the phasic firing pattern of rat hypothalamic magnocellular vasopressin neurons. J Neurosci 18:1879–1885

Kandel ER (1964) Electrical properties of hypothalamic neuroendocrine cells. J Gen Physiol 47:691–717

Leng G, Brown D (1997) The origins and significance of pulsatility in hormone secretion from the pituitary. J Neuroendocrinol 9:493–513

Leng G, Ludwig M (2008) Neurotransmitters and peptides: whispered secrets and public announcements. J Physiol 586:5625–5632

Lockard MA, Ebert MS, Bargmann CI (2017) Oxytocin mediated behavior in invertebrates: an evolutionary perspective. Dev Neurobiol 77:128–142

Loehlin DW, Carroll SB (2016) Expression of tandem gene duplicates is often greater than two-fold. Proc Natl Acad Sci USA 113:5988–5992

Ludwig M, Leng G (2006) Dendritic peptide release and peptide-dependent behaviours. Nat Rev Neurosci 7:126–136

MacGregor DJ, Leng G (2012) Phasic firing in vasopressin cells: understanding its functional significance through computational models. PLoS Comput Biol 8:e1002740

Maicas Royo J, Brown CH, Leng G, MacGregor DJ (2016) Oxytocin neurones: intrinsic mechanisms governing the regularity of spiking activity. J Neuroendocrinol 28

Manning M, Stoev S, Bankowski K, Misicka A, Lammek B, Wo NC, Sawyer WH (1992) Synthesis and some pharmacological properties of potent and selective antagonists of the vasopressor (V1-receptor) response to arginine-vasopressin. J Med Chem 35:382–388

Mayasich SA, Clarke BL (2016) The emergence of the vasopressin and oxytocin hormone receptor gene family lineage: clues from the characterization of vasotocin receptors in the sea lamprey (Petromyzon marinus). Gen Comp Endocrinol 226:88–101

Moos F, Poulain DA, Rodriguez F, Guerne Y, Vincent JD, Richard P (1989) Release of oxytocin within the supraoptic nucleus during the milk ejection reflex in rats. Exp Brain Res 76:593–602

Moran S, Moenter SM, Khadra A (2016) A unified model for two modes of bursting in GnRH neurons. J Comput Neurosci 40:297–315

Nozaki M, Gorbman A (1983) Immunocytochemical localization of somatostatin and vasotocin in the brain of the Pacific hagfish, Eptatretus stouti. Cell Tissue Res 229:541–550

Pang PKT, Griffith RW, Atz JW (1977) Osmoregulation in Elasmobranchs. Am Zool 17:365–377

Pekarek BT, Kochukov M, Lozzi B, Wu T, Hunt PJ, Tepe B, Hanson Moss E, Tantry EK, Swanson JL, Dooling SW, Patel M, Belfort BDW, Romero JM, Bao S, Hill MC, Arenkiel BR (2022) Oxytocin signaling is necessary for synaptic maturation of adult-born neurons. Genes Dev 36:1100–1118

Pickering AD, Morris R (1970) Osmoregulation of Lampetra fluviatilis L. and Petromyzon marinus (Cyclostomata) in hyperosmotic solutions. J Exp Biol 53:231–243

Rossoni E, Feng J, Tirozzi B, Brown D, Leng G, Moos F (2008) Emergent synchronous bursting of oxytocin neuronal network. PLoS Comput Biol 4:e1000123

Sabatier N, Leng G (2007) Bistability with hysteresis in the activity of vasopressin cells. J Neuroendocrinol 19:95–101

Sabatier N, Brown CH, Ludwig M, Leng G (2004) Phasic spike patterning in rat supraoptic neurones in vivo and in vitro. J Physiol 558:161–180

Saito D, Urano A (2001) Synchronized periodic Ca2+ pulses define neurosecretory activities in magnocellular vasotocin and isotocin neurons. J Neurosci 21:RC178

Saito D, Komatsuda M, Urano A (2004) Functional organization of preoptic vasotocin and isotocin neurons in the brain of rainbow trout: central and neurohypophysial projections of single neurons. Neuroscience 124:973–984

Sawyer WH (1970) Vasopressor, diuretic, and natriuretic responses by lungfish to arginine vasotocin. Am J Phys 218:1789–1794
Sawyer WH (1977) Evolution of neurohypophyseal hormones and their receptors. Fed Proc 36:1842–1847
Scharrer B (1978) Peptidergic neurons: facts and trends. Gen Comp Endocrinol 34:50–62
Seale AP, Breves JP (2022) Endocrine and osmoregulatory responses to tidally-changing salinities in fishes. Gen Comp Endocrinol 326:114071
Seale LA, Gilman CL, Zavacki AM, Larsen PR, Inokuchi M, Breves JP, Seale AP (2021) Regulation of thyroid hormones and branchial iodothyronine deiodinases during freshwater acclimation in tilapia. Mol Cell Endocrinol 538:111450
Sladek CD, Johnson AK (2013) Integration of thermal and osmotic regulation of water homeostasis: the role of TRPV channels. Am J Physiol Regul Integr Comp Physiol 305:R669–R678
Stafflinger E, Hansen KK, Hauser F, Schneider M, Cazzamali G, Williamson M, Grimmelikhuijzen CJ (2008) Cloning and identification of an oxytocin/vasopressin-like receptor and its ligand from insects. Proc Natl Acad Sci USA 105:3262–3267
Stinnakre J, Tauc L (1973) Calcium influx in active Aplysia neurones detected by injected aequorin. Nat New Biol 242:113–115
Takei Y, Hiroi J, Takahashi H, Sakamoto T (2014) Diverse mechanisms for body fluid regulation in teleost fishes. Am J Physiol Regul Integr Comp Physiol 307:R778–R792
Tessmar-Raible K, Raible F, Christodoulou F, Guy K, Rembold M, Hausen H, Arendt D (2007) Conserved sensory-neurosecretory cell types in annelid and fish forebrain: insights into hypothalamus evolution. Cell 129:1389–1400
Tobin V, Leng G, Ludwig M (2012) The involvement of actin, calcium channels and exocytosis proteins in somato-dendritic oxytocin and vasopressin release. Front Physiol 3:261
Tsuji T, Tsuji C, Ludwig M, Leng G (2016) The rat suprachiasmatic nucleus: the master clock ticks at 30 Hz. J Physiol 594:3629–3650
van Soest PF, Kits KS (1998) Conopressin affects excitability, firing, and action potential shape through stimulation of transient and persistent inward currents in mulluscan neurons. J Neurophysiol 79:1619–1632
Venkatesh B, Si-Hoe SL, Murphy D, Brenner S (1997) Transgenic rats reveal functional conservation of regulatory controls between the Fugu isotocin and rat oxytocin genes. Proc Natl Acad Sci USA 94:12462–12466
Vorster AP, Krishnan HC, Cirelli C, Lyons LC (2014) Characterization of sleep in Aplysia californica. Sleep 37:1453–1463
Wagenaar DA, Hamilton MS, Huang T, Kristan WB, French KA (2010) A hormone-activated central pattern generator for courtship. Curr Biol 20:487–495
Yagi K, Bern HA (1965) Electrophysiologic analysis of the response of the caudal neurosecretory system of Tilapia mossambica to osmotic manipulations. Gen Comp Endocrinol 5:509–526
Yagi K, Azuma T, Matsuda K (1966) Neurosecretory cell: capable of conducting impulse in rats. Science 154:778–779

Further Reading

Acher R, Chauvet J, Chauvet MT, Rouille Y (1999) Unique evolution of neurohypophysial hormones in cartilaginous fishes: possible implications for urea-based osmoregulation. J Exp Zool 284:475–484 *"Most bony vertebrate species display a great evolutionary stability of their two neurohypophysial hormones, so that two molecular lineages, isotocin-mesotocin-oxytocin and vasotocin-vasopressin, have been traced from bony fishes to mammals. Chondrichthyes, in contrast, show a striking diversity of their oxytocin-like hormones … It is suggested that [this] great diversity … expresses a release from an evolutionary receptor-binding constraint, so that amino-acid substitutions reflect neutral evolution. In contrast, the preservation of vasotocin suggests a selective pressure, which may be related to the regulation of renal urea transporter-recruitment mechanisms, as it has been shown for vasopressin in mammals."*

Eaton JL, Holmqvist B, Glasgow E (2008) Ontogeny of vasotocin-expressing cells in zebrafish: selective requirement for the transcriptional regulators *orthopedia* and *single-minded 1* in the preoptic area. Dev Dyn 237:995–1005

Murphy D, Si-Hoe SL, Brenner S, Venkatesh B (1998) Something fishy in the rat brain: molecular genetics of the hypothalamo-neurohypophysial system. Bioessays 20:741–749

Ocampo Daza D, Bergqvist CA, Larhammar D (2022) The evolution of oxytocin and vasotocin receptor genes in jawed vertebrates: a clear case for gene duplications through ancestral whole-genome duplications. Front Endocrinol (Lausanne) 12:792644 *"The neuronal and neuroendocrine peptides oxytocin (OT) and vasotocin (VT), including vasopressins, have six cognate receptors encoded by six receptor subtype genes in jawed vertebrates. These six receptor subtype genes arose from a syntenic ancestral gene pair, one VTR1/OTR ancestor and one VTR2 ancestor, through the early vertebrate whole-genome duplications."*

Sawyer WH (1967) Evolution of antidiuretic hormones and their functions. Am J Med 42:678–686 *"The importance of the vasopressins as antidiuretic hormones among mammals is established. Among the nonmammalian tetrapods arginine vasotocin may serve a similar function. In teleost fishes and in the lungfish, however, arginine vasotocin may actually be diuretic."*

Sawyer WH (1977) Evolution of neurohypophyseal hormones and their receptors. Fed Proc 36:1842–1847 *"It is possible that the most primitive neurohypophyseal functions were related to cardiovascular regulation and that the neurohypophysis acquired its osmoregulatory functions later in vertebrate evolution."*

Tessmar-Raible K, Raible F, Christodoulou F, Guy K, Rembold M, Hausen H, Arendt D (2007) Conserved sensory-neurosecretory cell types in annelid and fish forebrain: insights into hypothalamus evolution. Cell 129:1389–1400 *"We propose that cell types with dual sensory-neurosecretory properties were the starting point for the evolution of neurosecretory brain centers in Bilateria."*

Venkatesh B, Brenner S (1995) Structure and organization of the isotocin and vasotocin genes from teleosts. Adv Exp Med Biol 395:629–638 *"Genomic organisation of the Fugu vasotocin/isotocin locus suggests that the teleost vasotocin and isotocin genes have evolved from a common ancestor through tandem duplication and that this locus seems to have undergone a localised reorganisation during vertebrate evolution."*

How Does the Camel Survive in the Desert Without Drinking?

13

Panjiao Lin, Abdu Adem, David Murphy, and Michael Greenwood

Abstract

To live and thrive in arid environments place enormous evolutionary pressure on processes of water conservation. These are epitomized by the numerous adaptations manifested in those mammals specialized to limited water access, such as camels. At the level of the kidney, camels produce low volumes of highly concentrated urine, more so when water is scarce, to conserve body water. Two hormones, arginine vasopressin (AVP) and oxytocin (OXT), both produced in the supraoptic nucleus (SON), the core hypothalamic osmoregulatory control centre, are vital for this adaptive process. This chapter reviews what we know about the mechanisms that enable water-stressed mammals to cope with osmotic challenge. We describe the evolution of AVP-like molecules from invertebrates to mammals, the roles of AVP and OXT in osmoregulation, the structural/functional adaptations of the hypothalamus-kidney axis associated with AVP and the specific water conservation strategies employed by camels. Comparative studies on the molecular sequences of the AVP and OXT genes in mammals have revealed conserved and divergent elements across species. The homogeneity of these neu-

Supplementary Information The online version contains supplementary material available at https://doi.org/10.1007/978-3-031-80209-6_13.

P. Lin · D. Murphy · M. Greenwood (✉)
Molecular Neuroendocrinology Research Group, Translational Health Sciences, Bristol Medical School, University of Bristol, Bristol, UK
e-mail: mike.greenwood@bristol.ac.uk

A. Adem
Department of Pharmacology, College of Medicine and Health Sciences, Khalifa University, Abu Dhabi, United Arab Emirates

V. Grinevich, R. F. Oliveira (eds.), *Evolutionary and Comparative Neuroendocrinology*, Masterclass in Neuroendocrinology 17,
https://doi.org/10.1007/978-3-031-80209-6_13

ropeptides could be affected by their evolutionary relationships and the environmental stressors (such as aridity) that may have shaped their genetic structures. This chapter highlights the common elements between camels and other species that could be key factors of water homeostasis, as well as elements that appear to be unique to the camels that might be essential for adaptations necessary for life in the desert. For future studies, the identification of genetic variations associated with osmoregulation will provide insights for evaluating the possibility of evolutionary rescue of species currently challenged by increased desertification, especially the extremely endangered wild camels.

Keywords

Water conservation · Osmoregulation evolution · Vasopressin · Oxytocin · Camel · Hypothalamus · Kidney · Comparative studies · Genetic adaptations · Molecular variations

13.1 Introduction

In the context of the specific water conservation strategies employed by camels, this chapter describes the roles of arginine vasopressin (AVP) and oxytocin (OXT) in water balance regulation, focussing on the evolution of AVP/OXT peptides, the synthesis and transport of AVP and OXT and their functions in the hypothalamic neurohypophysial system (HNS)-kidney axis. Furthermore, we incorporate our recent comparative studies performed on the genetic variation of *AVP* and *OXT* between camels and other mammals, to explore how genetic differences may influence AVP and OXT in different species. We close this chapter with a summary of the main findings and discuss possible future research directions.

13.2 Hypothalamic Osmoregulation

The HNS is a crucial neuroendocrine system for controlling water homeostasis in mammals. The HNS is composed of specialized osmosensitive magnocellular neurons (MCNs) that are responsible for the synthesis of osmoregulatory neuropeptide hormones (de Souza Mecawi et al. 2015). MCNs are located in the supraoptic nucleus (SON) and paraventricular nucleus (PVN) in the hypothalamus. The SON MCNs are commonly classified as OXT- and AVP- synthesizing MCNs (cell type-specific expression) that account for a large proportion of the MCN population. But there is a small proportion of the SON MCNs that has been shown to co-express AVP and OXT at more comparable levels (Gainer 2012). The axons of MCNs course through the internal zone of the median eminence (ME) and terminate on blood capillaries of the pituitary gland pars nervosa (PN) (Bargmann 1966). The PN is a neurovascular interface by which the brain can release neuroendocrine hormones to maintain homeostasis. It constitutes the majority of the posterior pituitary gland (PP, also known as neurohypophysis) where AVP and OXT are stored before being released to the systemic circulation (Fig. 13.1). AVP and OXT are known to be multifunctional, acting at receptor targets throughout the body. As shown in Fig. 13.1, they are classically involved in defence against water limitation and in reproduction, respectively (Burbach et al. 2001; Leng and Russell 2019).

AVP and OXT are also released within the hypothalamic SON from MCN dendrites and cell bodies (Ludwig et al. 2022). This release mechanism is usually independent of the axonal release into the circulation. AVP and OXT in the extracellular

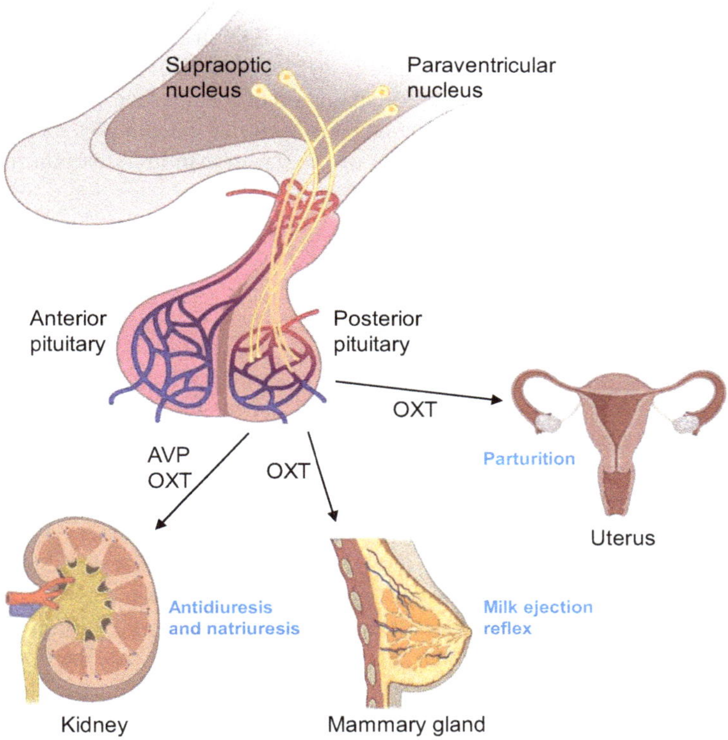

Fig. 13.1 Neurohypophysial control of vasopressin and oxytocin secretion. AVP- and OXT-synthesizing MCNs are located in the supraoptic (SON) and paraventricular (PVN) nuclei of the hypothalamus. Axons from these neurons project down the infundibular stalk to terminate along the veins and blood capillaries in the posterior lobe of the pituitary gland (PP). The neurosecretory products AVP and OXT are temporarily stored in axon terminals of the PP, where they are released into the circulation upon neural stimulation that activates the neuron's soma in the hypothalamus. AVP and OXT regulate the functions of organs including the kidney, mammary gland and uterus. AVP has antidiuretic actions to reduce water excretion via the kidneys. OXT stimulates natriuresis through actions on the kidney, uterine contractions during parturition and the milk ejection reflex (milk letdown), which is the ejection of milk from the mammary gland soon after suckling. AVP: arginine vasopressin, OXT: oxytocin. Figure adapted from Baribeau and Anagnostou (2015) and Borrow et al. (2020). Figure created using BioRender.com

space can exert both autocrine and paracrine effects by activating receptors expressed in the brain or reaching more remote targets by spilling over through the extracellular space into the cerebrospinal fluid.

It is also known that MCNs send collateral projections to brain nuclei regulating salt appetite (Sakai et al. 2000), food intake (Klockars et al. 2018), pain perception (Eliava et al. 2016), affective/emotional states (Wahis et al. 2021; Hernández-Pérez et al. 2019), social behaviour (Zhang et al. 2021; Tang et al. 2020) and fear (Hasan et al. 2019; Zhang et al. 2016).

There are extensive changes to the MCNs in response to dehydration (DH, when animals are restricted from adequate water intake) and salt loading (when animals are orally administered an excess amount of sodium chloride with food/water), as is apparent from the transcriptomic analyses of the SON in laboratory rats (Qiu et al. 2011; Johnson et al. 2015; Greenwood et al. 2015; Pauža et al. 2021). These changes include hypertrophy of the MCNs, reorganization of the neurone-glial architecture, increased synthesis of the MCN products, intracellular machineries and transcription factors and upregulated production of the gasotransmitters carbon monoxide and nitric oxide (Tasker et al. 2012; Reis et al. 2015).

13.3 AVP and OXT

13.3.1 The Evolution of AVP and OXT

The AVP/OXT neuropeptide signalling system has been identified in various organisms from invertebrates to mammals. Across evolutionary linages, it is functionally conserved in osmoregulation, reproduction, learning and memory (Murphy and Gainer 2016). Previous phylogenetic analysis suggested that AVP might share a common origin with cardioactive peptide, neuropeptide S and other AVP-like peptides, which could be products of ancient duplication events (Pitti and Manoj 2012; Mirabeau and Joly 2013). OXT originated from the duplication event of the ancestral vasotocin gene in a jawless fish species that lived around 450 million years ago. After gene duplication, usually one copy maintains the original function, leaving the other copy to mutate and diverge freely, as happened for AVP and OXT, respectively (Murphy and Gainer 2016; Leng and Russell 2019; Sparapani et al. 2021). Invertebrates, with few exceptions, are considered to have only one AVP/OXT-like peptide homolog. Related peptides of AVP and OXT are detectable in all vertebrate species (Baribeau and Anagnostou 2015). Within these lineages, the genes encoding these peptides are found in close proximity on the same chromosome (Murphy and Gainer 2016). The two hormones consist of nine amino acid residues in a cyclic structure (formed by a disulfide bond between two cysteines at positions 1 and 6) and differ by only two residues: phenylalanine and arginine in AVP, and isoleucine and leucine in OXT at position 3 and 8 (from N- to C-terminal) (Fig. 13.2a). When comparing AVP/OXT peptides between different organisms, certain positions are highly variable, whilst others are more conserved. For example, the conserved residues are located at positions 2 and 3 (hydrophobic or aromatic residues), positions

4 and 5 (polar or charged residues), position 7 (proline) and position 9 (glycine), whereas position 8 is highly variable (Murphy and Gainer 2016). The sequence variations of AVP/OXT homologs could contribute to the specificity of ligand binding, species-selective recognition and activation of different receptors (Murphy and Gainer 2016).

13.3.2 The Synthesis, Processing, Transport and Release of AVP and OXT

The biosynthesis of the mature AVP and OXT peptides are derived from the translation of the respective mRNAs into large precursor preprohormones (Fig. 13.2b). The organization of the two precursors is similar; both include a signal peptide, a hormone and a neurophysin, with the only exception being the presence of a glycoprotein (copeptin) moiety in the AVP precursor but not in the OXT precursor (Fig. 13.2b) (McEwen 2004). As established by Ando et al. (1988), McEwen (2004) and Qureshi et al. (2014), the AVP and OXT preprohormones undergo a series of post-translational processing events (Fig. 13.2b, c). After the cleavage of the signal peptide and the glycosylation of copeptin, the N-terminus of AVP fits into a binding pocket formed on the neurophysin II region. The remaining prohormone folds, dimerizes, forms multiple intrachain disulfide bonds and is then routed unidirectionally to the trans-Golgi network.

The protein is exported from the Golgi apparatus, targeted specifically into the large dense core granules of the regulated secretory pathway. The AVP prohormone is protected by neurophysin II from proteolytic cleavage during transport (Qureshi et al. 2014). Whilst these secretory granules are anterogradely transported towards the PP, the prohormone is processed by prohormone convertase to generate AVP hormone, neurophysin II and copeptin, which can occur either in the soma or the axon. These resulting secretory products are temporarily stored in the neuronal terminals in PP and released upon electrical/chemical stimulation of the cell into the circulation (McEwen 2004). In response to action potential invasion of the neurosecretory membrane, AVP is released by exocytosis of dense-core vesicles (Brown et al. 2013). OXT preprohormone follows almost the same process but involves the participation of a different carrier protein (neurophysin I) (Ando et al. 1988). The neurophysins remain bound to the hormones during storage, which may protect the hormones from diffusing away from the storage granule prior to secretion (Norman and Litwack 1987).

The release of AVP is primarily affected by altered blood volume, blood pressure and plasma osmolality (Zaffanello et al. 2012). The osmotic induction of OXT release includes increased plasma osmolarity and blood volume expansion (Caligioni and Franci 2002). In the SON, the pattern and rate of AVP and OXT release are determined by the pattern and frequency of action potentials (Fig. 13.2c), which results from the overall excitatory and inhibitory synaptic inputs on the MCN soma (Brunton and Russell 2015). AVP secretion is under the control of osmoreceptors in hypothalamic structures including SON and PVN (Sladek and Armstrong

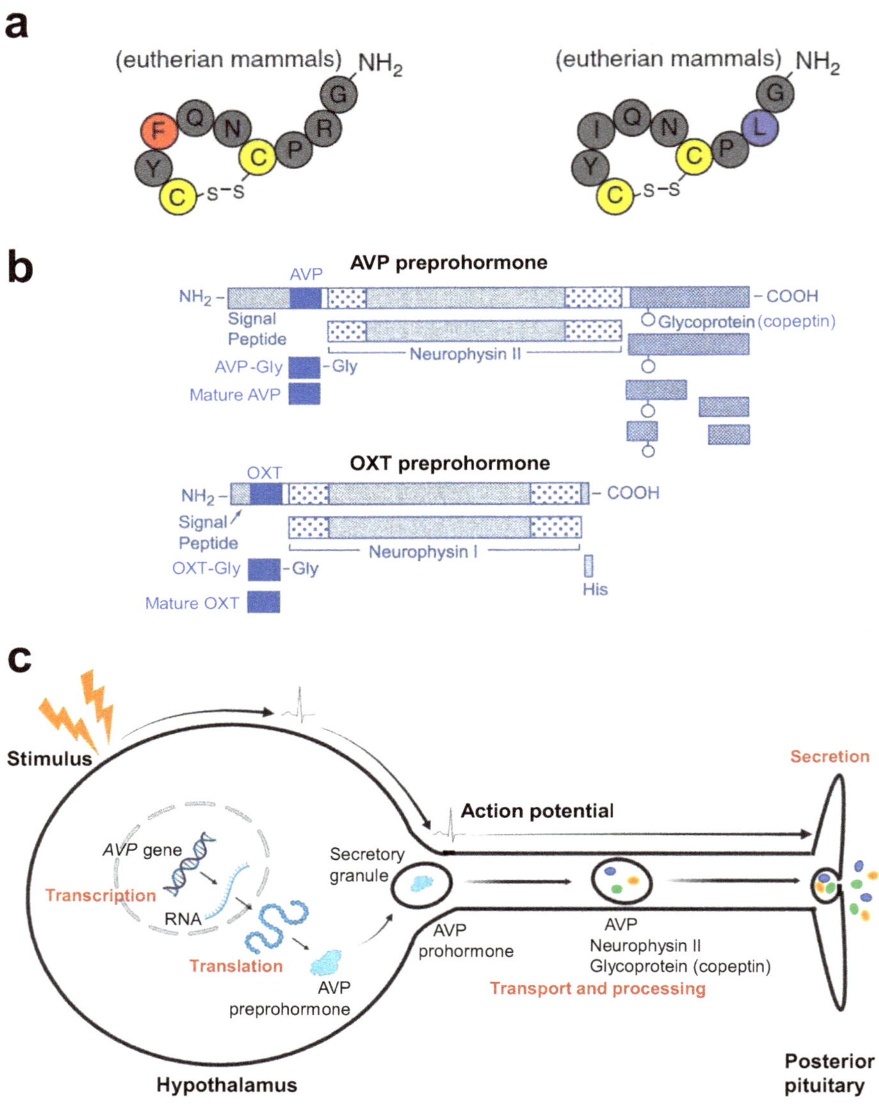

Fig. 13.2 The synthesis, processing, transport and release of AVP and OXT. (**a**) The peptide sequences are shown in one-letter amino acid code for AVP and OXT hormones in eutherian mammals. Disulfide bonds formed on the highly conserved cysteine residues (coloured in yellow) assist in the formation of a cyclic structure. Residues that are identical to vasotocin (the common ancestor of AVP and OXT) are coloured in dark grey. Residues that have changed during the evolution of AVP and OXT are coloured in red and dark blue, respectively. Figure source: Murphy and Gainer (2016). (**b**) Schematic illustrations of AVP and OXT preprohormones. For each precursor, the proteolytic maturation proceeds from top to bottom. The organizations of the two precursors are similar, the only exception being that a glycoprotein

2009). At the cellular level, osmotic stimulation of AVP release is associated with activation of the transient receptor potential cation channel subfamily V member channels—TRPV1 and TRPV4 (Bichet 2016; Janas et al. 2016). Inputs from multiple neurotransmitting and neuromodulating neurons secreting angiotensin II, ATP, glutamate, serotonin, histamine, acetylcholine, catecholamines, neuropeptide Y, substance P and orexins are received by the hypothalamic AVP-expressing neurons. AVP release is thought to be inhibited by gamma-aminobutyric acid (GABA), nitric oxide, natriuretic peptides, relaxin-3 and apelin (Szczepanska-Sadowska et al. 2018). Special attention is required for the functions of GABA on the AVP-releasing neurons when the organism is under different conditions, as it has been indicated that the inhibitory drive of GABA innervation might be excitatory under some pathological circumstances such as osmotic stress (Szczepanska-Sadowska et al. 2018).

13.3.3 AVP and OXT Functions in the Kidney

The modulation of water and sodium retention by the kidney is primarily mediated by the actions of AVP and OXT. The increase in plasma osmolality following DH is detected by MCNs (Bourque 2008) and by specialized osmoreceptive neurons located in the organum vasculosum of the lamina terminalis (OVLT) and the subfornical organ (SFO), which generate excitatory inputs to shape the firing activity of MCNs for hormone release into the circulation (McKinley et al. 2004). After release, AVP interacts with AVP type 2 receptor (V_2R) located in the kidney to promote water (Breyer and Ando 1994) and sodium reabsorption (Ares et al. 2011). OXT is thought to have natriuretic activity (Conrad et al. 1993) and antidiuretic functions (Li et al. 2008) in the kidney. The roles of OXT in the kidney appear to be species-specific (Rasmussen et al. 2004), since no osmotic regulation of OXT secretion and no clear renal actions have been observed in humans (Williams et al. 1986; Conrad et al. 1993; Rasmussen et al. 2004), but sodium excretion was found to be mediated

Fig. 13.2 (continued) (copeptin, glycosylation site labelled by a lollipop symbol) is included in the AVP precursor. For the neurophysins, the conserved regions and variable regions are represented by dark and light stippled bars, respectively. AVP-gly (Lote et al. 1986) and OXT-gly (Guillou et al. 1992) are maturation intermediates. (**c**) Synthesis, processing, transport and release of AVP in the MCN (similar manners for OXT). Stimuli triggered the propagation of an action potential to axonal terminals in the PP, leading the secretion. After transcription, AVP preprohormone is translated on rough ER and then transported and processed intracellularly. After the removal of signal peptide by signal peptidase, the resulting AVP prohormone is packaged into secretory granules. The neurophysins perform as carrier proteins that facilitate the neurosecretory granule packaging and transport of AVP prohormone via axons, protecting it from degradation. The secretory granule is the site of post-translational processing of the prohormone into AVP, neurophysin II and copeptin by prohormone convertases. This can occur either in the cell body or in the axon during axonal transport. These smaller products, including the mature AVP, are secreted from the neuron terminals by depolarization with the presence of extracellular Ca^{2+}. (**b**, **c**) Were adapted from McEwen (2004) and Fenske et al. (2018), respectively. (**c**) Was created using BioRender.com

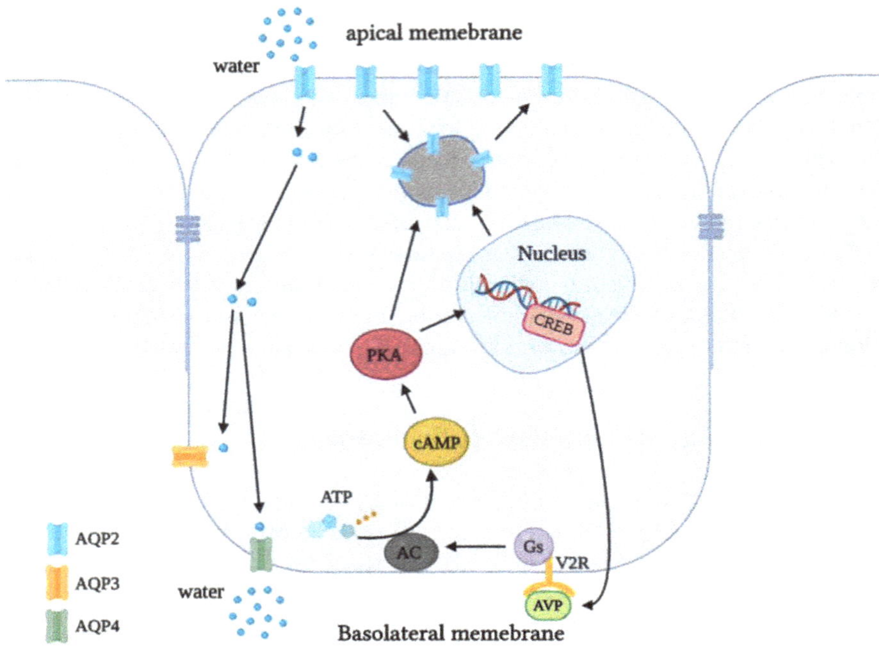

Fig. 13.3 Water reabsorption mediated by AVP in the kidney collecting duct. Model of water permeability mediated by AVP, AQP2, AQP3 and AQP4 in kidney collecting duct (principal) cells. Interaction of AVP with a V2/Gs-coupled receptor in the basolateral membrane activates adenylated cyclase (AC), which further upregulates the cyclic adenosine monophosphate (cAMP) levels. Protein kinase A (PKA) is activated downstream and promotes the trafficking of AQP2 storage vesicles to the apical membrane. The water permeability of the apical membrane of the principal cell is hence upregulated. PKA also phosphorylates the cAMP-responsive element-binding protein (CREB) that increases the *AQP2* transcription by binding to the *AQP2* promoter. Water from the tubular fluid (pre-urine) enters the principal cell through AQP2 and leaves via AQP3 and AQP4 on the basolateral membrane to return to the circulating blood. AVP, vasopressin; AQP, aquaporin; V2R, vasopressin V2 receptor; Gs, G stimulatory protein; AC, adenylated cyclase; cAMP, cyclic adenosine monophosphate; PKA, protein kinase A; CREB, cAMP-responsive element-binding protein. Figure adapted from Boone and Deen (2008). Figure created using BioRender.com

by OXT in dogs (Brooks and Pickford 1958), rats (Young and van Dyke 1968) and mice (Sinclair et al. 2010).

As shown in Fig. 13.3, water reabsorption in the collecting duct is mediated by the interaction between AVP-dependent V2/G_s-coupled receptor (V2R coupled with G stimulatory protein) signalling and the water channel protein, aquaporins (AQPs), in the kidney. Activation of V2R triggers the G_s signalling pathway that further activates cyclic adenosine monophosphate (cAMP) and protein kinase A (PKA) downstream (Holmes et al. 2003). The enhanced level of cAMP in renal cells induces the translocation and fusion of AQP2-bearing vesicles to the collecting duct principal cell membranes (Boone and Deen 2008; Thomsen et al. 2016). PKA phosphorylates the cAMP-responsive element-binding protein (CREB) which binds to the *AQP2* promoter and upregulates AQP2 synthesis (Boone and Deen 2008). Water from the tubular fluid (pre-urine) enters principal cells through AQP2 and leaves via AQP3 and AQP4 on the basolateral membrane to return to the circulation. During antidiuresis, when the circulating AVP levels are high, the water permeability of the collecting duct is upregulated and hence aids fluid reabsorption in the collecting duct system (Breyer and Ando 1994; Boone and Deen 2008). The expression levels of AQP2 and AQP3 (but not AQP4) increase in response to elevated circulating AVP (Terris et al. 1996).

OXT also has antidiuretic effects. A study (Pouzet et al. 2001) carried out in the Brattleboro rat, an AVP-deficient animal model that suffers from a recessive form of central diabetes insipidus, has shown that chronic infusion of OXT (3 μg/kg/h, intraperitoneal injection) via osmotic minipump for 6 days induced a marked antidiuretic effect (urine osmolality elevated to ~1500 mOsm/kg H_2O), which did not decline over time. This antidiuretic effect was abolished by the administration of a highly selective V2R antagonist, suggesting that this effect was due to the binding of V2R by the exogenous OXT. This study confirms that pharmacological dose of OXT given in vivo can exert powerful antidiuretic effects. Differently, the endogenous OXT can only exert a weak antidiuretic response, as indicated by DH experiments performed in the Brattleboro rats (Edwards and LaRochelle 1984). The constitutively synthesized OXT in vivo may occupy only a small proportion of V2Rs because their urine remained hypoosmotic (Pouzet et al. 2001).

13.4 Camel: A Desert-Adapted Species

Being well-known as the 'ship of the desert', camels have played important roles in human survival, especially in arid regions of African, Asian and Arabian deserts, due to its extraordinary adaptations to harsh conditions. Camels are adapted to a wide range of desert and semi-desert conditions. Historically, camels have played important roles in human culture. In recent years, they have regained interest as a production animal, mostly for dairy, with an increasing interest in the health benefits of camel dairy products (Tibary and El Allali 2020). Scientific studies on camels have thus progressively shifted their focus from considering this species as a mere biological curiosity to more thorough investigations into its biological adaptation (Tibary and El Allali 2020).

13.4.1 The Taxonomy of the Camelidae Family

Figure 13.4 demonstrates the most widely accepted classification of the Camelidae family (Wu et al. 2014). The origin of the camelids traces back to the Eocene (40–45 million years ago) when the first of their common ancestors was found in North America (Burger 2016). The camelids split into Camelini (the Old World camelids) and Lamini (the New World camelids), the former having migrated over the Bering land bridge to the eastern hemisphere (known as the Old World). The members of

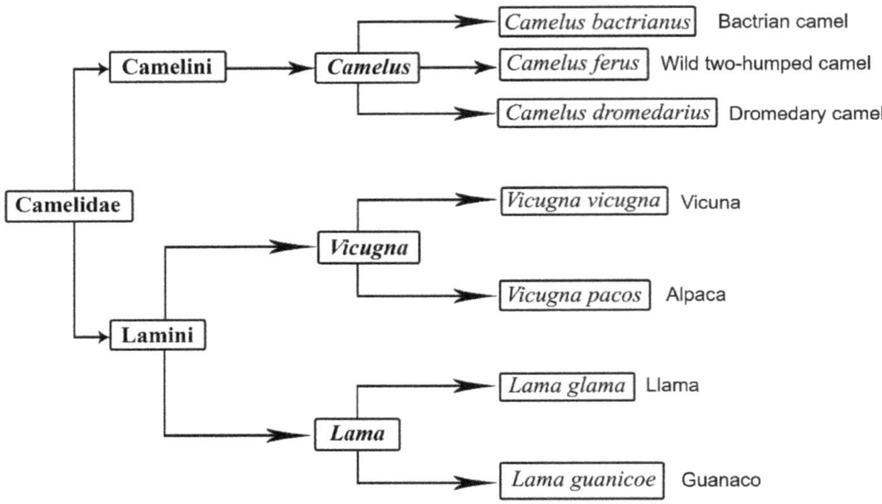

Fig. 13.4 Taxonomy of the Camelidae family. Figure adapted from Wu et al. (2014) and Ali et al. (2019)

the Camelini include two domesticated species: *Camelus dromedarius*—the dromedary camel (one-humped camel); *Camelus bactrianus*, the Bactrian camel (two-humped camel); and a wild species—*Camelus ferus*, the extremely endangered wild two-humped camel. According to the molecular evolutionary studies, the divergence between the dromedaries and the Bactrian camels occurred at about 5–8 million years ago (Ji et al. 2009; Wu et al. 2014). The Lamini consist of two wild species—vicuña (*Vicugna vicugna*) and guanaco (*Lama guanicoe*) and two domesticated species—alpaca (*Vicugna pacos*) and llama (*Lama glama*) (Kadwell et al. 2001). Notably, the term 'camel' used in this chapter represents the Camelini, not the Lamini.

13.4.2 Case Study: The Uses of the Dromedary Camel

Domestic camels had an important role in the development of the ancient civilizations of the Old World (Burger 2016). The dromedary camels (also known as Arabian camels) have evolved to withstand high temperatures, limited feed resources and long-term water deprivation, which makes them a prevalent sustainable livestock species throughout the Arab world (Gaughan 2011; Richard 2003). They have been used by humans as multipurpose animals for physical labour, transport, racing, tourism, trading and fieldwork and as a source of milk, meat, textile and leather (Bulliet 1975). Historically, the dromedary camels were first domesticated in the Arabian region around three to four thousand years ago (Almathen et al. 2016) and have been widely used in the trades across the vast deserts of Arabia for a very long time (Sapir-Hen and Ben-Yosef 2013). In the modern world, the demand for sustainable meat and milk production from the domestic dromedaries is continuously growing (Burger et al. 2019). Camel sport, which is an integral part of the Middle East region culture, has boosted the demand for special breeds of racing camels with outstanding resilience and endurance (Saadeldin et al. 2018). Moreover, the naturally occurring antibodies discovered in the camels can be further developed into recombinant nanobodies, suggesting a promising application of dromedaries in the medical field (Siontorou 2013).

Current environmental issues are influencing the survival of many mammalian species including traditional ruminant livestock such as cows, goats, camels, etc. (Burger et al. 2019). The camel is thus a promising animal model for studying the genetic and physiological adaptation to the drought weather that enables the survival of mammals in the harsh desert area. Such studies will provide insights for answering the ultimate questions that we are currently facing with the threat of heat stress and water shortage arising from global warming and desertification.

13.4.3 The Adaptations of the Camels to Desert Environment

13.4.3.1 Adaptations to Chronic Dehydration in Camels

In the desert environment, water is sparsely available from transient rainfed reservoirs and isolated oases, which makes it important for desert animals to reduce the amount of water loss and to tolerate significant levels of water loss (Ezcurra and Mellink 2013; Gebreyohanes and Assen 2017). Whilst all terrestrial mammals assiduously control the amount of water consumed and excreted (Antunes-Rodrigues et al. 2004), this is particularly important in the camels, which have a remarkable capacity to survive extended periods of DH (Bengoumi and Faye 2002; Gaughan 2011). Water loss is extremely well tolerated in the camel. Camels may go several months without drinking where green forage is available in mild climates or during the cold seasons (Gebreyohanes and Assen 2017). For instance, when the average ambient temperature reaches 30–35 °C, dromedary camels can go around 10–15 days without water. However, when the temperature is higher than 40 °C, more frequent drinking is necessary for survival (Kataria et al. 2002).

Whilst 12% water loss would be fatal to nondesert mammals due to cardiac failure resulting from circulatory disturbance (Ingram and Mount 1975), the camel can survive up to 30% (Schmidt-Nielsen 1959). However, it is water economy that is particularly vital for survival in the desert, and in the camel this is achieved by a number of mechanisms, such as minimal evaporative cooling (camels rarely sweat), water extraction from undigested food residues and variation in body temperature from 34 °C at night up to 42 °C during the day (Schmidt-Nielsen et al. 1967; Grigg et al. 2009). These mechanisms prevent the loss of about 5 litres of water per day by sweating.

Perhaps the most important water conservation mechanism in the camel is centred on the SON-kidney axis (Ben Goumi et al. 1993). In camels, water economy is achieved by the highly efficient reabsorption of water (Ben Goumi et al. 1993). This process is mediated by the actions of the antidiuretic hormone AVP to promote fluid reabsorption in the collecting duct (Breyer and Ando 1994) and sodium reabsorption in the thick ascending limb of the loop of Henle (Ares et al. 2011). In the dromedary camel, the level of circulating AVP increases following DH (Ben Goumi et al. 1993; Ali et al. 2012). The camel kidney reduces water loss by increasing water reabsorption in the collecting duct and decreasing the glomerular filtration rate (GFR), leading to the production of a low volume of highly concentrated urine following DH (Ben Goumi et al. 1993; Kataria et al. 2002).

Other unique features contributing to the survival of camels in the desert have been observed in their blood, thermoregulation capacity and digestive system. The life span of an erythrocyte in a DH camel is longer than that of hydrated camels, which saves the water and energy needed for erythrocyte turnover (Ouajd and Kamel 2009). The selective brain cooling by blood flow and bi-phasic respiratory patterns in camels helps to maintain a stable brain temperature, which is critical for survival under heat stress (Gaughan 2011; Gebreyohanes and Assen 2017). The digestive system of camel is also specialized for water economy. Cattle (*Bos taurus*) and camels are both large grazing ruminant livestock with similar digestive system structures and more comparable body weights than small ruminant farm animals (e.g. sheep, goats) (Breulmann et al. 2007). The end part of the camel intestine absorbs more fluid from food residues to produce faeces with less water content compared to cattle (Breulmann et al. 2007).

The genetic basis underlying the desert adaptations in camels has been widely studied using modern molecular genomic tools (Rocha et al. 2021). Evidence from a study identifying the nonsynonymous:synonymous substitutions ration (dN/dS) and gene family evolution revealed complex characteristics of adaptation in both dromedary and Bactrian camels, including strong selection for genes associated with the stress responses to heat, fat and water metabolism, insulin-signalling pathway, UV radiation and airborne dust (Wu et al. 2014). Moreover, differences in copy number of *CYP2* genes (encoding the cytochrome P450 CYP2 family of enzymes) between camels, cattle and other mammals were shown by the Bactrian Camels Genome Sequencing and Analysis Consortium (2012). This may contribute to the synthesis of a vasodilator of kidney preglomerular vessels that induce water reabsorption in the kidneys (Bactrian Camels Genome Sequencing and Analysis Consortium 2012). The extra copies of the *CPY2J2* gene (encoding the cytochrome P450 2J2) that is related to high-salt diets and blood pressure regulation (Bactrian Camels Genome Sequencing and Analysis Consortium 2012; Ali et al. 2019) found in camels may explain the capacity of camels to tolerate high salt ingestion from halophyte plants, pastures and fluid (Faye et al. 2011) without developing hypertension (Ali et al. 2019).

13.4.3.2 Adaptations to Rapid Rehydration in Dehydrated Camels

The astonishing adaptations that allow camels to thrive in desert environments include not only their endurance to survive for weeks of DH but also the capacity to quickly uptake vastly large amounts of water (rehydration (RH)) when available (such as from an oasis) to compensate for pre-existing water deficiency, without inducing apparently adverse effects (Irwin 2010; Bekele et al. 2013; Etzion et al.

1984; Etzion and Yagil 1986; Young and Schmidt-Nielsen 1985; Ali et al. 2020). The rapid water intake in camels reduces exposure to predators during drinking and allows the camel to leave the water source to browse over long distances, searching for a food source (Etzion and Yagil 1986). In a single session, a DH camel can drink up to 110 litres of water within 10 min (Schmidt-Nielsen et al. 1956; Irwin 2010) and by doing so resume its hydromineral balance and replenish 30%–40% loss of its body weight (Schmidt-Nielsen 1997; Ben Goumi et al. 1993). Such rapid consumption of large quantities of water following DH would lead to over-hydration and death in nondesert species (Gebreyohanes and Assen 2017).

Rehydration (RH) in camels is characterized by many remarkable physical and physiological adaptations. For example, the foregut of a RH camel can retain a large volume of water, approximately equal to 20% of its body weight, which buffers any short-term water deprivation that the animal has experienced/may experience later (Willmer et al. 2006; Cain et al. 2006). After a drinking session, the ingested water is rapidly absorbed into the bloodstream and tissue to restore the kidney function to pre-DH level within 2 h (Etzion and Yagil 1986; Crawford 1978). The rumen then retains the ingested water during the first few days, until the body tissues are fully rehydrated and water homeostasis is restored (Willmer et al. 2006). During these processes, it is important to prevent haemolysis and osmotic shock of tissues, which might be induced by the generation of a large osmotic gradient between the rumen fluid and the circulation due to the rapid RH (Cain et al. 2006).

Here comes the question: upon the rapid RH, how can camels manage to avoid severe adverse effects that could be caused by the rapid dilution of blood? The special features displayed by camel erythrocytes may provide solutions to this question. Camel erythrocytes are oval in shape and non-nucleated (Cohen and Terwilliger 1979), which helps to resist osmotic variation without rupturing. These cells are also enormously expandable and can swell to twice their initial volume following RH (Gebreyohanes and Assen 2017). Previous studies showed that camel erythrocytes are considerably more osmotically resistant than those of human and mesic ruminants, which might be attributed to the shorter and less saturated fatty acid chains and the unique lipoprotein structure identified in the membrane of camel erythrocytes that increases their stretching capacity (Livne and Kuiper 1973). The cytoskeleton of camel erythrocytes also differs from that of human erythrocytes and allows cell expansion with distilled water to 400% before rupturing. Compared to other domestic mammals like cattle and sheep, the maximum degree of swelling of camel erythrocytes in hypotonic sodium chloride solutions prior to haemolysis is much higher (Perk 1966). Indeed, the dilution of camel blood upon rapid RH does not lead to haemolysis, nor haemoglobinuria, a common occurrence caused by excessive intravascular haemolysis in cattle living in hot environments (Perk 1966).

Interestingly, the secretion of hypotonic saliva in RH camels recycles water back to the alimentary tract and transfers large amount of sodium from the circulation, assisting the return of elevated plasma osmolality (by DH) to normal levels (Silanikove 1994). Storage of consumed water in the gut and saliva together buffers the amount of rumen fluid in RH camels (Siebert and Macfarlane 1975; Silanikove 1994).

13.4.3.3 Comparison of Water Conservation Strategies Between Camels and Other Mammals

Water economy strategies in desert mammals show evidence of conservation and/or convergence at multiple levels under the same type of evolutionary pressure—the osmotic challenge (Ezcurra and Mellink 2013). Similarly to camels, eland (also known as antelope, capable of living in savanna, montane grassland, woodlands and fringes of deserts) also exhibits adaptive hyperthermia that allows them to reach 44 °C during the day with no harmful consequences and to preserve body water by not sweating (Ezcurra and Mellink 2013). Both camels and the desert kangaroo rats can acquire water metabolically by fat combustion (Holzapfel 2008). Presented with water, camels and wild asses (whose primary habitat is desert/semi-desert environment) both show remarkable traits of RH with large amounts of water in a very short period (Ezcurra and Mellink 2013). The desert-adapted Bedouin goat is another mammal that can drink large amounts of water as quickly as camels do, but the restoration of its kidney function is slower (not returning to normal within 48 h) because of the slow absorption of water from lumen (Etzion and Yagil 1986; Cain et al. 2006). Such buffering capability of rumens is apparently lost in cattle living in hot environments, since they often suffer from haemolysis after consumption of large volumes of water (Perk 1966).

The rock hyrax (*Procavia capensis*) is a desert species well-known for its resistance to desiccation. With arid environment (deserts, savannas, scrub forests and rocky terrains) as their preferred habitat (Sale 1966), rock hyraxes have kidneys with highly efficient functions in concentrating urine, where the maximum recorded urine osmolality (3088 mOsm/kg) is comparable to that of the camels (3178 mOsm/kg) (Louw et al. 1972).

Upper Galilee mountain blind mole rats (*Nannospalax galili*) (UGMBMR) inhabiting the cool and humid Upper Galilee mountains in northern Israel (Mediterranean climate, characterized by hot and dry summers and rainy winters) are defined as a mesic species (Nevo et al. 1989; Šklíba et al. 2016). It evolved from a semi-desert species, the Middle East blind mole rat (*Spalax ehrenbergi*), through adaptive speciation driven by the geographical radiation of the Middle East blind mole rats into different climates. There are four chromosomal species ($2n = 52$, 54, 58 and 60) displaying progressive stages of this speciation. Their adaptive radiation was closely associated with increasing savannization, deforestation and aridity. UGMBMR ($2n = 52$) is challenged the least, whilst $2n = 60$ is under the highest aridity stress, compared to other cryptic species derived from the common ancestor (Nevo et al. 1989), which makes the UGMBMR an engrossing model in which to study the adaptive variation of water conservation mechanisms after the transition from mesic to dry environment. Differing from the camels, the four chromosomally different blind mole rat species generally showed clear upper limits in the structural and functional capacities of the kidney, which prevented them from colonizing the real desert (Nevo et al. 1989). The highest resistance to hyperosmotic stress (salt-loading) was observed in the xeric species $n = 60$, whilst UGMBMR exhibits lower kidney concentrating capacity and diuresis under salt stress.

Lamini, the nearest relatives of camels, are adapted to living in the arid Andean high plateau of South America (Wu et al. 2014). These areas are characterized by high altitude, low oxygen level and a range of different biomes including deserts, semi-deserts, dry broadleaf forests, temperate grasslands, savannas and shrublands (Marin et al. 2013). Some traits are shared between the Lamini and the camels with regard to adaptation to water scarcity. The Lamini exhibit high efficiency in the metabolic use of water (Vilá and Arzamendia 2022). Similarly to the camels, voluminous drinking (15–20% of body weight) was also found in guanacos (Rosenmann and Morrison 1963). A low drinking frequency was described in llama and vicuña (once per day or twice on very dry days for vicuña) (Enke et al. 2022). Anatomically, the llama and guanaco's stomachs were found to contain sacculated structures that might function as temporary water reservoirs, whilst similar structures in the camel are less numerous but larger (Vallenas et al. 1971). In terms of the water reabsorption capacity of kidney, the Lamini are able to produce highly concentrated urine (3190 mOsm/L maximum urine concentration) and relatively dry, pelleted feces (Engelhardt et al. 1974). In addition, the similar elliptical erythrocyte structure shared across the Camelids allows the erythrocytes to continue circulating even during times of increased blood viscosity caused by DH (Fowler 1989).

Unlike the Camelini, Lamini are unable to adapt to elevated ambient temperatures by increasing body temperature and avoiding evaporative cooling (Davis 2005). The Lamini have an increased water demand during high ambient temperature. Therefore, they are considered to be less resistant to hot, arid environments, given the evidence that they are more susceptible to hyperthermia/heat stress than the Camelini (Agnew 2018). By considering the salt metabolism that has the main effect on water balance, a comparative study on the genomes of dromedary camel, Bactrian camel and alpaca revealed that the gene category associated with the voltage-gated potassium channel complex evolved rapidly in both camels compared to the alpaca (Wu et al. 2014). The *NR3C2* and *IRS1* genes that play important roles in sodium reabsorption and water balance in the kidney have two copies in the Bactrian camel but only a single copy in the alpaca, suggesting that the salt and water transport/reabsorption could be less efficient in alpaca than camels (Wu et al. 2014).

13.4.3.4 The Role of AVP in DH and RH Camels

The important neuropeptide regulator of water economy, AVP, has been measured in the plasma of camels subjected to DH and subsequent RH. The plasma AVP level increased sharply after 4 days of DH in non-lactating camels (Ben Goumi et al. 1993), or 8 days DH in lactating camels (Bekele et al. 2013). In DH lactating camels, hypovolaemia was found to occur as indicated by the elevated plasma protein concentration after 8 days (Bekele et al. 2013). It is acknowledged that hypovolaemia performs as the baroreceptor for low blood pressure that signals directly to the hypothalamus and/or indirectly through the renin-angiotensin system to stimulate AVP release and thirst during DH (Andersson 1977; McKinley and Johnson 2004). Hypovolaemia may contribute to the increase of plasma AVP concentration in DH camels to deliver both antidiuretic and vasoconstrictor effects (Bekele et al. 2013).

When RH begins, the camel plasma AVP concentration decreases after 2 h and reaches basal values after 7 days of RH (Ben Goumi et al. 1993). In camels rehydrated for 2 h, even though large volumes of water are absorbed, a significant fall of AVP is not accompanied by decreased plasma osmolality, suggesting a non-osmotic inhibition of AVP release, which could be mediated through gut receptors (Etzion and Yagil 1986; Ben Goumi et al. 1993).

At the level of hypothalamic osmoregulation, it is worth mentioning our recent transcriptomic study on the SON of dromedary camel (control: 20 days, chronic DH: 20 days, rapid RH: 3 days) (Lin et al. 2022). In this study, the expression of steady-state mature *AVP* and *OXT* transcripts as well as the precursor (intron-containing) hnRNA transcription products of these genes (as a proxy measure of transcription) in the camel SON was detected by qRT-PCR. The relative expression of the mature *AVP* mRNA was significantly increased following DH and remained significantly elevated following RH compared to controls. A similar trend was observed for the *hnAVP* RNA, whereas the expression level of *hnAVP* in DH dropped during RH (Lin et al. 2022).

13.5 Interspecific Comparison of *AVP/OXT* Genes and AVP/OXT Peptides

Given the important roles played by AVP and OXT in osmoregulation, identification of the sequence variability of the *AVP* and *OXT* genes in camels and other desert/nondesert species will help us to understand the regulation of the expression of these two hormones at the genetic level. Previous studies have characterized regulatory and functional regions of the *OXT* gene in the dromedary camel (Pauciullo et al. 2018) and the *AVP* gene in other species (Greenwood et al. 2014, 2016; Coulson et al. 1999), but the variations between these genes in the camels and other desert/nondesert species have not been investigated. To investigate the genetics underlying the regulation of two key genes involved in adaptation to deserts in the camels, the gene structures were predicted by carrying out multiple alignments between the camel *AVP* and *OXT* genes and their well-characterized orthologs in other species.

In this comparative study, phylogenetic analyses were performed on the *AVP* and *OXT* genes as well as their encoded products in dromedary camel, Bactrian camel, wild camel (*Camelus ferus*), vicuña, alpaca, guanaco, llama, rock hyrax, UGMBMR and mesic model organisms including rat (*Rattus norvegicus*), mouse (*Mus musculus*) and human (*Homo sapiens*), in order to test the hypothesis that these genes and their products are more conserved between desert species and to predict and compare ways in which gene expression is being regulated in animals in different environments with different levels of water availability. Though modern genetic tools such as whole genome sequencing have been applied to various non-model organisms, the genetic architectures and the gene annotations of *AVP* and *OXT* have been solely/both neglected for the vicuña, guanaco, llama, rock hyrax and UGMBMR. To bridge this knowledge gap, their *AVP* and *OXT* genetic sequences were retrieved from the corresponding genome assemblies for making comparisons with the camel species.

13.5.1 Phylogenetic Analysis Methodology

Sequences of *AVP* and *OXT* genes from different species (references gene sequences/genome assemblies are catalogued in Table 13.1) were aligned by inbuilt MUSCLE (Multiple Sequence Comparison by Log-Expectation) method (Edgar 2004) in software MEGA X (version 10.2.4) (Kumar et al. 2018), and the phylogenies of the genes between camels and other species were analysed by the inbuilt Maximum Likelihood (ML) method and Tamura-Nei model (Tamura and Nei 1993) of the same software. During the ML tree inference analysis, the tree topology and branch lengths that maximize the likelihood of the data given to the Tamura-Nei model were selected. The resulting branch lengths represent the expected number of base substitutions per site along each branch of the tree. Sequences of encoded products by *AVP* and *OXT* genes from different species were predicted by the 'translation' function of software ApE (A Plasmid Editor, version 3.1.0) (Davis and Jorgensen 2022) based on the coding sequences (CDSs) of the genes (Table 13.1). The sequences of AVP and OXT preprohormones were aligned and studied for phylogeny using the ML methods and JTT matrix-based model (Jones et al. 1992).

13.5.2 Structures of the Camel *AVP* and *OXT* Genes

Schematic diagrams of the camel *AVP* and *OXT* genes are shown in Fig. 13.5a. Similar structures are shared by the two genes, both containing three exons and two introns. In the dromedary camel, the two genes are linked together in a tail-to-tail orientation with an intergenic sequence of approximately 11.28 kbp. The two genes are transcribed from opposite DNA strands, strongly suggesting that these two structurally similar genes were derived from a common ancestor through gene duplication and inversion events. The same gene structures, orientations and the relative locations of the dromedary *AVP* and *OXT* are found in other species such as human (Yoshida 2008), rat (Ivell and Richter 1984) and mouse (Hara et al. 1990).

13.5.3 Interspecific Comparison of *AVP/OXT* Genes and AVP/OXT Peptides

Phylogenetic trees were built to compare genomic sequences [from 5′ to 3′: 350 bp upstream (proximal promoter region) of the start codon to 100 bp downstream of the 3′-end of exon3] of *AVP* (Fig. 13.5b) and *OXT* (Fig. 13.5c) genes between different species: the desert mammals including the Camelini and rock hyrax, the Lamini that are adapted to xeric cold-temperate environments (Marin et al. 2013), the mesic rodent UGMBMR and temperate species (also model organisms)

Table 13.1 Accession numbers for *AVP* and *OXT* genes in different species. Sequences and structural information of the genes were derived from NCBI Nucleotide, Gene and Assembly databases (NCBI Resource Coordinators 2018)

Species	*AVP*	*OXT*
Camelus dromedarius	OM963135 GCF_000803125.2 GCA_001640815.1	OM963134 GCA_000803125.2 MF464533.1
Camelus bactrianus	GCA_935104035.1	MF464532.1
Camelus ferus	GCF_009834535.1 XM_032461957.1	GCF_009834535.1 XM_032461171.1
Vicugna vicugna	GCA_013265495.1	GCA_013265495.1
Vicugna pacos	GCF_000164845.5 XM_031674975.1	GCF_000164845.5 MF464535.1
Lama guanicoe	GCA_013239625.1	GCA_013239625.1
Lama glama	GCA_013239585.1	GCA_013239585.1 MF464534.1
Procavia capensis	PVIO010020694.1 GCA_004026925.2	GCA_004026925.2
Homo sapiens	GCF_000001405.39 GenBank: M25647.1 NM_000490.5	GCF_000001405.39 NM_000915.4
Nannospalax galili	GCF_000622305.1	GCF_000622305.1
Rattus norvegicus	GCF_015227675.2 GenBank: AF112362.1 NM_016992.2	GCF_015227675.2 NM_012996.3
Mus musculus	GCF_000001635.27 GenBank: M88354.1 NM_009732.2	GCF_000001635.27 NM_011025.4

including mouse, rat and human. Sequences of the preprohormones encoded by the two genes were also compared across species (Figs. 13.6 and 13.7). Very low divergence was observed for the *AVP* and *OXT* genes and their encoded products among the Camelids (Figs. 13.6 and 13.7), suggesting strict conservations between closely related camel species, regardless of the long-distance migration across continents and long histories of domestications during their speciation/evolution (Burger et al. 2019). This attaches further importance to the roles of AVP and OXT in the adaptations of camels to different habitats. No divergence was found for the *OXT* genes between alpaca and llama (Fig. 13.7). In general, the dromedary *AVP* and *OXT* genomic sequences are closely related to those of two-humped camels and then with Lamini, human/rock hyrax and the rodents (UGMBMR, rat, and mouse), following a 'close-to-far' order in terms of the genetic relationship (Fig. 13.5b, c). Similar patterns are also observed in the regulatory regions and CDSs of *AVP* and *OXT* genes, and the AVP precursors, whereas the rodent OXT precursors are more closely related to the Camelids compared to human and rock hyrax (Supplementary Figs. S13.1, S13.2, S13.3, S13.4, 13.6 and 13.7).

Taken together, the hypothesis that the *AVP* and *OXT* genes and their encoded products are more homologous among desert species was not followed across the

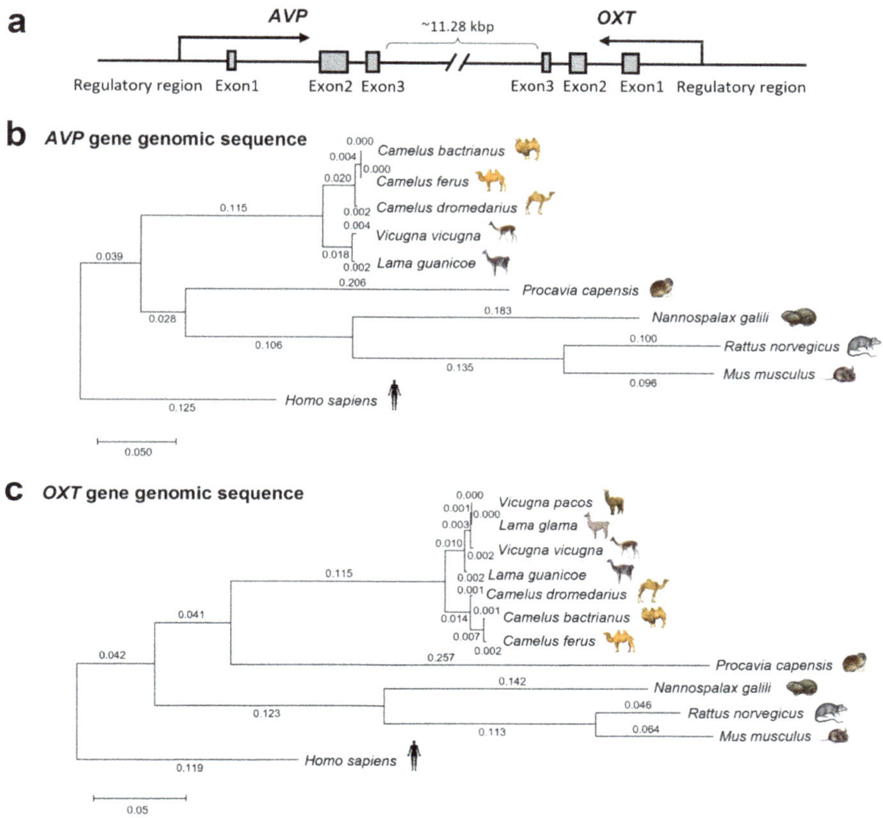

Fig. 13.5 Schematic diagram and phylogeny of *AVP* and *OXT* genes. (**a**) Schematic diagram of the camel *AVP* and *OXT* genes. (**b, c**) Maximum Likelihood trees of genomic sequence of *AVP* (**b**) and *OXT* (**c**) genes across species. *AVP* genes of *Lama glama* and *Vicugna pacos* are not included for comparison in (**b**) because complete sequences are not available. The evolutionary trees were generated using the Maximum Likelihood method and Tamura-Nei model. The trees with the highest log likelihood are shown. The evolutionary distances between species are represented by the number labelled on each branch (branch length). The trees were drawn to scale with branch lengths measured in the number of substitutions per site. Larger branch length represents more divergence. All trees were rooted with *Homo sapiens*

species studied here. The homogeneity of these sequences between the camels and other species could result from a combination of both evolutionary relationship and environmental stress. Evolutionary relationships are important factors especially for the Camelini and Lamini that belong to the same family and share a common ancestor. However, it is also possible that the similar selective pressures associated with heat and aridity may have favoured similar genetic features of *AVP* and *OXT* across different desert/xeric species (Bijlsma and Loeschcke 2005).

Two ATG start codons (AUG in mRNA) are observed at the 5′ end of *AVP* CDS in the camels and other species except rock hyrax. By comparing the *AVP* CDSs and

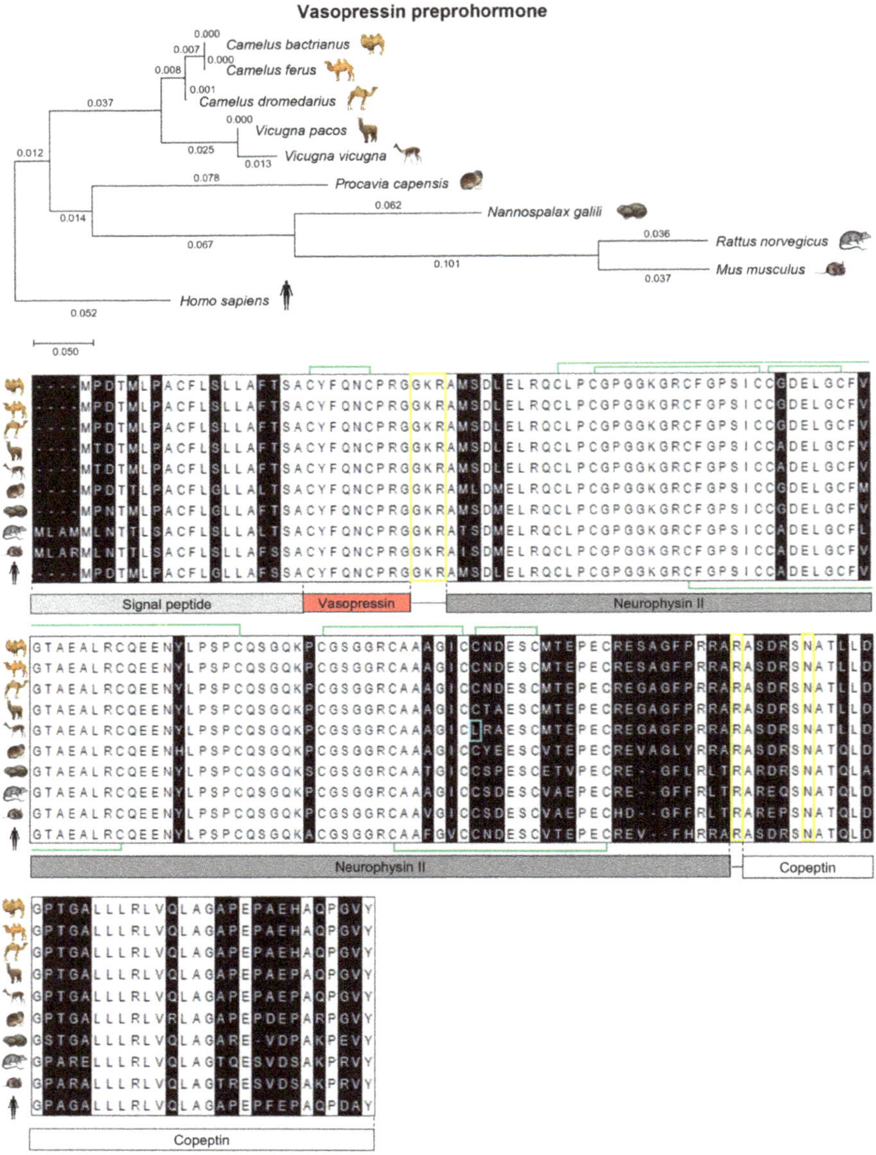

Fig. 13.6 Phylogeny of AVP preprohormones. Maximum Likelihood trees and alignment of the amino acid sequences of AVP preprohormones across species. Schematic diagram of AVP preprohormones are denoted underneath the alignments. In alignments, regions that are completely conserved across species are toggled together with white colour, for those not, with black colour; predicted post-translational sites are highlighted by yellow boxes; predicted disulfide bonds are marked by green semi-frames; a cysteine (C) substituted by leucine (L) at position 105 (a putative disulfide bond forming site in the dromedary camel) in the *Vicugna vicugna* AVP preprohormone is highlighted by a blue box. The AVP preprohormone of *Lama glama* is not included for

their encoded products (Fig. 13.6, Supplementary Fig. S13.1), the translation start site of *AVP* in these species was determined to be the first AUG codon (closer to the 5′ end) to give rise to a fully functional protein. Translation from the second AUG codon could generate a truncated protein yielding a shorter signal peptide. It is common to have multiple start codons in an mRNA sequence. Normally, the ribosomal scanning mechanism allows the AUG codon located closest to the 5′ end to be accessed by the translation machinery, since the specific motifs and structures of the mRNA sequence must be integrated correctly into the translation machinery to initiate the translation process (de Mayolo et al. 2006). However, mutations in the normal start codon might activate initiation from the downstream AUG codon. In this case, the encoded shorter signal peptide of AVP would not be recognized by the signal peptidase (Beuret et al. 1999); the resulting uncleaved AVP precursor folds incorrectly, fails the subsequent processing steps and eventually causes AVP retention in the endoplasmic reticulum. The hydromineral regulation is hence disrupted, leading to a disease known as diabetes insipidus in human (Beuret et al. 1999). As in humans, the feature of having two AUG codons in the *AVP* CDSs of camels, UGMBMR, rat and mouse demonstrates a conserved genetic risk in these organisms to suffer from such a genetic disorder caused by mutations in the upstream AUG codon.

13.5.4 Structure of the Camel AVP and OXT Preprohormones

The precursor hormones encoded by *AVP* and *OXT* genes resemble each other in structure (Figs. 13.6 and 13.7). From N- to C-terminal, the AVP preprohormone contains a signal peptide, the nonapeptide hormone itself, a 3-amino acid spacer 'GKR' (highlighted by yellow box), a carrier protein known as neurophysin, a monobasic site 'R' (highlighted by yellow box) and a glycoprotein named copeptin (Qureshi et al. 2014). The OXT preprohormone has almost the same composition but lacks a glycoprotein. The 'GKR' spacer in AVP and OXT preprohormones and the monobasic site 'R' in the AVP preprohormone are known to be cleavage sites (Lin et al. 2020; Hardiman et al. 2005; Qureshi et al. 2014) for post-translational processing and are completely conserved across all species studied. Copeptin of the AVP precursor contains a single N-linked glycosylation site (Balanescu et al. 2011) at an asparagine residue ('N' highlighted by yellow box) that is also conserved across species. The OXT precursor is not glycosylated (Gainer 1983). Compared to

Fig. 13.6 (continued) comparison because a complete sequence is not available. The AVP preprohormone of *Lama guanicoe* is not included for comparison because the 3′-end of exon2 is not aligned with other Lamini, which could be due to mutation, or an error in reference genome assembly. The evolutionary tree was generated using the Maximum Likelihood method and JTT matrix-based model. The tree with the highest log likelihood is shown. The evolutionary distances between species are represented by the numbers on each branch (branch length). The tree was drawn to scale with branch lengths measured in the number of substitutions per site. Larger branch length represents more divergence. The tree was rooted with *Homo sapiens*

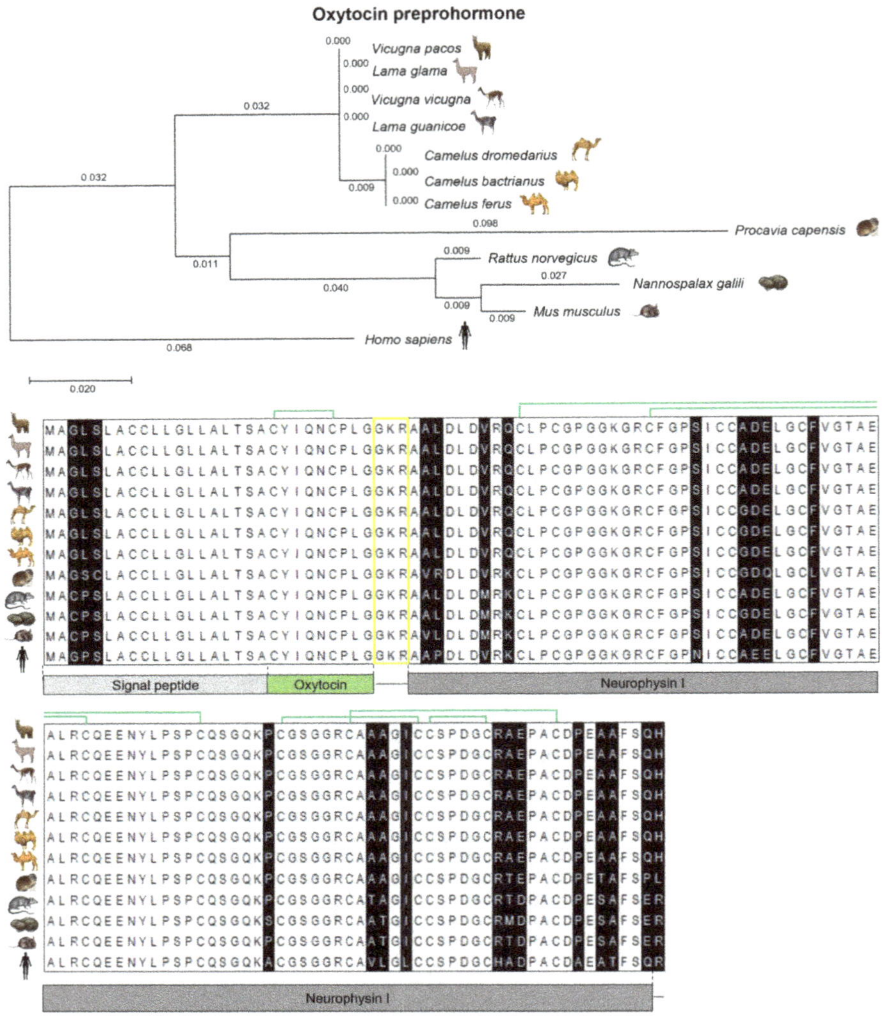

Fig. 13.7 Phylogeny of OXT preprohormones. Maximum Likelihood trees and alignment of the amino acid sequences of OXT preprohormones across species. Schematic diagram of OXT preprohormones are denoted underneath the alignments. In alignments, regions that are completely conserved across species are toggled together with white colour, for those not, with black colour; predicted post-translational sites are highlighted by yellow boxes; predicted disulfide bonds are marked by green semi-frames. The evolutionary tree was generated using the Maximum Likelihood method and JTT matrix-based model. The tree with the highest log likelihood is shown. The evolutionary distances between species are represented by the number on each branch (branch length). The tree was drawn to scale with branch lengths measured in the number of substitutions per site. Larger branch length represents more divergence. The tree was rooted with *Homo sapiens*

the dromedary *AVP* gene, there is an extra 12 bp sequence flanking the 5′ end of the *AVP* CDS in mouse and rat (Supplementary Fig. S13.1), encoding 4 more amino acids at the N-terminus of the AVP preprohormone (Fig. 13.6). This pattern is not observed in the OXT preprohormones.

All AVP/OXT-related peptides have a very similar primary structure, which features a cyclic nonapeptide with one disulfide bond and a glycylamide C-terminal residue (Hoyle 1998). The formation of cyclic hexapeptide ring structure is based on the disulfide bond between cysteine residues 1 and 6 of the nonapeptide hormone (Möller and Marí 2007), which could also exist in the dromedaries, as predicted according to the sequences of the highly conserved (identical across all species compared in this study) AVP and OXT nonapeptide hormones that contain two cysteines at the same locations (marked by green semi-frames in Figs. 13.6 and 13.7). In addition, seven more disulfide bond-forming sites are predicted according to previous studies of bovine neurophysin II (Schlesinger et al. 1972) and AVP/OXT in several arthropods including a model species water flea (*Daphnia pulex*) (Lin et al. 2020). The predicted disulfide bond-forming sites in both genes are conserved across all species studied here, the only exception being a putative disulfide bond site in the vicuña AVP precursor, where the cysteine residue is substituted by a leucine (labelled by a blue box in Fig. 13.6).

As established by others (Qureshi et al. 2014; Ando et al. 1988), the AVP preprohormone undergoes a series of post-translational processing steps (described in Sect. 13.3.2), which could be affected by the peptide sequence. The N-linked glycosylation site, although not described before for its function, might be important in assisting the refolding of misfolded AVP prohormone (Qureshi et al. 2014; Barat et al. 2004). The main function of the disulfide bonds in AVP and OXT nonapeptide hormones is to form and keep the cyclic conformations that enable the noncovalent interaction with their receptors (Rabenstein and Yeo 1994). Moreover, the seven predicted putative disulfide bonds in the camel neurophysin II, as previously identified in the cow, may contribute to the formation of a highly specific topography (globular, with minimal alpha-helical structure) to facilitate the AVP hormone-carrying function by reducing the number of biofunctional conformers of neurophysin II (Schlesinger et al. 1972). The transportation of camel OXT likely benefits in a similar way, since the disulfide bond features are shared by camel neurophysin I and II. Consequently, the homology of the N-linked glycosylation sites, the predicted disulfide bond formation sites and the cleavage sites of AVP/OXT precursors observed in this study suggest that the post-translational processing of AVP and OXT hormone is likely conserved in the camel.

13.5.5 Synonymous Substitution Sites in *AVP* and *OXT* Genes

Some genetic divergence is regarded as a synonymous mutation where the base substitution in the exon of a protein-coding gene may result in no change in the amino acid sequence, due to the degeneracy of codons (Chamary et al. 2006). Such mutations may have occurred in the *AVP* and *OXT* genes. By collectively comparing the gene and protein sequences across species, some synonymous substitution sites were observed in the CDSs that encode highly conserved amino acid blocks including the AVP and OXT nonapeptide hormones, the 'GKR' spacer in AVP precursor and the central evolutionarily conserved region of neurophysins. This may contribute to structural conservation, and potentially the functional conservation, of the two hormones in different lineages.

Despite the perception in earlier studies that synonymous mutations are 'silent', there is evidence showing that this type of mutation causes changes in the expression, conformation and function of protein (Sauna and Kimchi-Sarfaty 2011). Codon usage bias in relation to synonymous codons found in many organisms suggests that even the synonymous codons are under evolutionary pressure (Chamary et al. 2006; Plotkin and Kudla 2011). Synonymous mutations can lead to aberrant mRNA splicing (Cartegni et al. 2002) or altered mRNA stability (Nackley et al. 2006) and thus affect the production and function of proteins. Therefore, it is also possible that the synonymous single-nucleotide polymorphisms (SNPs) occurring in *AVP* and *OXT* genes have effects on the expression, conformation and function of the encoded preprohormones.

13.5.6 Regulatory Motifs of the Camel *AVP* and *OXT* Genes

The proximal regulatory regions of the dromedary *AVP* and *OXT* genes (Supplementary Figs. S13.3 and S13.4) are homologous to their orthologs in the Bactrian and wild camels. Both genes in the Camelini are highly homologous to that of the Lamini, in contrast to higher divergence compared to other species. Apart from the Lamini, the regulatory region of *AVP* of the Camelini species is more homologous to human, whilst that of *OXT* is more homologous to rock hyrax for the comparisons performed. The regulatory regions of *AVP* and *OXT* genes in the three nondesert rodents (mouse, rat and UGMBMR) are more distantly related to that of the camels.

By comparing *AVP* and *OXT* regulatory regions across species (Supplementary Figs. S13.3 and S13.4), many highly conserved regions were found, which could indicate the locations of functional regulatory motifs that play roles in regulating gene expression. It is well-known that the core promoter of a gene is the minimal DNA sequence adjacent to the transcription start site (TSS) that directs accurate transcription initiation at basal levels by the RNA polymerase transcription machinery (Roeder 1996). The TATA box (with TATAAA as the most common sequence) is a functional motif located in the core promoter. It can be bound with the TATA-binding protein and other transcription factors to recruit RNA polymerase. As a noncanonical form of the TATA box, CATA boxes (with CATAAA as the representative sequence) were discovered in many eukaryotic genes including *AVP* and *OXT* (Richter 1988). We hence predict that the CATAAA regions located 77 bp upstream of the start codon in dromedary camel *AVP* (Supplementary Fig. S13.3) and 64 bp in dromedary camel *OXT* (Supplementary Fig. S13.4), with perfect matches in other species (with one exception: the AATA box in the rock hyrax *OXT*), are the CATA boxes of the *AVP* and *OXT* genes in the camels, respectively.

Other regulatory motifs in the camel *AVP* were predicted based on previous studies on rat, human, mouse and cattle (Greenwood et al. 2014, 2016; Coulson et al. 1999). Those for camel *OXT* were based on a study on the Camelids (Pauciullo et al. 2018). The dromedary *AVP* is predicted to contain two CAAT boxes, a CRE1 element, a CRE2 element (CREB3L1 binding site), three E boxes (including a canonical CACGTC E box), a G box (CREB3L1 binding site), three AP binding sites and 10 potential CpG methylation sites upstream of the CATA box (Fig. 13.8a, Supplementary Fig. S13.3). The dromedary *OXT* gene is predicted to contain three SP1 binding sites, a NF1 binding site, a CHRE binding site, four ERE half motifs (two located in the CHRE binding site) and 26 potential CpG methylation sites upstream the CATA box (Fig. 13.8b, Supplementary Fig. S13.4). All predicted motifs are mostly conserved in Camelids. Two E boxes, an AP1 site and an AP2 site in *AVP*, and three ERE half motifs in *OXT* are completely conserved. In the *OXT* gene, one of the main regulatory elements CHRE is conserved among Camelids, human and rat. No authentic palindromic ERE elements (GGTCANNNTGACC) were identified in *OXT* of all species studied here (Supplementary Fig. S13.4).

The putative CpG sites were found to be highly diverse across different species. Only one predicted CpG site in the *AVP* gene is conserved across all species here, suggesting high variability and potentially large evolutionary divergence in the CpG sites in *AVP* and *OXT*. DNA methylation at CpG residues can inhibit transcription factor interactions with DNA. This is regarded as an epigenetic regulation of gene transcription and typically inhibits transcription. The CpG sites in the core promoter region of the rat *AVP* form a CpG island (Greenwood et al. 2016) that is defined as a ~200 bp DNA region with a GC content >50% and an observed CpG versus expected CpG ratio ≥ 0.6 (Gardiner-Garden and Frommer 1987). A study in rats (Greenwood et al. 2016) revealed a strong negative correlation between *AVP* promoter methylation and *AVP* hnRNA abundance in the rat SON. Hypermethylation of the proximal *AVP* promoter is induced by water deprivation in the SONs of water-deprived rats. The disturbance of normal DNA methylations and induced mutation in individual CpG sites in the *AVP* promoter in vitro resulted in altered levels of AVP biosynthesis (Greenwood et al. 2016). Taken together, it is speculated that putative CpG sites in the camel *AVP* proximal promoter could form a CpG island that regulates *AVP* transcription to coordinate plastic change of the SON during osmoregulation. Many putative CpG sites are found in the functional motifs such as CRE2, G box, SP1, CHRE and ERE half sites. Whether the methylation of these sites is altered by osmotic challenge, or if methylation alters *AVP*/*OXT* promoter activities, is an interesting topic for future studies.

Interestingly, more CpG sites were observed in the proximal promoter of the dromedary *OXT* than *AVP* (26 in *OXT*, 10 in *AVP*). In the proximal promoter of the *OXT* gene, multiple CpG sites (unique to camels) were found in or flanking regulatory elements including three SP1 sites, one CHRE site, two ERE half sites. Methylation of these CpG sites may hinder interactions between the SP1 motifs and the SP1 transcription factor, the CHRE site and carbohydrate response

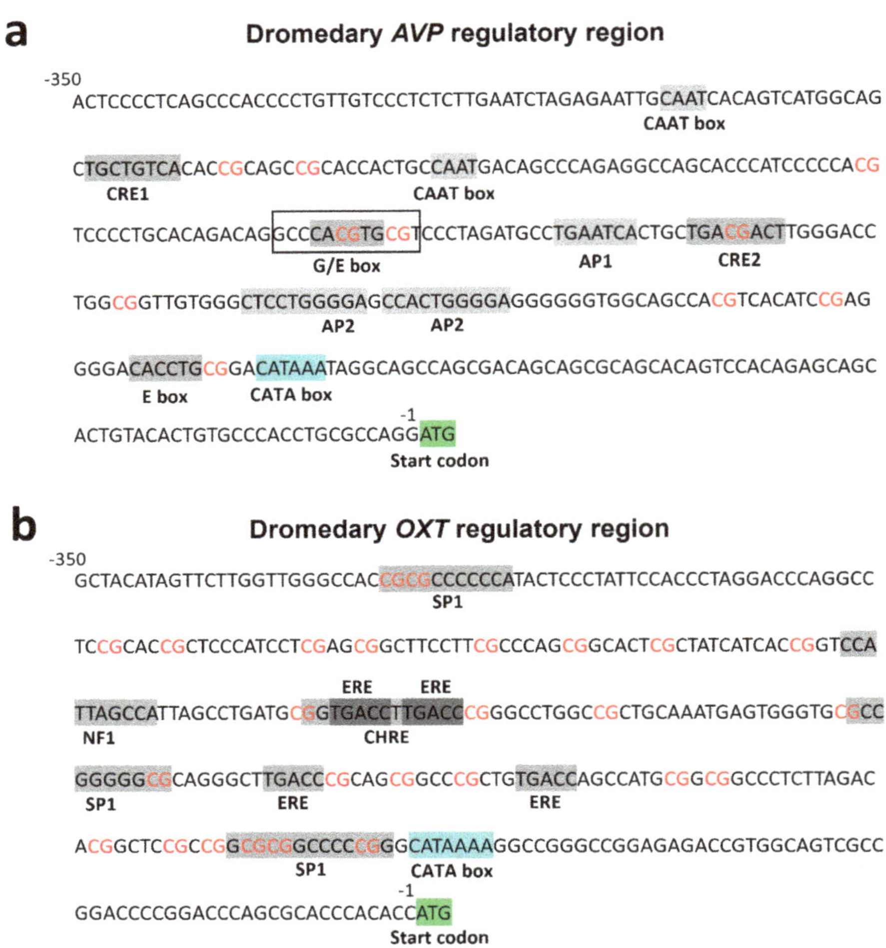

Fig. 13.8 Proximal regulatory regions of the dromedary AVP and OXT genes. Transcriptional factor binding sites/motifs (highlighted by grey or boxes) and CpG sites (denoted as red text) are predicted within the 350 bp regulatory regions upstream of the start codon of *AVP* (**a**) and *OXT* (**b**) genes. CAAT, CAAT box; CRE, cAMP response element; G/E-box, G-box or enhancer box; AP1, activator protein 1 binding site; AP2, activator protein 2 binding site; SP1, transcription factor SP1 binding site; NF1, nuclear factor 1 binding site; CHRE, carbohydrate response element-binding protein; ERE, oestrogen-responsive element motif (only two ERE half sites are found here); CG, CpG (cytosine-phosphate-guanine) site

element-binding protein, as well as the ERE half sites and oestrogen. As previously reported in the rat, the ERE half sites sit within the CHRE element of the camel *OXT* promoter may act synergistically with the full CHRE site located further upstream to promote *OXT* expression (Pauciullo et al. 2018).

CREB3L1 is a transcription factor of the *AVP* gene. It interacts with CRE2 and G box elements in the proximal promoter region of *AVP*. In comparison to the two CREB3L1 binding sites (the CRE2 element and the G box) in rat *AVP* gene, the putative 12-nucleotide G box in dromedary camel has a point mutation at residue 10 (rat: GCCCACGTGTGT, dromedary camel: GCCCACGTGCGT), and the putative CRE2 motif exhibits a mutation at residue 5 (rat: TGACAGCT, dromedary camel: TGACGACT), each introducing a new potential CpG site to the dromedary *AVP* for possible gene regulation (Supplementary Fig. S13.3). Moreover, the CRE1 motif in rat *AVP* was shown to be methylated at this CpG site to inhibit CREB-mediated transcription (Greenwood et al. 2016), whilst the camel *AVP* CRE1 motif contains no CpG sites.

Although the whole regions of the *AVP* and *OXT* introns are highly divergent, the predicted intron splicing sites GT...AG of the *AVP* and *OXT* genes are completely conserved across almost all species. Notably, the splicing sites of *AVP* intron 2 in vicuña and guanaco are GC...AG, which is known as one of the major non-canonical splice site combinations (Frey and Pucker 2020) (Supplementary Figs. S13.5 and S13.6).

13.6 Conclusions and Future Directions

The AVP-like mechanisms and water conservation adaptations in mammals, with a particular focus on the camels, are an exciting field of research that continues to unravel the mysteries of how organisms have adapted to challenges of limited water access throughout evolutionary history. Our comparative study on the *AVP* and *OXT* and their encoded products between camels and other species revealed shared gene/protein sequences between animals adapting to variable osmotic stress exerted by different habitats, as well as elements that appear to be unique to the camels and that may be essential for adaptations necessary for life in the desert. These understandings will be important for evaluating the possibility of evolutionary rescue of species currently challenged by increased desertification, especially the extremely endangered wild camels.

Whilst significant progress has been made in understanding the evolution and function of AVP-like mechanisms, there are still many promising directions for future research. First, the prediction of regulatory elements located in the intergenic sequences between the *AVP* and *OXT* genes can be further explored. Second, further genomic and transcriptomic studies can help to elucidate additional genetic variations and regulatory networks associated with water balance regulation. Third, comparative analyses across a broader range of water-stressed mammals in future may uncover novel genetic adaptations and shed light on the evolutionary trajectories of AVP-like mechanisms. Moreover, the desert adaptations observed in the camels can inspire innovative approaches for sustainable water management practices for animal husbandry in water-scarce regions, addressing the growing global demand for water resources. Last but not the least, understanding the behavioural adaptations, such as drinking patterns and foraging strategies, alongside the physiological mechanisms, can provide a more comprehensive picture of water conservation strategies in water-stressed mammals. As we delve deeper into the molecular, physiological and environmental aspects of AVP/OXT signalling and water balance regulation, such research will inform the development of therapeutic interventions for water balance disorders in both human and animal populations.

References

Agnew D (2018) Camelidae. In: Pathology of wildlife and zoo animals. Elsevier, pp 185–205. https://doi.org/10.1016/B978-0-12-805306-5.00007-9

Ali MA, Adem A, Chandranath IS, Benedict S, Pathan JY, Nagelkerke N, Nyberg F, Lewis LK, Yandle TG, Nicholls GM, Frampton CM, Kazzam E (2012) Responses to dehydration in the one-humped camel and effects of blocking the renin-angiotensin system. PLoS One 7(5):e37299. https://doi.org/10.1371/journal.pone.0037299

Ali A, Baby B, Vijayan R (2019) From desert to medicine: a review of camel genomics and therapeutic products. Front Genet 10. https://doi.org/10.3389/fgene.2019.00017

Ali MA, Abu Damir H, Ali OM, Amir N, Tariq S, Greenwood MP, Lin P, Gillard B, Murphy D, Adem A (2020) The effect of long-term dehydration and subsequent rehydration on markers of inflammation, oxidative stress and apoptosis in the camel kidney. BMC Vet Res 16(1):458. https://doi.org/10.1186/s12917-020-02628-5

Almathen F, Charruau P, Mohandesan E, Mwacharo JM, Orozco-terWengel P, Pitt D, Abdussamad AM, Uerpmann M, Uerpmann H-P, de Cupere B, Magee P, Alnaqeeb MA, Salim B, Raziq A, Dessie T, Abdelhadi OM, Banabazi MH, Al-Eknah M, Walzer C et al (2016) Ancient and modern DNA reveal dynamics of domestication and cross-continental dispersal of the dromedary. Proc Natl Acad Sci 113(24):6707–6712. https://doi.org/10.1073/pnas.1519508113

Andersson B (1977) Regulation of body fluids. Annu Rev Physiol 39:185–200

Ando S, Murthy AS, Eipper BA, Chaiken IM (1988) Effect of neurophysin on enzymatic maturation of oxytocin from its precursor. J Biol Chem 263(2):769–775

Antunes-Rodrigues J, de Castro M, Elias LLK, Valença MM, McCann SM (2004) Neuroendocrine control of body fluid metabolism. Physiol Rev 84(1):169–208. https://doi.org/10.1152/physrev.00017.2003

Ares GR, Caceres PS, Ortiz PA (2011) Molecular regulation of NKCC2 in the thick ascending limb. Am J Physiol Renal Physiol 301(6):F1143–F1159. https://doi.org/10.1152/ajprenal.00396.2011

Bactrian Camels Genome Sequencing and Analysis Consortium (2012) Genome sequences of wild and domestic bactrian camels. Nat Commun 3(1):1202. https://doi.org/10.1038/ncomms2192

Balanescu S, Kopp P, Gaskill MB, Morgenthaler NG, Schindler C, Rutishauser J (2011) Correlation of plasma copeptin and vasopressin concentrations in hypo-, iso-, and hyperosmolar states. J Clin Endocrinol Metab 96(4):1046–1052. https://doi.org/10.1210/jc.2010-2499

Barat C, Simpson L, Breslow E (2004) Properties of human vasopressin precursor constructs: inefficient monomer folding in the absence of copeptin as a potential contributor to diabetes insipidus. Biochemistry 43(25):8191–8203. https://doi.org/10.1021/bi0400094

Bargmann W (1966) Neurosecretion. Int Rev Cytol 19:183–201. https://doi.org/10.1016/S0074-7696(08)60567-7

Baribeau DA, Anagnostou E (2015) Oxytocin and vasopressin: linking pituitary neuropeptides and their receptors to social neurocircuits. Front Neurosci 9:335. https://doi.org/10.3389/fnins.2015.00335

Bekele T, Olsson K, Olsson U, Dahlborn K (2013) Physiological and behavioral responses to different watering intervals in lactating camels (Camelus dromedarius). Am J Physiol Regul Integr Comp Physiol 305(6):R639–R646. https://doi.org/10.1152/ajpregu.00015.2013

Ben Goumi M, Riad F, Giry J, de la Farge F, Safwate A, Davicco M-J, Barlet J-P (1993) Hormonal control of water and sodium in plasma and urine of camels during dehydration and rehydration. Gen Comp Endocrinol 89(3):378–386. https://doi.org/10.1006/gcen.1993.1045

Bengoumi M, Faye B (2002) Adaptation du dromadaire à la déshydratation. Sécheresse 13(2):121–129

Beuret N, Rutishauser J, Bider MD, Spiess M (1999) Mechanism of endoplasmic reticulum retention of mutant vasopressin precursor caused by a signal peptide truncation associated with diabetes insipidus. J Biol Chem 274(27):18965–18972. https://doi.org/10.1074/jbc.274.27.18965

Bichet DG (2016) Vasopressin at central levels and consequences of dehydration. Ann Nutr Metab 68(Suppl 2):19–23. https://doi.org/10.1159/000446200

Bijlsma R, Loeschcke V (2005) Environmental stress, adaptation and evolution: an overview. J Evol Biol 18(4):744–749. https://doi.org/10.1111/j.1420-9101.2005.00962.x

Boone M, Deen PMT (2008) Physiology and pathophysiology of the vasopressin-regulated renal water reabsorption. Pflugers Arch 456(6):1005–1024. https://doi.org/10.1007/s00424-008-0498-1

Borrow AP, Stover SA, Bales NJ, Handa RJ (2020) Posterior pituitary hormones. In: Hormonal signaling in biology and medicine. Elsevier, pp 203–226. https://doi.org/10.1016/B978-0-12-813814-4.00010-9

Bourque CW (2008) Central mechanisms of osmosensation and systemic osmoregulation. Nat Rev Neurosci 9(7):519–531. https://doi.org/10.1038/nrn2400

Breulmann M, Boer B, Wernery U, Wernery R, El-Shaer H (2007) A proposal towards combating desertification via establishment of camel farms based on fodder production from indigenous plants and halophytes. UNESCO-Doha Office, Qatar, United Arab Emirates

Breyer MD, Ando Y (1994) Hormonal signaling and regulation of salt and water transport in the collecting duct. Annu Rev Physiol 56:711–739. https://doi.org/10.1146/annurev.ph.56.030194.003431

Brooks FP, Pickford M (1958) The effect of posterior pituitary hormones on the excretion of electrolytes, in dogs. J Physiol 142(3):468–493. https://doi.org/10.1113/jphysiol.1958.sp006031

Brown CH, Bains JS, Ludwig M, Stern JE (2013) Physiological regulation of magnocellular neurosecretory cell activity: integration of intrinsic, local and afferent mechanisms. J Neuroendocrinol 25(8):678–710. https://doi.org/10.1111/jne.12051

Brunton PJ, Russell JA (2015) Maternal brain adaptations in pregnancy. In: Knobil and Neill's physiology of reproduction. Elsevier, pp 1957–2026. https://doi.org/10.1016/B978-0-12-397175-3.00044-2

Bulliet RW (1975) The camel and the wheel. Columbia University Press, New York

Burbach JP, Luckman SM, Murphy D, Gainer H (2001) Gene regulation in the magnocellular hypothalamo-neurohypophysial system. Physiol Rev 81(3):1197–1267. https://doi.org/10.1152/physrev.2001.81.3.1197

Burger PA (2016) The history of Old World camelids in the light of molecular genetics. Trop Anim Health Prod 48(5):905–913. https://doi.org/10.1007/s11250-016-1032-7

Burger PA, Ciani E, Faye B (2019) Old World camels in a modern world – a balancing act between conservation and genetic improvement. Anim Genet 50(6):598–612. https://doi.org/10.1111/age.12858

Cain JW, Krusman NPR, Turner JC (2006) Mechanisms of thermoregulation and water balance in desert ungulates. Wildl Soc Bull 34(3):570–581

Caligioni CS, Franci CR (2002) Oxytocin secretion induced by osmotic stimulation in rats during the estrous cycle and after ovariectomy and hormone replacement therapy. Life Sci 71(24):2821–2831. https://doi.org/10.1016/s0024-3205(02)02139-2

Cartegni L, Chew SL, Krainer AR (2002) Listening to silence and understanding nonsense: exonic mutations that affect splicing. Nat Rev Genet 3(4):285–298. https://doi.org/10.1038/nrg775

Chamary JV, Parmley JL, Hurst LD (2006) Hearing silence: non-neutral evolution at synonymous sites in mammals. Nat Rev Genet 7(2):98–108. https://doi.org/10.1038/nrg1770

Cohen WD, Terwilliger NB (1979) Marginal bands in camel erythrocytes. J Cell Sci 36(1):97–107. https://doi.org/10.1242/jcs.36.1.97

Conrad KP, Gellai M, North WG, Valtin H (1993) Influence of oxytocin on renal hemodynamics and sodium excretion. Ann N Y Acad Sci 689:346–362. https://doi.org/10.1111/j.1749-6632.1993.tb55559.x

Coulson JM, Fiskerstrand CE, Woll PJ, Quinn JP (1999) E-box motifs within the human vasopressin gene promoter contribute to a major enhancer in small-cell lung cancer. Biochem J 344(Pt 3):961–970

Crawford CS (1978) Environmental physiology of animals. J Arid Environ 1:195–196

Davis MW, Jorgensen EM (2022) ApE, A Plasmid Editor: a freely available DNA manipulation and visualization program. Front Bioinform 2:818619. https://doi.org/10.3389/fbinf.2022.818619

de Mayolo AA, Lisby M, Erdeniz N, Thybo T, Mortensen UH, Rothstein R (2006) Multiple start codons and phosphorylation result in discrete Rad52 protein species. Nucl Acids Res 34(9):2587–2597. https://doi.org/10.1093/nar/gkl280

de Souza Mecawi A, Ruginsk SG, Elias LLK, Varanda WA, Antunes-Rodrigues J (2015) Neuroendocrine regulation of hydromineral homeostasis. Comp Physiol 5(3):1465–1516. https://doi.org/10.1002/cphy.c140031

Edgar RC (2004) MUSCLE: multiple sequence alignment with high accuracy and high throughput. Nucl Acids Res 32(5):1792–1797. https://doi.org/10.1093/nar/gkh340

Edwards BR, LaRochelle FT (1984) Antidiuretic effect of endogenous oxytocin in dehydrated Brattleboro homozygous rats. Am J Physiol 247(3 Pt 2):F453–F465. https://doi.org/10.1152/ajprenal.1984.247.3.F453

Eliava M, Melchior M, Knobloch-Bollmann HS, Wahis J, da Silva Gouveia M, Tang Y, Ciobanu AC, Triana del Rio R, Roth LC, Althammer F, Chavant V, Goumon Y, Gruber T, Petit-Demoulière N, Busnelli M, Chini B, Tan LL, Mitre M, Froemke RC et al (2016) A new population of parvocellular oxytocin neurons controlling magnocellular neuron activity and inflammatory pain processing. Neuron 89(6):1291–1304. https://doi.org/10.1016/j.neuron.2016.01.041

Engelhardt W, Becker G, Engelhardt W, Hauffe R, Hinderer S, Rübsamen K, Schneider W (1974) Energy, water, and urea metabolism in the llama. In: Tracer studies on nonprotein nitrogen for Ruminants II. IAEA

Enke N, Brinkmann L, Runa RA, Südekum K-H, Tholen E, Gerken M (2022) Drinking behaviour of llamas (Lama glama) in choice tests for fresh or saline water. Small Rumin Res 216:106806. https://doi.org/10.1016/j.smallrumres.2022.106806

Etzion Z, Yagil R (1986) Renal function in camels (Camelus dromedarius) following rapid rehydration. Physiol Zool 59(5):558–562

Etzion Z, Meyerstein N, Yagil R (1984) Tritiated water metabolism during dehydration and rehydration in the camel. J Appl Physiol 56(1):217–220. https://doi.org/10.1152/jappl.1984.56.1.217

Ezcurra E, Mellink E (2013) Encyclopedia of biodiversity, vol 2, 2nd edn. Academic Press, Waltham

Faye B, Bengoumi M, Seboussi R (2011) The mineral nutrition and imbalances in camel: a constraint in pastoral areas. In: Bediye S, Tilahun S, Animut G, Egie M, Getahun T (eds) The international conference on camel research and development bridging knowledge and technology gaps in camel production. ICCED

Fenske W, Refardt J, Chifu I, Schnyder I, Winzeler B, Drummond J, Ribeiro-Oliveira A, Drescher T, Bilz S, Vogt DR, Malzahn U, Kroiss M, Christ E, Henzen C, Fischli S, Tönjes A, Mueller B, Schopohl J, Flitsch J et al (2018) A copeptin-based approach in the diagnosis of diabetes insipidus. New Engl J Med 379(5):428–439. https://doi.org/10.1056/NEJMoa1803760

Fowler ME (1989) Medicine and surgery of South American Camelids, 2nd edn. Iowa State University Press

Frey K, Pucker B (2020) Animal, fungi, and plant genome sequences harbor different non-canonical splice sites. Cells 9(2):458. https://doi.org/10.3390/cells9020458

Gainer H (1983) Precursors of vasopressin and oxytocin. Prog Brain Res 60:205–215. https://doi.org/10.1016/S0079-6123(08)64388-5

Gainer H (2012) Cell-type specific expression of oxytocin and vasopressin genes: an experimental Odyssey. J Neuroendocrinol 24(4):528–538. https://doi.org/10.1111/j.1365-2826.2011.02236.x

Gardiner-Garden M, Frommer M (1987) CpG Islands in vertebrate genomes. J Mol Biol 196(2):261–282. https://doi.org/10.1016/0022-2836(87)90689-9

Gaughan JB (2011) Which physiological adaptation allows camels to tolerate high heat load and what more can we learn? J Camelid Sci 4:85–88

Gebreyohanes MG, Assen AM (2017) Adaptation mechanisms of camels (Camelus dromedarius) for desert environment: a review. J Vet Sci Technol 08(06). https://doi.org/10.4172/2157-7579.1000486

Greenwood M, Bordieri L, Greenwood MP, Rosso Melo M, Colombari DSA, Colombari E, Paton JFR, Murphy D (2014) Transcription factor CREB3L1 regulates vasopressin gene expression in the rat hypothalamus. J Neurosci 34(11):3810–3820. https://doi.org/10.1523/JNEUROSCI.4343-13.2014

Greenwood MP, Mecawi AS, Hoe SZ, Mustafa MR, Johnson KR, Al-Mahmoud GA, Elias LLK, Paton JFR, Antunes-Rodrigues J, Gainer H, Murphy D, Hindmarch CCT (2015) A comparison of physiological and transcriptome responses to water deprivation and salt loading in the rat supraoptic nucleus. Am J Physiol Regul Integr Comp Physiol 308(7):R559–R568. https://doi.org/10.1152/ajpregu.00444.2014

Greenwood MP, Greenwood M, Gillard BT, Loh SY, Paton JFR, Murphy D (2016) Epigenetic control of the vasopressin promoter explains physiological ability to regulate vasopressin transcription in dehydration and salt loading states in the rat. J Neuroendocrinol 28(4). https://doi.org/10.1111/jne.12371

Grigg G, Beard L, Dörges B, Heucke J, Coventry J, Coppock A, Blomberg S (2009) Strategic (adaptive) hypothermia in bull dromedary camels during rut; could it increase reproductive success? Biol Lett 5(6):853–856. https://doi.org/10.1098/rsbl.2009.0450

Guillou MD, Barre N, Bussenot I, Plevrakis I, Clamagirand C (1992) COOH-terminally-extended processing forms of oxytocin in human ovary. Mol Cell Endocrinol 83(2–3):233–238. https://doi.org/10.1016/0303-7207(92)90163-z

Hara Y, Battey J, Gainer H (1990) Structure of mouse vasopressin and oxytocin genes. Mol Brain Res 8(4):319–324. https://doi.org/10.1016/0169-328X(90)90045-F

Hardiman A, Friedman TC, Grunwald WC, Furuta M, Zhu Z, Steiner DF, Cool DR (2005) Endocrinomic profile of neurointermediate lobe pituitary prohormone processing in PC1/3- and PC2-Null mice using SELDI-TOF mass spectrometry. J Mol Endocrinol 34(3):739–751. https://doi.org/10.1677/jme.1.01812

Hasan MT, Althammer F, Silva da Gouveia M, Goyon S, Eliava M, Lefevre A, Kerspern D, Schimmer J, Raftogianni A, Wahis J, Knobloch-Bollmann HS, Tang Y, Liu X, Jain A, Chavant V, Goumon Y, Weislogel J-M, Hurlemann R, Herpertz SC et al (2019) A fear memory engram and its plasticity in the hypothalamic oxytocin system. Neuron 103(1):133–146.e8. https://doi.org/10.1016/j.neuron.2019.04.029

Hernández-Pérez OR, Hernández VS, Nava-Kopp AT, Barrio RA, Seifi M, Swinny JD, Eiden LE, Zhang L (2019) A synaptically connected hypothalamic magnocellular vasopressin-locus coeruleus neuronal circuit and its plasticity in response to emotional and physiological stress. Front Neurosci 13. https://doi.org/10.3389/fnins.2019.00196

Holmes CL, Landry DW, Granton JT (2003) Science review: vasopressin and the cardiovascular system part 1--receptor physiology. Critical Care (London, England) 7(6):427–434. https://doi.org/10.1186/cc2337

Holzapfel C (2008) Deserts. In: Encyclopedia of ecology. Elsevier, pp 879–898. https://doi.org/10.1016/B978-008045405-4.00326-8

Hoyle CHV (1998) Neuropeptide families: evolutionary perspectives. Regul Pept 73(1):1–33. https://doi.org/10.1016/S0167-0115(97)01073-2

Ingram D, Mount L (1975) Man and animals in hot environment. Springer, Berlin

Irwin R (2010) Camel. Reaktion Books

Ivell R, Richter D (1984) Structure and comparison of the oxytocin and vasopressin genes from rat. Proc Natl Acad Sci 81(7):2006–2010. https://doi.org/10.1073/pnas.81.7.2006

Janas S, Seghers F, Schakman O, Alsady M, Deen P, Vriens J, Tissir F, Nilius B, Loffing J, Gailly P, Devuyst O (2016) TRPV4 is associated with central rather than nephrogenic osmoregulation. Pflugers Archiv 468(9):1595–1607. https://doi.org/10.1007/s00424-016-1850-5

Ji R, Cui P, Ding F, Geng J, Gao H, Zhang H, Yu J, Hu S, Meng H (2009) Monophyletic origin of domestic bactrian camel (Camelus bactrianus) and its evolutionary relationship with the extant wild camel (Camelus bactrianus ferus). Anim Genet 40(4):377–382. https://doi.org/10.1111/j.1365-2052.2008.01848.x

Johnson KR, Hindmarch CCT, Salinas YD, Shi Y, Greenwood M, Hoe SZ, Murphy D, Gainer H (2015) A RNA-Seq analysis of the rat supraoptic nucleus transcriptome: effects of salt loading on gene expression. PLoS One 10(4):e0124523. https://doi.org/10.1371/journal.pone.0124523

Jones DT, Taylor WR, Thornton JM (1992) The rapid generation of mutation data matrices from protein sequences. Bioinformatics 8(3):275–282. https://doi.org/10.1093/bioinformatics/8.3.275

Kadwell M, Fernandez M, Stanley HF, Baldi R, Wheeler JC, Rosadio R, Bruford MW (2001) Genetic analysis reveals the wild ancestors of the llama and the alpaca. Proc Biol Sci 268(1485):2575–2584. https://doi.org/10.1098/rspb.2001.1774

Kataria N, Kataria AK, Agarwal VK, Garg SL, Sahni MS (2002) Effect of long-term dehydration on serum constituents in extreme climatic conditions in camel (Camelus dromedarius). Indian J Physiol Pharmacol 46(2):218–222

Klockars OA, Klockars A, Levine AS, Olszewski PK (2018) Oxytocin administration in the basolateral and central nuclei of amygdala moderately suppresses food intake. Neuroreport 29(6):504–510. https://doi.org/10.1097/WNR.0000000000001005

Kumar S, Stecher G, Li M, Knyaz C, Tamura K (2018) MEGA X: molecular evolutionary genetics analysis across computing platforms. Mol Biol Evol 35(6):1547–1549. https://doi.org/10.1093/molbev/msy096

Leng G, Russell JA (2019) The osmoresponsiveness of oxytocin and vasopressin neurones: mechanisms, allostasis and evolution. J Neuroendocrinol 31(3):e12662. https://doi.org/10.1111/jne.12662

Li C, Wang W, Summer SN, Westfall TD, Brooks DP, Falk S, Schrier RW (2008) Molecular mechanisms of antidiuretic effect of oxytocin. Journal of the American Society of Nephrology 19(2):225–232. https://doi.org/10.1681/ASN.2007010029

Lin D, Wei Y, Ye H (2020) Role of oxytocin/vasopressin-like peptide and its receptor in vitellogenesis of mud crab. Int J Mol Sci 21(7):2297. https://doi.org/10.3390/ijms21072297

Lin P, Gillard BT, Pauža AG, Iraizoz FA, Ali MA, Mecawi AS, Alim FZD, Romanova EV, Burger PA, Greenwood MP, Adem A, Murphy D (2022) Transcriptomic plasticity of the hypothalamic osmoregulatory control centre of the Arabian dromedary camel. Commun Biol 5(1):1008. https://doi.org/10.1038/s42003-022-03857-0

Livne A, Kuiper PJC (1973) Unique properties of the camel erythrocyte membrane. Biochim Biophys Acta 318(1):41–49. https://doi.org/10.1016/0005-2736(73)90334-9

Lote CJ, McVicar AJ, Smyth DG (1986) Effects of vasopressin-glycine and vasopressin-glycine-lysine-arginine on renal function in the rat. J Endocrinol 108(2):255–260. https://doi.org/10.1677/joe.0.1080255

Louw E, Louw GN, Retief CP (1972) Thermolability, heat tolerance and renal function in the Dassie or Hyrax, *Procavia Capensis*. Zool Afr 7(2):451–469. https://doi.org/10.1080/00445096.1972.11447455

Ludwig M, Sabatier N, Bull PM, Landgraf R, Dayanithi G, Leng G (2022) Intracellular calcium stores regulate activity-dependent neuropeptide release from dendrites. Nature 418(6893):85–89. https://doi.org/10.1038/nature00822

Marin JC, González BA, Poulin E, Casey CS, Johnson WE (2013) The influence of the arid Andean high plateau on the phylogeography and population genetics of guanaco (*Lama guanicoe*) in South America. Mol Ecol 22(2):463–482. https://doi.org/10.1111/mec.12111

McEwen BB (2004) General introduction to vasopressin and oxytocin: structure/metabolism, evolutionary aspects, neural pathway/receptor distribution, and functional aspects relevant to memory processing. Adv Pharmacol 50:1–50. https://doi.org/10.1016/S1054-3589(04)50001-7

McKinley MJ, Johnson AK (2004) The physiological regulation of thirst and fluid intake. Physiology 19(1):1–6. https://doi.org/10.1152/nips.01470.2003

McKinley MJ, Mathai ML, McAllen RM, McClear RC, Miselis RR, Pennington GL, Vivas L, Wade JD, Oldfield BJ (2004) Vasopressin secretion: osmotic and hormonal regulation by the lamina terminalis. J Neuroendocrinol 16(4):340–347. https://doi.org/10.1111/j.0953-8194.2004.01184.x

Mirabeau O, Joly JS (2013) Molecular evolution of peptidergic signalling systems in bilaterians. Proc Natl Acad Sci U S A 110(22):E2028–E2037. https://doi.org/10.1073/pnas.1219956110

Möller C, Marí F (2007) A vasopressin/oxytocin-related conopeptide with γ-carboxyglutamate at position 8. Biochem J 404(3):413–419. https://doi.org/10.1042/BJ20061480

Murphy D, Gainer H (2016) Molecular neuroendocrinology: from genome to physiology. Wiley

Nackley AG, Shabalina SA, Tchivileva IE, Satterfield K, Korchynskyi O, Makarov SS, Maixner W, Diatchenko L (2006) Human catechol-*O*-methyltransferase haplotypes modulate protein expression by altering mRNA secondary structure. Science 314(5807):1930–1933. https://doi.org/10.1126/science.1131262

NCBI Resource Coordinators (2018) Database resources of the National Center for Biotechnology Information. Nucleic Acids Res 46(D1):D8–D13. https://doi.org/10.1093/nar/gkx1095

Nevo E, Simson S, Beiles A, Yahav S (1989) Adaptive variation in structure and function of kidneys of speciating subterranean mole rats. Oecologia 79(3):366–371. https://doi.org/10.1007/BF00384316

Norman AW, Litwack G (1987) Posterior pituitary hormones. In: Hormones. Elsevier, pp 133–169. https://doi.org/10.1016/B978-0-12-521440-7.50007-8

Ouajd S, Kamel B (2009) Physiological particularities of dromedary (Camelus dromedarius) and experimental implications. Scand J Lab Anim Sci 36(1)

Pauciullo A, Ogah DM, Iannaccone M, Erhardt G, di Stasio L, Cosenza G (2018) Genetic characterization of the oxytocin-neurophysin I gene (OXT) and its regulatory regions analysis in domestic old and new world camelids. PLoS One 13(4):e0195407. https://doi.org/10.1371/journal.pone.0195407

Pauža AG, Mecawi AS, Paterson A, Hindmarch CCT, Greenwood M, Murphy D, Greenwood MP (2021) Osmoregulation of the transcriptome of the hypothalamic supraoptic nucleus: a resource for the community. J Neuroendocrinol 33(8):e13007. https://doi.org/10.1111/jne.13007

Perk K (1966) Osmotic hemolysis of the camel's erythrocytes. I. A microcinematographic study. J Exp Zool 163(3):241–246. https://doi.org/10.1002/jez.1401630303

Pitti T, Manoj N (2012) Molecular evolution of the neuropeptide S receptor. PLoS One 7(3):e34046. https://doi.org/10.1371/journal.pone.0034046

Plotkin JB, Kudla G (2011) Synonymous but not the same: the causes and consequences of codon bias. Nat Rev Genet 12(1):32–42. https://doi.org/10.1038/nrg2899

Pouzet B, Serradeil-Le Gal C, Bouby N, Maffrand J, le Fur G, Bankir L (2001) Selective blockade of vasopressin V2 receptors reveals significant V2-mediated water reabsorption in Brattleboro rats with diabetes insipidus. Nephrol Dial Transplant 16(4):725–734. https://doi.org/10.1093/ndt/16.4.725

Qiu J, Hindmarch CCT, Yao ST, Tasker JG, Murphy D (2011) Transcriptomic analysis of the osmotic and reproductive remodeling of the female rat supraoptic nucleus. Endocrinology 152(9):3483–3491. https://doi.org/10.1210/en.2011-1044

Qureshi S, Galiveeti S, Bichet DG, Roth J (2014) Diabetes insipidus: celebrating a century of vasopressin therapy. Endocrinology 155(12):4605–4621. https://doi.org/10.1210/en.2014-1385

Rabenstein DL, Yeo PL (1994) Kinetics and equilibria of the formation and reduction of the disulfide bonds in arginine-vasopressin and oxytocin by thiol/disulfide interchange with glutathione and cysteine. J Org Chem 59(15):4223–4229. https://doi.org/10.1021/jo00094a039

Rasmussen MS, Simonsen JA, Sandgaard NCF, Høilund-Carlsen PF, Bie P (2004) Effects of oxytocin in normal man during low and high sodium diets. Acta Physiol Scand 181(2):247–257. https://doi.org/10.1111/j.1365-201X.2004.01286.x

Reis WL, Biancardi VC, Son S, Antunes-Rodrigues J, Stern JE (2015) Carbon monoxide and nitric oxide interactions in magnocellular neurosecretory neurones during water deprivation. J Neuroendocrinol 27(2):111–122. https://doi.org/10.1111/jne.12245

Richard S (2003) Near eastern archaeology: a reader. Eisenbrauns

Richter D (1988) Molecular events in expression of vasopressin and oxytocin and their cognate receptors. Am J Physiol Renal Physiol 255(2):F207–F219. https://doi.org/10.1152/ajprenal.1988.255.2.F207

Rocha JL, Godinho R, Brito JC, Nielsen R (2021) Life in deserts: the genetic basis of mammalian desert adaptation. Trends Ecol Evol 36(7):637–650. https://doi.org/10.1016/j.tree.2021.03.007

Roeder RG (1996) The role of general initiation factors in transcription by RNA polymerase II. Trends Biochem Sci 21(9):327–335

Rosenmann M, Morrison PR (1963) The physiological response to heat and dehydration of the guanaco, Llama guanicoe. Physiol Zool 36:45–51

Saadeldin IM, Abdel-Aziz Swelum A, Alzahrani FA, Alowaimer AN (2018) The current perspectives of dromedary camel stem cells research. Int J Vet Sci Med 6(Suppl):S27–S30. https://doi.org/10.1016/j.ijvsm.2018.01.002

Sakai RR, Mcewen BS, Fluharty SJ, Ma LY (2000) The amygdala: site of genomic and nongenomic arousal of aldosterone-induced sodium intake. Kidney Int 57(4):1337–1345. https://doi.org/10.1046/j.1523-1755.2000.00972.x

Sale JB (1966) The habitat of the rock hyrax. J East Afri Nat Hist Soc Natl Mus 25(3):205

Sapir-Hen L, Ben-Yosef E (2013) The introduction of domestic camels to the southern Levant: evidence from the Aravah Valley. Tel Aviv 40(2):277–285. https://doi.org/10.1179/033443513X13753505864089

Sauna ZE, Kimchi-Sarfaty C (2011) Understanding the contribution of synonymous mutations to human disease. Nat Rev Genet 12(10):683–691. https://doi.org/10.1038/nrg3051

Schlesinger DH, Frangione B, Walter R (1972) Covalent structure of bovine neurophysin-II: localization of the disulfide bonds. Proc Natl Acad Sci 69(11):3350–3354. https://doi.org/10.1073/pnas.69.11.3350

Schmidt-Nielsen K (1959) The physiology of the camel. Sci Am 201(6):140–151. https://doi.org/10.1038/scientificamerican1259-140

Schmidt-Nielsen K (1997) Animal physiology: adaptation and environment. Cambridge University Press

Schmidt-Nielsen B, Schmidt-Nielsen K, Houpt TR, Jarnum SA (1956) Water balance of the camel. Am J Physiol Legacy Content 185(1):185–194. https://doi.org/10.1152/ajplegacy.1956.185.1.185

Schmidt-Nielsen K, Crawford EC, Newsome A, Rawson K, Hammel H (1967) Metabolic rate of camels: effect of body temperature and dehydration. Am J Physiol Legacy Content 212(2):341–346. https://doi.org/10.1152/ajplegacy.1967.212.2.341

Siebert BD, Macfarlane WV (1975) Dehydration in desert cattle and camels. Physiol Zool 48:36–48

Silanikove N (1994) The struggle to maintain hydration and osmoregulation in animals experiencing severe dehydration and rapid rehydration: the story of ruminants. Exp Physiol 79(3):281–300. https://doi.org/10.1113/expphysiol.1994.sp003764

Sinclair MS, Perea-Martinez I, Dvoryanchikov G, Yoshida M, Nishimori K, Roper SD, Chaudhari N (2010) Oxytocin signaling in mouse taste buds. PloS One 5(8):e11980. https://doi.org/10.1371/journal.pone.0011980

Siontorou CG (2013) Nanobodies as novel agents for disease diagnosis and therapy. Int J Nanomed 8:4215. https://doi.org/10.2147/IJN.S39428

Šklíba J, Lövy M, Koeppen SCW, Pleštilová L, Vitámvás M, Nevo E, Šumbera R (2016) Activity of free-living subterranean blind mole rats *Spalax galili* (Rodentia: Spalacidae) in an area of supposed sympatric speciation. Biol J Linn Soc 118(2):280–291. https://doi.org/10.1111/bij.12741

Sladek CD, Armstrong WE (2009) Neurohypophyseal system. In: Encyclopedia of neuroscience. Elsevier, pp 449–458. https://doi.org/10.1016/B978-008045046-9.00466-6

Sparapani S, Millet-Boureima C, Oliver J, Mu K, Hadavi P, Kalostian T, Ali N, Avelar CM, Bardies M, Barrow B, Benedikt M, Biancardi G, Bindra R, Bui L, Chihab Z, Cossitt A, Costa J, Daigneault T, Dault J et al (2021) The biology of vasopressin. Biomedicines 9(1). https://doi.org/10.3390/biomedicines9010089

Szczepanska-Sadowska E, Czarzasta K, Cudnoch-Jedrzejewska A (2018) Dysregulation of the Renin-Angiotensin system and the vasopressinergic system interactions in cardiovascular disorders. Curr Hypertens Rep 20(3):19. https://doi.org/10.1007/s11906-018-0823-9

Tamura K, Nei M (1993) Estimation of the number of nucleotide substitutions in the control region of mitochondrial DNA in humans and chimpanzees. Mol Biol Evol. https://doi.org/10.1093/oxfordjournals.molbev.a040023

Tang Y, Benusiglio D, Lefevre A, Hilfiger L, Althammer F, Bludau A, Hagiwara D, Baudon A, Darbon P, Schimmer J, Kirchner MK, Roy RK, Wang S, Eliava M, Wagner S, Oberhuber M, Conzelmann KK, Schwarz M, Stern JE et al (2020) Social touch promotes interfemale communication via activation of parvocellular oxytocin neurons. Nat Neurosci 23(9):1125–1137. https://doi.org/10.1038/s41593-020-0674-y

Tasker JG, Oliet SHR, Bains JS, Brown CH, Stern JE (2012) Glial regulation of neuronal function: from synapse to systems physiology. J Neuroendocrinol 24(4):566–576. https://doi.org/10.1111/j.1365-2826.2011.02259.x

Terris J, Ecelbarger CA, Nielsen S, Knepper MA (1996) Long-term regulation of four renal aquaporins in rats. Am J Physiol 271(2 Pt 2):F414–F422. https://doi.org/10.1152/ajprenal.1996.271.2.F414

Thomsen ARB, Plouffe B, Cahill TJ, Shukla AK, Tarrasch JT, Dosey AM, Kahsai AW, Strachan RT, Pani B, Mahoney JP, Huang L, Breton B, Heydenreich FM, Sunahara RK, Skiniotis G, Bouvier M, Lefkowitz RJ (2016) GPCR-G protein-β-Arrestin super-complex mediates sustained G protein signaling. Cell 166(4):907–919. https://doi.org/10.1016/j.cell.2016.07.004

Tibary A, El Allali K (2020) Dromedary camel: a model of heat resistant livestock animal. Theriogenology 154:203–211. https://doi.org/10.1016/j.theriogenology.2020.05.046

Vallenas A, Cummings JF, Munnell JF (1971) A gross study of the compartmentalized stomach of two new-world camelids, the llama and guanaco. J Morphol 134(4):399–423. https://doi.org/10.1002/jmor.1051340403

Vilá B, Arzamendia Y (2022) South American Camelids: their values and contributions to people. Sustain Sci 17(3):707–724. https://doi.org/10.1007/s11625-020-00874-y

Wahis J, Baudon A, Althammer F, Kerspern D, Goyon S, Hagiwara D, Lefevre A, Barteczko L, Boury-Jamot B, Bellanger B, Abatis M, Da Silva GM, Benusiglio D, Eliava M, Rozov A, Weinsanto I, Knobloch-Bollmann HS, Kirchner MK, Roy RK, Wang H, Pertin M, Inquimbert P, Pitzer C, Siemens J, Goumon Y, Boutrel B, Lamy CM, Decosterd I, Chatton JY, Rouach N, Young WS, Stern JE, Poisbeau P, Stoop R, Darbon P, Grinevich V, Charlet A (2021) Astrocytes mediate the effect of oxytocin in the central amygdala on neuronal activity and affective states in rodents. Nat Neurosci 24(4):529–541. https://doi.org/10.1038/s41593-021-00800-0

Williams TD, Abel DC, King CM, Jelley RY, Lightman SL (1986) Vasopressin and oxytocin responses to acute and chronic osmotic stimuli in man. J Endocrinol 108(1):163–168. https://doi.org/10.1677/joe.0.1080163

Willmer P, Ston G, Johnston I (2006) Environmental physiology of animals, 2nd edn. Blackwell Publishing

Wu H, Guang X, Al-Fageeh MB, Cao J, Pan S, Zhou H, Zhang L, Abutarboush MH, Xing Y, Xie Z, Alshanqeeti AS, Zhang Y, Yao Q, Al-Shomrani BM, Zhang D, Li J, Manee MM, Yang Z, Yang L et al (2014) Camelid genomes reveal evolution and adaptation to desert environments. Nat Commun 5(1):5188. https://doi.org/10.1038/ncomms6188

Yoshida M (2008) Gene Regulation system of vasopressin and corticotoropin-releasing hormone. Gene Regul Syst Biol 2:GRSB.S424. https://doi.org/10.4137/GRSB.S424

Young SR, Schmidt-Nielsen K (1985) Animal physiology: adaptation and environment. J Appl Ecol 22:291

Young TK, van Dyke HB (1968) Repletion of vasopressin and oxytocin in the posterior lobe of the pituitary gland of the rat. J Endocrinol 40(3):337–342. https://doi.org/10.1677/joe.0.0400337

Zaffanello M, Dotta A, Emma F (2012) Renal modulation. In: Nephrology and fluid/electrolyte physiology: neonatology questions and controversies. Elsevier, pp 179–198. https://doi.org/10.1016/B978-1-4377-2658-9.00011-X

Zhang L, Hernández VS, Vázquez-Juárez E, Chay FK, Barrio RA (2016) Thirst is associated with suppression of Habenula output and active stress coping: is there a role for a non-canonical vasopressin-glutamate pathway? Front Neural Circuits 10:13. https://doi.org/10.3389/fncir.2016.00013

Zhang L, Hernandez VS, Gerfen CR, Jiang SZ, Zavala L, Barrio RA, Eiden LE (2021) Behavioral role of PACAP signaling reflects its selective distribution in glutamatergic and GABAergic neuronal subpopulations. ELife 10. https://doi.org/10.7554/eLife.61718

The Evolution of Oxytocin Receptor Signalling

14

Alina I. Sartorius, Constantina Theofanopoulou,
and Daniel S. Quintana

Abstract

Oxytocin is a neuromodulator and hormone that helps coordinate a variety of physiological and behavioural functions. Oxytocin exerts its diverse effects by binding to oxytocin receptors throughout the brain and the periphery, which initiates various signalling cascades. The location and distribution of oxytocin receptors can differ between species, which can facilitate species-specific behaviours that promote survival and reproduction. Yet, despite the increasing interest in the oxytocin signalling system, the extent of its functions in human health and well-being are currently not well understood. Pinpointing the evolutionary trajectory of the oxytocin receptor is one approach that can help unravel the current purpose of the oxytocin signalling system. Here, we demonstrate three analytical approaches that help uncover the evolutionary history of the oxytocin receptor: sequence homology searches, synteny mapping, and positive selection analysis. Our non-technical guide includes suggestions for software packages and previ-

A. I. Sartorius
Department of Psychology, University of Oslo, Oslo, Norway

C. Theofanopoulou
Rockefeller University, New York, NY, USA

New York University, New York, NY, USA

D. S. Quintana (✉)
Department of Psychology, University of Oslo, Oslo, Norway

KG Jebsen Centre for Neurodevelopmental Disorders, University of Oslo and Oslo University Hospital, Oslo, Norway

NevSom, Department of Rare Disorders & Disabilities, Oslo University Hospital, Oslo, Norway
e-mail: daniel.quintana@psykologi.uio.no

V. Grinevich, R. F. Oliveira (eds.), *Evolutionary and Comparative Neuroendocrinology*, Masterclass in Neuroendocrinology 17,
https://doi.org/10.1007/978-3-031-80209-6_14

ously published analysis scripts that can be used for preprocessing and performing the analysis. These complementary analytical approaches have helped uncover the evolutionary history of the oxytocin receptor, suggesting that oxytocin receptor orthologs can be traced back to the origin of vertebrates and that the oxytocin receptor has been positively selected for in an ancient marsupial branch.

Keywords

Oxytocin · Gene expression · Oxytocin receptor · Evolution

14.1 Introduction

Oxytocin is a neuromodulator and hormone that is mainly synthesized in the hypothalamus and secreted both centrally and peripherally (Jurek and Neumann 2018). Oxytocin was first discovered over a century ago for its role in childbirth (Dale 1906) and milk provisioning (Ott and Scott 1910). In the 2000s, there was a surge of interest in oxytocin regarding its role in social behaviour (Kenkel 2019; Leng and Leng 2021) and its potential to assist social behaviour challenges, which are associated with many psychiatric disorders and developmental conditions (Kennedy and Adolphs 2012). While early results from oxytocin administration trials were encouraging, more recent results have been mixed (Alvares et al. 2017). Various reasons for these conflicting results have been proposed, including questionable research practices (Lane et al. 2016) that may also have inflated effect sizes (Audunsdottir et al. 2023) and study designs with small sample sizes (Walum et al. 2016) that cannot detect or reject a wide range of effect sizes (Quintana 2020). More recent research indicates that oxytocin's role in cognition goes beyond the processing of social stimuli, which suggests that the scope of prevailing theories has been too narrow (Quintana and Guastella 2020). For example, oxytocin administration has been found to increase one's ability to delay gratification in a non-social context (Kapetaniou et al. 2021) and rule adherence in an economic task (Gross and De Dreu 2017). Beyond behaviour, more recent research also revealed several new physiological processes that oxytocin has been linked to, such as bone remodelling (Colaianni et al. 2012) and metabolic regulation (Chaves et al. 2013).

Despite these more recent discoveries, the full extent of oxytocin's roles in human physiology and behaviour has yet to be established. A key assumption of oxytocin administration studies is that oxytocin levels in the central nervous system can be experimentally increased using exogenous oxytocin administration. While the intranasal route is by far the most popular, some studies in recent years have also used an oral route (e.g. Zhuang et al. 2022). The upper and posterior regions of the nasal cavity are innervated by both trigeminal and olfactory nerve fibres (Djupesland et al. 2014; Quintana et al. 2018), which can transport oxytocin molecules directly to the brain (Lee et al. 2018; Striepens et al. 2013). Thus, intranasal oxytocin administration can be used to causally investigate oxytocin's role in various neurocognitive and behavioural functions.

Although randomized controlled trials are a common approach for investigating the role of oxytocin in various cognitive functions, understanding the evolutionary history of the oxytocin system in non-human species can shed light on the shared functions of oxytocin across species, as well as on species-specific functions. From this perspective, evaluating the evolutionary history of a gene or a gene family helps better understand its purpose in human physiology and survival (Bateson and Laland 2013; Tinbergen 1963). In this chapter, we will provide an overview of the evolution of the oxytocin receptor via three analytical approaches: sequence homology, synteny and positive selection analysis. We will also provide a non-technical walk-through of a novel analytical pipeline (Sartorius et al. 2024; Theofanopoulou et al. 2021) that can be used to help determine the evolutionary history of genes and, by extension, signalling gene pathways.

14.2 Methods for Investigating Oxytocin Receptor Evolution

Oxytocin receptor molecules have been found in all vertebrate classes and oxytocin receptor-like homologs in some invertebrate classes. Various approaches have been used to identify the presence or absence of this receptor gene across different species and lineages, which we will cover below. We will review a combination of these approaches that we have used to determine, as conclusively as possible, the distribution of the oxytocin receptor across species (Sartorius et al. 2024; Theofanopoulou et al. 2021).

14.2.1 Sequence Homology

An approach that is often used to understand the evolutionary distribution of genes across species is to search for sequence homology, which means whether a genetic sequence in one species can be found to share above-chance similarity to a sequence in another (or several other) species. Research using sequence homology analyses has successfully shown that homologous sequences for the gene coding for the oxytocin receptor (*OXTR*) in humans can be found throughout all major vertebrate lineages, including non-human primate species, rodents, birds, amphibians and teleost fish, extending as far back as to the most ancient group of vertebrates, jawless fish (Sartorius et al. 2024; Theofanopoulou et al. 2021).

There are a variety of available tools that can be used to assess sequence homology. The most widely applied is the Basic Local Alignment Search Tool (BLAST) (Altschul et al. 1990), while there are alternative or adjacent methods, such as DIAMOND (Buchfink et al. 2015) or the BLAST-like alignment tool (BLAT; Kent et al. 2002). BLAST is available both as an online tool (accessible via the NCBI website, https://ncbi.nlm.nih.gov/) and a command line program, and can provide an initial overview of cross-species sequence similarity. It can be run both with nucleotide (BLASTn) and protein/amino acid (BLASTp) sequences. The general mode of function for BLAST is to search for sequence similarity between a reference or query sequence of a species of interest and the genomes or proteomes of target species. This can facilitate the evaluation of sequence similarity between genomic sequences of closely related or distant species and, thus, to homologous sequences and the evolution of a sequence.

We have previously used BLASTp to identify the *OXTR* sequence across 13 species representing all vertebrate lineages (Sartorius et al. 2024). We chose protein-based sequences instead of nucleotide-based sequences because of their availability in several invertebrate species; however, both a nucleotide-based and a protein-based approach are associated with unique benefits that can yield valuable information. The first step for this process was to identify the protein sequence for *OXTR* in humans (*Homo sapiens*) as the reference sequence, and then to 'BLAST' this sequence against the entire proteomes of different species along the vertebrate phylogenetic tree, such as the rhesus macaque (*Macaca mulatta*, primate lineage), chicken (*Gallus gallus*, reptilia lineage), western clawed frog (*Xenopus tropicalis*, tetrapode lineage), zebra fish (*Danio rerio*, actinopterygii lineage) and sea lamprey (*Petromyzon marinus*, jawless vertebrate lineage). This way, BLASTp searches for sequence similarity between the *OXTR* protein sequence in humans and the proteomes of the other species provided protein sequences from protein-coding genes that displayed meaningful similarity with the query sequence of *OXTR*.

A BLAST analysis provides nine different outcome metrics. These outcome metrics include the description of each putative sequence similarity (each 'hit'), the scientific name of the species the hit was found in, the Max Score, Total Score, Query Cover, E value, Percent Identity, Accession Length and Accession. These metrics quantify the strength and credibility of a hit, in descending order of quality. The stronger the hit, the higher the likelihood of true sequence homology. Two of these measures are particularly relevant as they are most often used to assess the strength of a hit. The E value ('expect value') indicates whether two sequences share a higher degree of similarity than that to be *expected* by chance, where a lower E value stands for a higher likelihood of non-chance similarity. Importantly, the magnitude of the E value is dependent on the length of the query sequence (i.e. a hit resulting from the comparison of two shorter sequences on average tends to yield weaker evidence) and its complexity, as well as the completeness of the database. There is no established consensus E value threshold that separates stronger from weaker sequence similarities. This means the value is to be understood as a continuum from stronger to weaker evidence. The Max Score ('maximum score') is the second most relevant value and indicates the highest alignment or 'bit' score, with a higher Max Score indicating a better alignment. The output for the BLAST query for *OXTR* across vertebrate species is displayed in Table 14.1 (compare to findings reported in Sartorius et al. 2024). A common practice is to select a specific threshold (e.g. for Max Score, E value, or both), based on which sequences are considered homologous. However, there are caveats that warrant close attention regarding the selection of hits. One caveat is due to gene nomenclature: sequences with high similarity scores that would be assessed as homologous might carry a name that is ambiguous and misleading (Theofanopoulou et al. 2021; Theofanopoulou and Jarvis 2023). For instance, in a BLASTp query for *OXTR*, homologous hits in

Table 14.1 First ten rows of BLASTp query output for human *OXTR* sequence against different vertebrate species

Description	Scientific name	Max score	Total score	Query cover	E value	Per. Ident	Acc. Len	Accession
Oxytocin receptor [*Pan troglodytes*]	Pan troglodytes	781	781	100%	0.0	98.97%	389	XP_016795834.1
Oxytocin receptor [*Macaca mulatta*]	*Macaca mulatta*	769	769	100%	0.0	97.43%	389	NP_001038197.1
Oxytocin receptor [*Macaca mulatta*]	*Macaca mulatta*	768	768	100%	0.0	97.17%	389	EHH16201.1
Oxytocin receptor [*Bos taurus*]	*Bos taurus*	689	689	99%	0.0	89.49%	391	NP_776559.1
Oxytocin receptor isoform 1 [*Mus musculus*]	*Mus musculus*	659	659	95%	0.0	91.96%	388	NP_001074616.1
Unnamed protein product [*Mus musculus*]	*Mus musculus*	643	643	93%	0.0	92.03%	380	BAE25346.1
Oxytocin receptor [*Ornithorhynchus anatinus*]	*Ornithorhynchus anatinus*	607	607	98%	0.0	79.79%	392	XP_028905593.1
Oxytocin receptor [*Sarcophilus harrisii*]	*Sarcophilus harrisii*	592	592	98%	0.0	76.94%	395	XP_003762470.1
Oxytocin receptor [*Pan troglodytes*]	*Pan troglodytes*	573	573	74%	0.0	98.96%	289	AAM76566.1
Oxytocin receptor [*Gallus gallus*]	*Gallus gallus*	568	568	98%	0.0	74.15%	391	NP_001026740.1

Note. Max score = maximum score, E value = expect value, Per. Ident = percent identity, Acc. Len = accession length

different fish or shark species (not included in this report) could be labelled 'isotocin receptor' or 'glumitocin receptor' instead of 'oxytocin receptor'. At first glance, they appear to not be related to *OXTR*, yet, based on different types of analyses that we will explain below (Sartorius et al. 2024; Theofanopoulou et al. 2021), they are true oxytocin receptors, but have been named according to a different nomenclature tradition. As a result, the same gene that appears as *OXTR* in mammalian species, appears, for instance, as the isotocin receptor in different species such as fish.

Another caveat is due to the fact that BLASTp is not optimally equipped to distinguish similar genes that are orthologous versus paralogous. Following the phylogeny in Fig. 14.1, two or more genes are orthologous when they are the result of speciation of species, rather than gene duplication, leading to distinct species carrying the same gene (Altenhoff et al. 2019). When two genes are paralogous, they resulted from a gene duplication event. Due to their common history, paralogous genes often share similar nucleotide or amino acid sequences and thus appear in BLAST searches, where they can add noise to the results. An example where this is likely the case is for the vasotocin and oxytocin receptors, where, during genomic

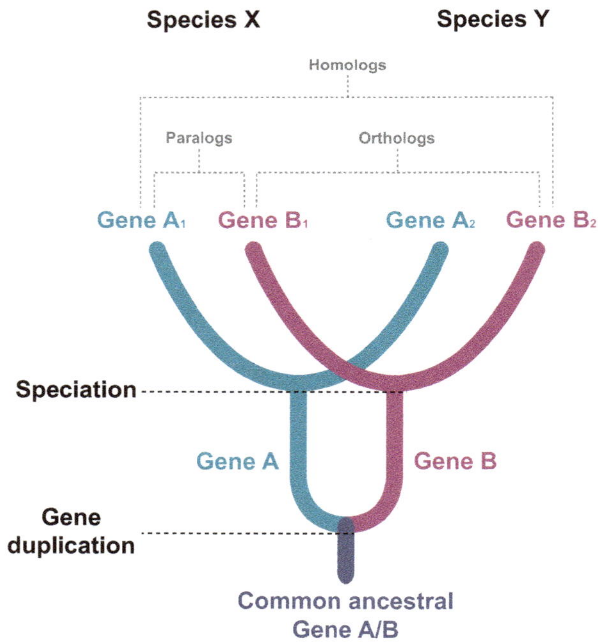

Fig. 14.1 The emergence and relationship of orthologs, paralogs and homologs. From a common ancestral gene A/B in an ancestral species, two distinct genes, A and B, are produced by a gene duplication event. During a later speciation event, in which two new and distinct species have emerged, gene A and gene B continue to be present in the two new species X and Y, as gene A_1 and B_1, and A_2 and B_2, respectively. In the new species X, gene A_1 and gene B_1 are paralogs, and gene A_1 and A_2 between species X and Y are orthologs, as they are the same gene but in different species. All genes A/B, A, B, A_1, A_2, B_1, and B_2 are homologous

duplication around the evolution of vertebrates (Gemmer et al. 2022), two new and discrete sibling genes—vasotocin and oxytocin—emerged from a single common ancestral vasotocin-oxytocin-*like* receptor gene (Mayasich and Clarke 2020; Theofanopoulou et al. 2021). All orthologs and paralogs are, and relate to each other as, homologs. When evaluating BLAST hits (e.g. those reported in Table 14.1), there is not yet an established threshold or criterion that can distinguish orthologous from paralogous *OXTR* sequences. As a result, highly similar sequences (based on BLASTp comparisons) can only be reliably considered to reflect homologous relationships. This was also the case for our BLASTp search. Although the hits in each searched organism were deemed strong, based on common thresholds we encountered across relevant studies (e.g. Liutkeviciute et al. 2016) and the natural limitations of BLAST, we refrained from using the analyses to make any statements on the orthology status of the hits. Thus, another approach is required to disentangle the BLAST results and to uncover orthologous relationships.

Various tools are available to help determine orthologous relationships from BLAST search results, which adopt different methods and assumptions. Commonly used software packages are fully automated, like OrthoFinder (Emms and Kelly 2015, 2019) and OrthoMCL (Li et al. 2003) for large-scale automated orthologous group ('orthogroup') identification. Tools for automated and manual synteny searches are also available and will be discussed next.

14.2.2 Synteny-Based Orthology Identification

Synteny is another approach for gene orthology identification. In the context of comparative evolutionary genomics, when genes 'reside' in conserved gene territories, meaning when they are surrounded by a high number of orthologous genes in the genome across two or more species, they are in synteny (but see the original definition in Passarge et al. 1999). A 'syntenic gene block' can be conserved over several millions of years across diverged species. These syntenic blocks can be of interest on a micro level (microsynteny, i.e. a genomic area focussing on 'only' a few genes around the gene of interest), or on a macro level (macrosynteny, i.e. entire 'segments' of chromosomes containing thousands of protein-coding genes).

Both microsynteny and macrosynteny approaches can serve different research purposes. Macrosynteny, for instance, can be applied to identify the translocation of a specific genomic segment from its chromosome of origin to a target chromosome, while microsynteny can be more suitable for tracing individual genes. Genes in conserved microsyntenic blocks are likely to be orthologous across species. Comparing a manually tractable number of genes (e.g. ten genes) flanking a focal gene against the same gene window around a potential homolog (previously identified with BLAST) may lead to a more robust estimation of gene orthology. In other words, if the genes located in the gene windows under search share homologous genes, namely, genes of the same gene family with high sequence identity, then it is much more likely that the focal genes they are flanking are orthologous, that is, the exact same gene in different species, rather than 'just' homologous. Synteny, especially microsynteny, can therefore be a useful addition to filter BLAST results. Thus, we applied a manual microsynteny analysis to expand our results from the *OXTR* BLASTp queries to be able to make informed statements about the presence of *OXTR* orthologs in the tested vertebrate lineages (Sartorius et al. 2024). The manual mapping is relatively straightforward and does not require coding or computational skills; however, depending on the size of the gene window, the number of species included and the number of genes or genetic loci of interest, the procedure can become time intensive, which should be considered when planning a study. Any established genome browser (e.g. NCBI, UCSC, Ensembl) can be used for microsyntenic searches. Using a ten-gene-window microsynteny analysis in our study, we found a genomic block of the genes surrounding *OXTR* in humans that is relatively conserved across vertebrates up to the sea lamprey, which is among the most ancient vertebrates (Sartorius et al. 2024; Theofanopoulou et al. 2021; Fig. 14.2).

In Fig. 14.2, the ten genes surrounding *OXTR* in the modern human at the top depict a genomic 'microblock' that is almost perfectly conserved across mammalian species (7–310 mya; Sartorius et al. 2024; Theofanopoulou et al. 2021). The microsynteny is slightly less conserved when the relevant mammalian territory is compared with the one present in amniotes (chicken/red junglefowl) and tetrapodes (western clawed frog) and even less conserved when compared with the zebrafish (480 mya) and other species. Traces of the mammalian syntenic block can be found even in the first vertebrate species (i.e. sea lamprey, 540 mya), suggesting that the microsyntenic gene block that can be observed in humans and other non-human mammals dates back to the origin of vertebrates. Here, microsynteny was used to complement the previous BLASTp search and to make informed estimations of the orthology relationship regarding the *OXTR* sequences in humans, the rhesus macaque, chicken, the great white shark and other vertebrates (Sartorius et al. 2024; Theofanopoulou et al. 2021). With microsynteny, one can assume with more

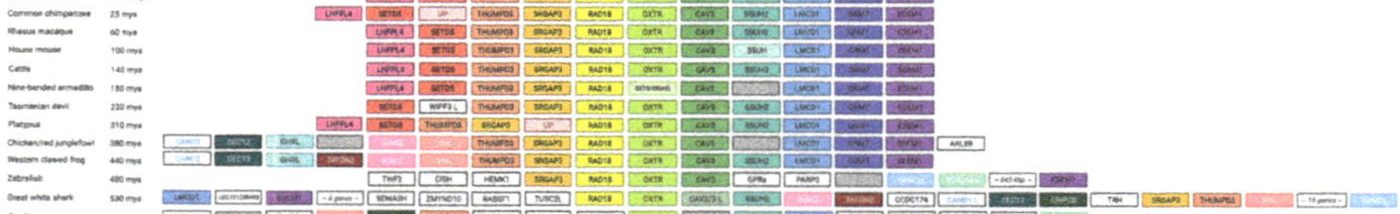

Fig. 14.2 **Vertebrate microsynteny of the oxytocin receptor gene *OXTR*.** The microsynentic block in modern humans consisting of five genes to the left and right flanking *OXTR* is conserved throughout mammalian species (7–310 mya). In amniotes (e.g. chicken, 380 mya) and tetrapodes (e.g. frog, 440 mya), the microsyntenic block is more scattered. Only in the zebrafish (teleost fish, 480 mya) and more ancient species (530 mya and older) the syntenic block visibly less conserved, likely due to whole-genome duplication events. Finally, in the most ancient vertebrates (e.g. sea lamprey, 540 mya), traces of the microsyntenic block are still discernible but markedly less coherent. Common names of the species are given in the outer left column, followed by the approximate branch split times in mya, and the genes in the microsyntenic block. A rectangle represents one gene (abbreviated gene label provided). Grey, empty rectangles indicate potential missing genes. Non-coloured rectangles indicate no reoccurrence of a gene in the displayed species within microsyntenic distance. mya = Million years ago; UP = Uncharacterized protein. Figure modified from Sartorius et al. (2024)

confidence than from using BLASTp alone that orthologs of *OXTR* are present in most vertebrates and that the receptor gene is, hence, not specific to gnathostomes but arose earlier, presumably at least before jawless vertebrate evolution.

One limitation of synteny analysis worth noting is that it depends on the sequencing status and the annotation of the genomes analysed. While the identification of synteny in species like the rhesus macaque or the great white shark relies on genomes that are complete and well annotated, in other species, poor genome quality or annotation can influence the quality of synteny analyses. Such genomes are often not sequenced at the chromosome level; hence, genes or genetic blocks are scattered across incoherent scaffolds, rendering placing genes in the correct order difficult. Thus, (micro-)synteny to bolster BLASTp query results can only be used meaningfully in species whose genomes are well-enough sequenced and annotated.

Other than the manual synteny methodology we reviewed here, there are also tools and programs for more automated synteny analyses, which can be accessed online. We refer the reader to Synima (Farrer 2017), SynFind (Tang et al. 2015) and SynMap (Haug-Baltzell et al. 2017) for macrosynteny and Simple Synteny (Veltri et al. 2016) or Genomicus (Muffato et al. 2010) for microsynteny. Theofanopoulou et al. (2021) and Theofanopoulou and Jarvis (2023) used a combination of manual and automated tools (SynFind, SynMap and Genomicus) in their identification of the evolutionary history of the oxytocin and vasopressin/vasotocin ligands and receptors. Since most of these automated tools are based on a combination of BLAST, phylogenetic and gene order conservation analyses, in our experience, they can also lead to paralogous (not orthologous) gene hits and often require manual confirmation. Until robust automated tools are in place, we recommend that these automated tools are used with caution when it comes to understanding the evolutionary history of single genes.

14.2.3 Positive Selection Analysis

According to the above two criteria, orthologs of *OXTR* can be found throughout the vertebrate animal kingdom, using combined BLAST and microsynteny analyses (Sartorius et al. 2024; Theofanopoulou et al. 2021). In addition to identifying homologous or orthologous genes, examining whether genes were positively selected during a specific time period in evolution can improve our understanding of the evolutionary trajectory of genes. When a gene is under positive selection, a common interpretation is that the trait(s) the gene encodes could have been beneficial, or at least functional, during that time in evolution. Likewise, when a gene is under positive selection in a species or population with known phenotypic differences compared to others, speculations can be drawn regarding the involvement of the positively selected genes in those phenotypes.

There are several positive selection identification tools that can serve more hypothesis-driven or exploratory purposes. An established tool with which to perform positive selection analyses is the software package PAML (codeml; Yang 1997, 2007). By providing preprocessed multiple sequence alignments of a gene and a phylogenetic tree, PAML can provide estimates on whether a gene of interest was under positive selection in an a priori specified foreground evolutionary branch (e.g. the jawed vertebrate ancestral branch) based on a likelihood ratio test. PAML is, thus, a hypothesis-driven phylogenetic analysis tool that requires the user to have a hypothesis regarding the branch where positive selection signals are expected. We refer the interested reader to specific studies that highlight the variety of PAML's application possibilities (e.g. Álvarez-Carretero et al. 2023; Jeffares et al. 2015).

When studies are more exploratory in nature, positive selection analysis using aBSREL (Smith et al. 2015) is generally more suitable compared to PAML. aBSREL, which is included in the HyPhy software package (Kosakovsky Pond et al. 2020), stands for 'adaptive branch-site random effects likelihood' and is a model that allows one to test for episodic diversifying positive selection without the specification of a foreground branch. Briefly, it compares a full alternative model against a null model and subsequently, like codeml from PAML, runs a likelihood ratio test. aBSREL performs multiple tests for several branches and nodes at the same time, making it more time-efficient. However, this comes at the cost of reduced statistical power and the need to control strictly for multiple comparisons. aBSREL automatically corrects raw *p*-values for multiple comparisons with the Holm-Bonferroni correction method. Since this correction method tends to be rather conservative, with an increasing number of branches tested, it is worth considering whether another correction method (e.g. false discovery rate) would be more suitable.

Using aBSREL on 13 different vertebrate species, representing the major lineages and branches during vertebrate evolution, and 10 nodes (a branching point splitting a branch into 2 novel species lineages from a last common ancestor (Baum 2008)) we showed that *OXTR* was under positive selection in the ancestral marsupial branch leading to the extant Tasmanian devil, approximately 220 mya (Sartorius et al. 2024). This is an interesting finding considering that marsupials are known for having a particularly complex and highly adaptive lactation system, since milk composition can drastically change in some marsupial species throughout the lactation cycle (Brennan et al. 2007; Lefèvre et al. 2010). Indeed, oxytocin plays a fundamental role in milk let-down (Ruis et al. 1981). Before aBSREL can be run on a set of genes and branches, the input data must undergo several preprocessing steps, which we describe next. Performing these steps is critical for the final outcome of the analysis; thus it is briefly mentioned here, with a more detailed description available elsewhere (Sartorius et al. 2024). In short, for downloading the orthologous protein *and* nucleotide sequences for the gene(s) of interest in the species of interest, any recognized gene bank can be used, such as NCBI or Ensembl. The grouping or concatenation of the downloaded protein and nucleotide sequences can be done manually or using a script (see Álvarez-Carretero et al. 2023). For the alignment of the concatenated protein sequences, we recommend using MUSCLE (Edgar 2004), a software package for the alignment of multiple sequences. Other software packages include the Clustal suite (i.e. ClustalOmega, ClustalW, ClustalX; Chenna et al. 2003; Higgins and Sharp 1988) or MAFFT (Katoh et al. 2002).

The resulting multiple protein sequence alignment must be translated into a codon-based alignment file. PAL2NAL (Suyama et al. 2006) is widely applied software for translation from protein into codon-based alignment; otherwise, MACSE can be an alternative approach (Ranwez et al. 2011, 2018). Once a codon-based multiple alignment file is in place, it needs to undergo two further processing steps—adjusting the sequence identifiers so that they match the branch labels of the phylogenetic tree later and the conversion to PHYLIP format. For the former, this again can be done manually or with a script (see Sartorius et al. 2024). For the latter, the software TriFusion (Silva 2018) can be used. Finally, a phylogenetic tree needs to be provided. Sartorius et al. (2024) utilized TimeTree (Kumar et al. 2022) for this purpose. Another commonly used program package is OrthoFinder which, next to identifying orthogroups as mentioned above, also generates phylogenetic trees (Emms and Kelly 2015, 2019). Before using any phylogenetic tree for the analysis, the branch lengths should be removed, and the species names should match the names in the codon-based multiple alignment file. It is important to note that, here again, the results of the positive selection analysis depend on the sequences that are being used. The results are only as good as the quality of the input sequences. Furthermore, the results may change depending on which species are included and on the number of included species.

14.3 Summary

Here, we described three methods and relevant software packages (sequence homology via BLAST, manual microsynteny, positive selection analysis via aBSREL (preprocessing with MUSCLE, PAL2NAL, TriFusion, TimeTree)) to illustrate one approach for quantifying the evolution of *OXTR*, which we have used in previous work (Sartorius et al. 2024). With sequence similarity searching using BLASTp, we first found that there is meaningful sequence homology between the human *OXTR* sequence and sequences in different vertebrate species up to cyclostomes. However, it was unclear whether the homologous sequences could be considered orthologous. Thus, the BLASTp analysis results were narrowed down with microsynteny, which revealed robust evidence for which homologs were true orthologs, confirming that orthologs of the human *OXTR* were likely to be already present in the ancestor of jawless vertebrates, as represented by the sea lamprey (Sartorius et al. 2024; Theofanopoulou et al. 2021). Lastly, to better understand the history of the vertebrate *OXTR* orthologs, an exploratory positive selection analysis was performed, investigating whether *OXTR* was 'favoured' by evolution at any point in time during the evolution of vertebrates. This revealed, among other things, significant episodic diversifying positive selection in the ancestral branch leading to the Tasmanian devil (Sartorius et al. 2024).

This type of analysis pipeline can be performed for entire signalling systems by repeating the analysis steps for all genes in a signalling system, thus gaining a better understanding of the evolution of an entire signalling gene pathway. We do not claim that the three methods described here are exhaustive, nor are the software programs detailed above. Wherever possible and relevant, common alternatives were mentioned with an explanation of the complementary questions they can address.

References

Altenhoff AM, Glover NM, Dessimoz C (2019) Inferring orthology and paralogy. In: Anisimova M (ed) Evolutionary genomics: statistical and computational methods, methods in molecular biology. Springer, New York, pp 149–175

Altschul SF, Gish W, Miller W, Myers EW, Lipman DJ (1990) Basic local alignment search tool. J Mol Biol 215:403–410. https://doi.org/10.1016/S0022-2836(05)80360-2

Alvares GA, Quintana DS, Whitehouse AJ (2017) Beyond the hype and hope: critical considerations for intranasal oxytocin research in autism spectrum disorder. Autism Res 10:25–41. https://doi.org/10.1002/aur.1692

Álvarez-Carretero S, Kapli P, Yang Z (2023) Beginner's guide on the use of PAML to detect positive selection. Mol Biol Evol 40:msad041. https://doi.org/10.1093/molbev/msad041

Audunsdottir K, Sartorius AM, Kang H, Glaser BD, Boen R, Nærland T, Alaerts K, Kildal ES-M, Westlye LT, Andreassen OA, Quintana D (2023) The effects of oxytocin administration on social and routinized behaviors in autism: a preregistered systematic review and meta-analysis. PsyArXiv. https://doi.org/10.31234/osf.io/nk2rh

Bateson P, Laland KN (2013) Tinbergen's four questions: an appreciation and an update. Trends Ecol Evol 28:712–718. https://doi.org/10.1016/j.tree.2013.09.013

Baum D (2008) Reading a phylogenetic tree: the meaning of monophyletic groups. Nat Educ 1:190

Brennan AJ, Sharp JA, Lefevre C, Topcic D, Auguste A, Digby M, Nicholas KR (2007) The tammar wallaby and fur seal: models to examine local control of lactation. J Dairy Sci 90:E66–E75. https://doi.org/10.3168/jds.2006-483

Buchfink B, Xie C, Huson DH (2015) Fast and sensitive protein alignment using DIAMOND. Nat Methods 12:59–60. https://doi.org/10.1038/nmeth.3176

Chaves VE, Tilelli CQ, Brito NA, Brito MN (2013) Role of oxytocin in energy metabolism. Peptides 45:9–14. https://doi.org/10.1016/j.peptides.2013.04.010

Chenna R, Sugawara H, Koike T, Lopez R, Gibson TJ, Higgins DG, Thompson JD (2003) Multiple sequence alignment with the Clustal series of programs. Nucleic Acids Res 31:3497–3500. https://doi.org/10.1093/nar/gkg500

Colaianni G, Sun L, Benedetto AD, Tamma R, Zhu L-L, Cao J, Grano M, Yuen T, Colucci S, Cuscito C, Mancini L, Li J, Nishimori K, Bab I, Lee H-J, Iqbal J, Young WS, Rosen C, Zallone A, Zaidi M (2012) Bone marrow oxytocin mediates the anabolic action of estrogen on the skeleton. J Biol Chem 287:29159–29167. https://doi.org/10.1074/jbc.M112.365049

Dale HH (1906) On some physiological actions of ergot. J Physiol 34:163–206. https://doi.org/10.1113/jphysiol.1906.sp001148

Djupesland PG, Messina JC, Mahmoud RA (2014) The nasal approach to delivering treatment for brain diseases: an anatomic, physiologic, and delivery technology overview. Ther Deliv 5:709–733. https://doi.org/10.4155/tde.14.41

Edgar RC (2004) MUSCLE: multiple sequence alignment with high accuracy and high throughput. Nucleic Acids Res 32:1792–1797. https://doi.org/10.1093/nar/gkh340

Emms DM, Kelly S (2015) OrthoFinder: solving fundamental biases in whole genome comparisons dramatically improves orthogroup inference accuracy. Genome Biol 16:157. https://doi.org/10.1186/s13059-015-0721-2

Emms DM, Kelly S (2019) OrthoFinder: phylogenetic orthology inference for comparative genomics. Genome Biol 20:238. https://doi.org/10.1186/s13059-019-1832-y

Farrer RA (2017) Synima: a Synteny imaging tool for annotated genome assemblies. BMC Bioinformatics 18:507. https://doi.org/10.1186/s12859-017-1939-7

Gemmer A, Mirkes K, Anneser L, Eilers T, Kibat C, Mathuru A, Ryu S, Schuman E (2022) Oxytocin receptors infuence the development and maintenance of social behavior in zebrafsh (Danio rerio). Sci Rep 12:4322. https://doi.org/10.1038/s41598-022-07990-y

Gross J, De Dreu CKW (2017) Oxytocin conditions trait-based rule adherence. Soc Cogn Affect Neurosci 12:427–435. https://doi.org/10.1093/scan/nsw138

Haug-Baltzell A, Stephens SA, Davey S, Scheidegger CE, Lyons E (2017) SynMap2 and SynMap3D: web-based whole-genome synteny browsers. Bioinformatics 33:2197–2198. https://doi.org/10.1093/bioinformatics/btx144

Higgins DG, Sharp PM (1988) CLUSTAL: a package for performing multiple sequence alignment on a microcomputer. Gene 73:237–244. https://doi.org/10.1016/0378-1119(88)90330-7

Jeffares DC, Tomiczek B, Sojo V, dos Reis M (2015) A beginners guide to estimating the non-synonymous to synonymous rate ratio of all protein-coding genes in a genome. In: Peacock C (ed) Parasite genomics protocols, methods in molecular biology. Springer, New York, pp 65–90. https://doi.org/10.1007/978-1-4939-1438-8_4

Jurek B, Neumann ID (2018) The oxytocin receptor: from intracellular signaling to behavior. Physiol Rev 98:1805–1908. https://doi.org/10.1152/physrev.00031.2017

Kapetaniou GE, Reinhard MA, Christian P, Jobst A, Tobler PN, Padberg F, Soutschek A (2021) The role of oxytocin in delay of gratification and flexibility in non-social decision making. eLife 10:e61844. https://doi.org/10.7554/eLife.61844

Katoh K, Misawa K, Kuma K, Miyata T (2002) MAFFT: a novel method for rapid multiple sequence alignment based on fast Fourier transform. Nucleic Acids Res 30:3059–3066. https://doi.org/10.1093/nar/gkf436

Kenkel WM (2019) Corpus colossal: a bibliometric analysis of neuroscience abstracts and impact factor. Front Integr Neurosci 13. https://doi.org/10.3389/fnint.2019.00018

Kennedy DP, Adolphs R (2012) The social brain in psychiatric and neurological disorders. Trends Cogn Sci 16:559–572. https://doi.org/10.1016/j.tics.2012.09.006

Kent WJ, Sugnet CW, Furey TS, Roskin KM, Pringle TH, Zahler AM, Haussler D (2002) The human genome browser at UCSC. Genome Res 12:996–1006. https://doi.org/10.1101/gr.229102

Kosakovsky Pond SL, Poon AFY, Velazquez R, Weaver S, Hepler NL, Murrell B, Shank SD, Magalis BR, Bouvier D, Nekrutenko A, Wisotsky S, Spielman SJ, Frost SDW, Muse SV (2020) HyPhy 2.5—a customizable platform for evolutionary hypothesis testing using phylogenies. Mol Biol Evol 37:295–299. https://doi.org/10.1093/molbev/msz197

Kumar S, Suleski M, Craig JM, Kasprowicz AE, Sanderford M, Li M, Stecher G, Hedges SB (2022) TimeTree 5: an expanded resource for species divergence times. Mol Biol Evol 39:msac174. https://doi.org/10.1093/molbev/msac174

Lane A, Luminet O, Nave G, Mikolajczak M (2016) Is there a publication bias in behavioral intranasal oxytocin research on humans? Opening the file drawer of one lab. J Neuroendocrinol 28. https://doi.org/10.1111/jne.12384

Lee MR, Scheidweiler KB, Diao XX, Akhlaghi F, Cummins A, Huestis MA, Leggio L, Averbeck BB (2018) Oxytocin by intranasal and intravenous routes reaches the cerebrospinal fluid in rhesus macaques: determination using a novel oxytocin assay. Mol Psychiatry 23:115–122. https://doi.org/10.1038/mp.2017.27

Lefèvre CM, Sharp JA, Nicholas KR (2010) Evolution of lactation: ancient origin and extreme adaptations of the lactation system. Annu Rev Genomics Hum Genet 11:219–238. https://doi.org/10.1146/annurev-genom-082509-141806

Leng G, Leng RI (2021) Oxytocin: a citation network analysis of 10 000 papers. J Neuroendocrinol 33:e13014. https://doi.org/10.1111/jne.13014

Li L, Stoeckert CJ, Roos DS (2003) OrthoMCL: identification of ortholog groups for eukaryotic genomes. Genome Res 13:2178–2189. https://doi.org/10.1101/gr.1224503

Liutkeviciute Z, Koehbach J, Eder T, Gil-Mansilla E, Gruber CW (2016) Global map of oxytocin/vasopressin-like neuropeptide signalling in insects. Sci Rep 6:39177. https://doi.org/10.1038/srep39177

Mayasich SA, Clarke BL (2020) Vasotocin and the origins of the vasopressin/oxytocin receptor gene family. In: Litwack G (ed) Vitamins and hormones, vasopressin. Academic Press, pp 1–27. https://doi.org/10.1016/bs.vh.2019.08.018

Muffato M, Louis A, Poisnel C-E, Crollius HR (2010) Genomicus: a database and a browser to study gene synteny in modern and ancestral genomes. Bioinformatics 26:1119–1121. https://doi.org/10.1093/bioinformatics/btq079

Ott I, Scott JC (1910) The action of infundibulin upon the mammary secretion. Proc Soc Exp Biol Med 8:48–49. https://doi.org/10.3181/00379727-8-27

Passarge E, Horsthemke B, Farber RA (1999) Incorrect use of the term synteny. Nat Genet 23:387–387. https://doi.org/10.1038/70486

Quintana DS (2020) Most oxytocin administration studies are statistically underpowered to reliably detect (or reject) a wide range of effect sizes. Compr Psychoneuroendocrinol 4:100014. https://doi.org/10.1016/j.cpnec.2020.100014

Quintana DS, Guastella AJ (2020) An allostatic theory of oxytocin. Trends Cogn Sci 24:515–528. https://doi.org/10.1016/j.tics.2020.03.008

Quintana DS, Smerud KT, Andreassen OA, Djupesland PG (2018) Evidence for intranasal oxytocin delivery to the brain: recent advances and future perspectives. Ther Deliv 9:515–525. https://doi.org/10.4155/tde-2018-0002

Ranwez V, Harispe S, Delsuc F, Douzery EJP (2011) MACSE: multiple alignment of coding SEquences accounting for frameshifts and stop codons. PLoS One 6:e22594. https://doi.org/10.1371/journal.pone.0022594

Ranwez V, Douzery EJP, Cambon C, Chantret N, Delsuc F (2018) MACSE v2: toolkit for the alignment of coding sequences accounting for frameshifts and stop codons. Mol Biol Evol 35:2582–2584. https://doi.org/10.1093/molbev/msy159

Ruis H, Rolland R, Doesburg W, Broeders G, Corbey R (1981) Oxytocin enhances onset of lactation among mothers delivering prematurely. BMJ 283:340–342. https://doi.org/10.1136/bmj.283.6287.340

Sartorius AM, Rokicki J, Birkeland S, Bettella F, Barth C, de Lange A-MG, Haram M, Shadrin A, Winterton A, Steen NE, Schwarz E, Stein DJ, Andreassen OA, van der Meer D, Westlye LT, Theofanopoulou C, Quintana DS (2024). An evolutionary timeline of the oxytocin signaling pathway. Commun Biol 7, 471. https://doi.org/10.1038/s42003-024-06094-9

Silva DNPR (2018) Phylogenomic and population genomic insights on the evolutionary history of coffee leaf rust within the rust fungi (doctoral thesis)

Smith MD, Wertheim JO, Weaver S, Murrell B, Scheffler K, Kosakovsky Pond SL (2015) Less is more: an adaptive branch-site random effects model for efficient detection of episodic diversifying selection. Mol Biol Evol 32:1342–1353. https://doi.org/10.1093/molbev/msv022

Striepens N, Kendrick KM, Hanking V, Landgraf R, Wüllner U, Maier W, Hurlemann R (2013) Elevated cerebrospinal fluid and blood concentrations of oxytocin following its intranasal administration in humans. Sci Rep 3:3440. https://doi.org/10.1038/srep03440

Suyama M, Torrents D, Bork P (2006) PAL2NAL: robust conversion of protein sequence alignments into the corresponding codon alignments. Nucleic Acids Res 34:W609–W612. https://doi.org/10.1093/nar/gkl315

Tang H, Bomhoff MD, Briones E, Zhang L, Schnable JC, Lyons E (2015) SynFind: compiling syntenic regions across any set of genomes on demand. Genome Biol Evol 7:3286–3298. https://doi.org/10.1093/gbe/evv219

Theofanopoulou C, Jarvis ED (2023) Reply to: the case for standardizing gene nomenclature in vertebrates. Nature 614:E33–E36. https://doi.org/10.1038/s41586-022-05634-9

Theofanopoulou C, Gedman G, Cahill JA, Boeckx C, Jarvis ED (2021) Universal nomenclature for oxytocin–vasotocin ligand and receptor families. Nature 592:747–755. https://doi.org/10.1038/s41586-020-03040-7

Tinbergen N (1963) On aims and methods of ethology. Ethology 20:410–433

Veltri D, Wight MM, Crouch JA (2016) SimpleSynteny: a web-based tool for visualization of microsynteny across multiple species. Nucleic Acids Res 44:W41–W45. https://doi.org/10.1093/nar/gkw330

Walum H, Waldman ID, Young LJ (2016) Statistical and methodological considerations for the interpretation of intranasal oxytocin studies. Biol Psychiatry 79:251–257. https://doi.org/10.1016/j.biopsych.2015.06.016

Yang Z (1997) PAML: a program package for phylogenetic analysis by maximum likelihood. Bioinformatics 13:555–556. https://doi.org/10.1093/bioinformatics/13.5.555

Yang Z (2007) PAML 4: phylogenetic analysis by maximum likelihood. Mol Biol Evol 24:1586–1591. https://doi.org/10.1093/molbev/msm088

Zhuang Q, Zheng X, Yao S, Zhao W, Becker B, Xu X, Kendrick KM (2022) Oral administration of oxytocin, like intranasal administration, decreases top-down social attention. Int J Neuropsychopharmacol 25:912–923. https://doi.org/10.1093/ijnp/pyac059

Nonapeptide Evolution and the Regulation of Social Behaviour in Teleost Fish: From Molecules to Sociality

15

Kyriacos Kareklas, Pol Sorigue, and Rui F. Oliveira

Abstract

The evolution of the nonapeptide system in vertebrates is characterised by gene duplication events that gave rise to a separation of behavioural functions between two signalling pathways, oxytocin and vasotocin. This is particularly marked in fish, which present the greatest diversity in social behaviour among vertebrates. Notably, the conservation of nonapeptide activity across homologous brain areas offers an opportunity to identify the mechanisms underlying social evolution in this group of animals. Here we describe the evolutionary history of the nonapeptide ligands and their receptors in fish and discuss up-to-date evidence of their role in social phenotypes. We present evidence for a broad implication of nonapeptides in the variation in social behaviour both across and within species. Much of the evidence is derived from pharmacological manipulations, but recently genetic tools have also been developed for model organisms, such as zebrafish and medaka. The time is now right to expand genetic approaches from model organisms in more ecologically relevant models in the field.

Keywords

Oxytocin · Vasotocin · Nonapeptide · Vertebrates · Ligands

Kyriacos Kareklas and Pol Sorigue contributed equally to this study and are listed by alphabetical order.

K. Kareklas · P. Sorigue
Gulbenkian Institute for Molecular Medicine, Oeiras, Portugal

R. F. Oliveira (✉)
ISPA - University Institute for Psychological Social and Life Sciences, Lisbon, Portugal

GIMM - Gulbenkian Institute for Molecular Medicine, Oeiras, Portugal
e-mail: ruiol@ispa.pt

V. Grinevich, R. F. Oliveira (eds.), *Evolutionary and Comparative Neuroendocrinology*, Masterclass in Neuroendocrinology 17,
https://doi.org/10.1007/978-3-031-80209-6_15

15.1 Introduction

Ray-finned fish are the largest and most phenotypically diverse group of vertebrates, consisting of more than 30,000 species and exhibiting a range of adaptive social behaviours across species or genera (Ravi and Venkatesh 2018; Facey et al. 2022). For example, some African cichlids exhibit mouthbrooding, where eggs are collected and nurtured in the female's mouth until their free-swimming larval stage (Solomon-Lane et al. 2022), while Siamese fighting fish males (*Betta splendens*) aggressively defend territory where they construct bubble-nests for their offspring (Kareklas et al. 2022; Oliveira et al. 2022). Some Central American convict cichlids are monogamous and perform biparental care, and some African cichlids exhibit cooperative breeding by third parties (Oldfield and Hofmann 2011). In some species dense aggregations are sometimes favoured (Almeida et al. 2012), while in others smaller groups are defined by social hierarchies (Huffman et al. 2015), or by aggressive males maintaining territories for polygynous mating (Oldfield et al. 2013). Further still, some fish species exhibit interspecific cooperative interactions, like the cleaning mutualism between cleaner wrasse and their client reef fishes (Soares et al. 2012), or like joint predator inspection in guppies (Dimitriadou et al. 2022). This diversity of social phenotypes in fish is paralleled by the evolution of specific neural circuits, as indicated by specific patterns of localised brain activity and functional connectivity across brain regions and neuroendocrine systems controlling downstream physiological outputs (Ogawa et al. 2021). Notably, all these examples of social behaviours are associated with one such mechanism that is highly conserved across vertebrates—the nonapeptide signalling system. Here, we consider the evolution and functions of nonapeptides, from central neuromodulatory functions to neuroendocrine control of peripheral body functions (Banerjee et al. 2017). The system is composed of two peptides, vasotocin- and oxytocin-like peptides, and respective receptors distributed in the brain and the periphery and whose diversity has been extended by gene duplication events in fish (Ocampo Daza et al. 2022). In the following sections, we discuss the evolution of the ligands, their receptors and their binding dynamics in fish and how these contribute to the neural circuitry of teleost social behaviour. We also examine evidence of how this circuitry has been functionally defined by its involvement in affiliative and aggressive social phenotypes across reproductive, parental, grouping and interspecific contexts. In doing so, we adopt the nomenclature proposed by Theofanopoulou et al. (2021) and supported by other studies (e.g. Ocampo Daza et al. 2022), where vasotocin (*VT*) refers to vasotocin orthologues (vasotocin and vasopressin) and oxytocin (*OT*) refers to all oxytocin orthologues (mesotocin, isotocin and other vertebrate lineage-specific terms). Similarly, receptors of OT-like and VT-like peptides are generally named *OTR* and *VTR*, respectively.

15.2 Evolution of Nonapeptide Signalling Mechanisms

15.2.1 Structure and Evolution of the Ligands

Nonapeptides are encoded in larger precursor genes that typically span over 1000–2000 base pairs in the genome. The protein precursor is usually slightly larger than 100 amino acids long, and it undergoes post-translational processing and cleavage before generating the final nonapeptide hormone. Generally, these precursors comprise a signal peptide, which is removed from the complex during the translation in the endoplasmic reticulum, the hormone moiety and a neurophysin domain (Ivell and Richter 1984). In vertebrate VT, as well as in teleost OT, the precursor gene also contains a copeptin domain.

The precursor genes of both *VT* and *OT* are found in tandem in most vertebrate species, with a head-tail orientation, except in mammals where they exhibit a head-head orientation. However, in teleosts the genes are often found in different chromosomes, or, when they are in the same chromosome, they are separated by other genes (Gwee et al. 2009). It is widely accepted that the second nonapeptide arose after a segment duplication, which explains the tandem distribution in most vertebrates, and later it would have undergone a translocation in a basal teleost, supported by the fact that most teleost species have both nonapeptide genes in different chromosomes (Banerjee et al. 2017; Gwee et al. 2009).

Nonapeptide genes generally contain three exons. The first one comprises the 5'UTR region, the signal peptide, hormone moiety and N-terminal neurophysin. The second is the most conserved part of the neurophysin. The third exon produces the copeptin. Although this is the most common structure, several teleost species have a slightly different arrangement of the oxytocin gene, with a fourth exon coding for a part of the 5'UTR in the oxytocin precursor gene, and in other species such as the white sucker (*Catostomus commersonii*), containing a single exon (Figueroa et al. 1989; Kanda et al. 2003). The amino acid substitution rates differ substantially between each protein domain, indicating a different selective pressure likely due to their functionality.

The origin of the VT/OT system dates to 600 million years ago, before vertebrates appeared. VT/OT-like peptides have been found in several nonchordate phyla: Tunicata (Kawada et al. 2008), Echinodermata (Elphick and Rowe 2009), Nematoda (Beets et al. 2012; Garrison et al. 2012), Mollusca (Cruz et al. 1987; Reich 1992), Annelida (Oumi et al. 1996) and Arthropoda (Gruber 2014; Gruber and Muttenthaler 2012; Rémy et al. 1979). In most cases, the neuropeptide has a slightly different amino acid sequence, and different names have been attributed to it according to phylogenetic characterisations. However, there is a predominantly conserved structure across clades (Elphick et al. 2018). Invertebrates have only one copy of a VT/OT-like peptide and its cognate receptor (Elphick et al. 2018). It was not until the emergence of vertebrates that duplications of the neuropeptide and receptors genes occurred. All major vertebrate clades have two nonapeptides except for the cyclostomes (Gwee et al. 2009). It has been commonly hypothesised that *VT* is the ancestral gene that underwent a gene duplication, followed by the accumulation of mutations and eventually giving rise to *OT* (Banerjee et al. 2017; Hoyle 1999; Theofanopoulou et al. 2021). This is supported by the presence of a VT-like peptide and absence of OT-like peptide at the base of the gnathostome lineage, in the jawless hagfishes and lampreys (Suzuki et al. 1995; Theofanopoulou et al. 2021).

Nonapeptides possess a cyclic structure formed by 6 amino acids and a three-amino acid tail (Fig. 15.1a, b). Positions 1 and 6 are cysteine residues that form a disulphide bridge, conferring the cyclic structure, while position 9 contains a glycine residue in all vertebrates (Banerjee et al. 2017; Grigor'eva and Golubeva 2010; Wircer et al. 2016). Positions 3, 4 and 8 are the more variable ones across all clades, while 1, 5, 6, 7 and 9 present almost no substitution (Banerjee et al. 2017; Fig. 15.1a). The nature of the amino acid at position 8 has led to the current classification of nonapeptide families, namely, basic nonapeptides (VT family) and neutral nonapeptides (OT family). In teleosts, amino acid sequences of the VT and OT nonapeptides are homologous, except for positions 4 and 8 (Gimpl and Fahrenholz 2001; Fig. 15.1a).

In line with the two ligand families, two groups of receptors have evolved, suggesting that position 8 might be key to ligand selectivity. From an ancient VT peptide, several new amino acid combinations have been found across vertebrates, with the neutral family being the most diverse, with the accumulation of several lineage-specific mutations providing a wider range of sequences (Banerjee et al. 2017; Fig. 15.1a). However, all of them keep the abovementioned conserved amino acids, suggesting a key role in the nonapeptide structure and function.

VT is very conserved across nonmammalian vertebrates and most species studied so far have the same sequence, whereas some substitutions in positions 3 or 8

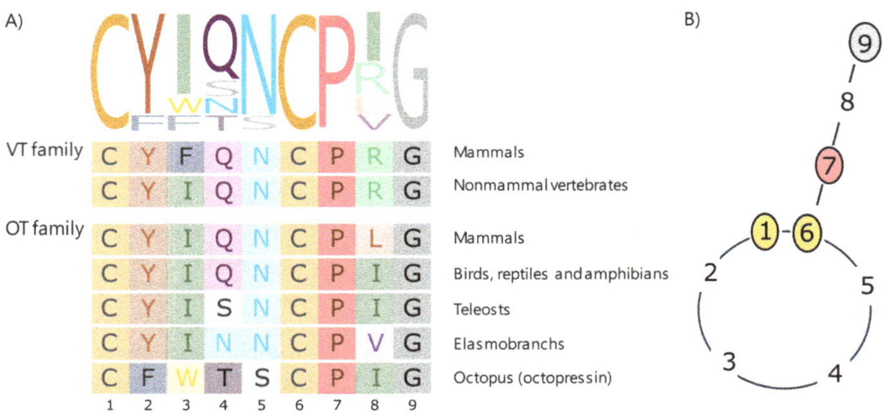

Fig. 15.1 (**a**) Nonapeptide sequences (OT: oxytocin; VT: vasotocin) of representatives of the major vertebrate clades. Only the most common sequence in each phylum is shown. However, some species or subclades have a different sequence. This is particularly the case of elasmobranch (cartilaginous fish), in which more than one OT copy per species can be found with a different amino acid sequence. (**b**) Schematic secondary structure of the nonapeptides. The ring structure is locked by a disulphide bond between the two conserved cysteine residues. The highlighted amino acids show the most conserved positions across species in both nonapeptide families

occurred in the mammalian clade. The most known mammalian VT-like sequence, named vasopressin (VP), underwent a mutation from isoleucine to phenylalanine in position 3 (Richter and Schmale 1983; Fig. 15.1). OT shows higher variability than VT across clades. Many different names have been attributed to the OT-like peptide across clades. Within vertebrates, teleost OT-like peptide has been named isotocin (IT), while mesotocin (MT) has been used for amphibians, reptiles, birds and some marsupials. The term oxytocin was first only used for eutherian mammals, although some marsupials have MT (Banerjee et al. 2017). All these different peptides differ mainly in the amino acids in positions 3, 4 and 8. Typically most species have only two nonapeptides (one of the neutral family and one of the basic family), although some elasmobranchs (cartilaginous fish) have been reported to have more than one neutral nonapeptide (Chauvet et al. 1994; Hyodo et al. 2004; Yves et al. 1988). In this case, the appearance of a new urea-based osmoregulation system, as opposed to the common ion-based, might have released the pressure on nonapeptides to control of ionic homeostasis, leading to different nonapeptide sequences (Acher 1996).

The neurophysin shows conservation across the animal kingdom, being present both in invertebrates and vertebrates (Van Kesteren et al. 1995). The neurophysin has 14 cysteine residues that are very conserved and is linked to the hormone by a glycine-lysine-arginine (GKR) residue, which is a key site for a three-step enzymatic cleavage and amidation of the hormone (Fig. 15.1). Two different neurophysins have been described, type I in the oxytocin precursor and type II in the vasotocin/vasopressin precursor. These carrier proteins are in charge to bind, translocate and store the nonapeptides in the granules before their release. Moreover, their multiple cysteine residues are responsible for the correct folding of the precursor, favouring the post-translational processing and cleavage (De Bree and Burbach 1998; Grigor'eva and Golubeva 2010).

In addition, vasotocin, as well as the oxytocin precursors in teleost, contain a copeptin domain at the carboxyl terminal, the function of which remains unclear. It has been suggested to be a prolactin-releasing factor (Flores et al. 2007; Hyde et al. 1989; Nagy et al. 1988). Copeptin normally has a conserved leucine-rich core and conserved glycosylation and cleavage sites, which facilitate the removal of copeptin from the neurophysin (Acher 1996; De Bree and Burbach 1998). However, teleosts seem to have lost the glycosylation and cleavage sites, presumably leading to longer neurophysin peptides in both neurohypophysial hormones (Banerjee et al. 2015).

15.2.2 Structure and Evolution of the Receptors

Oxytocin and vasotocin receptors (OTRs, VTRs) are G-protein coupled receptors with seven transmembrane α-helical domains, with three intracellular loops and three extracellular loops (Fig. 15.2). The transmembrane domains form a ring structure that results in an extracellular pocket-like conformation, which confers the binding site for the circular part of the nonapeptide (Mouillac et al. 1995; Waltenspühl et al. 2022). OTR and VTR show high sequence and structural similarity, but subfunctionalisation of duplicated receptors led to a broader range of functions, often

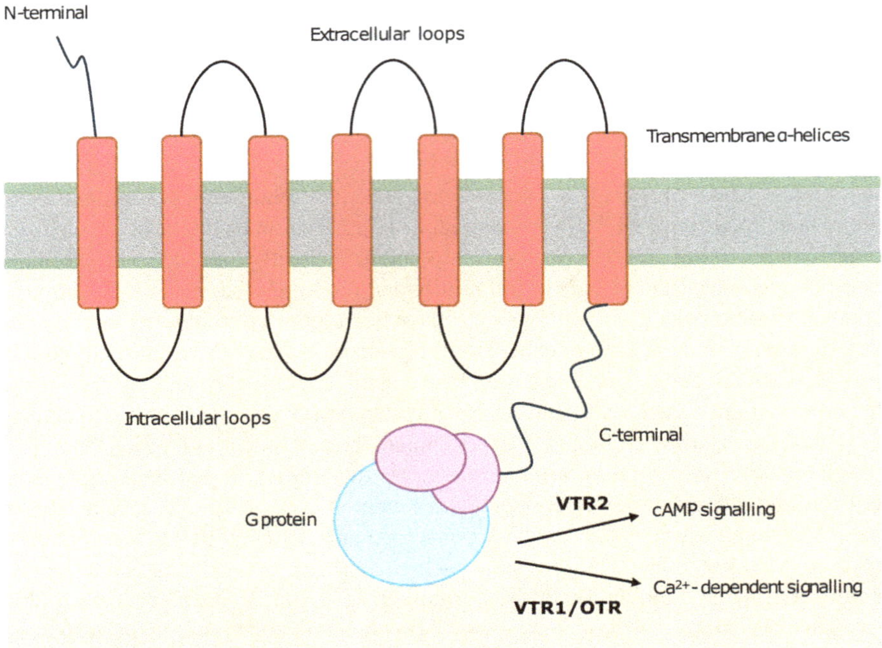

Fig. 15.2 General structure of the nonapeptide receptors. The configuration follows the typical of a G-protein coupled receptor: seven transmembrane domains are intercalated with three intracellular loops and three extracellular loops. The N-terminal is exposed to the extracellular space, whereas the C-terminal is physically linked to the G-protein, which operates the signal transduction. In the cell membrane, the transmembrane domains form a circular structure that confers the pocket-like space for the nonapeptide binding

shared across species but also sometimes species-specific. In general, OTR and VTR receptors are encoded in two exons separated by an intron. In teleost species, this is not the case for both copies of *VTR2A*, which have gradually accumulated three additional exons, before the third whole genome duplication (WGD) event occurred (Ocampo Daza et al. 2012): one in basal ray-finned fishes, a second one before teleost divergence and a third one in the early teleost evolution (Ocampo Daza et al. 2022).

Among vertebrates, six canonical nonapeptide receptors have been described, one for OT (*OTR*) and five for VT (*VTR1A*, *VTR1B*, *VTR2A*, *VTR2B*, *VTR2C*). However, further lineage-specific gene losses resulted in most vertebrate species having a maximum of five receptors (Fig. 15.3). How these genes originated is still a topic of discussion. It is widely accepted that the ancestral VTR underwent an initial gene duplication ca. 600 million years ago, likely before the ligand gene duplication (Gwee et al. 2009), resulting on VTR1 and VTR2 being present at the base of the vertebrate lineage (Ocampo Daza et al. 2022; Theofanopoulou et al. 2021). The subsequent evolutionary history of these receptors is still uncertain. One prevailing hypothesis is that four copies of a receptor pair resulted from two rounds

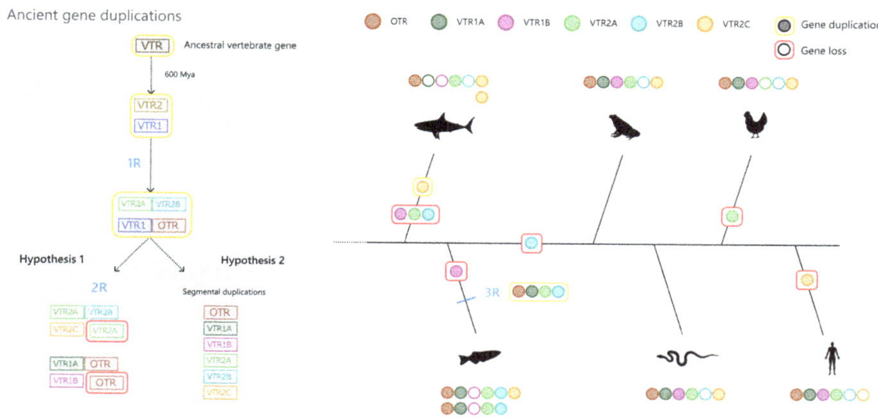

Fig. 15.3 Evolution of the six receptors (OTR and VTRs) from an ancestral *VTR* gene. The two different hypotheses are shown: hypothesis 1 states that the emergence of all duplicates was due to a second WGD event (2R), whereas the second supports the idea that segmental duplications led to the current set of receptors. On the right, the phylogenetic distribution of the OTRs and VTRs across vertebrates is shown. The distribution shows the most common set of receptors in each clade. However, many species- or subclade-specific gene gains or losses are not considered. Major vertebrate clades are represented as follows: elasmobranchs (shark), teleost (fish), amphibians (frog), reptiles (snake), birds (chicken) and mammals (human). Yellow boxes indicate gene duplication and red boxes indicate gene deletion. Filled circles indicate presence of the gene in the clade and empty circle means absence

of WGD events, followed by the loss of an *OTR/VTR1* and a *VTR2* duplicate before the radiation of the jawed vertebrates (Ocampo Daza et al. 2022). An alternative hypothesis proposes that only one WGD followed by segmental duplications lead to the present distribution of *OTR* and *VTR*s. In this case, two segmental translocated duplications of *VTR1A* and *VTR2A* gave rise to *VTR1B* and *VTR2B*, respectively (Mayasich and Clarke 2016; Theofanopoulou et al. 2021). Subsequently, the teleost-specific WGD produced extra copies of *OTR* (*OTRa* and *OTRb*), *VTR1A* (*VTR1Aa* and *VTR1Ab*) and *VTR2B* (*VTR2Ba* and *VTR2Bb*; Fig. 15.3).

In addition, the teleost *VTR1B* copy was lost at the base of their divergence from other ray-finned fishes, as evidenced by the presence of this receptor in the cartilaginous fish species elephant shark (*Callorhinchus milii*) but not in the teleost lineage (Gwee et al. 2009; Ocampo Daza et al. 2012; Theofanopoulou et al. 2021). Therefore, teleosts usually have two *OTRs* (*OTRa*, *OTRb*) and six to seven *VTRs* (*VTR1Aa*, *VTR1Ab*, *VTR2Aa*, *VTR2Ab*, *VTR2Ba*, *VTR2Ba*, *VTR2C*), although some clade-specific gene losses or duplications have occurred (Fig. 15.3). The two copies of teleost *OTRs* are usually found in the same chromosome, suggesting a tandem duplication event origin. Similarly, the two copies of *VTR2As* are also found in the same chromosome. Interestingly, the intracellular loop 3 of the receptor is substantially longer in both copies of *VTR2A* compared to any other vertebrate. This, together with the particular exon-intron structure of these two genes, supports the idea of a common origin of the two copies by a segmental duplication (Ocampo Daza et al. 2012).

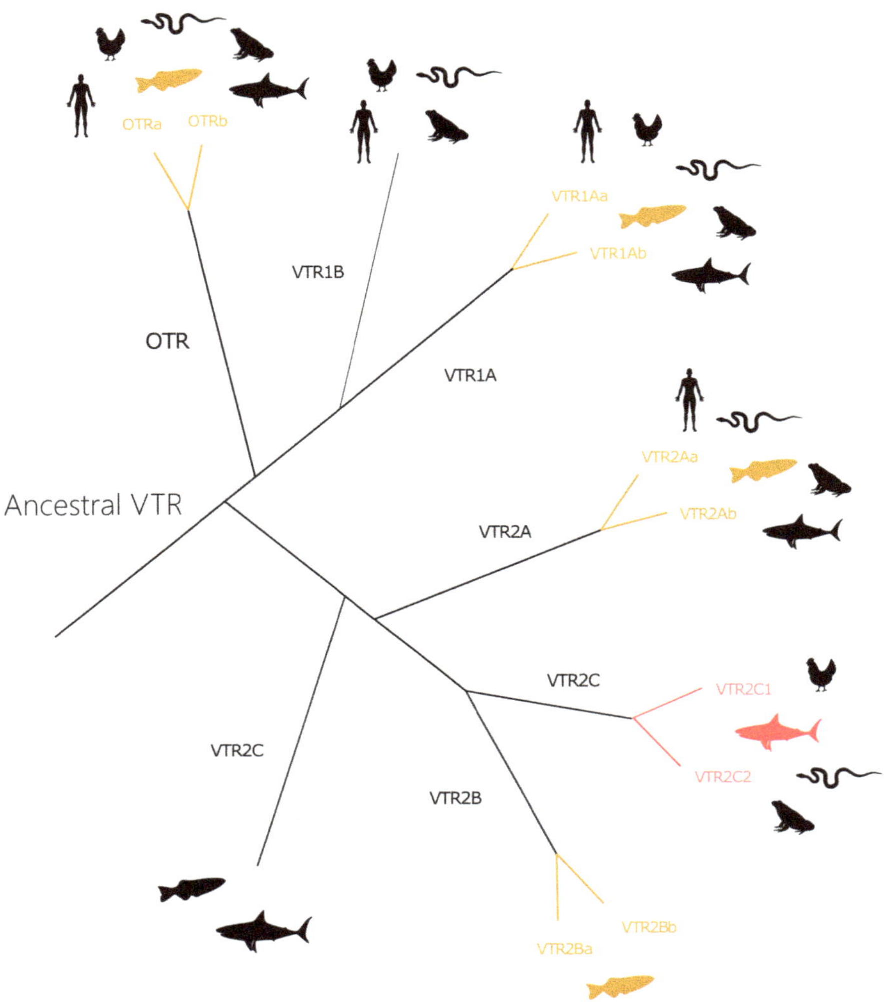

Fig. 15.4 Schematic representation of the phylogenetic relationships between the nonapeptide receptors. Duplication events are highlighted in elasmobranchs (red) and teleost (yellow). Branch lengths are not informative of the phylogenetic distance. Major vertebrate clades are represented as follows: elasmobranchs (shark), teleost (fish), amphibians (frog), reptiles (snake), birds (chicken) and mammals (human)

A phylogenetic analysis of receptor sequences shows that the *VTR1* family is more similar to *OTR* than to *VTR2*, which, together with a similar signalling mechanism of VTR1 and OTR (Brinton et al. 1994; Gimpl and Fahrenholz 2001; Zhang et al. 1997), suggests a common origin and a more recent divergence of these two genes (Ocampo Daza et al. 2022; Fig. 15.4). Only one copy of *VTR2C* has been found in teleost species studied up to date, suggesting that if *VTR2C* was duplicated during the teleost WGD, most clades lost one of the copies. Phylogenetic analysis separates two subgroups of *VTR2Cs*, with one being closer to *VTR2B* and the other forming a separate group from the base of *VTR2s* (Ocampo Daza et al. 2022; Fig. 15.4). This paraphyletic scenario can be explained either by fast evolution of *VTR2Cs* in only some species or by a distinct origin of the two *VTR2Cs*, followed by convergent evolution. Only some cartilaginous fish have two copies of *VTR2Cs*, each one belonging to one of the *VTR2C* subgroups (Fig. 15.4). Notably, cartilaginous fishes lost *VTR1B* and both *VTR2* copies. In addition, lineages of small ray-finned fish, such as the sterlet (*Acipenser ruthenus*) and the American paddlefish (*Polyodon spathula*), experienced a recent duplication of *VTR2C* (Ocampo Daza et al. 2022).

15.2.3 Cross Talk Between Nonapeptides and Their Receptors

Both OT and VT bind the receptor pocket in a similar conformation. However, during binding to VTR2, the tail peptide of VT faces the extracellular space, favouring contact with extracellular loops 1 and 3 and the N-terminus of VTR2, whereas during binding to OTR, the tail of OT stays in the plane of the nonapeptide ring. In addition, the helix I (transmembrane domain I) of OTR is pushed, favouring the insertion of both the ring and the tail (Waltenspühl et al. 2022). Due to these structural constraints, OT is less compatible to binding VTR2 but VT appears to have affinity for OTR (Chini et al. 1995). The second structural difference between OTR and VTR2 relies on the intracellular region, given that both receptors interact with a different region of the G-protein (Waltenspühl et al. 2022). Thus, both nonapeptides bind in a slightly different arrangement/configuration in the pocket, therefore modulating the downstream signalling and leading to changes in the outcome of the ligan-receptor interaction. OTRs, VTR1s, VTR2Bs and VTR2C share the calcium-mobilising mechanism as a second messenger for downstream signalling through phospholipase C pathway (Brinton et al. 1994; Ocampo Daza et al. 2022; Yamaguchi et al. 2012). Instead, VTR2A activates the adenylate cyclase/cAMP pathway (Zhang et al. 1997; Fig. 15.2).

Growing evidence shows that both OT and VT can bind each other's canonical receptors. In several cases, both nonapeptides have been shown to activate the same receptor, leading to similar effects (Song and Albers 2018). Although each nonapeptide system has been associated to particular and distinct behaviours, their roles are often intertwined, and both can modulate each other's functions (Hicks et al. 2014; Kawamata et al. 2003; Ramos et al. 2016). However, between-pathway affinity seems to depend on species and tissues type (Albers 2015; Lyu et al. 2021). In humans, VT (named AVP) appears to have a similar affinity for OTR and VTRs, whereas OT has higher affinity for its canonical receptor than VTRs (Chini et al. 1995). Although similar studies in teleosts are lacking, this may also be the case.

Given that the two ligands can bind to each other's receptors, the cellular colocalisation of receptors can lead to competition for binding. However, OTRs and VTRs are frequently anatomically separated; hence the effects of nonapeptides likely rely on which receptor is activated rather than which ligand is responsible for it. This is particularly relevant considering that often OT and VT produce different or even opposite phenotypes (Thompson and Walton 2004; Watanabe et al. 2007). Considering the idea that OTRs and VTRs rarely co-occur in the same regions, this cross talk would largely occur in situations of neuropeptide spill-over beyond synaptic regions, rather than the classic synaptic release (Song and Albers 2018). These different patterns of receptor activation could then contribute to the complex regulation of behaviours and physiological processes. Finally, it is worth noting that receptor selectivity has largely been considered under the bases of OT/VT affinity for the receptor. However, as discussed by Albers (2015), other means could have a similar effect, such as receptor up- or down-regulation, ligand-specific degrading enzymes, local ligand production or receptor clustering.

15.3 Nonapeptide Regulation of Neural Substrates of Social Behaviour

Across teleosts, nonapeptide neurons have been found in the preoptic area (POA), where distinct OT and VT cell populations can overlap, and direct contact between them has been reported (Hatton and Yang 1996). Within the POA, subpopulations of VT and OT cells are classified according to their soma size as parvo-, magno- and giganto-cellular and are differentially involved in different behaviours (Sawchenko and Swanson 1982; Greenwood et al. 2008; Knobloch and Grinevich 2014). While parvocellular neurons occupy the ventral part of the neurosecretory POA, magnocellular and gigantocellular cells are present in the dorso-caudal area (Saito et al. 2004). Different POA nonapeptide neuronal populations project to different targets, namely, the neurohypophysis, for peptide release to the systemic circulation, the hindbrain and the ventral forebrain (e.g. ventral telencephalon and the ventral thalamus) (Saito et al. 2004; Gonçalves and Oliveira 2011; Goodson 2008; Zhang et al. 2021; Mennigen et al. 2022). The projections to distinct brain regions define the neuromodulatory roles of the nonapeptides (Almeida et al. 2012; Goodson 2008; Saito et al. 2004).

The local expression of receptors largely determines the localised activity of nonapeptides across brain regions but also determines functional connectivity across regions. Thus, the spatial and temporal distribution of receptors is crucial for modulating behaviour. Receptor expression can be regulated at the gene expression level (e.g. transcription factors, epigenetic markers) or at the translational level via non-coding RNAs (Jurek and Neumann 2018). In general, nonapeptide receptors are widely distributed throughout the fish forebrain, with OTR and VTR1 receptors reported across areas of the telencephalon and diencephalon (Huffman et al. 2012), as well as in the midbrain, cerebellum and hindbrain (Lema 2010). Notably, nonapeptide receptors have been found in all the brain areas of the teleost social decision-making network, an evolutionarily conserved network, consisting of the mesolimbic reward system, the social behaviour network and shared areas connecting the two, that together regulate social behaviour (Newman 1999; Goodson 2005; O'Connell and Hofmann 2011, 2012; Huffman et al. 2012). As mentioned above, the receptors characterised in the fish brain and generally associated to behaviour regulation belong to five gene families: OTRa, OTRb, VTR1Aa, VTR1Ab, as well as VTR1B in some ray-finned fish, but not in teleosts or elasmobranchs (Yang et al. 2018; Mayasich and Clarke 2020; Ocampo Daza et al. 2022). VTR2 transcripts have a limited localisation, and receptor activity is largely reserved to osmoregulation and in peripheral tissues such as the liver, kidney, eye, gonads and intestine (Lema 2010; Singh and Joy 2009; Mayasich and Clarke 2020).

The downstream effects of nonapeptides on behaviour have been defined across two axes: neuromodulatory and neuroendocrine (Banerjee et al. 2017). Across vertebrates, nonapeptide neuromodulatory effects are prescribed to (i) excitatory activity, via depolarisation changes, (ii) regulation of inhibitory transmission by controls on release and interneuron spiking and (iii) the modulation of other modulators, particularly monoamine pathways such as dopamine and serotonin (Froemke and Young 2021; Rigney et al. 2022). For example, OT regulates the social contagion of alarm in zebrafish (*Danio rerio*), where animals replicate the distress behaviour they observe in others, via localised controls of excitatory and inhibitory neurons in the ventral telencephalon but also by excitatory and inhibitory functional connectivity changes across a network of social decision areas (Akinrinade et al. 2023). Furthermore, an interaction between serotonin and OT signalling has been revealed in goldfish during feeding and reproduction (Mennigen et al. 2017).

The primary neuroendocrine activity of nonapeptides is integrated in the regulation of the hypothalamic-pituitary-interrenal stress axis (HPI; Aguilera and Rabadan-Diehl 2000; Gesto et al. 2014). In particular, VT initiates the hypothalamic release of corticotropin-releasing hormone (CRH), which leads to a series of downstream hormonal triggers, from the release of adrenocorticotropic hormone in the anterior pituitary to the release of glucocorticoids (cortisol in fish) from the interrenal glands (Denver 2009; Lowry and Moore 2006; Wendelaar Bonga 1997). This is often equated with an anxiogenic role for VT as an initiator of the neuroendocrine stress response, whereas OT release has been found to reduce glucocorticoid release through an inhibitory effect, such as colocalised controls on CRH receptor 2, which also carries out anxiolytic functions (Dabrowska et al. 2011; Winter and Jurek 2019). This has not been extensively studied in fish, but OT neurons have been found to project to CRH receptor sites in the anterior pituitary across three species with distinctly varied behavioural repertoires: two oviparous fish, the marine European seabass (*Clarias gariepinus*) and the freshwater African sharp tooth catfish (*Dicentrarchus labrax*) and a live-bearing freshwater fish, the sailfin molly (*Poecilia latipinna*) (Batten et al. 1999). Recent accounts also characterise nonapeptides as key regulators of the hypothalamic-pituitary-gonadal (HPG) axis, implicated in reproduction and sex differences, and with their further interaction with the HPI axis considered to elicit stress-induced fertility disorders (Gołyszny et al. 2022). In some fish species, OT is colocalised with gonadotropin releasing hormone (GnRh) in parvocellular neurons (e.g. dwarf gourami, *Colisa lalia*; Maejima et al. 1994), while in others it colocalises with secretoneurin which stimulates the HPG axis (e.g. goldfish, *Carassius auratus*; Canosa et al. 2011). In addition, oestrogen has been shown to increase OTR expression in the ovary of guppies (Li et al. 2024), playing an important role in reproduction and maternal care. Finally, in the gilthead sea bream (*Sparus aurata*) VT treatments elicited upregulated expression of thyroid hormone RNA (*trh*) and an up-regulation of hypothalamic *VTR2* expression, which indicates interactions with the hypothalamic-pituitary-thyroid (HPT) axis and a potential role in HPT metabolic functions (Martos-Sitcha et al. 2019).

15.4 Nonapeptide Regulation of Behaviour Across Social Contexts

The central neuromodulatory functions of nonapeptides in the fish brain have been implicated in cognitive phenotypes—such as perception, appraisal and memory—responsible for the organisation of social behaviour across two main components: affiliation and aggression. For example, genetic zebrafish models, show that OTRa receptors modulate motivated approach and interaction with others by regulating the perception of biological form and movement (Nunes et al. 2020) and that the conditional ablation of OT neurons early in development in transgenics elicits adult deficits in social affiliation, which are mediated by a reduction in the number of dopaminergic cells in specific clusters and concomitant changes in functional connectivity across social decision-making areas (Nunes et al. 2021). Furthermore, using zebrafish mutants of either of the two OT receptors, as well as of the OT ligand, Akinrinade et al. (2023) demonstrated the oxytocinergic control of recognition, memorisation and recall of behavioural states in others, for organising later approach towards those in distress. This evidence implicates OT in zebrafish affiliative behaviour. On the other hand, the appraisal of aggressive interactions and contest outcomes implicates both nonapeptides. For instance, in male zebrafish, aggression with or without resolution (live animals vs. mirror fights) resulted in elevated diencephalic VT concentrations, while elevated VT in the telencephalon was related to winning and losing experiences, whereas OT concentrations presented outcome-specific effects, being elevated in the diencephalon for losers and reduced in the olfactory bulbs for winners (Teles et al. 2016). Across fish species, these nonapeptide neuromodulatory effects are joined with nonapeptide neuroendocrine controls in regulating contrasting affiliative and aggressive social phenotypes across contexts, including dyadic contests, reproduction, parental care, social grouping and interspecific interactions (Fig. 15.5).

15.4.1 Aggressiveness and Agonistic Behaviour

Most of the evidence on the role of nonapeptides in fish aggressiveness is related to the contribution of VT to male contest behaviour, which is generally found to increase aggression, while evidence on the role of OT suggests that it is either unrelated or that it inhibits aggression (Godwin and Thompson 2012; Lema et al. 2015). Nonetheless, the effects vary considerably across species and often within species, based on sex, social phenotype or context. For example, in male territorial

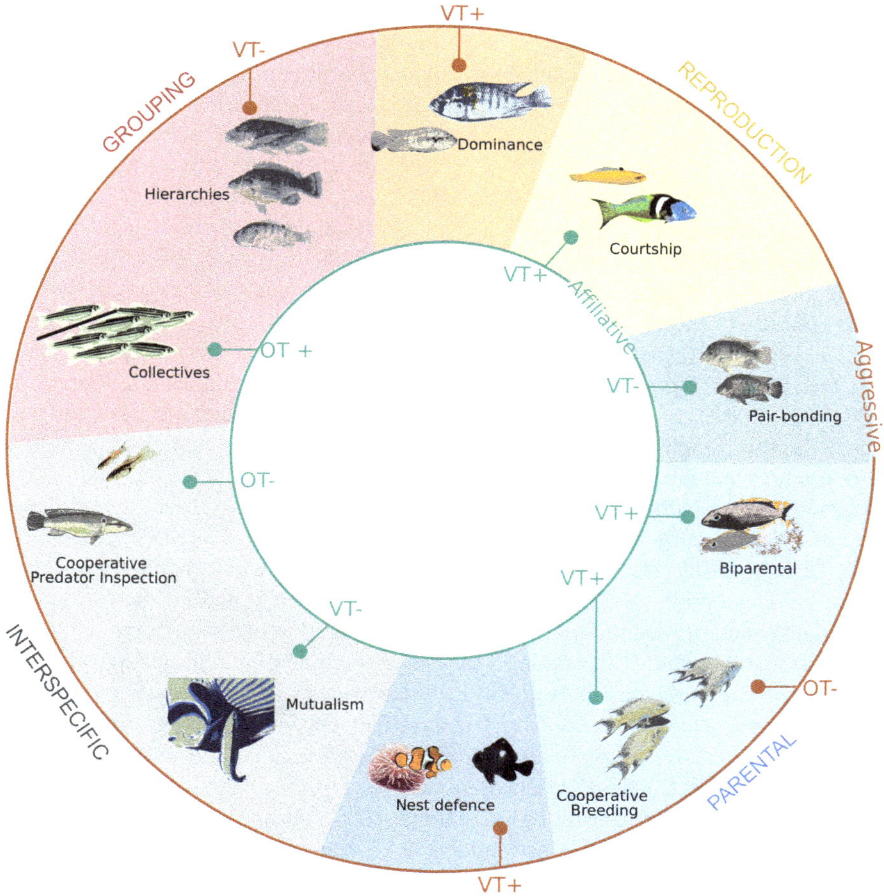

Fig. 15.5 Representation of the contribution of vasotocin and oxytocin pathways to different social phenotypes across the fish phylogeny. The diverse phenotypes are characterised by affiliative and aggressive components represented by behaviours across overlapping ecological contexts, including: Reproduction, defined by dominance in males (*Astatotilapia burtoni*), sexual morphs and courtship behaviour (*Thalassoma bifasciatum*), and monogamous pair-bonding versus polygyny (*Herichthys* spp.); Parental care, defined by parental nurture, such as biparental care systems (*Neolamprologus caudopunctatus)*, cooperative breeding systems, with dominant paired breeders and subordinate and punishable helpers (*Neolamprologous pulcher*), and by offspring/ nest defence against intruders, including other species (*Amphiprion ocellaris*); Interspecific interactions, defined by defensive behaviour, mutualistic repertoires, such as the cleaning of skin parasites as a food source (*Labroides dimidiatus*), and cooperative antipredator functions (*Poecilia reticulata*); Social grouping, defined by synchronised and cohesive collective responses (*Danio rerio*) and hierarchical dynamics (*Oreochromis mossambicus*). Major contribution to each phenotype by nonapeptides is noted by the pathway, OT or VT, and whether it positively or negatively drives the behaviour

beaugregory damselfish, *Stegastes leucostictus*, VT elicited higher aggressive bite rates towards intruders but only at medium dosages, while OT injections had no noticeable effect (Santangelo and Bass 2006). In contrast, in the plainfin midshipman (*Porichthys notatus*), VT inhibited aggressive vocalisations in males of the courting morph but had no effects in the sneaker morph or females, where decreased aggression was instead elicited by OT treatments (Goodson and Bass 2000). VT treatments also reduced territorial aggression in the Amargosa pupfish (*Cyprinodon nevadensis amargosae;* Lema and Nevitt 2004) and the juvenile rainbow trout (*Oncorhynchus mykiss;* Backström and Winberg 2009) and inhibited the production of electric organ signals during fights in the weakly electric *Apteronotus leptorhynchus* (Bastian et al. 2001), whereas, in the sex-changing bluehead wrasse (*Thalassoma bifasciatum*), VT pharmacological treatments decreased aggression in territorial males but increased aggression in nonterritorial males (Semsar et al. 2001). Similarly, in cichlids *Astatotilapia burtoni*, VT treatment also reduced aggression in established dominant males and increased their chance of becoming subordinates and increased dominance in otherwise subordinate males under unstable hierarchies (Huffman et al. 2015). Notably, dominant *A. burtoni* males had higher VT expression levels than subordinates (Greenwood et al. 2008; Renn et al. 2008), and short-term increases in these levels were noted when encountering a bigger dominant opponent (Chen and Fernald 2011). The signalling pathways and modulatory activity behind these effects also varies, often as a function of the differential involvement of the three POA subpopulations of VT projecting neurons. For instance, VT expression in gigantocellular neurons is higher in *A. burtoni* territorial males, while expression in parvocellular cells is higher in nonterritorial males, which may relate to the distinction between dominant and subordinate traits (Greenwood et al. 2008). Further, dominant *A. burtoni* males showed activity in magnocellular neurons during fights, compared to courting males (Loveland and Fernald 2017). The role of nonapeptide POA neuronal subpopulations may also present species differences. In the South American cichlids *Cichlasoma dimerus*, subordinates have larger parvocellular cells (Ramallo et al. 2012), but in the tilapia (*Oreochromis mossambicus*), subordinates have larger magnocellular and gigantocellular neurons (Almeida and Oliveira 2015). However, this may also relate to where these neuronal subpopulations project and which receptors are activated at the receiving end of the signal. For example, between closely related North American cichlids, the more polygynous *Herichthys minckleyi* exhibits no VT signalling differences from the monogamous *Herichthys cyanoguttatus*, other than higher VT1A receptor levels in its territorial males, suggesting a role of these receptors specific to territoriality (Oldfield et al. 2013). For OT, the evidence is largely restricted to effects from pharmacological treatments. For example, in *Betta splendens* males, where aggression is related to paternal care and the territoriality of bubble-nest sites, OT treatments reduce aggressive display and attack (Oliveira et al. 2022). However, there is often no relationship between aggressiveness and OT expression or concentration levels in the brain, such as when comparing dominant and subordinate males in *O. mossambicus* and in *Neolamprologus pulcher* (Almeida et al. 2012; Reddon et al. 2015). This may rely on OT having indirect effects mediated by interaction with other neuroendocrine systems. In goldfish (*Carassius auratus*), OT injections increased plasma levels of oestradiol, which in fish inhibits aggressive behaviour (Mennigen et al. 2008; Gonçalves et al. 2007). In zebrafish winners and losers have a similar VT profile across brain regions, but compared to controls, losing implicated elevated testosterone and greater diencephalic OT levels, whereas winning implicated elevated cortisol and lower OT levels in the olfactory bulbs (Teles et al. 2016; Teles and Oliveira 2016). In turn, in the cooperatively breeding *N. pulcher*, OT mRNA levels were positively correlated with affiliative and submissive behaviours (O'Connor et al. 2016) but correlate negatively with affiliation levels in subordinates (Reddon et al. 2015).

Beyond the control of individual aggression levels, the role of nonapeptides extends to how aggressive behaviour is modulated during agonistic interactions in response to the assessment of opponents. For instance, in simulated territorial fights, OT-injected *N. pulcher* exhibited greater submission and responded to their opponent's size, exhibiting higher aggression towards bigger opponents, while the behaviour of control fish was only predicted by their own aggression levels (from mirror tests), suggesting that OT modulates signal assessment during fights (Reddon et al. 2012). Response to opponent signals is considered an improved contest strategy but can impose time and energy costs for information processing and distract from an individual's own contest performance, which may explain its absence under native signalling (Arnott and Elwood 2009). Alternatively, the lack of assessment without OT treatments may be due to sensitisation to social signals, as suggested by Maruska et al. (2022). Another possibility is that matching the aggression levels of opponents is a display strategy related to motivational shifts via the nonapeptide modulation of monoamine reward signalling. For instance, Silva et al. (2013) found that territorial contests in the solitary weakly electric fish *Gymnotus omarorum* implicate downmodulation in serotonergic activity (i.e. aggression diminishes with receptor agonist) and that VT treatments increase aggressive motivation (first-attack speed) and electric signal rates, such that treated submissive animals were comparable to non-treated dominant ones. In the same study, during contests over reproductive opportunity in the gregarious *Brachyhypopomus gauderio*, treatment with a VT antagonist instead led dominant animals to match the electric discharge rates of subordinates.

15.4.2 Mating and Reproduction

Nonapeptides have long been implicated in mating behaviour and reproduction. One of the earliest pieces of evidence comes from the induction of the spawning reflex in killifish (*Fundulus heteroclitus*) by the administration of neurohypophysial extracts (Wilhelmi et al. 1955), which was subsequently shown to be more efficiently elicited by VT than OT (Pickford and Strecker 1977). Although nonapeptides could not elicit a similar response in other species (e.g. *Misgurnus anguillicaudatus*, Egamii and Ishii 1962; *C. auratus*, Macey et al. 1974), in seahorses (*Hippocampus hippocampus*) OT induced male parturition reflexes (Fiedler

1970). This earlier evidence indicates a regulatory role of nonapeptides in behaviours directly related to the stimulated release of gametes, but in later work we can find a series of examples where VT and OT also modulate pre-spawning courtship and mating phenotypes. For example, VT administration increases male sexual display in white perch (*Morone americana* Salek et al. 2002) and male electric signals in the weakly-electric fish *A. leptorhynchus* (Bastian et al. 2001), during their courtship of females. The role of VT also extends to female courtship, such as in the grass puffer (*Takifugu niphobles*), where VT mRNA expression in the brain was elevated in females during the pre-spawning period, while OT mRNA expression remained stable (Motohashi et al. 2008). However, in most cases the effect of nonapeptides is sex-specific, either influencing behaviour in one sex or differently between males and females. For instance, in three-spined stickleback (*Gasterosteus aculeatus*), high-performance liquid chromatography (HPLC) demonstrated that brain OT levels reach a peak during the reproductive season in females but not males (Gozdowska et al. 2006). In turn, VT administration to both territorial and nonterritorial bluehead wrasse in their terminal change from female to male promotes courtship behaviour (Semsar et al. 2001), but it does not induce these effects in females or sex-changing males at their initial phase (Semsar and Godwin 2004). In contrast, in the peacock blenny (*Salaria pavo*), VT mRNA increases more in females and female-like sneaker males than in nesting males (Grober et al. 2002). Recently, Altmieme et al. (2019) found a joint VT and OT contribution to zebrafish fertility, where pharmacological inhibition with antagonists of either pathway led to reduced numbers of fertilised eggs. The authors also characterised deficits in male courtship behaviour for animals treated with the VT and not the OT antagonist but also found that exposure of males to the female prostaglandin pheromone PGF2a-induced whole brain increases in the abundance of OT mRNA, but not VT mRNA. This suggests that contributions to fertility are derived from response in males during mating and that OT is involved more strongly in the perception of social signals and VT in the control of behavioural responses. However, the OT and VT pathways may also have antagonistic effects on reproductive phenotypes, across sexes, such as mate preference in the nonmonogamous Japanese medaka (*Oryzias latipes*). On the one hand, mutations affecting VT and VTR1A elicited deficits in dominant-male guarding behaviour during contests over a female (Yokoi et al. 2015). On the other hand, OT and OTRa mutations elicited sex-specific effects: in mutant females, preference for familiar mates was eliminated, whereas in mutant males mate-guarding of familiar females increased and courtship display to unfamiliar mates decreased (Yokoi et al. 2020).

Notably, there is evidence that females of some species exhibit mechanosensory channels that enable intraovarian pressure to induce hypertrophy in neurosecretory POA neurons (e.g. walking catfish, *Clarias batrachus*: Subhedar et al. 1987, 1996; Deshmukh and Subhedar 1993). This suggests feedback between reproductive state and nonapeptide activity, which could also be a result of interactions with the HPG axis. For example, in rainbow trout, GnRH in vitro treatments to VT neurons of the POA elicited an electrical response (Saito et al. 2003). In single sex groups of the black molly (*Poecilia sphenops*), female masculinisation was related to decreased levels of both nonapeptides, but also increased 11-ketotestosterone (11-KT) and lower 17β-oestradiol (E2) levels, compared to control females (Kulczykowska et al. 2014). Conversely, same-sex sexual behaviour in males, such as copulation displays and attempts, were related with increased VT and OT levels, and both lower E2 and 11-KT. The relationship between nonapeptides and sex hormone activity may also rely on reproductive strategies, which can vary across species. For instance, between closely related North American cichlids of the genus *Herichthys*, the monogamous *H. cyanoguttatus* maintain a small territory with its pair-bonded female partner, while males of the polygynous *H. minckley* maintain larger territories with more females. Oldfield et al. (2013) identified that reproductive males of both species exhibited greater androgen levels than nonreproductive males, but the monogamous species expressed overall more testosterone and the polygynous species more 11-KT, and that VTR1A2 was markedly more strongly expressed in the polygynous species. In turn, where the reproductive state elicited increases in hypothalamic VTR1A2 expression of the polygynous *H. minckley* males, it instead elicited decreases in VTR1A2 expression in the monogamous *H. cyanoguttatus*. In three-spined sticklebacks, the presence of males increased VT and OT concentrations in the brains of females, but experimental manipulations that decreased the probability of access to males from 0.33 to 0.1 reduced OT levels to that of single females, potentially a response to lower reproductive opportunity (Kulczykowska and Kleszczyńska 2014). The role of nonapeptides in preparing females for reproduction is also evident in the decrease in VT and OT concentrations after egg disposition. In general, VT and OT are involved in mating behaviour, the perception and response to courtship signals, the stimulation of sperm and egg release and to ultimate reproductive outcomes, such as fertility rates. Further, the effects can be sex-specific and often involve interactions with the HPG axis and sex hormone activity.

15.4.3 Parental Care

The involvement of nonapeptides in parental care is characterised by three components: (i) the phenotype, whether it is the organisation of direct offspring nurture, e.g. feeding or nest building, or aggressive defence from intruders; (ii) the parental strategy, whether mono- or biparental and with shared or distinct contributions from either parent; and (iii) cooperative breeding strategies, extending to help from third parties and the social dynamics between breeders and helpers. Regarding phenotype and parental strategy, in the false clown anemonefish (*Amphiprion ocellaris*), males exhibit lower aggression than females during nest defence but exhibit higher parental care than females in the absence of territorial intruders, and this trade-off is predicted by the contrasting effects of OT and VT (DeAngelis et al. 2020). Namely, comparing antagonist treatments with saline controls, the inhibition of VT signalling elicited deficits in aggressive defence and a trend to elevated egg care, while inhibition of OT signalling elicited increased defence and decreased egg care. Furthermore, male clownfish treated with an OT antagonist exhibited reduced paternal behaviour, including nips and egg fanning, and shorter nest visits, but a VT antagonist had no effect (DeAngelis et al. 2017). This suggests that OT drives parental care phenotypes and VT parental defence phenotypes and that their different implication in males or females depends on parental phenotype. This also suggests that in biparental systems, with stronger pair bonding, nonapeptide parental strategy effects may derive from intra-pair dynamics. In the monogamous convict cichlid *Amatitlania nigrofasciata*, treatments with a combined OT and VT antagonist elicited immediate short-term deficits in male affiliative behaviour towards females and aggression towards third parties, during pair formation, while in formed pairs the effects of the antagonist on males relied strongly on the female partner's behaviour and the presence of offspring (Oldfield and Hofmann 2011). In the same monogamous cichlids, single fathers and those involved in biparental care show equal investment in aggression, which, although elevated compared to unpaired males, coincided with decreases in 11-KT (O'Connell et al. 2012). However, single fathers elevated their paternal behaviour in an apparent compensation for their missing partner (experimental removal), and this implicated greater OT neuronal activity in parvocellular POA cells and neuronal responses in the ventral telencephalon, while OT antagonist treatments elicited deficits in paternal care. This demonstrates that in parental contexts, phenotype-specific effects of OT and VT are dependent on interactions between parental pairs and the needs of their offspring. However, VT and OT may not always serve the same role across species. For instance, whereas in the South American cichlid *A. nigrofasciata*, OT drives affiliative phenotypes and VT aggressive ones, in the also monogamous and biparental Lake Tanganyika cichlid, *Neolamprologus caudopunctatus*, only VT brain concentration was elevated in post-spawning pairs, and it positively predicted both pair-bonding (i.e. greater proximity and lower aggression between pairs) and nest maintenance (Cunha-Saraiva et al. 2019). Furthermore, while paternal care is positively predicted by VT levels in pipefish (*Syngnathus* spp.; Ripley and Foran 2010), it is promoted rather than inhibited by VTR1A antagonist treatments in anemonefish (*A. ocellaris*; DeAngelis et al. 2017). Notably, the involvement of nonapeptides in offspring care also extends to cooperative breeding systems, which suggests a likely prosocial function defined by the investment to improve survival in others without direct fitness benefits. For instance, another Lake Tanganyika cichlid, *N. pulcher*, is a cooperative breeder where male and female breeding pairs defend their territory and dominate over helpers (Balshine-Earn et al. 1998). Breeding females exhibit more aggression than helpers of either sex and show higher testosterone and VT mRNA levels (Aubin-Horth et al. 2007). Conversely, dominant males of *N. pulcher* that more often visited their brood expressed lower OT and OTRa expression in the POA (Culbert et al. 2024). A comparison of eight Lake Tanganyika cichlids also identified that those involved in cooperative breeding had fewer parvocellular OT neurons than independent breeders and that classification of OT neuronal counts could discriminate cooperative from independent breeders (Reddon et al. 2017).

15.4.4 Social Groups: Collective Behaviour and Dominance Hierarchies

Organisation in social groups implicates diverse phenotypes across fish, which include the drive to approach and join groups, dynamic interactions leading to collective responses or dominance hierarchies and the transmission, facilitation and buffering of responses to external conditions. Nonapeptides have been implicated across these phenotypes, often with combined neuromodulatory effects from OT and VT signalling and the implication of other neuroendocrine systems. For instance, in the Trinidadian guppy, *Poecilia reticulata*, treatments with OT increased social affiliation tendencies but also did exposure to higher-predation levels which also led to greater VT levels in gigantocellular subpopulations of the POA (Mehr 2022). This is supported by separate work on wild populations of this species, where animals from high-predation areas exhibited elevated VT gene expression levels in the brain (Reddon et al. 2022). The association of both nonapeptide pathways to social grouping may relate to the neurophysiological components that promote this response under different condition. For instance, the baseline control of OT may relate to affiliative group formation, while group formation under predation threat may implicate VT release during the activation of HPI stress-coping responses. The basal regulation of affiliative phenotypes by OT was illustrated by genetic zebrafish models, where mutations eliminating either of the OT receptors elicited early life deficits in approach towards social groups and in their cohesion within groups (e.g. interindividual and nearest neighbour distances) at 8 weeks post-fertilisation (Gemmer et al. 2022). However, zebrafish group cohesion was also deficient in mutations linked with reduced VT release, such as mutation to the endothelin receptor type aa, which present decreased VT concentrations in the brain, as well as apparent compensatory increases in the expression of VTR1 and dopamine synthesis genes (Carreño Gutiérrez et al. 2019). Paradoxically, in a separate study, treatments with VT decreased interaction times with a conspecific shoal, while VT antagonist treatments imposed similar deficits, as well as reduced shoaling behaviour (Lindeyer et al. 2015). The mixed results may derive from contributions by both pathways, a cross talk between the two based on between-pathway receptor binding affinity, or simply on interactions with other neuroendocrine systems, which may also implicate sex-specific effects. For instance, in zebrafish, treatments with the oestrogen steroid hormone E2 elicited deficits in male and female shoal cohesion and group preferences, and these effects implicated sex-specific changes in the mRNA expression of nonapeptide system genes, including decreased VTR and elevated VT expression in females and increased OTR expression in males (Salahinejad et al. 2020).

Group responses may also represent internal dynamics derived from dominance hierarchies. In *O. mossambicus*, dominant males that form dense territorial aggregations exhibited higher OT concentrations in the hindbrain, while subordinate non-territorial males that mimic the colouration and phenotype of females to remain inside aggregations exhibited higher VT concentrations in the pituitary (Almeida et al. 2012). In *N. pulcher*, social groups typically consist of a single highly territorial dominant female, which was found to express higher VT and VTR1A2 expression in the POA than her dominant male breeding partner or the up to 20 subordinate helpers (Culbert et al. 2024). Dominant females also exhibit higher levels of the nonapeptide inactivating enzyme leucyl-cystinyl aminopeptidase (*lnpep*), which reflects greater VT activity from production to binding and turnover. The role of nonapeptides in social hierarchies may also implicate controls on other associated systems and particularly androgens which have long been implicated in the aggressive interactions that determine dominance. For instance, in the monogamous cichlid *C. dimerus*, dominance hierarchies are dependent on male territoriality. Compared to territorial males, nonterritorial males presented larger soma sizes in their VT neurons, and VT had a positive control on gonadotropin secretion, eliciting response in both the luteinising (β-LH) and follicle-stimulating (β-FSH) hormones and a concentration-dependent increase in testicular androgen synthesis and secretion (Ramallo et al. 2012). The effects may also relate to the social role of individuals and how their behaviour impacts the perception of how they perform that role by other group members. For instance, in the cooperatively breeding cichlid, *N. pulcher*, the temporary removal of subordinates that take the role of helpers elicits aggression towards them and potential eviction from the group. OT treatments of these removed subordinates increased their submissive behaviour upon their return, and this increased the aggression directed towards them by others (Hellmann et al. 2015). Reddon et al. (2015) noted that *N. pulcher* helpers show greater VT levels in the brain than breeders and similar levels of affiliation, which negatively predicted OT levels. Indeed, in a separate experiment in *N. pulcher*, peripheral treatment with exogenous OT inhibited affiliation with others, while OT antagonist treatment in males increased grouping tendency (Reddon et al. 2014). However, the role of nonapeptides in social behaviour in groups does not only extend to grouping tendency, collective responses and within-group dynamics but also to situational responses to external conditions. In *N. pulcher*, behavioural and physiological stress imposed by removal from the water was buffered by recovery in a social group, compared to being alone, and this elicited decreases in OT transcription (via mRNA quantitative real-time polymerase chain reaction -qPCR; Culbert et al. 2019). In zebrafish, OT is a necessary and sufficient signalling mechanism for the transmission of alarm responses within social groups (Akinrinade et al. 2023). Finally, the maintenance of within-group affiliation, communication and hierarchical structures enables the aggressive defence of groups during intergroup conflict, which may involve tend and defend roles of OT that can predict "in-group" "out-group" dynamics (see conceptual model by Triki et al. 2022).

15.4.5 Interspecific Interactions

Affiliative phenotypes may also extend to interspecific interactions. For example, the obligate cleaner wrasse *Labroides dimidiatus*, which engages in mutualistic interactions with client reef fish, exhibited less and smaller VT gigantocellular neurons in the POA compared to the closely related non-cleaner wrasse *Labrichthys unilineatus* (Mendonça et al. 2013). Indeed, in the same species, VT treatments increased intraspecific and partner-oriented affiliation and decreased cleaner-oriented interaction times, rates and motivation, while treatments with Manning compound (VT antagonist) increased cleaning rates (Soares et al. 2012). In addition, VT increased cases where this cleaning fish would cheat and eat the client's mucus instead of ectoparasites, something they will typically supress to maintain clients and avoid punishing responses (Cardoso et al. 2015). This joint evidence points to the specificity of VT in controlling reward drives that are disruptive to interspecific interactions. This may also relate to specific local activity in different brain areas, which is devoted for such mutualistic behaviour. For example, obligatory cleaners *L. dimidiatus* and *Labroides bicolor* exhibit higher VT concentrations in the cerebellum, compared to the facultative cleaner *Labropsis australis* and the non-cleaner *L. unilineatus*, while higher OT levels in the non-cleaner *L. unilineatus* may rely on the sexual dimorphism occurring only in this species (Kulczykowska et al. 2015).

Aggressive phenotypes during interspecific interactions may extend to competition over resources, such as territory or food, or protection from predators. In the false clown anemonefish *A. ocellaris*, aggressive defence against territorial intrusion by the damselfish *Dascyllus trimaculatus* is driven by VT signalling and elevated by inhibited OT signalling (DeAngelis et al. 2020). In turn, guppies (*P. reticulata*) from high-predation environments in the wild exhibit a sex-independent increase in VT expression levels, but no differences in OT expression, compared to fish from low-predation areas (Reddon et al. 2022). In a separate study, guppies from low-predation conditions exhibited closer predator inspection when having a cooperating partner, but fish from high-predation environments exhibited close inspection irrespective of having a cooperating partner or not (Dimitriadou et al. 2022). However, fish from high-predation conditions also exhibited lower OTR transcription when inspecting predators with partners and higher OTR response when inspecting plants with partners, compared to no-partner controls. The combined results suggest that VT is involved in threatening interspecific interactions and that OT maintains vigilance based on the level of threat. In particular, lower OTR levels towards neutral stimuli or when inspecting predators with a cooperative partner and higher OTR levels when with a partner without threat or alone and under threat are possibly a dichotomy between the affiliative and anxiolytic functions of OT signalling (Dimitriadou et al. 2022).

15.5 Conclusions and Perspectives

The evolution of nonapeptides in vertebrates is marked by the ligand duplication of the ancestral VT gene at the base of gnathostome radiation and again in fish, at the base of the actinopterygian lineage, which have led to the subsequent separation of functions between the VT and OT (Banerjee et al. 2017). A number of receptors have been described for VT and OT, some of which present varied expression across the phylogeny ostensibly linked to functional localisation and derived from isolated and whole genome duplication events in fish (Theofanopoulou et al. 2021; Ocampo Daza et al. 2022). The conservation of the ligand structure, the maintenance of receptor expression across a network of social behaviour and decision-making areas and the broader involvement of the system in neuromodulatory and neuroendocrine functions emphasise its role across the diverse social phenotypes of fish (Huffman et al. 2012; O'Connell and Hofmann 2012). Yet, the precise circuitry that drives the effects of nonapeptides on fish behaviour is still poorly understood. This is because the existing work which we discuss here is largely dependent on the use of exogenous pharmacological treatments, which cannot distinguish between peripheral and central effects in the brain. The pharmacological approach is also limited by the availability of receptor-specific agonists and antagonists, which cannot disentangle cross-talk effects whereby the ligand of one nonapeptide binds to the receptors of the other. More recent approaches, using HPLC and qPCR techniques, have improved capacities to quantify ligand concentrations and receptor expression and localisation, respectively, in wild populations, but the availability or even the feasibility of some techniques in wild-caught animals is limited. On the one hand, genetic mutants and transgenic lines still offer better opportunities to target precisely each pathway and its receptors and even to conditionally eliminate the expression of either OT or VT so that developmental outcomes can be assessed (Akinrinade et al. 2023; Nunes et al. 2020, 2021). On the other hand, burgeoning laboratory models, such as zebrafish, benefit from all necessary genetic, molecular and cellular tools and from automated, high-throughput and controlled phenotyping (Kareklas et al. 2023) but are also less representative of natural selection and ecological pressures. Therefore, in better characterising the role of nonapeptides in the evolution of sociality in fish, such that we can identify conserved components and variations linked to adaptive phenotypes, there is space to develop integrative and interdisciplinary approaches that facilitate comparisons across species. To do so, we propose a two-way approach. First, the development of phenotypic concepts that define nonapeptide social functions across species can enable the identification of conserved components. One example is the simplification of social behaviours to decisions, which are more effective in defining the direction of a response—approach or avoid, affiliate or aggress—than any observable repertoire of movements, sounds or signals encompassing the behaviour. Decisions are a common component of behaviour that is independent of sensory-motor adaptations and underlie phenotypic plasticity by defining response based on current information and motivational state (Wallace and Hofmann 2021). Another example is social homeostatic systems that act to maintain social conditions at desired levels and which define behaviour based on these internal needs, against external pressures, and open to redefinition based on the phenotype of each individual, population or species (see Kareklas and Oliveira 2024). Second, the incorporation of ecological context in laboratory experiments and the development of protocols incorporating reliability and validity tests against alternative functional explanations (Voelkl et al. 2020; Holleman et al. 2020), enables the development of genetic neurobiological models that expand mechanistic-study capacities such that they offer relevant comparative value to wild species. Overall, there is a major opportunity to develop research on nonapeptide social functions, and there is a particular value of doing this in fish, including the abundance of diverse phenotypes and the parallel genomic variation that can explain it. Finally, identifying conserved social neurocognitive constructs in fish can reveal the bases on which further adaptations have occurred in more recent vertebrates, either at the mechanistic or behavioural level.

Acknowledgements This review was supported by a PhD Studentship to PS and as part of a project grand (https://doi.org/10.54499/PTDC/BIA-COM/3068/2020) awarded to RFO by the Portuguese Foundation for Science and Technology (FCT), and by an FCT Research Fellowship awarded to KK (https://doi.org/10.54499/2021.01659.CEECIND/CP1674/CT0003).

References

Acher R (1996) Molecular evolution of fish neurohypophysial hormones: neutral and selective evolutionary mechanisms. Gen Comp Endocrinol 102(2):157–172. https://doi.org/10.1006/gcen.1996.0057

Aguilera G, Rabadan-Diehl C (2000) Vasopressinergic regulation of the hypothalamic–pituitary–adrenal axis: implications for stress adaptation. Regul Pept 96(1–2):23–29

Akinrinade I, Kareklas K, Teles MC, Reis TK, Gliksberg M, Petri G et al (2023) Evolutionarily conserved role of oxytocin in social fear contagion in zebrafish. Science 379(6638):1232–1237

Albers HE (2015) Species, sex and individual differences in the vasotocin/vasopressin system: relationship to neurochemical signaling in the social behavior neural network. Front Neuroendocrinol 36:49–71. https://doi.org/10.1016/j.yfrne.2014.07.001

Almeida O, Oliveira RF (2015) Social status and arginine vasotocin neuronal phenotypes in a cichlid fish. Brain Behav Evol 85(3):203–213

Almeida O, Gozdowska M, Kulczykowska E, Oliveira RF (2012) Brain levels of arginine–vasotocin and isotocin in dominant and subordinate males of a cichlid fish. Horm Behav 61(2):212–217

Altmieme Z, Jubouri M, Touma K, Coté G, Fonseca M, Julian T, Mennigen JA (2019) A reproductive role for the nonapeptides vasotocin and isotocin in male zebrafish (Danio rerio). Comp Biochem Physiol B: Biochem Mol Biol 238:110333

Arnott G, Elwood RW (2009) Assessment of fighting ability in animal contests. Anim Behav 77(5):991–1004

Aubin-Horth NADIA, Desjardins JK, Martei YM, Balshine S, Hofmann HA (2007) Masculinized dominant females in a cooperatively breeding species. Mol Ecol 16(7):1349–1358

Backström T, Winberg S (2009) Arginine–vasotocin influence on aggressive behavior and dominance in rainbow trout. Physiol Behav 96(3):470–475

Balshine-Earn S, Neat FC, Reid H, Taborsky M (1998) Paying to stay or paying to breed? Field evidence for direct benefits of helping behavior in a cooperatively breeding fish. Behav Ecol 9(5):432–438

Banerjee P, Chaube R, Joy KP (2015) Molecular cloning, sequencing and tissue expression of vasotocin and isotocin precursor genes from ostariophysian catfishes: phylogeny and evolutionary considerations in teleosts. Front Neurosci 9(APR):1–16. https://doi.org/10.3389/fnins.2015.00166

Banerjee P, Joy KP, Chaube R (2017) Structural and functional diversity of nonapeptide hormones from an evolutionary perspective: a review. Gen Comp Endocrinol 241:4–23. https://doi.org/10.1016/j.ygcen.2016.04.025

Bastian J, Schniederjan S, Nguyenkim J (2001) Arginine vasotocin modulates a sexually dimorphic communication behavior in the weakly electric fish Apteronotus leptorhynchus. J Exp Biol 204(11):1909–1923

Batten TF, Moons L, Vandesande F (1999) Innervation and control of the adenohypophysis by hypothalamic peptidergic neurons in teleost fishes: EM immunohistochemical evidence. Microsc Res Tech 44(1):19–35

Beets I, Janssen T, Meelkop E, Temmerman L, Suetens N, Rademakers S, Jansen G, Schoofs L (2012) Vasopressin/oxytocin-related signaling regulates gustatory associative learning in C. elegans. Science 338(6106):543–545. https://doi.org/10.1126/science.1226860

Brinton RD, Gonzalez TM, Cheung WS (1994) Vasopressin-induced calcium signaling in cultured hippocampal neurons. Brain Res 661(1–2):174–180. https://doi.org/10.1016/0006-8993(94)91194-0

Canosa LF, Lopez GC, Scharrig E, Lesaux-Farmer K, Somoza GM, Kah O, Trudeau VL (2011) Forebrain mapping of secretoneurin-like immunoreactivity and its colocalization with isotocin in the preoptic nucleus and pituitary gland of goldfish. J Comp Neurol 519(18):3748–3765

Cardoso SC, Paitio JR, Oliveira RF, Bshary R, Soares MC (2015) Arginine vasotocin reduces levels of cooperative behaviour in a cleaner fish. Physiol Behav 139:314–320

Carreño Gutiérrez H, Colanesi S, Cooper B, Reichmann F, Young AM, Kelsh RN, Norton WH (2019) Endothelin neurotransmitter signalling controls zebrafish social behaviour. Sci Rep 9(1):3040

Chauvet J, Rouille Y, Chauveau C, Chauvet MT, Acher R (1994) Special evolution of neurohypophysial hormones in cartilaginous fishes: Asvatocin and phasvatocin, two oxytocin-like peptides isolated from the spotted dogfish (Scyliorhinus caniculus). Proc Natl Acad Sci USA 91(23):11266–11270. https://doi.org/10.1073/pnas.91.23.11266

Chen CC, Fernald RD (2011) Visual information alone changes behavior and physiology during social interactions in a cichlid fish (Astatotilapia burtoni). PLoS One 6(5):e20313

Chini B, Mouillac B, Ala Y, Balestre MN, Trumpp-Kallmeyer S, Hoflack J, Elands J, Hibert M, Manning M, Jard S, Barberis C (1995) Tyr115 is the key residue for determining agonist selectivity in the V1a vasopressin receptor. EMBO J 14(10):2176–2182. https://doi.org/10.1002/j.1460-2075.1995.tb07211.x

Cruz LJ, de Santos V, Zafaralla GC, Ramilo CA, Zeikus R, Gray WR, Olivera BM (1987) Invertebrate vasopressin/oxytocin homologs. Characterization of peptides from Conus geographus and Conus straitus venoms. J Biol Chem 262(33):15821–15824. https://doi.org/10.1016/s0021-9258(18)47661-2

Culbert BM, Gilmour KM, Balshine S (2019) Social buffering of stress in a group-living fish. Proc R Soc B 286(1910):20191626

Culbert BM, Ligocki IY, Salena MG, Wong MY, Hamilton IM, Bernier NJ, Balshine S (2024) Social regulation of arginine vasopressin and oxytocin systems in a wild group-living fish. Horm Behav 161:105521

Cunha-Saraiva F, Balshine S, Gozdowska M, Kulczykowska E, Wagner RH, Schaedelin FC (2019) Parental care and neuropeptide dynamics in a cichlid fish Neolamprologus caudopunctatus. Horm Behav 116:104576

Dabrowska J, Hazra R, Ahern TH, Guo JD, McDonald AJ, Mascagni F et al (2011) Neuroanatomical evidence for reciprocal regulation of the corticotrophin-releasing factor and oxytocin systems in the hypothalamus and the bed nucleus of the stria terminalis of the rat: implications for balancing stress and affect. Psychoneuroendocrinology 36(9):1312–1326

De Bree FM, Burbach JPH (1998) Structure-function relationships of the vasopressin prohormone domains. Cell Mol Neurobiol 18(2):173–191. https://doi.org/10.1023/A:1022564803093

DeAngelis R, Gogola J, Dodd L, Rhodes JS (2017) Opposite effects of nonapeptide antagonists on paternal behavior in the teleost fish Amphiprion ocellaris. Horm Behav 90:113–119

DeAngelis R, Dodd L, Rhodes J (2020) Nonapeptides mediate trade-offs in parental care strategy. Horm Behav 121:104717

Denver RJ (2009) Structural and functional evolution of vertebrate neuroendocrine stress systems. Ann N Y Acad Sci 1163(1):1–16

Deshmukh MK, Subhedar N (1993) Stretch sensory ovarian-preoptic pathway in the teleost, Clarias batrachus (Linn.): transection studies. J Exp Zool 267(4):359–364

Dimitriadou S, Santos EM, Croft DP, van Aerle R, Ramnarine IW, Filby AL, Darden SK (2022) Social partner cooperativeness influences brain oxytocin transcription in Trinidadian guppies (Poecilia reticulata). Behav Brain Res 423:113643

Egamii N, Ishii S (1962) Hypophyseal control of reproductive functions in teleost fishes. Gen Comp Endocrinol 1:248–253

Elphick MR, Rowe ML (2009) NGFFFamide and echinotocin: structurally unrelated myoactive neuropeptides derived from neurophysin-containing precursors in sea urchins. J Exp Biol 212(8):1067–1077. https://doi.org/10.1242/jeb.027599

Elphick MR, Mirabeau O, Larhammar D (2018) Correction: evolution of neuropeptide signalling systems (doi:10.1242/jeb.151092). J Exp Biol 221(19). https://doi.org/10.1242/jeb.193342

Facey DE, Bowen BW, Collette BB, Helfman GS (2022) The diversity of fishes: biology, evolution and ecology. Wiley

Fiedler K (1970) Hormonale auslösung der geburtsbewegungen beim seepferdchen (Hippocampus, Syngnathidae, Teleostei). Z Tierpsychol 27(6):679–686

Figueroa J, Morley SD, Heierhorst J, Krentler C, Lederis K, Richter D (1989) Two isotocin genes are present in the white sucker Catostomus commersoni both lacking introns in their protein coding regions. EMBO J 8(10):2873–2877. https://doi.org/10.1002/j.1460-2075.1989.tb08435.x

Flores CM, Muñoz D, Soto M, Kausel G, Romero A, Figueroa J (2007) Copeptin, derived from isotocin precursor, is a probable prolactin releasing factor in carp. Gen Comp Endocrinol 150(2):343–354. https://doi.org/10.1016/j.ygcen.2006.09.005

Froemke RC, Young LJ (2021) Oxytocin, neural plasticity, and social behavior. Annu Rev Neurosci 44:359–381

Garrison JL, Macosko EZ, Bernstein S, Pokala N, Albrecht DR, Bargmann CI (2012) Oxytocin/vasopressin-related peptides have an ancient role in reproductive behavior. Science 338(October):540–543. https://doi.org/10.1016/B978-0-12-819770-7.00052-9

Gemmer A, Mirkes K, Anneser L, Eilers T, Kibat C, Mathuru A et al (2022) Oxytocin receptors influence the development and maintenance of social behavior in zebrafish (Danio rerio). Sci Rep 12(1):4322

Gesto M, Soengas JL, Rodríguez-Illamola A, Míguez JM (2014) Arginine vasotocin treatment induces a stress response and exerts a potent anorexigenic effect in rainbow trout, Oncorhynchus mykiss. J Neuroendocrinol 26(2):89–99

Gimpl G, Fahrenholz F (2001) The oxytocin receptor system: structure, function, and regulation. Physiol Rev 81(2):629–683. https://doi.org/10.1152/physrev.2001.81.2.629

Godwin J, Thompson R (2012) Nonapeptides and social behavior in fishes. Horm Behav 61(3):230–238

Gołyszny M, Obuchowicz E, Zieliński M (2022) Neuropeptides as regulators of the hypothalamus-pituitary-gonadal (HPG) axis activity and their putative roles in stress-induced fertility disorders. Neuropeptides 91:102216

Gonçalves DM, Oliveira RF (2011) Hormones and sexual behavior of teleost fishes. In: Hormones and reproduction of vertebrates. Academic Press, pp 119–147

Gonçalves D, Alpedrinha J, Teles M, Oliveira RF (2007) Endocrine control of sexual behavior in sneaker males of the peacock blenny Salaria pavo: effects of castration, aromatase inhibition, testosterone and estradiol. Horm Behav 51(4):534–541

Goodson JL (2005) The vertebrate social behavior network: evolutionary themes and variations. Horm Behav 48(1):11–22

Goodson JL (2008) Nonapeptides and the evolutionary patterning of sociality. Prog Brain Res 170:3–15

Goodson JL, Bass AH (2000) Forebrain peptides modulate sexually polymorphic vocal circuitry. Nature 403(6771):769–772

Gozdowska M, Kleszczyńska A, Sokołowska E, Kulczykowska E (2006) Arginine vasotocin (AVT) and isotocin (IT) in fish brain: diurnal and seasonal variations. Comp Biochem Physiol B Biochem Mol Biol 143(3):330–334

Greenwood AK, Wark AR, Fernald RD, Hofmann HA (2008) Expression of arginine vasotocin in distinct preoptic regions is associated with dominant and subordinate behaviour in an African cichlid fish. Proc R Soc B Biol Sci 275(1649):2393–2402

Grigor'eva ME, Golubeva MG (2010) Oxytocin: structure, synthesis, receptors, and basic effects. Neurochem J 4(2):75–83. https://doi.org/10.1134/S1819712410020017

Grober MS, George AA, Watkins KK, Carneiro LA, Oliveira RF (2002) Forebrain AVT and courtship in a fish with male alternative reproductive tactics. Brain Res Bull 57(3-4):423–425

Gruber CW (2014) Physiology of invertebrate oxytocin and vasopressin neuropeptides. Exp Physiol 99(1):55–61. https://doi.org/10.1113/expphysiol.2013.072561

Gruber CW, Muttenthaler M (2012) Discovery of defense- and neuropeptides in social ants by genome-mining. PLoS One 7(3):1–12. https://doi.org/10.1371/journal.pone.0032559

Gwee PC, Tay BH, Brenner S, Venkatesh B (2009) Characterization of the neurohypophysial hormone gene loci in elephant shark and the Japanese lamprey: origin of the vertebrate neurohypophysial hormone genes. BMC Evol Biol 9(1):1–15. https://doi.org/10.1186/1471-2148-9-47

Hatton GI, Yang QZ (1996) Synaptically released histamine increases dye coupling among vasopressinergic neurons of the supraoptic nucleus: mediation by H1 receptors and cyclic nucleotides. J Neurosci 16(1):123–129

Hellmann JK, Reddon AR, Ligocki IY, O'Connor CM, Garvy KA, Marsh-Rollo SE et al (2015) Group response to social perturbation: impacts of isotocin and the social landscape. Anim Behav 105:55–62

Hicks C, Ramos L, Reekie T, Misagh GH, Narlawar R, Kassiou M, McGregor IS (2014) Body temperature and cardiac changes induced by peripherally administered oxytocin, vasopressin and the non-peptide oxytocin receptor agonist WAY 267,464: a biotelemetry study in rats. Br J Pharmacol 171(11):2868–2887. https://doi.org/10.1111/bph.12613

Holleman GA, Hooge IT, Kemner C, Hessels RS (2020) The 'real-world approach' and its problems: a critique of the term ecological validity. Front Psychol 11:529490

Hoyle CHV (1999) Neuropeptide families and their receptors: evolutionary perspectives. Brain Res 848(1–2):1–25. https://doi.org/10.1016/S0006-8993(99)01975-7

Huffman LS, O'Connell LA, Kenkel CD, Kline RJ, Khan IA, Hofmann HA (2012) Distribution of nonapeptide systems in the forebrain of an African cichlid fish, Astatotilapia burtoni. J Chem Neuroanat 44(2):86–97

Huffman LS, Hinz FI, Wojcik S, Aubin-Horth N, Hofmann HA (2015) Arginine vasotocin regulates social ascent in the African cichlid fish Astatotilapia burtoni. Gen Comp Endocrinol 212:106–113

Hyde JF, Ben-Jonathan N, North WG (1989) The vasopressin-associated glycopeptide is not a prolactin-releasing factor: studies with lactating brattleboro rats. Endocrinology 125(1):35–40. https://doi.org/10.1210/endo-125-1-35

Hyodo S, Tsukada T, Takei Y (2004) Neurohypophysial hormones of dogfish, Triakis scyllium: structures and salinity-dependent secretion. Gen Comp Endocrinol 138(2):97–104. https://doi.org/10.1016/j.ygcen.2004.05.009

Ivell R, Richter D (1984) Structure and comparison of the oxytocin and vasopressin genes from rat. Proc Natl Acad Sci U S A 81(7):2006–2010. https://doi.org/10.1073/pnas.81.7.2006

Jurek B, Neumann ID (2018) The oxytocin receptor: from intracellular signaling to behavior. Physiol Rev 98(3):1805–1908

Kanda A, Takuwa-Kuroda K, Iwakoshi-Ukena E, Minakata H (2003) Single exon structures of the oxytocin/vasopressin superfamily peptides of octopus. Biochem Biophys Res Commun 309(4):743–748. https://doi.org/10.1016/j.bbrc.2003.08.061

Kareklas K, Oliveira RF (2024) Oxytocin and social isolation: nonapeptide regulation of social homeostasis. In: Neuroendocrinology of behavior and emotions: environmental and social factors affecting behavior. Springer, Cham, pp 195–215

Kareklas K, Kunc HP, Arnott G (2022) Complex strategies: an integrative analysis of contests in Siamese fighting fish. BMC Zool 7(1):59

Kareklas K, Teles MC, Nunes AR, Oliveira RF (2023) Social zebrafish: Danio rerio as an emerging model in social neuroendocrinology. J Neuroendocrinol 35(9):e13280

Kawada T, Sekiguchi T, Itoh Y, Ogasawara M, Satake H (2008) Characterization of a novel vasopressin/oxytocin superfamily peptide and its receptor from an ascidian, Ciona intestinalis. Peptides 29(10):1672–1678. https://doi.org/10.1016/j.peptides.2008.05.030

Kawamata M, Mitsui-Saito M, Kimura T, Takayanagi Y, Yanagisawa T, Nishimori K (2003) Vasopressin-induced contraction of uterus is mediated solely by the oxytocin receptor in mice, but not in humans. Eur J Pharmacol 472(3):229–234. https://doi.org/10.1016/S0014-2999(03)01914-9

Knobloch HS, Grinevich V (2014) Evolution of oxytocin pathways in the brain of vertebrates. Front Behav Neurosci 8:31

Kulczykowska E, Kleszczyńska A (2014) Brain arginine vasotocin and isotocin in breeding female three-spined sticklebacks (Gasterosteus aculeatus): the presence of male and egg deposition. Gen Comp Endocrinol 204:8–12

Kulczykowska E, Kalamarz-Kubiak H, Nietrzeba M, Gozdowska M (2014) Brain nonapeptide and gonadal steroid responses to deprivation of heterosexual contact in the black molly. Biol Open 4(1):69–78

Kulczykowska E, Cardoso SC, Gozdowska M, André GI, Paula JR, Ślebioda M et al (2015) Brain levels of nonapeptides in four labrid fish species with different levels of mutualistic behavior. Gen Comp Endocrinol 222:99–105

Lema SC, Nevitt GA (2004) Exogenous vasotocin alters aggression during agonistic exchanges in male Amargosa River pupfish (Cyprinodon nevadensis amargosae). Horm Behav 46(5):628–637

Lema SC (2010) Identification of multiple vasotocin receptor cDNAs in teleost fish: sequences, phylogenetic analysis, sites of expression, and regulation in the hypothalamus and gill in response to hyperosmotic challenge. Mol Cell Endocrinol 321(2):215–230

Lema SC, Sanders KE, Walti KA (2015) Arginine vasotocin, isotocin and nonapeptide receptor gene expression link to social status and aggression in sex-dependent patterns. J Neuroendocrinol 27(2):142–157

Li J, Lyu L, Wen H, Li Y, Wang X, Yao Y, Qi X (2024) Estrogen regulates the transcription of guppy isotocin receptors. Comp Biochem Physiol B: Biochem Mol Biol 269:110895

Lindeyer CM, Langen EM, Swaney WT, Reader SM (2015) Nonapeptide influences on social behaviour: effects of vasotocin and isotocin on shoaling and interaction in zebrafish. Behaviour 152(7-8):897–915

Loveland JL, Fernald RD (2017) Differential activation of vasotocin neurons in contexts that elicit aggression and courtship. Behav Brain Res 317:188–203

Lowry CA, Moore FL (2006) Regulation of behavioral responses by corticotropin-releasing factor. Gen Comp Endocrinol 146(1):19–27

Lyu LK, Li JS, Wang XJ, Yao YJ, Li JF, Li Y, Wen HS, Qi X (2021) Arg-vasotocin directly activates isotocin receptors and induces COX2 expression in ovoviviparous guppies. Front Endocrinol 12(April):1–15. https://doi.org/10.3389/fendo.2021.617580

Macey MJ, Pickford GE, Peter RE (1974) Forebrain localization of the spawning reflex response to exogenous neurohypophysial hormones in the killifish, Fundulus heteroclitus. J Exp Zool 190(3):269–279

Maejima K, Oka Y, Park MK, Kawashima S (1994) Immunohistochemical double-labeling study of gonadotropin-releasing hormone (GnRH)-immunoreactive cells and oxytocin-immunoreactive cells in the preoptic area of the dwarf gourami, Colisa lalia. Neurosci Res 20(2):189–193

Martos-Sitcha JA, Cádiz L, Gozdowska M, Kulczykowska E, Martínez-Rodríguez G, Mancera JM (2019) Arginine vasotocin and cortisol co-regulate vasotocinergic, isotocinergic, stress, and thyroid pathways in the gilthead sea bream (Sparus aurata). Front Physiol 10:443407

Maruska KP, Anselmo CM, King T, Mobley RB, Ray EJ, Wayne R (2022) Endocrine and neuroendocrine regulation of social status in cichlid fishes. Horm Behav 139:105110

Mayasich SA, Clarke BL (2016) The emergence of the vasopressin and oxytocin hormone receptor gene family lineage: clues from the characterization of vasotocin receptors in the sea lamprey (Petromyzon marinus). Gen Comp Endocrinol 226:88–101. https://doi.org/10.1016/j.ygcen.2016.01.001

Mayasich SA, Clarke BL (2020) Vasotocin and the origins of the vasopressin/oxytocin receptor gene family. Vitam Horm 113:1–27

Mehr BA (2022) Effects of the brain Nonapeptides arginine-vasotocin and isotocin on shoaling behaviour in the guppy (Poecilia Reticulata). Doctoral dissertation, The University of Western Ontario, Canada

Mendonça R, Soares MC, Bshary R, Oliveira RF (2013) Arginine vasotocin neuronal phenotype and interspecific cooperative behaviour. Brain Behav Evol 82(3):166–176

Mennigen JA, Martyniuk CJ, Crump K, Xiong H, Zhao E, Popesku J et al (2008) Effects of fluoxetine on the reproductive axis of female goldfish (Carassius auratus). Physiol Genomics 35(3):273–282

Mennigen JA, Volkoff H, Chang JP, Trudeau VL (2017) The nonapeptide isotocin in goldfish: evidence for serotonergic regulation and functional roles in the control of food intake and pituitary hormone release. Gen Comp Endocrinol 254:38–49

Mennigen JA, Ramachandran D, Shaw K, Chaube R, Joy KP, Trudeau VL (2022) Reproductive roles of the vasopressin/oxytocin neuropeptide family in teleost fishes. Front Endocrinol 13:1005863

Motohashi E, Hamabata T, Ando H (2008) Structure of neurohypophysial hormone genes and changes in the levels of expression during spawning season in grass puffer (Takifugu niphobles). Gen Comp Endocrinol 155(2):456–463

Mouillac B, Chini B, Balestre M-N, Elands J, Trumpp-Kallmeyer S, Hoflack J, Hibert M, Jard S, Barberis C (1995) The binding site of neuropeptide vasopressin V1a receptor. J Biol Chem 270(43):25771–25777. https://doi.org/10.1074/jbc.270.43.25771

Nagy G, Mulchahey J, Smith D, Neill J (1988) The glycopeptide moiety of Vasopressin-neurophysin precursor is neurohypophysial prolactin releasing factor. Biochem Biophys Res Commun 151(1):524–529

Newman SW (1999) The medial extended amygdala in male reproductive behavior a node in the mammalian social behavior network. Ann N Y Acad Sci 877(1):242–257

Nunes AR, Carreira L, Anbalagan S, Blechman J, Levkowitz G, Oliveira RF (2020) Perceptual mechanisms of social affiliation in zebrafish. Sci Rep 10(1):3642

Nunes AR, Gliksberg M, Varela SA, Teles M, Wircer E, Blechman J et al (2021) Developmental effects of oxytocin neurons on social affiliation and processing of social information. J Neurosci 41(42):8742–8760

O'Connell LA, Hofmann HA (2012) Evolution of a vertebrate social decision-making network. Science 336(6085):1154–1157

Ocampo Daza D, Lewicka M, Larhammar D (2012) The oxytocin/vasopressin receptor family has at least five members in the gnathostome lineage, inclucing two distinct V2 subtypes. Gen Comp Endocrinol 175(1):135–143. https://doi.org/10.1016/j.ygcen.2011.10.011

Ocampo Daza D, Bergqvist CA, Larhammar D (2022) The evolution of oxytocin and vasotocin receptor genes in jawed vertebrates: a clear case for gene duplications through ancestral whole-genome duplications. Front Endocrinol 12(February):1–21. https://doi.org/10.3389/fendo.2021.792644

O'Connell LA, Hofmann HA (2011) The vertebrate mesolimbic reward system and social behavior network: a comparative synthesis. J Comp Neurol 519(18):3599–3639

O'Connell LA, Matthews BJ, Hofmann HA (2012) Isotocin regulates paternal care in a monogamous cichlid fish. Horm Behav 61(5):725–733

O'Connor CM, Marsh-Rollo SE, Aubin-Horth N, Balshine S (2016) Species-specific patterns of nonapeptide brain gene expression relative to pair-bonding behavior in grouping and non-grouping cichlids. Horm Behav 80:30–38

Ogawa S, Pfaff DW, Parhar IS (2021) Fish as a model in social neuroscience: conservation and diversity in the social brain network. Biol Rev 96(3):999–1020

Oldfield RG, Hofmann HA (2011) Neuropeptide regulation of social behavior in a monogamous cichlid fish. Physiol Behav 102(3–4):296–303

Oldfield RG, Harris RM, Hendrickson DA, Hofmann HA (2013) Arginine vasotocin and androgen pathways are associated with mating system variation in North American cichlid fishes. Horm Behav 64(1):44–52

Oliveira JA, da Silva Souza JG, de Jesus Paula DA, Virote BDCR, Murgas LDS (2022) Oxytocin reduces the frequency of aggressive behaviours in male betta fish (Betta splendens). Behav Process 200:104689

Oumi T, Ukena K, Matsushima O, Ikeda T, Fujita T, Minakata H, Nomoto K (1996) Annetocin, an annelid oxytocin-related peptide, induces egg-laying behavior in the earthworm, Eisenia foetida. J Exp Zool 276(2):151–156. https://doi.org/10.1002/(SICI)1097-010X(19961001)276:2<151::AID-JEZ8>3.0.CO;2-N

Pickford GE, Strecker EL (1977) The spawning reflex response of the killifish, Fundulus heteroclitus: isotocin is relatively inactive in comparison with arginine vasotocin. Gen Comp Endocrinol 32(2):132–137

Ramallo MR, Grober M, Cánepa MM, Morandini L, Pandolfi M (2012) Arginine-vasotocin expression and participation in reproduction and social behavior in males of the cichlid fish Cichlasoma dimerus. Gen Comp Endocrinol 179(2):221–231

Ramos L, Hicks C, Caminer A, Couto K, Narlawar R, Kassiou M, McGregor IS (2016) MDMA ('ecstasy'), oxytocin and vasopressin modulate social preference in rats: a role for handling and oxytocin receptors. Pharmacol Biochem Behav 150–151:115–123. https://doi.org/10.1016/j.pbb.2016.10.002

Ravi V, Venkatesh B (2018) The divergent genomes of teleosts. Annu Rev Anim Biosci 6:47–68

Reddon AR, O'Connor CM, Marsh-Rollo SE, Balshine S (2012) Effects of isotocin on social responses in a cooperatively breeding fish. Anim Behav 84(4):753–760

Reddon AR, Voisin MR, O'Connor CM, Balshine S (2014) Isotocin and sociality in the cooperatively breeding cichlid fish, Neolamprologus pulcher. Behaviour 151(10):1389–1411

Reddon AR, O'Connor CM, Marsh-Rollo SE, Balshine S, Gozdowska M, Kulczykowska E (2015) Brain nonapeptide levels are related to social status and affiliative behaviour in a cooperatively breeding cichlid fish. R Soc Open Sci 2(2):140072

Reddon AR, O'Connor CM, Nesjan E, Cameron J, Hellmann JK, Ligocki IY et al (2017) Isotocin neuronal phenotypes differ among social systems in cichlid fishes. R Soc Open Sci 4(5):170350

Reddon AR, Aubin-Horth N, Reader SM (2022) Wild guppies from populations exposed to higher predation risk exhibit greater vasotocin brain gene expression. J Zool 316(2):118–127

Reich G (1992) A new peptide of the oxytocin/vasopressin family isolated from nerves of the cephalopod Octopus vulgaris. Neurosci Lett 134(2):191–194. https://doi.org/10.1016/0304-3940(92)90514-8

Rémy C, Girardie J, Dubois MP (1979) Vertebrate neuropeptide-like substances in the suboesophageal ganglion of two insects: Locusta migratoria R. and F. (Orthoptera) and Bombyx mori L. (Lepidoptera) immunocytological investigation. Gen Comp Endocrinol 37(1):93–100. https://doi.org/10.1016/0016-6480(79)90050-9

Renn SC, Aubin-Horth N, Hofmann HA (2008) Fish and chips: functional genomics of social plasticity in an African cichlid fish. J Exp Biol 211(18):3041–3056

Richter D, Schmale H (1983) The structure of the precursor to arginine-vasopressin: a model preprohormone. Prog Brain Res 60(C):227–233. https://doi.org/10.1016/S0079-6123(08)64390-3

Rigney N, de Vries GJ, Petrulis A, Young LJ (2022) Oxytocin, vasopressin, and social behavior: from neural circuits to clinical opportunities. Endocrinology 163(9):bqac111

Ripley JL, Foran CM (2010) Quantification of whole brain arginine vasotocin for two Syngnathus pipefishes: elevated concentrations correlated with paternal brooding. Fish Physiol Biochem 36:867–874

Saito D, Hasegawa Y, Urano A (2003) Gonadotropin-releasing hormones modulate electrical activity of vasotocin and isotocin neurons in the brain of rainbow trout. Neurosci Lett 351(2):107–110

Saito D, Komatsuda M, Urano A (2004) Functional organization of preoptic vasotocin and isotocin neurons in the brain of rainbow trout: central and neurohypophysial projections of single neurons. Neuroscience 124(4):973–984

Salahinejad A, Naderi M, Attaran A, Meuthen D, Niyogi S, Chivers DP (2020) Effects of chronic exposure to bisphenol-S on social behaviors in adult zebrafish: disruption of the neuropeptide signaling pathways in the brain. Environ Pollut 262:113992

Salek SJ, Sullivan CV, Godwin J (2002) Arginine vasotocin effects on courtship behavior in male white perch (Morone americana). Behav Brain Res 133(2):177–183

Santangelo N, Bass AH (2006) New insights into neuropeptide modulation of aggression: field studies of arginine vasotocin in a territorial tropical damselfish. Proc R Soc B Biol Sci 273(1605):3085–3092

Sawchenko PE, Swanson LW (1982) The organization of noradrenergic pathways from the brainstem to the paraventricular and supraoptic nuclei in the rat. Brain Res Rev 4(3):275–325

Semsar K, Godwin J (2004) Multiple mechanisms of phenotype development in the bluehead wrasse. Horm Behav 45(5):345–353

Semsar K, Kandel FL, Godwin J (2001) Manipulations of the AVT system shift social status and related courtship and aggressive behavior in the bluehead wrasse. Horm Behav 40(1):21–31

Silva AC, Perrone R, Zubizarreta L, Batista G, Stoddard PK (2013) Neuromodulation of the agonistic behavior in two species of weakly electric fish that display different types of aggression. J Exp Biol 216(13):2412–2420

Singh V, Joy KP (2009) Relative in vitro seasonal effects of vasotocin and isotocin on ovarian steroid hormone levels in the catfish Heteropneustes fossilis. Gen Comp Endocrinol 162(3):257–264

Soares MC, Bshary R, Mendonça R, Grutter AS, Oliveira RF (2012) Arginine vasotocin regulation of interspecific cooperative behaviour in a cleaner fish. PLoS One 7(7):e39583

Solomon-Lane TK, Butler RM, Hofmann HA (2022) Vasopressin mediates nonapeptide and glucocorticoid signaling and social dynamics in juvenile dominance hierarchies of a highly social cichlid fish. Horm Behav 145:105238

Song Z, Albers HE (2018) Cross-talk among oxytocin and arginine-vasopressin receptors: relevance for basic and clinical studies of the brain and periphery. Front Neuroendocrinol 51(2017):14–24. https://doi.org/10.1016/j.yfrne.2017.10.004

Subhedar N, Krishna NR, Deshmukh MK (1987) The response of nucleus preopticus neurosecretory cells to ovarian pressure in the teleost, Clarias batrachus (Linn.). Gen Comp Endocrinol 68(3):357–368

Subhedar N, Deshmukh MK, Jain MR, Khan FA, Krishna NR (1996) Activation of hypothalamic neurons by intraovarian pressure signals in a teleost fish, Clams batrachus: role of mechanosensitive channels. Brain Behav Evol 47(4):179–184

Suzuki M, Kubokawa K, Nagasawa H, Urano A (1995) Sequence analysis of vasotocin cDNAs of the lamprey, Lampetra japonica, and the hagfish, Eptatretus burgeri: evolution of cyclostome vasotocin precursors. J Mol Endocrinol 14(1):67–77. https://doi.org/10.1677/jme.0.0140067

Teles MC, Oliveira RF (2016) Androgen response to social competition in a shoaling fish. Horm Behav 78:8–12

Teles MC, Gozdowska M, Kalamarz-Kubiak H, Kulczykowska E, Oliveira RF (2016) Agonistic interactions elicit rapid changes in brain nonapeptide levels in zebrafish. Horm Behav 84:57–63

Theofanopoulou C, Gedman G, Cahill JA, Boeckx C, Jarvis ED (2021) Universal nomenclature for oxytocin–vasotocin ligand and receptor families. Nature 592(7856):747–755. https://doi.org/10.1038/s41586-020-03040-7

Thompson RR, Walton JC (2004) Peptide effects on social behavior: effects of vasotocin and isotocin on social approach behavior in male goldfish (Carassius auratus). Behav Neurosci 118(3):620–626. https://doi.org/10.1037/0735-7044.118.3.620

Triki Z, Daughters K, De Dreu CK (2022) Oxytocin has 'tend-and-defend' functionality in group conflict across social vertebrates. Philos Trans R Soc B 377(1851):20210137

Van Kesteren RE, Smit AB, De Lange RPJ, Kits KS, Van Golen FA, Van Der Schors RC, De With ND, Burke JF, Geraerts WPM (1995) Structural and functional evolution of the vasopressin/oxytocin superfamily: vasopressin-related conopressin is the only member present in Lymnaea, and is involved in the control of sexual behavior. J Neurosci 15(9):5989–5998. https://doi.org/10.1523/jneurosci.15-09-05989.1995

Voelkl B, Altman NS, Forsman A, Forstmeier W, Gurevitch J, Jaric I et al (2020) Reproducibility of animal research in light of biological variation. Nat Rev Neurosci 21(7):384–393

Wallace KJ, Hofmann HA (2021) Decision-making in a social world: integrating cognitive ecology and social neuroscience. Curr Opin Neurobiol 68:152–158

Waltenspühl Y, Ehrenmann J, Vacca S, Thom C, Medalia O, Plückthun A (2022) Structural basis for the activation and ligand recognition of the human oxytocin receptor. Nat Commun 13(1):1–9. https://doi.org/10.1038/s41467-022-31325-0

Watanabe Y, Sakihara T, Mukuda T, Ando M (2007) Antagonistic effects of vasotocin and isotocin on the upper esophageal sphincter muscle of the eel acclimated to seawater. J Comp Physiol B 177(8):867–873. https://doi.org/10.1007/s00360-007-0184-1

Wendelaar Bonga SE (1997) The stress response in fish. Physiol Rev 77(3):591–625

Wilhelmi AE, Pickford GE, Sawyer WH (1955) Initiation of the spawning reflex response in Fundulus by the administration of fish and mammalian neurohypophysial preparations and synthetic oxytocin. Endocrinology 57(2):243–252

Winter J, Jurek B (2019) The interplay between oxytocin and the CRF system: regulation of the stress response. Cell Tissue Res 375:85–91

Wircer E, Ben-Dor S, Levkowitz G (2016) Non-mammalian models for neurohypophysial peptides. In: Molecular neuroendocrinology: from genome to physiology. Wiley, pp 301–328. https://doi.org/10.1002/9781118760369.ch14

Yamaguchi Y, Kaiya H, Konno N, Iwata E, Miyazato M, Uchiyama M, Bell JD, Toop T, Donald JA, Brenner S, Venkatesh B, Hyodo S (2012) The fifth neurohypophysial hormone receptor is structurally related to the V2-type receptor but functionally similar to V1-type receptors. Gen Comp Endocrinol 178(3):519–528. https://doi.org/10.1016/j.ygcen.2012.07.008

Yang W, Zhang N, Shi B, Zhang S, Zhang L, Zhang W (2018) Isotocin regulates growth hormone but not prolactin release from the pituitary of ricefield eels. Front Endocrinol 9:350426

Yokoi S, Okuyama T, Kamei Y, Naruse K, Taniguchi Y, Ansai S et al (2015) An essential role of the arginine vasotocin system in mate-guarding behaviors in triadic relationships of medaka fish (Oryzias latipes). PLoS Genet 11(2):e1005009

Yokoi S, Naruse K, Kamei Y, Ansai S, Kinoshita M, Mito M et al (2020) Sexually dimorphic role of oxytocin in medaka mate choice. Proc Natl Acad Sci 117(9):4802–4808

Yves R, Chauvet M-T, Chauvet J, Archer R (1988) Dual duplication of neurohypophysial hormones in an Australian marsupial: mesotocin, oxytocin, lysine vasopressin and arginine vasopressin in a single gland of the northern bandicoot (Isoodon macrourus). Biochem Biophys Res Commun 154(1):346–350

Zhang J, Sato M, Duzic E, Kubalak SW, Lanier SM, Webb JG (1997) Adenylyl cyclase isoforms and vasopressin enhancement of agonist-stimulated cAMP in vascular smooth muscle cells. Am J Physiol Heart Circ Physiol 273(2:42–42. https://doi.org/10.1152/ajpheart.1997.273.2.h971

Zhang B, Qiu L, Xiao W, Ni H, Chen L, Wang F et al (2021) Reconstruction of the hypothalamo-neurohypophysial system and functional dissection of magnocellular oxytocin neurons in the brain. Neuron 109(2):331–346

Insights and Future Perspectives of OT-Like Peptides in Birds

16

Lubrano Saverio and Canoine Virginie

Abstract

Oxytocin (OT) is a widely studied hormone since it plays a central role in shaping and regulating many endogenous and behavioural mechanisms important for social behaviour in mammals. The avian OT, formerly called Mesotocin (MT) appears to have a similar influence. However, its function and control mechanisms in birds are still unclear as research on avian OT was scarce and only recently received more attention thanks to the development of novel and less-invasive methods. OT in birds is associated with parental care and social behaviours such as mating, nest-building and chick-rearing, as well as physiological dynamics such as water balance, oviposition and diuresis. In this review, we try to describe the role and control mechanisms of OT in regulating behaviour and physiology in birds, summarizing information from over 60 years of literature. First, we will elucidate the state-of-the-art findings about OT evolution across taxa and report recent discussions regarding nomenclature. Then, we will describe the hormone action in the central and peripheral system, including its behavioural influences and its interaction with another closely related peptide, vasotocin. To conclude, we will describe novel methods used to quantify OT, underlining their pros and their cons.

L. Saverio
Centre for the Advanced Study of Collective Behaviour, Max-Planck Institute for Animal Behaviour, Konstanz, Germany

Department of Animal Behaviour, University of Konstanz, Konstanz, Germany
e-mail: saverio.lubrano@uni-konstanz.de

C. Virginie (✉)
Department of Behavioural and Cognitive Biology, University of Vienna, Vienna, Austria
e-mail: virginie.canoine@univie.ac.at

V. Grinevich, R. F. Oliveira (eds.), *Evolutionary and Comparative Neuroendocrinology*, Masterclass in Neuroendocrinology 17,
https://doi.org/10.1007/978-3-031-80209-6_16

Keywords

Neuropeptides · Avian-oxytocin · Mesotocin · Hypothalamus · Social behaviour

List of Abbreviations

BNSTm	Bed nucleus of the stria terminalis
HPA	Hypothalamus-pituitary-adrenal
HPG	Hypothalamus-pituitary-gonadal
LS	Lateral septum
OT	Oxytocin
OT/VT-ir	Oxytocin/vasotocin immunoreactive
PVN	Paraventricular nucleus of the hypothalamus
RIA	RadioImmuno Assay
SON	Supraoptic nucleus of the hypothalamus
VP	Vasopressin
VT	Vasotocin

16.1 Introduction

Since the nonapeptide oxytocin (OT) was crystallized and described by Du Vigneaud in 1953 (du Vigneaud et al. 1953), many studies have demonstrated its importance in regulating social, maternal and reproductive behaviours. In mammals, the dual importance of OT has been well documented: first, as a neuroendocrine hormone modulating various social behaviours (Froemke and Young 2021; Menon and Neumann 2023) and, second, in the periphery stimulating labour during childbirth and in the regulation of lactation. Extensive research by Young and colleagues has unfolded the importance of OT in regulating social monogamy by comparing the monogamous prairie vole (*Microtus ochrogaster*) with other vole species which differ in their mating systems (Donaldson and Young 2008; Froemke and Young 2021). Since then, the evolutionary significance of OT in regulating social monogamy and bi-parenting behaviours has been marked (Carter 2010). OT is also involved in regulating mate preference, trust, pair- and maternal bonding and can have opposite/aversive behaviours depending on the familiarity of the opponent (Goodson et al. 2012; Neumann and Slattery 2016). Therefore, it is not surprising that OT is often known as the "love" or "empathy" hormone (Carter 2020).

Together with another neuropeptide, vasopressin (VP), a homologue of the phylogenetically ancient vasotocin (VT), OT has been shown to regulate aggressive behaviour, anxiety and depression (Neumann and Landgraf 2012). However, our current knowledge on OT derives mostly from studies conducted in classical study models such as rats, apes, and humans—all mammals. It is comprehensible that focussing research on classical mammalian models was aimed at better understanding the regulation of OT in humans. However, this approach offers only a snapshot on the function of OT in animals as both nonapeptides, OT-like and VT-like, are evolutionarily conserved and have existed for more than 600 million years. Even though the endocrine mechanisms of OT-like peptides might be quite similar across species, its neuroendocrinological regulation remains still obscure in non-classical study species. Yet, the phylogenetic history and evolutionary processes that led to OT function in mammals are strictly linked to the peculiarity of their reproductive and parental physiology, i.e. complex social behaviours, lactation and labour.

How did this hormone evolve/diverge to its present specialized forms? What were its original functions? Which evolutionary constraints have led to such complex divergent and specific regulatory mechanisms? Is its role in modulating social behaviour a novel function, or have OT-like peptides always been involved in prosocial behaviour? To answer these questions, it is essential to study this peptide from a broader perspective, namely, by studying the behavioural neuroendocrinology of other non-mammalian taxa such as invertebrates, insects, reptiles or birds. In this review we provide a general overview of OT-like peptides in birds. We will present general insights but also discuss limitations and difficulties in this field, which might explain why little research has still been done in other vertebrates compared to mammals.

16.2 Evolution and (Bio)Chemical Structure, Homology in Birds

16.2.1 Why Birds?

Birds represent an ideal study model in which to investigate the behavioural endocrinology of OT-like peptides. Compared to mammals, more than 75% of bird species are socially monogamous, but they present a large variability of mating systems such as promiscuity or polygamy. Within those mating systems, various parenting systems can be found, such as bi-parenting, alloparenting or even sex-role reversal, e.g. Black coucals (*Centropus grillii*). Some species breed in large groups (e.g. colonial breeders), whereas others, such as many *Corvidae* and Fairy-Wrens (*Malurus cyaneus*) form complex social structures (Camerlenghi et al. 2023). This variability in the social, mating and breeding system, most of which is analogous across species of different families, offers the opportunity to select the appropriate model to study behavioural endocrinology of OT, with the possibility of analysing its variation and evolution across various environmental conditions/niches. In fact, *Neuroecology* is a new field emerging within the neurosciences that exploits the opportunity of a wider perspective to understand the evolution of regulation mechanisms.

16.2.2 Evolution and Homology of OT-Like Peptides

Nonapeptides are estimated to have existed since more than 600 million years ago and are evolutionarily well conserved. Indeed, all vertebrates have the ancient nonapeptide VT, homologous to mammalian VP (Acher and Chauvet 1995; Donaldson and Young 2008). Following a gene duplication, a second lineage emerged, resulting in two distinct but evolutionarily very close lineages: VT, the ancient peptide that is preserved and evolutionarily quite stable, and OT-like peptides, which appear less stable (Fig. 16.1). Although the sequences are quite similar, biochemical studies revealed minor changes in amino acids of these two nonapeptides and their respective receptors, leading to a variety of nomenclatures (Theofanopoulou 2021). Birds together with reptiles and amphibians share the same OT-like peptide, traditionally called mesotocin (MT), which is homologous to the mammalian OT. Birds have one OT receptor (OTR) and four VT receptors (VTR1A, VTR1B; VTR2A, VTR2B; (Leng and Ludwig 2016; Theofanopoulou 2021) Fig. 16.1) which all have a high affinity for both OT and VT. Birds and mammals are the only taxa with two less active receptors compared to jawed vertebrates. Differences in the nomenclature contributed to extensive confusion among scientists working on OT-and VT-like peptides and may be the reason why research in this field often remained confined within taxa. Recently, Theofanopoulou has shown in a large genetic analysis that both VT and OT-like peptides and their respective receptors are genetically identical to the human VP and OT, as well as their receptors (Theofanopoulou 2021). The

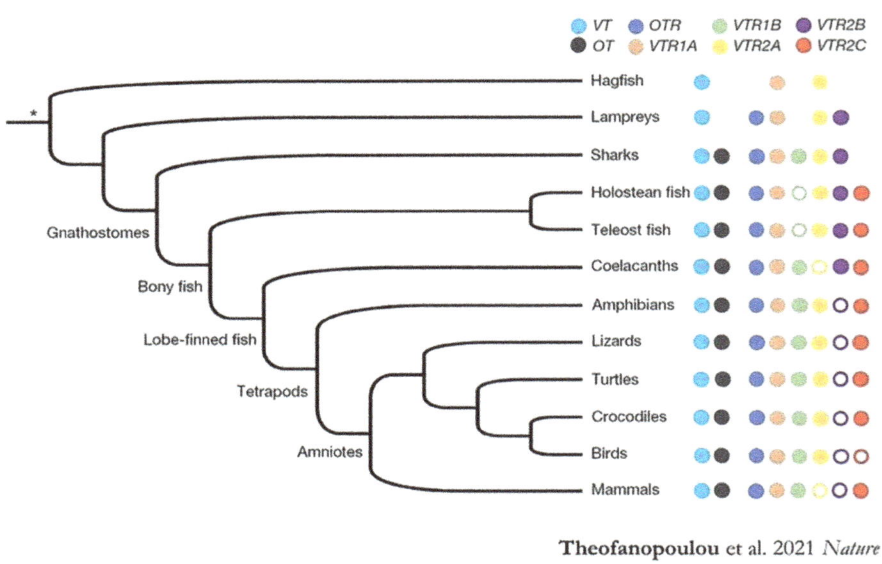

Fig. 16.1 From Theofanopoulou et al. (2021); "Phylogenetic distribution of OT, VT and OTR-VTR genes among vertebrates". OTR and VTR genes codes for the receptors. Birds don't code for two receptors; however, the remaining receptors show high affinity towards both VT and OT

authors thus proposed to simplify the nomenclature to facilitate unification of the research field (Theofanopoulou et al. 2021). In this review, we will follow the recommendation of Theofanopoulou and use the term OT also for avian MT.

16.3 Pathways, Regulation Mechanisms and Behavioural Influences

The study of the control mechanism of OT and VT neuropeptides in birds is quite complex as it presents several challenges. Unlike other hormones such as corticosterone or testosterone, the central pathway eliciting OT release in birds remains somehow obscure. We will try in this review to elucidate this pathway, tracing the journey from signal acquisition to integration within the central and peripheral systems.

Nonapeptides in birds share the same precursors as mammals, namely, VLDV-neurophysin for OT and MSEL-neurophysin for VT (Chauvet et al. 1986). While VT-producing neurons are distributed across the brain, OT is synthesized mainly in parvo- and magnocellular neurons located within the paraventricular (PVN) and supraoptic nuclei (SON) of the hypothalamus (Fig. 16.2). These neuro-sensitive

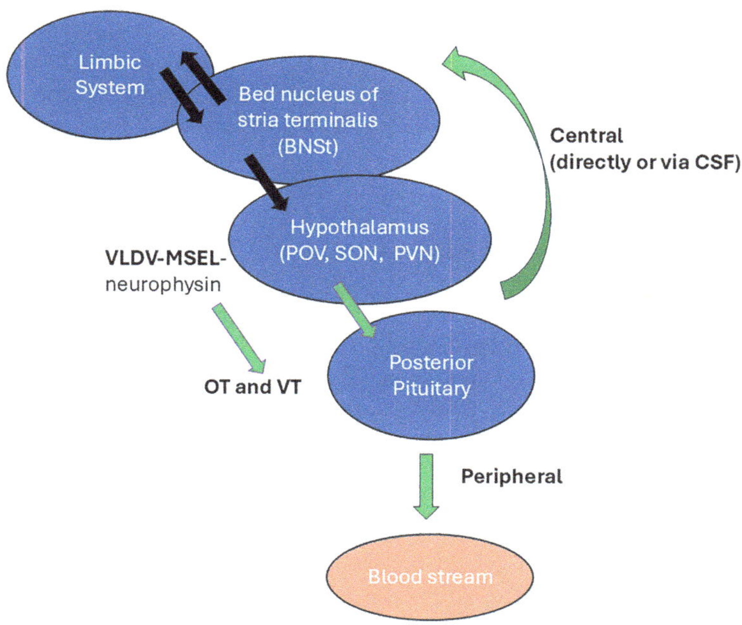

Fig. 16.2 Cycle of avian-OT and avian-VT production in the brain and secretion into the bloodstream. The afferent information spread centrally through the BNSTm, and it is processed in the POV, SON (only OT) and PVN of the hypothalamus. There the neurophysin VLDV and MSEL become namely OT and VT. The molecules, once ready, can be either released into the blood stream through the posterior pituitary, stored in vesicles for neurotransmission or be released into the cerebrospinal fluid

Table 16.1 Distribution and relative abundance of VT and OT immunoreactive neurons across the avian brain (Robinzon et al. 1988a) See also Haakenson et al. 2022.

Hormones	MT	MT > VP	MT = VP	VP > MT
Areas	Cerebellum	Medulla oblongata Medial and posterior hypothalamus	Septal area Archistriatum (arcopallium) Anterior hypothalamus	Paleostriatum Optic lobe Midbrain Anterior hypothalamus

neurons receive input from higher brain levels and via action potentials through their axons, projecting across the hypothalamus-hypophyseal tract. From here, the neuropeptides are released from neurosecretory vesicles into the blood system (Acher and Chauvet 1988; Chauvet et al. 1986; Goodson and Kingsbury 2011; Korf et al. 1988; Thayananuphat et al. 2011; Triki et al. 2022). Although hypothalamic parvo- and magnocellular neurons can synthesize both OT and VT, they show a specialization in producing either OT or VT (only a few neurons produce both at the same times (Goossens et al. 1977). OT-ir neurons and fibres have been highlighted in basically all areas of the brain such as the cerebellum, lateral septum, optic lobe and medulla oblongata (Table 16.1) (Robinzon et al. 1988b; Aste et al. 1998; Goodson et al. 2009c), consistently in many bird species such as the domestic fowl (*Gallus gallus*; (Kamkrathok et al. 2017; Robinzon et al. 1988a)), dark-eyed junco (*Junco hyemalis*; (Panzica et al. 1999), blue tit (*Cyanistes caeruleus*; (Montagnese et al. 2015) and zebra finch (*Taeniopygia guttata*; Haakenson et al 2022, (Goodson et al. 2009b) among others. OT and VT both act within the brain (centrally) through their projections to important brain-nuclei and in the periphery in various target tissues, where they have distinct functions.

16.3.1 Centrally Regulated Behaviours

Within the brain, OT plays an important role in regulating social, aggressive and sexual behaviour, among others (Goodson 2013). Differences in the quantity of OT and VT-ir in various brain areas underpin their complex involvement in a wide range of behaviour (Table 16.1). The nexus of complexity increases considering that birds have one OTR and four VT receptors, all of which have a high affinity to both OT and VT (Fig. 16.1). While the structures associated with VT are highly conserved across birds (Theofanopoulou et al. 2021), changes in the number of active OT- and VP-inergic neurons innervating the different areas of the brain appear to be species-specific and vary with sex, life history stages, across seasons and social preferences (Chokchaloemwong et al. 2013; Goodson et al. 2009b; Haakenson et al. 2022; Plumari et al. 2004). All main brain areas in which OTR and VTR are expressed and to which the fibres project are important for reproductive, social and aggressive behaviour, in which androgen, oestrogen and/or glucocorticoid receptors are colocalized. This underlies an interaction of OT with other endocrine systems such as the hypothalamus-pituitary-gonadal (HPG-) and/or adrenal (HPA-) axis.

Nonapeptides interact with other hormones such as steroid hormones (Jurkevich et al. 1999). In male domestic fowl, a decrease in testosterone concentrations due to castration or ageing results in reduced VT-ir (VT-immunoreactive) innervations of the lateral septum (LS), while administration of exogenous testosterone elicits an increase of such innervations (Viglietti-Panzica et al. 1994). Further, male zebra finches starting breeding activity express almost twice as many VT-ir neurons in the bed nucleus of the stria terminalis (BNSTm) as females (Kelly and Goodson 2014). These findings indicate that VP- and OT-inergic innervations are sexually dimorphic and are dependent on gonadal steroid levels. This is not surprising, considering that OT and VP play an important role in reproduction, which is often associated with seasonally increased sex steroids and during which individuals express sex-specific tasks and behaviour. Seasonal changes in OT might be more pronounced in seasonal breeders than in opportunistic breeders, which rely more on day length and have distinct reproductive windows. This is supported by a study in which female canaries, which are seasonal breeders, kept in short day conditions were found to express fewer OT-ir neurons in LS than male canaries and both female and male zebra finches, which are opportunistic breeders (Haakenson et al. 2022). However, it is likely that multiple factors are involved in the OT cycle stimulation: sex steroids increase the sensitivity to perceive and evaluate visual and acoustic stimuli of conspecific individuals, which are essential during courtship and territorial displays.

Recent studies have shown that OT receptors are present in nuclei of the song system, important for the performance and the perception of vocalization (Haakenson et al. 2022). In addition, it has been suggested that VT may act as a promoter of social motivation for vocal learning and language (Thayananuphat et al. 2011; Theofanopoulou et al. 2017). Visual and acoustic stimuli are essential for attachment not only in adults but also in juveniles. While OT promotes parental care and attachment of adults to their offspring, this seems to be true also for the offspring towards their parents: OT-ic injections in domestic fowl chicks significantly increased preferences for a stuffed hen, a potential mother (Loveland et al. 2019).

The birds' parenting cycle involves nest construction, egg laying, incubating and rearing (up to post-fledging parental care in some species) (Mota-Rojas et al. 2023). OT plays a role in almost all these life history stages, governing nest construction choice (Hall et al. 2015), egg laying (Takahashi and Kawashima 2008), incubation (Thayananuphat et al. 2011) and rearing (Aleksandrova 2019); (reviewed by Mota-Rojas et al. 2023; Fig. 16.3). Such governance, as mentioned before, is expressed neuronally and behaviourally with a sexual dimorphism.

During the nest construction phase in zebra finches, nest-building individuals have more OT-ir neurons in the BNSt compared to "non-builders", independent of the sex. In the same experiment, VT-ir neurons were found to be more numerous in males that were bringing more nest material and spending more time in proximity to the partner, suggesting a nested effect structure between OT and VT (Hall et al. 2015). The density of OT-ir neurons keeps on rising during the breeding season, peaking during incubation and/or rearing (Chokchaloemwong et al. 2013; Mota-Rojas et al. 2023). Such an increase of OT in the central area is reflected also in the periphery, as circulating hormonal concentration (Stocker et al. 2021). Across a

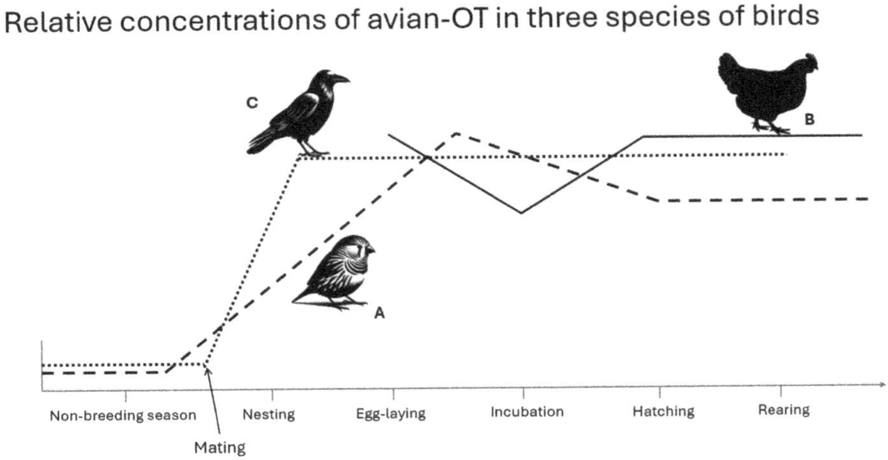

Fig. 16.3 Adapted from Mota-Rojas et al. (2023), OT relative concentrations during the breeding season in three bird species. (**a**) Zebra finch (long-dashed line) OT concentration peaks during the incubation, and then it decreases constantly. (**b**) Thai hens (continuous line) OT concentration decrease during the incubating time, to rise again once it is time to rear the chicks (post-hatching) (Chokchaloemwong et al. 2013). (**c**) Ravens (dotted line) salivary OT concentrations are higher during the mating and breeding season rather than outside that window (Stocker et al. 2021). OT concentrations fluctuate, finely tuning the breeding season in function of the species' specific exigencies, in hens only females incubate, and the lack of social interactions with other conspecifics correlates with OT decrease Vs biparental incubation in zebra finches. The stable high levels of OT post-hatching in hens (compared to the decrease in zebra finches) could be due to the different parenting style (altricial Vs precocial). Birds' images have been generated with Logo creator from OpenAI

breeding season, OT fluctuations appear to follow a species-specific pattern to adjust the social needs. For instance, while zebra finches' OT concentrations peak during egg-laying and then decrease constantly, in hens OT peaks again after hatching (Fig. 16.3). The authors hypothesize that in hens this reflects the social life involvement, since they exhibit mono-parental care during incubation and need to actively look for proximity with chicks after hatching, due to the precocial development pattern.

We want to emphasize that OT involvement is not exclusively within a "breeding context" or limited to seasonal specific time windows but also acts outside these periods. Interindividual social aspects or mate preference are also important outside the breeding context. In this framework, there is an overall consensus about the positive effect of central OT in promoting positive social behaviours, no matter whether such behaviours are directed towards a potential mate, related subordinates, e.g. parental cares (Chokchaloemwong et al. 2013; Thayananuphat et al. 2011) or other unrelated individuals (Duque et al. 2018; Goodson et al. 2009b). Following intracerebroventricular administration of OT, female zebra finches spend more time in close proximity to unrelated conspecifics and choose groups with more familiar individuals (Goodson et al. 2009b; Kabelik et al. 2009). In pinyon jays (*Gymnorhinus cyanocephalus*), intranasal administration of OT promotes prosocial behaviour—the behaviour benefits the actor and, without cost, also the receiver (Duque et al. 2018).

On the other hand, much less consensus exists about the effects of VT on behaviour; overall described as a counterpart of OT, studies show a huge variability in relation to the stimulated brain area (Goodson et al. 2009a; Kabelik et al. 2009). VT appears to have an influence very similar to the mammalian VP on social behaviours, at many levels and with due equivalences. In competitive situations, such as mate competition, zebra finches show VT dose-dependent increase in aggressiveness in both sexes (Goodson et al. 1998). VT antagonists reduce aggressions only in males, suggesting that endogenous VT plays, in this context, a more central role in shaping males' behaviour than females'. However, such a correlation does not appear when it is time to choose a mating partner based on familiarity (Goodson et al. 2009b). In birds, comparable to mammals, VT is involved in the stimulation of the HPA ("stress") axis (Kuenzel et al. 2013; Masunari et al. 2013; Petersson et al. 1999; Romero and Wingfield 2001). However, while in mammals OT interacts with the HPA in preventing stress-induced apoptosis in the hippocampus and exerting an anti-inflammatory effect (Latt et al. 2018), this is not evident in birds (Cornett et al. 2013; Nephew et al. 2005; Nordquist et al. 2020).

16.3.2 Peripheral Mechanisms

As previously mentioned, OT and VT act not only centrally but are secreted within the blood system to exert their functions on peripheral tissues. Technically OT and VT have been quantified successfully in urine (Seguchi et al. 2022), saliva (Stocker et al. 2021) and blood (Nouwen et al. 1984)—see the next section. The diffusion of OT and VT outside the brain seems to be through two main pathways: release into the blood system through the anterior pituitary and release from the third ventricle into the cerebrospinal fluid to the spinal chord (see Fig. 16.3). (Korf et al. 1988; Tennyson et al. 1985). Peripheral VT has been associated with water balance in birds as well as in other taxa. In Pekin ducks (*Anas platyrhynchos domesticus*), magnocellular neurosecretory cells can be inhibited by afferent connections from the vagal terminal nuclei. This connection is based on volume-sensitive receptors and regulates VT production, inducing an antidiuretic effect when the connection is blocked (Korf 1984; Korf et al. 1982). VT mainly controls osmoregulation (Nouwen et al. 1984; Panzica et al. 2002), whereas OT seems to be more involved in restoring the volumetric balance of water (Nouwen et al. 1984). Peripheral VT is also associated with egg laying (Saito et al. 1987) (Nouwen et al. 1984). On the other hand, OT binding affinity and activity to OT receptors within uterine tissues in hens is increased prior oviposition compared to non-egg-laying phase (Takahashi and Kawashima 2008. In turn, VTs stimulate the smooth muscle of the uterine tissues, allowing the egg to be laid (Acher and Chauvet 1995; Takahashi and Kawashima 2008). Overall, it appears that the functionality of OT and VT in non-mammalian vertebrates is quite similar, except perhaps for amphibians (Acher and Chauvet 1995).

While there is almost no evidence for a direct link of VT concentrations between the central area and the periphery in birds, recent research seems to underpin a link between peripheral injections and cognitive/social capabilities (Trigo et al. 2022). This suggests that in birds central and peripheral levels are correlated and peripheral administrations can reach the central nervous system, opening the door to non-invasive research about VT, which is so far mostly unexplored.

16.3.3 Reflections

To conclude, although we have a general understanding of how OT and VT modify various behaviours in birds, it remains unclear which external clues trigger MT and VT production. A few studies suggest that the transcription of VT-fos mRNA is triggered by visual stimuli (Goodson et al. 2009a, b). Also, visually naive domestic fowls increased their neuronal activation (quantified by increased c-fos marker in archipallium, LS and nucleus taeniae) following conspecific presentation (Mayer et al. 2019). However, more research is needed to explore the relationship between causes and consequences. Unfortunately, most work on OT refers to captive individuals, providing limited insights about the effects of OT on social behaviour in natural environments (Goodson 2013). It should be mentioned also that most avian studies were conducted on zebra finches, Thai hens, etc., all species that underwent high artificial selective pressure on breeding traits (Tobari et al. 2022).

16.4 Quantification

In mammals, OT concentration is commonly measured in matrices such as in plasma, saliva, cerebrospinal fluid, milk and urine (Bienboire-Frosini et al. 2017; Lürzel et al. 2020; Martins et al. 2020; Neumann and Slattery 2016; Rault 2016). In birds most studies exploring the behavioural neuroendocrine role of OT have been conducted by quantifying receptor density, by administering OT or by blocking OTR. A large body of work has been conducted by Goodson and colleagues using these techniques, unfolding valuable insights. In a comparative study across various bird species with different social structures, Goodson and colleagues showed species-specific distribution of OT-like binding sites that mirrored evolutionary convergence and divergences in flocking and territorial behaviour (Goodson and Kingsbury 2011). Quantification of VT and OT receptors in specific nuclei that are known to regulate specific social behaviours (such as PVN or nucleus accumbens) was by using immunohistochemistry (Haakenson et al. 2022), in situ-hybridization or receptor-binding techniques (Kingsbury et al. 2013). As previously mentioned, it should be noted that OTRs are expressed in nuclei within the hypothalamus and limbic areas, known for other important interconnective axis such as HPA or HPG axis and which are modulated by sex steroids to regulate behaviour (e.g. territoriality or reproduction). This makes it difficult to disentangle the concrete role of OT on specific behaviours. Consequently, another widely employed method was the administration of OT or its antagonist intracerebrally, subcutaneously (Goodson et al. 2009b) or intranasally (Duque et al. 2018); Lubrano et al. in prep.) to observe subsequent behavioural changes. Each of these techniques has advantages and disadvantages: intracerebral administration, a quite invasive technique, requires not only special ethical permits and training but is also difficult to perform in free-living birds. Although it is probably the most appropriate technique, considering that OT plays a major role within the brain, one ideally would need to inject OT within a given target nucleus to elicit specific behaviour. Modern microdialysis might be a good option to operate on a more fine-tuned level (Ikeda et al. 2015), yet to our knowledge, no experiments have been conducted with this method using OT (Loveland et al. 2019). Subcutaneous injections are not typically used, as the peripheral action of OT is not yet clarified and undoubtedly needs more investigations. Nevertheless, various studies indicate a possible role of OT in the periphery (Goodson et al. 2009b; Trigo et al. 2022).

The most common way to administrate OT is intranasally, a non-invasive and easy method. However, various inconveniences are associated with this method. First, it is difficult to administer an accurate and consistent physiological concentration of OT acting on the target tissues (Rault et al. 2017). Further, within an experiment the dosage between subjects might vary, depending on the bird's reaction, as some of the liquid often spills over when the experimenter applies it into the nostril located on the beak. Moreover, it is not yet clear whether intranasal administration of OT has mainly central or peripheral effects (Martins et al. 2020; Neumann and Landgraf 2012). Further studies are needed to fully understand the dose-response curve of this peptide via this path (Rault 2016).

Today, a major challenge is to quantify circulating levels of OT in bird species and relate them to expressed behaviours. Although in classical study models OT concentration is commonly quantified either by radioimmunoassay (RIA; (Martins et al. 2020) or ELISA (Bienboire-Frosini et al. 2017; Leng and Ludwig 2016), this is not the case in birds. The use of immunoassays for quantifying OT has been strongly debated for various reasons, such as low sensitivity, matrix interference resulting in high variation or cross-reaction with VT. In addition, a relatively large sample volume is required to reach the sensitivity appropriate to detecting OT (Bienboire-Frosini et al. 2017). Further, the efficiency of the antibodies has been often called in question, which makes useful interpretation of the data quite difficult (Neumann and Landgraf 2012). Contradictory results in the concentration of circulating OT nourished all of these doubts, including the use of the appropriate extraction procedure (Bienboire-Frosini et al. 2017; Brandtzaeg et al. 2016). Hence, recent validation studies by means of LCMS or HPLC, (which have higher sensitivity) have confirmed that circulating concentration of OT in the periphery might be higher than previously thought, due to the strong interaction of OT and binding proteins (humans: Brandtzaeg et al. 2016; in birds: Canoine, Fusani, Hykollari, in prep*). All of these points and the fact that plasma volumes obtained from birds are typically smaller than in classical study species may have dissuaded research from quantifying OT in birds. Nevertheless, a few studies working on large birds such as chicken used RIA successfully, as large plasma volumes can be obtained (Takahashi and Kawashima 2008). Lately, a few commercially available kits (e.g. Arbor Assay, Cayman) using an antibody which appears specific enough also for birds, revealed to be useful to quantify OT in birds (Stocker et al. 2021); Lubrano et al. in prep).

Theoretically, as in classical study models, various types of sample can be taken to quantify OT in birds. Plasma, which is obtained after centrifugation of blood samples, is probably the most common matrix to analyse. Blood sampling in birds is already well established in quantifying numerous other hormones, such as glucocorticoids and sex steroids, but not for OT. In larger birds, saliva samples have been successfully used to quantify OT and represent an excellent opportunity for non-invasive sampling (Stocker et al. 2021). Obviously, the biggest challenge and limitation of this method is that it requires training of individuals to chew on cotton swabs, but in a longer term, this approach might open up very interesting opportunities to study complex social or cognitive behaviour in species such as ravens or jays. This experimental procedure could be applied both in the laboratory and in nature.

16.5 Perspectives

We hope to have convinced readers that birds are excellent study models in which to investigate behavioural neuroendocrinology. Studies in both captive and free-living species are essential to gain a complete view of the neuroendocrinological mechanisms of OT (and VT) related to social behaviour. Nowadays several techniques offer the opportunity to explore the role of OT and VT and its evolutionary processes on a large spectrum of species. Overall, research on these peptide hormones would profit from an exchange of information between teams working with different species and the development of methodologies that can be applied to and compared between several taxa.

References

Acher R, Chauvet J (1988) Structure, processing and evolution of the neurohypophysial hormone-neurophysin precursors. Biochimie 70(9):1197–1207. https://doi.org/10.1016/0300-9084(88)90185-x

Acher R, Chauvet J (1995) The neurohypophysial endocrine regulatory cascade: precursors, mediators, receptors, and effectors. Front Neuroendocrinol 16(3):237–289. https://doi.org/10.1006/frne.1995.1009

Aste N, Balthazart J, Absil P, Grossmann R, Mülhbauer E, Viglietti-Panzica C, Panzica GC (1998) Anatomical and neurochemical definition of the nucleus of the stria terminalis in japanese quail (Coturnix japonica). Journal of Comparative Neurology, 396(2):141–157. https://doi.org/10.1002/(SICI)1096-9861

Aleksandrova Y (2019) Neuroendocrine control of maternal behaviour in birds

Bienboire-Frosini C, Chabaud C, Cozzi A, Pageat P (2017) Validation of a commercially available enzyme immunoassay for the determination of oxytocin in plasma samples from seven domestic animal species. Front Neurosci 11:290748

Brandtzaeg OK, Johnsen E, Roberg-Larsen H, Seip KF, MacLean EL, Gesquiere LR, Leknes S, Lundanes E, Wilson SR (2016) Proteomics tools reveal startlingly high amounts of oxytocin in plasma and serum. Sci Rep 6(1):31693

Camerlenghi E, Nolazco S, Farine DR, Magrath RD, Peters A (2023) Multilevel social structure predicts individual helping responses in a songbird. Curr Biol 33(8):1582–1587.e1583

Carter CS (2010) Love and monogamy: an endocrine story. J Androl

Carter CS (2020) Oxytocin, love and the COVID-19 crisis. Clin Neuropsychiatry 17(3):195

Chauvet MT, Levy B, Michel G, Chauvet J, Acher R (1986) Precursors of mesotocin and vasotocin in birds: identification of VLDV- and MSEL- neurophysins in chicken, goose, and ostrich. Biosci Rep 6(4):381–385. https://doi.org/10.1007/BF01116425

Chokchaloemwong D, Prakobsaeng N, Sartsoongnoen N, Kosonsiriluk S, El Halawani M, Chaiseha Y (2013) Mesotocin and maternal care of chicks in native Thai hens (Gallus domesticus). Horm Behav 64(1):53–69. https://doi.org/10.1016/j.yhbeh.2013.04.010

Cornett LE, Kang SW, Kuenzel WJ (2013) A possible mechanism contributing to the synergistic action of vasotocin (VT) and corticotropin-releasing hormone (CRH) receptors on corticosterone release in birds. Gen Comp Endocrinol 188:46–53. https://doi.org/10.1016/j.ygcen.2013.02.032

Donaldson ZR, Young LJ (2008) Oxytocin, vasopressin, and the neurogenetics of sociality. Science 322(5903):900–904

du Vigneaud V, Ressler C, Trippett S (1953) The sequence of amino acids in oxytocin, with a proposal for the structure of oxytocin. J Biol Chem 205(2):949–957

Duque JF, Leichner W, Ahmann H, Stevens JR (2018) Mesotocin influences pinyon jay prosociality. Biol Lett 14(4). https://doi.org/10.1098/rsbl.2018.0105

Froemke RC, Young LJ (2021) Oxytocin, neural plasticity, and social behavior. Annu Rev Neurosci 44:359–381

Goodson JL (2013) Deconstructing sociality, social evolution and relevant nonapeptide functions. Psychoneuroendocrinology 38(4):465–478. https://doi.org/10.1016/j.psyneuen.2012.12.005

Goodson JL, Kingsbury MA (2011) Nonapeptides and the evolution of social group sizes in birds. Front Neuroanat 5:13. https://doi.org/10.3389/fnana.2011.00013

Goodson J, Eibach R, Sakata J, Adkins-Regan E (1998) Effect of septal lesions on male song and aggression in the colonial zebra finch (Taeniopygia guttata) and the territorial field sparrow (Spizella pusilla). Behav Brain Res

Goodson JL, Rinaldi J, Kelly AM (2009a) Vasotocin neurons in the bed nucleus of the stria terminalis preferentially process social information and exhibit properties that dichotomize courting and non-courting phenotypes. Horm Behav 55(1):197–202. https://doi.org/10.1016/j.yhbeh.2008.10.007

Goodson JL, Schrock SE, Klatt JD, Kabelik D, Kingsbury MA (2009b) Mesotocin and nonapeptide receptors promote estrildid flocking behavior. Science 325(5942):862–866. https://doi.org/10.1126/science.1174929

Goodson JL, Rinaldi J, Kelly AM (2009c) Vasotocin neurons in the bed nucleus of the stria terminalis preferentially process social information and exhibit properties that dichotomize courting and non-courting phenotypes. Hormones and Behavior, 55(1):197-202. https://doi.org/10.1016/j.yhbeh.2008.10.007

Goodson JL, Wilson LC, Schrock SE (2012) To flock or fight: neurochemical signatures of divergent life histories in sparrows. Proc Natl Acad Sci 109(suppl 1):10685–10692

Goossens N, Blahser S, Oksche A, Vandesande F, Dierickx K (1977) Immunocytochemical investigation of the hypothalamo-neurohypophysial system in birds. Cell Tissue Res 184(1):1–13. https://doi.org/10.1007/BF00220523

Haakenson CM, Balthazart J, Madison FN, Ball GF (2022) The neural distribution of the avian homologue of oxytocin, mesotocin, in two songbird species, the zebra finch and the canary: a potential role in song perception and production. J Comp Neurol 530(13):2402–2414

Hall ZJ, Healy SD, Meddle SL (2015) A role for nonapeptides and dopamine in nest-building behaviour. J Neuroendocrinol 27(2):158–165. https://doi.org/10.1111/jne.12250

Ikeda MZ, Jeon SD, Cowell RA, Remage-Healey L (2015) Norepinephrine modulates coding of complex vocalizations in the songbird auditory cortex independent of local neuroestrogen synthesis. J Neurosci 35(25):9356–9368

Jurkevich A, Barth SW, Kuenzel WJ, Kohler A, Grossmann R (1999) Development of sexually dimorphic vasotocinergic system in the bed nucleus of stria terminalis in chickens. J Comp Neurol 408(1):46–60. https://doi.org/10.1002/(sici)1096-9861(19990524)408:1<46::aid-cne4>3.0.co;2-5

Kabelik D, Klatt JD, Kingsbury MA, Goodson JL (2009) Endogenous vasotocin exerts context-dependent behavioral effects in a semi-naturalistic colony environment. Horm Behav 56(1):101–107. https://doi.org/10.1016/j.yhbeh.2009.03.017

Kamkrathok B, Porter TE, El Halawani ME, Chaiseha Y (2017) Distribution of mesotocin-immunoreactive neurons in the brain of the male native Thai chicken. Acta Histochem 119(8):804–811. https://doi.org/10.1016/j.acthis.2017.10.004

Kelly AM, Goodson JL (2014) Social functions of individual vasopressin–oxytocin cell groups in vertebrates: what do we really know? Front Neuroendocrinol 35(4):512–529

Kingsbury MA, Miller KM, Goodson JL (2013) VPAC receptor signaling modulates grouping behavior and social responses to contextual novelty in a gregarious finch: a role for a putative prefrontal cortex homologue. Horm Behav 64(3):511–518

Korf HW (1984) Neuronal organization of the avian paraventricular nucleus: intrinsic, afferent, and efferent connections. J Exp Zool 232(3):387–395. https://doi.org/10.1002/jez.1402320303

Korf HW, Simon-Oppermann C, Simon E (1982) Afferent connections of physiologically identified neuronal complexes in the paraventricular nucleus of conscious Pekin ducks involved in regulation of salt- and water-balance. Cell Tissue Res 226(2):275–300. https://doi.org/10.1007/BF00218359

Korf H, Panzica G, Viglietti-Panzica C, Oksche A (1988) Pattern of peptidergic neurons in the avian brain: clusters--local circuitries--projections. Basic Appl Histochem 32(1):55–75

Kuenzel WJ, Kang SW, Jurkevich A (2013) Neuroendocrine regulation of stress in birds with an emphasis on vasotocin receptors (VTRs). Gen Comp Endocrinol 190:18–23

Latt HM, Matsushita H, Morino M, Koga Y, Michiue H, Nishiki T, Tomizawa K, Matsui H (2018) Oxytocin inhibits corticosterone-induced apoptosis in primary hippocampal neurons. Neuroscience 379:383–389. https://doi.org/10.1016/j.neuroscience.2018.03.025

Leng G, Ludwig M (2016) Intranasal oxytocin: myths and delusions. Biol Psychiatry 79(3):243–250

Loveland JL, Stewart MG, Vallortigara G (2019) Effects of oxytocin-family peptides and substance P on locomotor activity and filial preferences in visually naïve chicks. Eur J Neurosci 50(10):3674–3687

Lürzel S, Bückendorf L, Waiblinger S, Rault J-L (2020) Salivary oxytocin in pigs, cattle, and goats during positive human-animal interactions. Psychoneuroendocrinology 115:104636

Martins D, Gabay AS, Mehta M, Paloyelis Y (2020) Salivary and plasmatic oxytocin are not reliable trait markers of the physiology of the oxytocin system in humans. Elife 9:e62456

Masunari K, Khan MS, Cline MA, Tachibana T (2013) Central administration of mesotocin inhibits feeding behavior in chicks. Regul Pept 187:1–5. https://doi.org/10.1016/j.regpep.2013.10.004

Mayer U, Rosa-Salva O, Loveland JL, Vallortigara G (2019) Selective response of the nucleus taeniae of the amygdala to a naturalistic social stimulus in visually naive domestic chicks. Sci Rep 9(1):9849

Menon R, Neumann ID (2023) Detection, processing and reinforcement of social cues: regulation by the oxytocin system. Nat Rev Neurosci 24(12):761–777

Montagnese CM, Szekely T, Csillag A, Zachar G (2015) Distribution of vasotocin- and vasoactive intestinal peptide-like immunoreactivity in the brain of blue tit (Cyanistes coeruleus). Front Neuroanat 9:90. https://doi.org/10.3389/fnana.2015.00090

Mota-Rojas D, Marcet-Rius M, Domínguez-Oliva A, Buenhombre J, Daza-Cardona EA, Lezama-García K, Olmos-Hernández A, Verduzco-Mendoza A, Bienboire-Frosini C (2023) Parental behavior and newborn attachment in birds: life history traits and endocrine responses. Front Psychol 14:1183554

Nephew BC, Aaron RS, Romero LM (2005) Effects of arginine vasotocin (AVT) on the behavioral, cardiovascular, and corticosterone responses of starlings (Sturnus vulgaris) to crowding. Horm Behav 47(3):280–289. https://doi.org/10.1016/j.yhbeh.2004.11.007

Neumann ID, Landgraf R (2012) Balance of brain oxytocin and vasopressin: implications for anxiety, depression, and social behaviors. Trends Neurosci 35(11):649–659

Neumann ID, Slattery DA (2016) Oxytocin in general anxiety and social fear: a translational approach. Biol Psychiatry 79(3):213–221

Nordquist RE, Zeinstra EC, Dougherty A, Riber AB (2020) Effects of dark brooder rearing and age on hypothalamic vasotocin and feather corticosterone levels in laying hens. Front Vet Sci 7:19. https://doi.org/10.3389/fvets.2020.00019

Nouwen EJ, Decuypere E, Kuhn ER, Michels H, Hall TR, Chadwick A (1984) Effect of dehydration, haemorrhage and oviposition on serum concentrations of vasotocin, mesotocin and prolactin in the chicken. J Endocrinol 102(3):345–351. https://doi.org/10.1677/joe.0.1020345

Panzica GC, Plumari L, Garcia-Ojeda E, Deviche P (1999) Central vasotocin-immunoreactive system in a male passerine bird (Junco hyemalis). J Comp Neurol 409(1):105–117. https://www.ncbi.nlm.nih.gov/pubmed/10363714

Panzica GC, Bakthazart J, Pessatti M, Viglietti-Panzica C (2002) The parvocellular vasotocin system of Japanese quail: a developmental and adult model for the study of influences of gonadal hormones on sexually differentiated and behaviorally relevant neural circuits. Environ Health Perspect 110(Suppl 3):423–428. https://doi.org/10.1289/ehp.02110s3423

Petersson M, Hulting AL, Uvnas-Moberg K (1999) Oxytocin causes a sustained decrease in plasma levels of corticosterone in rats. Neurosci Lett 264(1–3):41–44. https://doi.org/10.1016/s0304-3940(99)00159-7

Plumari L, Plateroti S, Deviche P, Panzica GC (2004) Region-specific testosterone modulation of the vasotocin-immunoreactive system in male dark-eyed junco, Junco hyemalis. Brain Res 999(1):1–8. https://doi.org/10.1016/j.brainres.2003.10.037

Rault J-L (2016) Effects of positive and negative human contacts and intranasal oxytocin on cerebrospinal fluid oxytocin. Psychoneuroendocrinology 69:60–66

Rault J-L, van den Munkhof M, Buisman-Pijlman FT (2017) Oxytocin as an indicator of psychological and social well-being in domesticated animals: a critical review. Front Psychol 8:288410

Robinzon B, Koike TI, Neldon HL, Kinzler SL, Hendry IR, Halawani ME (1988a) Physiological effects of arginine vasotocin and mesotocin in cockerels. Br Poult Sci 29(3):639–652. https://doi.org/10.1080/00071668808417090

Robinzon B, Koike TI, Neldon HL, SL (1988b) Distribution of immunoreactive mesotocin and vasotocin in the brain and pituitary of chickens. Peptides 9(4):829–833. https://doi.org/10.1016/0196-9781(88)90129-5

Romero LM, Wingfield JC (2001) Regulation of the hypothalamic-pituitary-adrenal axis in free-living pigeons. J Comp Physiol B 171(3):231–235. https://doi.org/10.1007/s003600000167

Saito N, Shimada K, Koike TI (1987) Interrelationship between arginine vasotocin, prostaglandin, and uterine contractility in the control of oviposition in the hen (Gallus domesticus). Gen Comp Endocrinol 67(3):342–347

Seguchi A, Mogi K, Izawa EI (2022) Measurement of urinary mesotocin in large-billed crows by enzyme-linked immunosorbent assay. J Vet Med Sci 84(4):520–524. https://doi.org/10.1292/jvms.21-0635

Stocker M, Prosl J, Vanhooland LC, Horn L, Bugnyar T, Canoine V, Massen JJM (2021) Measuring salivary mesotocin in birds - seasonal differences in ravens' peripheral mesotocin levels. Horm Behav 134:105015. https://doi.org/10.1016/j.yhbeh.2021.105015

Takahashi T, Kawashima M (2008) Mesotocin increases the sensitivity of the hen oviduct uterus to arginine vasotocin. Poult Sci 87(10):2107–2111. https://doi.org/10.3382/ps.2008-00076

Tennyson VM, Hou-Yu A, Nilaver G, Zimmerman EA (1985) Immunocytochemical studies of vasotocin and mesotocin in the hypothalamo-hypophysial system of the chicken. Cell Tissue Res 239(2):279–291. https://doi.org/10.1007/BF00218005

Thayananuphat A, Youngren OM, Kang SW, Bakken T, Kosonsiriluk S, Chaiseha Y, El Halawani ME (2011) Dopamine and mesotocin neurotransmission during the transition from incubation to brooding in the Turkey. Horm Behav 60(4):327–335. https://doi.org/10.1016/j.yhbeh.2011.06.009

Theofanopoulou C (2021) Reconstructing the evolutionary history of the oxytocin and vasotocin receptor gene family: insights on whole genome duplication scenarios. Dev Biol 479:99–106

Theofanopoulou C, Boeckx C, Jarvis ED (2017) A hypothesis on a role of oxytocin in the social mechanisms of speech and vocal learning. Proc R Soc B Biol Sci 284(1861):20170988

Theofanopoulou C, Gedman G, Cahill JA, Boeckx C, Jarvis ED (2021) Universal nomenclature for oxytocin–vasotocin ligand and receptor families. Nature 592(7856):747–755

Tobari Y, Theofanopoulou C, Mori C, Sato Y, Marutani M, Fujioka S, Konno N, Suzuki K, Furutani A, Hakataya S (2022) Oxytocin variation and brain region-specific gene expression in a domesticated avian species. Genes Brain Behav 21(2):e12780

Trigo S, Silva PA, Cardoso GC, Soares MC (2022) A test of context-and sex-dependent dopaminergic effects on the behavior of a gregarious bird, the common waxbill, Estrilda astrild. J Exp Biol 225(6):jeb243861

Triki Z, Daughters K, De Dreu CKW (2022) Oxytocin has 'tend-and-defend' functionality in group conflict across social vertebrates. Philos Trans R Soc Lond Ser B Biol Sci 377(1851):20210137. https://doi.org/10.1098/rstb.2021.0137

Viglietti-Panzica C, Aste N, Balthazart J, Panzica GC (1994) Vasotocinergic innervation of sexually dimorphic medial preoptic nucleus of the male Japanese quail: influence of testosterone. Brain Res 657(1-2):171–184. https://doi.org/10.1016/0006-8993(94)90965-2

Key Literature

Chokchaloemwong D, Prakobsaeng N, Sartsoongnoen N, Kosonsiriluk S, El Halawani M, Chaiseha Y (2013) Mesotocin and maternal care of chicks in native Thai hens (Gallus domesticus). Horm Behav 64(1):53–69 A very straightforward experiment to prove the influence of OT in rearing and incubating

du Vigneaud V, Ressler C, Trippett S (1953) The sequence of amino acids in oxytocin, with a proposal for the structure of oxytocin. J Biol Chem 205(2):949–957 The authors crystallized the OT molecule for the first time

Goodson JL (2013) Deconstructing sociality, social evolution and relevant nonapeptide functions. Psychoneuroendocrinology 38(4):465–478 Review of OT and VT effects comparing all taxa

Goodson JL, Schrock SE, Klatt JD, Kabelik D, Kingsbury MA (2009) Mesotocin and nonapeptide receptors promote estrildid flocking behavior. Science 325(5942):862–866 The authors compare the effects of VT and OT intra-cranial administration revealing sex-dependent effects and an influence on the social behavior outside the breeding season. Further, activity in the BNSTm correlates positively with the complexity of group living, across 5 species

Haakenson CM, Balthazart J, Madison FN, Ball GF (2022) The neural distribution of the avian homologue of oxytocin, mesotocin, in two songbird species, the zebra finch and the canary: a potential role in song perception and production. J Comp Neurol 530(13):2402–2414 In this study, the authors describe neuronal OT-ir distribution comparing 2 different song bird species and showing similarities with non-oscine bird species, indicating a conserved OT-system also in birds. Further authors propose a possible OT involvement on mating and singing behaviour.

Korf HW (1984) Neuronal organization of the avian paraventricular nucleus: intrinsic, afferent, and efferent connections. J Exp Zool 232(3):387–395 Extensive description of the PVN and its magnocellular and parvocellular neurons' afferences and efferences

Loveland JL, Stewart MG, Vallortigara G (2019) Effects of oxytocin-family peptides and substance P on locomotor activity and filial preferences in visually naïve chicks. Eur J Neurosci 50(10):3674–3687 Shows for the first time that OT also shapes chicks' behaviour during the rearing phase

Stocker M, Prosl J, Vanhooland LC, Horn L, Bugnyar T, Canoine V, Massen JJM (2021) Measuring salivary mesotocin in birds - seasonal differences in ravens' peripheral mesotocin levels. Horm Behav 134:105015 The first study to show season-dependent changes in OT concentrations by means of salivary samples in Raven.

Theofanopoulou C, Gedman G, Cahill JA, Boeckx C, Jarvis ED (2021) Universal nomenclature for oxytocin–vasotocin ligand and receptor families. Nature 592(7856):747–755 The authors used an evolution based approach to provide a new nomenclature for nonapepties

Glossary

AgRP Agouti-related Protein

Akt Akt kinase/protein kinase B

Antidromic identification When an axon is stimulated electrically, an action potential is initiated that propagates both orthodromically, towards the axonal endings, and antidromically towards the cell body. This antidromic potential will arrive at the soma at a constant latency after the stimulus, unless it first meets a descending spontaneous action potential, in which event both action potentials will be extinguished in the collision. These two features—constant latency and collision—are used to determine whether a given neuron projects to the site at which a stimulating electrode is placed.

BAT Brown Adipose Tissue

Bilateria A clade of the animal kingdom comprising taxa that exhibit bilateral symmetry during at least one stage during their life cycle and that all evolved from a common 'urbilaterian' ancestor.

BPAE Bovine Pulmonary Artery Endothelial

cAMP Cyclic Adenosine Monophosphate

Carboxypeptidases A group of enzymes that cleave amino acids from the carboxyl terminal of a peptide or protein.

Central nervous system Brain and medulla.

Central Pattern Generator Neuronal circuits that when activated by a non-rhythmic input can produce rhythmic output patterns that drive specific behaviours.

CHO Chinese Hamster Ovarian

Chordates Animals in the phylum Chordata characterized by the presence of a notochord and a post-anal tail at some stage in their development. Chordates include vertebrates (e.g., lampreys, fish, amphibians, reptiles, birds and mammals) and some invertebrates (e.g., amphioxus and tunicates).

CRE cAMP Response Element

CREB cAMP Response Element Binding Protein

CRISPR/Cas9 Clustered Regularly Interspaced Short Palindromic Repeats—CRISPR associated Protein 9

V. Grinevich, R. F. Oliveira (eds.), *Evolutionary and Comparative Neuroendocrinology*, Masterclass in Neuroendocrinology 17,
https://doi.org/10.1007/978-3-031-80209-6

Deep homology A term proposed by Neil Shubin and collaborators to describe independent origin of similar genetic regulatory pathways in two or more unrelated lineages.

Deuterostomes A clade of the Bilateria that comprises chordates (including vertebrates), hemichordates, and echinoderms. The name deuterostome refers to the typical mode of embryonic development in these animals where the mouth is formed from the second invagination of the blastula.

Elasmobranch A subgroup of cartilaginous fish that emerged about 390 million years ago; extant members include sharks, rays, skates, and swordfish

ERE Estrogen Response Element

Evolutionary convergence the result of natural selection on two or more unrelated species that face a particular environmental pressure and come up with the same solution, in which case, different structures may have similar function or share similar materials of development (e.g. genes) without being homologous.

FSH Follicle-Stimulating Hormone

GH Growth Hormone

GHRH Growth Hormone Releasing Hormone

GHRH-R Growth Hormone Releasing Hormone Receptor

GLP-1 Glucagon Like Peptide-1

Gonopore A genital pore in many invertebrates. Most hexapods, including insects, have a single common gonopore.

GPCR G Protein Coupled Receptor

Hatschek's Pit Structure involved in filter feeding, allowing the organism to capture small food particles from the surrounding water. It is a feature found in certain chordates such as amphioxus.

Homologue Similar DNA sequences found in different species that share a common origin.

Hypothalamo-neurohypophysial system (HNS) In reptiles, birds, and mammals the magnocellular neurosecretory neurons of the HNS are aggregated in two large neuronal groups, a paraventricular nucleus and a supraoptic nucleus, but there are also many smaller aggregations (the 'accessory nuclei' of the HNS). Magnocellular neurons project to the neural lobe of the pituitary but collaterals of these axons innervate a diversity of forebrain structures. The paraventricular nucleus of mammals also contains several populations of 'parvocellular' oxytocin- and vasopressin-neurons which do not project to the neurohypophysis ('parvocellular' is a misnomer because there is little if any difference in soma size compared to magnocellular neurons).

IGF Insulin-like Growth Factor

Immunohistochemistry Use of antibodies to check for antigens. Used to localize proteins.

Immunoreactivity Immune reaction due to an interaction with an antigen.

In-situ-hybridisation Use of genetic material to localize precise nucleic acids in a DNA sequence.

INS1 Insulinoma 1

IRS2 Insulin Receptor Substrate 2

LH Luteinizing Hormone

Life histories All the organism's traits that affect fitness (survival, reproduction, and fecundity). For example, rate of growth, reproductive age, number of offspring, parental care, and so on.

LPS Lipopolysaccharide

Magnocellular preoptic nucleus A thin layer of neurosecretory neurons that flanks the walls of the third ventricle in teleosts, including the neurosecretory neurons that project to the neurohypophysis.

Median eminence Densely vascularised region at the base of the third ventricle in mammals where hypothalamic neurosecretory neurons release factors that regulate hormone secretion from the anterior pituitary into the hypophysial portal blood vessels. In teleosts, the region corresponding to the median eminence is traditionally referred to as the 'anterior neurohypophysis'. A specialized region of the hypothalamus that serves as a site where regulatory hormones are released into the bloodstream in order to travel to the anterior pituitary gland and control the secretion of various hormones that influence the neuroendocrine axis.

Monophyletic group A taxonomic group that shares a common ancestor and all of its evolutionary descendants.

mRNA messenger RNA

MVEC Microvascular Endothelial Cells

Neoteny A biological trait in which an organism retains juvenile or immature characteristics for their adult stage. These can include physical features, behaviours, or developmental traits that are typically associated with the earlier stages of an organism.

Neurohypophysis This term originally meant the median eminence and the neural lobe in all vertebrate classes, but now the terms 'neural lobe', 'neurohypophysis', and 'posterior pituitary' are used interchangeably. In teleosts, the brain region corresponding to the neural lobe is sometimes called the 'posterior neurohypophysis'.

Neuropeptide Peptide that can act as a neurotransmitter.

Neutral evolution A theory proposed by Moto Kimura and Tomoko Ohta that explains the molecular variation in both amino acids and nucleotide levels inside and between populations. Such variation cannot be eliminated or fixed in the population by negative or positive selection because it does not affect fitness, i.e. neutral, but is the result of the balance between genetic drift and mutation interaction.

NOD-SCID Non-Obese Diabetic—Severe Combined Immunodeficiency

Non-mammalian vertebrates Lampreys, fish, amphibians, reptiles, birds

Nonapeptide Nine-amino acid peptide

NPY Neuropeptide Y

NREM Non-Rapid Eye Movement

Orthologs Homologous genes (and their encoded proteins) in different species that have evolved from a common ancestral gene through the process of speciation. Genes in different species that have evolved from a common ancestral gene through speciation. These genes often have similar functions in their respective species.

PACAP Pituitary Adenylate Cyclase Activating Polypeptide

Paralogs Homologous genes (and their encoded proteins) in single species that have evolved from a common ancestral gene through the process of gene duplication.

PC1/3 Prohormone convertase 1/3

Peripheral nervous system Everything else than brain and medulla

Phylogeny Representation of the history of a group of organisms (e.g. animals) that reflects their evolutionary relationships.

PKA Protein Kinase A

Protostomes A clade of the Bilateria that includes ecdysozoans (e.g. arthropods, nematodes) and lophotrochozoans (e.g. annelids, molluscs). The name protostome refers to the typical mode of embryonic development in these animals where the mouth is formed from the first invagination of the blastula (the blastopore).

Tanycytes Specialized glial cells located in the mammalian hypothalamus. Tanycytes have long processes and are involved in various physiological functions, including regulating the blood–brain barrier, controlling the secretion of neurohormones, and the regulation of body weight and energy balance.

Teleost A class of modern bony fish that comprises most extant ray-finned fish.

Thyrocyte A cell type found in the thyroid gland that plays a main role in the production and secretion of thyroid hormones.

UCP1 Uncoupling Protein 1

VIP Vasoactive Intestinal Polypeptide

www.ingramcontent.com/pod-product-compliance
Lightning Source LLC
Chambersburg PA
CBHW051952031225
36163CB00026B/54
9783031802089